T0181644

Lecture Notes in Computer Science 12427

More information about this series at http://www.springer.com/series/7409

Constantine Stephanidis ·
Gavriel Salvendy · June Wei ·
Sakae Yamamoto · Hirohiko Mori ·
Gabriele Meiselwitz · Fiona Fui-Hoon Nah ·
Keng Siau (Eds.)

HCI International 2020 – Late Breaking Papers

Interaction, Knowledge and Social Media

22nd HCI International Conference, HCII 2020
Copenhagen, Denmark, July 19–24, 2020
Proceedings

Springer

Editors
Constantine Stephanidis
University of Crete and Foundation for Research and Technology – Hellas (FORTH)
Heraklion, Crete, Greece

June Wei
University of West Florida
Pensacola, FL, USA

Hirohiko Mori
Tokyo City University
Tokyo, Japan

Fiona Fui-Hoon Nah
Missouri University of Science and Technology
Rolla, MO, USA

Gavriel Salvendy
University of Central Florida
Orlando, FL, USA

Sakae Yamamoto
Tokyo University of Science
Tokyo, Japan

Gabriele Meiselwitz
Towson University
Towson, MD, USA

Keng Siau
Missouri University of Science and Technology
Rolla, MO, USA

ISSN 0302-9743 ISSN 1611-3349 (electronic)
Lecture Notes in Computer Science
ISBN 978-3-030-60151-5 ISBN 978-3-030-60152-2 (eBook)
https://doi.org/10.1007/978-3-030-60152-2

LNCS Sublibrary: SL3 – Information Systems and Applications, incl. Internet/Web, and HCI

This Springer imprint is published by the registered company Springer Nature Switzerland AG
The registered company address is: Gewerbestrasse 11, 6330 Cham, Switzerland

Foreword

The 22nd International Conference on Human-Computer Interaction, HCI International 2020 (HCII 2020), was planned to be held at the AC Bella Sky Hotel and Bella Center, Copenhagen, Denmark, during July 19–24, 2020. Due to the COVID-19 pandemic and the resolution of the Danish government not to allow events larger than 500 people to be hosted until September 1, 2020, HCII 2020 had to be held virtually. It incorporated the 21 thematic areas and affiliated conferences listed on the following page.

A total of 6,326 individuals from academia, research institutes, industry, and governmental agencies from 97 countries submitted contributions, and 1,439 papers and 238 posters were included in the volumes of the proceedings published before the conference. Additionally, 333 papers and 144 posters are included in the volumes of the proceedings published after the conference, as "Late Breaking Work" (papers and posters). These contributions address the latest research and development efforts in the field and highlight the human aspects of design and use of computing systems.

The volumes comprising the full set of the HCII 2020 conference proceedings are listed in the following pages and together they broadly cover the entire field of human-computer interaction, addressing major advances in knowledge and effective use of computers in a variety of application areas.

I would like to thank the Program Board Chairs and the members of the Program Boards of all Thematic Areas and Affiliated Conferences for their valuable contributions towards the highest scientific quality and the overall success of the HCI International 2020 conference.

This conference would not have been possible without the continuous and unwavering support and advice of the founder, conference general chair emeritus and conference scientific advisor, Prof. Gavriel Salvendy. For his outstanding efforts, I would like to express my appreciation to the communications chair and editor of HCI International News, Dr. Abbas Moallem.

July 2020 Constantine Stephanidis

HCI International 2020 Thematic Areas and Affiliated Conferences

Thematic Areas:

- HCI 2020: Human-Computer Interaction
- HIMI 2020: Human Interface and the Management of Information

Affiliated Conferences:

- EPCE: 17th International Conference on Engineering Psychology and Cognitive Ergonomics
- UAHCI: 14th International Conference on Universal Access in Human-Computer Interaction
- VAMR: 12th International Conference on Virtual, Augmented and Mixed Reality
- CCD: 12th International Conference on Cross-Cultural Design
- SCSM: 12th International Conference on Social Computing and Social Media
- AC: 14th International Conference on Augmented Cognition
- DHM: 11th International Conference on Digital Human Modeling & Applications in Health, Safety, Ergonomics & Risk Management
- DUXU: 9th International Conference on Design, User Experience and Usability
- DAPI: 8th International Conference on Distributed, Ambient and Pervasive Interactions
- HCIBGO: 7th International Conference on HCI in Business, Government and Organizations
- LCT: 7th International Conference on Learning and Collaboration Technologies
- ITAP: 6th International Conference on Human Aspects of IT for the Aged Population
- HCI-CPT: Second International Conference on HCI for Cybersecurity, Privacy and Trust
- HCI-Games: Second International Conference on HCI in Games
- MobiTAS: Second International Conference on HCI in Mobility, Transport and Automotive Systems
- AIS: Second International Conference on Adaptive Instructional Systems
- C&C: 8th International Conference on Culture and Computing
- MOBILE: First International Conference on Design, Operation and Evaluation of Mobile Communications
- AI-HCI: First International Conference on Artificial Intelligence in HCI

Conference Proceedings – Full List of Volumes

1. LNCS 12181, Human-Computer Interaction: Design and User Experience (Part I), edited by Masaaki Kurosu
2. LNCS 12182, Human-Computer Interaction: Multimodal and Natural Interaction (Part II), edited by Masaaki Kurosu
3. LNCS 12183, Human-Computer Interaction: Human Values and Quality of Life (Part III), edited by Masaaki Kurosu
4. LNCS 12184, Human Interface and the Management of Information: Designing Information (Part I), edited by Sakae Yamamoto and Hirohiko Mori
5. LNCS 12185, Human Interface and the Management of Information: Interacting with Information (Part II), edited by Sakae Yamamoto and Hirohiko Mori
6. LNAI 12186, Engineering Psychology and Cognitive Ergonomics: Mental Workload, Human Physiology, and Human Energy (Part I), edited by Don Harris and Wen-Chin Li
7. LNAI 12187, Engineering Psychology and Cognitive Ergonomics: Cognition and Design (Part II), edited by Don Harris and Wen-Chin Li
8. LNCS 12188, Universal Access in Human-Computer Interaction: Design Approaches and Supporting Technologies (Part I), edited by Margherita Antona and Constantine Stephanidis
9. LNCS 12189, Universal Access in Human-Computer Interaction: Applications and Practice (Part II), edited by Margherita Antona and Constantine Stephanidis
10. LNCS 12190, Virtual, Augmented and Mixed Reality: Design and Interaction (Part I), edited by Jessie Y.C. Chen and Gino Fragomeni
11. LNCS 12191, Virtual, Augmented and Mixed Reality: Industrial and Everyday Life Applications (Part II), edited by Jessie Y.C. Chen and Gino Fragomeni
12. LNCS 12192, Cross-Cultural Design: User Experience of Products, Services, and Intelligent Environments (Part I), edited by P.L. Patrick Rau
13. LNCS 12193, Cross-Cultural Design: Applications in Health, Learning, Communication, and Creativity (Part II), edited by P.L. Patrick Rau
14. LNCS 12194, Social Computing and Social Media: Design, Ethics, User Behavior, and Social Network Analysis (Part I), edited by Gabriele Meiselwitz
15. LNCS 12195, Social Computing and Social Media: Participation, User Experience, Consumer Experience, and Applications of Social Computing (Part II), edited by Gabriele Meiselwitz
16. LNAI 12196, Augmented Cognition: Theoretical and Technological Approaches (Part I), edited by Dylan D. Schmorrow and Cali M. Fidopiastis
17. LNAI 12197, Augmented Cognition: Human Cognition and Behaviour (Part II), edited by Dylan D. Schmorrow and Cali M. Fidopiastis

18. LNCS 12198, Digital Human Modeling & Applications in Health, Safety, Ergonomics & Risk Management: Posture, Motion and Health (Part I), edited by Vincent G. Duffy
19. LNCS 12199, Digital Human Modeling & Applications in Health, Safety, Ergonomics & Risk Management: Human Communication, Organization and Work (Part II), edited by Vincent G. Duffy
20. LNCS 12200, Design, User Experience, and Usability: Interaction Design (Part I), edited by Aaron Marcus and Elizabeth Rosenzweig
21. LNCS 12201, Design, User Experience, and Usability: Design for Contemporary Interactive Environments (Part II), edited by Aaron Marcus and Elizabeth Rosenzweig
22. LNCS 12202, Design, User Experience, and Usability: Case Studies in Public and Personal Interactive Systems (Part III), edited by Aaron Marcus and Elizabeth Rosenzweig
23. LNCS 12203, Distributed, Ambient and Pervasive Interactions, edited by Norbert Streitz and Shin'ichi Konomi
24. LNCS 12204, HCI in Business, Government and Organizations, edited by Fiona Fui-Hoon Nah and Keng Siau
25. LNCS 12205, Learning and Collaboration Technologies: Designing, Developing and Deploying Learning Experiences (Part I), edited by Panayiotis Zaphiris and Andri Ioannou
26. LNCS 12206, Learning and Collaboration Technologies: Human and Technology Ecosystems (Part II), edited by Panayiotis Zaphiris and Andri Ioannou
27. LNCS 12207, Human Aspects of IT for the Aged Population: Technologies, Design and User experience (Part I), edited by Qin Gao and Jia Zhou
28. LNCS 12208, Human Aspects of IT for the Aged Population: Healthy and Active Aging (Part II), edited by Qin Gao and Jia Zhou
29. LNCS 12209, Human Aspects of IT for the Aged Population: Technology and Society (Part III), edited by Qin Gao and Jia Zhou
30. LNCS 12210, HCI for Cybersecurity Privacy and Trust, edited by Abbas Moallem
31. LNCS 12211, HCI in Games, edited by Xiaowen Fang
32. LNCS 12212, HCI in Mobility, Transport and Automotive Systems: Automated Driving and In-Vehicle Experience Design (Part I), edited by Heidi Krömker
33. LNCS 12213, HCI in Mobility, Transport and Automotive Systems: Driving Behavior, Urban and Smart Mobility (Part II), edited by Heidi Krömker
34. LNCS 12214, Adaptive Instructional Systems, edited by Robert A. Sottilare and Jessica Schwarz
35. LNCS 12215, Culture and Computing, edited by Matthias Rauterberg
36. LNCS 12216, Design, Operation and Evaluation of Mobile Communications, edited by Gavriel Salvendy and June Wei
37. LNCS 12217, Artificial Intelligence in HCI, edited by Helmut Degen and Lauren Reinerman-Jones
38. CCIS 1224, HCI International 2020 Posters (Part I), edited by Constantine Stephanidis and Margherita Antona
39. CCIS 1225, HCI International 2020 Posters (Part II), edited by Constantine Stephanidis and Margherita Antona

40. CCIS 1226, HCI International 2020 Posters (Part III), edited by Constantine Stephanidis and Margherita Antona
41. LNCS 12423, HCI International 2020 – Late Breaking Papers: User Experience Design and Case Studies, edited by Constantine Stephanidis, Aaron Marcus, Elizabeth Rosenzweig, P.L. Patrick Rau, Abbas Moallem, and Matthias Rauterberg
42. LNCS 12424, HCI International 2020 – Late Breaking Papers: Multimodality and Intelligence, edited by Constantine Stephanidis, Masaaki Kurosu, Helmut Degen, and Lauren Reinerman-Jones
43. LNCS 12425, HCI International 2020 – Late Breaking Papers: Cognition, Learning and Games, edited by Constantine Stephanidis, Don Harris, Wen-Chin Li, Dylan D. Schmorrow, Cali M. Fidopiastis, Panayiotis Zaphiris, Andri Ioannou, Xiaowen Fang, Robert Sottilare, and Jessica Schwarz
44. LNCS 12426, HCI International 2020 – Late Breaking Papers: Universal Access and Inclusive Design, edited by Constantine Stephanidis, Margherita Antona, Qin Gao, and Jia Zhou
45. LNCS 12427, HCI International 2020 – Late Breaking Papers: Interaction, Knowledge and Social Media, edited by Constantine Stephanidis, Gavriel Salvendy, June Wei, Sakae Yamamoto, Hirohiko Mori, Gabriele Meiselwitz, Fiona Fui-Hoon Nah, and Keng Siau
46. LNCS 12428, HCI International 2020 – Late Breaking Papers: Virtual and Augmented Reality, edited by Constantine Stephanidis, Jessie Y.C. Chen, and Gino Fragomeni
47. LNCS 12429, HCI International 2020 – Late Breaking Papers: Digital Human Modeling and Ergonomics, Mobility and Intelligent Environments, edited by Constantine Stephanidis, Vincent G. Duffy, Norbert Streitz, Shin'ichi Konomi, and Heidi Krömker
48. CCIS 1293, HCI International 2020 – Late Breaking Posters (Part I), edited by Constantine Stephanidis, Margherita Antona, and Stavroula Ntoa
49. CCIS 1294, HCI International 2020 – Late Breaking Posters (Part II), edited by Constantine Stephanidis, Margherita Antona, and Stavroula Ntoa

HCI International 2020 (HCII 2020)

The full list with the Program Board Chairs and the members of the Program Boards of all thematic areas and affiliated conferences is available online at:

http://www.hci.international/board-members-2020.php

HCI International 2021

The 23rd International Conference on Human-Computer Interaction, HCI International 2021 (HCII 2021), will be held jointly with the affiliated conferences in Washington DC, USA, at the Washington Hilton Hotel, July 24–29, 2021. It will cover a broad spectrum of themes related to human-computer interaction (HCI), including theoretical issues, methods, tools, processes, and case studies in HCI design, as well as novel interaction techniques, interfaces, and applications. The proceedings will be published by Springer. More information will be available on the conference website: http://2021.hci.international/.

General Chair
Prof. Constantine Stephanidis
University of Crete and ICS-FORTH
Heraklion, Crete, Greece
Email: general_chair@hcii2021.org

http://2021.hci.international/

Contents

Interacting with Information and Knowledge

Software Crowdsourcing Design: An Experiment on the Relationship Between Task Design and Crowdsourcing Performance

Turki Alelyani[1](✉), Paul T. Grogan[2], Yla Tausczik[3], and Ye Yang[2]

[1] Najran University, Najran 61241, Saudi Arabia
tnalelyani@nu.edu.sa
[2] Stevens Institute of Technology, Hoboken, NJ 07030, USA
{pgrogan,ye.yang}@stevens.edu
[3] University of Maryland, College Park, MD 20740, USA
ylatau@umd.edu

Abstract. Software crowdsourcing platforms allow for wide task accessibility and self-selection, which require participants to interact with a wide range of task options. This increased accessibility enables crowd workers to freely choose tasks based on their skills, experience and interests. However, the inconsistencies in crowd workers' comprehension and interpretation of the designs of different tasks and the platform interaction strategies may introduce task uncertainty into the selection process. This may result in low performance, which can diminish the quality of the crowdsourcing platform. Previous studies show that the quality of the outcome of these systems depends on different design parameters in both social and technical contexts. Therefore, through the lens of Socio-Technical Systems, our experimental study quantifies and compares 30 participants' behavior and performance throughout their interaction and selection of different software tasks via an online-based tool named "SoftCrowd," using different design strategies and task features. Results showed a significant relationship between task submission rate and platform design, including interaction strategies, task ambiguity and task instruction design. These findings suggest that our proposed design factors may have unanticipated effects on the quality of crowdsourcing, calling for more studies of crowdsourcing design to inform better strategies for various aspects of platform design.

Keywords: Crowdsourcing · Software development · Task design · Crowd performance · Socio-technical systems

1 Introduction

Numerous crowdsourcing platforms, such as TopCoder, Amazon Mechanical Turk, and Upwork, have successfully attracted talented individuals to participate in crowdsourcing and to produce different products. These platforms had

C. Stephanidis et al. (Eds.): HCII 2020, LNCS 12427, pp. 3–27, 2020.
https://doi.org/10.1007/978-3-030-60152-2_1

also expanded their respective scopes of software development, which could lead to fundamental and distributive changes in the way crowdsourcing software is developed [5]. The crowdsourcing platform's format allows individuals and organizations to outsource their software development tasks to an undefined group of networks so that developers from different backgrounds and countries with diverse expertise can produce high-quality software at reduced costs [18]. However, this format is subject to error, as the crowdsourcing system may have limited context and understanding but powerful contributors. This can be explained by applying design strategies that don't fully support the system goal to produce its desired outcome.

A current challenge in crowdsourcing is how to both engage participants and produce high quality work simultaneously [5]. It is particularly challenging to design engaging "requirements" that are specifiable in sufficient detail without onerous overhead [5]. Crowdsourcing platforms also showed that system design and participants both define and limit the interaction with different features within the system [6]. For example, the crowdsourcing platforms may not reliably support the goal of solving some available tasks, which may cause workers to abandon the system or exhibit low performance. That includes a lack of systematic task design that helps platforms avoid any ambiguity or cognitive loads in the posted tasks, so workers face no difficulty in the development process.

Quality assurance is key to software development, whether developed in house or by external parties [30]. This is particularly true in crowdsourcing, as task requestors know neither the developers nor the process to follow. Furthermore, the quality of the work is related to the design of the task, not merely the quality of the contributors [33]. Task design contains many features: incentives, description layout, complexity, workflow, task clarity, etc. Task descriptions, which must be clear to the requester, can be difficult for workers to understand, and overcomplex interfaces can diminish task result quality. Therefore, improving and changing the task design regarding instructions, clarity and incentives can lead to better results, which can improve the interaction between crowdsourcing tasks and assigned workers.

Nevertheless, the effects of different design techniques in a software crowdsourcing context are not clear. Several studies examined how the design quality in micro-task platforms may influence a participant's behavior, which can maximize or minimize the quality of the delivered work [30]. Kittur et al. show that crowd performance can be related to task design. Clear task description can engage participants, improve worker performance and satisfaction [9]. Upfront task design includes task decomposition, reward, workflow mechanisms, and instructions conducive to high-quality work.

Yet more studies are needed to investigate the influence of different task designs on the level of performance and which task design strategies that can maximize performance. Well-defined studies with specific objectives should be proposed so results can be easily compared for a more coherent understanding of crowdsourcing design [28]. As an initial step toward characterizing different socio-technical conditions in software crowdsourcing, this study quantifies and compares participants' behavior and performance throughout their interaction and selection of different software tasks posted on a web-based tool the authors of this study developed. Using empirical data from the conducted human subjects experiment, we can better understand the relationship between developers' choices of the task based on its instruction design, ambiguity level, and the offered reward and the achieved performance in submitted tasks. Our goal is to motivate and assess new strategies and design mechanisms in the context of software crowdsourcing.

To examine the possible effects of these designs on workers, we performed a human experiment with 30 subjects. Crowdsourcing participants were recruited to perform short, well-defined software tasks with a diversity of formats, reward, and technical and social conditions. More specifically, through a controlled environment we observed participants making their selection decisions under these conditions:

- Two task design themes were introduced: (1) narrative-based tasks (NBT), and (2) descriptive-based tasks (DBT). The proposed task design followed a proposal for open source software requirements [24].
- The proposed tasks varied in ambiguity, which we, again, called choice under uncertainty, in which information is known to be missing or instructions can be interpreted in many ways. All tasks were within a similar range of difficulty, small to medium based on the measurement by HackerRank Platform[1].
- We applied social features to encourage participants to contribute and to see how their contributions would influence their engagement levels. This included regular interaction with some subjects as well as some motivational notes.
- Finally, we assigned different rewards to participants to quantify the influence of rewards on the selection process as well as the participant's overall performance.

At the end of our study, a survey was completed by participants about their experience. This can help us to provide more reasoning about the observed behavior so further studies can be built upon our findings. The remaining of this paper is organized as follows: Sect. 2 discusses the background on software crowdsourcing design, incentives, behavior, and Socio-technical systems. Section 3 introduces our research objective and hypothesis. In Sect. 4, we describe experimental design and analysis. Sections 5 and 6 present results, discussion, and conclusion.

2 Background

2.1 Crowdsourcing Incentive Design

Crowdsourcing design has gained much interest in recent years and has been studied from different perspectives. One main question raised in virtual communities is how people are motivated to contribute and collaborate with others

[1] http://www.hackerrank.com.

voluntarily. Motivation has been extensively studied in virtual communities and found to be effective in contribution and work quality [23]. In terms of crowdsourcing, studies show how extrinsic and intrinsic motivations can improve output quality. Intrinsic motivation occurs when individuals have the curiosity to explore, learn, and understand [22]. When no extrinsic motivation was given (no payment), intrinsic incentive did not play any role in motivating people [37]. In a crowdsourcing context, individuals are intrinsically motivated to contribute and produce some works. Another type of intrinsic motivation is motivation toward accomplishment which occurs when engaging in an activity can result in feeling of achievement. For instance, individuals who feel a personal satisfaction while they are mastering certain coding skills are intrinsically motivated and would result in enjoyable experience [26].

Intrinsic motivation to experience stimulation occurs when someone engages in the experience for excitement or enjoys being engaged in the activity. In paid crowd work, workers are compensated for completing their required tasks [1]. This is often called reward or price; posted along with the task when it is created. Several crowdsourcing platforms give workers monetary incentives in exchange for their project participation: TopCoder (the largest software crowdsourcing platform), Amazon Mechanical Turk, Upwork, Utest etc.

Mason and Watts identified different incentives to improve the quality of results, such as workers being asked to evaluate their peers' work, while financial incentives did not affect the quality [19]. At the same time, Thalmann [31] shows that quality assessment feedback is well received by workers and can be useful for achieving better results. Despite the success of these platforms in attracting talented individuals to participate and produce different products, studies show that crowdsourcing platforms are still challenged to design effective incentive systems for their different tasks, sometimes due to requesters' constraints and the online market setting of a particular project [25]. Thus, calling for more studies to better understand crowd behavior so adequate mechanisms can be designed.

2.2 Crowdsourcing Participants Behavior

Several studies conducted in-depth research in different topics related to participants' behavior on crowdsourcing and its relationship to work quality. For instance, [2] studied dependability of crowd workers in selecting tasks for software crowdsourcing. Reliability ratings categorize workers in terms of the value of rewards and workers with high-reliability ratings are more specific in the types of rewards they select. The study shows evidence that the most reliable workers–the high-reliability group–tend to be more motivated to quickly register for a task and complete it within an optimal time.

Prior work used a game theoretic model to analyze the behavior of the contest outcome on crowdsourcing [4]. This study focused on the design of a single contest; how many prizes should be awarded and the reward value. Another related study by Archak studied the effects of the reputation system currently used by TopCoder on behavior of contestants [3]. The study found individual specific traits together with the project payment and the number of project requirements are significant predictors of the final project quality. Furthermore, the study found significant evidence of strategic behavior of contestants. High rated contestants face tougher competition from their opponents in the competition phase of the contest. To ease the competition, they move first in the registration phase of the contest, signing up early for particular projects. The study concluded that the reputation and cheap talk mechanisms employed by TopCoder have a positive effect on allocative efficiency of simultaneous all-pay contests and should be considered for adoption in other crowdsourcing platforms. A recent study used social network analysis to study the impact of developer behavior and community structure on software crowdsourcing practices. The study found there are temporal bursty patterns of online engagement of TopCoder users which leads to the similarity in the type and level of participation at TopCoder [36].

2.3 Crowdsourcing Task Design

Previous studies show the design of a data collection interface affects the quality and users' contribution [20,25]. While crowd workers perform some assigned tasks, the task interface was found to be an important factor for the delivered work quality. Another related study represented crowdsourcing task processes in a model called task choice decision [9]. One part of the study investigated the relationship between work intention and task presentation. While task can be presented in different formats in terms of its wording style and the length description, the study suggests that different design patterns should be adapted depending on the task type and worker group. However, these influences have not yet been studied.

A study on relatedness and growth in crowdsourcing looked at task design from a different perspective. Moussawi and Koufaris investigated how the tasks' nature and size determine the level of innovation, complexity, and proficiency required to complete them [20]. This study found the primary criterion used for choosing tasks to be completed was a pay vs. time trade off. The study suggests more investigations for the type of incentive mechanisms are needed when designing tasks. Another related work studied the task design, in terms of its instructions and user interface, and how that can also affect the workers' perception of the task [29]. Based on a user experiment, the study shows that complex user interface gives worse results than the simple user interface. Drommi et al. reviewed the issues of cognitive load theory and its relevance for developing interface systems [10]. The study suggests one aspect of cognitive load can address the concept of memory load when users interact with the interface. One way to manage cognitive load is considering users' expertise and avoiding any redundant contents. In crowdsourcing, specifically, Hass et al. defined task quality as a set of components that can measure the quality of a task [13]. These components include time spent on a task, task length, task resources, and price options.

After applying a regression model on reviewed tasks, the results show that previously mentioned factors contribute in measuring the task quality.

2.4 Socio-Technical System

Socio-Technical System (STS) design approach states that system is composed of two main factors that influence its business performance namely technical and social [32]. A socio-technical system is structured mostly around the interdependencies among its different components. A successful system should take into account the properties and the interaction of these two components. Bostrom and Heinen (1977) demonstrated the need to frame any management information system design within the Socio-Technical Systems (STS) design approach. The STS approach is used for redesigning existing work systems as well as for new site designs. In a crowdsourcing setting, a platform is made up of two jointly interacting system – the social and the technical. The technical system is concerned with tasks, technology and design to transform input into output. The social system is concerned with people, skills, values, relationships and reward systems. Thus, any design or redesign of system must deal with both systems in an integrated format. Cartelli (2007) shows that the cornerstone of the socio-technical system approach is that the fit is achieved by a design process aiming at the joint optimization of the subsystems. In other words, any organizational system maximizes performance only if the interdependency of the subsystems is explicitly recognized.

Previous studies described software crowdsourcing as socio-technical system [15]. In the crowdsourcing context, a set of relationships connects individuals, technologies and work activities to constitute the crowdsourcing platform. Specifically, the interaction among task requestors, developers and platforms is key to the overall system's success. These components increase interdependency between system designs and the people who interact and connect throughout the crowdsourcing process.

Prior studies explored five dimensions of risk in a self-reporting study to investigate their effect on crowdsourcing performance. Some of these risks were associated with workers, requestors, relationship, requirements, and complexity of tasks [11].

As a system design can determine participant behavior, we must study the impact of these components on the process, a topic surprisingly unexplored. Using questionnaires to study how social subsystem risk influences crowdsourcing performance, existing research found that social subsystems diminish crowdsourcing performance [35], also revealing that different types of risks interact with each other to influence performance.

Much of the literature shows a relationship among crowdsourcing performance, task design, price, workers' experience, skills, motivation and other crowdsourcing system components [25,38]. Crowdsourcing performance refers to the degree of effective task completion [39]. Based on previous studies, this paper hypothesizes that, given the high technological and social uncertainties in crowdsourcing, crowdsourcing risk is associated with different design aspects in task description, reward, communication, and engagement strategies.

2.5 Research Objectives

One goal of the proposed study is to fill the research gap on how the social and technical conditions that crowd workers encountered–which may have confronted them with some uncertainties–affected crowdsourcing performance. Key elements of the social system are people, social relationships, values and attitudes associated with humans [7], while the technical system includes technology as well as the proposed tasks. Related studies included two factors in the technical system–requirements and complexity [16]–which we explored in our study. Studying social and technical contexts in a controlled setting can improve our understanding of the effect of different design strategies in crowdsourcing platforms, thus help crowdsourcing managers in their decision-making on designs.

To more fully address the aforementioned two components in crowdsourcing systems, our study includes a human subjects experiment using a simplified crowdsourcing platform called "SoftCrowd" to control technical and social conditions among software developers. Our applied social and technical conditions varied among the study participants and were solely based on previous studies' measures to better quantify subjects' performance with regard to each condition. Our study in a controlled environment is expected to reveal how different crowdsourcing design strategies–including task design features, ambiguity level, reward, and developers' interaction–affect developers' actions and performance. Improving our understanding of crowdsourcing design strategies will allow future studies to build upon our findings in improving the existing software crowdsourcing systems, allowing managers to act accordingly. Figure 1 illustrates our research model. There are two main conditions that we hypothesize affect crowdsourcing performance: Technical Conditions and Social Conditions. In Technical conditions, we introduce three factors: instructions design, ambiguity level and task variety. Where in Social Condition, we introduce three factors: interaction level, recognition and reward. The two factors inside the dashed line were only collected through the survey.

To better conceptualize this problem, our study builds upon recent studies of crowdsourcing as a socio-technical system [16,35]. Those studies emphasized risk mechanisms to capture crowd behavior and proposed incentives and designs to produce effective strategies for a better performance. To address the aforementioned research problem, this paper asks:

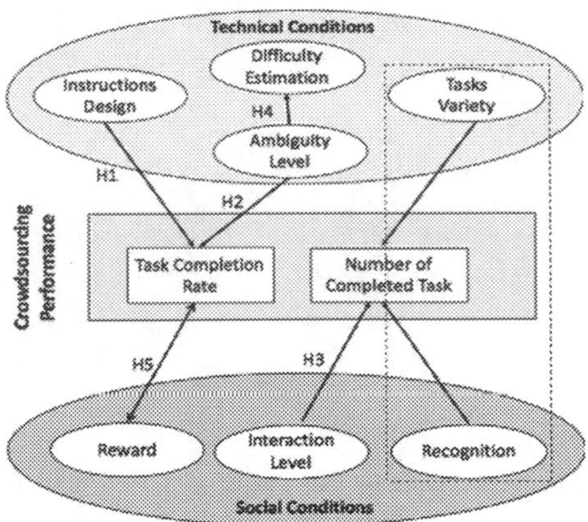

Fig. 1. The research model.

What are the effects of social and technical design conditions on software crowdsourcing performance?

3 Hypotheses

To develop reasonable hypotheses of task design and workers' performance, we draw on Socio-Technical System (STS) design approach to understand and interpret what may influence crowdsourcing performance. Previous studies referred to task design as a technical condition key to crowdsourcing design. In our experiment, we presumed that lack of the systematic design necessary to ensure the posted task's consistency and competency would adversely affect the delivered work, whereas simple, clear instructions may improve participant engagement. A major issue in software requirements is ambiguity [27]; ambiguous requirements can be interpreted in multiple ways. We hypothesize that:

H1. Narrative-based task increases task submission rate where descriptive-based task decreases task submission rate.
H2. High ambiguity level in software task decreases task submission rate.

Complex software crowdsourcing tasks tend to be larger in requirements with more dependencies. This explains why previous research shows motivating workers to accomplish such tasks can be very challenging, as the posted task's design could affect their level of engagement [11]. In addition, prior studies show that mechanisms to provide better interaction with crowdsourcing workers should be addressed as one way to motivate and engage crowdsourcing participants. Therefore:

H3. Interaction with workers during the experiment time increases task submission rate.

Task choice decision in crowdsourcing is the essential self-selection process in which workers engage in the first step of task selection. Potential workers must choose among a number of other tasks, i.e., build their own estimations regarding task worth or complexity, which will affect the accuracy of their decisions. So, we assumed that a well-designed task would give subjects a more accurate estimation of its difficulty, compared to an ambiguous, unclear task. Therefore:

H4. High ambiguity level in software task decreases subjects' accurate estimation of the potential task's difficulty.

Different inferences workers use in choosing tasks–task title, payment, description, task user interface, etc.–affect their decisions as well as the task's cost [25]. A worker's assessment can be misled by lack of careful design in the posted task. Worker experience and skills may also play a role in the selection process. Previous studies show that more experienced and reliable workers tend to choose tasks with higher prices [2]. Accordingly, the selected task will be affected by workers' experience and the posted reward.

H5. The offered reward on software crowdsourcing platforms is positively associated with the task submission rate.

4 Experimental Methodology

To understand the effects of different social and technical conditions on crowdsourcing workers' selection and submission rate of tasks, we conducted a human subjects experiment in which we crowdsourced 15 software engineering tasks via our *SoftCrowd* platform. Our two-week software crowdsourcing experiment was perceived as a series of competitive-based tasks where workers competed with each other to deliver most of the proposed tasks.

SoftCrowd provides tasks to subjects and lets them freely decide on a task. Subjects were not required to do installation. The platform was hosted on our server, which enabled subjects to access the experiment through a web link. There were 15 data structure and design Java programming tasks.

Participants were introduced to the experiment, and to software crowdsourcing in general, in a 20-min session. The following day they received an email with instructions and were asked to sign a consent form. They could choose any task, and each task had certain points associated with it. These points represent range of gift cards for which subjects were entered in a drawing upon completing the experiment if they accumulated 20 points. This procedure gave us sufficient data for each subject for us to draw a meaningful conclusion about their behavior in selection and contribution.

At the end of the experiment, subjects were asked to complete a survey of questions about their experiences and the challenges they encountered during the experiment. The survey comprised 30 questions in three categories: (1) personal information, (2) background information on their experience with software engineering and crowdsourcing, and (3) their experience in and perception of the experiment, including selection motivation, effort, task features, and strategies employed when selecting and solving problems. All survey questions were formed on a five-point Likert scale that ranged from "Strongly disagree" (1) to "Strongly agree" (5). These measurements were adapted from prior study [38].

Survey responses were also used to draw conclusions about subjects' actions during the experiment. Besides the objective data we collected from the experiment, this subjective data will be used to investigate what task features might matter most when selecting a task and what challenges it may present.

4.1 Experimental Design

In this section, we explain the design of our experiment for both social and technical conditions. To examine subjects' selections among different tasks, we considered three factors: (1) the level of ambiguity (2) the offered reward, and (3) the task type. Thus, we could understand subjects' decisions under different levels of either ambiguity, reward, or the task description type and how those decisions related to the recorded performance. Prior studies concluded that crowdsourcing developers tend to make consequential decisions based on relatively little information, which may result in unsuccessful tasks.

Social Conditions. In our experiment we applied one social condition called Interaction, where the survey has two social factors: Task Variety and Recognition. For interaction, we regularly sent notifications to 11 subjects. These notifications contained specific texts that included motivational messages about how rewarding programming can be or following up on their progress during the experiment. These notifications were not designed to help subjects by giving hints, but to engage subjects while they were working on the platform. Variety represents the range of skills, domain knowledge and actions that subject should apply to complete the selected tasks. Tasks with wide range of options may make subjects experience more exciting and engaging. Recognition was designed in alignment with crowdsourcing platforms where public record is published so that crowd workers can be recognized for their achievement.

Technical Conditions. Each task $T_1, \ldots\ldots, T_{15}$ is associated with number of Technical Features:

$$T = [x_1, x_2, x_3]$$

x_1 represents the task ambiguity level, a concept borrowed from traditional software engineering, where ambiguity in requirements can be categorized into two types: (1) linguistic ambiguity; (2) software engineering ambiguity [28].

In our experiment, we manually inject "linguistic ambiguity" to some of the tasks instruction. Some keywords were added within task instruction including:

o *Much smaller, larger*
o *As much time as possible*
o *If you think it would help*

We used a binary measure to describe the ambiguity: low L and high H. Low means there was no intention to include ambiguity in the description where high ambiguity means ambiguity were intentionally added to the description.

$$(x_1) = \begin{cases} 1 \\ 0 \end{cases} \tag{1}$$

x_2 is the reward we assign per each task. Rewards in this experiment represent points between 2 and 6, which represent range of gift cards in the amount of \$40, \$30, and \$20. Subjects who complete the experiment will be entered to win a gift card based on the points they collect. In our experiment, when the task ambiguity level is high, it gets the following points: (4, 5, 6). High score means higher ambiguity.

We mapped rewards to points as follows:

- \$40: Subjects who complete and submit more than 30 points.
- \$30: Subjects who complete and submit points from 21 to 30.
- \$20: Subjects who submit total of 20 points but with some incomplete tasks. x_3 represents how each task's instructions were designed. In this study we employed the two aforementioned conditions for task description design based on prior study [24], narrative-based task and descriptive-based task. In NBT, the task description is more succinctly specified and uses a formal "how to" description that explicitly declares the task requirements. It has three parts: introduction, body, and conclusion.

In contrast, DBT represents an informal task instruction that appears as composition or decomposition of any of the proceeding. Developers may have to try to make sense of it to elicit the proposed requirements.

$$(x_3) = \begin{cases} NBT \\ DBT \end{cases} \tag{2}$$

The description of the study main variables including Dependent Variable (DV) and Independent Variable (IV) is as follows:

1. Subject performance (DV): Subject performance is measured by number of completed tasks per subject.
2. Task submission (DV): This variable measures the successful submission rate each task receives: Number of submitted tasks/Number of available tasks.
3. Task ambiguity (x_1)(IV): Language ambiguity; semantic ambiguity. Binary codes (0,1) represent low and high level of ambiguity per each task description.
4. Reward x_2 (IV): Points 2–6 assigned to different gift cards.
5. Narrative-based tasks (NBT) and Descriptive-based tasks (DBT) x_3 (IV):
   - NBT: Explicit declaration of the requirements as formal HowTo format. Introduction, body, and conclusion.
   - DBT: Informal HowTo formats appear as a selection or composition of the narrative-based tasks.

6. Interaction: (IV) Binary codes (0,1) as low and high interactions with subjects.
7. Task Variety (IV): Posting similar tasks in terms of domain knowledge, the required actions and skills at the same time versus different tasks which require variety of skills.
8. Recognition (IV): The importance of recognition as a motivating factor for each subject.
9. Gender (IV): Whether a subject is male or female.
10. Participant Education Degree (IV): Whether a subject is an undergraduate student or master student.
11. Industry Experience (IV): Subject years of experience as a software engineer.

4.2 Subjects

We recruited participants from software engineering and computer science departments. All were working on undergraduate or graduate degrees. All participants (N = 30, female: 10, male: 20, aged between 20–30) had prior experience in programming in Java through school projects or industry experience. Table 1 summarizes the subjects' demographics.

4.3 Study Metrics

We evaluated the subjects' selection, performance, task submission rate, effort estimation and reward range. Subjects were also asked to complete a short survey based on well-established measures from previous research. This section explains the way we calculate the study variables.

Tasks Submission Rate. The task submission rate $\boldsymbol{r_i}$ is the number of successful submissions per task $\boldsymbol{T}$ divided by the number of available subjects $\boldsymbol{S}$. Where T_c^i denotes number of submissions for task $\boldsymbol{i}$.

$$r_i = \frac{T_c^i}{s} \tag{3}$$

Subjects Performance. Given the small range of difficulty of the proposed tasks, we measured subject $\boldsymbol{s}$ performance by the sum of their submitted tasks T_c^i, as Eq. 4 shows.

$$\boldsymbol{T_c^s} = \sum_{i=1}^{T} z_i^s \tag{4}$$

Where

$$Z_i^S = \begin{cases} 1, & \text{if } s \text{ submites task } i \\ 0, & \text{otherwise} \end{cases} \tag{5}$$

Effort Estimation. Effort estimation evaluates a task's difficulty level and the effort required to complete it. Subjects were asked to enter an initial difficulty estimation E_{start} before they began working on the task and after they finished it E_{end}. Thus, we could measure their ability to evaluate any potential tasks, given controlled variables such as level of ambiguity. The difficulty estimation ranges between 1 to 10. On a software crowdsourcing platform, its developers' selection process can be influenced by both a task's clarity and their ability to effectively measure its difficulty level. In addition, we presumed that developers who chose to opt out from any task on any platform–TopCoder, for instance–after agreeing to work on the task did so because of the task's level of clarity. We applied the following formula to capture our subjects' accuracy estimation:

$$E = E_{end} - E_{start} \tag{6}$$

Reward Selection. A different task reward R_i, ranging from 2 to 6, was set for each task. Subjects were asked to solve no fewer than 20 points' worth of tasks in total. This allowed us to glean more meaningful findings about subjects' behavior in choosing among different tasks. We applied the following formula to capture the reward selection R_s for each subject based on the submitted task.

$$R_S = \frac{\sum_{i=1}^{T} R_i \cdot z_i^s}{\sum_{i=1}^{T} z_i^s} \tag{7}$$

Subjects Interaction. To encourage subjects to contribute and maybe increase their engagement levels, some subjects (N = 11) were randomly selected to receive some motivational messages during the experiment. Equation 7 shows a binary measure: 0 (low) if the subject received no messages, and 1 (High) if the subject received a message. The experimenter interacted with the selected 11 subjects 4 to 6 times. During these interactions *I*, subjects were asked about their progress and whether their tasks were unclear to them in any way. Subjects also received motivational texts to encourage them to continue their efforts to complete their tasks.

$$I(i) = \begin{cases} 1 \\ 0 \end{cases} \tag{8}$$

Tasks Variety. Variety represents the range of skills, domain knowledge and actions that subject should apply to complete and submit the selected tasks. Tasks with wide range of options may make subjects experience more exciting and engaging. Prior study argues that if crowd task was designed in such a way that a crowd worker needs to perform repetitive actions and fewer skills, that will negatively affect theory intrinsic motivation [38]. For this metric, we only analyzed the qualitative data which we received from the survey.

Recognition. Recognition was designed in alignment with crowdsourcing platforms where crowd records are publicly available so that crowd workers can be recognized for their achievement. This metric used to measure motivation in crowdsourcing platforms.

5 Analysis Method

Our analysis method follows two steps:

1. Based on collected experimental data, we conducted statistical tests to determine whether socio-technical conditions affected crowdsourcing performance. Based on the experimental data, we examined four independent variables: (1) narrative or descriptive, (2) ambiguity level, (3) reward, (4) level of interaction. The dependent variables were the task submission rate and subjects' performance; the number of completed tasks. To do so, we applied mixed effects logistic regression model using glmer method in lme4 package in R. We used this method because our dependent variable, task completion, is binary and we were using a fully crossed design in which multiple observations were taken per participant and per task. Thus, in our first model we controlled for both participantID and taskID. To further examine the effects of participants features such as Gender, Graduate Vs. Undergraduate and Experience, our second model controlled for participantID. This allowed us to address H1, H2, H3, H4, H5. In addition, we tested for possible interactions effects between the proposed variables.
2. We collected the subjects' responses in a survey format, which was designed on social and technical measures borrowed from previous study [38]. As Fig. 1 shows, two variables in the dashed line were added to the previously mentioned model to make 6 factors collected through survey. The open-ended questions asked subjects for their subjective opinions about the relationship among task design, communication, reward, and their efforts to submit different tasks.

To analyze the collected survey data, we applied the Partial Least Squares (PLS) structural modeling equation (SME) method to study the relationship among the variables in the proposed model. This approach was used in relevant studies and was proven suitable for our small size sample [12].

We assessed the proposed model variables by calculating the composite reliability (CR), the average variance explained (AVE), the average shared variance (ASV), discriminant validity (DV), and Cronbach's α for each of the variables in the model. Finally, we reported path significance, p-values, and R^2 to determine whether we should accept or reject the proposed hypothesis.

Table 1. Demographic characteristics of study sample (N = 30)

	Characteristics	Frequency
Gender	Male	20
	Female	10
Age	20–30 years	30
Current level of study	Master	11
	Undergraduate	19
Field of study	Computer Science	15
	Software Engineering	15
Is english your first language?	Yes	14
	No	16
Lines of code (prior experience)	1000–2000	18
	2000–4000	4
	4000–6000	4
	6000–8000	4

6 Results

6.1 Effect Task-Participant Characteristics and Requester Interaction on Task Completion

On average participants submitted 6.1 out of 15 tasks (Median = 6, SD = 1.2). In order to understand how technical and social characteristics of the task design and platform affected task completion, we constructed a mixed effects logistic regression model, controlling for participant and task. We first constructed a model with main effects for: task instructions (narrative vs. descriptive), task ambiguity (low vs. high), requester interaction (no messages vs. messages) and reward amount (2–6). Table 2 shows the results of the model. We found a significant effect of task instructions on task completion (Coef. $= 1.44, \mathrm{SE} = 0.25, \mathrm{p} < 0.0001$). Participants completed more narrative-based instructions (M = 0.51; SD = 0.14) than a descriptive-based instruction (M = 0.16; SD = 0.10). These results suggest that individuals preferred tasks with explicit requirements and/or found them easier to complete (Table 3).

Next, we examined the effect of task ambiguity. We found a significant difference in the task completion based on the degree of ambiguity in the task requirements (Coef. = −1.27, SE = 0.40, p = 0.002). Tasks which had low ambiguity in their requirements were completed at higher rates (M = 0.46; SD = 0.21), than those with higher ambiguity in their requirements (M = 0.21; SD = 0.11). These results suggest individuals preferred and/or found tasks easier to complete when the requirements were less ambiguous.

Surprisingly we found no significant effect of reward on task completion, whether reward was included in a model with or without the other variables

Table 2. Mixed effect logistic regression model predicting task completion controlling for participant and task (Observations = 450, Users = 30, Tasks = 15)

Factor	Coefficient	SE	z	p
Intercept	−2.00	0.60	−3.36	<0.0001
Task instructions	1.44	0.25	5.86	<0.0001
Task ambiguity	−1.27	0.40	−3.16	0.002
Reward	0.27	0.17	1.56	0.12
Requester interaction	0.69	0.22	3.14	0.002

Table 3. Proportion of tasks completed grouped by variable

Variable	Level	Mean	SD
Task instructions	Descriptive-based	0.161	0.102
	Narrative-based	0.512	0.140
Ambiguity	Low	0.466	0.219
	High	0.214	0.118
Requester interaction	No messages	4.58	1.774
	Messages	7.27	1.191

(Coef. = 0.27, SE = 0.17, p = 0.12). We expected that reward might be a motivating factor and perhaps compensate for other characteristics of tasks that might make them more difficult, such as more ambiguity, however there was no evidence that reward increased task completion.

We examined the impact of requester interaction on task completion. We also found a significant effect of requester interaction on task completion (Coef. = 0.69, SE = 0.22, p = 0.002). Individuals were more likely to submit tasks when they received encouraging messages from the requester (M = 0.49, SD = 0.50) than when they received no messages (M = 0.35, SD = 0.48). This suggests that interacting with an encouraging requester can motivate more work.

We also investigated whether there were any significant interactions between the technical conditions–task instructions, task ambiguity, reward–and task completion. We constructed mixed effects logistic regression models predicting task completion controlling for participant and task, in these models we considered all possible two-way and three-way interactions. We found no significant three-way interaction (Coef. = 0.28, SE = 0.78, p = 0.72) and no significant two-way interactions between task instructions and ambiguity (Coef. = −0.07, SE = 0.49, p = 0.89), task instructions and reward (Coef. = 0.03, SE = 0.22, p = 0.89), ambiguity and reward (Coef. = −0.48, SE = 0.34, p = 0.16).

Finally, we constructed a model with main effects as well as subjects' characteristics including gender, experience, as well as education degree (Undergraduate VS. Graduate). Table 4 shows the results of the model. We found no significant effect of participants' gender (Coef. = −0.004, SE = 0.33, p = 0.984), industry experience, and education degree (Coef. = −0.08, SE = 0.22, p = 0.712) on task completion. These results suggest that individuals' characteristics did not significantly affect their performance. Future research on real crowdsourcing with more data may investigate this further.

Table 4. Mixed effect logistic regression model predicting task completion controlling for participant (Observations = 450, Users = 30, Tasks = 15)

Factor	Coefficient	SE	z	p
Intercept	−1.73	0.55	−314	0.001
Task instructions	1.42	0.21	6.51	<0.0001
Task ambiguity	−1.21	0.36	−3.37	0.0007
Reward	0.21	0.15	1.42	0.15
Requester interaction	0.63	0.23	2.67	0.007
Gender	−0.004	0.33	−0.02	0.984
Industry experience	−0.008	0.193	−0.046	0.963
Education degree	−0.08	0.22	−0.36	0.712

6.2 Effects of Task Characteristics on Perception of Task Difficulty

Participants assessed the difficulty of each task before and after completion, which helped us to understand their assessment dynamics during task selection. Their accuracy in estimation may affect their decision whether to proceed on a particular task. In this experiment, we tried to examine how the ambiguity level per each task may affect level of accuracy when subjects assess different tasks. After applying an independent sample t-test on the two groups as Table 5 shows, we found that ambiguity does play a role in the accuracy estimation for low ambiguous tasks (M = 1.87, SD = 2.42) and high ambiguous tasks (M = −4.71, SD = 3.73). These results suggest that ambiguity level increases the level of uncertainty or biases during the selection decision (t (10) = 3.99, p = 0.001).

Table 5. Ambiguity and subjects' estimation accuracy of task difficulty

Ambiguity	N	Mean	SD	P-value
L	8	1.87	2.42	2* 0.001
H	7	−4.71	3.73	

6.3 Effects of Self-reported Task Perceptions on Task Effort

As mentioned, subjects' responses were collected after experiment completion. The questionnaire was constructed according to the socio-technical system approach that adapts measures to assist results from experimental studies. The measures we used in the questionnaire–task design features, reward, communication–fell into two categories: technical condition (TC) and social condition (SC).

To analyze the data, we applied Partial Least Squares (PLS) to study the effect among the variables in the proposed model. To establish the reliability and validity of our construct measurements before their use in the model, we assessed the model variables by examining cross-loadings, composite reliability (CR), Cronbach's α, average variance extracted (AVE), internal consistence, discriminant validity, and convergent validity.

Item loading was greater than 0.7 in most measures. Few cases experienced low loadings, but due to the size of the collected data we did not remove the low loadings from the data. Table 6 represents the results of CR, Cronbach's α, and average variance (AV) for each variable in the model. Regarding AV, our model variables met the AVE threshold, which was greater than (0.500), for all proposed variables. To establish the reliability, our model variables also met the threshold of CRs, as all factors were greater than 0.700. After conducting these analyses, we concluded that our model variables were sufficiently valid to be constructed in the model.

Table 6. Descriptive statistics, reliability, and AE

Variable	Cronbach' α	CR	AE
Design	0.820	0.874	0.589
Social	0.618	0.802	0.587
Effort	0.873	0.940	0.888

After we examined the collected measures and their validity, we constructed our model, which was aligned well with the initial one. Two factors affected subjects' efforts to contribute to the crowdsourcing experiment: TC and SC. In TC, we considered three measures: (1) task clarity, (2) presentation of instructions, and (3) variety of available tasks. In SC, we considered three measures: (1) the interaction during the experiment, (2) the reward, and (3) the recognition from the community (Table 7).

After ensuring validity, we examined that path significance of the hypothesized relationship in our proposed model, using PLS. Figure 2 presents the results of the structural model that includes the path coefficients, their significant level, and R2. R2 indicates the explained variance, which was 75%. As Fig. 2 shows, TC and SC positively influenced subjects' task-completion efforts. The path coefficient of TC was (0.50) ($P < 0.01$), whereas that of SC was (0.35) ($P < 0.1$). These findings suggest that both TC and SC influenced subjects' behavior, selection decisions and performance.

Table 7. Mixed effect logistic regression model predicting task completion controlling for participant (Observations = 450, Users = 30, Tasks = 15)

Variables	Coefficient	SE	Z	P
Intercept	−1.73	0.55	−314	0.001
Task instructions	1.42	0.21	6.51	<0.0001
Task ambiguity	−1.21	0.36	−3.37	0.0007
Reward	0.21	0.15	1.42	0.15
Requester interaction	0.63	0.23	2.67	0.007
Gender	−0.004	0.33	−0.02	0.984
Industry experience	−0.008	0.193	−0.046	0.963
Education degree	−0.08	0.22	−0.36	0.712

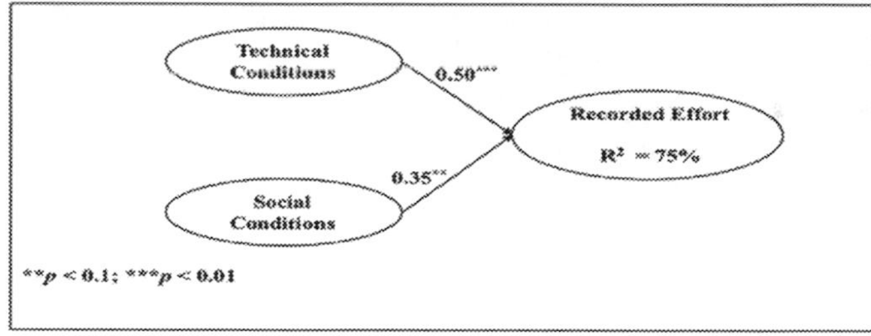

Fig. 2. The structure model.

7 Discussion

7.1 Interpretation of Results

Our research established five hypotheses based on the socio-technical System (STS) design approach. We sought to examine how different social and technical conditions may influence subjects' selections and then task submission rates in a real experiment. Four hypotheses out of five were supported at different significant levels.

Our analysis suggests distinct differences in the effectiveness of diverse task instruction designs. The narrative-based tasks gave subjects a clear understanding of task requirements by clarifying the instructions and offering clear outputs and constraints. The applied test on the collected data indicates a significant difference between the submission rates for narrative and those for descriptive tasks revealing that narrative tasks may have influenced subjects' decisions.

Our findings are consistent with previous studies of the design of project requirements in open-source platforms [24]. Our study denoted two types of task instructions: formal and informal design. The formal design includes explicit declarations of its purpose as a "how to" and may be called a system tutorial. Subjects found this type of format more engaging and helpful in their decision whether or not to complete a task. This is significant, given some major challenges software crowdsourcing workers face in crowdsourcing platforms: how to make unbiased decisions in task selection. This challenge is evident when workers start a task but withdraw later because of issues beyond their control [38].

One of the subjects provided the following feedback regarding the challenge he or she faced:

"clearer descriptions, test code or input to ensure its working right would be helpful, and the super long-winded ones just confused me more than helped."

Another subject provided the following comment regarding their experience with some of the tasks they were not able to solve:

"designer can provide more clearly description of the questions."

Clearly there was less satisfaction with some of the descriptions which contributed to less engagement with those tasks. This calls for more thorough research into crowdsourcing design regarding techniques for more systematic reasoning of internal consistency, completeness or correctness of task instructions before the task is posted to the crowd.

Our findings demonstrate that most subjects are risk-averse when considering each task's ambiguity level. This conclusion is evident where tasks with low ambiguity received 63% of the submissions. We denote tasks with higher ambiguity as riskier tasks. Therefore, subjects who choose to complete these tasks may take more risks, especially if the subject did not ask for clarification beforehand. The applied independent t-test shows significant difference between the average submission rates for low-ambiguity tasks and high-ambiguity tasks. This observation indicates that tasks with unclear or inconsistent requirements contribute to less engagement in completing them. We also found that increasing incentive for unclear tasks did not increase the quality of their submission. This finding is consistent with previous studies showing that increasing task costs did not affect the quality of the submitted work [34]. Subjects who undertook unclear tasks tended to experience deficiency in their solutions, by either introducing more requirements or missing some required inputs. Previous studies showed how complex crowdsourcing task instruction can increase cognitive demands on workers, thus hampering their performance [11]. Future research can explore different pricing mechanisms to improve developers' decision during the selection process, so it could reduce any potential issue.

Our results also found that ambiguity contributed to less accurate estimation of task difficulty. Subjects who tackled tasks with more ambiguity tended to show a big variance between their first and final estimations. This often happens when, during the selection process, inaccurate estimation of a task may lead to withdrawal from it. The applied test shows a significant difference in the average estimations of difficulty in low-ambiguity tasks and high-ambiguity tasks. Findings from our experimental data and subsequent survey show that the design of task instructions, including their wording, affects subjects' engagement in tasks, hence their success. This also suggests the need for further work on the design and implementation of tasks in crowdsourcing.

Our results further show that our communication and interaction with crowd participants significantly influenced their engagement and contribution levels.

Notifications, including motivational statements and encouragement of more participation via messages sent during the experiment, further increased participant engagement. This finding is consistent with prior studies on communication and relationship as mitigators of social and technical risks [35]. This phenomenon may also be attributed to different practices in traditional software engineering development, where project managers ensure that the software development process functions as intended. This suggests that effective interaction with crowd workers would help as a design strategy for more effective task outcome.

Finally, our results indicate a negative correlation between tasks with high reward and their respective submission rates. As mentioned, tasks with high ambiguity were given high rewards in our experiment, thus subjects were more risk-averse in their selection decisions. The applied correlation found a negative correlation between the task rewards and submission rates. This result is consistent with previous findings that reward increases did not improve the quality of crowd work, suggesting that future studies should examine effective pricing mechanisms that can overcome subjects' biases during their task selection process [19].

Implications for Design. Our results show that socio-technical system design approach explains different crowdsourcing ecosystem conditions that may affect subjects' behavior in the task selection process as well as their engagement in and contribution to their tasks. Technical risks, specifically in task instructions and overall design, significantly affect crowd decision in task selection or even pursuit. This finding suggests that social and technical factors need to be framed as part of crowdsourcing platform design. Applying these two factors should also ensure the validity of the developed platform.

Our findings may explain some challenges in crowdsourcing platforms, where some workers opt out from registered tasks, thus sometimes fail them. Also, our exploration of the two types of instruction design–descriptive and narrative–shows that the way instruction is designed and worded engages subjects, thus contributes to task success. This calls for more effort to find ways to prepare crowdsourcing tasks, from the decomposition process to preparing a clear task.

The effect of ambiguity level per each task significantly contributes to task submission rate, as a lower level of description accuracy in crowd tasks could undermine their success, because ambiguous requirements lead to confusion, wasted effort, and rework. So task requirements must be preprocessed to identify ambiguous sentences in each task description and thus improve description accuracy. To achieve this, previous research suggested natural language processing (NLP) to extract requirements from the document, tag their sentences, find duplicate requirements, translate the machine, and extract ambiguous requirements [27]. A related study suggests two ways to deal with software requirement ambiguity: (1) use a constrained natural language (a subset of natural language ruled by an underlying ontology) to write specifications; and (2) use a linguistic-driven inspection of the requirements [11]. So the question remains: How could software crowdsourcing platforms improve their current design?

Our crowdsourcing research problem can also draw on concepts from organizational behavior research [14], specifically the formal-informal coordination. In crowdsourcing context, most workflows are akin to the formal coordination processes and artifacts used in organizations. However, from our findings and previous studies, we argue that crowdsourcing systems should support formal and informal coordination to respond to the challenges crowd workers face while working on tasks [21]. For instance, in our experiment we show that interaction with subjects helped to clarify some of their confusions and encourage them to make more effort to complete their tasks. An experiment designer in this study who interacted with a group of participants is analogous to a software engineering development project manager that can solve any problem during development. In this way future crowdsourcing platforms should consolidate the gap between software crowdsourcing and aspects of traditional software engineering development. This would integrate aspects of software engineering and organizational behavior into a more effective crowdsourcing design.

Recent research into crowdsourcing workflow and coordination suggests that a small design change could give the crowd "panic button" anxiety when the workflow or task design is preventing them from completing the task [21]. From our study and this finding, we conclude that establishing a communication channel between crowd workers and their project manager would mitigate potential risk. An early interaction with the platform may allow for quick inspection and repair of the task description or workflow.

A prior study [15] showed that some crowdsourcing challenges could be mapped to coordination dependencies identified by [17] that also apply to human organizations. The study discusses two categories of overlap among coordination dependencies, their analogs in distributed computing, and their implications to crowd work: (1) managing shared resources that ensure task completion regardless of any opt-out during the development time, and (2) managing producer/consumer and task/sub-task relationships. These conditions can be observed in the execution of complex and interdependent tasks.

The coordination theory justifies some of our findings, suggesting that software crowdsourcing platforms must adapt more transparent workflow in a team format. This can be applied to both competitive and collaborative crowdsourcing models in which the project manager can monitor task-completion progress in a more transparent mode that includes the current workers in the project, their progress, the level of interdependency among their tasks, and the timeline for the entire project. Such a format would add more responsibility on crowd workers while they interact with the rest of the group and allow a project manager to proactively respond to any risk.

Hierarchy, specifically how subjects' interaction with the experiment designer facilitated their contribution, is the primary management strategy in traditional organizations. It benefits coordination, decision-making and quality control, and assigns incentives and sanctions [8]. As discussed, the task design can allow workers to act more like teams, e.g., in accountability standard development, decision-making, and conflict resolution [15].

8 Limitations

This study has a few limitations which must be considered while interpreting and generalizing results. First, the study was conducted in a university setting rather than a crowdsourcing platform, and subjects had limited experience in freelancing or crowdsourcing compared to most software developers. Nonetheless, analysis of this study's data reveals interesting results conducive to further understanding of software crowdsourcing system design. Future research should carefully implement these findings in a larger crowdsourcing setting. Second, the specific type of programming problems, reward mechanisms, and task difficulty levels likely influence the resulting behavior of subjects in the study. To improve the generalization of results, future studies should apply the existing study method on real crowdsourcing platform data, including task descriptions and selections and their results. Third, despite our attempt to recruit subjects within similar experience ranges, unknown factors may have introduced some variances we did not control. Future studies would apply different metrics in controlling the level of experience for crowdsourcing subjects. Finally, regardless of our effort to recruit more subjects, the study's sample is relatively small, which may have limited the generalizability of the results. However, this study closely followed the software crowdsourcing platform format.

9 Conclusion

This study examined the relationship between task design in both social and technical conditions and crowd performance in the context of software crowdsourcing. By integrating socio-technical system design approach, the study investigated the influence of different design conditions on subjects' behaviors. It revealed empirical evidence that social and technical conditions influence crowd workers' task selection, engagement and contribution. These findings reveal that different types of risks affect crowdsourcing selection and performance in different ways. Our study was one of the few that applied socio-technical system in a real experiment within the context of software crowdsourcing. Our contribution would encourage future crowdsourcing researchers and practitioners to develop more systematic approaches to cope with risks in crowdsourcing based on pertinent organizational behavior and coordination theories.

Several extensions of this study are left for future research, which can investigate the effects of social and technical conditions on a more granular level. This can be done by using data from software crowdsourcing platforms with diverse workers, problems and difficulty levels. Future studies could also examine the influences of different user interfaces on crowd engagement and contribution, including gamification and unified modeling language techniques for task description. Furthermore, more qualitative studies of real crowdsourcing workers should be considered, including interviews and focus groups. Such studies would reveal important information regarding these workers' motivation, selection, and contribution. Future study could also propose collaborative models to investigate how they differ from our method; for example, team progress shared among the members in a more transparent model.

References

1. Alelyani, T., Mao, K., Yang, Y.: Context-centric pricing: early pricing models for software crowdsourcing tasks. In: Proceedings of the 13th International Conference on Predictive Models and Data Analytics in Software Engineering (2017)
2. Alelyani, T., Yang, Y.: Software crowdsourcing reliability: an empirical study on developers behavior. In: Proceedings of the 2nd International Workshop on Software Analytics, pp. 36–42 (2016)
3. Archak, N.: Money, glory and cheap talk: analyzing strategic behavior of contestants in simultaneous crowdsourcing contests on TopCoder.com. In: Proceedings of the 19th International Conference on World Wide Web, pp. 21–30 (2010)
4. Archak, N., Sundararajan, A.: Optimal design of crowdsourcing contests. In: ICIS 2009 Proceedings, p. 200 (2009)
5. Begel, A., Herbsleb, J.D., Storey, M.A.: The future of collaborative software development. In: Proceedings of the ACM 2012 conference on Computer Supported Cooperative Work Companion, pp. 17–18 (2012)
6. Bernstein, M.S.: Crowd-powered systems. KI-Künstliche Intelligenz **27**(1), 69–73 (2013)
7. Bostrom, R.P., Heinen, J.S.: MIS problems and failures: a socio-technical perspective, Part II: the application of socio-technical theory. MIS Q. **1**, 11–28 (1977)
8. Coase, R.H.: The nature of the firm. Economica **4**(16), 386–405 (1937)
9. Dow, S., Kulkarni, A., Klemmer, S., Hartmann, B.: Shepherding the crowd yields better work. In: Proceedings of the ACM 2012 Conference on Computer Supported Cooperative Work, pp. 1013–1022 (2012)
10. Drommi, A., Ulferts, G.W., Shoemaker, D.: Interface design: a focus on cognitive science. In: The Proceedings of ISECON 2001, vol. 18 (2001)
11. Finnerty, A., Kucherbaev, P., Tranquillini, S., Convertino, G.: Keep it simple: reward and task design in crowdsourcing (2013)
12. Gefen, D., Rigdon, E.E., Straub, D.: Editor's comments: an update and extension to SEM guidelines for administrative and social science research. MIS Q. **35**, iii–xiv (2011)
13. Haas, D., Greenstein, M., Kamalov, K., Marcus, A., Olszewski, M., Piette, M.: Reducing error in context-sensitive crowdsourced tasks. In: First AAAI Conference on Human Computation and Crowdsourcing (2013)
14. Hackman, J.R., Lorsch, J.: Handbook of Organizational Behavior. Prentice-Hall, Englewood Cliffs (1987)
15. Kittur, A., et al.: The future of crowd work. In: Proceedings of the 2013 Conference on Computer Supported Cooperative Work, pp. 1301–1318 (2013)
16. Liu, S., Xia, F., Zhang, J., Wang, L.: How crowdsourcing risks affect performance: an exploratory model. Manage. Decis. **54**, 2235–2255 (2016)
17. Malone, T.W., Crowston, K.: The interdisciplinary study of coordination. ACM Comput. Surv. (CSUR) **26**(1), 87–119 (1994)
18. Mao, K., Yang, Y., Wang, Q., Jia, Y., Harman, M.: Developer recommendation for crowdsourced software development tasks. In: 2015 IEEE Symposium on Service-Oriented System Engineering, pp. 347–356. IEEE (2015)
19. Mason, W., Watts, D.J.: Financial incentives and the "performance of crowds". In: Proceedings of the ACM SIGKDD Workshop on Human Computation, pp. 77–85 (2009)
20. Moussawi, S., Koufaris, M.: Working on low-paid micro-task crowdsourcing platforms: an existence, relatedness and growth view (2015)

21. Retelny, D., Bernstein, M.S., Valentine, M.A.: No workflow can ever be enough: how crowdsourcing workflows constrain complex work. Proc. ACM Hum.-Comput. Interact. **CSCCSCW**(1), 1–23 (2017)
22. Rogstadius, J., Kostakos, V., Kittur, A., Smus, B., Laredo, J., Vukovic, M.: An assessment of intrinsic and extrinsic motivation on task performance in crowdsourcing markets. In: Fifth International AAAI Conference on Weblogs and Social Media (2011)
23. Rymill, S.J., Dodgson, N.A.: A psychologically-based simulation of human behaviour. In: TPCG, pp. 35–42. Citeseer (2005)
24. Scacchi, W.: Understanding the requirements for developing open source software systems. IEE Proc. Softw. **149**(1), 24–39 (2002)
25. Schulze, T., Krug, S., Schader, M.: Workers' task choice in crowdsourcing and human computation markets (2012)
26. Schulze, T., Seedorf, S., Geiger, D., Kaufmann, N., Schader, M.: Exploring task properties in crowdsourcing-an empirical study on mechanical turk (2011)
27. Shah, U.S., Jinwala, D.C.: Resolving ambiguities in natural language software requirements: a comprehensive survey. ACM SIGSOFT Softw. Eng. Notes **40**(5), 1–7 (2015)
28. Slivkins, A., Vaughan, J.W.: Online decision making in crowdsourcing markets: theoretical challenges. ACM SIGecom Exchanges **12**(2), 4–23 (2014)
29. Stewart, O., Lubensky, D., Huerta, J.M.: Crowdsourcing participation inequality: a scout model for the enterprise domain. In: Proceedings of the ACM SIGKDD Workshop on Human Computation, pp. 30–33 (2010)
30. Stol, K.J., Fitzgerald, B.: Two's company, three's a crowd: a case study of crowdsourcing software development. In: Proceedings of the 36th International Conference on Software Engineering, pp. 187–198 (2014)
31. Thalmann, D.: The foundations to build a virtual human society. In: de Antonio, A., Aylett, R., Ballin, D. (eds.) IVA 2001. LNCS (LNAI), vol. 2190, pp. 1–14. Springer, Heidelberg (2001). https://doi.org/10.1007/3-540-44812-8_1
32. Trist, E.: The evolution of socio-technical systems: a conceptual framework and an action research program. Ontario Ministry of Labour (1981)
33. Vaish, R., Organisciak, P., Hara, K., Bigham, J.P., Zhang, H.: Low effort crowdsourcing: Leveraging peripheral attention for crowd work. In: Second AAAI Conference on Human Computation and Crowdsourcing (2014)
34. Wu, H., Corney, J., Grant, M.: Relationship between quality and payment in crowdsourced design. In: Proceedings of the 2014 IEEE 18th International Conference on Computer Supported Cooperative Work in Design (CSCWD), pp. 499–504. IEEE (2014)
35. Xia, F., Liu, S., Zhang, J.: How social subsystem and technical subsystem risks influence crowdsourcing performance. In: PACIS, p. 222 (2015)
36. Zhang, H., Wu, Y., Wu, W.: Analyzing developer behavior and community structure in software crowdsourcing. In: Kim, K.J. (ed.) Information Science and Applications. LNEE, vol. 339, pp. 981–988. Springer, Heidelberg (2015). https://doi.org/10.1007/978-3-662-46578-3_117
37. Zhao, Y.C., Zhu, Q.: Effects of extrinsic and intrinsic motivation on participation in crowdsourcing contest. Online Inf. Rev. **38**, 896–917 (2014)
38. Zheng, H., Li, D., Hou, W.: Task design, motivation, and participation in crowdsourcing contests. Int. J. Electron. Commer. **15**(4), 57–88 (2011)
39. Zhu, H., Djurjagina, K., Leker, J.: Innovative behaviour types and their influence on individual crowdsourcing performances. Int. J. Innov. Manage. **18**(06), 1440015 (2014)

Implementation of Descriptive Similarity for Decision Making in Smart Cities

Maryna Averkyna[1,2(✉)]

[1] Estonian Business School, A. Lauteri, 3, Tallinn, Estonia
maryna.averkyna@oa.edu.ua
[2] The National University of Ostroh Academy, Seminarska, 2, Ostroh, Ukraine

Abstract. The paper deals with forming the descriptive similarity based on algorithm and a computer program for the decision making support in order to select the suitable solution for implementation from portfolio of the existing experiences related to public transport from various cities, especially in Smart Cities. It helps to satisfy needs in six fields of Smart City and to form rapid decisions for the problems solving. The deeper focus of the work is to develop tools for supporting a decision-making process in which computer systems and people inevitability to participate together. People will not be able to process and analyze the required amounts of data within the required time. However, computers cannot, in principle, decide for humans what to consider as equivalent, what is appropriate and inappropriate for humans. This work focuses on those aspects, which are related to the numerical evaluation of similarity that are needed to make decisions based on analogy, higher prevision of descriptions.

Keywords: Descriptive similarity · Smart City · Information system · Algorithm · Application · Software

1 Introduction

Implementation of information technologies for cities' management is important. The United Nations (UN) forecasts that 6.5 billion people will live in cities by 2050 [7]. Rapid urbanization process due to uptake peri-urban land and towns by cities. Peri-urban land and towns require the relevant management of urban transportation system. One of the areas of implementation making decisions is based on the similarity of situations. Solutions are made of possible consequences. In other words, the local government first tries so far to find out situations and developments similarity and the make a decision to implement or not an approach or solution. Solutions used in similar situations elsewhere are then examined to eliminate "bad solutions" and the "best one" is chosen. Therefore it is crucial to use experience of towns in order to manage transportation in cities. Also, that is why it is relevant to use information technologies and concept Smart city with the purpose decision-making process. Li, Zhu, and Wang pointed out that 'many cities are focusing their efforts to become 'smarter' by employing Information and Communication Technologies (ICT) to improve various aspects of city operation and management, including: local economy, transport, traffic management, environment, quality of life for citizens, and electronic delivery of public

C. Stephanidis et al. (Eds.): HCII 2020, LNCS 12427, pp. 28–39, 2020.
https://doi.org/10.1007/978-3-030-60152-2_2

services' [12]. Elvira Ismagilova, Laurie Hughes, Yogesh K. Dwivedi, K. Ravi Raman emphasized that 'smart cities employ information and communication technologies to improve the quality of life for its citizens, the local economy, transport, traffic management, environment, and interaction with government. Due to the relevance of smart cities to various stakeholders and the benefits and challenges associated with its implementation, the concept of smart cities has attracted significant attention from researchers within multiple fields, including information systems' [3]. Smart city can help to solve the problem in specific areas such as transport and mobility, building technologies, cities' management.

Nevertheless, it is important to investigate how Decision Making Process and Descriptive Similarity can be implemented in the concept of Smart City.

The aim of the paper is to create application (algorithm) for Smart City in order to calculate Systems' Similarity. Research question of the paper is how the formal (Logic) approach can be used in order to the creation table of statements (as predicate calculus formulas) and table comparisons for application of Smart City.

In the present paper, the author will discuss first of all the Smart City concept and importance of the Descriptive Similarity for Decision Making Process implementation. Then we will study application of the logic approach for calculation of the descriptive similarity between small towns and their public transport systems. We will present logic approach for preparation table of statements and table of comparison. Then we will provide the algorithm as application Smart City for calculation coefficient similarity, as well as discussion and conclusions.

2 The Smart City Concept

The concept of the smart city dates back to the 1960s as a cybernetically planned city [11]. Hsi-Peng Lu, Chiao-Shan Chen, Hueiju Yu Hsi-Peng Lu, Chiao-Shan Chen, Hueiju Yu pointed out that 'the smart city concept began in 2010, with increased academic research and policy discussion, particularly in the EU' [6]. Boyd Cohen proposed the Smart City Wheel as an indicator framework used to provide a broad and deep assessment of smart city attributes. The framework encompasses the six fields of smart city development, namely Smart Government, Smart Economy, Smart Environment, Smart Mobility, Smart People, and Smart Living [1].

In the report 'Mapping smart cities in the UE, European Parliament. Directorate-General for Internal Policies. Policy Department: Economic and Scientific Policy A' authors proposed six Smart City characteristics as well as Boyd Cohen [2]. They emphasized that:

1. Smart Governance means joined up within-city and across-city governance using ICT and e-government in participatory decision-making and co-created e-services, for example apps.
2. Smart Economy refers to e-business and e-commerce, increased productivity, ICT-enabled and advanced manufacturing and delivery of services, ICT-enabled innovation, as well as new products, new services and business models. It also establishes smart clusters and eco-systems (e.g. digital business and entrepreneurship).
3. Smart Mobility means ICT supported and integrated transport and logistics systems.

4. Smart environment includes smart energy including renewables, ICT enabled energy grids, metering, pollution control and monitoring, renovation of buildings and amenities, green buildings, green urban planning, as well as resource use efficiency, re-use and resource substitution which serves the above goals.
5. Smart People mean e-skills, working in ICT-enabled working, having access to education and training, human resources and capacity management, within an inclusive society that improves creativity and fosters innovation.
6. Smart Living refers to ICT-enabled life styles, behaviour and consumption [2].

It is necessary to point out that there are plenty of scientists' approaches, which define smart city. Ortiz-Fournier, Márquez, Flores, Rivera-Vázquez, and Colon defined 'smart cities in the context of their smart inhabitants, educational degree, quality' [8]. Peng, Nunes, and Zheng defined 'Smart Cities as cities developed to utilize a set of advanced technologies including: smart hardware devices, e.g. wireless sensors, smart meters, smart vehicles, smart phones, mobile networks, data storage technologies and software' [5]. Manville et al. pointed out that 'the idea of Smart Cities is rooted in the creation and connection of human capital, social capital and information and Communication technology (ICT) infrastructure in order to generate greater and more sustainable economic development and a better quality of life' [2]. Giuseppe D'Aniello, Matteo Gaeta, Francesco Orciuolib, considered that 'the Smart City as an adaptive system that, starting from data gathered from different sources in several domains, is able to process and integrate information (and infer new knowledge) to achieve two main goals: support the decision making processes and enrich the city domain knowledge' [4]. We can see that Smart City concept is concentrated on management, operational, sustainable and technological aspects in order to satisfy needs in six fields of Smart City.

We suppose that Smart City should not only provide IT projects implementation in six fields of Smart City but also support to form rapidly decisions for the problems solving. Since decision-making depends on time limits and the precedent solutions which are embedded in the information system, it is important to implement and build the appropriate decision making support systems that base it and whether a precedent solution is embedded in the information system, so it is important to implement and create the suitable algorithms that based on a descriptive similarity.

Peeter Lorents and Maryna Averkyna in the paper published in 2019 presented the application of descriptive similarity in the process of managing the development of urban transport systems in small towns in Estonia and Ukraine [9]. This approach allows managers to understand towns' similarity. In this case, if similarity coefficient is closer to one, managers can use information obtained in one city for urban transportation management during Decision-Making Process. Thus, managers can form system of precedent solution based on the real experience.

It is necessary to remember that assessment of the systems' similarity is time-consuming. We have obtained information from 10 and it took plenty of time to calculate coefficient similarity. And it rather hard to imagine labor resources to be spend for the same calculation in the scale of 100 cities. Therefore, we propose to form an approach forming the basis of a computer program for the automated processing of cities' statements in order to identify precedent decisions. Such a program should be formulated as an application and implemented in the Smart City.

3 Algorithmic Approach for Calculation Descriptive Similarity

3.1 Preparation of the Table of Statements

Let us have a set of statements H with at least two statements. We agree that the number of elements of the final set K will be E(K). In this case, $E(H) \geq 2$. Let's make a step-by-step the *prepared set* Prep(H).

Step 0. Let that $H_0 = H, \mathrm{Prep}_0(H) = \varnothing$.

1. Let us take the first element from H_0, the first statement marked with the symbol h_0.
2. We create a set $\mathrm{Equ}(h_0)$ which elements are h_0 and all of these elements of H_0 (i.e., in this case, all such claims) that we have decided to equate with the element h_0 and (NB!) also equate with each other. We choose one such a statement that suits us, which we think is equivalent to all the claims from $\mathrm{Equ}(h_0)$ and denote it by the symbol $\mathrm{Sequ}(h_0)$.
3. We form the sets $H_0 - \mathrm{Equ}(h_0)$ and $\mathrm{Prep}_0(H) \cup \{\mathrm{Sequ}(h_0)\}$.

4.1. If $H_0 - \mathrm{Equ}(h_0) = \varnothing$, the formation of Prep(H) ends in step 0 and then

$$\mathrm{Prep}\,(H) = \mathrm{Prep}_0(H) \cup \{\mathrm{Sequ}(h_0)\} \tag{1}$$

Comment. Note that in this case

$$\mathrm{Prep}\,(H) = \{\mathrm{Sequ}(h_0)\} \tag{2}$$

4.2. If $H_0 - \mathrm{Equ}(h_0) \neq \varnothing$, then the formation of Prep (H) proceeds to the next step. We agree that:

$$H_1 = H_0 - \mathrm{Equ}(h_0) \tag{3}$$

$$\text{and} \qquad \mathrm{Prep}_1(H) = \mathrm{Prep}_0(H) \cup \{\mathrm{Sequ}(h_0)\} \tag{4}$$

$$\textbf{NB}!\,\text{Note that} \qquad E(H_1) \leq m - 1 \tag{5}$$

Suppose now that the formation did not end in stage e, where:

$$1 \leq e \leq m - 1 \tag{6}$$

Step e + 1. Here we have two options:

$$\mathbf{(I)}\; \mathbf{e + 1 < m} \tag{7}$$

or

$$(\mathbf{II})\ \mathbf{e+1=m} \tag{8}$$

$$\text{Let} \qquad H_{e+1} = H_e Equ(h_e) \tag{9}$$

$$\text{and let} \qquad Prep_{e+1}(H) = Prep_e(H) \cup \{Sequ(h_e)\} \tag{10}$$

Note that in this case:

$$Prep_{e+1}(H) = \{Sequ(h_0), \ldots, Sequ(h_e)\} \tag{11}$$

$$\text{Also note that:} \qquad E(H_{e+1}) \leq m - (e+1) \tag{12}$$

The first option: e + 1 < m.

Let us take the first element of He + 1, which we denote by the symbol he + 1.

We create a set of claims Equ(he + 1) with elements of he + 1 and all of the elements from He + 1 that we have decided to equate with the element he + 1 and (NB!) equate with each other. We choose one such statement that suits us, which we think can be equated with all the claims from Equ(he + 1) and mark it with the symbol Sequ(he + 1).

We form the sets He + 1–Equ (he + 1) and Prepe + 1(H) ∪ {Sequ(he + 1)}.

4.1. If $H_{e+1} - Equ(h_{e+1}) = \varnothing$, then the formation of Prep(H) ends in step e + 1 and Prep(H) = $Prep_{e+1}(H) \cup \{Sequ(h_{e+1})\}$.

In doing so – we see that:

$$Prep(H) = \{Sequ(h_0), \ldots, Sequ(h_e),\ Sequ(h_{e+1})\} \tag{13}$$

4.2. If $H_{e+1} - Equ(h_{e+1}) \varnothing$, then the formation of Prep (H) proceeds to the next step.

Here we agree that $H_{e+2} = H_{e+1} - Equ(h_{e+1})$ and $rep_{e+2}(H) = Prep_{e+1}(H) \cup \{Sequ(h_{e+1})\}$.

Comment. Note that in this case:

$$Prep_{e+2}(H) = \{Sequ(h_0), \ldots, Sequ(h_e),\ Sequ(h_{e+1})\} \tag{14}$$

The second options: e + 1 = m. If e + 1 = m, then that $E(H_{e+1}) \leq m - (e+1)$, follows $E(H_{e+1}) = 0$ and $H_{e+1} = H_e - Equ(h_e) = \varnothing$, which in its turn leads to the conclusion that the formation of Prep(H) already ended in step e.

This approach can be used in order to determinate town's statements (see Table 1).

Table 1. Determination of Town's Statements

Number of the predicate	Town's Statements
1	Small town
2	Minimum population
3	////////////////////////////
4	///////////////////////////
……	///////////////////////////
n	Local council's manager does not calculate the number of passengers who get off in each bus station

It is the first step for structuring qualitative information, which describes town and its urban transportation system.

Then we proposed to form a table of equalized statements of the towns in order to estimate the Lorents coefficient, which was presented by Peeter Lorents and Maryna Averkyna in the work published in 2019. In this case, we propose to use logic approach for creation table comparisons (see Table 2).

3.2 Algorithmic Approach for Creation Table Comparisons

Let us have at least two-element sets P and Q. We assume that P and Q are already prepared and E (P) $\leq$ E(Q). We form stepwise sets of Spec(P), Spec(Q), Com(P, C). At these sets, we have considered that Spec(P) contains those elements which are so-called specific for P and not equalizable by any of the elements Q; Spec(Q) contains those elements that are 'specific' to Q and are not equalizable with any of the elements P; Com(P, Q) is composed of pairs $\langle p,q \rangle$, where p is the element of the set P, q is the element of Q, and p and q are considered to be equalized.

Step 0.

Suppose now that $\mathrm{Spec}_0(P) = \varnothing$, $\mathrm{Com}_0(P,Q) = \varnothing$, $\mathrm{Spec}_0(Q) = Q$.

Step 1.

Let's take the first element of P – $\mathrm{Spec}_0(P)$, which is p_1.

If Q contains such an element (NB, only one, because Q is prepared) q_1, which we equalize with p_1, then

$$\mathrm{Spec}_1(P) = \mathrm{Spec}_0(P) \tag{14}$$

$$\mathrm{Com}_1(P,Q) = \mathrm{Com}_0(P,Q) \cup \{\langle p_1, q_1 \rangle\} \tag{15}$$

$$\mathrm{Spec}_1(Q) = \mathrm{Spec}_0(Q) - \{q_1\} \tag{16}$$

If there are no such elements in Q that we would like to equalize with p_1, then:

$$\mathrm{Spec}_1(P) = \mathrm{Spec}_0(P) \cup \{p_1\} \tag{17}$$

$$\mathrm{Com}_1(P,Q) = \mathrm{Com}_0(P,Q), \mathrm{Spec}_1(Q) = \mathrm{Spec}_0(Q) \tag{18}$$

Step e + 1.

Let's take the first element of P – $\mathrm{Spec}_e(P)$, let's p_{e+1}.

If Q contains such an element (NB, only one, because Q is prepared) q_{e+1}, which we equalize with p_{e+1}, then

$$\mathrm{Spec}_{e+1}(P) = \mathrm{Spec}_e(P) \tag{17}$$

$$\mathrm{Com}_{e+1}(P,Q) = \mathrm{Com}_e(P,Q) \cup \{\langle p_{e+1}, q_{e+1}\rangle\} \tag{18}$$

$$\mathrm{Spec}_{e+1}(Q) = \mathrm{Spec}_e(Q) - \{q_{e+1}\} \tag{19}$$

If there are no such elements in Q that we would like to equalize with p_{e+1}, then:

$$\mathrm{Spec}_{e+1}(P) = \mathrm{Spec}_e(P) \cup \{p_{e+1}\} \tag{20}$$

$$\mathrm{Com}_{e+1}(P,Q) = \mathrm{Com}_e(P,Q), \mathrm{Spec}_{e+1}(Q) = \mathrm{Spec}_e(Q) \tag{21}$$

Next, there are two possibilities:

$$(\mathrm{I})\, e + 1 < E(P), \tag{22}$$

$$(\mathrm{II})\, e + 1 = E(Q) \tag{23}$$

(I) If $e+1<E(P)$, then the formation of the sets Spec(P), Spec(Q), Com(P, C) proceeds to step e + 2 in exactly the same way as in step e + 1.

(II) If e + 1 = E(Q), the formation of the set Spec(P), Spec(Q), Com(P, C) ends with step e + 1, where we take:

$$\mathrm{Spec}(P) = \mathrm{Spec}_{e+1}(P), \tag{24}$$

$$\mathrm{Com}(P,Q) = \mathrm{Com}_{e+1}(P,Q), \mathrm{Spec}(Q) = \mathrm{Spec}_{e+1}(Q) \tag{26}$$

Note. Calculating is an estimation of similarity as follows:

$$\begin{aligned} \mathrm{Sim}(P,Q) &= E(\mathrm{Com}(P,Q)) : [E(P) - E(\mathrm{Com}(P,Q)) + E(Q)] \\ &= E(\mathrm{Com}(P,Q)) : [E(\mathrm{Spec}(P)) + E(\mathrm{Com}(P,Q)) + E(\mathrm{Spec}(Q))] = \ldots\ldots \end{aligned} \tag{27}$$

Then we propose an example table comparisons of towns' statements (see Table 2). As far as we know, that assessment of the towns' similarity required plenty of time we propose to create a computer program, which follows to the next algorithm (Fig. 1). It allows managers to more quick calculate coefficient similarity the number of towns.

Table 2. Town's Statements

Town P (Ostroh) describes the statements	Equivalent statements			Town Q (Valga) describes the statements
	The wording from the first	The wording from the second	The wording for both	
1.	Small town	Small town	Small town	
2.	Min. pop. 5000	Min. pop. 5000	Population	
3.	Density at 300 inhab.	Density at 300 inhab.	Density	
4.	Less than 50% lives in high-density clusters	Less than 50% lives in high-density clusters	Concentration	
5.	Intermediate area (towns and suburbs)	Intermediate area (towns and suburbs)	Intermediate area (towns and suburbs)	
n. Local council's manager does not calculate the number of passengers who get off in each bus station				
n + 1				Local council's manager calculates the number of passengers who get off in each bus station

This algorithm (flowchart is depicted in Fig. 1) is the basis for writing an RStudio-based software that can be used as an application for the Smart City. It will also simplify the coefficient similarity procedure calculation for managers. In order to implement the algorithm, we propose to create Google forms with statements describing the towns and their transportation system. The using of Google forms is convenient for estimating the similarity factor, which will be calculated using the RStudio software. Google forms and code written for RStudio software will simplify the algorithm itself, since the number and order of claims is the same for each set, and usage of the embedded loops become obsolete (see the flowchart in Fig. 2 and Fig. 3).

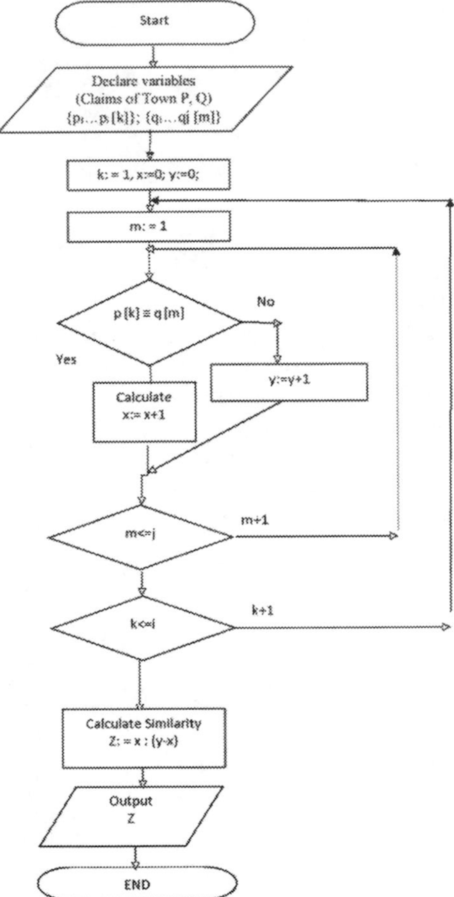

Fig. 1. Flowchart for general size of claims

This application is quite flexible and does not require a plenty of time for its implementation in Smart City. As far as we know RStudio is freeware, so it is not necessarily to spend money for implementation code in Smart City by local councils' managers. Also, this code can be easily updated along with changes made to Google forms. It is crucial for managers because every city has features, which do not similar with other ones. When managers fill in Google forms they can add new information as question. This option is considered in written code as 'N/A'.

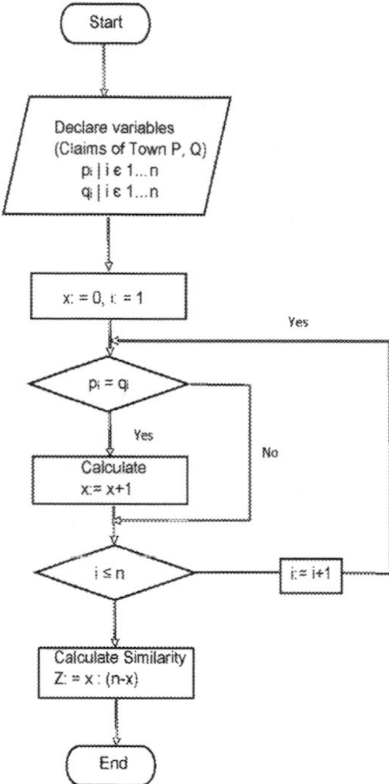

Fig. 2. Flowchart for Google-form based similarity analysis

```
data = read.csv("Similarity Questionnaire.csv", header = TRUE)

similarity <- function(indices) {
   city1 = data[indices[1], ]
   city2 = data[indices[2], ]

   data1 = city1[, -c(1, 2)]
   data2 = city2[, -c(1, 2)]

   x = sum(data1 == data2 & data1 != "N/A" & data2 != "N/A" &
!is.na(data1) & !is.na(data2))
   y = length(data1)

   city1_name = as.character(city1[, 2])
   city2_name = as.character(city2[, 2])

   print(city1_name)

   list("City 1" = city1_name, "City 2" = city2_name, "Score" = x / (y - x))
}

combn(rownames(data), 2, FUN = similarity)
```

Fig. 3. Fragment of R code for similarity analysis

4 Conclusion

In this paper we researched algorithms, which are related to descriptive similarity estimates. They are to support implementation decisions, which are necessary for small town's governance as to transport management. We discussed the Smart City concept and importance of the Descriptive Similarity for Decision Making Process implementation. We pointed out that Smart City allows local councils managers to form decisions rapidly for problem solving. We emphasized that it is necessary to use experience of towns in order to manage transportation in cities. That is why is crucial to use the Descriptive Similarity approach in order to form the precedent solutions which can be embedded in the information system of the city. We presented approach for preparation table of statements and table of comparison. It helps to create algorithm and to write code for implementation in the Smart City. It was presented the algorithm and written code as flexible application of Smart City.

Acknowledgments. The author of the given work expresses profound to professor Peeter Lorents for assistance in a writing of given clause and to professor Yaroslav Pasternak, Roman Suiatinov for the help with improving the flowchart for writing an RStudio-based software.

References

1. Cohen, B.: The Smart City Wheel. https://www.smart-circle.org/smartcity/blog/boyd-cohen-the-smart-city-wheel
2. Manville, C., et al.: Mapping smart cities in the UE, European Parliament. Directorate-General for Internal Policies. Policy Department: Economic and Scientific Policy A (2014)
3. Ismagilova, E., Hughes, L., Dwivedi, Y.K., Raman, K.R.: Smart cities: advances in research —an information system perspective. Int. J. Inf. Manag. **47**, 88–100 (2019)
4. D'Aniello, G., Gaeta, M., Orciuolib, F.: An approach based on semantic stream reasoning to support decision processes in smart cities. Telemat. Inform. **35**, 68–81 (2018)
5. Peng, G.C.A., Nunes, M.B., Zheng, L.: Impacts of low citizen awareness and usage in smart city services: the case of London's smart parking system. Inf. Syst. E-Bus. Manag. **15**(4), 845–876 (2017)
6. Lu, H.-P., Chen, C.-S., Yu, H., Lu, H.-P., Chen, C.-S., Yu, H.: Technology roadmap for building a smart city: an exploring study on methodology. Future Gener. Comput. Syst. **97**, 727–742 (2019)
7. Streitz, N.: Citizen-centered design for humane and sociable hybrid cities. Hybrid City, 17–20 (2015)
8. Ortiz-Fournier, L.V., Márquez, E., Flores, F.R., Rivera-Vázquez, J.C., Colon, P.A.: Integrating educational institutions to produce intellectual capital for sustainability in Caguas, Puerto Rico. Knowl. Manag. Res. Pract. **8**(3), 203–215 (2010)
9. Lorents, P., Averkyna, M.: Some mathematical and practical aspects of decision-making based on similarity. In: Stephanidis, C. (ed.) HCII 2019. LNCS, vol. 11786, pp. 168–179. Springer, Cham (2019). https://doi.org/10.1007/978-3-030-30033-3_13
10. Keegan, S., O'Hare, G., O'Grady, M.: Retail in the digital city. Int. J. E-Bus. Res. **8**(3), 18–32 (2012)

11. Roque-Cilia, S., Tamariz-Flores, E.I., Torrealba-Meléndez, R., Covarrubias-Rosales, D.H.: Transport tracking through communication in WDSN for Smart cities. Measurement **139**, 205–212 (2019)
12. Li, X., Zhu, Y., Wang, J.: Efficient encrypted data comparison through a hybrid method. J. Inf. Sci. Eng. **33**(4), 953–964 (2017)

The Conformity Utilization on Community Resources on Base of Urban Renewal—Taking Xinhua Community of Shaoyang City as the Case

Wei Bi[1,3(✉)], Yang Gao[2,3], and Zidong He[1,3]

[1] School of Art and Design, Guangdong University of Finance and Economics, Guangzhou, People's Republic of China
{beebvv,35290984}@qq.com

[2] Graduate School of Creative Industry Design, National Taiwan University of Arts, Taipei, Taiwan
lukegao1991@gmail.com

[3] Guangzhou Creative Intelligent Manufacturing, Culture Communication Co. Ltd., Guangzhou, People's Republic of China

Abstract. Since the year of 2011 when urbanization of China reached 50%, "Urban Renewal" has become the hot topic among scholars from all industries of the society. Meanwhile, there appeared relevant discussion and researches against the methods of urban renewal. Compared with the previous developing mode which takes city expansion as the main stream, the comprehensive programme in large scale is preferred. At present, on facing the urban renewal situation switching from increment to stock. Small scale, local and progressive renewal method is gradually becoming the main stream. The community unit was one of the important parts of city structure, it is unique and could not be replaced on historical value and social meaning. In this way, when considering to melt it into current urban development, multi intervening is more required. It's not only about the physical fusion in view of traditional meaning, but also includes the fusion of social environment. This thesis takes Xinhua Community of Shaoyang City as the case, regards the optimization and adjustment of all plots, as well as the coordination of benefit of all aspects as the entry point, with the grand background of "wiping off unitization", to reconsider the relation between community units and all hierarchies of the city, as well as the way how community units blend into city.

Keywords: Community units · Transformation of resources · Resources activation · Optimized space module

1 Introduction

Xinhua Community of Shaoyang City is located in the Second Xinhua Printing House of Hunan Province (hereinafter referred to as the 2nd Printing House). The location is on south side of Dongda Rd, Shuangqing District, Shaoyang City, Hunan Province, on north side it is next to Dongda Rd, the east side and west side is respectively close to

C. Stephanidis et al. (Eds.): HCII 2020, LNCS 12427, pp. 40–50, 2020.
https://doi.org/10.1007/978-3-030-60152-2_3

Xinhua Rd and Shuangpo South Rd. Around it, there exist a lot of large scaled living areas and residence zones. The comprehensive condition is surrounded. On gate of the 2nd Printing House, there are all kinds of public transportations which reach the downtown directly, however as the geographic position is quite remote. The distance from the 2nd Printing House to downtown is 30 min by driving. There are small scaled supermarkets and living facilities around the 2nd Printing House, basic demands for living could be satisfied.

2 Basic Research on Environmental Resources of Xinhua Community

Hunan Xinhua 2nd Printing House was founded in 1964 by the national government for the tertiary construction, which was a large scaled basic facility construction for national defense, science and technology, industrial and transportation launched by the government of the P.R.C for middle and west area of China, which covered 13 provinces and municipalities.

The 2nd Printing House was applied by Hunan Provincial Bureau of Culture on Feb 3rd. 1966, approved by the Hunan Provincial Party Committee that Hunan Xinhua Printing House was allowed to build up the Shaoyang Branch. After which, the basic facility construction team was established. The plant was started on construction in Sep 1966 in Shuangpoling of Shaoyang City. In year of 1969, the printing house was accomplished, divided into 2 major areas including production area and residence area (the current Xinhua Community). The total area of the factory was 190,000 m^2, including which production area was on west side of the 2nd Printing House, with production shops, warehouses, back-mountain reservoirs and other functional areas, the overall floor space was around 118,000 m^2, shared 60% of the total area of Xinhua Second Printing House. Xinhua Community was on east side of the 2nd Printing House, which included residential buildings, shops, snack shops, dinning tall, kindergarten and other functional areas. The overall floor space was around 72,000 m^2, which was the 40% of the total area of Xinhua the Second Printing House. The research on in-house environment is majorly against Xinhua Community.

2.1 Spatial Arrangement

As the 2nd Printing House was constructed in 1960's, so the overall arrangement is quite monotonous. All buildings in Xinhua Community are in row layout which the rows are flat and the lines are straight. Meanwhile, the public service facilities and playgrounds are quite centralized. As the most popular vehicle in 1960's was bike, so there is only one main road from the community to the outside world. Also, at the stage when factory constructed, all buildings were limited within 10 m. By reading the former plan, it could be realized that, the 2nd Printing House is remaining very outstanding independent unit features since the construction accomplished. All areas of factory are entirely different, all boundaries between the unit and the city are also outstanding by walls or terrain differences.

2.2 Road System

The road system in the 2nd Printing House can be divided into 2 zones. Among which, there is only one main road from the community to gate of the factory, with width of 6.5 m. All others roads in the factory are limited with width of 4.5 m. When the unit developed fast, the in-community road conformed to the common principle for unit residence zone in the 1980's that: "Currently, the travel mode of our nation mainly relies on foot and bike. In some large scaled residence areas, bus is rcquired. Some special economic zones with highly developed economic shall consider the transportation of private motorbikes and cars. 1" So, though the former road design in Xinhua Community could satisfy the demands for walking and low speed path, as more and more residents in Xinhua Community own cars, the roads in community are hard to divide foot passengers from vehicles. What's worse, as the community was not designed with specific area for parking, most cars have to be parked on playground or grass land. In this way public playground is shared, and it's obviously bringing negative impacts to the roads which were good for foot passengers. On the other hand, sometimes some huge cargo trucks have to cross the production area, and park on the public road from factory gate to community. The cargo trucks have no area to park, so sometimes it's causing block against internal roads (Figs. 1 and 2).

Fig. 1. Road in front of residential building in Xinhua Community (image from: author's shoot)

2.3 Landscaping and Public Activity Area

In Xinhua Community, grown trees and lawns are the major constituent parts of landscaping. There are also some mini pot cultures in which, planted by residents in front of their residential buildings. There are a few small and short green plants. Most existing plants in the factory were planted when the factory was constructed, after growing over 50 years, they are very strong. Besides, there is a pristine green area in the factory. However, due to lack on relevant plan, it is still idle till now. on the other hand, though the percentage of landscaping in Xinhua Community is not low, the relevant public space for activity is obviously lacking, so systematic space for public activities could not be established. Instead, there are some public spaces spreading around residential buildings in small area. But, with the growing number of cars, some

Fig. 2. The unique main road in Xinhua Community (Image from: author's shoot)

exiting facilities for fitness are cancelled for cars to park. In this way, area of public space for activities is decreased again. Some other zones in the community are short for corresponding facilities either, with very monotonous facilities for rest and interaction (Figs. 3, 4, 5 and 6).

Fig. 3. Landscaping space in Xinhua Communit (image from: author's shoot)

2.4 Public Service Facility

Same to other communities, on initial stage of construction Xinhua Second Printing House was equipped with clinic, card room, senior citizens activity center, outdoors facilities, kindergarten and primary school. But till now, the community is operating clinic and card room only. All other facilities were dismantled or distributed to other institutions so they do not belong to Xinhua Community any more. On one hand, the previous outdoor facilities and part of activity rooms for the olds were removed due to construction of new buildings. On the other hand, though the production area and community are existed with kindergartens, but they do not belong to Xinhua Second Printing House, and most of the students are not from the factory or community. In this way, on aspect of public facility, Xinhua Community owns corresponding facilities in field of medical care and education. However the relevant facilities in fields of culture,

Fig. 4. Landscaping area with little application (image from: author's shoot)

Fig. 5. Landscaping with a few applicants (image from: author's shoot)

Fig. 6. Public space with low upgrades (image from: author's shoot)

physical education, social welfare and commerce are quite short. The outstanding conflict is on aspect of public space, due to lack of physical education, people of different groups are getting less and less satisfied with this.

In summary, according to basic research on Xinhua Community, it can be realized that the environment of Xinhua Community is mainly displayed with the following features: First, Xinhua Community is highly closed to the society, the high wall and unique entrance is stopping the communication between the city and the community. Thus the development of Xinhua Community is quite off-grid to the urban renewal process to a certain extent. Secondly, traffic in the community is quite chaotic, which is caused by the lack of concerning on parking area and the design for pedestrian system separated from vehicle system. It even has some hidden safety risk for residents during daily travels. Thirdly, the community is lack of rich and hierarchy on designing internal landscaping and public space, nor necessary maintenance, renewal. So there appeared a lot of useless spaces in the community. In this way, the applicants attending public space is inversely proportional to the utility rate of space. Fourthly, the service facility in community is undeveloped, and even decreasing, which makes residents in community very inconvenient to apply which.

3 Resources Integration of the Community—Driving Force

3.1 Suitable Space Environment

Xinhua Community is a typical community unit, it owns space environment which is good for alternation. Which appears via two aspects: 1. The spatial scale in Xinhua Community is commonly designed for people's daily communication and walking, so in community there are a lot of scatted spaces for people to chat, walk and rest, which conforms to the application which small groups can gather. 2. There are a lot of idle spaces in Xinhua Community, such idle spaces are lack of necessary maintenance and upgrades so they are never concerned. However due to residents nearby have demands on public space, so they have huge exploring potentials. With different thematic plans and designing methods, such spaces could be involved into community upgrading.

3.2 The Space Alternation Supported by Government

At present there are 23 residential buildings in Xinhua Community. Among which, 10 of them are red brick buildings constructed when the 2nd Printing House established. According to the governmental document of Shaoyang City <The Implementation Opinions about Speeding Up the Alternation Works in Shanty Town of Shaoyang People's Government> , which plans to process the alternation works against shanty town from the year of 2013 to 2017, and the 10 ancient residential buildings of Xinhua Community are also included. The original buildings with 3 to 5 floors are to be rebuilt to a new residential building with 24 floors. Part of the property right belongs to the original residents who provisionally moved out because of removing, and the rest part belongs to the community. The new built high-raise building has advantage on attracting fresh people into Xinhua Community. On the other hand, it provides perfect chance for the community to upgrade and alter the original idle public spaces.

3.3 Residents' Demands on Optimizing Spaces

At present, in public spaces of the factory, there are only two types of spaces for walking and ball type games. According to author's observation and interview, the residents in community need at least 4 types of spaces including: 1. All types of ball type games—basketball, ping-pong ball, badminton. 2. Spaces for dancing and excises. 3. Spaces for walking and resting, which belongs to relaxing activities. 4. All types of outdoors appliances and low-speed running, which may consume a lot of power.

When supplements are obviously lower than demands, all types of applicants are hard to use spaces with their desires. Some spaces even have to play a lot of roles. For example, Mr. Lei, a worker of the factory and his friends, there is only one basketball court in office area in the community, which is also a open space with middle size. Mr. Lei and his friends have to play basketball in the afternoon, as when evening comes, some middle and old aged ladies would come up to share the only basketball court of the community, to dance and do exercises there. "Anyway we are used to it already, but sometimes when we work overtime, we would have no chance to play basketball in the afternoon. Then in the evening if we want play, no way at all. However, this is a play for playing basketball." (Mr. Lei, 24 years old).

Just as Mr. Lei's words, basketball court is the place for playing basketball. However for middle and old aged ladies, this is a perfect stage for them to dance and do exercises. "It's big enough, so more and more people can join us to dance. Previously when we danced in front of the residential building, we were disliked as it's too noisy. Now we are not disturbing anyone" (Mrs. Wang, 48 years old). Meanwhile, as the basketball court is surrounded by the paths which a lot of people may cross, so it's more like a "stage", while dancing on the basketball court, the ladies may attract more attentions so they are more encouraged to dance. In this way, Mr. Lei and his friends, middle and old aged ladies, these two groups are gathering more and more conflicts. With this situation, all activity groups of Xinhua Community are more desired to their independent activity spaces.

3.4 Residents' Demands on Service Modes and Employments

As the community is hard to upgrade the service facilities, so in Xinhua Community there is little service. So at present, Xinhua Community is sharing living facilities together with other communities along Dongda Rd. "Previously there are restaurants in the factory, now when I want to have dinner with friends in some delicious restaurant, I have to go to Hongqi Rd to eat." (Mr. Cao, 39 years old). "There is only one barber's here, every time when I come, I have to wait. So when I am urgent I prefer having my haircut at home. And when Spring Festival comes, I may have to wait for a day long." (Miss. Mao, 43 years old). It can be easily realized that, due to the disappearing of original services, residents of the community are getting more and more inconvenient. On the other hand, a lot of residents of the community are not satisfied with their salary so they wish to work more, or do some part time jobs. There are also some laid-off workers in the community, they run to different places for odd jobs for one day. Everyone wish to have stable employment, however the residents have limited skills, and also the nearby environment could not provide enough job opportunities, so most of the residents could just wish.

4 Integration Strategy on Community Resources: Standardize and Alter in Orders, Open Source and Attract Out Sources

Public spaces and landscaping are the most attractive internal exclusive resources for residents nearby. The original walking paths in Xinhua Community connect scattered public spaces with landscaping, which also promotes the optionality and accessibility for public activity participants by the crossed and floor orders. However, as the unit had never processed essential upgrading the public spaces, landscaping, outdoors facilities in community, which makes the public spaces are using the public spaces too centralized, while the others are idle. Thus, at initial stage of alternation against Xinhua Community, based on view of resource complementarity, to take advantage of and upgrade exiting internal resources would be preferred. By integrating and developing the existing landscaping and public spaces in community to create an outdoors playground which would be shared by the residents from and out of the community. This is also making foundation for Xinhua Community to bring in new form of industries, and bring in more employment opportunities (Fig. 7).

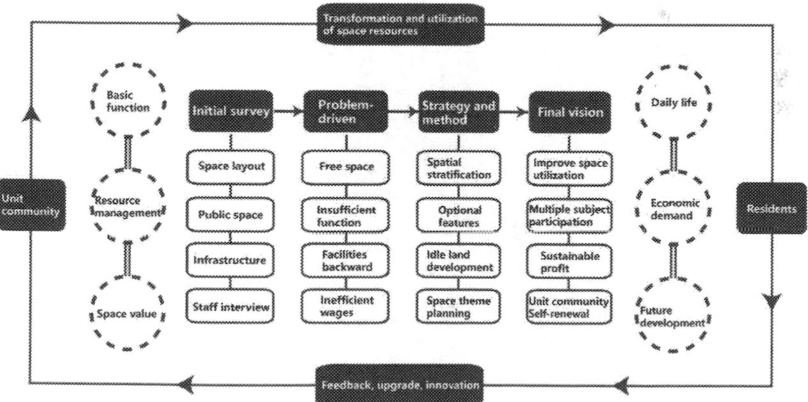

Fig. 7. Frame of community resource integration strategy (image from: author's drawing)

4.1 Renovate Current Public Spaces and Place into Relevant Activity Facilities

For the renovation of current public spaces in Xinhua Community, 2 major parts as below: The first part is the public spaces on opposite side of north gate of Xinhua Second Printing House, which is divided into 4 major areas. Including area A and B are not well designed due to lack of relevant exercises facilities, so they are in idle situation. Area C is the communal basketball court for young people to play basketball and the middle – old aged ladies t for dancing. Area D is the path for walking and running, surrounding the office building. According to applicants' demands and conflicts, area A and B are considered to be re-designed, aiming to realizing the application pressure because of concentrated application and enhance the attraction of the idle spaces. From which, functional areas for different applicants can be divided, so the comprehensive applying situation is more reasonable. On one hand, it's supplying "stage" for the middle and old aged ladies, which they could be concerned by passengers, and the basketball court could be returned to the young people. On the other hand, it also partakes the pressure for area D due to too many applicants, and the walkers, runners could have more choices. The second part of altering public space is mainly to active the idle areas in east Xinhua Community, which is rarely applied. With the re-plan to idle area and the design on activity facilities, interaction devices according to application demands. The service facility and rest facility in space are enriched, the recognition and playability is also enhanced. In this way more applicants could be attracted to use the public spaces. After which, when the new form of industry is ready, the population and application rate of community space can be well improved (Fig. 8).

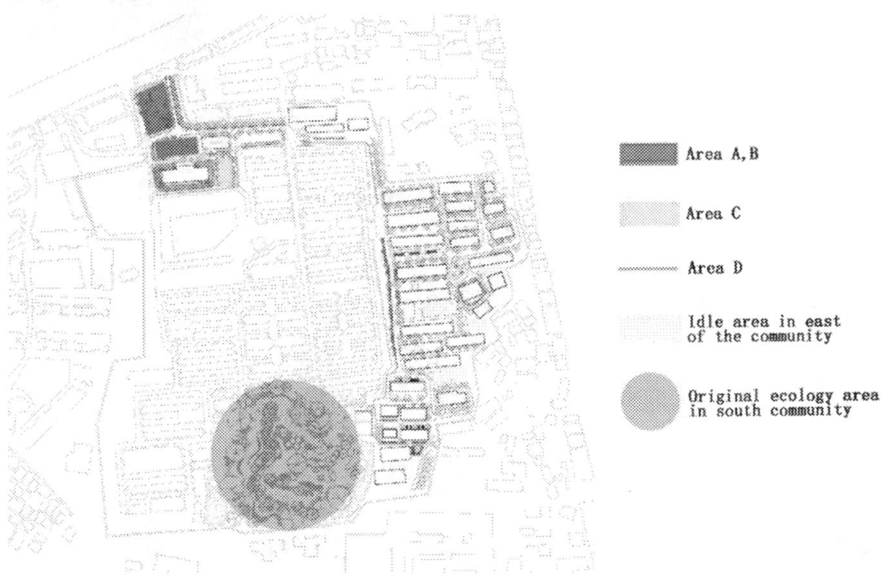

Fig. 8. Layout overview on all types of public spaces in Xinhua Community (Image from: author's drawing)

4.2 Develop Idle Landscaping and Create Sharing Park

With a great amount of landscaping, the community is always attracting a lot of residents nearby to come into Xinhua Community. However, the 6,000 m^2' original ecology area in south part of Xinhua Community is always lack of development, so it's totally "wild". In this way, people in and out of community are hard to use it. So when altering the landscaping in Xinhua Community, this "wild" area could be taken as a landmark of Xinhua Community to develop, as it's always idle. With the following 3 aspects of planning to design it into a open city park. First, retain part of the existing green plants, and take this advantage to arrange space paths, and reasonable activity areas. Second, arrange different space types for different applicants, and enhance the interactivity, attraction by interaction devices. Third, reasonably bring in a certain of retailing industries in different areas to try the best effort on improving service on all aspects. When the original ecology area in Xinhua Community is transferred into city park, the community could be well connected to the outside world, and also it's a great chance to improve internal sources of Xinhua Community.

Thus, when developing shared park, on concerning on topics about the attraction to nearby residents of the comprehensive spaces, required time period for advertising and the sustainability of capital, this ecology area shall be divided into different stages for arrangement. The first stage shall be forced on the relatively outer spaces. And when the terrain is satisfied, the wilderness area shall be initially altered, some functional areas such as low-speed running paths, biking paths, ball game areas and other spaces for residents' activities shall be arranged as well. The second stage shall be based on the activity areas in stage one, to bring in part of retailing industries including light food shops, sports goods shops, gift shops and so on. On one hand it's forming a complementary experience relation with the activity areas, and on the other hand it's providing ideal and comfortable place for applicants by combining with the to be developed spaces. With the first and second stage developments, in the landscaping areas there would be some population and a certain of commercial atmosphere, more consumptive display zones could be brought in, such as the outdoors manual workshops for parent – child interaction, planting display shops for mini plants observing, and so on. When improving popularity, some of the areas could also be used as

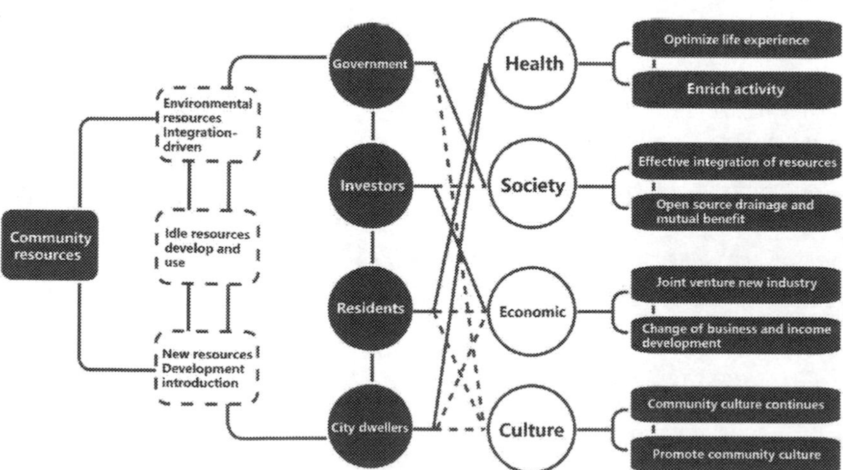

Fig. 9. Frame of community resource transformation (image from: author's drawing)

commercial playground for enterprises to hold activity, which brings more economical benefits (Fig. 9).

5 Conclusion and Suggestion

The development on original ecology landscaping area issued by integration resources strategy, on one hand it is providing all applicants from and out of Xinhua Community activity spaces with more choices, and more comfortable. On the other hand it is providing a period of employment opportunities for residents of Xinhua Community, as well as the nearby residents. Meanwhile, the commercial system assorted with the activity spaces can also supply economical stands for the construction of Xinhua Community. With the situation of "the community has now power to admin, and the government admins little", the community is gradually seizing the ability of "self management and construction". Also, with the multi choice on space exploration to form up sustainable economical benefits step by step. All which are aiming to supply feasible conditions for the development of Xinhua Community, and to improve the living standard of residents in Xinhua Community. Anyway, during the actual alternation, there would be more complicated matters including ownership of land, relations of people and so on. So there would be more comprehensive elements to be concerned before actual development starts.

References

1. Xie, S.: The Development and Arrangement of Urban Community. China Substance Press, Beijing (2007)
2. Chang, T.: The Discussion on New Community. China Society Press, Beijing (2005)
3. Zhao, M., Zhao, W.: Theory and Practice on Arrangement of Community Development. China Building Industrial Press, Beijing (2006)
4. Gail, Y.: Communication and Space (trans: He, R.). China Building Industrial Press, Beijing (2002)
5. Xu, Y.: The Discussion on Development of Community. East China University of Science and Technology Press, Shanghai (2000)
6. Zhang, J.: The Organization and Management of Urban Community in China. Southeast University Press, Nanjing (2004)
7. Li, D.: Generality on Environment Behavior Studies. Tsinghua University Press, Beijing (1999)
8. Gu, C.: Urban Sociology. Southeast University Press, Nanjing (2002)
9. Zhou, L.: Arrangement Principle on Urban Residential Area. Tongji University Press, Shanghai (1999)

Tools with Histories: Exploring NFC-Tagging to Support Hybrid Documentation Practices and Knowledge Discovery in Makerspaces

Daragh Byrne[1(✉)] and Marti Louw[2]

[1] School of Architecture, Carnegie Mellon University, Pittsburgh, PA 15213, USA
daraghb@andrew.cmu.edu

[2] Human Computer Interaction Institute, Carnegie Mellon University, Pittsburgh, PA 15213, USA
martil@andrew.cmu.edu

Abstract. We present the design research process towards a novel learning technology to improve instructional documentation in makerspaces. Our focus is on the ways in which smart tools can better support learning practices, with a particular emphasis on role of documentation plays. We first describe a co-design process with educational stakeholder that generated concepts for NFC-enabled forms of hybrid documentation. This solution developed offers a process for coupling physical tools, parts and materials to online resources in order to help make documentation ready-at-hand for learners. Findings from a subsequent focus group led to a refined implementation. Our preliminary evaluation in a high school settings highlighted the value of integrating online documentation formats to support youth navigating a broad array of fabrication tools and parts. Educators valued the solutions ability to support self-directed learning and to increase student access and agency with instructional resources.

Keywords: Makerspace · Education · Learning technology · NFC · Smart tools

1 Introduction and Background

Much work has explored the role of documentation in making learners' thinking visible [3, 5, 24] and revealing process and accomplishments [3, 29]. By externalizing and sharing concepts and craft [26], documentation prepared by others also plays a significant role in learning [7, 22], particularly in maker-based contexts; as evidenced by the popularity of digital guides like Instructables and Hackster.

Makerspaces contain a wide array of tools, materials and components visible and ready-to-use for learners [17]. While the tools are ready-at-hand, the knowledge and practices needed to use them is not. This is particularly challenging for novices who need to successfully navigate many choices to identify appropriate tools for their projects [11, 13]. Each tool often requires safety, training and imposes a large number of practical, logistical and instructional overheads on facilities and staff [4]. Recent studies have noted the many complexities present for learner and practitioner alike: the

C. Stephanidis et al. (Eds.): HCII 2020, LNCS 12427, pp. 51–67, 2020.
https://doi.org/10.1007/978-3-030-60152-2_4

tools in makerspaces require much training and guidance; they are often not suited to youth learners without training and guidance; they require alignment with instructional methods and outcomes to be successful; and they require learners to successfully and regularly transition from physical work spaces to virtual spaces to manage documentation and knowledge discovery [21, 23].

In this paper, we address these challenge and present a mechanism for distributing knowledge into the makerspace: to more closely integrate tools, components and physical assets with supportive digital instructional resources. Recognizing the same problem, Knibble et al. [13] created an augmented workbench to overlay contextually-relevant instructional guides and augment project-work in makerspaces. Schoop et al. [25] similarly proposed a mixed physical-digital environment to build confidence with tools and technical skills. These two approaches highlight how technology augmentation can positively enhance the practices within a makerspace. Both examples, however, employing expensive and technically intensive strategies that are unlikely to be replicable or affordable for most makerspaces. We instead use NFC technology to offer a low-cost, scalable solution.

We outline the design research process behind this solution. We first describe a multisite co-design process to generate concepts for technology-enhanced making. This process is used to closely with educators and identify technology-driven concepts that aligns with the cultures, values and practices of educational stakeholders. We discuss how this process was leveraged to synthesize and identify solution for hybrid documentation - the coupling of a physical object to online digital content like a blog-post, portfolio or Instructable. We describe the system prepared, its operation and how it offers a low cost, scalable alternative to existing and emerging approaches. Finally, this system is evaluated both through an early stage focus group with two experts and through a preliminary evaluation with seven participants at a high school makerspace. In so doing, we illustrate an effective approach to introducing NFC technology within educational contexts and additionally demonstrate the value and applicability of NFC in creative and technical documentation to support learning within a makerspace.

2 Design Research and Concept Development

In this section, we describe our design process for documentation in educational makerspaces.

2.1 Co-design Workshops

We emphasize a design research and participatory approach in how we examine new technology supports for maker- and project-based learning. In so doing, we seek to include educational stakeholders in identifying and responding to instructional practices, learning goals and outcomes, and cultures of documentation valued in their spaces. To do this, we developed a series of co-design workshops to solicit concepts and possible design enactments. This four-hour workshop was conducted with stakeholders at across three contexts which emphasize maker based learning: an undergraduate degree program (four participants), a high school (eleven participants) and a youth summer program (six participants). The curricula at all three sites was organized around hands-on project-based learning and the instructors all explicitly included documentation activities within their instructional practice. Participants included instructors, program coordinators and staff. Initial activities in the workshop encouraged educators to develop a shared understanding of documentation, how it benefits their curricular programming and the pain-points encountered when facilitating documentation in educational settings. Pain points were actively discussed and prioritized as they related to each program. This elicitation of challenges was central to supporting generative design activities later. The workshop then transitioned to introduce technologies and emerging approaches that could be used to scaffold documentation practices. After which, the remainder of the workshop was dedicated to a series of participatory design activities. These encouraged educators to speculate on the ways in which technology (e.g. cameras and timelapse/video capture, live-scribe pens, smart home products, etc.) could be deployed to augment existing or foster new documentation practices in their classroom. In this, we encouraged participants to consider technologies that are situated in learning activities and routines, would improve documentation skills and outcomes, would be valued by educators and learners and would address the pain points and challenges identified by the group. At the end of the activity, participants reported out on their concepts and as a group prioritized concepts for supporting learning in technology enhanced makerspaces (Fig. 1).

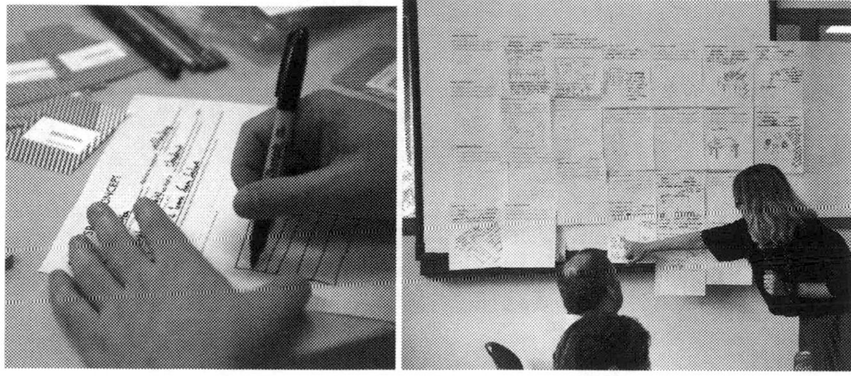

Fig. 1. Participants brainstorm, discuss and prioritize concepts.

Across all sites, participants converged on a cluster of related concepts that identified the need to trace the history of interactions, and to support peer-to-peer learning and knowledge discovery with the objects, tools and parts that are typically found in a makerspace. Several related concepts suggested that tracking patterns of use would encourage students to explore new materials and tools. Another set of concepts valued augmenting tools with digital instructional guides through spoken audio or projection; while others considered how the tools themselves could gather feedback and reflections from students while in use.

For example, at the high school, participants prioritized three concepts relating to material and tool usage, and ways in which the tools and materials themselves could enable students to learn how to use them effectively by their documenting past interactions with (other) learners. A concept from one participant imagined an augmented parts bin where "[a]s a user comes to materials, [the device] tracks materials they look at and take/put back. Students then "get a list of considered/used/returned materials" to help them develop a bill of materials, identify relevant online tutorials and guides, and to support the creative exploration of parts as they apply to assigned project work. A second, and similar, concept saw a design of a materials or parts bin that would help students to "understand the purpose of complex parts and their effectiveness". It would do this by allowing a student to place materials or parts in front of a camera, and then it would begin "displaying text and describing use of parts and materials through video and images". This notional concept also included the idea that students could add to these descriptions based on how they had used the tool, giving richer resources to the next student to use the same material or part. In the same vein, the third use case, imagined a storage area or cubby that would actively track tool use and encourage a student to "articulate how they used tools to support their learning". During use or on returning a tool to the storage area, this solution might prompt them with messages like "How did you use this tool today?" to verbally record their experience with a tool, encourage reflection on skills and project development and provide knowledge resources to other students. These tools were organized and described by the group as *"tools with histories."* They are illustrated in Fig. 2.

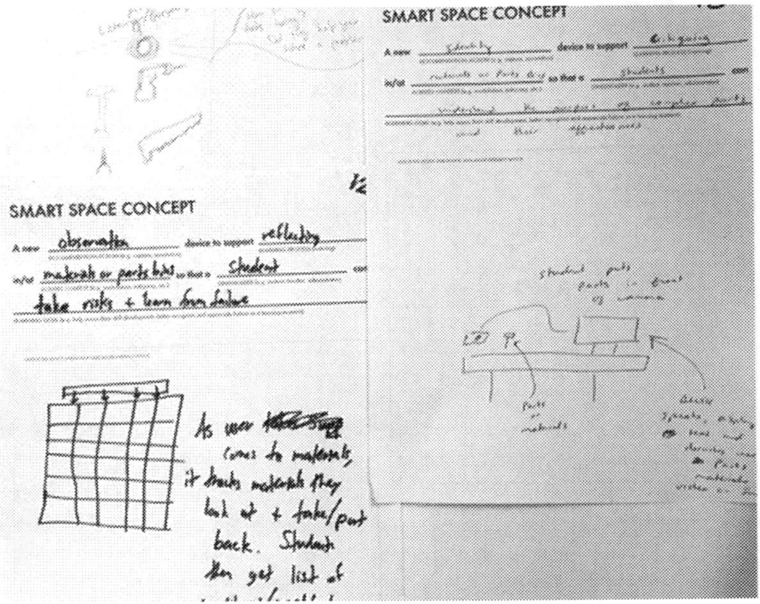

Fig. 2. Concepts for 'Tools with Histories' generated by participants.

2.2 Concept Development and Iterative Prototyping

Concepts were analyzed and key features elicited by participants drawn out. This was synthesized and conceptualized into an underlying solution for enabling hybrid documentation – the coupling of a physical object to online digital content like a blog-post, portfolio or instructable, brief surveys and reflective prompts. At the core of the concept was the need to enable proximity-based interactions with a wide variety of tools, materials and components that were of an array of shapes, sizes, arrangements, etc. Additional constraints on scalability, deployment and use were set: namely it must create the least amount of effort to infrastructure a space for a resource bound educator, it must be a broadly accessible solution to students with the least amount of overhead to adopt/participate with the solution, and it must provide agency to educators and students alike in developing knowledge resources.

These requirements guided our survey of ubiquitous proximity-based technologies for a suitable solution including include printed markers (barcodes and QR codes), RFID/NFC, BLE or indoor localization strategies, and computer vision techniques; all of which could be used in our approach. Computer vision (object detection and recognition) has been used in identifying parts in electronics [12] and tools within spaces, however it requires extensive visual examplars to be collected for training purposes making it impractical as a broadly scalable solution. Furthermore and as Gong et al. [9] note: "While these computer vision based methods are powerful, we deliberately excluded cameras in this work since the spaces that fabrication can take place in can be large, transitory and dynamic, the issue of occlusion can cause loss of information from camera streams." As an alternative, Bluetooth LE beacons could be applied to tools, but they are more costly than other solutions (e.g. QR codes, RFID/NFC), require power and are generally more bulky making them impractical to apply to varied surfaces and small components. Indoor localization strategies could be used to detect where a learner is in the makerspace and operating in tandem with a virtual model of the space could allow a learner to be presented with relevant (augmented) information. As parts, tools and materials are regularly moved in active makerspaces and not reliably returned to designated locations, this was also deemed impractical. In addition, both BLE and indoor localization would require the development and installation of a custom mobile application(s), adding overhead to deployment and immediate use by students. This left printed markers and NFC technologies as candidate technologies. Both were low-cost, and scalable mechanisms that had seen recent integration into modern smartphones (QR codes can be automatically detected in camera applications [30] while background tag reading for NFC is included in latest Android and Apple models [1]), removing the need for installed applications. QR codes offers a more adopted and ubiquitous technology, and one that shows past success in educational settings [8, 19]. In assessing the technology two issues were presented: readability issues could present [28], especially for small tools, or tools with irregular surfaces and while it would provide a visual signal to augmented content, it would also require extensive relabeling in established educational makerspaces – such makerspaces typically are well labeled, especially parts bins, componentry and storage. Relabeling to accommodate the inclusion of a QR code would be a resource intensive effort. While, NFC is less widely used in educational makerspaces, its potential has been suggested as an avenue for future exploration [16]. Making it an interesting technology to study further in this context. It has, however, has been extensively used in industry (e.g. payments, logistics, healthcare) [2, 14, 15, 18, 27], is proven reliable and recent iterations of smartphones have incorporated background NFC-tag scanning capabilities making it more broadly accessible [1]. In addition, as a flexible sticker, it can be widely applied to a variety of surfaces and can be placed on or behind existing labeling.

As such we identified and adopted NFC tags as a low-cost, easy-to-use and scalable mechanism to enable hybrid documentation. Thus, and by using NFC, in our scenario a learner would tap their phone on a soldering iron, for example, to reveal step-by-step guides curated by their instructor, prompts on past students experiences with the device and the opportunity for students to contribute and evolve the documentation. This would make learning guides ready-at-hand and contextually relevant to a student's activities.

2.3 Focus Group and Concept Validation

An initial prototype of a NFC-device was prepared. Two representatives from the co-design workshops were invited to take part in a short, 2-h focus group. This was intended to validate the conceptual design, assess relevance to their makerspaces and solicit further requirements and needs. The participants included one lead for a physical computing educationally-focused makerspace and a program coordinator for a youth summer program that focused on making and entrepreneurship. During the focus group, the prototype was introduced and the two participants were asked to consider scenarios and uses cases, share feedback on the initial implementation, discuss refinements necessary and consider its relevance and applicability at their sites.

During the session, participants immediately drew comparisons to QR codes but P2 noted "my experience has been that they've been kind of a little bit cumbersome for users in that they depending on the angle and the lighting and like how close you get and the camera resolution, it can be a little bit difficult" and "one of the things that I like about NFC is that with the latest of phones that they're passively scanning for NFC, that it's in the background and it's really easy to tap and get it." They also noted the drawbacks of QR codes: "I don't have to open the camera app." "I don't have to then tell it what to do when it gets a URL" and "it's like a four step process."

Both participants noted potential accessibility issues with an NFC based solution. They stressed it was "absolutely essential for some kind of other way of accessing [the resources]" as "we want to have everyone to be able to use" any solution deployed at their site. Participants suggested "having [another dedicated] device that's local, that we're not dependent on people's individual phones". Similarly they noted providing an additional device would allow students who did not have access to a smartphone to still benefit from the platform and gain access to its content easily. They recommended a tablet with NFC capabilities would be ideal as "[t]here's something nice about a tablet size for scrolling through."

The educators also voiced concern at the level of effort it might take to completely outfit an existing makerspace. They offered suggestions as to how the system might practically manage and minimize this effort: "bulk labeling could be super interesting" and linking with or supporting import from existing inventory spreadsheets or database software was seen as advantageous. Both recommendations were later incorporated into the platform.

Both participants saw immediate opportunities for any analytics of student interaction with NFC-tags as being beneficial. These comments centered on the potential of usage data to help instructors monitor and better support student progression through analytics of documentation (and material/part) use: "It would be an easy way at a glance to see if they needed help or resources in a certain area." P2 illustrated this further: "if I could see students as they scan one thing, and I knew what their next steps would be and they stopped or they started over or they got the wrong thing next…I could easily touch base with them and say, Oh, you seem like you're stuck on this one part or you're continually revisiting this." It was also noted that these analytics would be equally valuable to the learners themselves: "they're creating their own interactive history of learning."

Reflecting on this further, P2 remarked "I like the idea of the history, but I also like the idea *of them* creating the history (P2)." Both participants noted analytics would beneficial but fostering an active community that contributed to the continued development of associated tool documentation would be much more advantageous; they also suggested the system should become "a commons for technical components." P1 noted that the system currently repeated the existing model: "[w]hat I have to do now is either point students towards a resources that I wrote… or I point them to another tutorial I found on the internet that I think is a good tutorial." However, P1 preferred if the system could give more agency to the students in building knowledge resources: "I want a user editable page, for every single thing in our inventory, which is akin to a Wikipedia page" as it would be "great to be able to say here's a resource developed mainly in part by faculty and teachers of the space, and maybe in part by other students who's project used this part… and other student can build on it.

The participants concluded the session by considering additional ways this system could be deployed in their space beyond documentation. Tags could be added "in each cubby where students leave their work behind having a link to their process work", as a way of "the door having a feedback survey, that they can scan on the way out, on what they learned that day", and suggested it should allow "these tags to connect to any other network device in the space", for example "we can have a tag that sends a signal to the printer and it prints a recipe for parts and everything that they should go through."

To recap, we validated the "tools with history" concept with two educational stakeholders in a short focus group. They emphasized the needs for: supporting alternative means of accessing the documentation linked to a NFC tag; providing dedicated NFC-enabled devices in the classroom to allow students without access to a smartphone to participate with the platform; revealing analytics to enable learners to trace their tool use and development, in tandem with providing visibility of their learning experience so that instructors can better support them; to enable wiki-like editing of the documentation to give voice and agency to students in supporting their peers and in decentralizing the process of documentation development; and finally, enable tag activations to trigger actions with networked devices and resources. These requirements informed the development of a refined version of the platform described next.

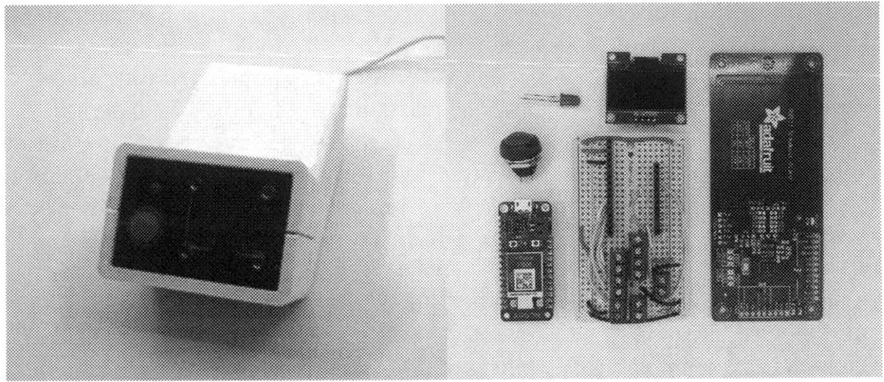

Fig. 3. *Left:* final prototype of NFC tag writer for hybrid documentation. *Right*: depiction of components used.

3 Implementation and System Description

The underlying framework for hybrid documentation comprises a custom NFC-writer device and a cloud platform. The NFC writer is a key component of the infrastructure (see Fig. 3); as it is used to write information to an NFC tag before it is applied to a tool, component or asset within the space. It is designed to be replicable with hardware and components commonly available in makerspaces. It contains a wifi-connected microcontroller, a PN532 NFC board, an OLED screen, a pushbutton, a LED and a 3D printed enclosure.

Figure 4 depicts the relationship of the NFC-writer to the larger framework. The NFC-tag writer is responsible for encoding a URL to each NFC tag. When a new tag is placed on the tag writer, it will be detected and prompt the user to set it up. During the setup process, the device connects with our middleware, which generates a unique URL per tag; comprised of middleware's domain and a tag specific shortcode. Thus, each time the tag is scanned, it first visits our server and is then redirected to the resource, allowing analytics to be gathered.

Once configured, the tag can be scanned with a recent model smartphone. The first time the tag is scanned, it will prompt the user to specify how it is mapped to documentation. Each tag can be set to either redirect to a web resource (e.g. an instructable or guide), or to a wiki-like webpage that can be collaboratively edited by the learning community. This process is illustrated in Fig. 5.

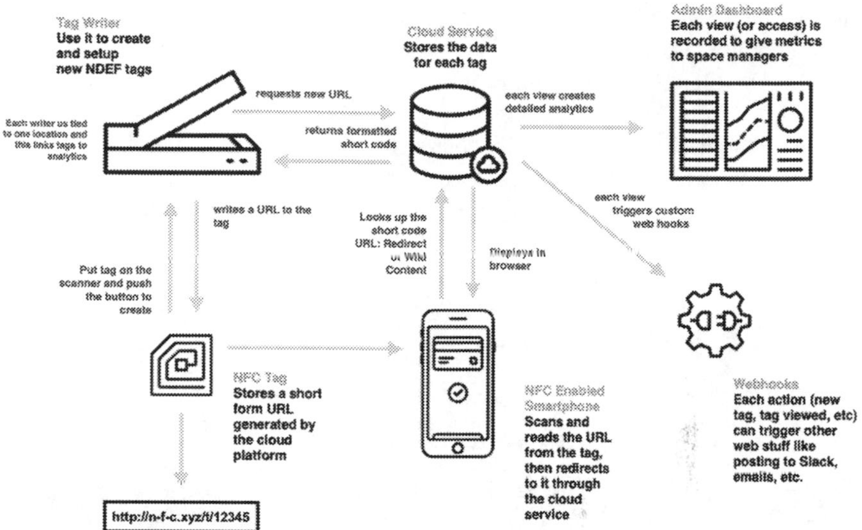

Fig. 4. An overview of the hybrid documentation framework.

The server provides a full featured web application for makerspace managers and authorized users. The user interface provides a dashboard to view analytics, manage tags, and connect webhooks. Webhooks are triggered when a tag is scanned, modified or setup.

Modern smartphones passively scan for NFC tags, so once the phone is proximal to a NFC tag it will immediately load the associated URL. Thereby making documentation ready-at-hand and convenient to access.

This approach is flexible, allowing multiple forms of documentation to be associated with a variety of tools and components. For example, a soldering iron might be linked to an Instructable to guide novices on its use and through a webhook activate its outlet connected with a smart socket. Similarly, and as suggested by focus group participants, storage areas or student project work might be linked to their online portfolio or process blog, while a component bin could be linked to a wiki that distributes the work to the community of describing how to use the resource in their projects and example use cases.

The novelty of this approach is threefold. Firstly, it provides a low-cost, easy-to-adopt, scalable framework that can be overlaid in existing makerspaces without requiring significant technical knowledge, redesign of labeling, signage or componentry's organization. Secondly, this solution offers an integrated platform to support learning in an educational makerspace. While RFID/NFC is already widely used in makerspaces for access control, here, it deployed to solves two common key challenges for learners in makerspaces: identifying and distributing instruction in-situ and ready-at-hand while providing agency to the learning community to continually gather and evolve learning resources through collaborative editing. Thirdly, and by gathering analytics of the documentation accessed by learners, it provides educators and instructors increased visibility on what knowledge resources are being frequently accessed and how this changes over the course of a semester. This can enable them to direct effort to digital documentation and make curricular moves to better support novices.

There are some limitations to this approach. Most notably, it is reliant on smartphones that support NFC technology, and this may limit the number of students who can adopt and interact with such an approach. As noted previously, this can be overcome by providing a dedicated NFC-enabled tablet or smartphone for students without access to a NFC-enabled device to access the content conveniently. In addition, the documentation resources can be accessed via a web frontpage for the makerspaces to provide another accessible mode of access.

1. Turn on the box by plugging it in. It may take a moment for the screen to turn on.

2. Place the tag on top of the box, and follow the instructions on the screen.

3. Scan the tag with your phone to set up the URL.

4. To change the URL on the tag again, repeat steps 2 and 3.

5. Peel off the tag and stick it somewhere!
Notes on placing tag:
The tags don't work on metal surfaces. The closer the phone can get to the tag, the better, but the tag can work up to 1 cm through wood, paper or plastic. The tag can be placed on electronic screens, and can be written on with a sharpie, pen or pencil.

Fig. 5. A step-by-step guide to set-up of a NFC-tag with the custom tag reader

4 Evaluation

As a formative evaluation of the hybrid documentation framework, we conducted a two-hour workshop at a high school with five educators and two administrators. The high school, and participants, had previously taken part in the previously described co-design workshop. As such, all participants were familiar with the concepts generated from this work and valued documentation in supporting student learning. The workshop was held in the school's makerspace, a large dedicated room situated in the school's library.

Table 1. A summary of concepts generated by the educators.

Physical item	Digital content	Learning goal or purpose
Wire stripper	Tool information and demo- how is it used and named; Pictures and video of students using the tool	Help familiarize students with tools and skill development
Computer project kit	Dis/assembly guide through rich media (text and video)	Set stage for students to understand what's inside the box and how to use the components
Soldering area	Tutorial on soldering, safety video, tools associated with soldering	Use of equipment; safe use of equipment
Roland mill	Tutorial on how the tool is used, a safety tutorial, student work produced on the tool	To help student to know how to use a piece of equipment
VR googles	Video introducing VR equipment and how to use google earth; guidance on setting it up for use in the classroom	Introduce students to the concepts of VR equipment for a project
Past project (Game Console)	Video of student work; reflections on construction; highlights video	Provide reflections on the creation of a product; highlights student contributions

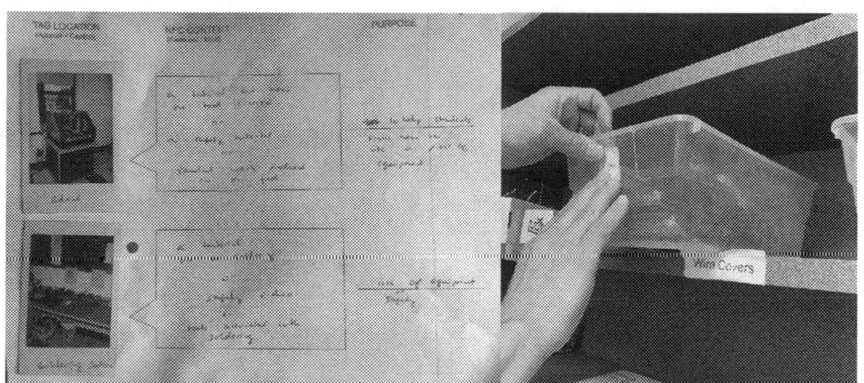

Fig. 6. *Left*: NFC tag brainstorming. *Right*: Instructors deploying tags in the makerspace

4.1 Procedure

The workshop began by reviewing the co-design process outcomes and discussing the 'tools with histories' concepts generated. The hybrid documentation framework was then introduced and demonstrated using a series of staged artifacts including linking a multimeter to a tutorial on its use and a clipboard linked to an online collaboratively editable to-do list.

After introducing the system, participants toured the makerspace using Polaroid photo elicitation activity to help identify scenarios that could be supported by the system [4]. As part of this process, individuals were asked to capture a polaroid photo of an object, tool, part, material or area in the space that could be augmented with NFC-enabled documentation. Using the worksheet (see Fig. 6), they recorded what content they would like to associate with it and the educational goal or purpose that it would serve. This exercise lasted 15 min, after which participants gathered to share out their use-cases. A total of unique 14 objects and areas were marked-up. This included an interactive white board and digital signage, equipment (green screen, thermal printer, Roland mill, 3D printer), tools (wire stripper, 3D Pen), storage bins and areas (soldering station, entry way), a project kit (a computer to be disassembled/reassembled) as well as one past student project (a gaming console). While more creative applications were covered (attendance tracking, gathering documentation), these initial scenarios emphasized providing introductory content to "introduce students to makerspace and equipment" and increase their recognition of the type and use of tools. They favored being able to help students name specific tools, increase understanding of tool affordances, encouraging safe tool and equipment use, providing curated introductory guides, and revealing procedures and setup processes for proper use. A subset of participant-generated scenarios is provided in Table 1. Working with the educators, they were then invited to use the framework to enact these scenarios within the makerspace, as illustrated in Fig. 6. This was followed by open-ended discussion with the stakeholders. The discussion was audio recorded, transcribed and synthesized. Findings are next described.

4.2 Discussion and Findings

Flexibility: Participants appreciated the system's potential to increase familiarity and contextual know-how with tools without instructor support, remarking that this approach to instructional resources is about "making it more accessible. It's more direct." They also valued its adaptability to a variety of uses including administering surveys, linking portfolios, and gathering documentation. Of most importance to the participants was the fact that it builds on, rather than replacing, existing digital content at their disposal. This school uses Google products (Drive and Sites to gather project documentation) so it would be convenient for both educators and students to work with: "Google drive is something that we're comfortable using", "we could write step-by-step guides for some of these scenarios…using Google drive", "probably that would [link to] a separate webpage of the student would build." By linking to online resources, they felt it could easily integrate with existing digital platforms in the school and layer on instructional practices already used in their curricula. This could save time and effort in deploying the system.

Effort and Coordination: However, educators shared that they did not have enough suitable content to begin building the use cases they envisioned. One teacher noted "I'd love to do [this] but I don't have the student reflections. I don't have the videos and that content created yet." Another agreed but noted "the real goal is getting the students to get content in there." Stakeholders further discussed that getting students to cooperatively prepare documentation and resources for the makerspace would be helpful but would need to be situated within a curricular process to be most effective. Coupling a class project with developing documentation for tools and part, would be most beneficial to students as "its a really nice use beyond just content or information… but if we're looking for students create content. By the end of this term, [a student] will have created two tutorials. The creating of the tutorial is kind of a bonus. Somebody else gets to learn from viewing the tutorial, but…there's a lot of evidence of learning in the process of creating the tutorial." As such, the suggestion was to have students document outcomes that contribute to the situated knowledge of the makerspace; this would allow the effort in developing resources to be shared, and mutually benefit student by creating shared resources for others in the space, while being a vehicle to evidence their own learning on projects.

Issues Surfaced: In considering a curricular framework where students cooperatively develop content, educators raised concern over broad access to edit the linked content. Instructors wanted greater oversight over content to avoid digital vandalism or misuse. Additionally, the group was cautious about the analytics, not wanting to be "too heavy handed with tracking" as that could have "raised other kinds of concerns about how the data is being used and that kind of thing" for students. Finally, they suggested the need to "develop some sort of visual convention, like a sticker, an icon or something. So that students start to recognize this as an NFC tag." This led to a discussion comparing the approach to QR codes, which have a more prominent visual signifier.

Knowledge Discovery and Credibility: During this comparison of the two approaches, one participant raised the question "why shouldn't they [the students] just Google?" to find the resources they need. This led to a productive conversation. One participant responded: "You can Google anything and we don't really want to test kids on things that they can Google. So if they can Google how to use this tool, why should we not just make it accessible for them?" Another noted that by making the content available in-situ and "on demand, that changes instructionally what can happen and what you can do because [the teacher]'s gained back that time" from formally introducing each tool in class. Discussion also noted that by curating recommended resources, "[t]he teacher is directing to where you're going. So it's a credible source. It's what you've selected as the training" It was noted that students who are newer to projects involving tools and technology, often don't know how independently identify suitable sources of knowledge for their project work. This solution overcomes those challenges.

Value in Learning: Throughout the discussion, the educators noted several opportunities to create value in their educational experiences, specifically in personalizing the experience, supporting reflection and inquiry, in leaving knowledge behind for others and in helping to gather multiple forms of evidence of learning.

First, this framework, the analytics gathered and the ability for students to contribute knowledge resources to the makerspace were all seen as new and interesting mechanisms to evidence student learning: by accessing the guidance and by tracking activity students are indirectly "documenting their ability to be able to use the tool." In offering ways for students to contribute to the documentation too, this could be more actively demonstrated. Discussing the soldering station, one educator remarked "I would love to have students literally sit here and take a 30-second video of the evidence" that they have practiced soldering and add it to the associated tag's documentation. In so doing, it was seen as a way for students to "start to learn how to articulate what they know and learn"

Second, in making details about tools, components and materials accessible in-situ and in the context of hands-on project work, it would "have a role in helping [students] reflect on what's possible for their projects" This was seen to be of benefit not only to increase familiarity with tools but to allow students to reflect on the forms of inquiry that the tools support.

Third, it was seen as an opportunity to enhance personalized learning and self-directed discovery within their makerspace: "the tools themselves, they provide training and guidance" and by virtue of this it would "be personalized to the students as well"; they can find the knowledge they need to work with a tool in the context of the activity. This was further elaborated: "some of the time that you spend in class explaining [tools] such as a hammer. This now becomes on demand learning…It's almost like the flipped classroom concept where you give them the basics ahead of time for them to view, or when they need to know how to do it." Educators noted that this is creates a means for students "to *own their learning*. And that helps them… It's not an easy shift but that's a very valuable and a meaningful shift"

Summary of Findings: The workshop yielded the following findings. The advantage of the hybrid documentation approach is to make existing guides, tutorials and resources more accessible and physically colocated with the tools, materials and parts they relate to. Educators noted that this has the additional advantages of being able to work within existing educational technology and platforms and the content they house. Educators recognized that by curating digital content within this framework, they were both reducing the friction for novice makers to identify knowledge and resources but also explicitly guiding students to credible, educator-validated resources. Finally, educators noted the approach to create multiple educational values, and importantly, the potential to increase student agency for self directed learning in a makerspace and to gather new forms of evidence of learning. These preliminary findings will be the subject of future investigation and validation.

5 Future Work

We have discussed NFC-enabled tags as a low-cost strategy for improving learning and present a framework to guide rapidly augmenting maker-based educational contexts. Initial evaluations demonstrated this approach as an accessible, ready-at-hand solution. There remains much further work to conduct to improve and validate this approach with students. An immediate next step for this project will be to conduct a longitudinal analysis where the tool is deployed in educational makerspace and coordinated with a curriculum over a full semester experience.

In addition, we plan to platform's support for hybrid documentation. Specifically, we plan to improve the features for knowledge discovery and organization. As the platform intends to provide agency to learners in iterating and gathering documentation relevant to their needs, we intend to explore user-centered mechanisms to incentivize peer-to-peer, collaborative documentation authoring and maintenance.

6 Conclusion

With growing interest in the maker movement and project-based learning in education, there has been much recent research on this space. This has largely coordinated around studying educational benefits, fostering documentation and portfolio practices, and identifying technological augmentations the tools offered to support novice learners and skilled work. In this work, we bridge all three areas. Towards this, we describe a co-design process with educators involved in maker- and project-based settings. This process identifies high value technological concepts to enhance education and support educator's learning goals. Outcomes were synthesized into concept for hybrid documentation and delivered a prototype that leverages NFC technology in educational makerspaces. This solution provides a low-cost, scalable alternative to recent research in augmented tools and spaces for making. We conducted a preliminary evaluation with educators to demonstrate the value of adapting NFC to educational contexts, as well as, the applicability of NFC in maker- and project-based learning documentation.

Acknowledgements. We would like to thank our educational partners in the Pittsburgh region for supporting this research and exploration. Additionally, we would like to acknowledge the work of the following students in supporting the work on this project: Alice Fang, Emily Chan, Jean Zhang and Wei Wei Chi. This work is supported by the National Science Foundation (Grant No. 1736189).

References

1. Apple. What's New in Core NFC (2018). https://developer.apple.com/videos/play/tech-talks/702/. Accessed 10 Mar 2019
2. Benyo, B., Vilmos, A., Kovacs, K., Kutor, L.: NFC applications and business model of the ecosystem. In: 2007 16th IST Mobile and Wireless Communications Summit, pp. 1–5. IEEE, July 2007
3. Brown, J.O.: Know thyself: the impact of portfolio development on adult learning. Adult Educ. Q. **52**(3), 228–245 (2002)
4. Byrne, D., et al.: MakeSchools higher education alliance state of making report (2015). http://makeschools.org/week_of_making/report. Accessed Jun 2015
5. Chang, S., Keune, A., Peppler, K., Maltese, A., McKay, C., Regalla, L.: Open portfolio project, phase 1: research brief series (2015). http://makered.org/opp/publications/
6. Collins, A., Brown, J.S., Holum, A.: Cognitive apprenticeship: making thinking visible. Am. Educ. **15**(3), 6–11 (1991)
7. Cross, N.: Designerly Ways of Knowing. Birkhäuser, Basel (2007)

8. Durak, G., Özkeskin, E., Ataizi, M.: QR codes in education and communication. Turk. Online J. Distance Educ. **17**, 42–58 (2016)
9. Gong, J., Anderson, F., Fitzmaurice, G., Grossman, T.: Instrumenting and analyzing fabrication activities, users, and expertise. In: Proceedings of the 2019 CHI Conference on Human Factors in Computing Systems, pp. 1–14, May 2019
10. Hall, L., Jones, S., Hall, M., Richardson, J., Hodgson, J.: Inspiring design: the use of photo elicitation and Lomography in gaining the child's perspective. In: Proceedings of the 21st British HCI Group Annual Conference on People and Computers: HCI… but not as we know it, vol. 1, pp. 227–236. British Computer Society, September 2007
11. Hira, A., Joslyn, C.H., Hynes, M.M.: Classroom makerspaces: identifying the opportunities and challenges. In: 2014 IEEE Frontiers in Education Conference (FIE) Proceedings, pp. 1–5. IEEE, October 2014
12. Huang, R., Gu, J., Sun, X., Hou, Y., Uddin, S.: A rapid recognition method for electronic components based on the improved YOLO-V3 network. Electronics **8**(8), 825 (2019)
13. Knibbe, J., Grossman, T., Fitzmaurice, G.: Smart makerspace: an immersive instructional space for physical tasks. In: Proceedings of the 2015 International Conference on Interactive Tabletops & Surfaces, pp. 83–92, November 2015
14. Islam, S.R., Kwak, D., Kabir, M.H., Hossain, M., Kwak, K.S.: The internet of things for health care: a comprehensive survey. IEEE Access **3**, 678–708 (2015)
15. Lazaro, A., Villarino, R., Girbau, D.: A survey of NFC sensors based on energy harvesting for IoT applications. Sensors **18**(11), 3746 (2018)
16. Lensing, K., Schwuchow, B., Oehlandt, S., Haertel, T.: How Makerspaces help to participate in technology: results of a survey to gain data about learners' activities in makerspaces. In: 2018 World Engineering Education Forum-Global Engineering Deans Council (WEEF-GEDC), pp. 1–5. IEEE, November 2018
17. Litts, B.K.: Resources, facilitation, and partnerships: three design considerations for youth makerspaces. In: Proceedings of the 14th International Conference on Interaction Design and Children, pp. 347–350, June 2015
18. Madakam, S., Lake, V., Lake, V., Lake, V.: Internet of Things (IoT): a literature review. J. Comput. Commun. **3**(05), 164 (2015)
19. Massis, B.E.: QR codes in the library. New Libr. World **112**(9/10), 466–469 (2011). https://doi.org/10.1108/03074801111182058
20. McHugh, S., Yarmey, K.: Near Field Communication: Recent Developments and Library Implications. Morgan & Claypool Publishers, San Rafael (2014)
21. Peppler, K., Maltese, A., Keune, A., Chang, S., Regalla, L.: The maker ed open portfolio project: survey of makerspaces, part II. In: Open Portfolios: Maker Education Initiative Research Brief Series, pp. 47–53 (2015)
22. Peppler, K., Keune, P.: "It helps create and enhance a community": Youth motivations for making portfolios. Mind Cult. Act. **26**(3), 234–248 (2019). https://doi.org/10.1080/10749039.2019.1647546
23. Open portfolios maker education initiative. Research Brief Series (2015). http://makered.org/opp/research-briefs/
24. Ritchhart, R., Perkins, D.: Making thinking visible. Educ. Leadersh. **65**(5), 57–61 (2008)
25. Schoop, E., Nguyen, M., Lim, D., Savage, V., Follmer, S., Hartmann, B.: Drill sergeant: supporting physical construction projects through an ecosystem of augmented tools. In: Proceedings of the 2016 CHI Conference Extended Abstracts on Human Factors in Computing Systems, pp. 1607–1614. ACM, May 2016
26. Schön, D.A.: The Design Studio: An Exploration of its Traditions and Potentials. RIBA Publications Limited, London (1985)

27. Shobha, N.S.S., Aruna, K.S.P., Bhagyashree, M.D.P., Sarita, K.S.J.: NFC and NFC payments: a review. In: 2016 International Conference on ICT in Business Industry & Government (ICTBIG), pp. 1–7. IEEE, November 2016
28. Tarjan, L., Šenk, I., Tegeltija, S., Stankovski, S., Ostojic, G.: A readability analysis for QR code application in a traceability system. Comput. Electron. Agric. **109**, 1–11 (2014)
29. Tseng, T., Yang, M., Ruthmann, S.: Documentation in progress: challenges with representing design process online. In: ASME 2014 International Design Engineering Technical Conferences and Computers and Information in Engineering Conference, pp. V003T04A027–V003T04A027. American Society of Mechanical Engineers, August 2014
30. Wired. The Curious Comeback of the Dreaded QR Code (2017). https://www.wired.com/story/the-curious-comeback-of-the-dreaded-qr-code/. Accessed 18th June 2020

The Adoption of Mobile Technologies in Healthcare: The Perceptions of Healthcare Professionals Regarding Knowledge Management Practices in Developing Countries

Avijit Chowdhury[1(✉)], Abdul Hafeez-Baig[1], Raj Gururajan[1], and Mirza Akmal Sharif[2]

[1] University of Southern Queensland, Toowoomba, QLD 4350, Australia
chowdhury.avijit@gmail.com

[2] University Medical and Dental College, Madina Teaching Hospital, Faisalabad, Pakistan

Abstract. The purpose of this paper is to explore the perceptions of healthcare professionals (HCPs) in the countries of India and Pakistan, that impede knowledge management practices by adopting mobile technologies in healthcare. An exploratory study was conducted among healthcare professionals in India and Pakistan to assess their perceptions regarding knowledge management practices using mobile applications in their working environment. This cross-sectional qualitative study used semi-structured interviews with healthcare professionals in both India and Pakistan. The interviews were analyzed using a thematic analysis, which revealed three major themes. These themes being: Medical education and training, Collaboration between HCPs, and Patient health education. Findings indicate that the HCPs associated with mobile technologies in healthcare in India and Pakistan generally view it in a functional manner, rather than in an explicit manner. The mobile healthcare technologies in use are looked upon as an alternative solution for clinical purposes such as viewing diagnostic reports or providing initial information to the patients, and many more. The recognition of the idea of explicit and tacit knowledge is not usual among the HCPs. Therefore, the knowledge-sharing ability of mobile healthcare technologies is understood below par and often underutilized.

Keywords: Healthcare · Developing countries · Adoption of mobile technologies in healthcare

1 Introduction

The purpose of this paper is to explore the perceptions of healthcare professionals (HCPs) in the countries of India and Pakistan, regarding knowledge management practices using mobile technologies. Various authors have defined mobile technologies in healthcare as, smartphone applications (West 2012), health monitoring wearable devices (Gao 2015; Li et al. 2016), infrared and radio-based locator badges

C. Stephanidis et al. (Eds.): HCII 2020, LNCS 12427, pp. 68–76, 2020.
https://doi.org/10.1007/978-3-030-60152-2_5

(Varshney 2007), wireless monitors for blood oxygen saturation, remote heart monitoring, digitized ECG, and EEG epilepsy monitoring (Wu et al. 2011). The mentioned studies considered intention, attitude, and factors influencing the adoption of mobile and wireless devices. Consequently, few studies focused on integrating knowledge management practices with adoption of mobile technologies by the HCPs (Räisänen et al. 2009; Standing et al. 2018, 2011). Further, there are fewer studies focused on developing countries (Chowdhury et al. 2019; Hafeez-Baig et al. 2009). As the adoption of mobile technologies in the healthcare environment will possibly rise (Koumpouros and Georgoulas 2019; Latif et al. 2018; Moghaddam and Lowe 2019; West 2012), the perception regarding knowledge management practices of the HCPs can provide an insight how the HCPs can collaborate and learn from interaction among themselves (Standing et al. 2018). Further, any knowledge on the perceptions of the HCPs regarding knowledge management practices will help to make standardized policies at governmental and non-governmental levels. Therefore, the research query was focussed on finding the perceptions of the HCPs regarding knowledge management practices while adopting mobile technologies in healthcare environments in India and Pakistan as scant studies have been undertaken on this topic, which was evident from the literature review. Further, the literature review suggests many similarities between the healthcare environment in India and Pakistan (Nishtar et al. 2015).

2 Literature Review

Mobile devices and mobile applications are in use in the healthcare domain for a considerable period, especially in developed countries (Trmcic et al. 2016). Efficient management of knowledge can enhance the health outcomes using such technologies (Eysenbach 2001). Nevertheless, managing knowledge can be a complicated affair to accomplish in healthcare practice (Dwivedi et al. 2001). The literature presents no standardized or transparent definitions of knowledge. One of the studies alludes to knowledge, as information retained in the human mind, with some meaning attributed to it (Alavi and Leidner 2001). Further, some authors defined knowledge as a belief which justifies an increase in organizational efficacy (Kelp 2016). The forms of knowledge can be codified, as found in systems software, or dynamic conversations may yield such knowledge (Bock et al. 2005; Chen et al. 2012; Nonaka 1994; Nonaka and Takeuchi 2007; Sveiby 1996). Thus, tacit knowledge is the personal knowledge stored in the human mind, but its expression entails complication. On the contrary, explicit knowledge is expressible in codes or written modes (Nonaka and Takeuchi 2007). Ideas, experiences, difficulties faced, requirements are included in tacit knowledge whereas ideas in written forms such as reports, articles, books are known to be explicit knowledge (Panahi et al. 2016). Recent studies, increasingly are focusing on the conversion of tacit knowledge to explicit knowledge (Shepherd and Cooper 2020). The role of information and communications technology (ICT) in converting tacit knowledge to explicit knowledge is ambiguous enough, though recent studies show the potential of its role in sharing knowledge and codifying the same in healthcare domain (Dagenais et al. 2020). ICT in healthcare can be used for sharing information and knowledge, although there are not enough standards (Standing et al. 2014).

3 Methodology

3.1 Design

The aim of this research was to gain an understanding of the perceptions of the HCPs in India and Pakistan, regarding how knowledge management practices influence the adoption of mobile technologies in healthcare. This exploratory qualitative research was conducted using in-depth semi-structured interviews with a range of HCPs in both India and Pakistan, to collect their perceptions about training and knowledge requirements in using mobile technologies. Doctors and other allied HCPs were interviewed with their responses being recorded. A manual thematic analysis (Braun and Clarke 2006) was conducted to analyze the healthcare professionals narratives so as to identify, evaluate and report themes regarding knowledge management practices within healthcare mobile technologies framework. Ethics approval to conduct this research was obtained from the relevant authorities prior to conducting the interviews.

3.2 Qualitative Data Collection

The researchers conducted in-depth, open-ended, semi-structured interviews in a few states of India and Pakistan to gain an understanding of the perceptions of HCPs in developing and emerging economies. The participants were HCPs in the capacity of doctors, dietitians, and healthcare administrators. The researchers utilized cross-sectional data collection for this research, with seventeen interviews being conducted between February 2019 to June 2019 with HCPs located in India and Pakistan. The participant HCPs were situated in the Punjab state (eight participants) in Pakistan, and the Indian states of Uttar Pradesh (three participants), Maharashtra (two participants), Karnataka (one participant) and Tamil Nadu (three participants) and all were acquainted with mobile technologies in healthcare. Random convenience sampling was used for selecting the participants based on ease, accessibility and low costs (Acharya et al. 2013; Creswell 2014). Ten participants were males and seven were females. Among the HCPs, twelve were doctors, four were dietitians, and one was a healthcare administrator. Six of the participants were involved in organized mobile technology set-up, and eleven were involved in online consultation using mobile technologies in a private capacity. A pilot study was conducted by considering the initial five interviews to test and confirm the reliability and validity of the questions and respective dataset. The interview questions were refined accordingly after the pilot study. The interviews continued till a final saturation point is reached wherein successive responses were similar, and theme saturation was achieved (Bowen 2008). The participants were asked for their opinion on sharing knowledge using mobile technologies and applications. The interview hovered around their perceptions on knowledge management, including sharing, willingness, usefulness, training requirements in the healthcare environment.

3.3 Qualitative Data Analysis

The initial phase of data analysis involved the transcription of the recorded data to written data by the researchers. The researchers elected not to use computer aided software to assist with the transcription and analysis as, although it is faster, the researchers wanted to become immersed in the data and get a sense of the whole interview prior to commencing the initial search for themes (Creswell 2014; Dohan and Sanchez-Jankowski 1998). An inductive thematic analysis was applied to the transcribed data as it enabled the researchers to identify and develop themes via iterative readings of the transcripts (Elo and Kyngäs 2008; Vaismoradi et al. 2013).

4 Findings and Discussion

Three main themes, associated with mobile technologies in healthcare, emerged from the analysis of the interview transcripts. These themes were as follows:

- Medical education and training
- Collaboration between HCPs
- Patient health education

4.1 Medical Education and Training

The theme emerged from the collected data when the participants identified that medical education and training facilities could be enhanced using mobile applications in healthcare. A participant from Pakistan (P1) identified that video recordings of his surgery could, later on, provide an opportunity for the students to recapitulate the critical steps performed by him.

> *"When I am teaching my students to examine the patients, many are making videos at the same time. Now if that video is seen afterwards they can see it many times, they can correct their methods and they can learn a lot, and that can be uploaded online for viewing by other persons."* (P1)

Several other Pakistani participants agreed.

> *"We are doing OPD…………particularly in dermatology and surgery as well and we are guiding the clinical attendant there……….medical officer or medical graduates…………they consult us using video link and we advise them what to do regarding minor surgery."* (P3)

The Indian participants expressed similar views on medical education and training both for the healthcare students and continuing medical education for healthcare professionals. One of the Indian participants (I7) expressed his views on intellectual debates on mobile platforms useful both for the healthcare students and professionals.

> *"Intellectual debates open up the platform for the doctors and students, and those platforms can be accredited by the medical council and doctors get points for it because it is part of their practice.....continuing medical education for doctors."* (I7)

Another Indian participant (I1) expressed affirmative views on using mobile technologies in providing training to the health workers.

> *"We also educate and train the health workers in various aspects of telemedicine.....using mobile applications."* (I1)

Interestingly it can be seen that a hint of using tacit knowledge to explicit knowledge has been demonstrated by participants in both countries, though the answers also reflect the lack of understanding of effective management of the same. Further, the participants instead showed a willingness to share their knowledge, but there was a lack of emphasis and strategy (Standing et al. 2018).

4.2 Collaboration Between HCPs

The theme emerged when several participants both from India and Pakistan opined about the possibility of collaboration between the HCPs using mobile technologies in the respective healthcare environment. An Indian participant (I6) shared her views on knowledge sharing with several others agreeing as well, which reflected collaboration among the fellow healthcare professionals.

> *"I am willing to share knowledge about online consulting as a dietitian. Webinars would be a good platform to share knowledge with fellow healthcare professionals."*(I6)

Another Indian participant agreed (I10):

> *"It is not teleconsultation, but it is the doctor to doctor, the specialist talks to the primary care physician and empowers him to manage the cases."* (I10)

The Pakistani participants shared similar views reflecting collaboration when asked about their perceptions on knowledge sharing in a healthcare environment using mobile technologies. One Pakistani participant (P4) expressed her views on knowledge sharing as:

> *"Mainly if the radiologist is sitting far away from the main city and he is sitting in the darkroom he can't report………he cannot report the clinical diagnosis of the person. By seeing or by taking the expert opinion of the radiologist the doctor can exactly take the professional opinion from the expert from the teaching hospitals by using this technology."*(P4)

Another Pakistani participant (P7) agreed on the same in as much as several other Pakistani doctors.

> *"Doctor-doctor interaction is going on……we collaborate and share our cases…..I upload some notorious cases……..seeking opinion of other doctors……..I discuss all over Pakistani doctors……everybody provides their opinion and sometimes the patient is seen by multiple people. That's a great use of technology."* (P7)

The similarity between the Indian and Pakistani participants on their views on knowledge sharing and collaboration is noted but, as with the first theme of medical education and training, a lack of strategy is evident. Further, the knowledge is poorly managed as the HCPs does not take a resource-based view where the creation of knowledge is considered as value addition in capability (Ferlie et al. 2012). Further, the participants from both India and Pakistan pointed out the challenges, including not having standard guidelines, improper ICT infrastructure, hindering the usage of mobile technologies in healthcare.

4.3 Patient Health Education

The theme of patient health education emerges in the course of the interviews when the participants view reflected on the lack of patient awareness of health when asked about knowledge sharing using mobile technologies. All the participants from both India and Pakistan agreed that mobile technologies serve an useful purpose in educating patients and health awareness among them. An Indian participant (I7) opines:

> *"The platform can be used to educate the doctors as well as the patients. It is absolutely important."* (I7)

The opinion was agreed upon by another Indian participant (I3) as well as others.

> *"Very useful. Can reach out to the globe. Skype, Whatsapp, facebook to educate patients."* (I3)

The Pakistani counterparts of the participants expressed similar views about using mobile technologies in knowledge sharing. One Pakistani participant (P5) expressed her views as:

> *"We can give awareness to………now this social media is known to everyone can contact…….. but if we can give them some knowledge on these areas that you can get facilities……..but same thing is……it will take time……they will take time to understand."* (P5)

Another Pakistani participant (P8) along with several others agreed on the need for patient education using mobile technologies.

> *"Teleconsultations are a very good option but the usage of these telehealth services are beneficial in those societies which are very highly educated and patient education on the grass root level is advisable."* (P8)

The interview excerpts reflect that the perceptions of the healthcare professionals both in India and Pakistan are similar in a way that both agreed on the usage of mobile technologies and applications in healthcare to spread and promote health awareness among patients.

A key finding of this research study is that the HCPs associated with mobile technologies in healthcare in India and Pakistan generally view it in a functional manner, rather than in an explicit manner. The mobile healthcare technologies in use are looked upon as an alternative solution for clinical purposes such as viewing diagnostic reports or providing initial information to the patients, and many more. The recognition of the idea of explicit and tacit knowledge is not usual among the HCPs. Therefore, the knowledge-sharing ability of mobile healthcare technologies is understood below par and often underutilized (Räisänen et al. 2009).

5 Limitations

As this research is based on a limited scale healthcare professionals' opinions, the findings cannot be generalized unless by conducting a wider scale quantitative survey. This paper provided valuable insight and encourages such future research endeavours as to the knowledge management practices involving mobile technologies in healthcare domain of the Indian subcontinent. The present cutting edge technologies can be used to solve the challenges, only if there is a robust adoption process (Bartolini and McNeill 2012). Developing patient awareness and increasing healthcare expenditure to address such issues is the need of the hour.

6 Conclusion

Mobile technologies in healthcare have become much more apparent especially in developed countries where the healthcare infrastructure is adequate and supportive (Standing et al. 2018). In contrast, there are necessary steps which can be undertaken for the effective sharing of knowledge and collaborative purposes for the HCPs in the developing countries. The support from top management in healthcare institutions, a vision plan for sharing knowledge, adequate ICT infrastructure, and an awareness of the latest approaches in knowledge management practices are some of the prerequisites for successful adoption of mobile technologies by HCPs in developing countries.

Acknowledgement. The authors acknowledge the support received by Avijit Chowdhury from the University of Southern Queensland in the form of fees research scholarship for Ph.D studies.

References

Acharya, A.S., Prakash, A., Saxena, P., Nigam, A.: Sampling: why and how of it. Indian J. Med. Spec. **4**(2), 330–333 (2013)

Alavi, M., Leidner, D.E.: Knowledge management and knowledge management systems: Conceptual foundations and research issues. MIS Q. **25**(1), 107–136 (2001)

Bartolini, E., McNeill, N.: Getting to Value: Eleven Chronic Disease Technologies to Watch. NEHI, Boston (2012). http://www.nehi.net/publications/30-getting-to-value-eleven-chronic-disease-technologies-to-watch/view. Accessed 12 Nov 2019

Bock, G.-W., Zmud, R.W., Kim, Y.-G., Lee, J.-N.: Behavioral intention formation in knowledge sharing: Examining the roles of extrinsic motivators, social-psychological forces, and organizational climate. MIS Q. **29**(1), 87–111 (2005)

Bowen, G.A.: Naturalistic inquiry and the saturation concept: a research note. Qual. Res. **8**(1), 137–152 (2008)

Braun, V., Clarke, V.: Using thematic analysis in psychology. Qual. Res. Psychol. **3**(2), 77–101 (2006)

Chen, S.-S., Chuang, Y.-W., Chen, P.-Y.: Behavioral intention formation in knowledge sharing: Examining the roles of KMS quality, KMS self-efficacy, and organizational climate. Knowl.-Based Syst. **31**, 106–118 (2012)

Chowdhury, A., Hafeez-Baig, A., Gururajan, R., Chakraborty, S.: Conceptual framework for telehealth adoption in Indian healthcare. In: Paper presented at the 24th Annual Conference of the Asia Pacific Decision Sciences Institute: Full Papers (2019)

Creswell, J.W.: Research Design: Qualitative, Quantitative, and Mixed Methods Approaches. SAGE, Thousand Oaks (2014)

Dagenais, C., Dupont, D., Brière, F.N., Mena, D., Yale-Soulière, G., Mc Sween-Cadieux, E.: Codifying explicit and tacit practitioner knowledge in community social pediatrics organizations: Evaluation of the first step of a knowledge transfer strategy. Eval. Program Plann. **79**, 101778 (2020)

Dohan, D., Sanchez-Jankowski, M.: Using computers to analyze ethnographic field data: theoretical and practical considerations. Ann. Rev. Sociol. **24**(1), 477–498 (1998)

Dwivedi, A., Bali, R.K., James, A.E., Naguib, R.N.G.: Telehealth systems: considering knowledge management and ICT issues. In; Paper presented at the 2001 Conference Proceedings of the 23rd Annual International Conference of the IEEE Engineering in Medicine and Biology Society, 25–28 October 2001 (2001)

Elo, S., Kyngäs, H.: The qualitative content analysis process. J. Adv. Nurs. **62**(1), 107–115 (2008)

Eysenbach, G.: What is e-health? J Med Internet Res **3**(2), e20 (2001) https://doi.org/10.2196/jmir.3.2.e20

Ferlie, E., Crilly, T., Jashapara, A., Peckham, A.: Knowledge mobilisation in healthcare: a critical review of health sector and generic management literature. Soc. Sci. Med. **74**(8), 1297–1304 (2012)

Gao, Y.: An empirical study of wearable technology acceptance in healthcare. Ind. Manage. Data Syst. **115**(9), 1704–1723 (2015). https://doi.org/10.1108/imds-03-2015-0087

Hafeez-Baig, A., Gururajan, R., Mula, J.M., Lin, M. K.: Study to investigate the determinants for the use of wireless technology in healthcare setting: a case of Pakistan. In: Paper presented at the 2009 Fourth International Conference on Cooperation and Promotion of Information Resources in Science and Technology, 21–23 November 2009 (2009)

Kelp, C.: Justified belief: Knowledge first-style. Philos. Phenomenol. Res. **93**(1), 79–100 (2016)

Koumpouros, Y., Georgoulas, A.: The rise of mHealth research in Europe: a macroscopic analysis of EC-funded projects of the last decade. In: Mobile Health Applications for Quality Healthcare Delivery, pp. 1–29. IGI Global (2019)

Latif, S., et al.: Mobile technologies for managing non-communicable diseases in developing countries. In: Mobile Applications and Solutions for Social Inclusion, pp. 261–287. IGI Global (2018)

Li, H., Wu, J., Gao, Y., Shi, Y.: Examining individuals' adoption of healthcare wearable devices: an empirical study from privacy calculus perspective. Int. J. Med. Informatics **88**, 8–17 (2016). https://doi.org/10.1016/j.ijmedinf.2015.12.010

Moghaddam, G. K. and Lowe, C.R.: Mobile healthcare. In: Health and Wellness Measurement Approaches for Mobile Healthcare, pp. 1–11. Springer, Cham. https://doi.org/10.1007/978-3-030-01557-2_1

Nishtar, S., et al.: Pak-India collaborations in health: insights and way forward. Global Public Health **10**(7), 794–816 (2015). https://doi.org/10.1080/17441692.2015.1035301

Nonaka, I.: A dynamic theory of organizational knowledge creation. Organ. Sci. **5**(1), 14–37 (1994)

Nonaka, I., Takeuchi, H.: The knowledge-creating company. Harvard Bus. Rev. **85**(7/8), 162 (2007)

Panahi, S., Watson, J., Partridge, H.: Information encountering on social media and tacit knowledge sharing. J. Inf. Sci. **42**(4), 539–550 (2016)

Räisänen, T., Oinas-Kukkonen, H., Leiviskä, K., Seppänen, M., Kallio, M.: Managing mobile healthcare knowledge: Physicians' perceptions on knowledge creation and reuse. In: Mobile Health Solutions for Biomedical Applications, pp. 111–127. IGI Global (2009)

Shepherd, A., Cooper, J.: Knowledge management for virtual teams. Issues Inf. Syst. **21**(1), 62–68 (2020)

Standing, C., Gururajan, R., Standing, S., Cripps, H.: Making the most of virtual expertise in telemedicine and telehealth environment. J. Organ. Comput. Electron. Commer. **24**(2–3), 138–156 (2014)

Standing, C., Standing, S., Gururajan, R., Fulford, R., Gengatharen, D.: Coming to terms with knowledge management in telehealth. Syst. Res. Behav. Sci. **35**(1), 102–113 (2018)

Standing, C., Volpe, I., Standing, S., Gururajan, R.: Making the most of virtual expertise in telemedicine and telehealth environments. In: Paper presented at the IEEE Ninth International Conference on Dependable, Autonomic and Secure Computing (2011)

Sveiby, K.-E.: Transfer of knowledge and the information processing professions. Eur. Manag. J. **14**(4), 379–388 (1996)

Trmčić, B.R., Labus, A., Bogdanović, Z., Babić, D., Dacić-Pilčević, A.: Usability of m-Health services: a health professional's perspective. Management **80**, 45–80 (2016). https://doi.org/10.7595/management.fon.2016.0022

Vaismoradi, M., Turunen, H., Bondas, T.: Content analysis and thematic analysis: implications for conducting a qualitative descriptive study. Nurs. Health Sci. **15**(3), 398–405 (2013)

Varshney, U.: Pervasive healthcare and wireless health monitoring. Mob. Netw. Appl. **12**(2–3), 113–127 (2007)

West, D.: How mobile devices are transforming healthcare. Issues Tech. Innov. **18**(1), 1–11 (2012)

Wu, I.-L., Li, J.-Y., Fu, C.-Y.: The adoption of mobile healthcare by hospital's professionals: an integrative perspective. Decis. Support Syst. **51**(3), 587–596 (2011). https://doi.org/10.1016/j.dss.2011.03.003

Examination of Communication Tools for "The Left-Behind Children"

Minzhi Deng(✉), Keiko Kasamatsu, and Takeo Ainoya

Tokyo Metropolitan University, 6-6 Asahigaoka, Hino, Tokyo, Japan
dengminzhi0717@yahoo.co.jp

Abstract. The left-behind children in China are children who remain in rural regions of China while their parents leave to work in urban areas. The lack of infrastructure and parental support have led to a host of additional challenges for left-behind children like quality education, physical well-being, and healthy social relationships. As a previous study, we analyze the psychological needs of specific age groups based on the educational methodology of school-age children. Children who are overpraised, on the other hand, might develop a sense of arrogance. Clearly, balance plays a major role at this point in development. The idea is that human psychology grows through interaction with the surrounding people. The psychological problems of special children called "The left-behind children" are extracted. We propose communication tools to address these issues. This method will be used to solve psychological problems that "The left-behind children" have. We make prototypes from ideas and conduct experiments and evaluations. You can repeat such a review process many times and use effective communication tools. He thinks that "absent children" can solve psychological problems caused by growing parents because they do not have time to attend.

Keywords: User experience · Human-Centered · Product design · Design process first section

1 Research Background

There are many children called "The left-behind children" in China. Children whose parents are migrating to remote areas and not living with them. Children remain in their parents' homes, such as mountain villages, where parents live in the city to work. There are 61 million "The left-behind children" nationwide, of which 10 million have been with their parents for more than a year. The situation where "The left-behind children" feel lonely in such an environment has been cited as a social problem in China. Resolving the loneliness of "The left-behind children" is one of the important issues in psychological support for them.

In order to maintain the emotional stability of "The left-behind children", parents are advised to meet with their children at least once every three months and talk with their children 1 to 2 times a week. Especially girls are prone to emotional confusion and insecurity. The reason is that girls mature faster than boys, and their feelings are sensitive and complex. In addition, there were differences in children's emotions

C. Stephanidis et al. (Eds.): HCII 2020, LNCS 12427, pp. 77–87, 2020.
https://doi.org/10.1007/978-3-030-60152-2_6

between parents who went out to earn money alone and parents who went out to earn money together. At least, it is ideal for children's psychological development that mothers are around their children. Moreover, these children also want to know about the big cities where their parents live and work. Therefore, it is hoped that the tool will provide information to children living in mountain villages and parents living in cities, so as to eliminate children's loneliness. The purpose of this study is to investigate and verify how to convey information about parents' living conditions, so as to further reduce the loneliness of the left-behind children, and propose new communication tools.

Although China's rapid economic development has increased the number of migrant workers migrating from rural to urban areas, it is difficult for them to live with their children in urban areas due to their family register system and economics, and they have placed their children in rural homes. Have left. These children are called "absent children." In recent years, the government has implemented a series of support measures to mitigate the problem of absence of children, and various support activities and welfare services have been seen. However, drastic measures are not easy due to restrictions on family register. No understanding or analysis of the problem has been made. In this paper, the current situation and problems of the out-of-home children in rural China were grasped, the background and the factors of formation were arranged, and from the viewpoint of child and family welfare, the problems of out-of-home children and the way of social support were examined.

2 From the Theory of Education

As a preliminary study, the psychological needs of specific age groups are analyzed according to the educational methodology of school-age children. According to a theory advocated by Erickson called "psychosocial development", Erickson's stage 4 of development was chosen: the choice of childhood. About 5–12 years old is the value for modifying the corresponding parameter as described in "School age". The psycho-social crisis that should be overcome is "industry vs. inferiority". Now is the time to start elementary school and learn about the joys of learning. Because the assignments that should be completed during the semester and summer vacation are coming up one after another, remember to "complete the project in a planned way and submit it". By doing this over and over again, you can increase your confidence and understand that you have "Competence".

Therefore, the formation of adolescent problems is formed by the feeling of trust in infants and young children. The feeling of trust is twofold. First, trust is from the accumulation of experience so far, learning to give the necessary external existence is always the same, as well as the continuity of trust. Trust in this sense is to trust others. Secondly, at the same time as the first aspect, that is, external existence becomes external predictability, it also brings internal certainty. Trust in this sense refers to trusting oneself. Moreover, the feeling of trust is interlinked with the feeling of "ir-relevancy", the feeling of being oneself associated with responsibilities and hopes, and the feeling that one expects and acknowledges to be others associated with knowledge and one's own trust. The next two stages are, after the basic trust stage, the self-discipline, the spontaneity, and the industriousness stage, which supports the development of feelings. In order to form self-identity in youth, it is necessary to cultivate a basic sense of trust, of self-discipline, of spontaneity, of diligence. Erikson himself recorded the following. The amount of trust an infant derives from the earliest experiences does not depend on the absolute amount of food and love expressed, but on the quality of the relationship with the mother. The mother's sensitivity to the infant's every need and the fact that, within the trustworthiness of the culture's way of life, she herself is imbued with the qualities of trust and confidence, are evidence of a method of upbringing that generates a feeling of trust in the child's mind. This lays the foundation for a sense of identity. This feeling of identity combines the subsequent feeling of "it's okay" with the feeling of being oneself, and the feeling of being someone else's trust in oneself. The two aspects of trust, trust relationship with others and trust relationship with oneself, are also related to the formation of self-identity. The relationship with others is the pillar of the sense of consistency in space and continuity in time, and the relationship with oneself is related to consistency from beginning to end. When the basic trust becomes a problem, the spatial uniformity and temporal continuity of the external being, the parent, is an important issue, but when the sense of self-identity becomes a problem, how to recognize the spatial uniformity and temporal continuity of the self-relationship between oneself and the larger external being is a problem. Here, the sense of personal identity and the sense of self-identity must be distinguished from each other.

2.1 Education in Rural China

Elaborated our country countryside education policy historical evolution, as well as the Chinese countryside elementary and middle schools' present situation and the challenge which faces. In the early days of the People's Republic of China (1949–1950s), the popularization of education was an important issue, and the improvement of adult literacy and children's school attendance rate was emphasized. Great Leap Forward and National Economic Adjustment Period (1958–1965). So, the exemption of tuition fees from primary school to high school has greatly promoted the school attendance of rural children, but the problems of incomplete school facilities, insufficient teachers and teachers' qualifications are too serious. By the time of the Cultural Revolution (1966–1976), the expansion of coverage was given priority over the improvement and enrichment of rural education levels. There has been little improvement in quality, and the quality of education is in question. Since the reform and opening up (1978-now), the Chinese government has indeed continued to make efforts in both the quantity and quality of education in rural areas. However, due to the rapid economic development, there is a big gap between urban and rural areas, and the new problems of rural education occur faster than the promotion of policies to improve education. On this historical basis, the current situation and issues of rural primary and secondary schools in China are mainly four aspects: education funds and education gap, education quality and substitute teachers, school merger (school merger), education of migrant workers' children (migrant children and left-behind children). The shortage of education funds can be said to be the root of the backwardness of rural primary and secondary education. It not only hinders the development of rural education, but also causes the education gap between urban and rural areas. Substitute teachers, as non-formal teachers, have a bad impact on the quality of education in rural primary and secondary schools because they do not have teacher qualifications and professional skills. On the other hand, the salary of substitute teachers is extremely low, and there is no health care and pension. Their treatment is also a problem. The purpose of the merged schools was originally to achieve the sustainable development of rural primary education, but in the process of implementation, there are too many students, too concentrated schools, incomplete boarding system and other issues. The education of migrant workers' children is a new problem with the change of the times, which has attracted people's attention in recent years. Both out-of-home children who stay away from their parents in rural areas and migrant children who move to cities with their parents face many challenges in education.

2.2 Submit from Education

According to the theory of "psychosocial development" put forward by Erickson, this paper analyzes the psychological state of corresponding age and puts forward various requirements. Combining with the educational theory and the educational situation of "The left-behind children", children lack the basic sense of trust. This report takes the sense of trust as the starting point and studies the ways to make up for the lack of trust between "The left-behind children" and their parents. First of all, in order to build the trust relationship between parents and children, the research points. There are many parents who want to build a strong trust relationship with their children. There are various reasons why a relationship of trust is not established.

(1) Didn't notice the child's troubles.
Suppose that parents usually do not care about their children, nor carefully observe their expressions, nor observe their actions. As a result, it is difficult to detect changes, it is difficult to detect the children's troubles. For the guardian of this attitude, the children can not say trouble, and so aware of the time, it is likely to deteriorate the situation. In order to establish a relationship of trust, it is not necessary to ask your child about the situation in detail, but it is important to keep abreast of the situation.
(2) Second, children lack affection. If the guardian is always angry, or does not listen to the child's speech, gives priority to the guardian's own mood, and can not establish a good trust relationship, the child can not act coquettishly when he wants to rely on the guardian. For guardians, children will feel that "it is impossible to understand their feelings because they can not communicate regularly". If children think that guardians are best friends, they can form dependence on guardians. Nothing can replace that sense of peace of mind. Whether they can feel the love of their guardians will directly or indirectly affect whether their children can maintain their mental health.
(3) Communication between parents and children becomes weak.

The depth of trust relationship is also related to the communication between parents and children. Importantly, it is important for children to think that parents want to understand their own affairs.

However, the guardian always seems very busy, it is difficult to meet the situation and do not attach importance to the conversation time and other do not care about the situation, communication will become weak. In such family life, because of the lonely mood and sense of loneliness, it may lead to over-reliance on the co-dependence of the relationship with specific objects.

In general, in order to compensate for the lack of dependency between parents and children, it is proposed to increase the opportunities for parent-oriented communication between "children who are not at home".

3 Creating Personas (Children and Parents)

3.1 Persona 1

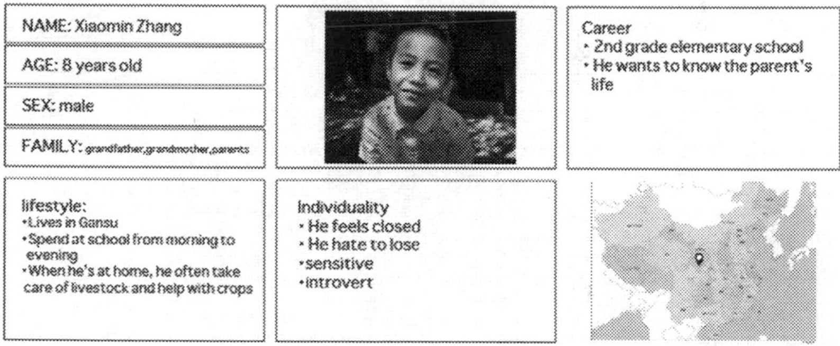

Fig. 1. Persona 1 for child

Note:

Xiaoming Zhang is strong, but really wants to be spoiled. I can only meet my parents once a year. He has no family register and cannot move to the city or attend school in the city. School is the only channel that knows the right answer outside (Fig. 1).

3.2 Persona 2

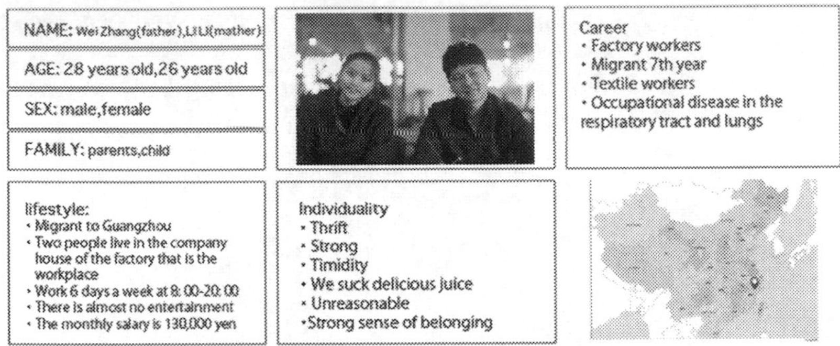

Fig. 2. Persona 2 for parents

Note:

The couple can only return to their hometown once a year during the Lunar New Year. Recently I have lost my understanding of how to interact with children (Fig. 2).

4 Human Centered Design

Human Centered Design (HCD) is a concept of designing products, services, websites, apps, etc., focusing on the ease of use of users who use the products. We will depart from the conventional idea that humans will adapt to the usage presented by the product developer, and pursue a design that is easy to use without stress from the perspective of the user (Fig. 3).

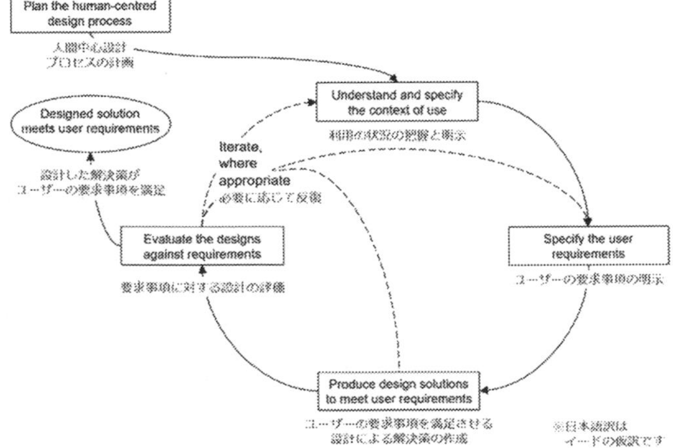

Fig. 3. Human Centered Design

According to this design theory, it was decided to discuss the process of communication tools for The left-behind children (Fig. 4).

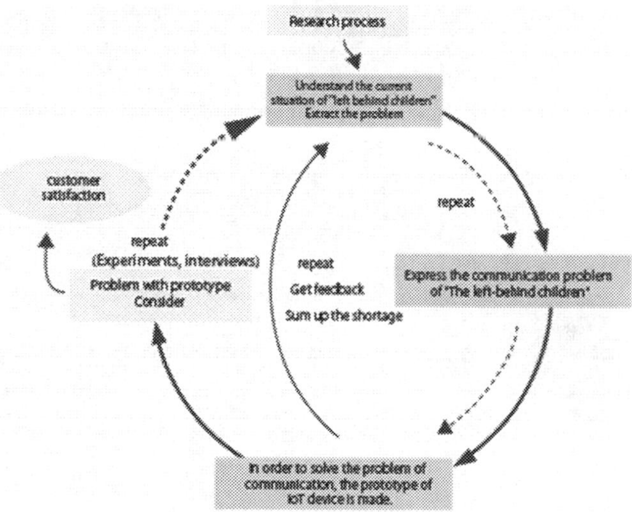

Fig. 4. Research flow

5 To Propose a Communication Tool for the Left-Behind Children and Parents

I would like to design communication tools from the design considerations of information communication. In order to solve the problem of children's loneliness, we should focus on the effect of non-verbal information such as expression and voice, and consider the means of conveying information both inside and outsidc thc village.

"The left-behind children" can not contact their parents and volunteers every time they feel negative emotions such as anxiety and loneliness. In my opinion, not only a lot of information is prompted, but also the negative emotions can be eliminated by conveying the feelings of caring parents to their children when they have such negative emotions.

As a design direction, consider the information provided to make it easier to imagine the situation of the parents when the "The left-behind children" feel lonely. As a result, it can satisfy children's mood and alleviate loneliness.

In addition, when taking the left-behind children and their families as the object, the following problems are listed.

① Parents and volunteers who work in big cities are very busy every day, can not communicate instantly, and have little time to communicate.

② “The left-behind children” in poor villages have almost no instant communication tools such as mobile phones and computers, and the cost of communication and data is also a great burden for poor families.

Based on the above points, considering the time constraints of parents and the possibility of cheap import, consider the convenient communication tools for families of “The left-behind children”.

Prototyping, experimenting and evaluating ideas. Such a discussion process can be repeated in several times, and effective communication tools can be used. I think “The left-behind children” can solve the psychological problems caused by the growing parents who have no time to accompany them (Figs. 5, 6 and 7).

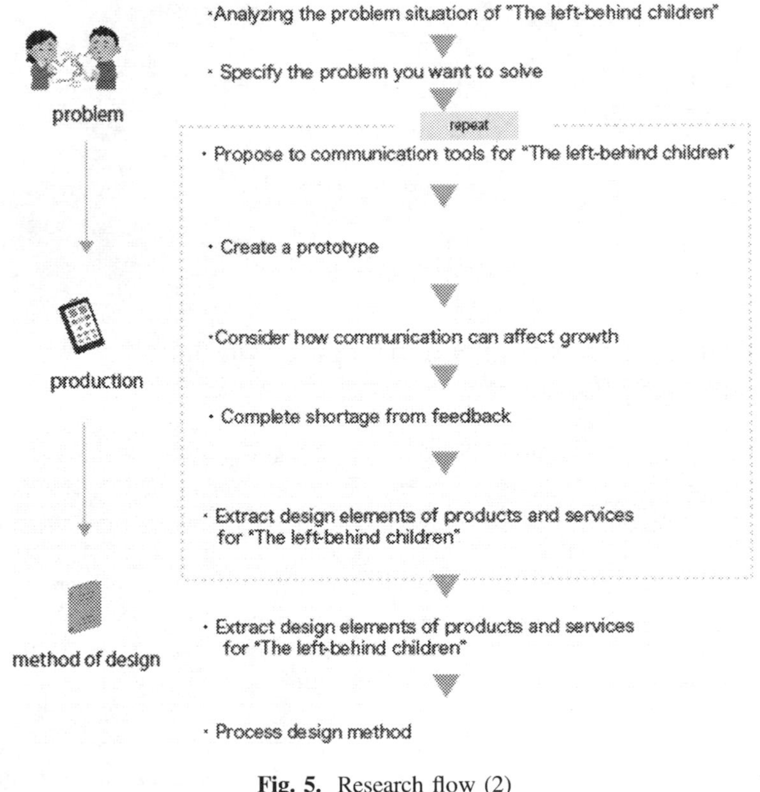

Fig. 5. Research flow (2)

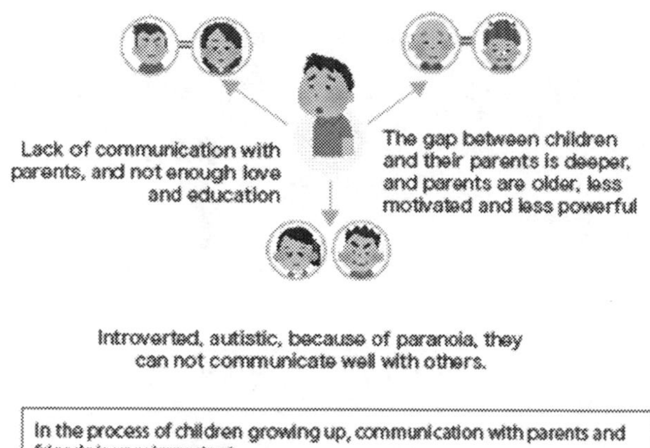

Lack of communication with parents, and not enough love and education
The gap between children and their parents is deeper, and parents are older, less motivated and less powerful
Introverted, autistic, because of paranoia, they can not communicate well with others.
In the process of children growing up, communication with parents and friends is very important.

Fig. 6. Communication between children, parents, and grandparents

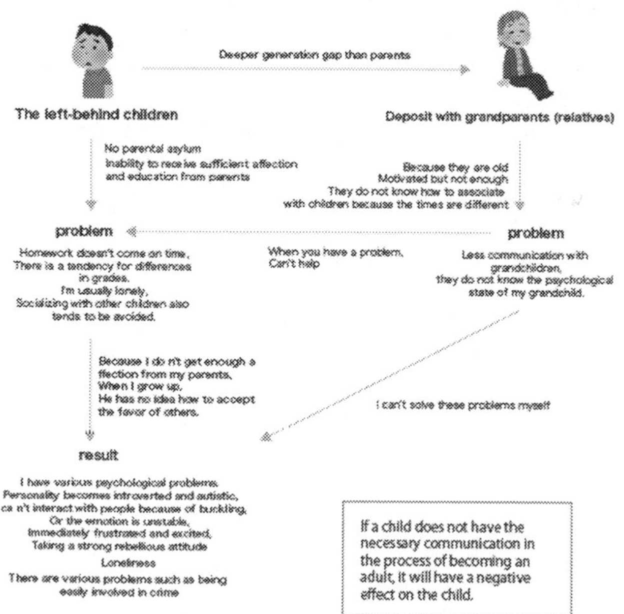

Fig. 7. Problems in children's communication with grandparents

Design of ioT device design of information transmission and reconstruction

It not only needs to present a lot of information, but also to convey to the children how parents care about them, so as to eliminate this when the children have this negative emotion (Fig. 8).

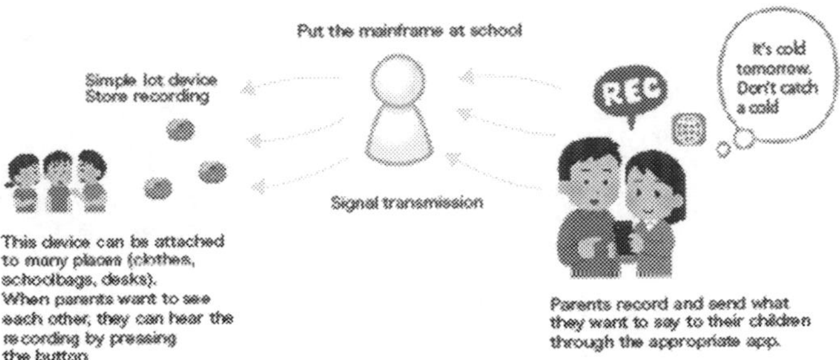

Fig. 8. ioT device design

If you want to solve communication problems, communication equipment is an important part. Therefore, it is necessary to investigate the popularity of rural infrastructure and people's communication equipment in China. In October 2015, the State Council introduced the compensation system for general telecommunications services in order to speed up the narrowing of the digital divide between urban and rural areas. Before that, the preparation of communication infrastructure in rural areas of China was entrusted to the project of "Every Village has a Telephone Link" which started in 2004. This project, as the name implies, is aimed at telephone calls in the village. In the process of implementation, three major communication operators (China Mobile, China Telecom, China Unicom) are forced to allocate sharing areas.

According to the information released by the Ministry of Industry and Information Technology in August 2019, in the three years since the promotion of the general service compensation system in 2015, the amount of central government subsidies to provide compensation for general services has exceeded 10 billion yuan, and the total amount of funds used by three major telecommunication operators for related infrastructure construction has exceeded 40 billion yuan. In the case of China Mobile, since 2004, it has invested a total of 55.2 billion yuan, opened 163,000 telephones and 74,000 broadband lines, and increased the broadband coverage to 98% and 96% in administrative villages and poor villages, respectively. Since 2013, China Telecom has invested more than 100 billion yuan in rural telecommunications infrastructure, achieving 100% coverage of 4G network in villages and towns and 90% coverage of optical broadband. In the past three years, it has undertaken 40% of all assigned tasks, invested more than 10 billion yuan in construction funds and opened 52000 administrative villages with optical fiber broadband. As a result, the number of rural areas with a communications network is increasing because infrastructure is not available in all areas. The products we designed can be purchased by the families of The left-behind children, and the cost of use is very low. Do not add excessive financial burdens to their lives.

6 Future Research

In line with this objective, there is a need to improve the previous programme. For hardware, propose an improved solution by investigating the price of the required equipment and the range of available networks. Communication methods (e.g.

Language and images) In this regard, further analysis is necessary in order to improve through the amount of communication required by the target child. In the subsequent investigation and research, we believe that the design of communication tools has been improved, and according to the above needs of the study. For the discussion part, make a model from this design and use that model as an experiment (Fig. 9).

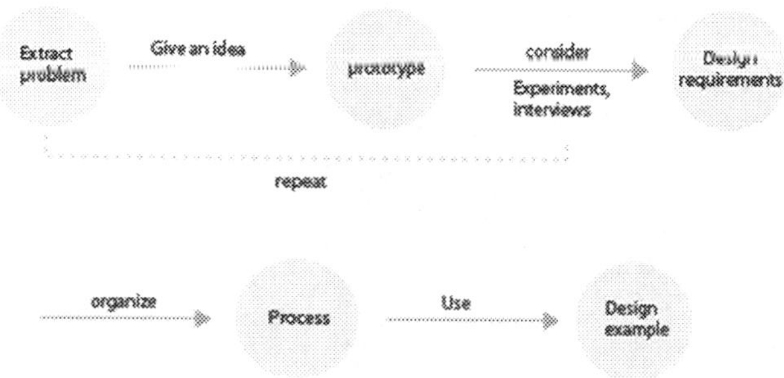

Fig. 9. Future research plans

References

1. Miyashita, K.: E.H. Erikson's identity theory and education. Faculty of Education, Chiba University, vol. 67, pp. 1–6 (2019)
2. Fukaya, T.: Learning and Development Studies Course: Development and evaluation of interventions to enhance learning and social emotional skills. For students and students
3. Higuchi, S.: A development of the theory of aesthetic education: expression. Bull. Hiroshima Univ. Graduate School Educ. Part 1 **68**, 11–20 (2019)
4. Tosaka, M.: The left behind children in rural districts in China. J. Kyushu Univ. Health Welf. **10**, 67–77 (2009)

Model for the Optimization of the Rendering Process, the Reduction of Workflow and Carbon Footprint

Felipe González-Restrepo(✉), Jorge Andrés Rodríguez-Acevedo(✉), and Sara B. Ibarra-Vargas(✉)

Facultad de Producción y Diseño - Departamento de Diseño, Institución Universitaria Pascual Bravo, Medellín, Colombia
{f.gonzalezre, jorgea.rodriguez, s.ibarrava}@pascualbravo.edu.co

Abstract. Renders or digital images of high quality and visual effect based on a 3D model or scenes made in a specialized computer program are generated along the chain of realization of animation products. A render is generated from a series of calculations between formulas and algorithms on the model that simulate the bounces of light rays on 3D objects in the scene and software that performs a Raytracing or tracking of the surfaces generated through a virtual camera that is integrated into the process.

A 3D superior or production animation product would imply extensive rendering times, which is expressed in project costs per machine use. The carbon footprint of CO2 emissions imply on animation projects has been poorly or non explored, so this document presents an empirical experience that puts on the table a free and efficient alternative to optimize rendering processes in digital animation projects based on local machine adjustment and measurement techniques aimed at reducing carbon footprint, process flow and labor costs.

Keywords: Render optimization · Digital animation · Carbon footprint

1 Introduction

The cultural and creative industries have gained an important place in the economic dynamics of developing countries through the incorporation of various management activities in sophisticated production processes and large-scale production chains. Oriented to provide content with high artistic, cultural and/or heritage impact to almost every market in the world (Felipe Buitrago Restrepo - Presidential Adviser on Economic Affairs et al. 2019). These products - goods or services - that the cultural and creative industries generate can appear in different formats, but for the purposes of this research the interest is focused on audiovisual, interactive products that involve digital animation processes such as rendering.

Along the chain of realization of animation products Renders or digital images of high quality and visual effect are generated based on a 3D model or scenes, which are made in a specialized computer program. A render is generated from a series of calculations between formulas and algorithms on the model (Kajiya 1986) that simulate

C. Stephanidis et al. (Eds.): HCII 2020, LNCS 12427, pp. 88–94, 2020.
https://doi.org/10.1007/978-3-030-60152-2_7

the bounces of light rays on 3D objects in the scene and software that performs a Raytracing or tracking of surfaces generated through a virtual camera that is integrated into the process (Levoy and Hanrahan 1996; McMillan and Bishop 1995; Saito and Takahashi 1990).

Based on calculations on the properties of the objects in the 3D scene, formal attributes such as textures and surface roughness are simulated, which, together with computations around physical effects of reflection, refraction, occlusion, dispersion and caustics, are arranged so that It gives the appearance of a real-photorealistic image - either by movements or attributes of the scene - which is finally the most common purpose of 3D artists. The rendering process is one of the most important and complex, since the quality of the final image depends not only on attributes related to software and hardware capabilities but also on the time that is defined for the system to execute calculations on the scene. In general terms, a 3D animation product of superior quality would imply extensive rendering times, which is expressed in project costs per machine use (Morcillo and Jiménez 2007) and CO2 generation.

However, on the issue of time in relation to the latter issue, which carbon footprint and energy impact in the realization of animation projects has been little explored. Each regular computer-hour on emits between 52 and 234 grams of CO2 equivalent considering a power of between 80 and 360 W depending on the carbon footprint consumption, that is, a computer can generate equivalent emissions of 180 kg of CO2 per year, which represents more than 1,000 km of travel by private vehicle on the road (Planeta 2016).

In digital animation projects, a short movie has a duration of 7 min, which decomposes in 420 s at 24 frames per second, resulting in a total of 10,080 rendered frames (Alzate Castelblanco 2018; Renderizado et al. n.d.). Thus, the rendering process of a digital animation scene can take weeks since each frame can take several days per frame, which implies leading the machines that process the images to prolonged wear and higher energy consumption of the processor as well as of the cooling system of the computers that are running the task.

The digital rendering softwares allow you to adjust the quality values of the final rendered image based on the product display needs. These adjustments are inversely proportional to the rendering time, in other words it means that the lower the rendering time, the lower the quality of the 3D model (Dally et al. 2003; Morcillo and Jiménez 2007). In this sense, by ensuring the visualization requirements of the image in a digital animation project it is possible to define the rendering times for each scene. Even more in which chase is integrated a specialized rendering software tools as well as the correct manipulation of their variables would imply a substantial improvement in the time-quality image relationship in a digital animation project, since high-level results with wear values are ensured of machine and runtime below standard (Amanatides 1992; Osorio 2010).

Globally it is possible to recognize alternatives for the optimization of the rendering process such as Zyncrender (Google n.d.) and Optix Denoiser. The first, proposed by Google, offers users to perform the rendering task not on the local server or on the home machine (Buck et al. 2004; Buck and Hanrahan 2003) but this processing is done in the cloud (Baharon et al. 2013; Carroll et al. 2012; Cho et al. 2014) - which means the use of slightly more powerful machines - with a view to decreasing times of rendering and therefore the energy consumption is reduced, this omitting the consumption per download of the final product. Although this alternative may be attractive in projects of free licensing or small independent processes, the option has been little received by the animation industries mainly because of the implications on copyright and intellectual property registrations.

Denoiser, on the other hand, performs the rendering process on the local server and allows modifying parameters around the number of samples through an occlusion algorithm, making use of render passes such as Normal Noise, Diffuse Albedo, Z-Depth, Variance, Specular and Transmission. These Render passes calculate the proximity between objects and soften the noise generated by the lack of samples in the lighting calculations between objects. This attribute has a direct relationship not only with the quality of the final scene but also with machine wear and usage time. Precisely on these attributes is that this research is carried out, the implications in the quality-time relationship to modify the number of samples and the use of a support software system to optimize the process and reduce energy consumption.

2 Methodology

The empirical explorations that this document presents were developed in the creation workshop "Optimizes the render-contributes to the environment", within the framework of the eighth edition of the International Symposium on Sustainable Design carried out by the Pascual Bravo University Institution. This workshop was attended by 20 students of the first and second semester of Digital Animation Technology of the Institution convened for their interest in optimizing the rendering process within the profession of 3D animation. All students with basic knowledge in the use of Autodesk Maya 2018® rendering software and the Arnold® rendering engine supported by the Arnold native Denoiser system and the NVIDIA brand OPTIX denoiser system. The group of assistants was accompanied by two teachers and an assistant laboratory of Digital Animation with advanced knowledge in modeling, texturing, lighting and 3D rendering.

The classroom where the workshop was held was equipped with 21 HP Z440 workstations, with INTEL XEON E5-1620 v4 3.5 GHz processors, with 16 Gb DDR3 RAM, 4 GB NVIDIA QUADRO M2000 video card, 24 HP Z24n IPS screen 24"And resolution of FullHD (1920 × 1200), and WACOM Cintiq PRO 24 24" digitizer tablet and 4 K resolution: Ultra HD (3840 × 2160).

The modeling and rendering experiments were carried out on the HP Z440 workstation, as a local server, guiding the scene configuration parameters to the reduction of rendering times, which consequently achieves a lower consumption of KW/H, which translates into a direct reduction of the carbon footprint of the rendering process.

Within the rendering tests, the power consumption in Watts of each workstation is verified and recorded by means of a DC power-voltage panel meter, Elegant ac 100A digital led.

Each student did an independent job on the machine throughout the instruction. individually each assistant created a 3D scene to which they should assign material and render lights. It also made the recording of the parameters of the scene and included aspects such as number of polygons, types and quantity of lights, materials used and the rendering configuration specifying the parameters of Sampling, and use of the Denoiser in the scene. The control scene used in the tests is presented in the Table 1.

Table 1. Control scene 1

Attribute	Description
Geometry	– 3d text "RAW" (2978 verts, 4892 faces) Material: Ai standarSurface preset glass – 3d Dino (270724 verts, 270656 faces) Material: Ai standarSurface preset honey – 3d Plane (121 verts, 100 faces) Mat: Lambert
Lights	– aiSkyDomeLight (intensity: 1, Exposure: 0, Samples: 1 Max bounces: 999)
Image format Size	1920 1080 (format: EXR)
Camera	– Deep of focus: OFF, Motion blur: OFF, f.Stop: 5.6, focal length 35, focus distance: 5, shutter angle 144, angle view: 54.43

This record also included rendering times in each of the tests (attribute modification/render) and quality obtained in each frame. During this last exercise of contrast between Denoiser attributes, quality and rendering time, the measurements related to the consumption in watts of each workstation were taken (set CPU + Screen + Digitizer Tablet).

The calculation on the KW/H consumption of each frame was performed following the equation presented below:

```
KwH per Frame =((render Seconds/360)*KwH Workstation on rendering)
```

The carbon footprint for the city of Medellin delivers an equivalence rate in which 1 kwh is equivalent to 500 kg of CO2, that is 0.360 kwh is equivalent to 180 kg of CO2, this rate allows to establish the carbon footprint of a rendered animation frame in the industry of the animation for this context.

The data obtained in this exercise of experimentation-participatory action were included in an excel document shared online with the attending students.

3 Results and Discussion

The data recorded by the students during the creation workshop described above are presented in Table 2. For the analysis of the render quality obtained in each of the tests, it is taken as a reference an Fig. 1, which defines grain quality levels and allows the results to be assessed for later analysis.

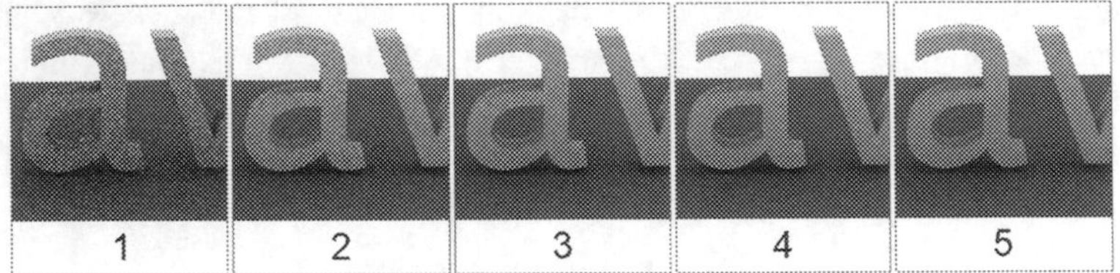

Fig. 1. Render result visible grain. Imagen quality level

Table 2. Data recorded during scene rendering.

A: Scene; B: # render test; C: Render time, *C1:* Value (sec), *C2:* Value (mm:ss); D: Render Result Visible Grain (image quality level); E: Render config. Parameters, *E1:* Camera AA, *E2:* Diffuse, *E3:* Specular, *E4:* Transmission, *E5:* SSS, *E6:* Volume Indirect, *E7*: Progressive Render; F: Adapting Sampling, *F1:* Addapting Sampling, *F2:* Max AA, *F3:* Adaptive Threshold; G: Denoiser (Nvidia Optix); H: Render process Energy consumption, *H1:* Watts (rendering), *H2:* KW/h PC Rendering, *H3:* KW/h per Render Frame.

A	**B**	**C**		**D**	**E**							**F**			**G**	**H**		
		C1	C2		E1	E2	E3	E4	E5	E6	E7	F1	F2	F3		H1	H2	H3
											YES/NO	**YES/NO**			**YES/NO**	**269,5**	**0,25**	**Value**
	D1	0:07	7	2	0	0	0	0	0	0	NO	NO	–	–	YES	269,5	0,25	0,0049
	D2	0:27	27	2	0	1	1	1	1	1	NO	NO	–	–	YES	269,5	0,25	0,0188
	D3	1:13	73	3	0	2	2	2	2	2	NO	NO	–	–	YES	269,5	0,25	0,0507
	D4	2:41	161	5	0	3	3	3	3	3	NO	NO	–	–	YES	269,5	0,25	0,1118
	D5	0:07	7	2	1	0	0	0	0	0	NO	NO	–	–	YES	269,5	0,25	0,0049
	D6	0:26	26	2	1	1	1	1	1	1	NO	NO	–	–	YES	269,5	0,25	0,0181
	D7	1:11	71	3	1	2	2	2	2	2	NO	NO	–	–	YES	269,5	0,25	0,0493
	D8	2:30	150	5	1	3	3	3	3	3	NO	NO	–	–	YES	269,5	0,25	0,1042
	D9	0:19	19	2	2	0	0	0	0	0	NO	NO	–	–	YES	269,5	0,25	0,0132
	D10	1:30	90	3	2	1	1	1	1	1	NO	NO	–	–	YES	269,5	0,25	0,0625
	D11	4:29	269	4	2	2	2	2	2	2	NO	NO	–	–	YES	269,5	0,25	0,1868
	D12	9:27	567	5	2	3	3	3	3	3	NO	NO	–	–	YES	269,5	0,25	0,3938
	D13	0:39	39	3	3	0	0	0	0	0	NO	NO	–	–	YES	269,5	0,25	0,0271
	D14	3:09	189	4	3	1	1	1	1	1	NO	NO	–	–	YES	269,5	0,25	0,1313
	D15	9:58	598	5	3	2	2	2	2	2	NO	NO	–	–	YES	269,5	0,25	0,4153
	D16	21:16	1276	5	3	3	3	3	3	3	NO	NO	–	–	YES	269,5	0,25	0,8861
	D17	0:07	7	2	0 (def value)	0	0	0	0	0	NO	YES	0	N/A	YES	269,5	0,25	0,0049
	D18	0:08	8	2	0 (def value)	0	0	0	0	0	NO	YES	1	0.015	YES	269,5	0,25	0,0056
	D19	0:25	25	4	0 (def value)	0	0	0	0	0	NO	YES	2	0.015	YES	269,5	0,25	0,0174
	D20	0:35	35	4	0 (def value)	0	0	0	0	0	NO	YES	3	0.015	YES	269,5	0,25	0,0243
	D21	0:24	24	2	0 (def value)	1	1	1	1	1	NO	YES	0	N/A	YES	269,5	0,25	0,0167
	D22	0:25	25	2	0 (def value)	1	1	1	1	1	NO	YES	1	0.015	YES	269,5	0,25	0,0174
	D23	1:51	111	3	0 (def value)	1	1	1	1	1	NO	YES	2	0.015	YES	269,5	0,25	0,0771
	D24	5:11	311	5	0 (def value)	1	1	1	1	1	NO	YES	3	0.015	YES	269,5	0,25	0,2160
	D25	1:14	74	4	0 (def value)	2	2	2	2	2	NO	YES	0	N/A	YES	269,5	0,25	0,0514
	D26	1:14	74	4	0 (def value)	2	2	2	2	2	NO	YES	1	0.015	YES	269,5	0,25	0,0514
	D27	6:24	384	5	0 (def value)	2	2	2	2	2	NO	YES	2	0.015	YES	269,5	0,25	0,2667
	D28	14:22	862	5	0 (def value)	2	2	2	2	2	NO	YES	3	0.015	YES	269,5	0,25	0,5986
	D29	2:39	159	5	0 (def value)	3	3	3	3	3	NO	YES	0	N/A	YES	269,5	0,25	0,1104
	D30	2:41	161	5	0 (def value)	3	3	3	3	3	NO	YES	1	0.015	YES	269,5	0,25	0,1118
	D31	15:04	904	5	0 (def value)	3	3	3	3	3	NO	YES	2	0.015	YES	269,5	0,25	0,6278
	D32	32:31	1951	5	0 (def value)	3	3	3	3	3	NO	YES	3	0.015	YES	269,5	0,25	1,3549

For the analysis of this exploration exercise, images with quality 3 (acceptable), quality 4 (good) and quality 5 (superior) are of particular interest since a digital animation project is valued for this technical-artistic aspect.

According to the results obtained, the tests with quality values 3 have an average rendering execution of 30 s, while those of superior quality 5 are in the 240 s. This confirms the quality/time relationship in the animation projects. Thus, the rendering times change in an approximate ratio of 1 to 100 to obtain results rendered in quality 4 and approximately 1 to 50 for quality results 5 making use of the Denoiser.

By modifying the rendering conditions and applying the Denoiser component to tests 3 of acceptable quality. It is possible to find that the render improves to quality 5 and the power consumption of the machine does not yet exceed the values delivered by a render of superior quality in the first execution. Although a simple task can be added in the rendering process (Activation of the Denoiser within the render elements), this action has no major implications on energy consumption and instead a higher quality of the final image is obtained, which finally represents reduction of costs and CO2 emissions.

4 Conclusions

This research puts on the table a free and efficient alternative to optimize rendering processes in digital animation projects for creative and cultural industries worldwide since the proposal that this document develops, based on machine adjustment and measurement techniques local, it is aimed at reducing carbon footprint, process flow and labor costs as well as security on your digital animation projects, because working in the cloud still involves some risks expressed in legal gaps on copyright and security over the files that are hosted there that many animation studios are not willing to take. Moreover, the integration of renders elements in the rendering process on the scene allows to achieve jumps of improvement of image quality of up to two levels without sacrificing the time of use of the machine or increasing the carbon footprint.

References

Alzate Castelblanco, D.F.: Desarrollo de un plug-in para limpiar curvas de animación 3dD y configuración de granja de render para la herramienta de autor 3DSMAX (2018)

Amanatides, J.: Algorithms for the detection and elimination of specular aliasing. In: Proceedings - Graphics Interface, pp. 86–93 (1992)

Baharon, M.R., Shi, Q., Llewellyn-Jones, D., Merabti, M.: Secure rendering process in cloud computing. In: 2013 Eleventh Annual Conference on Privacy, Security and Trust, pp. 82–87 (2013)

Buck, I., et al.: Brook for GPUs: stream computing on graphics hardware. ACM Trans. Graph. **23**(3), 777–786 (2004). https://doi.org/10.1145/1015706.1015800

Buck, I., Hanrahan, P.: Data Parallel Computation on Graphics Hardware. Elements (2003)

Carroll, M.D., Hadžić, I., Katsak, W.A.: 3D rendering in the cloud. Bell Labs Tech. J. **17**(2), 55–66 (2012). https://doi.org/10.1002/bltj.21544

Cho, K., et al.: Render verse: hybrid render farm for cluster and cloud environments. In: 7th International Conference on Control and Automation, pp. 6–11 (2014)

Dally, W.J., et al.: Merrimac: supercomputing with streams. In: Proceedings of the 2003 ACM/IEEE Conference on Supercomputing, SC 2003, vol. 35 (2003). https://doi.org/10.1145/1048935.1050187

Felipe Buitrago Restrepo -Consejero Presidencial en Asuntos Económicos, P., Manuel Restrepo Abondano -Ministro Saúl Pineda Hoyos -Viceministro de Desarrollo Empresarial, J., Victoria Angulo González -Ministra Ministerio de Hacienda Crédito Público Alberto Carrasquilla Barrera -Ministro, M., Patricia Gutiérrez Castañeda -Ministra, N., Cristina Constaín Rengifo - Ministra Ministerio de Trabajo Alicia Victoria Arango Olmos -Ministra, S., Daniel Oviedo Arango -Director Ricardo Valencia Ramírez -Subdirector Dirección Nacional de Derecho de Autor Carolina Romero Romero -Directora, J., … Santoro Trujillo COLCIENCIAS Diego Fernando Hernández Losada, F.: CONSEJO NACIONAL DE ECONOMÍA NARANJA Presidencia de la República Iván Duque Márquez-Presidente Entidades invitadas 56 (2019)

Google: Zync (n.d.). Zync Render https://doc.zyncrender.com/

Kajiya, J.T.: The rendering equation. In: Proceedings of the 13th Annual Conference on Computer Graphics and Interactive Techniques, pp. 143–150 (1986). https://doi.org/10.1145/15922.15902

Levoy, M., Hanrahan, P.: Light field rendering. In; SIGGRAPH 1996 (1996)

McMillan, L., Bishop, G.: Plenoptic modeling: an image-based rendering system. In: SIGGRAPH 1995 (1995)

Morcillo, C.G., Jiménez, F.L.: Optimización del proceso de render 3D distribuido con software libre. Revista de La Asociaciónde Técnicos de Informática (2007)

Osorio, J.D.: Evaluación de una solución de renderizado distribuido en las salas de cómputo de la Universidad Icesi para los estudiantes de Diseño Industrial y de Medios Interactivos. Sistemas & Telemática. **8**(15), 79–87 (2010). https://doi.org/10.18046/syt.v8i15.1023

Planeta: Planeta. Retrieved from Los ordenadores también emiten CO2 (2016). https://www.ecoembes.com/es/planeta-recicla/blog/los-ordenadores-tambien-emiten-co2

Renderizado, A., Leonardo, A., Rodríguez, C.: Renderizado en Tiempo Real de Modelos de Terrenos 3D : una Revisión del Estado del Arte (n.d.)

Saito, T., Takahashi, T.: Comprehensible rendering of 3-D shapes. In; SIGGRAPH 1990 (1990)

Development of Empowered SPIDAR-Tablet and Evaluation of a System Presenting Geographical Information Using it

Yuki Hasumi[1(✉)], Keita Ueno[1], Sakae Yamamoto[1], Takehiko Yamaguchi[2], Makoto Sato[3], and Tetsuya Harada[1]

[1] Tokyo University of Science, 6-3-1 Niijuku, Katsushika-ku, Tokyo, Japan
hasumi.yuki.hrlb@gmail.com

[2] Suwa University of Science, Toyohira, Chino-City, Nagano 5000-1, Japan

[3] Tokyo Institute of Technology, 4259 Nagatsuta-cho, Midori-ku, Yokohama, Kanagawa, Japan

Abstract. Nowadays, there are many opportunities to see maps on smartphones and tablet devices. The maps we use every day have no information about the slope, and we don't know if a road is uphill or downhill. In particular, in hazard maps that map the predicted damage caused by natural disasters and the extent of the damage, slope information is important for understanding which route to follow during evacuation. Therefore, we thought that it is possible to grasp the inclination information when tracing the road by feeding back the force sense to the touch panel operation. In order to present the force sense on tablet devices, we employed SPIDAR-tablet. In our previous research, this device had a weak force sensation. Therefore, in this study, we strengthened the force by replacing the motor and rebuilding the frame. In addition, we have developed a new map application that can output the slope information of the traced road as an easy-to-understand force sense. Then, using this application, we conducted an evaluation experiment of the effectiveness of presenting the inclination angle of the road by force. As a result of this experiment, it was found that the difference in the inclination angle of the road could be detected by force. As a future perspective, we would like to present more disaster information to the hazard map by adding not only force information but also sound and temperature.

Keywords: Force feedback · Hazzard map · Tablet device

1 Introduction

1.1 Background

Recently, it is possible to easily view a map using a smartphone or tablet device. The maps we use every day have no information about the slope, and we don't know if a road is uphill or downhill. In particular, in hazard maps, which predict damage caused by natural disasters and map the extent of the damage, it is necessary to avoid locations where secondary disasters occur and display appropriate evacuation routes and locations. Therefore, it is necessary to be able to read various map information quickly when

C. Stephanidis et al. (Eds.): HCII 2020, LNCS 12427, pp. 95–103, 2020.
https://doi.org/10.1007/978-3-030-60152-2_8

using a hazard map. The elevation information can be obtained from contour lines, but it is difficult to display it on a map that presents normal route information. Therefore, we decided to present such information by a method other than visual sense, and to express the information of the road inclination angle by a force sense presentation. For this purpose, we used SPIDAR-tablet [1], a force display device for tablet devices.

1.2 Purpose

The force sense is strengthened by changing the motor of the conventional SPIDAR-tablet that we developed, and a new frame is manufactured fitting to the motor. In addition, we will recreate the previously developed map application to match the improved SPIDAR-tablet. Then, we confirm that it is possible to present the road inclination information by the force presentation.

2 Empowerment of the Force Sense Presentation of SPIDAR-Tablet and Improving of Map Application

2.1 SPIDAR-Tablet

Figure 1 shows the former SPIDAR-tablet that we developed. The frame consists of four motors, strings, pulleys, and a control circuit using PIC. By controlling the tension by winding the strings with pulleys attached to the shaft of the motors, it is possible to present a force sense in any direction on a two-dimensional plane on the tablet display. In addition, in order to present the force sensation to the fingertips that operate the tablet device, a hook-and-loop fastener is attached to the ends of the four threads and it is wrapped around the finger. The tablet device and PIC controller are connected via USB, and the output force information is sent from the tablet device to the PIC controller. Based on the received information, the PWM function of the motor is used to present a force sense to the user. Microsoft's Surface Pro4 is employed for the tablet device.

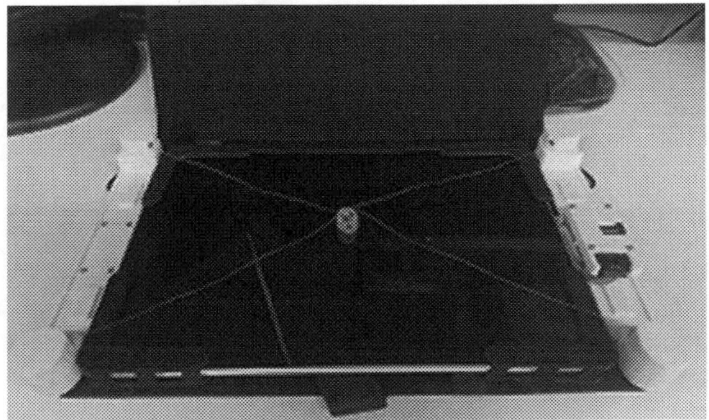

Fig. 1. SPIDAR-tablet

2.2 Empowerment of Force Sense of SPIDAR-Tablet

To improve the weakness of force sense, which is a problem with the conventional SPIDAR-tablet, we aimed to change the motors and create a new frame. The tablet device and the PIC control board were not changed.

In the change of the motors, the DC motor RF-300FA-12350 [2] manufactured by MABUCHI MOTOR Co., Ltd. was replaced to the FAULHABER coreless DC motor 1724T006SR [3]. Table 1 shows the specifications of the motor.

Table 1. Motor specifications

	RF-300FA-12350	1724T006SR
Rated voltage [V]	3	6
Rated torque [mNm]	0.48	4.2
Stall torque [mNm]	2.51	11.5
Weight [g]	22	27

When making a new SPIDAR-tablet frame, we focused on not changing the position of the thread hole on the pulley cover.

In consideration of the above point, a new frame was designed using 3DCAD based on the conventional frame, and the frame was manufactured using a 3D printer. Figure 2 shows the SPIDAR-tablet parts, and Fig. 3 shows the improved SPIDAR-tablet frame.

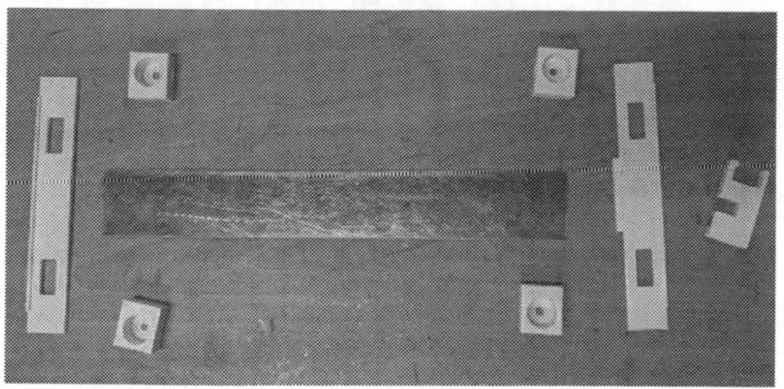

Fig. 2. SPIDAR-tablet parts

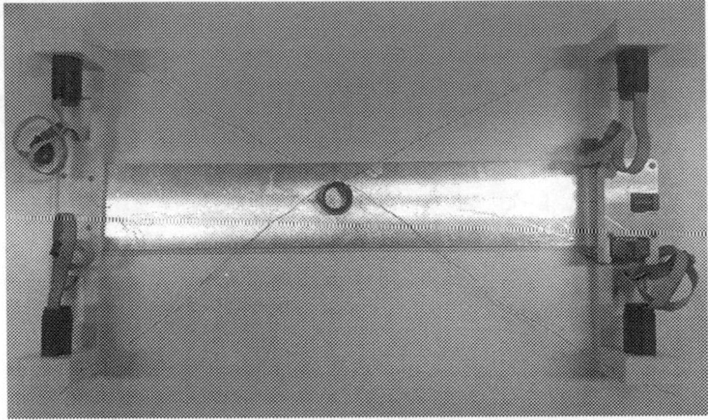

Fig. 3. Improved SPIDAR-tablet frame

Fig. 4. Horizontal installation of the motor

The main change from the conventional type is that the length of the motor has become longer, making it difficult to mount it vertically as before. Therefore, motor mounting parts were changed for horizontal motor mount as shown in Fig. 4.

The maximum tension of the string of the conventional and improved frames were measured. The application was set to drive one motor with the maximum output of the control board, and measurement was performed using Digital Force Gauge RZ-2 [4] manufactured by Aikoh Engineering Co., Ltd. As a result, the maximum tension of the conventional type was 0.47 N, while the maximum tension of the improved type was 1.36 N.

2.3 Structure of Map Application

The conventional map application [1] created by Unity3D displays a map on the screen of SPIDAR-tablet shown in Fig. 5, and when the user traces the road, the information of the slope is presented as a force sense at the fingertip. However, by changing the motor to strengthen the sense of force, there occurred a problem that strong vibrations were felt frequently during the tracing and on the curved part of the road. The main causes are that the force sensation suddenly becomes 0 when the finger comes off the road, and because the slope was defined by dividing the road into short blocks, the force sensation suddenly changed at the joint. Therefore, a new map application was developed based on the used map image and altitude data.

The main change is the output method of haptics. In previous map application, nodes containing elevation data were created at intersections, corners, or intermediate points of roads, and rectangular areas for presenting the force sense were defined between specified nodes. We considered that the map was constructed by connecting the rectangular areas, and that the vibrations were caused by the overlapping of them in the curve. Therefore, in order to eliminate the joints in the rectangular area and to smoothly trace the curve, we considered generating a force sensation according to the grayscale shading that reflects the elevation data.

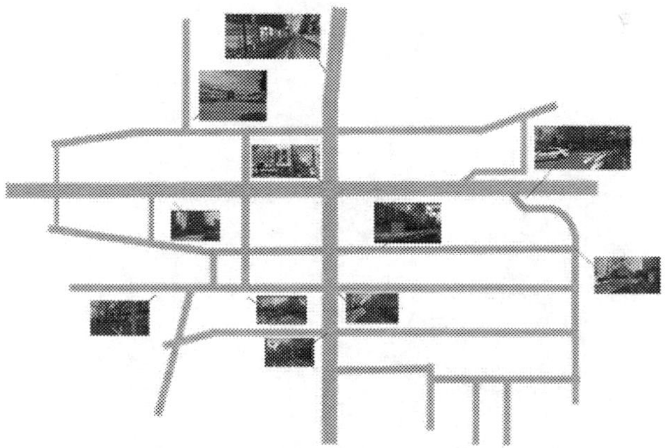

Fig. 5. Map image used in the conventional map application

In the same way as before, the application was developed using the integrated development environment Unity3D. First, the 8 bits grayscale gradient map shown in Fig. 6 was created. Next, the following processes were performed using this map. As shown in Fig. 7, the pixel values of 15 pixels in the positive and negative four directions of the x-axis and the y-axis on the two-dimensional plane centered on the point touched by the finger of the screen of the tablet device are read from the map and the average values x_1, x_2, y_1, y_2 are calculated. Using $x = x_1 - x_2$ and $y = y_1 - y_2$ as the slope in the lateral direction and the slope in the longitudinal direction, respectively, using the following Eqs. (1) and (2), the force sensations in the x-axis direction and the y-axis direction are calculated.

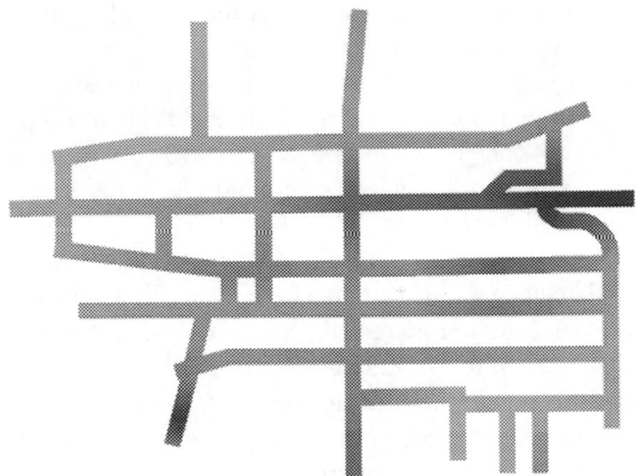

Fig. 6. Grayscale gradient map

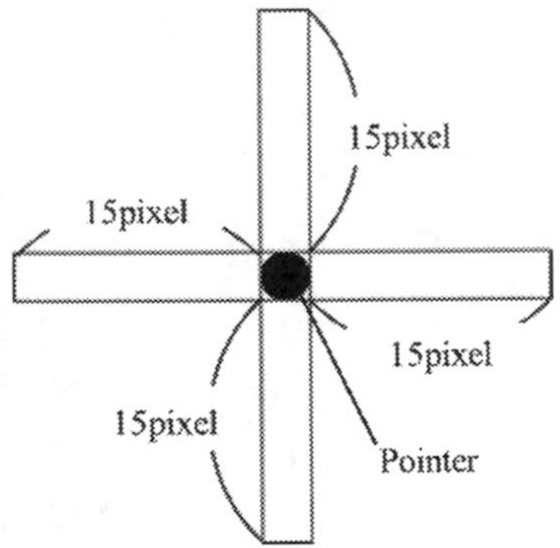

Fig. 7. Areas used for calculating the slopes

$$\begin{cases} x' = x \cdot 18.6 \\ y' = y \cdot 30.1 \end{cases} \tag{1}$$

$$\begin{cases} X = \frac{F}{\log_{10}(A+1)} \log_{10}\left(A \frac{|x'|}{|x'|_{max}} + 1\right) \cdot \frac{x'}{|x'|} \\ Y = \frac{F}{\log_{10}(A+1)} \log_{10}\left(A \frac{|y'|}{|y'|_{max}} + 1\right) \cdot \frac{y'}{|y'|} \end{cases} \tag{2}$$

x: Normalized slope value in the x-axis direction
y: Normalized slope value in the y-axis direction
X: Force sense value in x-axis direction
Y: Force sense value in y-axis direction
A: log curve adjustment parameter (>0)
F: Upper limit of force sense value

In Eq. (1), the aspect ratio of the string exit distance is 18.6:30.1, so the aspect ratio is normalized to 1:1. Also, since the area other than the road is set to the maximum grayscale value, when the finger comes off the road, a force is generated to return it to the road. These processes are performed every frame (about 4 ms). Due to the above processing, the occurrence of vibration was suppressed.

3 Evaluation Experiment of Road Inclination Expression by Force Sense

3.1 Experimental Method

In the hands-on lesson at Aijitsu Elementary School, we had 56 pupils in 6th grade (11–12 years old) experience our device. We also conducted an evaluation experiment on 12 students aged 22 to 24 at Tokyo University of Science. Hereinafter, for the sake of convenience, they are referred to as a child participant and an adult participant, respectively. We asked the participants to attach the hook-and-loop fastener to the index finger of the dominant hand and trace the way. Figure 8 shows the maps.

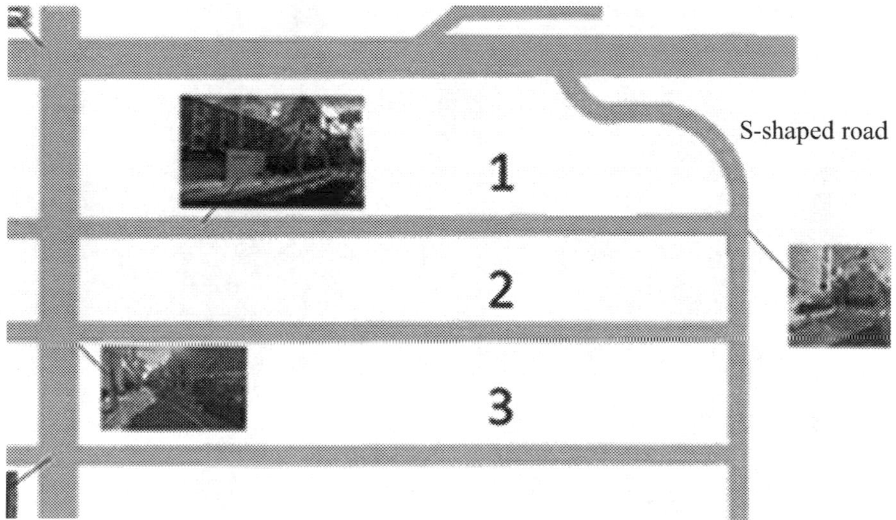

Fig. 8. Map used in the evaluation experiments

In the hands-on lesson for children, they traced the S-shaped road for 20 s, and then traced the roads numbered 1, 2, and 3 for 30 s. At this time, we considered the order of experience to prevent order effects. After the experience, we asked them to answer the questionnaire.

In the experiment for adults, they traced the S-shaped road three times and then traced the roads numbered 1, 2, and 3 three times. Also, we considered the order of experience to prevent order effects. After the experience, we asked them to answer the questionnaire. We summarized the results of children and adults and evaluated whether the haptic presentation of SPIDAR-tablet was appropriate.

The order of inclination was S-shaped > No. 1 > No. 2 > No. 3. The road No. 3 was almost flat.

3.2 Experimental Results and Discussion

First, we summarize whether the magnitude of force sense was correctly recognized for three numbered paths, which are straight lines of the same length (see Table 2 and Fig. 9). About 75% of children and almost all adults could recognize the difference in force sense on slopes. From this, it can be said that the difference in the magnitude of force sense i.e. the difference of the slope was able to be presented. However, some people did not feel the difference, so it should be added a function that can customize the haptic dynamic range for each user.

Table 2. Number of people who correctly recognized the magnitude of force sense

	Children	Adults (beige map)
Correctly recognized	41	11
Incorrectly recognized	14	1

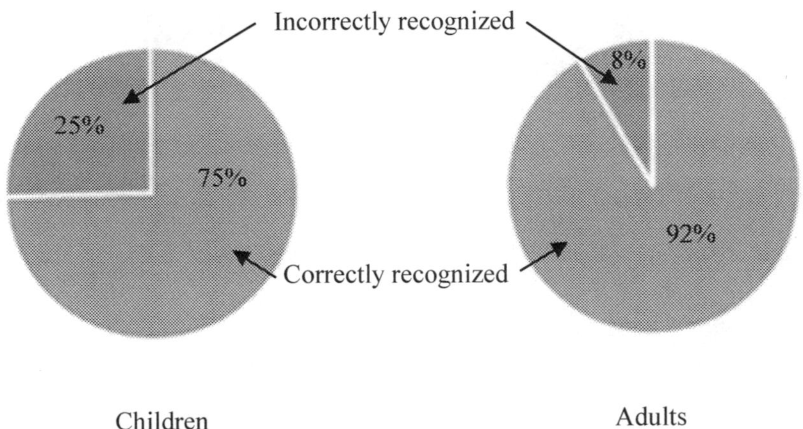

Children Adults

Fig. 9. Percentage of people who were able to correctly recognize the magnitude of haptics

In addition, when asked which of the four roads, including the S-shaped road that had the largest force sensation, more than 75% of the participants answered that the S-shaped road had the largest. From this result, the force sensation was clearly produced even on the curve road (see Table 3).

Table 3. The road that felt the greatest force sense

	Children	Adults
S-shaped road	50	9
No. 1 road	6	3
No. 2 road	0	0
No. 3 road	0	0

4 Summary and Outlook

In this study, the frame was improved by changing the motor to enhance the force sense of SPIDAR-tablet. In addition, a new map application was created, and sense of force could be obtained by reading the grayscale map. The evaluation experiments showed that the slope information was successfully presented by force sense.

However, some people did not feel the difference, so it should be added a function that can customize the haptic dynamic range for each user.

In addition, in order to use it as a hazard map, we would like to make it possible to display disaster prevention geographic information such as topographic features and ground characteristics that cause disasters, past disaster history, and evacuation routes, etc. adding sound and temperature presentation on SPIDAR-tablet.

Acknowledgments. We would like to thank all the research participants. This work was supported by JSPS KAKENHI Grant Number JP17H01782.

References

1. Tasaka, Y., et al.: Development of frame for SPIDAR tablet on windows and evaluation of system-presented geographical information. In: Yamamoto, S., Mori, H. (eds.) HIMI 2018. LNCS, vol. 10904, pp. 358–368. Springer, Cham (2018). https://doi.org/10.1007/978-3-319-92043-6_30
2. MABUCHI MOTOR CO., LTD. https://product.mabuchi-motor.com/detail.html?id=74. Accessed 8 June 2020
3. FAULHABER MINIMOTOR SA. https://www.faulhaber.com/en/products/series/1724sr/. Accessed 8 June 2020
4. Aikoh Engineering Co., Ltd. http://www.aikoh.co.jp/en/forcegauge/rz/. Accessed 8 June 2020

Jarvis: A Multimodal Visualization Tool for Bioinformatic Data

Mark Hutchens[1], Nikhil Krishnaswamy[1(✉)], Brent Cochran[2], and James Pustejovsky[1]

[1] Brandeis University, Waltham, MA, USA
{mhutchens,nkrishna,jamesp}@brandeis.edu
[2] Tufts University School of Medicine, Boston, MA, USA
brent.cochran@tufts.edu

Abstract. In this paper we present Jarvis, a multimodal explorer and navigation system for biocuration data, from both curated sources and text-derived datasets. This system harnesses voice and haptic control for a bioinformatic research context, specifically manipulation of data visualizations such as heatmaps and word clouds showing related terms in the dataset. We combine external speech systems with Clustergrammer [1] for the generation of bioinformatic queries, the BoB interface [2] for answering queries in that domain, and the VoxML framework [12] for manipulating the results and semantic grounding. We deploy the resulting system to iOS on an iPad for use by researchers over a test dataset of gene expression in tumor samples. The intent is to integrate multimodal control (here voice and haptics), so as to facilitate interaction with and analysis of data, taking advantages of using both modalities.

Keywords: Multimodal · Bioinformatics · Haptics · Visualization

1 Motivations

Due to the size of current curated biological datasets, such as protein-protein interaction networks, navigating and exploring the data in such collections can be a challenge. Information visualization is a valuable technique to navigate and cogently understand it, and the visualization should have the ability to smoothly manipulate large quantities of data in a variety of ways, including interactions between different visual techniques [16,22].

At the same time, the naming schemes and underlying ontologies used in bioinformatics datasets, such as those for genes/proteins, discourage pure voice-based interactions for understanding queries (for instance, "angiotensin-converting enzyme 2" is usually referred to by its acronym, ACE2, which sounds like the phrase "ace two," a term difficult to ground semantically directly from a domain-independent speech model). Additional modalities, e.g., haptics indicating regions in which these terms appear, simplifies grounding these entities to the data, hence multimodal methods for manipulating the data can be helpful.

C. Stephanidis et al. (Eds.): HCII 2020, LNCS 12427, pp. 104–116, 2020.
https://doi.org/10.1007/978-3-030-60152-2_9

Voice commands can include demonstratives such as "this" and "that", while a haptic interface can specify intended targets.

Our system, named Jarvis, combines speech and haptic controls to encourage more robust and flexible question and answer interactions over biocuration data. More generally, this platform enables a user to navigate and explore any dataset using multimodal queries, with more traditional language input as well through pointing, swiping, tapping, and other haptic gestures. As we demonstrate, the modality of expression for a query or part of a query varies considerably, depending on the content and context. That is, sometimes it is more appropriate (and easier) to simply point to a region on a heatmap rather than try to describe it in language; in other contexts, the exact term is needed, where there would be no recourse to a gesture. By allowing the user to interact through both modalities of language and haptics (individually and together), we hope to enhance the navigability and potential for discovery of results returned through complex queries.

2 Prior Work

With the advent of tablets and smartphones, haptic interfaces have become a common method of interacting with such devices. As speech recognition technology has improved, voice commands have also increased in prevalence, particularly in contexts where touching the device is prohibited or ill-advised, such as driving. However, despite the co-occurrence of these two modal interfaces on the same device, they rarely overlap in the same use case. Some notable exceptions are in work to integrate conversational with situational context in request fulfillment, where users (for example) can ground to locations on maps using the touchscreen interface and make requests that are localized within those locations [8]. The underlying data, regardless of its domain or format, and its visualization provides the user with background knowledge when making a relevant request, and integrating multiple interactive modalities should also allow computer systems to take advantage of such background knowledge [9,20].

While there has been considerable work on the analysis of both biological and biomedical literature (e.g., [10,13]), there has been less research devoted to natural language dialogue-based interactions over biocuration datasets. Of particular note in this area is the work presented in [4,6,14], where computational agents with deep knowledge of biological processes are queried through natural language expressions. More recently, [23] extends this to allow for dialogue-based interactions over biological knowledge bases (http://pathwaymap.indra.bio).

Visualization of dense data for bioinformatics has been accomplished through clustergrams [19], and dimensionality reductions have been performed through manifold approximation [15] and principal component analysis [3,22]. Other visualization tools include Cytoscape [21] and ProViz [7]. We bring together the integration of multiple modalities, the background data derived from bioinformatic literature, and the aforementioned visualization techniques in the Jarvis interface.

3 Architecture

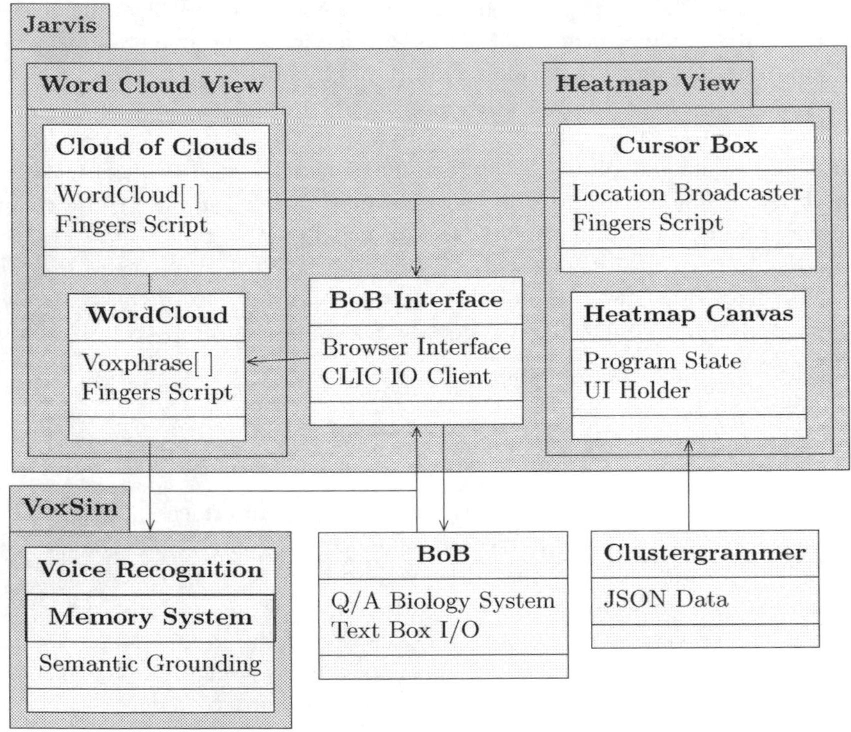

Fig. 1. The main components of Jarvis

Jarvis uses the Unity game engine [5] to render graphics and process I/O. The rendered environment is developed on top of the VoxSim platform [12], a Unity-based semantic event simulator that facilitates the manipulation of the visualized objects. This allows for movement of the objects by voice command. It also provides semantic grounding to know what words like "this" may refer to in context of previous inputs and in each modality.

VoxSim is built on the modeling language VoxML [18], which encodes the semantics of *voxemes* or visual instantiations of lexical items. This allows the visualized objects to be manipulated in 3D space based on concrete properties like concavity or symmetry, or abstract properties like location in 3D space or graspability. Jarvis exploits the abstract properties of voxemes to render elements from the underlying dataset as manipulable objects to facilitate data exploration.

Objects implemented as voxeme-derived instances[1] in VoxSim include both words displayed to the user and the groupings of them in clouds. Objects are

[1] "Voxphrases" in Fig. 1 are voxemes representing words as manipulable 3D objects. These are always billboarded to display facing the camera.

made interactable by touch on the iPad with scripts from the publicly-available Unity asset *Fingers*, which parallels native iOS gestures. The Unity object has access to both the properties of the voxeme class and the gestures accessible through Fingers.

The data for heatmaps is generated through Clustergrammer [1], which groups hierarchically-clustered heatmaps from gene expression data and saves them to JSON files. These are visualized on the Heatmap Canvas. The heatmap visualization algorithm is also adapted from Clustergrammer.

Through Unity APIs, Jarvis can be configured to consume recognized speech from a variety of services, including Google SR, IBM Watson, or custom speech models. The VoxSim platform also facilitates external parsers through TCP sockets and REST connections.

Natural language processing and question answering is done via a connection to the BoB biocuration system [2]. An example of an exchange with BoB is given in Table 1. Jarvis uses BoB's associated CLIC IO Client (also [2]) to format requests relating to larger datasets such that BoB may parse them, and Jarvis may read the results. It attaches lists of data to phrases passed into BoB and receives the desired results.

Figure 1 shows the architecture and components of Jarvis.

Table 1. A typical exchange with BoB

USER	Create the gene set [*This also passes a JSON structure containing all of the genes selected from the visual heatmap interface*]
BoB	Okay
BoB	I created the gene-set selection with 7 items
BoB	What would you like to do next?
USER	Which of these are transcription factors?
BoB	Of those 7 genes, PLAGL1 is a transcription factor

4 Haptic Control

The user can interact with Jarvis via a combination of voice and haptic control. Table 2 shows the gestures available in the Jarvis interface and what their associated function. Use of gestures depends on the visualized context and the technique currently being used (e.g., heatmap or word cloud). Some gestures may mean different things in different context and not all gestures have a use in all contexts.

In Fig. 2 we see a selector box and the heatmap we wish to zoom on. The heatmap shown represents data on gene expression over tumor samples, that is run through Clustergrammer [1] to group the presence of proteins along rows and source tissue samples along columns.

Table 2. Gestures available in Jarvis

Gesture	Heatmap interpretation	Word cloud interpretation
Tap	–	Semantically ground word
Swipe	Swap view	Swap view
Pan	Move selection box	–
Scale	Resize selection box	Zoom camera
Rotate	–	Rotate word cloud
Long press	–	Semantically ground cloud

The colors in the heatmap are salient to relationships between proteins and gene expression, so analyzing the redder area in the lower left that displays higher associated activity is likely to be useful. These few dozen proteins are best grabbed by selecting a region, which we do by dragging the selection box.

Haptic commands through touch on the iPad are used to drag and resize the selector box, and saying "Zoom in here" is used to zoom in. This simultaneously grounds the selection for transmission to BoB. In the absence of voice commands, such as in the event of speech recognition failure, a corresponding text input box is provided to facilitate the same natural language functionality.

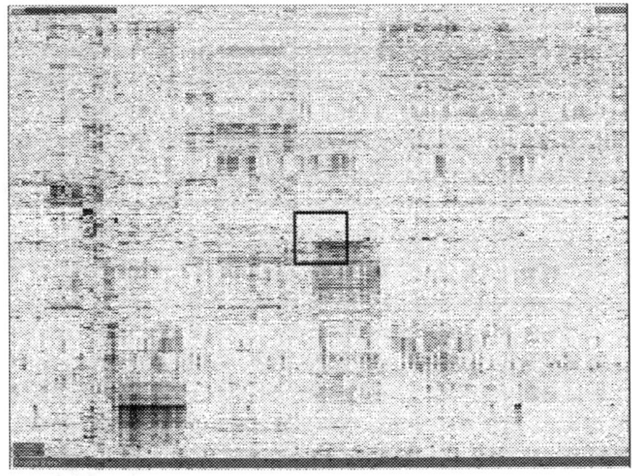

Fig. 2. A heatmap of proteins vs. tissue samples with user interface overlaid

5 Voice Interaction

As mentioned above, speech-to-text is handled by packages connected to the VoxSim platform [12], such as IBM Watson or Google Speech Recognition.

When the user asks a question like "Which of these are transcription factors?" the system reads the currently selected genes and passes the list and question to BoB. The resultant list is passed back to Jarvis for visualization.

The two lists are then visualized as word clouds, with each word in the cloud individually interactable to encourage item-to-item juxtaposition. The full list of genes selected initially corresponds to one word cloud, while the subset of transcription factors can be separated. Each word's position in the clouds is determined by factors in the underlying data, such as frequency of occurrence or similarity to a data point represented by another selected word.

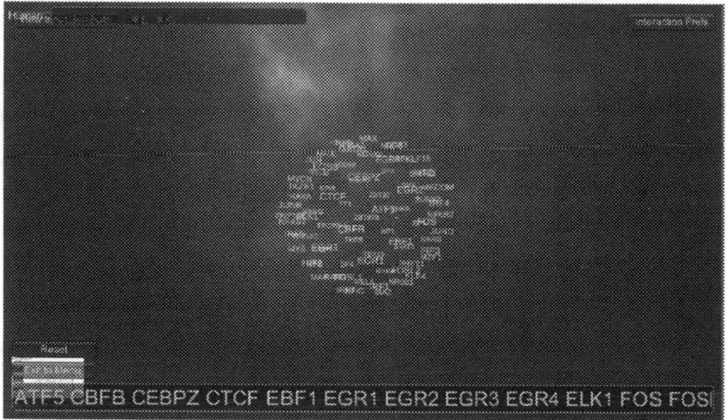

Fig. 3. A cloud of proteins for manipulation

In Fig. 3 we see a cloud of protein names that BoB returned. These proteins can be organized in the cloud based on a number of attributes, and the cloud itself responds to haptic commands as well. A subset of the cloud, e.g. which proteins are transcription factors, may be pulled out and manipulated independently.

When a word is selected, the cloud can be reordered to correspond to relationships with that word. In future, the 3-dimensional display would be useful for groupings related to up to three different criteria simultaneously.

6 Multimodal Integration

Certain modalities are better suited to grounding certain information than others. For example, in human-to-human conversation, deictic gesture—i.e., pointing—grounds naturally to locations, while language may be better at grounding concept labels or attribute descriptions. In a tablet environment, such as that on which Jarvis is deployed, projective gesture maps neatly to haptics and touch feedback, and the gesture set listed in Table 2.

The nature of bioinformatic data, particularly protein and gene names, poses a problem for purely speech-based systems. In the parlance used in biology, gene names which may be initialisms or collections of characters that are not easily pronounced, such as "MAPK" nonetheless have conventionalized pronunciations in common use (e.g., "map-kay"). These pronunciations are not likely to be covered by the phonetic model of a speech recognition system; a user saying "map-kay" is likely to result is a transcription of "Map K[ay]" (or as might be the case on a smartphone, an inferred request for directions to the house of someone named "Kay").

Therefore, navigating this domain via speech alone is exceedingly difficult. Providing text input may alleviate the speech recognition issues but is time consuming and still does not solve the problem if the user does not know quite how to phrase their request. In addition, the typical ways in which large bioinformatic datasets are typically presented (heatmaps, graphs, word clouds) do not necessarily lend themselves well to being explored purely using language. These kinds of data presentations are inherently visual as well.

Including another modality helps with this. Using haptics to indicate regions or entities of interest allows the user to ground their actions and requests to specific entities using easily recognizable demonstratives ("this," "that," "these," etc.), obviating the need to try to pronounce or spell out the entity references in the data. This makes navigation easy and more tractable for a large dataset.

In principle, it would be possible to navigate and interact with the data using only haptic gesture, by linking defined actions to other haptic gestures (e.g., double or triple tap, two-fingered pan, long press, etc.—see Table 2), but the limit in the number of gestures allowed limits the available vocabulary. Therefore, the functionality that is provided by speech and language input is also crucial. The mixture of haptics and language allows for less discursive language to ask the same question due to entities being focused using haptics and passed along using language.

7 Evaluation

Because Jarvis is under active development, evaluation of its capabilities is still in the early stages. Nonetheless, since the goal of Jarvis in this particular use case is well-defined—to enable biologists to accomplish novel discoveries in large datasets—we can at least evaluate the usability of the system in accomplishing this task. In addition, we can evaluate its interactive capabilities for accomplishing data exploration over large datasets of arbitrary provenance (see Sect. 8); strictly biological data is not a requirement for useful multimodal exploration, it is simply an illustrative use case.

7.1 Interactive Usability

We propose a method for evaluating the multimodal capabilities of the interaction using simple metrics and object and event semantics, one agnostic to the precise modalities in use in the interaction [11], making it ideal to assess particular areas where the Jarvis system needs improvement.

When defining a multimodal interaction, it is necessary to specify the vocabulary expressible in each modality. The system discussed in [11], also built on VoxSim, uses speech and projective gesture recognition in a Blocks World environment. To evaluate Jarvis, we swap out the projective gesture recognition (pointing, pushing, grasping, etc.) for the available haptic gestures (see Table 2), and instead of specific objects in the interaction, we may instead use delineated regions of the data in the view presented if the presentation allows (e.g., heatmap).

A robust multimodal evaluation scheme should be able to be applied to a human-computer interaction on a system and return a result that is representative of the system's coverage of the total possible interactions within the system's domain (e.g., exploration of bioinformatic data using a variety of visualization techniques).

Previous multimodal evaluation using this schema presented assumed that a user must be truly naive, having very little to no knowledge of exactly what the system understands. However, evaluating in this manner defeats the purpose of a system like Jarvis, whose assumed users must be domain experts. Therefore, evaluating the system's coverage should represent how easy it is for the domain expert, without much prior knowledge of the interface, to use that interface to accomplish their task.

An interaction consists of "moves" taken by each participant, which are logged live. These are timestamped and coded by participant and modality (S for speech, G for gesture, A for action; these are listed in the subscripts in the sample log in Table 3).

Usability metrics of the system can be conditioned on a particular modality. For instance, user studies in aggregate might find that speech input is a point of difficulty, with users struggling to figure out appropriate phrasings, where haptic input provides for an easier semantic grounding of entities (some of these possibilities are shown in the sample log in Table 3). Conversely, they may discover that haptic input is less precise than necessary, requiring the ability to select finer-grained regions than currently allowed.

The sample log in Table 3 is based on the BoB dialogue from Table 1. We model the USER and two "agents," BoB and JARVIS. This is to separate the biocuration backend functionality provided by BoB from the grounding interface provided by Jarvis. BoB normally delivers output through text, and so the speech output attributed to BoB here is actually delivered through Jarvis via text-to-speech. Therefore Jarvis and BoB are perceived as one from the perspective of the user, but BoB is nevertheless given attribution to focus on the semantic content. This allows us to evaluate difficulties due to the BoB output versus difficulties due to the Jarvis interface, giving Jarvis a way to evaluate both itself and its backend.

In the sample interaction, we see Jarvis making moves that facilitate moves by the user (e.g., moves 0–1, Jarvis presents the data in heatmap form, facilitating the user selecting a region by haptic panning). The data presentation in this fashion followed by a user action over it can be assumed to mean that the user

Table 3. Sample interaction log.

1	JARVIS$_A$	CREATE_HEATMAP(data[])	0.000000
2	USER$_G$	PAN_TO (<.14674;.24371>)	1.145281
3	USER$_S$	"Which of these are transcription factors?"	2.452981
4	BoB$_S$	"I am having trouble, possibly because I don't know what 'these' refers to"	5.803915
5	BoB$_s$	"I don't know what genes you mean"	7.818170
6	USER$_S$	"Create the gene set"	8.642095
7	BoB$_S$	"Okay"	10.041973
8	BoB$_S$	"I created the gene-set selection with 7 items"	12.803915
9	BoB$_S$	"What would you like to do next?"	14.500183
10	USER$_S$	"Which of these are transcription factors?"	15.661427
11	JARVIS$_A$	CREATE_WORDCLOUD(geneset[])	18.891054
12	JARVIS$_A$	CREATE_WORDCLOUD(subset[])	18.891054
13	BoB$_S$	"Of those 7 genes, PLAGL1 is a transcription factor"	18.891054

understands how to use this type of data presentation, along with the haptic gesture. This can be further quantified by looking at the length of the delay between when the data is shown to when the user selects an area. A long delay may signal confusion on the part of the user, e.g., in how to use the system and/or what to do with this type of data presentation.

We also see places where the dialogue breaks down or fails to proceed smoothly. At move 3, the user says something that BoB does not understand, and so BoB gives the user a reason why. The user responds with a new instruction that grounds the demonstrative "these" to a particular set, that allows BoB to complete the request. Since BoB prompts for a correction that later incorporates information gathered through the Jarvis interface, we can see that it needed more information at move 4, which is alleviated through use of an additional modality provided by Jarvis. This then becomes a way of validating intent detection within a dialogue systems (cf. [24]).

7.2 Fidelity of Data Transfer

Successful discovery using a multi-component system like Jarvis, particularly one intended for use with arbitrary backends, depends on the fidelity of data transferred between the subsystems. Therefore, we propose the following methods to evaluate fidelity of particular subcomponents:

Speech Recognition. One of the primary difficulties that arises in a multimodal system is poor recognition of one modality hamstringing the interaction in other modalities, often by forcing the interaction into a bad state from which it cannot recover. To assess quality of speech recognition specifically and figure out where it needs improvement, we can have users execute a scripted dialogue that provides a known ground truth, and then measure the accuracy of the recognized input compared to that reference using standard metrics, e.g., BLEU score [17].

Semantic Grounding. Correct semantic grounding is essential to ensure that the correct information extracted from the underlying data is passed to BoB along with the query. To assess this through the interaction, we look at "blocks" bounded by moves that negate a prior move and redirect the interaction to new focus objects or actions (e.g., move 5 in Table 3). We assume that within a block, information is being correctly grounded, thus enabling the user to make satisfactory requests, so the longer a block proceeds without redirection or correction, the better the grounding mechanism is performing. Thus, we can also assess grounding via speech vs. grounding via haptics.

Visualization Accuracy. Numerous aspects of the visualization can be assessed for accuracy. BoB will return gene subsets represented as word clouds, and the visualization must be sure to accurately represent all elements in the gene subset. Each voxeme object representing a gene in the subset has an underlying textual representation (to which speech and language input is grounded), so this can simply be compared to the string representing the same gene returned by BoB. A 100% match between the two sets means 100% coverage of Jarvis over the BoB data.

Region selection can also be assessed for consistency, by calculating the overlap when a user attempts to select the same region multiple times and calculating how the selections, determined from the selection box's size and location (see Fig. 2) overlap. This can be assessed using haptics with selection via mouse as a baseline to see how much precision is gained or lost from the use of haptics.

Region selection can also be correlated to selection in the underlying data by selecting the same indicated region multiple times and calculating the variance over the data subsets that are extracted by selecting that region.

All these component-specific evaluations can be combined with the general time-based usability evaluation (Sect. 7.1) to determine the errors in components that led to difficulties in the overall interaction, e.g., does a particular misrecognized word regularly lead to a command BoB fails to understand or does an inaccurate presentation of the data make it difficult to make a discovery based on the previous query? Therefore, different levels of functionality each have distinct kinds of evaluation criteria measuring performance and usability. Evaluations using domains experts are planned and ongoing.

8 Future Work

Using the results of our planned evaluation, we anticipate being able to fine-tune speech recognition language models to suit the biocuration domain and BoB's capability. However, the particular domain for a multimodal Jarvis interface is determined by the data source and the backend inference and query engine.

Hence, while bioinformatic data is our current test case, the Jarvis interface is not limited solely to the display of biological data. For future developments, we would like to implement the following additions to enable robust interaction and exploitation of arbitrary datasets:

- 3D visualizations of the heatmap view to add an additional dimension. This will be useful for representing time sequence data. In the biocuration use case, for example, this could represent single-cell or cell-cycle proliferation data. Such a visualization would be requested by a swipe to the left or right from the heatmap view, effectively rotating the 3D heatmap display, to view the temporal dimension side-on.
- More extended commands for data manipulation. For example, the Fingers package can recognize double and triple taps, which could be used on words for highlighting different kinds of relationships between them, or rearranging locations in clouds to show which genes regulate which proteins, etc.
- More in-depth gene-to-gene relationships in the word cloud visualization. For instance, grouping genes by known gene clusters, or genes that encode similar proteins or share generalized functions. For other types of data, such as protein-protein interaction, these relationship could also be proteins that interact with other proteins or classes or proteins in similar ways (e.g., "TNF-inhibitors," or proteins that collectively inhibit the expression of members of the tumor necrosis factor superfamily).
- Better integration between Fingers and VoxSim. There is a memory stack in VoxSim that tracks which objects the user has interacted with, and any interaction through haptics should be guaranteed to be tracked.
- The ability to "save state" in a visualization and revisit it (for example using the "swipe" gesture to swap views). This would allow users to conduct multiple searches that may lead them on long paths through the data, and then subsequently return "home" and compare their results to make higher-level discoveries.

9 Conclusion

In this paper, we have introduced a multimodal tool for visualizing and exploring bioinformatic datasets. This tool, Jarvis, combines speech and haptic control with a robust biocuration dialogue system in iOS on an iPad. These features encourage smooth interactions over complex data in this domain.

The underlying mechanism that enables the integration of the two distinct modalities is the transformation of data into a manipulable object. This allows domain specialists to navigate through the data using multiple grounding techniques. For large, complex datasets such as those containing biological entities and relations, multiple grounding techniques should not only allow users to have a less cumbersome user experience, but also allow them to switch modalities to achieve greater (or less) precision when desired, and allow one modality's strength to mitigate weaknesses of the other(s) when interacting with the specifics of a dataset. This also allows the same underlying interface to be used with multiple datasets potentially in multiple domains.

Acknowledgements. This work is supported in part by US Defense Advanced Research Projects Agency (DARPA), Contract W911NF-15-C-0238; and DTRA grant DTRA-16-1-0002; Approved for Public Release, Distribution Unlimited. The views expressed are those of the authors and do not reflect the official policy or position of the Department of Defense or the U.S. Government. We would like to thank everyone at the Boston office of Smart Information Flow Technologies, particularly Laurel Bobrow, Robert Bobrow, Mark Burstein, David McDonald, and Matthew McLure; and Benjamin Gyori and John Bachman at Harvard Medical School. All remaining errors are, of course, those of the authors alone.

References

1. Fernandez, N.F., et al.: Clustergrammer, a web-based heatmap visualization and analysis tool for high-dimensional biological data (2017). https://doi.org/10.1038/sdata.2017.151
2. Burstein, M., et al.: Using multiple contexts to interpret collaborative task dialogs. In: Advanced in Cognitive Systems (2019)
3. Clark, N.R., Ma'ayan, A.: Introduction to statistical methods for analyzing large data sets: gene-set enrichment analysis. Sci. Signal. **4**(190), p. tr4 (2011). ISSN 1945–0877. https://doi.org/10.1126/scisignal.2001966. https://stke.sciencemag.org/content/4/190/tr4.full.pdf. https://stke.sciencemag.org/content/4/190/tr4
4. Friedman, S., et al.: Learning by reading: extending and localizing against a model. Adv. Cogn. Syst. **5**, 77–96 (2017)
5. Goldstone, W.: Unity Game Development Essentials. Packt Publishing Ltd., Birmingham (2009)
6. Gyori, B.M., et al.: From word models to executable models of signaling networks using automated assembly. Mol. Syst. Biol. **13**(11), 954 (2017)
7. Jehl, P., et al.: ProViz–a web-based visualization tool to investigate the functional and evolutionary features of protein sequences. Nucl. Acids Res. **44**(W1), W11–W15 (2016). ISSN 0305-1048. https://doi.org/10.1093/nar/gkw265. https://academic.oup.com/nar/article-pdf/44/W1/W11/18787253/gkw265.pdf
8. Johnston, M.: Building multimodal applications with EMMA. In: Proceedings of the 2009 International Conference on Multimodal Interfaces, pp. 47–54 (2009)
9. Johnston, M.: Multimodal integration for interactive conversational systems. In: The Handbook of Multimodal-Multisensor Interfaces: Language Processing, Software, Commercialization, and Emerging Directions, vol. 3, pp. 21–76 (2019)
10. Kim, J.-D.: Biomedical Natural Language Processing (2017)
11. Krishnaswamy, N., Pustejovsky, J.: An evaluation framework for multimodal interaction. In: Proceedings of the Eleventh International Conference on Language Resources and Evaluation (LREC 2018) (2018)
12. Krishnaswamy, N., Pustejovsky, J.: VoxSim: a visual platform for modeling motion language. In: Proceedings of COLING 2016, the 26th International Conference on Computational Linguistics: System Demonstrations. Osaka, Japan: The COLING 2016 Organizing Committee, December 2016, pp. 54–58. https://www.aclweb.org/anthology/C16-2012

13. Lee, J., et al.: BioBERT: a pre-trained biomedical language representation model for biomedical text mining. Bioinformatics **36**(4), 1234–1240 (2020)
14. McDonald, D., et al.: Extending biology models with deep NLP over scientific articles. In: Workshops at the Thirtieth AAAI Conference on Artificial Intelligence (2016)
15. McInnes, L., Healy, J.: UMAP: uniform manifold approximation and projection for dimension reduction. ArXiv abs/1802.03426 (2018)
16. O'Halloran, K.L., et al.: A digital mixed methods research design: integrating multimodal analysis with data mining and information visualization for big data analytics. J. Mixed Methods Res. **12**(1), 11–30 (2018)
17. Papineni, K., et al.: BLEU: a method for automatic evaluation of machine translation. In: Proceedings of the 40th Annual Meeting on Association for Computational Linguistics. Association for Computational Linguistics, pp. 311–318 (2002)
18. Pustejovsky, J., Krishnaswamy, N.: VoxML: a visualization modeling language. In: Proceedings of the Tenth International Conference on Language Resources and Evaluation (LREC 2016), pp. 4606–4613 (2016)
19. Schonlau, M.: The clustergram: a graph for visualizing hierarchical and nonhierarchical cluster analyses. Stata J. **2**(4), 391–402 (2002)
20. Selfridge, E., Johnston, M.: Interact: tightly-coupling multimodal dialog with an interactive virtual assistant. In: Proceedings of the 2015 ACM on International Conference on Multimodal Interaction, pp. 381–382 (2015)
21. Shannon, P., et al.: Cytoscape: a software environment for integrated models of biomolecular interaction networks. Genome Res. **13**(11), 2498–2504 (2003). https://doi.org/10.1101/gr.1239303
22. Tao, Y., et al.: Information visualization techniques in bioinformatics during the postgenomic era. Drug Discov. Today BIOSILICO **2**(6), 237–245 (2004)
23. Todorov, P.V., et al.: INDRA-IPM: interactive pathway modeling using natural language with automated assembly. Bioinformatics **35**(21), 4501–4503 (2019)
24. Yan, Z., et al.: Building task-oriented dialogue systems for online shopping. In: Thirty-First AAAI Conference on Artificial Intelligence (2017)

The Research of Regional Cultural Image of China

Exemplifying with Guandong Culture Area

He Jiang[1(✉)], Keiko Kasamatsu[1], and Takeo Ainoya[2]

[1] Tokyo Metropolitan University, Tokyo, Japan
jhe02@sina.com
[2] serBOTinQ, Tokyo Metropolitan University, Tokyo, Japan

Abstract. The original intention of this research is to spread Chinese culture. When communicating with foreigners, I discover quite a lot of them hold a view that Chinese culture is not easy to get to know and further be taken into use of daily life. Then I believe the first step I need to do is to let foreigners feel the easy-going aspect of Chinese culture. In this research, we choose the direct points of people in different areas of China as our starting point to make introduction of Chinese culture rather than make utilization of traditional geographical, historical and cultural classification pattern. Thus, current and real Chinese culture can be directly opened out in front of those foreigners who have desire to get to know China through self-experience, daily life and emotional memories. I reckon it would be far better to introduce China by daily details description, because this way can lead to a better understanding of Chinese culture, and otherwise the getting-to-know process may be humdrum and rigid. Here we utilize Guandong culture area as a detailed example of Chinese regional cultural image, through the introduction of the differences and common characters of Heilongjiang Province, Jilin Province and Liaoning Province.

Keywords: Chinese regional culture · Cultural image · Cultural symbols

1 Researching Background

With the pace of internationalization, China is communicating with all human society through Chinese culture, at the same time providing an alternative access to get knowledge about Chinese culture type for the whole world. When doing intercultural communication on basis of traditional Chinese culture, there are two factors that should always be taken into consideration: the first one is cultural differences between countries and the second one is the differences between receivers.

However, special attention should always be lied on the stressing and emphasis of the communication subject - person. It is person that both launch and receive the communication. The feeling of the communication varies with individuals of distinctive nationalities and ages due to their experience, knowledge, skills and time [1]. Therefore, when Chinese culture is supposed to be effectively spread, analysis and designs unfurled on foundation of human communication pattern would definitely be the most effective.

C. Stephanidis et al. (Eds.): HCII 2020, LNCS 12427, pp. 117–132, 2020.
https://doi.org/10.1007/978-3-030-60152-2_10

2 Preliminary Investigation

This research focused on international communication of Chinese culture and cultural communication of different countries. In order to understand the current status of foreigners' recognition for Chinese culture, I researched 'Foreigners' impression of Chinese culture' and generalized three representative personal cards through multi-pronged channels such as onlinc videos, commentaries, reviews, etc.

It can be analyzed from the language description of personal card (see Fig. 1 Character 1) that foreigners' impression and recognition of Chinese culture is superficial and onefold. However, with vast territory, large population and 56 minorities, China has a prominent feature of cultural diversity. Different regions may have completely different cultures.

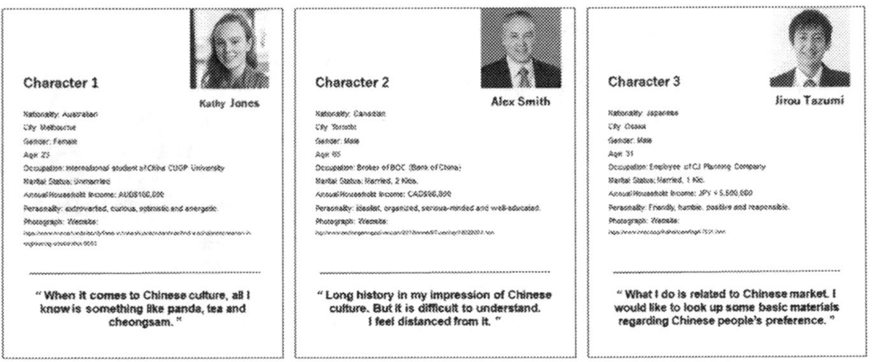

Fig. 1. Personal card

Here, I want to try and give you a point to understanding China. The point is that China is primarily a civilization instead of a nation state. Without understanding this notion of Chinese civilization, you cannot understand China who is the longest continuously existing polity in the world. The way we think of what China is, is not a function of the last hundred years of being a nation state. It is a function of more than 2000 years of being a civilization state. It is absolutely different from western countries. Western countries are constituted on the basis of national identity, like U.S. China is different; China is firstly and foremost constituted on the basis of civilization. I mean Chinese all regard ourselves to be multiracial multicultural. We have very unique relationship, confusion values, very unusual conception of families.

As mentioned above, because China is a civilization state which is different with nation state of western countries, it becomes more necessary to analyze and compare cultures between regions in China. Only then can foreigners be provided a comprehensive and accurate information reference of Chinese culture. Therefore, instead of comparing with cultures of other countries, this research only analyzes and compares cultures of regions in China (Here answers question 2 of comments).

To make foreigners understand the Chinese cultural diversity and the cultural images of different Chinese regions, the first thing is to divide regions by different culture. In recent years, Chinese scholars have divided Chinese culture regions from different perspectives (e.g. geography, history, politics, nationality, etc.). Qin Liangjie [2] and Wang Huichang [3] divided Chinese culture into 16 regions according to the region's customs. This compartmentalization is more suitable for my research. I made this map [4] as follows (see Fig. 2) to make it easier to understand base on this compartmentalization.

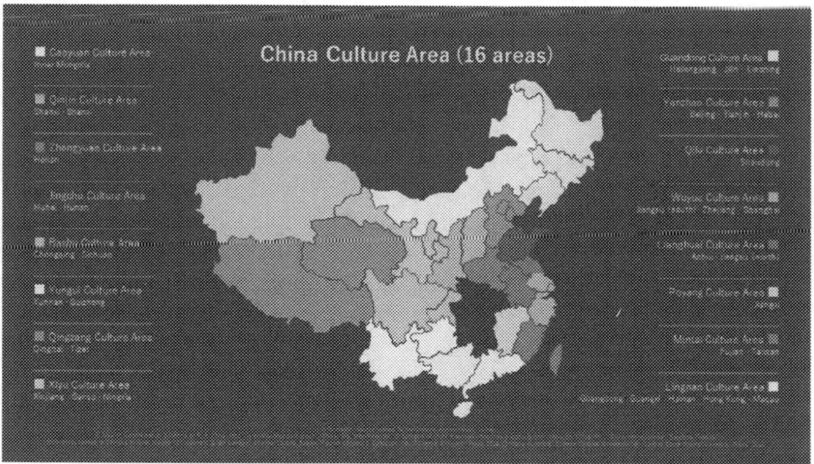

Fig. 2. China Culture Area (16 areas)

Furthermore, foreigners' impression of Chinese culture mostly comes from the signs of Chinese culture. As a result, it is non-negligible to extract cultural signs from different regions. Cultural impression and signs are closely related. People tend to form a set a culture impression through cultural impression. As shown in Fig. 3, this research makes all the culture signs either abstract of concrete, they belong to nature resources, lifestyle, celebrities, culture resources, artistic form, and philosophy. This kind way of classification to the greatest extent put all the cultural signs into consideration.

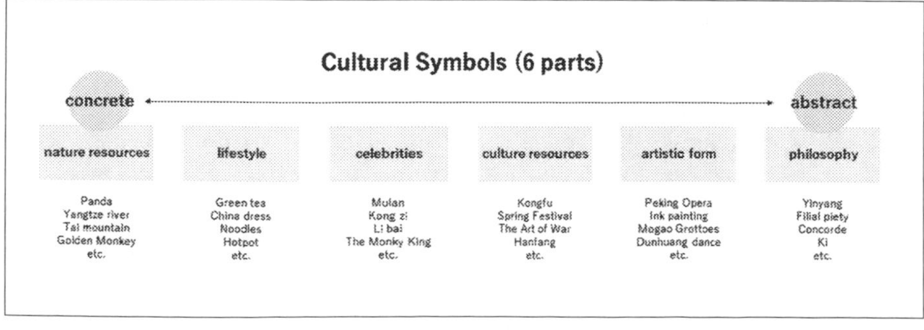

Fig. 3. Cultural Symbols (6 parts)

It can be analyzed from the language description of personal card (see Fig. 1 Character 2) that some foreigners are very interested in Chinese culture. However, they feel distant from Chinese culture because of language barrier and cultural difference. Real-life stories that are easy to understand and close to life are more attractive.

There is another example to support this point. Institute of cultural innovation and communication of Beijing Normal University is an institute that aims to international communication of Chinese culture. In 2014 and 2017, together with Survey Sampling International (SSI), this institute has issued *Foreigners' Cognition of Chinese Culture* and a series of reports [5]. According to the date results of *Foreigners' Cognition Status of Chinese Culture* (Fig. 4), among all the positive options, foreign respondents generally agree that Chinese culture is historic (91.4%), dynamic (74.9%) and attractive (71.8%). And intimate has the lowest identity with 59.7%. This can be verified from the data of negative impressions as well. There is a high proportion of respondents have chosen the options of mysterious (76.6%), conservative (74.9%), confusing (68.3%). Consequently, I have noticed an important problem point of international communication of Chinese culture - Inadequate intimacy.

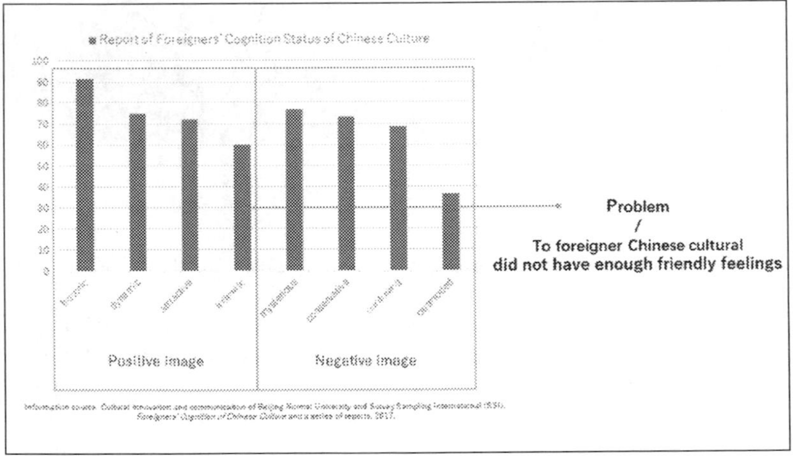

Fig. 4. Report of Foreigners' cognition status of Chinese culture

Trying to improve the intimacy of Chinese culture impression by specific practice becomes one of the problems we need to solve by this research. People's intimacy of another country should be gradually formed by detailed ways that close to real life and shorten the psychological gap. The intimacy also can be improved by other ways. For example, Chinese people may try to tell their personal experience with their own memories of culture image and impressions of hometown's culture.

It can be analyzed from the language description of personal card (see Fig. 1 Character 3) that some foreigners engage with Chinese culture proactively because of their interest. And some foreigners get to know about China and Chinese culture passively. In either case, it is necessary to conclude Chinese culture image, people's preference and regional cultural symbol. If my research proceeds smoothly, it not only will help to transmit Chinese culture images, but also will provide a reference for foreigners or overseas companies to learn the real cultural life of Chinese people.

3 Research Approaches

3.1 Research Model Design

In this research, we utilize psychological examination and distribute questionnaires to the interviewees. The questionnaire is all together divided into 2 parts: 12 basic personal information questions consisting of the first one, whereas 8 hometown culture impression questions consisting of the second one, all together 20 questions is in the questionnaires. The whole analysis process will be operated in center of data analysis and perspective description analysis (see Fig. 5).

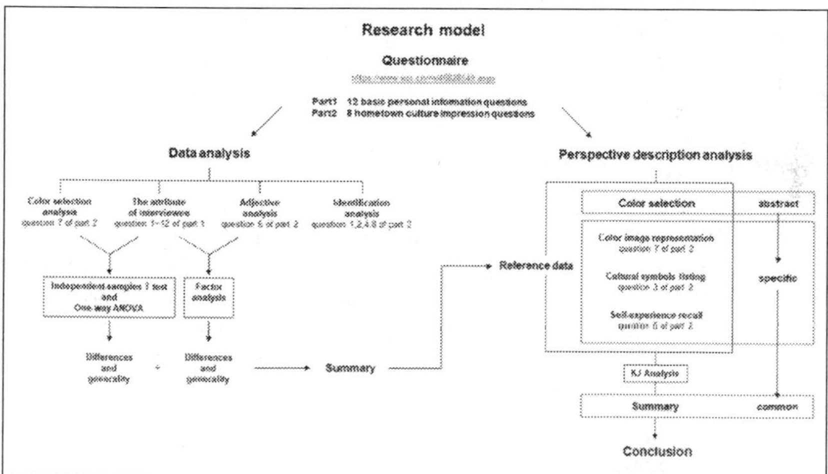

Fig. 5. Research model

As we all know, the main purpose of the research is to draw out the sharing common characters of culture impression through the interviewees perspective description of hometown colors, cultural symbols as well as memory and emotions, as a result, the data model design should serve for getting differences and generality reference data of perspective description. The research mainly uses independent-samples T test, one-way ANOVA, factor analysis.

The analysis of perspective description, with the inspiration of the Archetypal theory of C. G. Jung, color, a symbolic and abstract analysis factor, is here regarded as the access point. One step by another, we are trying to find out the specific connections between abstract descriptions and detailed things by color selection analysis, cultural symbols listing, self-experience recall and so on, therefore further conclude the collective cultural impression, also, this is an abstract-specific-common process. In the whole process, KJ Analysis pattern is in the main place.

To my point of view, the research mode above is suitable for investigation and analysis of all countries. Because in this research mode, the analysis of regional cultural images, mainly and basically by analyzing color, image, emotion, municipal impression of culture will further be drawn out. This research is human-centered and also centers on personal cognitive experience and mental feelings. In the end, I will also set about to do relating analysis of culture symbol on the aspect of personal memories and emotions, and then try to summarize the sharing common points. I believe that a collective and common cultural image is essential for every country and region (Here answers question1 of comments).

Of the other hand, we can also utilize modern information technology. We can put age, gender, income, profession, education into digital analysis system, so as to draw the conclusion whether people are interested in traditional or modern culture, what they are pursuing spirit layer of material layer, major focus point is natural environmental or social.

3.2 Summary of the Questionnaire

The Statistic Accounting of the Questionnaire Callback

According to the latest provincial population ranking report of China [6], the overall population of China is about 1,395,380,000. In order to improve the accomplishment ratio of the questionnaire, at the same time, ensure the analysis result which is on basis of the sample quantity, the survey is carried on by employing Equal reduction (ratio: 1/600,000). Macao, who has the smallest population, is here viewed as the reference value (set as 1 person), will collect back about 2386 questionnaires. The whole process, starting from this October while coming to an end today, has altogether collected 1182 ones, and 1204 ones are still needed (see Table 1).

Table 1. Demographics of China and the number of samples

*Rank	Administrative Divison	Population	Number of Samples	Collected Samples
1	Guangdong	113,460,000	189	19
2	Shandong	100,472,400	168	168
3	Henan	96,050,000	160	25
4	Sichuan	83,410,000	139	139
5	Jiangsu	80,293,000	134	85
6	Hebei	75,563,000	105	37
7	Hunan	68,602,000	115	70
8	Anhui	63,236,000	106	25
9	Hubei	59,170,000	99	14
10	Zhejiang	57,370,000	96	17
11	Guangxi	48,850,000	81	28
12	Yunnan	48,005,000	80	7
13	Jiangxi	46,221,000	77	77
14	Liaoning	43,593,000	73	73
15	Fujian	39,410,000	66	24
16	Shanxi	38,354,400	64	30
17	Heilongjiang	37,887,000	63	63
18	Shanxi	37,023,500	62	20
19	Guizhong	35,800,000	60	17
20	Chongqing	30,484,300	51	51
21	Jilin	27,174,300	46	46
22	Gansu	26,257,100	44	15
23	Inner Mongolia	25,340,000	42	16
24	Xinjiang	24,446,700	41	9
25	Shanghai	24,237,800	41	16
26	Taiwan	23,690,000	40	3
27	Beijing	21,707,000	36	36
28	Tianjin	15,596,000	27	21
29	Hainan	9,257,600	16	7
30	Hong Kong	7,482,500	13	0
31	Ningxia	6,817,900	12	6
32	Qinghai	5,983,800	10	10
33	Tibet	3,371,500	7	7
34	Macau	632,000	1	1
Total	China	1,395,380,000	2396	1192

As is shown in the forms (Fig. 6), Heilongjiang Province, Jilin Province and Liaoning Province are within Guandong culture area, Shandong Province within Qilu culture area, Jiangxi Province within Boyang Culture area, Sichuan Province and Chongqing Municipality within Bashu culture area, Qinghai Province and Xizang autonomous region within Qingzang culture area. The mentioned provinces within the five culture areas have already gathered enough questionnaires just as supposed to, therefore, the whole process can be moved forward to the stage of data disposal and analysis. The posted content of HCII 2020 takes Guandong culture area as an example and analyzes its area cultural images by using the research mode above.

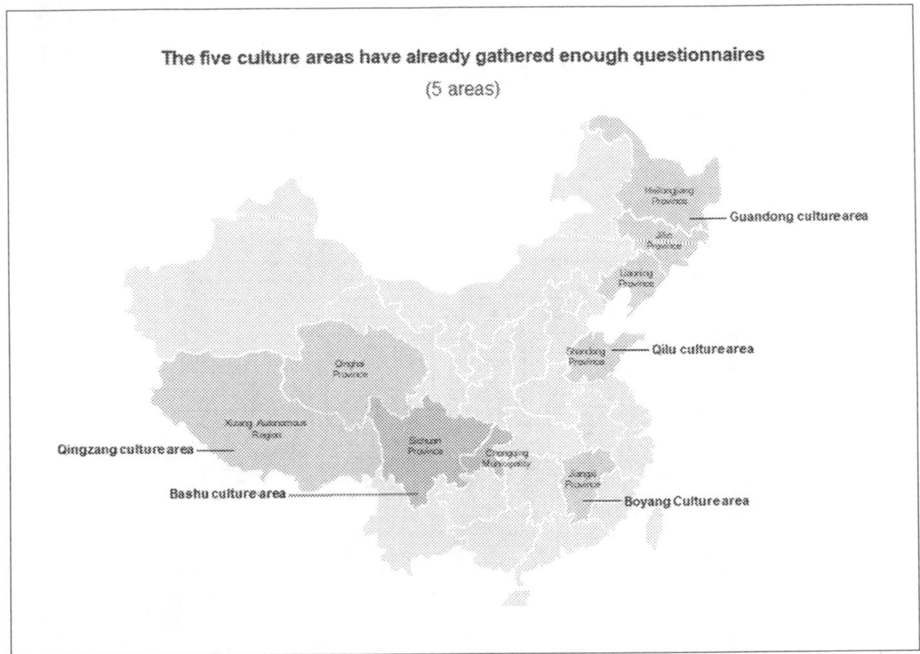

Fig. 6. The five areas have already gathered enough questionnaires

The Pictures of Gandong Culture Area

In order to provide a general idea of Guandong area culture before we elaborate the research, I found some typical photos (Fig. 7) according to the answers given by Respondents. These photos roughly show that Guandong is a frozen area with black land that covers by ice and snow. The people there are forthright. They love to eat meat and huge meals.

Fig. 7. The pictures of Guandong culture area (*The pictures stem from the network).

3.3 Data Analysis

The Attribute of Interviewee

Let's take Guandong culture area for instance, 63 from Heilongjiang, 46 from Jilin, and 73 from Liaoning (altogether182) have done the random questionnaires. The first part is unfurled around the interviewees' personal information, and there are 12 questions included.

Just as shown (Fig. 8), as for nationalities, the Han nationality takes the percentage of 84.1, while the minority takes 15.9, and the Manchu and the Korean nationality takes 8.8 and 3.8 respectively. As for those who have received higher education, undergraduate and above takes 72.5. In terms of personal monthly income, 73.7% of the interviewees earn over 3000 RMB monthly. When taking family monthly income per person as standard, 89.5% of the families are of medium and above level, while 59.3% of them belong to the group of high-level income people. Most interviewees come from cities (136 people), two-parent family (160 people), the number of single-child interviewees is 109 and the number of those who have siblings is as high as 73.

After the cross statistics (see Table 2), we draw the following conclusion. Among the interviewees, there are 84 males and 98 females. 106 of them are at the age of 18-28, taking the percentage of 58.2. When the number of the interviewees are set, 68 of them work for enterprises, 44 students, taking the place of No. 1 and No. 2. According

The questionnaire of hometown culture impression
Part 1 (12 questions)
Essential information

1 gender male / female

2 ethnic group (xx ethnic group) e.g. Han

3 age under 17 / 18-28 / 29-39 / 40-50 / 51-61 / over 62

4 educational degree
elementary school / middle school / high school / specialty / undergraduate / undergraduate above

5 occupation
enterprise / public institution / individuals / student / retire / wait for a job / other

6 personal monthly income (CNY)
3,000 RMB and under / 3,000-6,000 RMB / 6,000-10,000 RMB / 6,000-10,000 RMB and above

7 hometown (city name, xx Province) e.g. Jinan, Shandong Province

8 present residence (hometown or other provinces: xx province/ xx city)
(abroad: xx nation/ xx region)

9 one child of the family yes / no

10 hometown type city / country

11 family type single parent / parents

12 family monthly income per person (CNY)
under 500 RMB / 501-1,000 RMB / 1,001-2,000 RMB / 2,001-3,000 RMB / 3001-4,000 RMB / 4,001-5,000 RMB / over 5,000 RMB

The attribute of interviewee

item	category	frequency	percentage	cumulative percentage
gender	male	84	46.2	46.2
	female	98	53.8	100
ethnic group	Han	153	84.1	84.1
	Manchu	16	8.8	92.9
	Hui	2	1.1	94
	Korean	7	3.8	97.8
	Mongolian	2	1.1	98.9
	Xibe	2	1.1	100
age	under 17	1	0.5	0.5
	18-28	106	58.2	58.8
	29-39	28	15.4	74.2
	40-50	23	12.6	86.8
	51-61	24	13.2	100
	over 62	0	0	100
educational degree	elementary school	1	0.5	0.5
	middle school	5	2.7	3.3
	high school	10	5.5	8.8
	specialty	24	13.2	[illegible]
	undergraduate	87	47.8	75.3
	undergraduate above	45	24.7	100
occupation	enterprise	68	37.4	37.4
	public institution	25	13.7	51.1
	individuals	22	12.1	63.2
	student	44	24.2	87.4
	retire	8	4.4	91.8
	wait for a job	7	3.8	95.6
	other	4	2.2	100
personal monthly income	3,000 RMB and under	48	26.4	26.4
	3,000-6,000 RMB	72	39.6	68.2
	6,000-10,000 RMB	42	23.1	89
	10,000 RMB and above	20	11	100
hometown	Heilongjiang	63	34.6	34.6
	Jilin	46	25.3	59.9
	Liaoning	73	40.1	100
present residence	hometown	86	47.2	47.3
	other provinces	68	37.4	84.6
	abroad	28	15.4	100
one child of the family	yes	109	59.9	59.9
	no	73	40.1	100
hometown type	city	136	74.7	74.7
	country	46	25.3	100
family type	single parent	22	12.1	12.1
	parents	160	87.9	100
family monthly income per person	under 500 RMB	2	1.1	1.1
	501-1,000 RMB	6	3.3	4.4
	1,001-2,000 RMB	11	6	10.4
	2,001-3,000 RMB	25	13.7	24.2
	3001-4,000 RMB	30	16.5	40.7
	4,001-5,000 RMB	17	9.3	50
	over 5,000 RMB	91	50	100

Fig. 8. The attribute of interviewee

Table 2. The cross statistics (A ~ E)

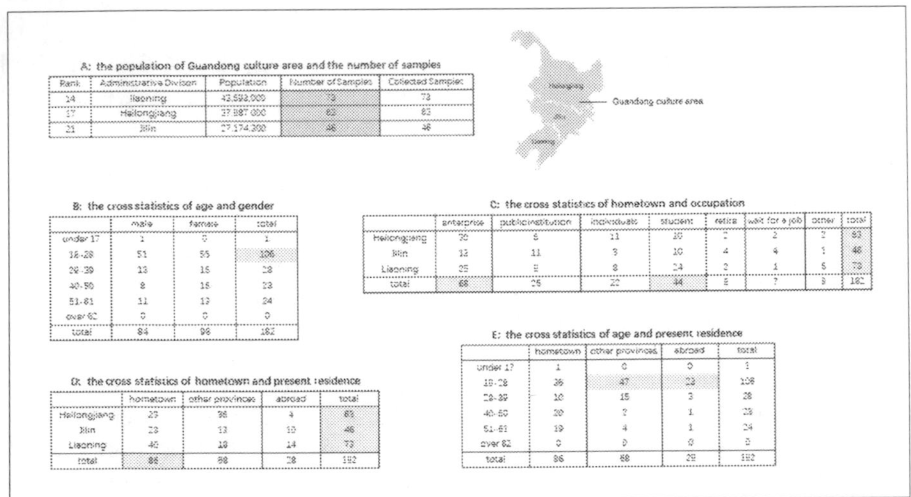

A: the population of Guandong culture area and the number of samples

Rank	Administrative Divison	Population	Number of Samples	Collected Samples
14	liaoning	43,598,000	73	73
17	Heilongjiang	37,887,000	83	83
21	Jilin	27,174,300	46	46

B: the cross statistics of age and gender

	male	female	total
under 17	1	0	1
18-28	51	55	106
29-39	13	15	28
40-50	8	15	23
51-61	11	13	24
over 62	0	0	0
total	84	98	182

C: the cross statistics of hometown and occupation

	enterprise	publicinstitution	individuals	student	retire	wait for a job	other	total
Heilongjiang	30	6	11	10	2	2	2	83
Jilin	13	11	3	10	4	4	1	46
Liaoning	25	8	8	24	2	1	5	73
total	68	25	22	44	8	7	9	182

D: the cross statistics of hometown and present residence

	hometown	other provinces	abroad	total
Heilongjiang	23	36	4	83
Jilin	23	13	10	46
Liaoning	40	18	14	73
total	86	68	28	182

E: the cross statistics of age and present residence

	hometown	other provinces	abroad	total
under 17	1	0	0	1
18-28	36	47	23	106
28-39	10	15	3	28
40-50	20	2	1	23
51-61	19	4	1	24
over 62	0	0	0	0
total	86	68	28	182

to the cross statistics of hometown and present residence, 86 of them are still in their hometown whereas 68 of them are in other provinces of China. 28 of them are now overseas. The number of those who are off their hometown is higher than those in, and the phenomenon is even more evident in Heilongjiang. And a majority of off-hometown interviewees are between 18–28.

Identification Analysis

Questions No. 1, No. 2 and No. 8 of the questionnaire are about interviewees' hometown culture interest degree, familiarity degree and spreading willingness. It employs Likert scale, positive answers are about five points, while the negative ones are about one point. According to statistics analysis, when asked whether interested in hometown culture, 43.4% of the interviewees' answers are very much, 35.2% degree are comparatively high, and the two statistics add up to 78.6%. When asked whether familiar with hometown culture, 41.2% answers are so-so or normal, however 33.5% of the answers are pretty high degree, and these two statistics add up to 74.7%. When asked whether have the willingness to introduce hometown culture to foreigners, 64.3% of them are of high willingness degree and the degree of 20.3% of them are of comparatively high, which two add up to 84.6%. In comparison to high degree of interest and spread willingness, the interviewers give out quite neutral answer in terms of familiarity degree (see Fig. 9).

The questionnaire of hometown culture impression
Part 2 (question 1,2,8)
Survey of hometown culture impression

1 Are you interested in the culture of the area where your hometown is? (multiple choice)
hate (1)
slightly hate (2)
not matter (3)
interested (4)
very much (5)

Interviewees' hometown culture interest degree

Likert scale	frequency	percentage	effective percentage	cumulative percentage
hate	0	0	0	0
slightly hate	2	1.1	1.1	1.1
not matter	37	20.3	20.3	21.4
interested	64	35.2	35.2	56.6
very much	79	43.4	43.4	100
total	182	100	100	100

2 Are you familiar with your hometown culture? (multiple choice)
not at all (1)
not familiar (2)
so-so (3)
familiar (4)
very much (5)

Interviewees' hometown culture familiarity degree

Likert scale	frequency	percentage	effective percentage	cumulative percentage
not at all	3	1.6	1.6	1.6
not familiar	6	3.3	3.3	4.8
so-so	75	41.2	41.2	46.2
familiar	61	33.5	33.5	79.7
very much	37	20.3	20.3	100
total	182	100	100	100

8 Are you willing to introduce your hometown culture to foreigners with pride? What's your attitude to your hometown culture? (multiple choice)
hate (1)
slightly hate (2)
not matter (3)
general acceptance (4)
high acceptance (5)

Interviewees' hometown culture spreading willingness

Likert scale	frequency	percentage	effective percentage	cumulative percentage
hate	2	1.1	1.1	1.1
slightly hate	2	1.1	1.1	2.2
not matter	24	13.2	13.2	15.4
general acceptance	37	20.3	20.3	35.7
high acceptance	117	64.3	64.3	100
total	182	100	100	100

Fig. 9. The questionnaire of hometown culture impression (Part 2 Questions No. 1, 2, 8)

Question No. 4 of the questionnaire is a multiple-choice question. According to the statistics, among all the 182 interviewees, people get to know their hometown culture via self-experience, parent-friend effect, they are respectively 144 people, 25.8% and 133 people, 23.8%. Combining the drawn-out conclusion and questions No. 3, No. 5 and No. 7 of the second part of the questionnaire, it is easy to discover interviewees' hometown culture impression description tends to operate on basis of daily and common living habits (see Fig. 10).

The questionnaire of hometown culture impression
Part 2 (question 4)
Survey of hometown culture impression

4 Where do you get the culture symbol? (Comprehensive choices)
personal experience / parents and relatives / network / broadcast / paper media / school curricula / other ______

Obtain channel survey of hometown culture symbols

	frequency	percentage	effective percentage	cumulative percentage
personal experience	144	25.8	25.8	25.8
parents and relatives	133	23.8	23.8	49.6
network	87	15.6	15.6	65.2
broadcast	90	16.1	16.1	81.3
paper media	55	9.8	9.8	91.1
school curricula	48	8.6	8.6	99.7
other	1	0.3	0.3	100
total	558	100	100	100

Fig. 10. The questionnaire of hometown culture impression (Part 2 Questions No. 4)

Color Selection Analysis

Question No. 7 of the questionnaire is the survey about the color psychological survey of hometown culture impression (see Fig. 11). When asked which color will you choose to represent the area where your hometown locates, 20.9% of the interviewees choose green, listing No. 1, and other following ratio are blue (15.4%) No. 2, white (13.2%) No. 3, red (12.6%)No. 4. In order to further study the relationship between interviewee attribution and color selection, cross statistics are made between interviewees' 12 basis information items and color selection. when the comparison target quantity is two (question 1, 9, 10, 11 of part 1), use independent-samples T test, and when the comparison target quantity is above two (question 2, 3, 4, 5, 6, 7, 8, 12 of part 1), use one-way ANOVA.

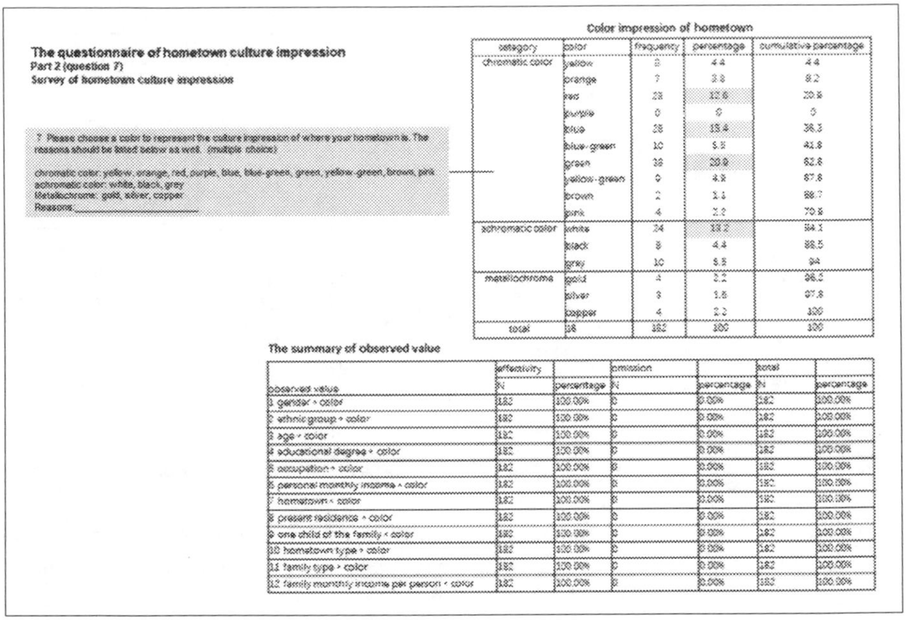

The questionnaire of hometown culture impression
Part 2 (question 7)
Survey of hometown culture impression

7. Please choose a color to represent the culture impression of where your hometown is. The reasons should be listed below as well. (multiple choice)

chromatic color: yellow, orange, red, purple, blue, blue-green, green, yellow-green, brown, pink
achromatic color: white, black, grey
Metallochrome: gold, silver, copper
Reasons:________________

Color impression of hometown

category	color	frequency	percentage	cumulative percentage
chromatic color	yellow	8	4.4	4.4
	orange	7	3.8	8.2
	red	23	12.6	20.8
	purple	0	0	0
	blue	28	15.4	36.3
	blue-green	10	5.5	41.8
	green	38	20.9	62.6
	yellow-green	9	4.9	67.6
	brown	2	1.1	68.7
	pink	4	2.2	70.9
achromatic color	white	24	13.2	84.1
	black	8	4.4	88.5
	grey	10	5.5	94
metallochrome	gold	4	2.2	96.2
	silver	3	1.6	97.8
	copper	4	2.2	100
total	18	182	100	100

The summary of observed value

observed value	effectivity		omission		total	
	N	percentage	N	percentage	N	percentage
1 gender * color	182	100.00%	0	0.00%	182	100.00%
2 ethnic group * color	182	100.00%	0	0.00%	182	100.00%
3 age * color	182	100.00%	0	0.00%	182	100.00%
4 educational degree * color	182	100.00%	0	0.00%	182	100.00%
5 occupation * color	182	100.00%	0	0.00%	182	100.00%
6 personal monthly income * color	182	100.00%	0	0.00%	182	100.00%
7 hometown * color	182	100.00%	0	0.00%	182	100.00%
8 present residence * color	182	100.00%	0	0.00%	182	100.00%
9 one child of the family * color	182	100.00%	0	0.00%	182	100.00%
10 hometown type * color	182	100.00%	0	0.00%	182	100.00%
11 family type * color	182	100.00%	0	0.00%	182	100.00%
12 family monthly income per person * color	182	100.00%	0	0.00%	182	100.00%

Fig. 11. The questionnaire of hometown culture impression (Part 2 Questions No. 7)

Independent-Samples T test

As for independent-samples T test, single-child, city and countryside origination, family type does not matter the selection of color. However, gender does matter the selection of color hometown culture impression.

When combining the two factors—gender and color selection and do the cross analysis, we can discover that the number of females who chose achromatic color and metallochrome is twice as many as males, and among those, women are even more likely to choose white and black than men (see Fig. 12). In contrast, the men who tend to colorful choose are far more likely to make a choice of green compared with women. As we all know, women basis carry twice cone cells as many as men, which lead to the result that women can see more colors, and thus being more sensitive to colors. And also, sometimes men and women may choose different colors to expressing their feelings toward one same thing. For instance, when required to describe the frozen snowy world, almost all men made the choice of white to do the expression, and none of the interviewees selected chose silver. In comparison, two women made use of silver to describe the same mentioned winter feelings. The color psychology differences and generality between men and women will be paid special attention in the study of next stage.

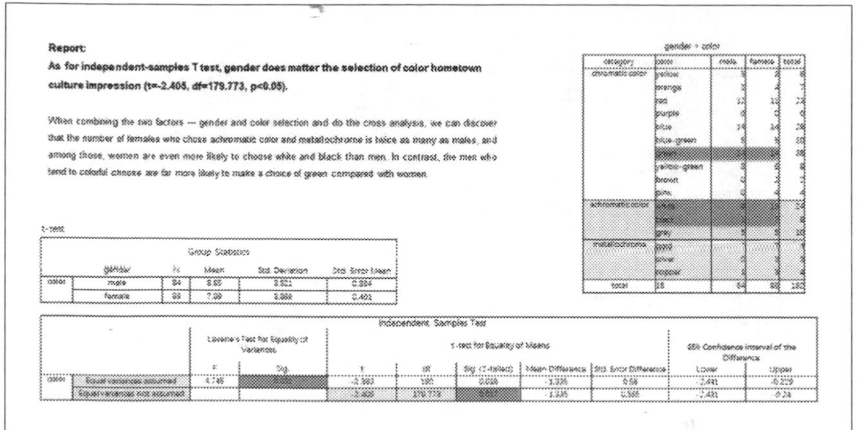

Report:

As for independent-samples T test, gender does matter the selection of color hometown culture impression (t=-2.405, df=179.773, p<0.05).

When combining the two factors — gender and color selection and do the cross analysis, we can discover that the number of females who chose achromatic color and metallochrome is twice as many as males, and among those, women are even more likely to choose white and black than men. In contrast, the men who tend to colorful choose are far more likely to make a choice of green compared with women.

gender * color

category	color	male	female	total
chromatic color	yellow	[illegible]	[illegible]	[illegible]
	orange	[illegible]	[illegible]	[illegible]
	red	[illegible]	[illegible]	22
	purple	[illegible]	[illegible]	[illegible]
	blue	[illegible]	[illegible]	[illegible]
	blue-green	[illegible]	[illegible]	[illegible]
	[illegible]	[illegible]	[illegible]	38
	yellow-green	[illegible]	[illegible]	[illegible]
	brown	[illegible]	[illegible]	[illegible]
	pink	[illegible]	[illegible]	[illegible]
achromatic color	[illegible]	[illegible]	[illegible]	24
	[illegible]	[illegible]	[illegible]	[illegible]
	grey	[illegible]	[illegible]	[illegible]
metallochrome	gold	[illegible]	[illegible]	[illegible]
	silver	[illegible]	[illegible]	[illegible]
	copper	[illegible]	[illegible]	[illegible]
total	18	84	98	182

t-test

Group Statistics

	gender	N	Mean	Std. Deviation	Std. Error Mean
color	male	84	8.65	3.521	0.384
	female	98	7.39	3.968	0.401

Independent Samples Test

		Levene's Test for Equality of Variances		t-test for Equality of Means					95% Confidence Interval of the Difference	
		F	Sig.	t	df	Sig. (2-tailed)	Mean Difference	Std. Error Difference	Lower	Upper
color	Equal variances assumed	4.746	[illegible]	-2.383	180	0.018	-1.336	0.56	-2.441	-0.229
	Equal variances not assumed			-2.405	179.773	[illegible]	-1.336	0.555	-2.431	-0.24

Fig. 12. The report of Independent-samples T test

One-Way ANOVA

As for the one-way ANOVA, those factors, such as ethnic group, age, educational degree, occupation, personal monthly income, family monthly income per person, present residence, don't matter with the color selection of hometown impression on the aspect of statistics. However, the hometown location does matter with the color selection of hometown impression on the aspect of statistics.

When combining the two factors - hometown location and color selection and do the cross analysis, we can discover that 16 of the interviewees who originate from Heilongjiang chose white, while none of those who originate from Liaoning make such a choice (Fig. 13). On contrast, more Liaoning-originating interviewees prefer to choose blue and red. Heilongjiang is a northernmost inland province of China, its summer is quite short while the freezing cold winter can be as long as 6 months, which is in line with the feelings that white can bring about. The south part of Liaoning is on the verge of the Bohai Sea and the Yellow sea, whose coast line is all together 2920 km, which is in line with the feelings that blue can bring about as well. Differing from the differences that following climate and geography, the two provinces have comparatively similar perspective description toward red, what usually means jubilation, liveliness, enthusiasm, forthright and outspoken. Therefore, the color psychological differences and generality between Liaoning and Heilongjiang interviewees will be the second noticeable reference in the study of next stage.

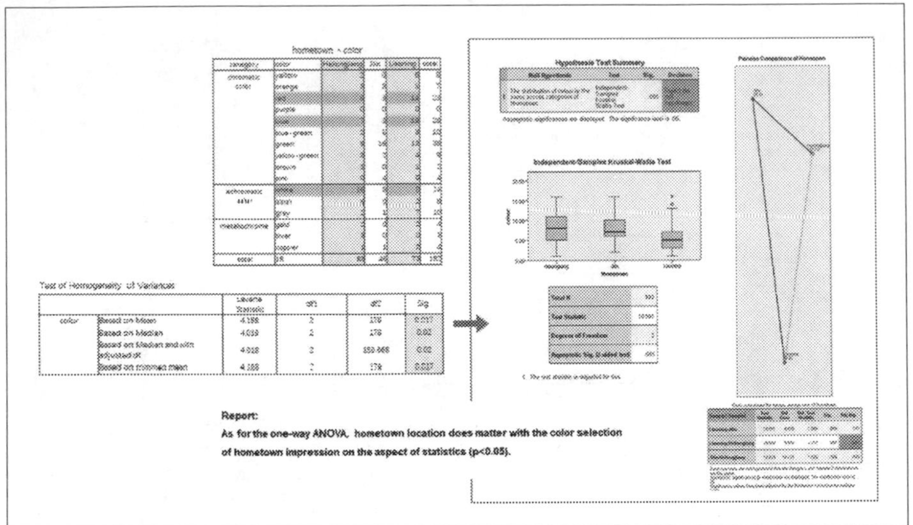

Fig. 13. The report of One-way ANOVA

4 Conclusion of These Parts

4.1 The Attribute of Interviewee and Summary

When it comes to the basis information of Guandong culture area investigation, the ratio of male against female is parallel to each other, and the Han nation, citizens and two-parent family samples cover the largest section. Quite a large majority of the interviewees have received higher education and have a considerable income. Quite a few of the interviewees even move home and abroad, and those who are away from their home are basically at the age from 18–28.

4.2 Identification Analysis and Summary

During the survey process of culture identification of Guandong culture area, over 70% of the interviewees are interested in hometown culture and also feel they have a good knowledge of hometown culture, and at the same time, they have the strong desire to introduce their hometown culture to foreigners. Most of the interviewees get to know all aspects of their hometown through self-experience and parent-friend effect, and they did the impression description on basis of daily surroundings and common living habits.

4.3 Color Selection Analysis and Summary

During the process of color impression of Guandong culture area, when studying the connection between interviewee attribution and color, we discover gender, Heilongjiang originating interviewees and Liaoning ones matter the differences and common characters of color psychology. And this part will be taken as two important reference factors in next study stage.

As the Fig. 5 shows, a part of the analysis has been finished analyzing currently. Next, I will summarize and analyze the adjective of Guandong culture area. The research result will be accomplished before the publishing of HCII meeting in July, 2020.

5 Anticipatory Result

Through the research, the PR plan that we want to present is about Chinese culture's experiential space. It is by five senses that people get in touch with new things and get knowledge of foreign culture. No matter it is language, word, food, dress, festival, religion, architecture and so on, they are all forms of expression on Chinese culture, and they all need to be transferred by those elementary senses of sight, sound, taste, smell and touch. Whether finally we are going to make the pattern like space of Blind restaurant which strengthens taste while weaken visual effect, or like space of Naked eye 3D restaurant which pay more attention on sight and sound, the conclusion needs to be drawn on foundation and basis of large amounts of investigation and study.

The advertising methods of PR meditation are different because of the occasions, therefore a large amount questions and data collection are required. We are going to combine the five senses in different ways (31 combinations are working out by the permutation and combination) to choose the most suitable advertising plans and approaches (see Fig. 14). In order to achieve a concrete application, I will summarize and analyze the data result of the 16 areas of China, conduct a targeted investigation of foreigner's "five senses" preference, combine the results of both sides, choose the best of 31 combinations, build up a human communication pattern. Finally, present the PR proposal. (Here answers question 3 of comments)

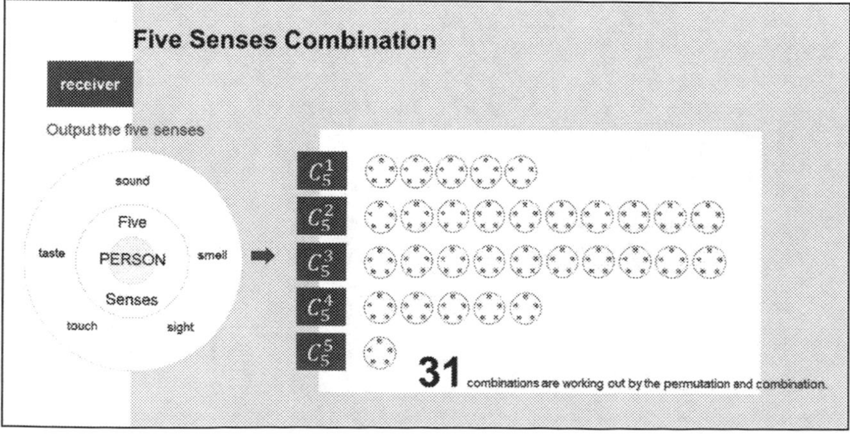

Fig. 14. Five senses combination

References

1. Hakuo Maruyama, T.: Science of the Five Senses: Special Issue. Komanii Corporation communication design research center, Japan (1993)
2. Qin Liangjie, T.: Chinese Culture Region Tourist Literature Collection. Tsinghua University Press, China (2014)
3. Wang Huichang, T.: Chinese Cultural Geography. Central China Normal University Press, China (2010)
4. Information source related to the Chinese administration frame: China City Map Homepage, http://www.ditu-map.com/gov, last accessed 2019/06/20. China Administration Divisions Homepage, http://www.xzqh.org/html/show/cn/37714.html. Accessed 20 June 2019
5. Beijing Institute of Culture Innovation and Communication (Cultural innovation and communication of Beijing Normal University) Homepage, https://bicic.bnu.edu.cn. Accessed 25 June 2019
6. National Bureau of Statistics of China Homepage, http://www.stats.gov.cn/tjsj/zxfb. Accessed 03 Nov 2019

Outside the Box: Contextualizing User Experience Challenges in Emergency Medical Technician (EMT) and Paramedic Workflows

Katelynn A. Kapalo[1(✉)], Joseph A. Bonnell[2], and Joseph J. LaViola Jr.[1]

[1] University of Central Florida, Orlando, FL, USA
kate.kapalo@knights.ucf.edu
[2] Phoenix Fire Department, Phoenix, AZ, USA

Abstract. Paramedics and Emergency Medical Technicians (EMTs) often serve dual roles in their communities as both emergency medical providers and firefighters. Therefore, the demands and needs of these providers are different than those working directly in a hospital emergency room or medical office. Medics are required to treat a patient in less than ideal conditions where seconds can mean the difference between life and death. The goal of this work is to better understand how research informing technology for the emergency room (ER) can be linked to improving the user experience of EMTs and paramedics in the field. Through a review of relevant literature, we capture lessons learned in conveying information quickly, linking necessary information from disparate sources, and giving these providers accurate information to successfully treat and transport patients during prehospital care.

Keywords: Human factors · Emergency medicine · Prehospital care · Electronic health records

1 Introduction

As technology in healthcare continues to advance, the field of health informatics shifted from a focus on simply managing patient medical records to the management of protected health information. The introduction of electronic medical records (EMRs) and electronic health records (EHRs) have changed the way providers and patients manage their care. However, some studies have demonstrated conflicting evidence regarding the successful implementation of EHRs in a variety of healthcare contexts [1].

One understudied area is the impact of these systems on prehospital emergency medical services (EMS). Prior to arriving at the emergency department, patients generally are transported in an ambulance. This portion of care, known as prehospital care, requires careful consideration. Prehospital care refers to the treatment patients receive prior to handoff at the hospital ED. Prehospital care varies depending on location, region, and funding. For example, in the US, prehospital care is generally performed by

C. Stephanidis et al. (Eds.): HCII 2020, LNCS 12427, pp. 133–150, 2020.
https://doi.org/10.1007/978-3-030-60152-2_11

trained and certified EMTs and paramedics [2]. In Europe and in other countries, physicians or nurses may accompany paramedics on an ambulance to provide care [3]. The majority of modern prehospital care systems are equipped to provide advanced life support (ALS) [3]. Breyer [4] detailed the importance of understanding the effect of the Patient Protection and Affordable Care Act (PPACA) on the current prehospital fire-based EMS systems operating in the United States (U.S.). The PPACA created the opportunity for fire-based EMS systems to transition from emergency response and patient transport, to a more integrated part of the healthcare system. Despite this transition, the introduction of EHRs in fire-based EMS is relatively new and poses a variety of unique challenges [4]. For example, some studies have demonstrated that patient satisfaction is directly correlated with perceived paramedic response times [5–7]. Results from a meta-analysis suggest that the total average time for prehospital care is approximately thirty minutes in urban areas, demonstrating that paramedics and EMTs must gather information quickly [8]. The goal of this work is to better understand the role of EHRs in prehospital care, as well as to outline areas where more research is required to afford EMTs and paramedics the information needed to deliver quality patient care.

We conducted a conceptual review of the literature, emphasizing the importance of lessons learned in the transition of paper medical records to electronic in emergency medicine. These lessons are applied to emergency medical services (EMS) to translate how these practices map to the needs of paramedics and EMTs in the field. Additionally, we identified gaps where further information is necessary to recommend solutions. To do this, we interviewed EMTs and medics to better understand the practical impact of best practices and the link to their current information needs to existing solutions. Drawing from human-computer interaction (HCI) methods, we conducted interviews with two separate fire-based EMS departments. We contend that the current technologies provided to medics and EMTs do not meet provide a positive user experience. Instead, these rescue units often face *distractions, interruptions, and issues with technologies, creating deficits related to patient care and response times*. This not only impacts the patient, but also has negative outcomes for emergency medical personnel [9]. The goal of this work is to better understand the current tools and technologies rescue units rely upon in order to provide patient care and to better understand interdependencies where the human user is required to intervene or interact with inadequate systems in prehospital care.

By capturing the information requirements of these emergency medical personnel in the context of a sociotechnical framework, we can map these needs to current or future technological capabilities. Ultimately the use of technology has an impact on patient safety and the safety of the medical professionals providing patient care [10], creating an opportunity for the human-computer interaction (HCI) and human factors communities to better assist the prehospital emergency medical domain through further analysis of their direct user experiences and the extant literature.

2 Related Work

Electronic health records (EHRs) recently became more prevalent due to the introduction of the Health Information Technology for Economic and Clinical Health (HITECH) Act [11]. Generally, there has been a focus on EHRs in traditional medical offices and hospitals. However, much of the literature does not focus on the impact of EHR use in ambulances by paramedics and EMTs. Seminal work in this area includes a systematic review of prehospital emergency care in England that examined the tensions between systems and providers [12]. Additionally work emphasizes the importance of systematic reviews for understanding the changing landscape as more hospitals transition to EHRs [13, 14]. Relatively few articles from the extant literature examine the impact of EHRs on prehospital care providers, and even fewer articles focus on understanding the impact of these systems on the stress and workload levels of these providers [15, 16]. Some of the core research conducted in this area includes an analysis of reports and handoffs to better support EMTs [15, 16]. Additional research in this area has focused on the perception of eHealth systems more broadly, from the perspective of multiple healthcare stakeholders [9, 17].

This lack of empirical data creates a critical gap in the literature since the role of paramedics, EMTs, and prehospital care providers are typically not included in discussions of EHRs. However, they are considered important stakeholders in the overall electronic health system, since they perform care prior to arrival in the ED. Newgard et al. [18] studied the impact of electronic data processing, providing evidence that the implementation of EHRs in ambulances positively impacted care. The use of EHR systems in prehospital care has largely been driven by a need to improve systems for billing, consequently enhancing quality assurance practices [12, 19]. However, these decisions are typically driven by administrative requirements and the end-users of the systems are generally not involved in the process of adoption, meaning paramedics and EMTs often *do not have a voice* in the process of selection or implementation of the EHR systems they rely upon to complete their daily job tasks.

2.1 Transition from Paper Records to Electronic Health Records (EHRs)

The concept of electronic health, known as eHealth, was first introduced in the late 1990s [20]. Moving into the early 2000s, advancements in computing made the possibility of electronic medical systems and records much more viable [21]. This transition made it possible to avoid clinical documentation errors such as illegible writing and allowed for the collection of other types of health care data, saving practitioners time. It also contributed to better information for insurance companies to file claims [21]. Despite this, some studies have identified deficiencies in the use of EHRs due to a lack of data on patient outcomes or the actual cost-effectiveness of these systems [22, 23].

The emergency department (ED), also known as the emergency room (ER), in modern US hospitals is an acute care facility. Emergency departments adopted EHRs primarily as a means of improving patient care through decision-making support on scene, continuity of care and access to patient information [24]. Studies have demonstrated that physicians in U.S. EDs complete over 100 tasks an hour, requiring a system that supports their ability to complete these tasks as quickly as possible [25].

Consequently, the impact of EHRs on ED efficacy and efficiency has been studied for over a decade [26–29]. Most studies suggest that EHRs generally enhance ED productivity and efficiency; however, there have been some studies that uncovered some issues associated with the use of commercial EHR systems in emergency medicine, such as increased task switching [30–32]. Additionally, some studies have emphasized the importance of addressing the impact of EHRs on provider well-being [33] and the importance of meeting the technology needs of first responders through user-centered design [34, 35].

This study contributes to the literature by identifying areas where further research could enhance user experience for *prehospital* emergency medical professionals who rely upon systems that do not meet their current needs or the needs of the patient due to outdated and unreliable technologies. Additionally, this work contributes to better understanding the needs of prehospital emergency providers who often lack access to historical health information to provide patients with better continuity of care.

3 Methods

We conducted a literature review on peer-reviewed publications related the implementation of EHRs between the years of 2000 and 2018. The year 2000 was chosen as a starting point since eHealth and EHRs were first introduced and implemented as a potential solution in the late 1990s [20]. The process involved three iterations and was conducted between November 2018 and January 2019. We conducted our search via the ISI Web of Knowledge and Science Direct. The search terms were derived from key words used in combination such as: "EHR and paramedic," "technology and emergency department," "user experience paramedic," "eHeath and emergency medicine," and "electronic medical records emergency." For the purposes of our literature review, articles must have met three criteria to warrant inclusion:

1. Peer reviewed and published work
2. Published between the years 2000 and 2018
3. Involved discussion of technology or eHealth systems in the context of emergency medicine or prehospital care.

We excluded articles and works that covered topics outside the scope of the emergency room care, such as disease genomics. We also excluded papers that used EHRs or EMRs as a source of archival analysis for a particular injury or condition. For example, some papers leveraged health records as a data source to identify whether emergency admissions generally peak at a certain period of time or on certain days of the year. Since this did not directly involve the use of technology for patient care, these papers were excluded from the review. Our search terms identified over 2,000 papers that were initially reviewed for relevance based on the criteria above. Two hundred articles were excluded due to the fact they were not peer-reviewed. Of the remaining articles, all were reviewed for relevancy, and 108 papers were included in this review. A representative sample of these articles are included in the reference list. We aggregated findings to communicate general trends related to the effective use of EMRs in an emergency medical setting, drawing from the sociotechnical framework originally proposed by Sittig and Singh [36] to guide our discussion. This framework conceptualizes key issues in the sociotechnical system that prehospital care providers operate in, providing the key dimensions of interest and scaffolding the coding scheme for understanding the cognitive and physical needs and demands placed upon prehospital emergency medical providers.

In addition to the literature review, we conducted interviews with four paramedics and two EMTs, (n = 6) from separate agencies that service different metropolitan areas in the Southeastern region of the United States. This process of data collection supplements the existing gaps in the literature with data from prehospital care providers who are not well-represented in the literature. To do this, we conducted interviews via phone and online conferencing systems. Our approach was grounded in the Applied Cognitive Task Analysis (ACTA) framework, a well-known method in the field of human factors that consists of three interview methods [37]. ACTA is designed to help inform HCI work for interface design and applied product development. The participants age (M = 33.67, SD = 4.32) and years of EMS job experience (M = 13, SD = 2.90) are captured in the table below (Table 1).

Table 1. Participant demographics

Participant	Age	Experience in EMS (in years)
1	36	17
2	27	9
3	34	15
4	40	14
5	33	12
6	32	11

4 Results

The Safety Assurance Factors for EHR Resilience (SAFER) guidelines were originally developed to help understand the importance of measurement and monitoring in the case of eHealth systems more broadly [38]. However, this framework emphasizes *self-assessment of risk*, allowing departments and organizations to take ownership of improving the safety and effectiveness of EHRs with the goal of improving the quality of patient care. These guidelines provide evidence that self-assessment of EHRs is an important component of ensuring that quality of care and satisfaction are not compromised in the implementation or use of EHRs. Despite the benefits, these guidelines generally apply to settings other than prehospital care. To better frame the risk to both patients and prehospital care providers, we categorized the results of our literature and interviews below in terms of the dimensions outlined in the sociotechnical model originally proposed by Sittig and Singh [36]. The table below is adapted from their original model [36]. We leveraged this model to create an *a priori* framework in order to analyze the qualitative data from interviews and to categorize the literature evaluated in our review. See Table 2 below for the code book that guided our analysis.

Table 2. Sociotechnical Model Proposed by Sittig and Singh for self-risk assessment of EHRs to improve the quality of patient care, adapted to prehospital care.

Sociotechnical dimension	Definition
Hardware and software	Computing foundation for applications and systems used in patient care
Clinical content	Data (text, images, etc.) contained in clinical documentation
Human-computer interface	How clinicians or other stakeholders interact with a computer system (including inputs and outputs)
People	All stakeholders involved in prehospital care
Workflow and communication	Procedures and protocols that ensure patient is being cared for
Internal organization features	The work environment surrounding the prehospital care provider
External rules and regulations	External limitations and regulations (e.g., accreditation, laws, etc.)
Measurement and monitoring	Evaluation methods used to determine the impact of computer systems on the provider and patient safety

4.1 Hardware and Software: Case Studies and Prototype Testing

A majority of studies focused on understanding the impact of new systems in specific locations or applied settings. Two studies in Crete, Greece demonstrated the importance of applying theoretical models, such as the Task-Technology Fit Model to better understand the role of EHRs in delivering better quality care and assisting prehospital care providers with technology to meet their needs [39, 40]. Additional work in this area evaluated a mobile system architecture for assisting emergency medical personnel rendering aid to victims of motor vehicle accidents and systems for triaging patients in the midst of a disaster or mass casualty incident [41–43]. These new systems have focused on understanding the impact of information availability on patient care, ultimately paving the way for future systems.

Some studies focused on the testing and evaluation of new software prototypes or architectures to better support both patients and healthcare providers [44–48]. For example, a large-scale project funded by the European Union demonstrated the efficacy of video-based information exchange to support prehospital care [49]. There have also been systems that use RFID technology to track care [50]. Majeed [51] introduced a new architecture for prehospital care reporting. Although these studies have demonstrated the efficacy of EHRs on prehospital care, it is important to note that the majority of them took place outside the United States, which may represent a different model of emergency care [52]. There is a paucity of literature on existing implementation of

EHRs in prehospital care, of those that do exist, most of them focus on ED physicians or other stakeholders rather than prehospital care providers [9, 53].

This dimension is difficult to measure due to the lack of consistency between departments and agencies. It is critical to note that each agency may have different requirements or interoperability standards, making it sometimes unfeasible or impractical for the paramedics or EMTs to have any input or influence on the system.

4.2 Clinical Content

Historically, fire-based EMS systems did not have access to hospital-based EHR systems [6]. However, this dynamic is changing. A number of papers we reviewed addressed the clinical data or content of charts to better understand trends in diagnosis and care of patients in the ED [54, 55]. Some recent studies have revealed negative provider perceptions on the use of EHRs for documentation of specific life-saving interventions, such as resuscitation in the ED [29]. This focus on clinical content is important for understanding some of the interaction design decisions that can negatively impact provider perceptions.

Continuity of Care. In our review, preliminary work focused on better understanding how information exchange leads to better clinical outcomes. Lammers et al. [56] demonstrated the importance of timely information exchange on the reduction of redundant imaging orders in the ED, saving both the patient and provider frustration, time, and unnecessary costs. Detailed work has been conducted on identifying some of the sociotechnical factors that contribute to the improvement of patient handover to [57]. Additionally work has focused on understanding how to more effectively chart patient information prior to conducting patient handover, also known as patient handoff [58–60]. However, it is important to note that paramedics do not typically have access to detailed health information or EHRs that are integrated with the patient's primary care doctor, resulting in potential information losses and reduced efficiency of care.

During an interview session, one paramedic succinctly captured the idea that while having access to prior emergency runs can be important in the case of opioid abuse or chronic health conditions, it also may bias prehospital care providers towards a particular treatment plan, as illustrated below:

> *"Sometimes you have a patient that calls us frequently. While it's important to understand that they most likely have a chronic condition or require treatment for a specific problem repeatedly, that is not always the case. You can go on several calls that don't necessarily give you the full history, but then they really do have an emergency. It's important to know and recognize when this happens. We can be biased without even recognizing it because we see them all the time for the same things over and over." -P2*

From this comment, it is important to note that patient outcomes can be inadvertently influenced by the existing information that is available to the provider, usually from previous hospital runs. Paramedics and EMTs are limited in the information provided to them and often have to make critical decisions about interventions in order to save a patient's life *without access to information about prior medical conditions or treatments plans*. However, for certain patients, this may not be the case as a department may run calls on a single patient multiple times per day depending upon their health and living conditions, etc. Balancing what little information is available with their own assessments can present challenges and increase paramedic workload. Moreover, research in this area has demonstrated that frequent users of urban EMS systems are typically treated for recurring health conditions that could be better managed outside the EMS system [61]. The accessibility of this information across providers has also been investigated as a means to improve patient care [62].

Managing Patient Engagement and Education. One area of emerging interest involves the use of mobile applications and technology to enhance patient engagement and education in their own treatment process [63]. For example, a mobile health application involving a text-based intervention seemed to demonstrate positive clinical outcomes and decreased the number of ER visits for patients with Type II Diabetes [64]. These applications of mobile health provide a more convenient way for patients and providers to interact. However, these solutions are often focused on longer term care and involve follow up with the patient's network of doctors, which is outside the scope of the prehospital care they receive in the rescue unit. However, in order to realize the goal of integrating fire-based EMS systems within the total healthcare ecosystem of a patient, this introduction of technology for education and outreach may have positive outcomes.

4.3 Human-Computer Interface

This dimension resulted in one of the most frequently discussed and rich areas of research in our review.

In describing the impact of interfaces and information systems on providers, we often conceptualize their functionality as something of importance. That is, without a fully functioning system, their job tasks are impeded. One paramedic we interviewed described the initial process of obtaining better devices. They were provided with tablets that featured a removable keyboard, as outlined below:

> *"We received these fancy new tablets last year with a detachable keyboard. When we need to get an elderly patient to provide a signature, the tablet is much less burdensome and less heavy than a Toughbook. However, now the problem is that the keyboard attachment has become loose and the keyboard will unsnap in the middle of typing up a report. It is frustrating and often just makes our tasking more difficult." -E1*

On one hand, this participant could name the benefits of the newly implemented system: lighter weight, easier for patients to hold, and the convenience of using a tablet in the rescue unit. However, the use of improperly functioning devices challenges providers who are already dealing with time-sensitive and critical patients. Clinical documentation and reporting are key features of many EHRs, but without a properly functioning system, EMTs and paramedics experience more stress and frustration than necessary. Some work in this area has also demonstrated the efficacy of an approach focusing on *non-acceptance* to better inform the design of new systems in disaster and emergency response [65]. The participant quote below illustrates the perspective of the end-user who sees impacts to productivity and time as critical:

"The system was not user friendly. Several features never worked such as populating patient information from prior runs or bring over CAD (Computer-Aided Dispatch) system information like addresses or times. Little things like that make a big difference for us in time management." -PM6

In the literature, clinician perceptions of EHRs were measured, but it was often in the context of the ED, thus excluding prehospital care [66–68]. More importantly, not all departments have swiftly transitioned from paper to electronic patient care reporting systems (ePCRs). Subsequently, there are challenges in the transition of paper to electronic health records. Studies comparing paper to electronic records demonstrated that more data elements were captured in the electronic documents. However, paper records were more likely to contain information about the amount of intravenous fluids administered before arrival to the ED [69]. Although this study did not highlight the role of prehospital care EMS, it did point to the idea that nurses were able to leverage existing EMS reports to better support patient care, thus improving the dialogue between paramedics and the hospital ED.

4.4 People

Along this dimension, our participants indicated that often they were required to complete tasks that may be outside their written job description. Additionally, although it was not directly captured in a participant quote, some of our participants indicated that their shift partner could make or break their performance.

Unclear or Ad-hoc Roles. One participant captured this issue well in his description of his daily job tasks. In addition to his duties as a firefighter/paramedic, he acknowledged the impact of budget cuts on department funds and operations. Two firefighters were appointed to order supplies and manage inventory. Instead of providing dedicated support or consultation personnel, they were required to learn this on the job. This added frustration and stress to the simple process of ordering supplies:

"Ordering supplies is an extremely frustrating process. I have to navigate to an internal website, download a form, fill out the form, save the form as a PDF, attach it to an email, and send it, just to get supplies. So instead of hiring a designated IT (information technology) person to handle this, they "promote" two firemen and expect them to just figure it out." -P1

Teammate Familiarity. Studies have demonstrated the importance of team composition and familiarity in reducing workplace injuries, increasing performance, and creating safer work environments, specifically in the context of EMS [70]. In considering what factors influence patient care, internal factors such as the composition of shifts and teams are critical. Although this is often not brought up in discussions of internal factors, this idea of teammate familiarity may also play a role in preventing unsafe practices, may enhance adherence to protocols, and could potentially impact the ability for providers to respond quickly to escalating situations, which we highlight in the discussion portion of this paper.

4.5 Workflow and Communication

Several studies focused on better understanding portions of workflow and communications within the ED are impacted by the use of EHRs [71, 72]. This is also critical for understanding the work environment of paramedics. One participant was able to capture his frustration with his current system's workflow in the excerpt from his interview below:

> *"There is so much repetitive information. You would document the same complaint in several different areas of the report. They tried to make everything "black and white" is [sic] the aspect of click boxes with predefined answers. This is not a "black and white" field." -PM1*

> *"Many of the options in the click boxes made no sense and did not fit the dynamic of the call we were documenting. All we wanted was a box to type in the exact issue. Not to have to find something that was the closest match which felt like lying on a report." -PM6*

Work in this area extended from understanding workflow more broadly, to more specific instances of workflow interruptions. For example, Madathil et al. [73] leveraged unified modeling language (UML) to demonstrate bottlenecks associated with patient consent processes. Interestingly, studies also indicated correlations between dissatisfaction of EHR usability and disruptions to clinical workflow [74]. Although these studies would need to be replicated in a prehospital care environment, this work does demonstrate that EHR user experience affected clinical workflow and direct patient care time. Further investigation is necessary to determine if this would have an impact on prehospital care perception as well. However, based upon limited study data, we contend that clinical workflow is interrupted in the case of the EMTs and paramedics interviewed during this study. More objective analysis is required to determine the magnitude and direction of the effect.

Ergonomics and Physical Environment. Paramedics identified that assessment and understanding of their work environment was one of the more critical areas requiring further analysis of needs. As illustrated in the quote below, some providers feel that there is little concern for the environment they work in when it comes to choosing EHR and information systems.

> *"I think the biggest thing people get wrong is the complexity of the situation. Sometimes I have to intubate a patient and he is lodged in between the bed and the wall because he fell. Sometimes I have a patient who is trying to jump out the back of the ambulance. The work we do is not always in a sterile operating room. It's uglier. We are given imperfect conditions and it's our job to do our best despite the circumstances." -PM1*

> *"The Phillips Monitor is so heavy it impedes my ability to do my job when I'm trying to hold it and take care of a patient." -PM2*

Because current systems place so many physical demands on the user, getting through a call efficiently can present a problem. This also creates additional strain on responding units who are running calls frequently during their shifts and frequently has implications for EMS-based standards. Despite the existing gaps in technology, EMS providers are still required to meet the demands of the department credentials and accreditation standards.

4.6 Internal Organization Factors

Training on EHR Systems. Studies have demonstrated evidence that peer-based instruction seemed to increase proficiency and satisfaction related to use of EHR systems for physicians [75]. When asked about the training received, most of the participants in our study indicated they were taught on the job or through "train the trainer courses." This is captured in the participant quote below:

> *"Most of it really is a learned on the job deal. The 10 rides with a seasoned preceptor give plenty of opportunity to learn the system and ask questions. There is no renewal or certification procedure for the EHR system. Once you got it, you got it for life." -PM6*

This lack of formalized training can create user frustration since this creates additional demands on the paramedics or EMTs who are training new staff. More importantly, there are no continuing training courses, so if providers struggle to pick up the system, they are typically required to manage any additional training on their own time.

4.7 External Organization Factors

Seminal work in the area of prehospital EMS focused on identifying measurable indicators of quality [76]. Since the late 1990 s, the landscape of prehospital care has vastly changed. Now more than ever, this changing landscape impacts EMS providers both directly and indirectly. Financial reimbursement, medical policy, insurance policy, and government legislation all contribute to the effectiveness of fire-based EMS, in addition to quality assurance measures of performance. Although paramedics and EMTs may not be directly involved in these processes, they are impacted by these changes. For example, as mentioned earlier in this paper, EMS response times are important for both patient satisfaction and quality assurance [77]. Paramedics are increasingly facing pressure to respond as fast as possible, despite the additional concerns outlined throughout this portion of our review.

Additionally, work in this area has focused on reviewing adherence to national and international prehospital emergency medical protocols [78]. Adherence to protocols is also correlated with quality assurance and performance measurements, directly connecting to the discussion of the Measurement and Monitoring dimension below.

4.8 Measurement and Monitoring

Due to a need for assessment and quality assurance, measurement and monitoring represent a key component of understanding the impact of EHRs on efficiency and improvement of patient care. However, because EHRs are relatively new to prehospital care, there is little information that exists on how to assess, maintain, monitor, and measure the impact of EHR systems on prehospital care providers, as well as patients [1, 79]. In looking at the literature reviewed, very few studies captured the importance of developing more robust or generalized frameworks for understanding the impact of these EHRs on emergency medical care outside of the hospital emergency department. Several papers emphasized a human-factors approach, but even these were targeted towards the clinicians in the hospital and not necessarily created to measure the workload of paramedics or EMTs [53].

5 Discussion

Based upon the findings of our literature review and study, we have identified areas where more empirical research is necessary to better understand the needs of prehospital care providers. From our literature review and our interview data, we found that paramedics and EMTs often deal with issues related to interoperability of systems, problems with functionality, and interface design. Towards this end we have identified key areas where immediate intervention could better support fire-based EMS providers. We recognize that this is not an exhaustive list of potential future research directions, but we have identified areas where the HCI community has the opportunity to better support medics and EMTs.

5.1 Recommendations for Improving End-User Experience

Designing for Multiple Users. Interestingly, in medicine it is not uncommon for a patient to use the same interface as the provider to provide consent for medical care, educational or discharge instructions, etc. However, through our study, we found that this challenge of designing for multiple stakeholders emphasized the idea that the current systems used in prehospital emergency medicine may require patient input, such as a signature, on the same computers prehospital care providers use to draft reports and retrieve health information. Due to the complexity and the nature of emergency prehospital care, we found that providers, in our study, struggled with systems that did not necessarily fit their long-term needs. While the introduction of the tablets outlined above solved one problem, it created several more. It is our hope that with this data as a foundation, agencies and departments can look to this research as a way to understand and mitigate similar technology risks, while also selecting and implementing technologies that support EMS personnel.

3D User Interfaces. Due to increasing workload and call volumes, many fire-based EMS systems are facing issues related to meeting EMS-based standards while also providing the best patient care possible. Based upon some of the literature we identified, it is possible that there may be opportunities for EMS providers to adopt solutions that other healthcare providers current use. For example, dictation software packages and voice-activated inputs could help reduce paramedic workload when drafting patient care reports. Although the data is limited to emergency physicians, previous work demonstrated that voice-input charting could also potentially reduce the workload and number of interruptions, thus this may be a solution for EMS preceptors who may be observing and monitoring the clinical work of interns and students while also providing patient care [80]. Furthermore, work has also focused on the use of biometrics to create a safe and more accessible method of obtaining patient records on the scene of an emergency [81]. More objective data is necessary to determine whether these systems would provide viable solutions for prehospital care providers, but from our data and the extant literature, further analysis is required to determine which configurations of interfaces best support fire-based EMS providers.

Potentially Violent Situations. Additionally, there is also the growing concern of potentially violent situations (PVS). Paramedics and EMTs may encounter belligerent patients or they may be ambushed while on the job. Recent studies have demonstrated that this risk has increased in the last four years and the risk of violence extends to international emergency care providers as well [82, 83]. This growing concern has created a need for these providers to be able to document situations, to call additional units and law enforcement for support, and to complete these calls for support without the need for another user interface or additional workload. This is where the need for interoperability and reliable communication systems becomes of utmost importance [84]. By providing support through interface design, we can assist prehospital care providers in keeping both themselves and their patients safer.

5.2 Approaches to Future Work

Although we found results consistent with previous studies, we must use caution when generalizing this information to other departments or agencies. In addition, some departments are moving towards alternative EMS models where non-emergency calls are handled differently to reduce call load [85]. For example, one department in Washington D.C. is implementing a new triage program in which first responders will assess the severity of calls to determine whether a patient needs routine care from a clinic or requires care from the emergency room [38]. The goal is to reduce the number of routine calls that do not require emergency medical attention to give rescue units the opportunity to treat the most critical patients. Similar programs have demonstrated success in other areas of the United States [39]. Additionally, some other work has focused on leveraging telemedicine and related solutions to better understand how to support paramedics as call loads increase [45].

Acknowledgments. The authors would like to thank the anonymous paramedics and EMTs who participated in this study. Additionally, Katelynn would like to thank Joel Youker, Stephen Cabrera, and Andy Long for comments that helped to improve the manuscript. This paper is dedicated to the memory of Christopher Asseff, Connor Kuschel, and David Dangerfield. Thank you for the lifesaving care you provided to so many. These views are the authors alone and do not represent the official views of the University of Central Florida, the U.S. Government, or any fire department.

References

1. Lluch, M.: Healthcare professionals' organisational barriers to health information technologies—a literature review. Int. J. Med. Inform. **80**(12), 849–862 (2011)
2. Pozner, C.N., Zane, R., Nelson, S.J., Levine, M.: International EMS systems: The United States: past, present, and future. Resuscitation **60**, 239–244 (2004)
3. Roudsari, B.S., et al.: International comparison of prehospital trauma care systems. Injury **38** (9), 993–1000 (2007)

4. Breyer, T.: An analysis of rules, regulations, and policies to identify opportunities and limitations for fire-based EMS systems to integrate into healthcare using a community paramedic model. Int. Fire Serv. J. Leadersh. Manage. **9**, 41–48 (2015)
5. Peyravi, M.R., Modirian, M.J., Ettehadi, R., Pourmohammadi, K.: Improving medical emergency services system by evaluating patient satisfaction: means for health management. J. Heal. Manag. Informatics **1**(1), 15–18 (2014)
6. Curka, P.A., Pepe, P.E., Zachariah, S., Gray, G.D.: Incidence, source, and nature of complaints received in a large, urban emergency medical services system. Acad. Emerg. Med. **2**(6), 508–512 (1995)
7. Persse, D.E., Jarvis, J.L., Corpening, J., Harris, B.: Customer satisfaction in a large urban fire department emergency medical services system. Acad. Emerg. Med. **6563**(03), 106–110 (2003)
8. Carr, C., et al.: When should ED physicians use an HIE? Predicting presence of patient data in an HIE. South. Med. J. **109**(7), 427–433 (2017)
9. Ratwani, R.M., Fairbanks, R.J., Zachary Hettinger, A., Benda, N.C.: Electronic health record usability: analysis of the user-centered design processes of eleven electronic health record vendors. J. Am. Med. Inform. Assoc. **22**(6), 1179–1182 (2015)
10. Bigham, B.L., et al.: Patient safety in emergency medical services: a systematic review of the literature. Prehosp. Emerg. Care **16**(1), 20–35 (2012)
11. Redhead, C.S.: The Health Information Technology for Economic and Clinical Health (HITECH) Act (2009)
12. Baird, S., Boak, G.: Leading change: introducing an electronic medical record system to a paramedic service. Leadersh. Health Serv. (Bradf. Engl.) **29**, 136–150 (2015)
13. Eden, R., Burton-Jones, A., Scott, I., Staib, A., Sullivan, C.: Effects of eHealth on hospital practice: synthesis of the current literature. Aust. Heal. Rev. **42**(5), 568–578 (2018)
14. Akhlaq, A., Sheikh, A., Pagliari, C.: Defining health information exchange: scoping review of published definitions. J. Innov. Heal. Inform. **23**(4), 684–764 (2016)
15. Cuk, S., Wimmer, H., Powell, L.M.: Problems associated with patient care reports and transferring data between ambulance and hospitals from the perspective of emergency medical technicians. Issues Inf. Syst. **18**(4), 16–26 (2017)
16. Cuk, S., Wimmer, H., Powell, L.M., Rebman, C.M.: Electronic emergency medical technician reports-testing a perception of a prototype. Issues Inf. Syst. **19**(3), 81–91 (2018)
17. Potter, L.E., Purdie, C., Nielsen, S.: The view from the trenches: satisfaction with eHealth systems by a group of health professionals. In: Proceedings of the 23rd Australasian Conference on Information Systems, ACIS 2012, pp. 1–9 (2012)
18. Newgard, C.D., Zive, D., Jui, J., Weathers, C.: Electronic versus manual data processing: evaluating the use of electronic health records in out-of-hospital clinical research. Acad. Emerg. Med. **19**, 217–227 (2012)
19. Landman, A.B., Lee, C.H., Sasson, C., Van Gelder, C.M., Curry, L.A.: Prehospital electronic patient care report systems: early experiences from emergency medical services agency leaders. PLoS ONE **7**(3), 1–8 (2012)
20. Eysenbach, G.: What is e-health? J. Med. Internet Res. **3**, 2–3 (2001)
21. Ahern, D.K., Patrick, K., Phalen, J.M., Neiley, J.D.: An introduction to methodological challenges in the evaluation of eHealth research: perspectives from the Health e-Technologies Initiative. Eval. Progr. Plan. **29**, 386–389 (2006)
22. Black, A.D., et al.: The impact of eHealth on the quality and safety of health care: a systematic overview. PLoS Med. **8**(1), e1000387 (2011)
23. Sidorov, J.: It ain't necessarily so: the electronic heath record and the unlikely prospect of reducing health care costs. Health Aff. **25**(4), 1079–1086 (2006)

24. Geisler, B.P., Schuur, J.D., Pallin, D.J.: Estimates of electronic medical records in U.S. emergency departments. PLoS ONE **5**(2), 3–6 (2010)
25. Abdulwahid, M.A., Booth, A., Turner, J., Mason, S.: Understanding better how emergency doctors work. Analysis of distribution of time and activities of emergency doctors: a systematic review and critical appraisal of time and motion studies. Emerg. Med. J. **35**(11), 692–700 (2018)
26. Furukawa, M.F.: Electronic medical records and the efficiency of hospital emergency departments. Med. Care Res. Rev. **68**, 75–95 (2010)
27. Desroches, B.C.M., et al.: Electronic health records' limited successes suggest more targeted uses. Health Aff. (Millwood) **29**, 639–646 (2009)
28. Hund, A.M., Nazarczuk, S.N.: The effects of sense of direction and training experience on wayfinding efficiency. J. Environ. Psychol. **29**, 151–159 (2009)
29. Sarangarm, D., Lamb, G., Weiss, S., Ernst, A., Hewitt, L.: Implementation of electronic charting is not associated with significant change in physician productivity in an academic emergency department. JAMIA Open **1**(2), 227–232 (2018)
30. Benda, N.C., Meadors, M.L., Hettinger, A.Z., Ratwani, R.M.: Emergency physician task switching increases with the introduction of a commercial electronic health record. Ann. Emerg. Med. **67**(6), 741–746 (2016)
31. Yamamoto, L.G., Khan, A.: Challenges of electronic medical record implementation in the emergency department. Pediatr. Emerg. Care **22**(3), 184–191 (2006)
32. Hill, R.G., Sears, L.M., Melanson, S.W.: 4000 Clicks: a productivity analysis of electronic medical records in a community hospital ED. Am. J. Emerg. Med. **31**(11), 1591–1594 (2013)
33. Ahmad, M., Kartiwi, M.: A model for measuring well-being of medical practitioners in EHR implementation. In: Proceedings of the 6th International Conference on Information and Communication Technology for the Muslim World, ICT4M 2016, pp. 148–153 (2017)
34. Choong, Y.-Y., Dawkins, S., Furman, S., Greene, K.K., Prettyman, S.S., Theofanos, M.F.: Voices of first responders – identifying public safety communication problems: findings from user-centered interviews, phase 1, volume 1 (2018)
35. Dawkins, S., et al.: Public safety communication user needs: voices of first responders. Proc. Hum. Factors Ergon. Soc. **1**, 92–96 (2018)
36. Sittig, D., Singh, H.: A new socio-technical model for studying health information technology in complex adaptive healthcare systems. Qual. Saf. Heal. Care **19**(Suppl. 3), 1–14 (2011)
37. Militello, L.G., Hutton, R.J.B.: Applied cognitive task analysis (ACTA): a practitioner's toolkit for understanding cognitive task demands. Ergonomics **41**, 1618–1641 (1998)
38. Sittig, D., Classen, D.: Safe Electronic Health Record use requires a comprehensive monitoring and evaluation framework. JAMA, J. Am. Med. Assoc. **303**(5), 450–451 (2010)
39. Tsiknakis, M., Kouroubali, A.: Organizational factors affecting successful adoption of innovative eHealth services: a case study employing the FITT framework. Int. J. Med. Inform. **8**, 39–52 (2008)
40. Tsiknakis, M., Spanakis, M.: Adoption of innovative eHealth services in prehospital emergency management: a case study. In: Proceedings of the 10th IEEE International Conference on Information Technology and Applications in Biomedicine (2010)
41. Lavariega, Juan C., Avila, A., Gómez-Martínez, Lorena G.: Software architecture for emergency remote pre-hospital assistance systems. In: Adibi, S. (ed.) Mobile Health. SSB, vol. 5, pp. 453–471. Springer, Cham (2015). https://doi.org/10.1007/978-3-319-12817-7_20
42. Acharya, S., Imani, O.V.: A novel resource management approach for paramedic triage systems, pp. 1–4 (2017)

43. Lenert, L.A., et al.: Design and evaluation of a wireless electronic health records system for field care in mass casualty settings. J. Am. Med. Inform. Assoc. **18**(6), 842–852 (2011)
44. Anantharaman, V., Lim, S.H.: HEAL (hospital & emergency ambulance link): using IT to enhance emergency pre-hospital care. Stud. Health Technol. Inform. **84**, 875 (2001)
45. Bergrath, S., et al.: Implementation phase of a multicentre prehospital telemedicine system to support paramedics: feasibility and possible limitations. Scand. J. Trauma. Resusc. Emerg. Med. **21**(1), 54 (2013)
46. Van Wynsberghe, A.: A method for integrating ethics into the design of robots. Ind. Robot Int. J. **40**(5), 433–440 (2013)
47. Koufi, V., Malamateniou, F., Vassilacopoulos, G.; Ubiquitous access to cloud emergency medical services. In: Proceedings of the IEEE/EMBS International Conference on Information Technology Application in Biomedicine (ITAB), pp. 19–22 (2010)
48. Elsaadany, A., Sedky, A., Elkholy, N.: A triggering mechanism for end-to-end IoT eHealth system with connected ambulance vehicles. In: 2017 8th International Conference on Information, Intelligence, Systems & Applications, IISA 2017, vol. 2018, pp. 1–6 (2018)
49. Metelmann, B., Metelmann, C.: M-Health in Prehospital Emergency Medicine: Experiences from the EU funded Project LiveCity. IGI Global, Pennsylvania (2016)
50. Turcu, C.E., Turcu, C., Popa, V.: An RFID-based system for emergency health care services. In: International Conference on Advanced Information Networking and Applications Workshops, pp. 624–629 (2009)
51. Majeed, R.W., Stöhr, M.R., Röhrig, R.: Architecture of a prehospital emergency patient care report system (PEPRS). Stud. Health Technol. Inform. **57**(7), 2013 (2013)
52. Pines, J.M., Abualenain, J., Scott, J., Shesser, R. (eds.) Emergency Care and the Public's Health (2014)
53. Benda, N.C., et al.: Human factors design in the clinical environment: development and assessment of an interface workload human factors design in the clinical environment: development and assessment of an interface for visualizing emergency medicine clinician workload. IISE Trans. Occup. Ergon. Hum. Factor. **6**(3–4), 225–237 (2018)
54. Shelton, D., Sinclair, P.: Availability of ambulance patient care reports in the emergency department. BMJ Qual. Improv. Rep. **5**(1), u209478.w3889 (2016)
55. Connelly, D.P., et al.: The impact of electronic health records on care of heart failure patients in the emergency room. J. Am. Med. Inform. Assoc. **19**(3), 334–340 (2012)
56. Lammers, E.J., et al.: Departments does health information exchange reduce redundant imaging? Evidence from emergency departments. Med. Care **52**(3), 227–234 (2020)
57. Balka, E., Tolar, M., Coates, S., Whitehouse, S.: Socio-technical issues and challenges in implementing safe patient handovers: insights from ethnographic case studies. Int. J. Med. Inform. **82**(12), e345–e357 (2013)
58. Cheung, D.S., et al.: Improving handoffs in the emergency department. Ann. Emerg. Med. **55**(2), 171–180 (2010)
59. Hilligoss, B., Zheng, K.: Chart biopsy: an emerging medical practice enabled by electronic health records and its impacts on emergency department-inpatient admission handoffs. J. Am. Med. Inform. Assoc. **20**(2), 260–267 (2013)
60. Collins, S.A., Stein, D.M., Vawdrey, D.K., Stetson, P.D., Bakken, S.: Content overlap in nurse and physician handoff artifacts and the potential role of electronic health records: a systematic review. J. Biomed. Inform. **44**(4), 704–712 (2011)
61. Norman, C., Mello, M., Choi, B.: Identifying frequent users of an urban emergency medical service using descriptive statistics and regression analyses. West. J. Emerg. Med. **17**(1), 39–45 (2016)

62. Hansagi, H., Olsson, M., Hussain, A., Öhlén, G.: Is information sharing between the emergency department and primary care useful to the care of frequent emergency department users? Eur. J. Emerg. Med. **15**(1), 34–39 (2008)
63. Kargl, F., Lawrence, E., Fischer, M.: Security, privacy, and legal issues in pervasive eHealth monitoring systems. In: International Conference on Mobile Business (2008)
64. Arora, S., Peters, A.L., Burner, E., Lam, C.N., Menchine, M.: Trial to examine text message – based mHealth in emergency department patients with diabetes (TExT-MED): a randomized controlled trial. Ann. Emerg. Med. **63**, 15–20 (2013)
65. Elmasllari, E., Reiners, R.: Learning from non-acceptance: design dimensions for user acceptance of E-triage systems. In: Proceedings of the International ISCRAM Conference, vol. 2017, pp. 798–813 (2017)
66. Chisolm, D.J., Purnell, T.S., Cohen, D.M., McAlearney, A.S.: Clinician perceptions of an electronic medical record during the first year of implementation in emergency services. Pediatr. Emerg. Care **26**(2), 107–110 (2010)
67. Hockstein, M.A., Pope, S.N., Donnawell, K., Chavez, S.A., Bhat, L.: Emergency medicine residents on electronic medical records: perspectives and advice. Cureus **11**(2), e4027 (2019)
68. Neri, P.M., et al.: Emergency medicine resident physicians' perceptions of electronic documentation and workflow: a mixed methods study. Appl. Clin. Inform. **6**(1), 27–41 (2015)
69. Coffey, C., et al.: A comparison of paper documentation to electronic documentation for trauma resuscitations at a Level I pediatric trauma center. J. Emerg. Nurs. **41**(1), 52–56 (2015)
70. Hughes, A.M., et al.: Teammate familiarity, teamwork, and risk of workplace injury in emergency medical services teams. J. Emerg. Nurs. **43**(4), 339–346 (2017)
71. Rosenfield, D., Harvey, G., Jessa, K.: Implementing electronic medical records in Canadian emergency departments. Can. J. Emerg. Med. **21**(1), 15–17 (2019)
72. Horsky, J., Gutnik, L., Patel, V.L.: Technology for emergency care: cognitive and workflow considerations. In: AMIA Annual Symposium Proceedings, pp. 344–348 (2006)
73. Madathil, K.C., Koikkara, R., Gramopadhye, A.K., Fryar, K.: An analysis of the general consenting process in an emergency department at a major hospital : challenges for migrating to an electronic health record (2011)
74. Denton, C.A., et al.: Emergency physicians' perceived influence of EHR use on clinical workflow and performance metrics. Appl. Clin. Inform. **9**(3), 725–733 (2018)
75. Dastagir, M.T., et al.: Advanced proficiency EHR training: effect on physicians' EHR efficiency, EHR satisfaction and job satisfaction. AMIA Annu. Symp. Proc. **2012**, 136–143 (2012)
76. Moore, L.: Measuring quality and effectiveness of prehospital EMS. Prehospital Emerg. Care **3**(4), 325–331 (1999)
77. Bernard, A.W., et al.: Postal survey methodology to assess patient satisfaction in a study. BMC Emerg. Med. **7**, 1–5 (2007)
78. Ebben, R.H.A., Vloet, L.C.M., Verhofstad, M.H.J., Meijer, S., de Groot, J.A.M., van Achterberg, T.: Adherence to guidelines and protocols in the prehospital and emergency care setting: a systematic review. Scand. J. Trauma. Resusc. Emerg. Med. **21**(1), 9 (2013)
79. Ben-Assuli, O.: Electronic health records, adoption, quality of care, legal and privacy issues and their implementation in emergency departments. Health Policy **119**(3), 287–297 (2015)
80. Dela Cruz, J.E., et al.: Typed versus voice recognition for data entry in electronic health records: emergency physician time use and interruptions. West. J. Emerg. Med. **15**(4), 541–547 (2014)

81. Díaz-Palacios, J.R., Romo-Aledo, V.J., Chinaei, A.H.: Biometric access control for e-health records in pre-hospital care. In: ACM International Conference on Proceeding Series, pp. 169–173 (2013)
82. Maguire, B.J., Browne, M., Neill, B.J.O., Dealy, M.T., Clare, D., Meara, P.O.: International survey of violence against EMS personnel: physical violence report. Prehosp. Disaster Med. **33**(5), 526–531 (2020)
83. Taylor, J.A., Barnes, Ã.B., Davis, A.L., Wright, J., Widman, S., Levasseur, M.: Expecting the unexpected: a mixed methods study of violence to EMS responders in an urban fire department. Am. J. Ind. Med. **163**, 150–163 (2016)
84. House, A., Power, N., Alison, L.: A systematic review of the potential hurdles of interoperability to the emergency services in major incidents: recommendations for solutions and alternatives. Cogn. Technol. Work **16**(3), 319–335 (2013)
85. Capp, R., et al.: Coordination program reduced acute care use and increased primary care visits among frequent emergency care users. Health Aff. **36**(10), 1705–1711 (2017)

Compogram: Development and Evaluation of ITS for Organizing Programming-Knowledge by Visualizing Behavior

Kento Koike[1(✉)], Tomohiro Mogi[1], Takahito Tomoto[2], Tomoya Horiguchi[3], and Tsukasa Hirashima[4]

[1] Graduate School of Engineering, Tokyo Polytechnic University, Kanagawa, Japan
k.koike@t-kougei.ac.jp
[2] Faculty of Engineering, Tokyo Polytechnic University, Kanagawa, Japan
[3] Graduate School of Maritime Sciences, Kobe University, Kobe, Japan
[4] Graduate School of Engineering, Hiroshima University, Hiroshima, Japan
https://www.koike.app/

Abstract. Currently, computing education, especially programming education has become more important. Meanwhile, programming education has many difficulties such as should learn many concepts and skills. Researches of Intelligent Tutoring System (ITSs) have been attempted to reduce these difficulties. ITSs are educational systems that able to adaptively pedagogical behavior and feedback and aim to supply adaptive tutoring to learner's profiles by alternate to human tutors. Despite there are much supports to programming education by ITSs, there no attempts for organizing knowledge in programming. Organizing knowledge is acquiring the systematized knowledge and its scalability which enabling to existing knowledge reuse to the same or similar problems that solved once by scaling knowledge. We considered that organizing knowledge is fosters problem-solving skills, and it gains Computational Thinking eventually. Therefore, we have been focused on supporting the process of solving problems by combining a bit of program. Then we selected the knowledge of "function" and "source program that achieves the function" as knowledge to be organized. And, we defined a pair of knowledge as a component. In this paper, we proposed and developed Compogram: an ITS for organizing knowledge by visualizing behavior in programming. Furthermore, for identifying learning gains, we conducted an evaluation compared to our conventional systems. Results were suggested that Compogram was fostering knowledge organizing skills that can apply to out of learning ranges.

Keywords: Programming education · Knowledge organization · Intelligent Tutoring System · Problem-solving process

C. Stephanidis et al. (Eds.): HCII 2020, LNCS 12427, pp. 151–162, 2020.
https://doi.org/10.1007/978-3-030-60152-2_12

1 Introduction

Computing education is widely known that necessity and importance. It is believed to leads gain Computational Thinking [20,23]. In recent years, also, many countries include subjects in elementary schools which related to computing education. In particular, it seems that programming education is the core of those.

Meanwhile, it becomes revealed that there are many difficulties in these programming education. In programming education, it is basic difficulties that understanding concepts such as variables, loops, recursions, arrays, and so on [17]. Moreover, there are advanced difficulties that achieving complex activities such as tracing, reading, writing, abstraction, and decomposition [7,14,15].

In research fields of computing education, many studies engaged in reducing these basic and advanced difficulties by tools or educational systems [16,19]. Notably, educational systems that able to adaptively pedagogical behavior and feedback by artificial intelligence technology is called Intelligent Tutoring Systems (ITSs) [18,22]. ITSs aims to supply adaptive tutoring to learner's profile by alternate to human tutors. And, ITSs which focusing on the domain of programming is also known as Intelligent Programming Tutors (IPTs) [3]. IPTs helps novice's overcome not only basic difficulties but also advanced difficulties.

Despite there are much supports to programming education by IPTs, there no attempts for organizing knowledge in programming. From this perspective, we aim to reveal an effective way to supports knowledge organization.

The purpose of knowledge organization is to produce well-organized knowledge. Well-organized knowledge is widely known as chunks in fields of cognitive psychology [2]. Also, it was indicated that chunks help to solve problems in programming [1]. Because, well-organized knowledge, chunks, is able to reuse existing knowledge to the same or similar problems that solved once by scaling knowledge. That is, organizing knowledge leads to acquiring systematized knowledge and its scalability. Therefore, we considered that organizing knowledge is fosters problem-solving skills, and thus it gains Computational Thinking eventually.

In this paper, we proposed and developed Compogram: an ITS for organizing knowledge by visualizing behavior in programming. Furthermore, for identifying learning gains, we conducted an evaluation compared to our conventional systems. Results were suggested that Compogram was fostering knowledge organizing skills that can apply to out of learning ranges.

2 Objective and Research Questions

To solve problems, it is important that learners understand component elements of problems, and acquire it as knowledge. However, there is a problem that difficulties to adequately reuse the same or similar solutions which have already been used by learners. For solving that, one of the ways is to organize the existing knowledge. That is, by organizing the existing knowledge, it is expected that existing knowledge can reuse the same or similar solutions by scaling knowledge as adapting to the grain size of problems.

The knowledge in programming involves many concepts such as grammar of specific languages and the object-oriented approach. Thus, it is necessary that select knowledge types for organizing accordingly the objective. In this research, we focused on supporting the process of solving problems by combining a bit of program. Then we selected the knowledge of the "function" and the "source program that achieves function" as knowledge to be organized. And, we defined a pair of knowledge as a component.

Therefore, we addressed the following research questions:

RQ1 *How to design/realize learning tasks/activities for bonding the "function" and the "source program that achieves function", acquiring as a component, and organizing as the hierarchy of components?*
RQ2 *Does provide such designed learning tasks/activities lead to learning gains?*
RQ3 *How to design/realize effective feedbacks/interventions in such learning tasks/activities?*

For this objective, we have been engaged in support of systematizing and organizing learners' existing knowledge as a component and have aimed at improving reusability [10–12].

First, we proposed a knowledge organization method called BROCs (Building method that Realizes Organizing Components) [10,11]. BROCs support to understand the relationship between components by presents problems stepwise and bottom-up. Next, we developed and evaluated a system that realized BROCs . We addressed RQ1 by designing BROCs as a learning method and developing a system of BROCs. Meanwhile, we addressed RQ2 by conducting comparison evaluations between the BROCs system with and without feedbacks, and textbook.

In [12], we addressed RQ3 by proposing a model of the problem-solving process in BROCs. Through discussion based on the model, it was pointed out that there is the "behavior" in the bonding process between the "function" and "source program that achieves function". Also, we pointed out that insufficient supports for understanding the "behavior". In the conventional system as stated above, however, the feedback only presented the excess or deficiency compared to the correct answer when learners mistake. Therefore, we conducted an evaluation to investigate the effectiveness of feedback included it related to the behavior, that produced by the constructed problem-solving process model. Results indicated the explanation based on the model seems to deepen the understanding was obtained.

In this paper, we developed a system that realizes support for organizing knowledge by visualizing program behavior, it called Compogram. Accordingly, we evaluated the system and addressed the following research question:

RQ4 *Does provide such designed feedbacks/interventions lead to more learning gains?*

3 System Design

3.1 Theory

BROCs: Acquiring Components Method for Knowledge Organization. In our research, as described in Sect. 2, we focused on acquiring the knowledge of "function" and "source program that achieves function", and defined a pair of knowledge as a component.

BROCs (Building method that Realizes Organizing Components) is a method of knowledge organization in programming that our proposed. In BROCs (Fig. 1), presenting tasks to learners according to follows two steps: (1) make small functions (such as swap) by combining primitive components (e.g.. assignment, if statement), (2) make larger functions (such as sort) by combining components that include already made it.

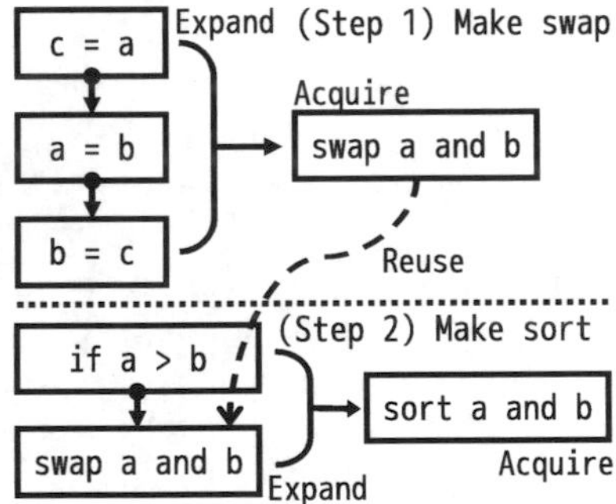

Fig. 1. An example of BROCs.

By repeat these steps, learners can structuralize and accumulate components stepwise. In fact, some studies claimed a learning approach is effective that makes small functions firstly, and then, combines these small functions to make larger functions [8,21]. Therefore, BROCs support learners' knowledge organization through repeat stepwise presenting these tasks. Due to this, BROCs situate as an important pedagogy in the proposed ITS.

Function, Behavior, and Structure in Problem-Solving Process. We have been explained the problem-solving process in activities of acquiring components accordingly Function-Behavior-Structure (FBS) Framework [12]. FBS Framework is a way of enabling clearly divide features of artifacts into the function, the behavior, and the structure. It helps the comprehension and utilization of artifacts by not only humans also computers. FBS Framework has been defined and utilized in fields of knowledge representation (e.g.. qualitative physics, applied ontology) [4–6,9].

We considered that understanding components from such a perspective provide more effective feedback to previously unclear features. As such, we mapped and considered FBS Framework into activities of acquiring components as follows (Fig. 2):

- Function: a function to be achieved, as a result of interpreting particular behavior,
- Behavior: a sequence of data states between initial-state (given) to end-state (result),
- Structure: a set of source program as operators that achieves a function by producing behavior.

In a similar way, Kruchten's study has the same perspectives as ours in the domain of software design [13].

In light of such problem-solving, we considered that learners need to think of relationships not only between function and structure but also among these and behavior, for acquiring components.

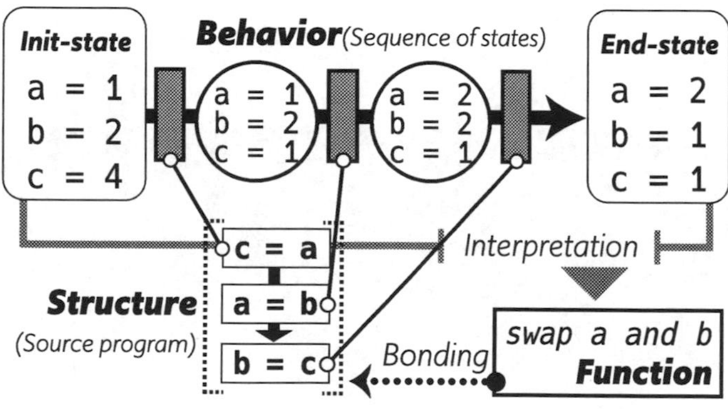

Fig. 2. The problem-solving process in programming.

3.2 Implementation

Based System Using BROCs. Heretofore, we developed an ITS using BROCs as a pedagogy [10,11]. This system is available in Windows as a desktop application and developed using `C#`. Figure 3 shows an user-interface (UI) of this system with BROCs. On the middle-top of UI shows the target component that learners should be acquired and a problem and constraints for acquiring it. To construct the target component, on the left of UI shows the component list for selecting components to add to the workspace. The workspace shows on the lower-right of UI. In the workspace, learners can be edit to the textbox, order, and level in the hierarchy, of components.

In addition, this system have following feedback functions for scaffolding learners' activities in BROCs:

- Simple Hint: give a hint that pointed out an error of textbox/order/level of the uppermost in the procedure,
- Detail Hint: give a correct answer to such error,
- Check Answer: tell learners whether their answers are correct.

However, these feedback functions only presented the excess or deficiency compared to the correct answer when learners mistake. In other words, learners were only given feedbacks related to the structure for achieving the function. Accordingly, there was no support for learners to understand the behavior that is considered to be important factors for acquiring components.

Table 1. Functional comparison of the conventional system and Compogram.

System	Simple hint	Detail hint	Check answer	Visualize behavior
Conventional system	Yes	Yes	Yes	-
Compogram	-	Yes	Yes	Yes

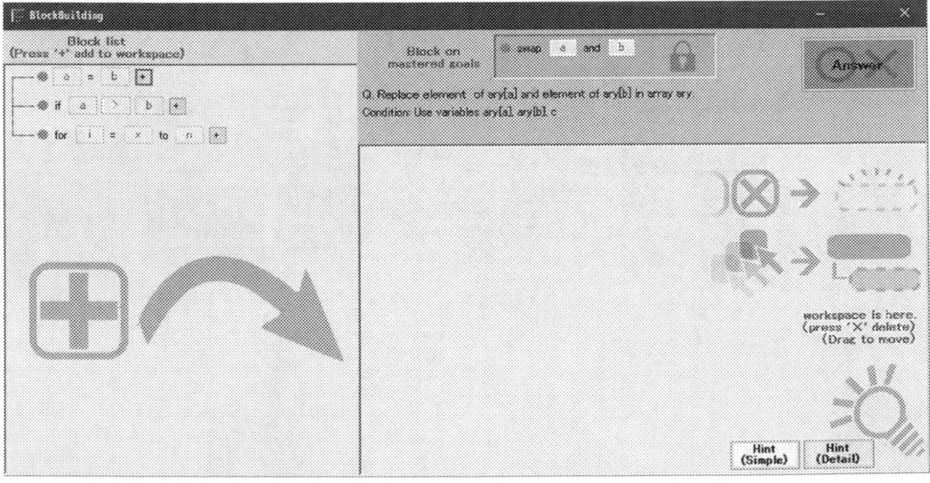

Fig. 3. An user-interface of our previous system.

Proposed System: Compogram. Here, we developed an ITS based on the conventional system, called Compogram. Compogram have a function of visualizing behavior by extended the conventional system. Due to this, as we stated the importance in Sect. 3.1, Compogram can promote that awareness of relationships not only between function and structure but also among these and behavior, in learners' activities of acquiring components. Like the conventional system, Compogram is available in Windows as a desktop application and developed using C#.

It is a mostly comprehensive extension compared to the conventional system's functionality (see Table 1). However, in order for learners to notice the errors by themselves, the simple hint function has been removed. On the other hand, the detail hint function was left in place to help learners clear up any impasse by themselves.

In Compogram, as stated above, we extended a function of visualizing behavior. Figure 4 shows an UI of Compogram. The added function in the right of UI presents the behavior and differences between learners' constructs and correct answers. And, the initial-state as input is changing per click the check button. Accordingly, if including control-flow effected by initial-state in constructs, also changing behavior.

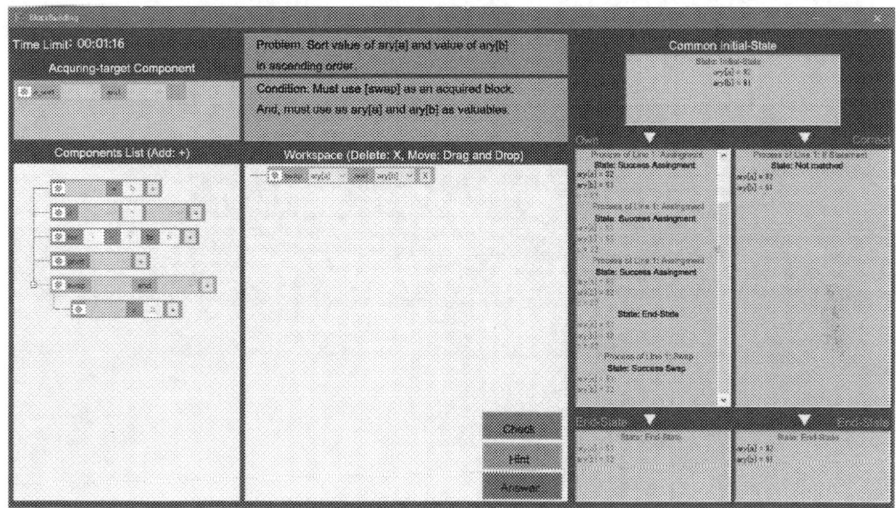

Fig. 4. An user-interface of Compogram.

4 Comparative Evaluation

4.1 Method

For addressing RQ4 (*Does provide our designed feedbacks/interventions lead to more learning gains?*), we conducted a comparative evaluation between the conventional system with/without feedback and Compogram.

Participants were 24 undergraduate and graduate students who have studied programming for at least three years in programming lectures. All participants

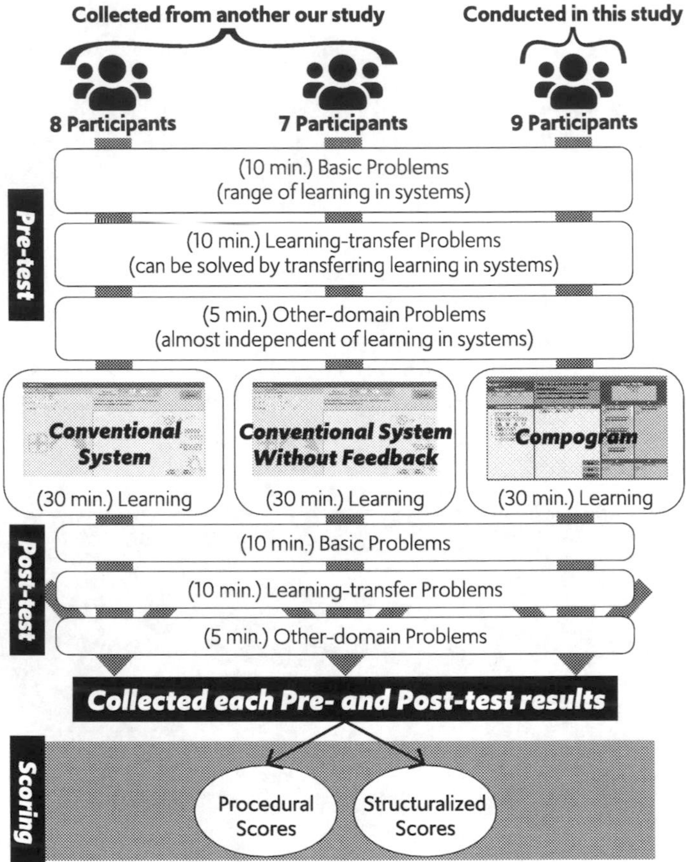

Fig. 5. Design of comparative evaluation.

had already learned basic concepts such as "for" and "if" statements, sorting algorithms, and functions. And, we divided participants into three groups: (CS) a group using the conventional system (8 participants), (CS-NF) a group using the conventional system without feedback (7 participants), and (CMPG) a group using Compogram (9 participants). As annotation, CS and CS-NF groups were already evaluated in [11]. In other words, CS and CS-NF evaluation data were collected from this another study.

Our design of comparative evaluation shows in Fig. 5. As a procedure, firstly, we conducted a pre-test in 25 min that contained three problem sets that can be developed stepwise to measure fundamental programming and design skills. These problem sets contain the following problems in stated time limits:

- Basic problems (10min.): it included 11 problems in the range of learning in systems,

- Learning-transfer problems (10 min.): it included 9 problems out of the scope of learning tasks, but that can be solved by transferring learning in systems,
- Other-domain problems (5 min.): it included simple 4 problems in the other domain, that are almost independent of learning in systems, for measuring the gain of skills.

Also, in the pre-tests (and post-tests), we instructed to do structuring such as functionalization or componentization in each problem to the extent possible, and reuse earlier answers in other problems. These evaluations made two types of evaluations: (1) a procedural score that evaluated whether the procedure is correct, and (2) a structuralized score that evaluated whether learners trying to functionalize and reuse in other problems, and it excluded the correctness of procedures. After that, we conducted learning tasks in divided groups with each system, for 30 min. Finally, we conducted a 25 min post-tests with the same contents and evaluation as the pre-tests.

4.2 Result and Analysis

Figure 6 and Fig. 7 show the results of means of procedural/structuralized scores in pre- and post-tests. In these figures, we normalized each score to percentages divided means by maximums which different per each problem set and evaluation method. To analyze learning gains, we conducted a 3 (CS, CS-NF, or CMPG) x 2 (pre- or post-test scores) ANOVA for each problem set per scoring.

In procedural scores of basic problems from Fig. 6, while results showed no significance among groups $[F(2,21) = 1.87, p > .10, \eta_p^2 = .15]$; it showed significant learning effects of each system within groups $[F(1,21) = 28.45, p < .01, \eta_p^2 = .58]$. Results of learning-transfer problems also showed no significance among groups $[F(2,21) = 0.16, p > .10, \eta_p^2 = .02]$ and significant learning effects of each system within groups $[F(1,21) = 13.78, p < .01, \eta_p^2 = .40]$.

In structuralized scores from Fig. 6, the results are similar to the results of procedural scores. In detail, the results of basic problems showed no significance among groups $[F(2,21) = 1.51, p > .10, \eta_p^2 = .13]$ and significant learning effects of each system within groups $[F(1,21) = 31.40, p < .01, \eta_p^2 = .60]$. Results of learning-transfer problems also showed no significance among groups $[F(2,21) = 0.03, p > .10, \eta_p^2 = .00]$ and significant learning effects of each system within groups $[F(1,21) = 13.86, p < .01, \eta_p^2 = .40]$.

In procedural scores of other-domain problems from Fig. 7, similar to the above, the results showed no significance among groups $[F(2,21) = 1.05, p > .10, \eta_p^2 = .09]$ and significant learning effects of each system within groups $[F(1,21) = 13.09, p < .01, \eta_p^2 = .38]$. On the other hand, the results in structuralized scores were different from the above. while the results showed no significance among groups $[F(2,21) = 1.28, p > .10, \eta_p^2 = .11]$; it showed significant learning effects of each system within groups $[F(1,21) = 3.51, p < .10, \eta_p^2 = .14]$. Furthermore, it showed significantly AxB interaction $[F(2,21) = 3.51, p < .05, \eta_p^2 = .25]$. In an analysis of AxB interaction, the results within groups were showed significant learning effects in only CMPG $[F(1,21) = 10.52, p < .01]$.

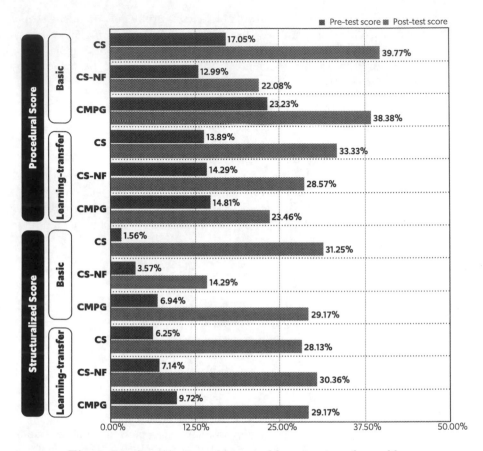

Fig. 6. Results of basic problems and learning-transfer problems.

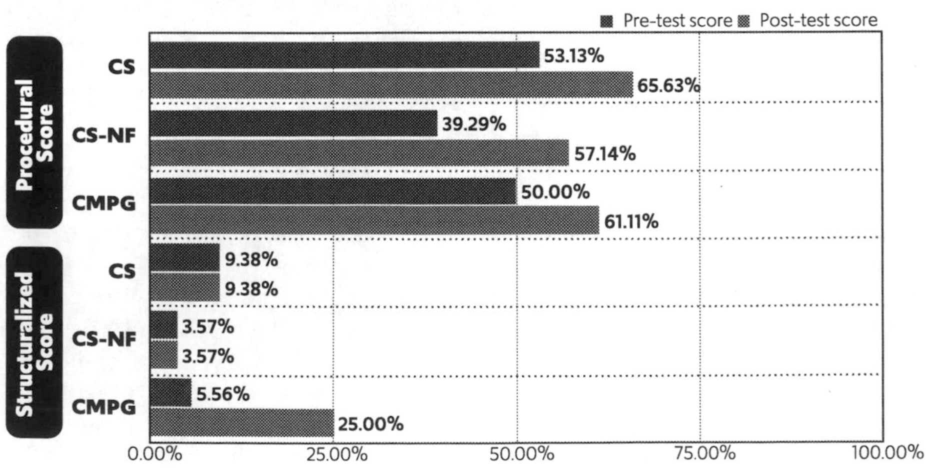

Fig. 7. Results of other-domain problems.

In consideration of the above results and analysis, we considered that the learning effects of Compogram are almost the same as the conventional system with feedback. On the other hand, as a distinguished point, Compogram demonstrated its ability for other-domain problems that almost not related to the learning range of systems. That is, results were suggested that Compogram was fostering knowledge organizing skills that can apply to out of learning ranges. As stated above, we addressed RQ4 (*Does provide our designed feedbacks/interventions lead to more learning gains?*).

5 Discussion and Future Work

In this paper, we pointed out that insufficient support for organizing knowledge in programming education. Therefore, based on our previous studies, we considered a new way of support for organizing knowledge in ITS. In this way, we developed and evaluated Compogram. Furthermore, evaluation results suggested that Compogram was fostering knowledge organizing skills that can apply to out of learning ranges.

Our future works are to conduct more large-scale experiments in classrooms, verify the learning effects, and realize more adaptive feedback in the system.

Acknowledgement. This work was supported by JSPS KAKENHI Grant Numbers JP18K11586, JP19H04227, and JP17H01839.

References

1. Adelson, B.: Problem solving and the development of abstract categories in programming languages. Mem. Cogn. **9**(4), 422–433 (1981). https://doi.org/10.3758/BF03197568
2. Chase, W.G., Simon, H.A.: Perception in chess. Cogn. Psychol. **4**(1), 55–81 (1973)
3. Crow, T., Luxton-Reilly, A., Wuensche, B.: Intelligent tutoring systems for programming education: a systematic review. In: ACM International Conference Proceeding Series, pp. 53–62. ACM (2018). https://doi.org/10.1145/3160489.3160492
4. De Kleer, J., Brown, J.S.: A qualitative physics based on confluences. Artif. Intell. **24**(1–3), 7–83 (1984). https://doi.org/10.1016/0004-3702(84)90037-7
5. Gero, J.S., Kannengiesser, U.: A function-behavior-structure ontology of processes. Artif. Intell. Eng. Des. Anal. Manuf. **21**(4), 379–391 (2007). https://doi.org/10.1017/S0890060407000340
6. Goel, A.K., Rugaber, S., Vattam, S.: Structure, behavior, and function of complex systems: the structure, behavior, and function modeling language. Artif. Intell. Eng. Des. Anal. Manuf. **23**(1), 23–35 (2009). https://doi.org/10.1017/S0890060409000080
7. Gomes, A., Mendes, A.J.N.: Learning to program-difficulties and solutions. In: International Conference on Engineering Education, pp. 1–5 (2007), http://ineer.org/Events/ICEE2007/papers/411.pdf
8. Hu, M., Winikoff, M., Cranefield, S.: A process for novice programming using goals and plans. In: Proceedings of the Fifteenth Australasian Computing Education Conference, pp. 3–12 (2013)

9. Kitamura, Y., Mizoguchi, R.: Ontology-based systematization of functional knowledge. J. Eng. Des. **15**(4), 327–351 (2004). https://doi.org/10.1080/09544820410001697163
10. Koike, K., Tomoto, T., Hirashima, T.: Proposal of a stepwise support for structural understanding in programming. In: ICCE 2017–25th International Conference on Computers in Education, Workshop Proceedings, pp. 471–481, December 2017
11. Koike, K., Tomoto, T., Horiguchi, T., Hirashima, T.: Proposal of the expandable modular statements method for structural understanding of programming, and development and evaluation of a learning support system. Trans. Japn. Soc. Inf. Syst. Educ. **36**(3), 190–202 (2019). https://doi.org/10.14926/jsise.36.190. in Japanese
12. Koike, K., Tomoto, T., Horiguchi, T., Hirashima, T.: Supporting knowledge organization for reuse in programming: proposal of a system based on function-behavior-structure models. In: ICCE 2019–27th International Conference on Computers in Education, Workshop Proceedings, vol. 2, pp. 388–398, December 2019
13. Kruchten, P.: Casting software design in the function-behavior-structure framework. IEEE Softw. **22**(2), 52–58 (2005). https://doi.org/10.1109/MS.2005.33
14. Lahtinen, E., Ala-Mutka, K., Järvinen, H.M.: A study of the difficulties of novice programmers. In: Proceedings of the 10th Annual SIGCSE Conference on Innovation and Technology in Computer Science Education, ITiCSE 2005, pp. 14–18. Association for Computing Machinery, New York (2005). https://doi.org/10.1145/1067445.1067453
15. Lopez, M., Whalley, J., Robbins, P., Lister, R.: Relationships between reading, tracing and writing skills in introductory programming. In: Proceedings of the Fourth International Workshop on Computing Education Research, pp. 101–112 (2008)
16. Luxton-Reilly, A., et al.: Introductory programming: a systematic literature review. In: Proceedings Companion of the 23rd Annual ACM Conference on Innovation and Technology in Computer Science Education, ITiCSE 2018 Companion, pp. 55–106. Association for Computing Machinery, New York (2018). https://doi.org/10.1145/3293881.3295779
17. Milne, I., Rowe, G.: Difficulties in learning and teaching programming - views of students and tutors. Educ. Inf. Technol. **7**(1), 55–66 (2002). https://doi.org/10.1023/A:1015362608943
18. Nwana, H.S.: Intelligent tutoring systems: an overview. Artif. Intell. Rev. **4**(4), 251–277 (1990)
19. Robins, A., Rountree, J., Rountree, N.: Learning and teaching programming: a review and discussion. Int. J. Phytorem. **21**(1), 137–172 (2003). https://doi.org/10.1076/csed.13.2.137.14200
20. Selby, C.C.: Relationships: computational thinking, pedagogy of programming, and Bloom's Taxonomy. In: Proceedings of the Workshop in Primary and Secondary Computing Education, pp. 80–87 (2015)
21. Shneiderman, B.: Software psychology. Winthrop, Cambridge, Mass **48**, 161–172 (1980)
22. Wenger, E.: Artificial intelligence and tutoring systems: computational and cognitive approaches to the communication of knowledge (1987)
23. Wing, J.M.: Computational thinking. Commun. ACM **49**(3), 33–35 (2006)

A Research and Development of User Centered Zongzi Leaves Cleaning Machine Design

Yann-Long Lee[1(✉)], Feng-Che Tsai[2], Tai-Shen Huang[1], Chuan-Po Wang[1], and Wei-Lun Lo[3]

[1] Chaoyang University of Technology, Wufeng District, Taichung 41349, Taiwan
yannlee@cyut.edu.tw
[2] National Formosa University, Huwei Township, Yunlin County 632, Taiwan
[3] Design Industry, Yilan County, Taiwan

Abstract. For the Chinese, Zongzi is a kind of food that often appears in our daily life. However, with the development of modern automation technology, the original hand-washed leaves were also cleaned by machines. As time experience accumulates, some of the problems that users find will help to improve the efficiency and cleaning effectiveness of the Zongzi leaves cleaning machine, which is worthy of further study.

Generally, it takes many steps to clean the leaves, which is quite time-consuming. Cleaning requires brushing back and forth to remove sand or bug eggs on the leaves. Because of the different thickness of the leaves, it is more likely to cause damage to the general machine, which greatly increases the cost. This study hopes to improve the problem and use the design of the human-machine interface to complete the development and design of the new machine.

This study used an in-depth interview to investigate the needs of the industry. Then, by the Systems engineering and User-centered design, the problems can be classified into four categories: filtering, cleaning, storage, and others. And for each function, find the suitable requirements and appropriate mechanism to design.

The results are suitable for small and medium-sized food factories, it can efficiently and quickly clean the Zongzi leaves and make consumers feel at ease eating.

Keywords: User centered design · Zongzi leaves cleaning machine · Mechanism design · Systems engineering

1 Introduction

1.1 Zongzi Culture in China

The Dragon Boat Festival is one of the three major festivals in China. In history, the most common saying is that in the Warring States period, the patriotic poet Qu Yuan plunge suicide in the Miluo River. In order to prevent the fish to bite the body, people used to put Zongzi (Also known as Rice dumpling) into the river to make fish eat rice without biting the body of Qu Yuan (Huang Yong, 2006) [1].

C. Stephanidis et al. (Eds.): HCII 2020, LNCS 12427, pp. 163–175, 2020.
https://doi.org/10.1007/978-3-030-60152-2_13

Year by year, the Zongzi has not only been enjoyed by certain festivals. It became a tabletop meal on a regular basis, and even be a high-priced food gift, and many well-known Zongzi shops have made.

1.2 Type of Zongzi Leaves

In different local customs Taiwan, the Zongzi is also divided into several major categories. The most common ones are Hakka Zongzi, Southern Taiwan Zongzi and Northern Taiwan Zongzi (Wu & Cai, 2009) [2]. Not only the difference in taste but also the materials and fillings used are also different. There are great differences in the leaves used. The common Zongzi leaves in Taiwan include "Muzhu Leaves", "Guizhu Leaves and "Shell ginger Leaves" (Hung, 2015) [3]. This study based on three kinds of most common Zongzi leaves in Taiwan to find the characteristics and differences.

1.3 Existing Zongzi Leaves Cleaning Machine Problems

This research mainly cooperates with an old-fashioned Zongzi shop in Tainan, Taiwan (TTNEWS, 2019) [4]. To assist in the design and development of a new type of Zongzi leaf cleaning machine, to improve the problem from the existing leaf cleaning process.

In general, cleaning the leaves requires many steps, which is quite time-consuming and labor-intensive. In addition, it is necessary to clean the sand or bug eggs by brushing them back and forth.

In response to a large number of Zongzi sales needs, many stores have gradually used the machine to assist in cleaning.

At present, the commercially available cleaning machine is mainly designed for single-leaf design and has problems such as iron plate shape, sediment accumulation, easy wear and tear, incomplete cleaning, etc., and the overall use is quite inconvenient.

Because of the different thickness of the leaves, on the machine that generally only has a single-size cleaning thickness, the leaves are more likely to cause damage during cleaning, resulting in an increase in cost.

This study will use the field interview method to sort out the demand for existing Zongzi cleaning machines and assist the industry to obtain higher benefits in the cleaning function.

1.4 Patent Analysis

Generally, the patent for the Zongzi leaves cleaning machine can be divided into two types: Single leaves cleaning machine and Automatic leaves cleaning device.

The single leaves cleaning machine has a mainly conveying device which is required to feed the leaves to the machine and use a sprinkler to clean the leaves. Its motor is exposed and its protection is not enough. It is a very simple product (Fig. 1).

The automatic leaves cleaning devices have bubbling and rinsing system to remove the dirt, hair, insects and other impurities on the leaves, and also the decomposition of pesticide residues. The leaves are washed again at the outlet by a spray system. It equipped with a speed control device and a filter device to recycle water could clean a large number of leaves at the same time (Fig. 2).

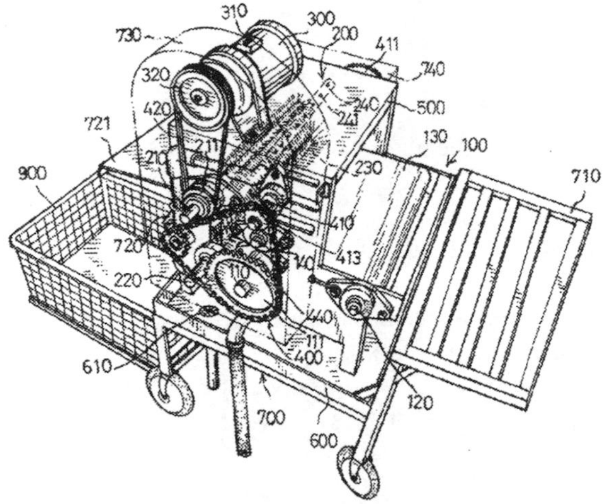

Fig. 1. Single leaves cleaning machine (Hong, 2002) [5]

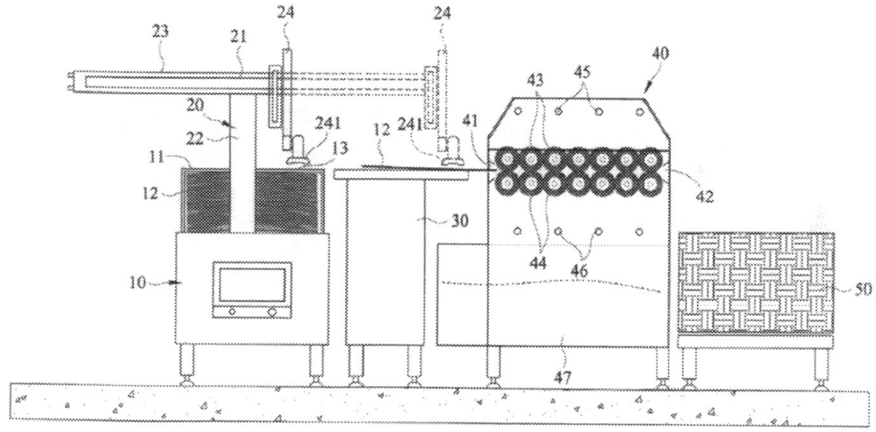

Fig. 2. Automatic leaves cleaning device (Liu, 2019) [6]

The single leaves cleaning machine has a small footprint and is suitable for small and medium-sized factories. The automatic leaves cleaning device is more suitable for large factories because of its large size and expensive. However, it is not specifically for Zongzi leaves, it also suitable for cleaning vegetables. Therefore, it is necessary to adjust the precision carefully, otherwise, the leaves are very easy to break or cannot be clean enough.

2 Methods

2.1 Systems Engineering

Lee (2008) [7] proposed that the main purpose of systems engineering is to standardize the regulation of engineering to ensure the successful development of the system and enhance product competitiveness.

Systems engineering includes: defining products, meeting and confirming system requirements, setting work plans, system optimization design, providing acceptance criteria, conduce the management system and providing the optimal support system. Indeed to meet the needs of the task, and fully use the expected functions. In this research, systems engineering was used to development and design of an innovative Zongzi Leaves Cleaning Machine (Fig. 3).

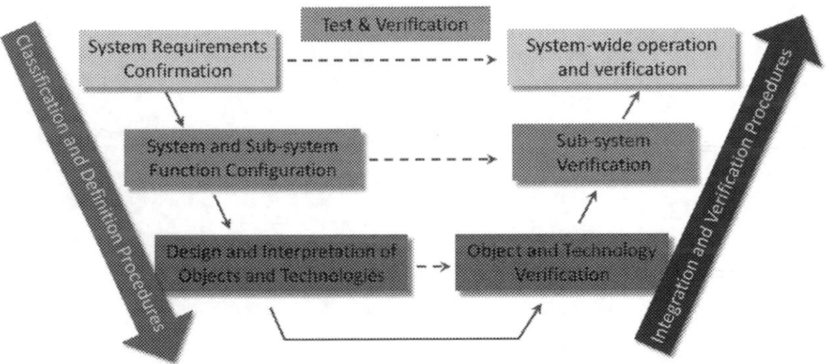

Fig. 3. Systems engineering procedure concept & steps

2.2 User Centered Design

Incorporating hidden information discovered from users in the design process is also one of the important issues of this research. Therefore, the study uses User Centered Design (UCD) to find the interaction between the user and the machine. From interviewing the suitable users, finding the purpose of the machine, to using reasonable design guidelines to create a rich experience, and thus making the design better.

Generally, good user experience is based on the concept of "User-centered design (UCD)". Most designs are based on design guidelines and add user-centered design for the design projects (Travis, 2014) [8]. The five focal points of UCD is shown in the Fig. 4.

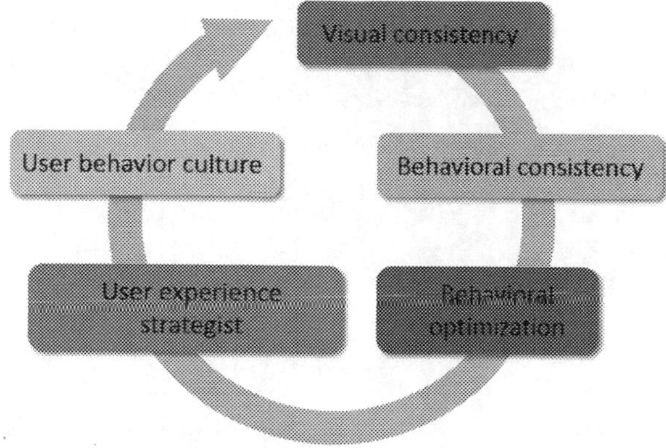

Fig. 4. The five focal points of "User-centered design (UCD)"

3 Results

3.1 Classification and Definition Procedures

a) System requirements confirmation – in-depth interview
b) System and Sub-system Function Configuration

This research uses the in-depth interview method to conduct interviews with the users and the person in charge of the cleaning machine. And clearly understand the existing problems and the items that need to be improved. Then, integrate and plan as the development goal of this design.

Based on the results of in-depth interviews, this study concludes the following points. According to the problems to be improved, they can be classified into four categories: Filtering system, cleaning system, storage system, and others system, as shown in the Fig. 5 below. It also analyze different functions separately, find out the appropriate function for each corresponding demand, to make a better design.

C) Design and Interpretation of Objects and Technologies - Mechanism engineering design

After the classification and definition procedures was completed, structure design was stated immediately. The design goal can be divided into the following: (see Table 1)

1. Steel main structure design
2. Fine-tuning and lifting structure design
3. Washing tank and washing structure design
4. Water circulation system and pipe design
5. Water proof sink design
6. Filtering drawer design
7. Working platform design

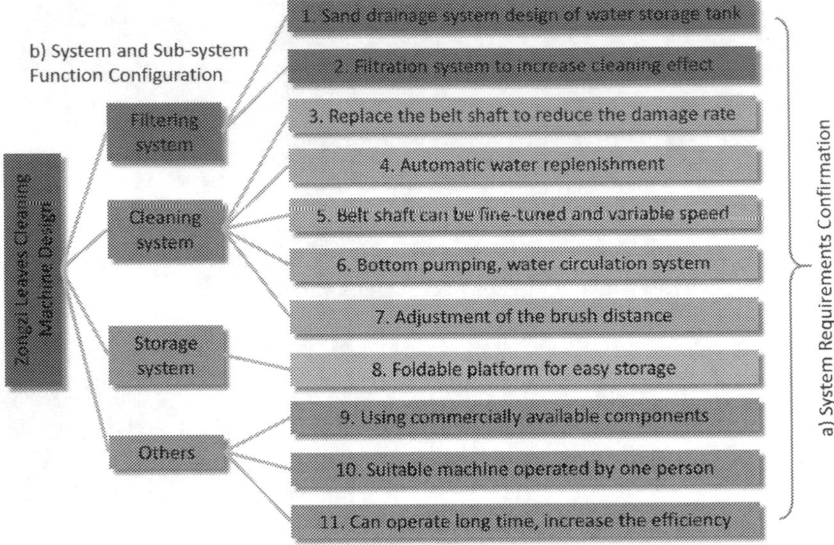

Fig. 5. The system requirements and function configuration

3.2 Integration and Verification Procedures

a) Object and Technology Verification – the overall design

In terms of the overall design, the cleaning machine effectively improves the existing problems and improves relative processing benefits. The design of the four systems is described in detail as follows (Fig. 6):

1. The filtering system

A mezzanine filter drawer that can be easily pulled out was set in the middle block of the cleaning machine for the preliminary impurity filtering (see Fig. 7). Next, water containing sediment was flowing into the structure of the water tank below. The water tank spacing makes the leaf debris float and causes sediment to accumulate, increasing the overall filtration efficiency. The position of the sand drainage hole is set directly front and rear of the water tank, and the valve can be quickly opened by turning.

2. The cleaning system

Replacing the original rubber belt with a timing belt will help reduce the rate of belt damage and slippage during machine operation. The water flow system is planned to pump its clean water through the submerged motor to the main cleaning position and designed two water outlet holes to increase the leaf cleaning efficiency. In addition, a waterproof layer is installed inside the machine to prevent rusting and slipping of components (see Fig. 8).

A speed controller is mounted on the motor, which can have a speed regulation function. It will be suitable for different types of leaves and prevent damage when washed. It's designed to adjust the cleaning interval by using a screw rod and a spiral

Table 1. Mechanism and structure design

Steel main structure and lifting structure design

Water proof sink design

Filtering drawer and Working platform design

slider, which can be adjusted about 3.7 cm up and down. The brush spacing not only helps to clean a variety of leaves but also avoids brush wear caused by prolonged friction.

3. The storage system

The workbench is divided into 3 parts, the middle, left and right parts, and the folding bracket design is used. When leaving work, the left and right parts of the workbench can be stowed to achieve the storage function, reducing collision danger and saving space (see Fig. 9).

4. Others

Except for the sheet metal parts, other components are assembled from commercially available parts, without the additional mold customization. The machine is designed for single-person operation, and only one person is needed for the cleaning job. The above design not only reduces labor costs, parts costs but also increases cleaning efficiency.

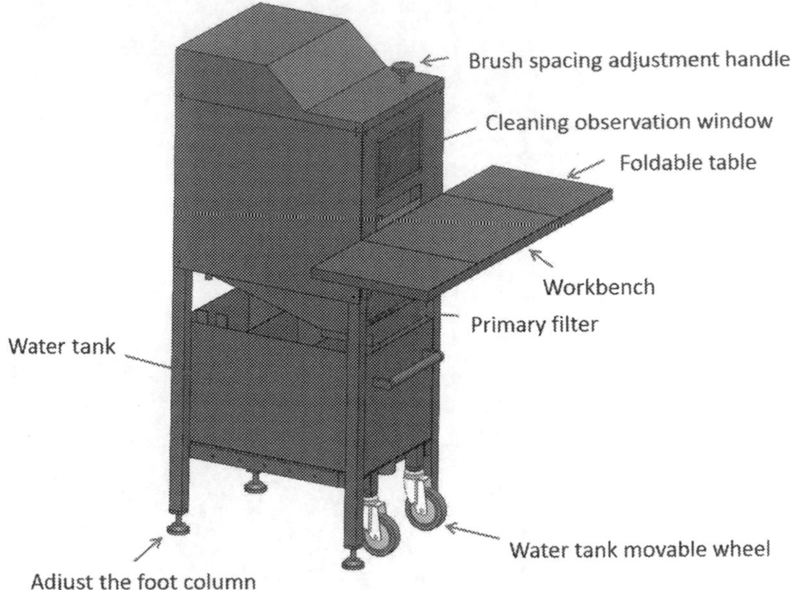

Fig. 6. The overall design of Zongzi Leaves Cleaning Machine

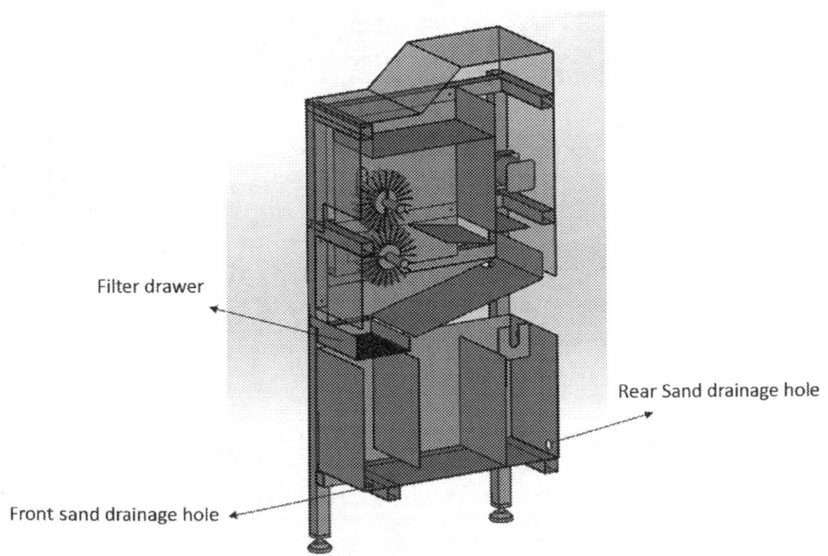

Fig. 7. The filtering system design

b) Sub-system Verification - Stress Analysis

In this research, stress analysis is used to analyze the assembly through the 3D simulation drawings and engineering drawings of various institutions. And for the main

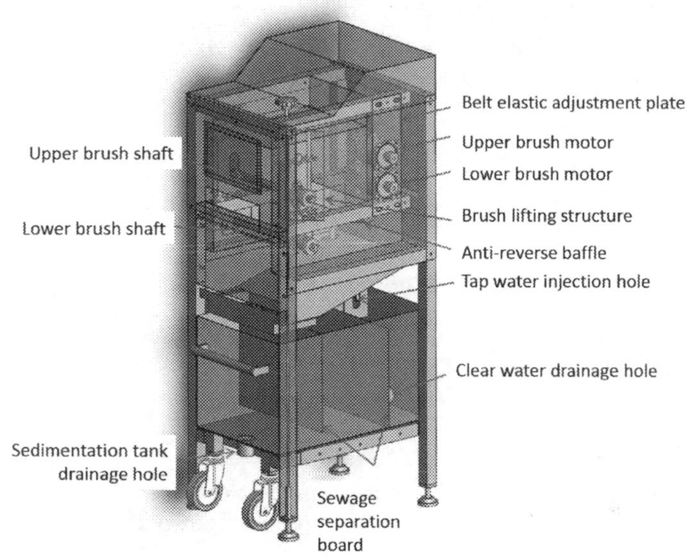

Fig. 8. Interior perspective of the cleaning system

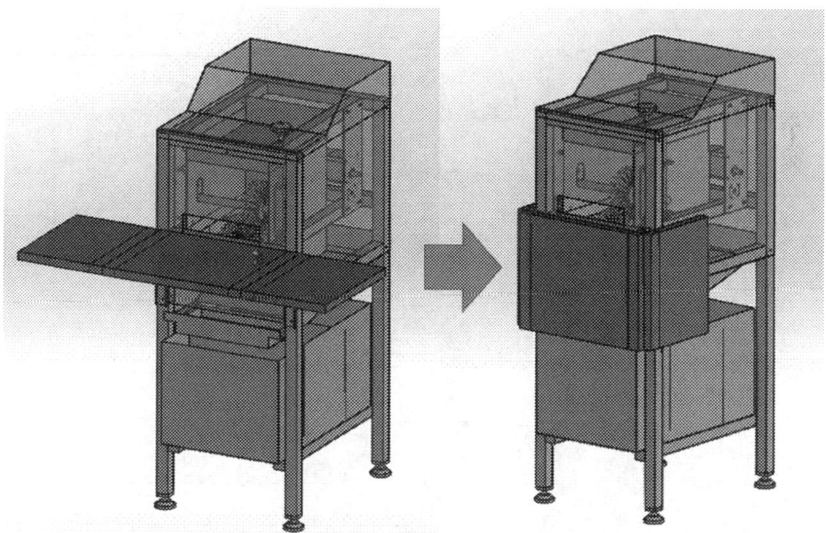

Fig. 9. The workbench storage system

structure and other mechanical stress analysis to confirm the feasibility analysis of the material, external load, deformation results, etc. The following uses the steel main structure as an example (Table 2).

By the Solidworks software, set the stress analysis according to parameter conditions. Because of the environment around the machine is wet and it will be touched to

Table 2. Description of analysis conditions for steel main structure

Material	Stainless steel (AISI304) Falling strength: 190 N/mm^2 Tensile strength: 500–700 N/mm^2 Elongation: 45% Vickers hardness: HV 150
Fixed surface	Steel main structure contact point with the ground
Force value	1.98 N (about 10 kg) 2.196 N (about 20 kg) 3.98 N (about 10 kg) 4.98 N (about 10 kg) 5. 1274 N (about 130 kg)
Mesh	Preset - Fine mesh

the water during operation, the material is set to stainless steel. Take the four contact points below the steel main structure as fixed points, and sets each force value. The components connected to the steel main structure are 5 application points, and the downward application force is respectively setting (see Fig. 10).

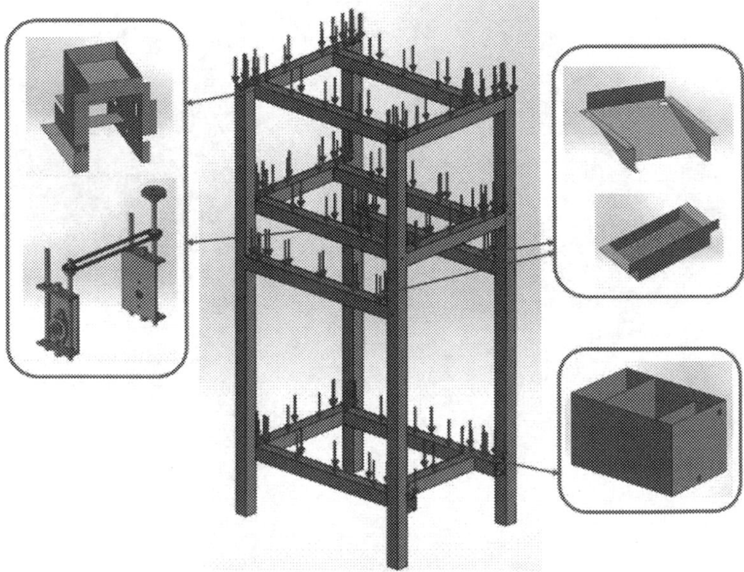

Fig. 10. The steel main structure application force setting

The analysis result of the steel main structure shows the highest stress value, the lowest stress value, and the yield strength. And through the safety factor analysis, it is

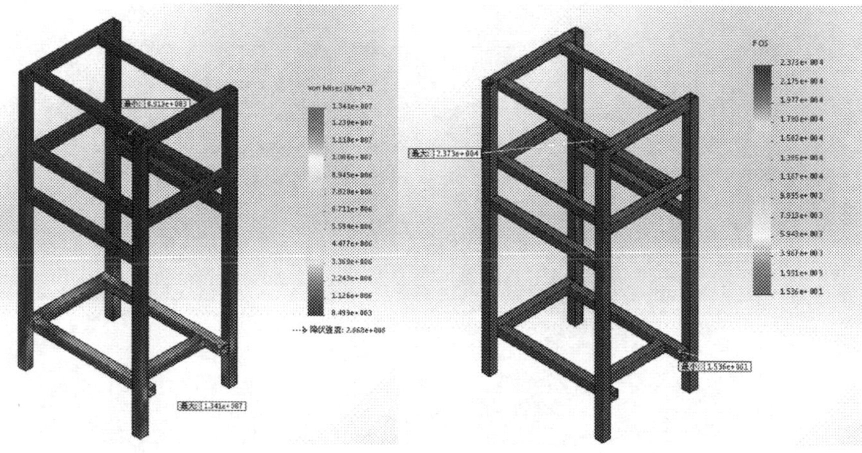

Fig. 11. The diagram of stress analysis and safety factor analysis

Table 3. The results of stress analysis and safety factor analysis

Yield strength	2.608×10^8
Minimum stress value	8.919×10^3
Maximum stress value	1.341×10^7
Minimum safety factor	15.36
Maximum safety factor	2.373×10^4

confirmed that the steel main structure is within the safety value range, which can indicate that this is an optimization mechanism (Fig. 11 and Table 3).

c) System-wide operation and verification - User-Centered Design

In the past cleaning process, the leaf washing machine replaced manual cleaning, but in the new cleaning process, the advancement and innovation of the machine have been improved. Considering the user-centered design, the water flow sped and temperature can be adjusted, and automatic feeding and brushing on both sides greatly improve the usability of the machine.

The following table (see Table 4) describes the five focal points of "User-centered design" and the design results of the Zongzi leaf washing machine (Fig. 12).

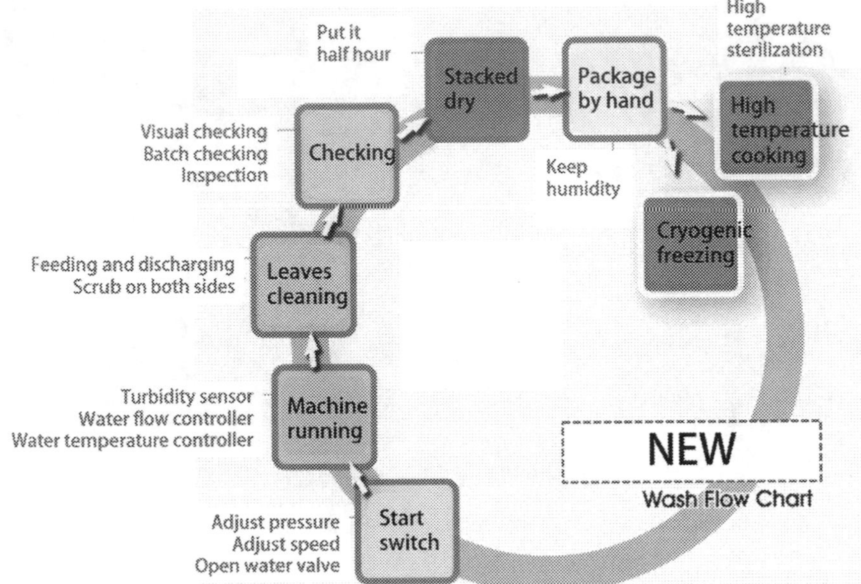

Fig. 12. The new cleaning process of the machine

Table 4. The five focal points of "User-centered design and the design results"

User-centered design: UCD	Design results
1. Visual consistency	The overall design is mainly made of food-grade stainless steel and sheet metal, which is simple and without excessive decoration and is obviously different from the previous product of exposed motor. The simple shape can give users a pleasant look
2. Behavioral consistency	All the design, content, design concepts and functions are approximately the same as the old model, making it easier for users to learn and use the product
3. Behavioral optimization	Through in-depth interviews and operation surveys, the new design optimizes the user's operating behavior, and several additional controllers are added to make the cleaning safer, cleaner and more efficient
4. User experience strategist	Through the analysis of users' actual operations, learned that users' perceptions, preferences, and behavioral responses and changes after contacting products, systems, and services, and unified the strategies for user experience
5. User behavior culture	The new design enables users to have better functions and operations, create new usage scenarios, and promote a new user culture

4 Conclusion

The study adopts the integration planning of Systems engineering and User-centered design, then completes the development and design of innovative Zongzi leaves cleaning machines through various functions such as lifting function, filtering system, and convenient operation. The results are as follows:

1. Developed and completed the innovative Zongzi leaves cleaning machine, which can be operated by one person to reduce the labor cost.
2. The new Zongzi leaves cleaning machine with lifting structure, and enhances the filtering and storage functions.
3. The different thickness of the Zongzi leaves can be cleaned and the quality of leaves cleaning can be improved.
4. The interaction, discussion, and sharing of student groups are very important.

References

1. Yong, H.: On QU Yuan's suicide. J. Zhuzhou Teach. Coll. **11**(1), 47–50 (2006)
2. Dai-En, W., Yi-Wen, C.: Taiwan local cuisine analysis – Zongzi and meatballs. The National Tainan Women's Middle School Press (2009)
3. Yu-Ying, H.: Origin, Evolution and Cultural Interpretation of Taiwanese Rice Dumpling, National Kaohsiung Normal University Thesis (2015)
4. TTNEWS: Snack food in Tainan. TraNEWS (2019). http://062745588.tw.tranews.com/
5. Mingqun, H.: Zongzi leaves cleaning machine, Patent announcement number: 491028. Intellectual Property Office, Ministry of Economic Affairs, Taiwan (2002)
6. Jiahui, L.: Automatic Zongzi leaf cleaning machine, Patent number: M583201. Intellectual Property Office, Ministry of Economic Affairs, Taiwan (2019)
7. Ling-Kuan, L.: A Built-up of Airplane Vertical Tail Manufacture Process. Chienkuo Technology University Thesis (2008)
8. Lowdermilk, T.: User-Centered Design. O'Reilly Books (2014)

Methodology of Controlling Subjective Speed While Watching CG Images

Yuki Motomura(✉), Hiroki Hashiguti, Takafumi Asao, Kentaro Kotani, and Satoshi Suzuki

Department of Mechanical Engineering, Kansai University, Osaka, Japan
{k402192, k671372, asao, kotani, ssuzuki}@kansai-u.ac.jp

Abstract. The technology of computer graphics (CG) has been rapidly developing, and is being utilized in various domains. Driving simulators (DSs) can provide driving experience to a user/driver by reproducing a virtual traffic environment using CG. In recent years, advanced driver-assistance systems have been actively researched, developed, and evaluated by automobile companies and ministries that hold the jurisdictions of transportation. Accordingly, the need to use DSs has become strong as they are being used instead of real vehicles. However, the subjective speed for driving a DS is lower than that for driving a real vehicle. Consequently, in the case of DSs, some participants aggressively drive at fairly high speeds, and thus some data collected from DSs become unreliable. Therefore, to increase the fidelity of the driver's behavior, a major challenge is to improve the subjective speed while using a DS to match the subjective speed in real driving. Two mainstream approaches employed to improve the subjective speed when using a DS are to spread the areas that display the images and to introduce a motion-based system. However, these approaches increase costs and require large space. In a previous study, the subjective speed was improved by correcting the CG image without the problems of high costs and large space. In this study, the objective is to control the subjective speed using CG image correction.

Keywords: Subjective speed · Driving simulators · Computer graphics

1 Introduction

Advanced driver-assistance systems have been developed by automobile manufacturers and suppliers considering an automated driving scenario in the near future [1, 2]. Accordingly, driving simulators (DSs) are being used to develop such systems [3]. DSs provide a sense of driving to users/drivers while being in a limited space and offer the following advantages [4]:

(1) Ease of ensuring driving safety,
(2) Ease of establishing a traffic environment and events with high repeatability,
(3) Several types of information can be measured more easily than a real automobile, and
(4) Independent of seasons, weather, and time of day.

C. Stephanidis et al. (Eds.): HCII 2020, LNCS 12427, pp. 176–186, 2020.
https://doi.org/10.1007/978-3-030-60152-2_14

Therefore, the demand for DSs is increasing [5]. The images used in DSs are live-action or computer graphics (CG) [6]. A live-action image is actually a video recorded on a real automobile, and a CG image is actually a virtual traffic scene [7]. Although the realistic sensation provided by CG images is lower than that provided by live-action ones, the former offers an advantage in that the traffic environment can be created without limitations. Therefore, CG images predominate in DSs. However, the subjective driving speed in DSs is lower than that in real automobiles [8–10]. Thus, the experimental results obtained using DSs become unreliable because some participants drive at fairly high speeds while using them [11]. Therefore, drivers must be provided with an environment as similar as possible to a real-automobile one to increase both the fidelity of their behavior and effectiveness of the experiments [12]. Therefore, increasing the subjective speed in DSs is a challenge. There exist two major approaches to improve the subjective speed: introducing a motion system and expanding the field-of-view (FOV) of the image. The motion system provides inertial acceleration to the drivers according to the real-automobile behavior [3]. A wide FOV is implemented using wide screens or displays. However, both the approaches are not preferred in the case of small research-and-development projects because they require high costs and large spaces [13, 14]. Drivers on a real automobile rely on their subjective speed based on visual cues from the optic flow of their surroundings, vestibular and somatosensory cues of acceleration and vibration, haptic cues of reaction force by a steering wheel and pedals, and auditory cues of engine sound and wind noise [15]. Particularly, the visual information gained by drivers comprises 90% of their driving information [11, 16]. In a previous study, the subjective speed was improved by correcting CG images with respect to the characteristics of a person's visual space [17]. In this study, empirical examinations were conducted to control the subjective speed to be the same as that on a real automobile by using an image-correction method.

2 Image Correction to Improve Subjective Speed

In a previous study [17], a correction coefficient f(z), defined using Eq. (1), was used to distort the CG image with respect to the human visual space to improve the subjective speed. One has

$$f(z) = \beta \times 10^{-5} \times |z| + (100 - \beta) \times 10^{-2} \tag{1}$$

where z denotes the depth coordinate of a CG object from a DS driver. The constant β denotes the degree of distortion and is determined how the position of a CG object becomes inward from its original location in the same depth as the driver is. The values of β were set at 0%, 40%, 60%, or 80% in following experiments.

The CG image was distorted using Eq. (2), where the coordinates of the position of the object before and after the correction are (x, y, z) and (X, Y, Z), respectively. One has

$$\begin{pmatrix} X \\ Y \\ Z \end{pmatrix} = \begin{pmatrix} f(z) & 0 & 0 \\ 0 & f(z) & 0 \\ 0 & 0 & 1 \end{pmatrix} \begin{pmatrix} x \\ y \\ z \end{pmatrix} \tag{2}$$

3 Experiment 1: Collection of Subjective Speed

The relation among the degree of distortion β, subjective speed Vp, and running speed Vg should be clarified. The objective of this experiment is to collect subjective speeds under certain conditions on both the degree of distortion and running speed.

3.1 Apparatus and Participants

Figure 1 shows the outline of the experimental environment. A computer (CERVO, Applied Corp.) equipped with a graphic board (Quadro K620, video memory: DDR3, 2 GB) was used to draw the CG images, which were projected using a liquid-crystal projector (EMP-1825, EPSON. Refresh rate: 60 Hz) on a screen (SRMS3D-100, Solidray). Each participant sat on a chair in front of the screen at a viewing distance of 1.5 m, and he/she put his/her head on a chin rest to match the line joining his/her eye points with the horizontal and vertical center of the image. The horizontal and vertical FOVs were 49.1° and 36.9°, respectively. The experimental apparatus was surrounded with partitions, and the light was turned off. The luminance at the eye point was 21.9 lx in average.

Figures 2 and 3 show the examples of CG images, which were created using OpenGL and depict a situation of running along a lone straight asphaltic road on a lawn. A dashed centerline was putted on the road with a length of 5 m and an interval of 5 m, and continuous side lines were also putted on right and left sides of the road. Moreover, mountains were placed at a significant distance and covered with a sky to augment the reality. A red circle is displayed on the middle of the image as a fixation point where participants were asked to look. The image had a resolution of 1024 × 768 pixels, and it was vertically synchronized with the projector.

Fifteen university students (20–25 years of age) participated after providing informed consent. Each of them had a valid driving license and normal or corrected-to-normal binocular vision.

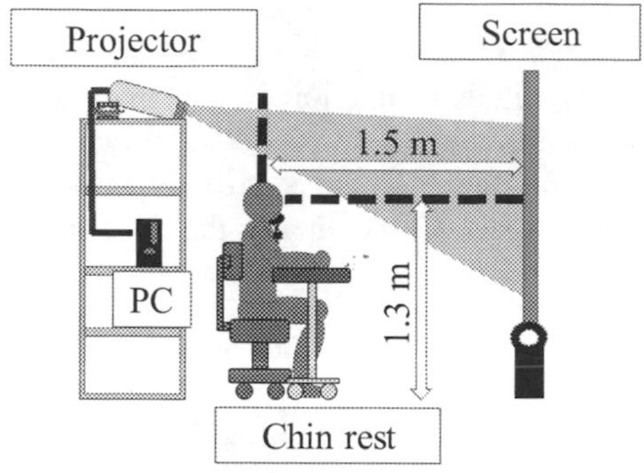

Fig. 1. Outline of the experimental environment.

Fig. 2. Example of a CG image with the degree of distortion β = 0%. (Color figure online)

Fig. 3. Example of a CG image with the degree of distortion β = 60%. (Color figure online)

3.2 Subjective Speed

The absolute method of magnitude estimation was adopted to measure the subjective speed. It is one of the methods used to construct a psychological scale of a physical stimulus [18]. This means that the participants verbally answered the subjective speed in the unit of km/h with respect to the image they viewed. Moreover, they were instructed that they could use any numbers that were integers, decimals, and fraction values.

3.3 Conditions

Table 1 lists the experimental conditions. The subjective speeds were not different between the images with 20% and 40% distortions, according to a previously conducted study [17]. Therefore, in this experiment, 80% distortion was adopted instead of 20%. The four distortion levels were 0%, 40%, 60%, and 80%. The values used for running speed Vg were 20, 40, 60, 80, and 100 km/h. Moreover, the speeds of 1 and 120 km/h were added as dummy conditions to eliminate the regression bias [19], which is a tendency to answer neither high nor low values. A block of 28 conditions (four image-distortions × seven running-speeds) was repeated five times.

Table 1. Conditions for Experiment 1

Degree of distortion β	0%, 40%, 60%, and 80%
Running speed V_g	1, 20, 40, 60, 80, 100, and 120 km/h
Repetition	5 times

3.4 Procedure

In one trial, a gray image was displayed for 5 s, and then stimulus images were presented for 5 s. This trial was repeated 28 times as one block of experiment. The participants were asked to verbally answer the subjective speed while the gray image was displayed, and they were instructed not to compare the stimulus image with the previous one. The order of the 28 conditions was completely randomized in each block for each participant.

At the beginning, the participants were instructed regarding the procedure of the experiment, following which they provided informed consent. Subsequently, they were tested for visual acuity. Next, they performed one block of training session. Furthermore, a test session, which comprised five blocks, was performed. The participants rested for 2 min between the blocks. After the test session, they answered questionnaires regarding the trials. The accuracy of the subjective speed that they answered was not fed back in either the training or test session.

3.5 Results

Figure 4 shows the relationship between subjective speed Vp and running speed Vg. Each plot represents the geometric mean of the subjective speeds among the participants, and the error bar denotes the standard error of the geometric mean. Evidently, the subjective speed becomes high with respect to the degree of distortion. This trend is the same as that in a previously conducted study [17].

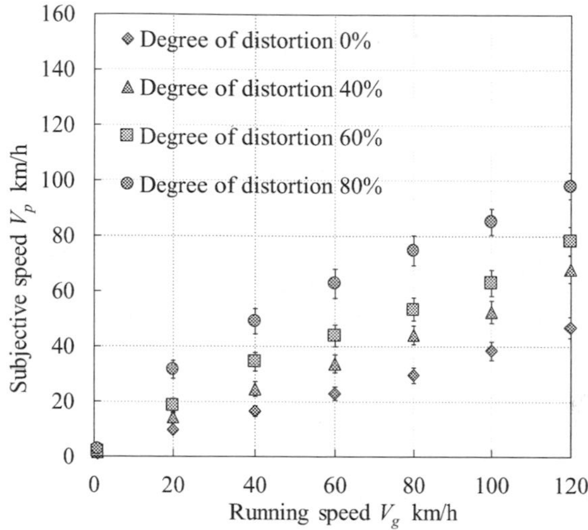

Fig. 4. Relationship between subjective and running speeds in Experiment 1.

The two-way analysis of variance was conducted to confirm the effect of the degree of distortion and running speed on the subjective speed. Consequently, the main effects of the running speed ($F(4, 280) = 80.29$; $p < 0.05$) and distortion ($F(3, 280) = 98.16$; $p < 0.05$) were significant. However, the interaction effect between the running speed and distortion was not significant ($F(12, 280) = 1.26$; $p = 0.244$). Certainly, the distortion of CG images affects the subjective speed.

4 Derivation of Appropriate Distortion to Control Subjective Speed

An equation that represents the relationships between the subjective speed, running speed, and degree of distortion is derived from the results of Experiment 1.

A power law, popularly called Stevens' power law, represents the relationship between physical and psychological quantities. The running speed is denoted as Vg, degree of distortion as β, and subjective speed as Vp. Accordingly, the following relationships hold:

$$V_p \propto V_g^A \tag{3}$$

$$V_p \propto \alpha^B \tag{4}$$

where A and B denote constants. The constant β is replaced by dis-distortion α defined as follows to avoid the value from being zero:

$$\alpha = (100 - \beta) \times 10^{-2} \tag{5}$$

Equations (3) and (4) can be transformed as follows by taking the common logarithm on both the sides of them:

$$\log_{10}\mathrm{V_p} \propto \mathrm{A}\log_{10}\mathrm{V_g} \tag{6}$$

$$\log_{10}\mathrm{V_p} \propto \mathrm{B}\log_{10}\alpha \tag{7}$$

Combining Eqs. (6) and (7), log10Vp can be expressed as follows:

$$\log_{10}\mathrm{V_p} = \mathrm{A}\log_{10}\mathrm{V_g} + \mathrm{B}\log_{10}\alpha + \mathrm{C} \tag{8}$$

where C denotes a constant.

The experimental data collected in Experiment 1 were approximated using Eq. (8). Thus, constant values A = 0.754, B = −0.617, and C = 0.0438 were obtained (R2 = 0.988). Figure 5 shows this three-dimensional approximation. Solving Eq. (8) for β using Eq. (5), an appropriate distortion with respect to running speed Vg and subjective speed Vp can be obtained as follows:

$$\beta = \left(1 - 10^{-\frac{C}{B}} \cdot V_p^{\frac{1}{B}} \cdot V_g^{-\frac{A}{B}}\right) \times 10^2 \tag{9}$$

A desired subjective-speed value is assigned to Vp in Eq. (9). Thus, distorted images would be generated, as a driver would feel as though he/she is running at a speed equal to Vp. In the next experiment, it is validated that the subjective speed can be controlled using Eq. (9).

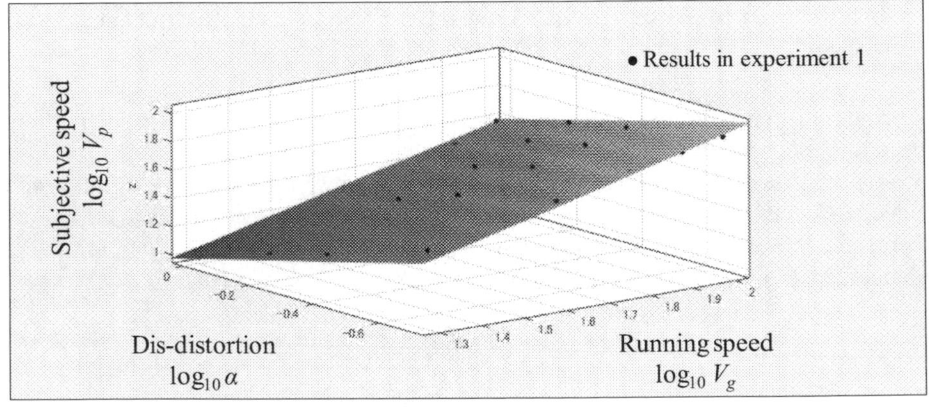

Fig. 5. Three-dimensional approximation of subjective speed using dis-distortion and running speed.

5 Experiment 2: Controlling the Subjective Speed

The objective of this experiment is to verify whether the subjective speed can be controlled, as mentioned in the previous section. The apparatus, measurement method of subjective-speed, and participants were the same as those in Experiment 1.

5.1 Conditions

Table 2 summarizes the experimental conditions. The running speed had five conditions, same as those in Experiment 1, namely, 20, 40, 60, 80, and 100 km/h. In this experiment, the objective was to reproduce the subjective speed as follows [21]:

$$V_p = 0.324\ V_g^{1.15} \tag{10}$$

where Vp denotes the subjective speed and Vg the driving speed. The degree of distortion β was expressed by equating Vp in Eq. (10) to Vp in Eq. (9).

$$\beta = \left(1 - 7.31\ V_g^{-0.634}\right) \times 10^2 \tag{11}$$

The CG-image distortion was changed for each condition of driving speed Vg according to Eq. (11). Moreover, two dummy trials were added to the conditions, and they had distortion β = 0% at the speed of 1 km/h and β = 80% at the speed of 120 km/h.

Table 2. Experimental conditions for Experiment 2.

Running speed V_g	20, 40, 60, 80, and 100 [km/h]
Degree of distortion β	$\beta = \left(1 - 7.31\ V_g^{-0.634}\right) \times 10^2$
Repetition	5 times

5.2 Procedure

The experiment was conducted similarly to Experiment 1. However, one block comprised seven trials.

5.3 Results

Figure 6 shows the relationship between subjective speed Vp and running speed Vg. Each plot represents the geometric mean of subjective speeds for the participants; the error bar denotes the standard error of the geometric mean; the red solid line represents the target speed to be perceived, as shown in Eq. (10). If the plot lies on the red line, it is indicated that the subjective speed can be controlled according to Eq. (10).

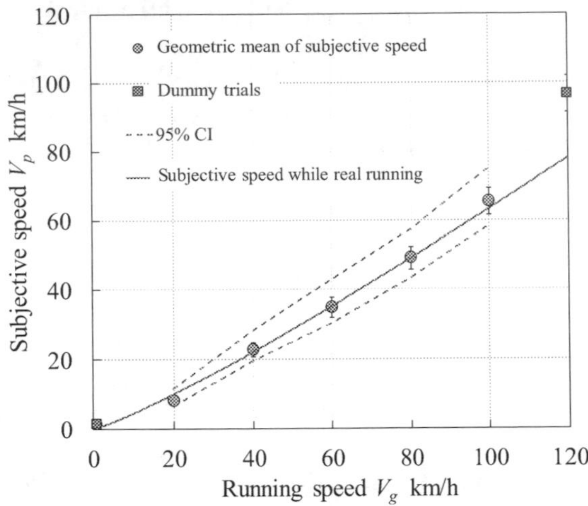

Fig. 6. Relationship between subjective and running speed in Experiment 2. The red solid line denotes target speed to control the subjective speed. (Color figure online)

The two-tailed t-test was conducted between the measured subjective speed and target speed Vp in Eq. (10) for each running speed. Consequently, the difference was not significant when the level of significance was 5% (Vg = 20 km/h: t(14) = 0.83, p = 0.420; Vg = 40 km/h: t(14) = 0.94, p = 0.364; Vg = 60 km/h: t(14) = 0.51, p = 0.621; Vg = 80 km/h: t(14) = 0.44, p = 0.664; Vg = 100 km/h: t(14) = 0.93, p = 0.366).

Moreover, the confidence interval (CI) was calculated at the level of significance of 95%. The results were as follows: Vg = 20 km/h: M = 9.04, CI = [3.47, 1.53];

Vg = 40 km/h: M = 24.33, CI = [2.48, 6.33]; Vg = 60 km/h: M = 36.70, CI = [4.81, 7.77]; Vg = 80 km/h: M = 48.96, CI = [5.56, 8.46]; Vg = 100 km/h: M = 66.84, CI = [4.70, 11.96]. The mean subjective speed for all running speeds was within a 95% CI.

5.4 Discussion

The difference between the subjective and target speeds was not significant. Moreover, the subjective speed while driving was within the 95% CI of the experimental results for all the running-speed conditions. Therefore, the subjective speed could be controlled through the adaptive distortion of the CG image. However, there are four considerations.

The first consideration is the personalization of the distortion. Although only the mean subjective speed among the participants was evaluated in this experiment, the individual subjective speed significantly varied among the participants. Therefore, a mechanism to consider individual variation must be proposed.

The second consideration is road alignment. Although only the scenes that depicted running along a straight road were examined, several types of road alignments exist in reality. Therefore, the validation of controlling the subjective speed must be conducted for a situation of running on a curved road.

The third consideration is the static change in image distortion. Although the distortion technique described in Sect. 4 can be applied when the running speed dynamically changes, the image actively changes according to the dynamic changes in distortion. Therefore, the subjective feeling of experiencing dynamic changes in image distortion must be evaluated.

The fourth consideration is the surrounding environment in a virtual scene. In these experiments, the scenes only simulated a rural-like simple field. Therefore, evaluations in various other surrounding environments must be performed.

6 Conclusions

Two experiments were conducted with the aim of controlling the subjective speed of DSs. The first experiment was conducted to collect the subjective speed with four conditions of distortion of CG image at five running-speed conditions. The results showed that the subjective speed became higher with larger image distortion at all the running-speed conditions. The second experiment was performed to control the subjective speed in a DS to be the same as that on a real vehicle. Consequently, it was indicated that the subjective speed could be controlled through CG-image distortion. This image-distortion technique can contribute to the development of DS images.

Acknowledgment. This research was supported in part by the Kansai University Grant in Aid for progress of research in graduate course for 2019 and in part by Kansai University's Overseas Research Program for 2014.

References

1. Asanuma, N., Kaseyama, H.: Circumstances of the vehicle prevention technologies for safety driving support. IATSS Rev. **31**(1), 56–61 (2006). (in Japanese)
2. Sagesaka, Y.: Advanced driver assistance system and the role of ITS communication system. J. Soc. Instrum. Control Eng. **54**(11), 845–848 (2015). (in Japanese)
3. Miki, K.: Real time high fidelity motion cueing technologies on driving simulator. J. Robot. Soc. Jpn. **10**(7), 878–884 (1992). (in Japanese)
4. Nerio, M., Chiku, Y.: Introduction to simulation technology for analysis of human factor. J. Soc. Instrum. Control Eng. **45**(8), 726–730 (2006). (in Japanese)
5. Oshima, D., et al.: A study on needs and advanced technologies for a driving simulator. Seisan Kenkyu: Bull. Inst. Ind. Sci. Univ. Tokyo **67**(2), 87–92 (2015). (in Japanese)
6. Tanba, Y.: Analysis of effects of difference between live-action and CG images on changes in cerebral blood flow and visual information, Summary of undergraduate thesis (2017). (in Japanese)
7. Ohta, H., Katakura, M., Oguchi, T., Shikata, S.: Effects of traffic safety devices on driver's behavior at a curve section. In: Proceedings of Infrastructure Planning of JSCE, vol. 26 (2002). (in Japanese)

8. Kuriyagawa, Y., Ohsuga, M., Kageyama, I.: Basic study on driver assistance system using driving simulator. Correspondences Hum. Interface **6**(2), 43–46 (2004). (in Japanese)
9. Ohta H., Komatsu, H.: A comparison of speed perception in driving with that in TV simulation. In: Proceedings of 11th World Congress of the International Association for Accident and Traffic Medicine, pp. 435–441 (1988)
10. Jamson, H.: Cross-platform validation issues. In: Fisher, D.L., Rizzo, M., Caird, J.F., Lee, J.D. (eds.) Handbook of Driving Simulation for Engineering, Medicine, and Psychology, pp. 12-1–12-13. CRC Press, Florida, US (2011)
11. Kuriyagawa, Y., Kageyama, I.: Study on using of driving simulator to measure characteristics of drivers. J. Coll. Ind. Technol. Nihon Univ. **42**(2), 11–18 (2009). (in Japanese)
12. Colombet, F., Paillot, D., Mérienne, F., Kemeny, A.: Impact of geometric field of view on speed perception. In: Proceedings of the Driving Simulation Conference Europe 2010, pp. 69–79 (2010)
13. Kien, H., Ideguchi, T., Okuda, T., Okuda, T.: Development and verification of an experimental platform (ARDS-Platform) for IVC-based driving safety support system. Trans. Inf. Process. Soc. Jpn. **53**(1), 223–231 (2012). (in Japanese)
14. Kihira, M., Itoh, Y.: Investigation of apparatus of simulated driving and design of exercise test on evaluating avoidance behavior. Trans. Soc. Automot. Eng. Jpn. **41**(2), 221–226 (2010). (in Japanese)
15. Kawashima, Y., Uchikawa, K., Kaneko, H., Fukuda, K., Yamamoto, K., Kiya, K.: Changing driver's sensation of speed applying vection caused by flickering boards placed on sides of road. J. Inst. Image Inf. Telev. Eng. **65**(6), 833–840 (2011). (in Japanese)
16. Sivak, M.: The information that drivers use: is it indeed 90% visual? Perception **25**(9), 1081–1089 (1996)
17. Fujiwara, K.: Proposition of method to develop speed feeling for VR images, B.E. Thesis (2015). (in Japanese)
18. Gescheider, G.A.: Psychophysics: The Fundamentals, vol. 1, 3rd edn., p. 166. Kyoto, Kitaohji (2002). (Translated version in Japanese)
19. Gescheider, G.A.: Psychophysics: The Fundamentals, vol. 1, 3rd edn., p. 138. Kyoto, Kitaohji (2002). (Translated version in Japanese)
20. Stevens, S.S.: On the psychophysical law. Psychol. Rev. **64**(3), 153–181 (1957)
21. Evans, L.: Speed estimation from a moving automobile. Ergonomics **13**(2), 219–230 (1970)

Improvement of SPIDAR-HS and Construction of Visual Rod Tracking Task Environment

Hiroya Suzuki[1(✉)], Ryuki Tsukikawa[1], Daiji Kobayashi[2], Makoto Sato[3], Takehiko Yamaguchi[4], and Tetsuya Harada[1]

[1] Tokyo University of Science, 6-3-1 Niijuku, Katsushika-Ku, Tokyo, Japan
8119534@ed.tus.ac.jp

[2] Chitose Institute of Science and Technology, 758-65 Bibi, Chitose, Hokkaido, Japan

[3] Tokyo Institute of Technology, 4259 Nagatsuta-Cho, Midori-Ku, Yokohama, Kanagawa, Japan

[4] Suwa University of Science, 5000-1 Toyohira, Chino-City, Nagano, Japan

Abstract. The final goal of this study is to elucidate the contribution of haptics and vision to the Sense of embodiment (SoE) in a virtual reality (VR) environment through physiological behavioral measurements. To achieve the objective, a rod-tracking task was employed in which a rod held in the hand is passed along a sinusoidal path in VR environment using HMD and haptic device. The problems identified in the previous system were the device positional accuracy and insufficiency of the force sensation. In order to solve these problems, the control circuit of the haptic device was modified. Then, the positional accuracy of the device and the force output characteristics of the device were evaluated. Through these evaluations, it was confirmed that the accuracy of the device was improved, and the strength of the force presentation was improved. In addition, we evaluated the experimental environment by measuring physiological behavior. The evaluation of the experimental environment by the physiological behavioral measurements was carried out by a questionnaire survey on the Rod Tracking Task, which included measurements of muscle potential and subjective evaluation. In the future, we will improve the accuracy of the device, improve the control circuit, and change the force-sensing model in consideration of the system construction.

Keywords: VR · SPIDAR · SoE · Haptics

1 Introduction

1.1 Background and Purpose

According to “Virtual Reality Science” [1] (2016), VR is a technology that constitutes a system that has three elements of virtual reality, namely, self-projectability (SP), three-dimensional spatiality (TS), and real-time interactivity (RI) and aims to allow humans to use an environment that is substantially equivalent to the real environment, and SP is the realization of a state that is consistent between different sensory

C. Stephanidis et al. (Eds.): HCII 2020, LNCS 12427, pp. 187–195, 2020.
https://doi.org/10.1007/978-3-030-60152-2_15

modalities that humans possess. It is said that SP is the only thing that is difficult to evaluate because of the user's sense. On the other hand, the "three-dimensional spatiality" and "real time interactivity" are regardless of the user and can be easily measured objectively.

In this study, we focus on the sense of agency. The sense of agency was originally studied in the field of cognitive neuroscience, and it is still being studied to elucidate the characteristics of the sense of agency in the real environment. [2, 3] In the field of HCI, researches on VR technology has been increasing in recent years. However, SP is a subjective sensation, and it is difficult to evaluate, and the index for quantitative evaluation has not yet been clarified. If the SP can be evaluated quantitatively, all three elements of virtual reality can be evaluated quantitatively, which will contribute to the development of VR technology.

The final goal of this study is to clarify the contribution of haptic and visual sensation to SoE in VR environment by physiological behavioral measurements and to develop quantitative evaluation indices. To achieve this goal, this study aims to construct a human-scale VR environment, to evaluate the environment by physiological behavioral measurements, and to elucidate the difference between the real and VR environments.

2 Construction of SPIDAR-HS

2.1 Spidar

SPIDAR is haptic devices consists by modules of a motor, a rotary encoder, and a pulley with a string. Generally, eight modules are combined to present the sense of force by the motors and measure the position and angle by the encoders. SPIDAR-HS [4] is an extension of the SPIDAR system to human scale. The SPIDAR-HS is shown in Fig. 1.

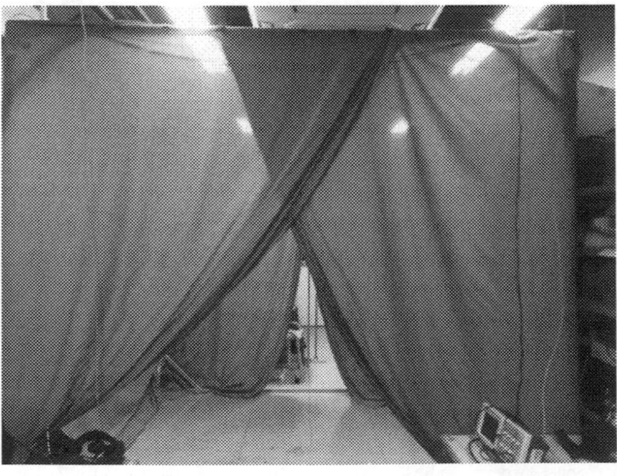

Fig. 1. SPIDAR-HS [4]

2.2 Improvement of SPIDAR-HS Controller

In this study, we improved the problem of SPIDAR-HS found in the previous experiment [5]. As a result of the previous experiment, inaccurate force presentation on end-effector was an issue. It is necessary to be able to present a large force in human scale large type SPIDAR. Although SPIDAR-HS was controlled by 5 V until previous, especially since this system used a 24 V motor (Table 1.), if the control is set to 24 V the motor's performance can be maximized. In addition, due to the large size of the frame, the wiring length is long and the switching noise of PWM may affect the performance. Therefore, analog control was used. This made it possible to present a larger force, moreover, has improved the position accuracy. Due to the limitation of the software environment, even the controller board do not support 24 V motor, it must be used. Thus, it was decided to expand it adding an analog circuit. The circuit is as follows. Figure 2 shows circuit diagram for one module. Figure 3 shows the overall circuit diagram. Until previous, the SPIDAR was controlled by 5 V PWM control, but in the new circuit, 24 V linear control is adopted. Replace the conventional PWM-controlled motor with a resistor and invert the source-drain voltage of the output FET from 0 to 12 V centered on 6 V and a low-pass filter using an operational amplifier. The MC33074AP single-supply general-purpose operational amplifier is used. The 0–12 V DC voltage was obtained. And using this voltage as a control voltage, 24 V linear control output was obtained using LM324 and a power MOSFET. Details of the motor and pulley are shown in Table 1 [6] and details of the encoder are shown in Table 2 [7].

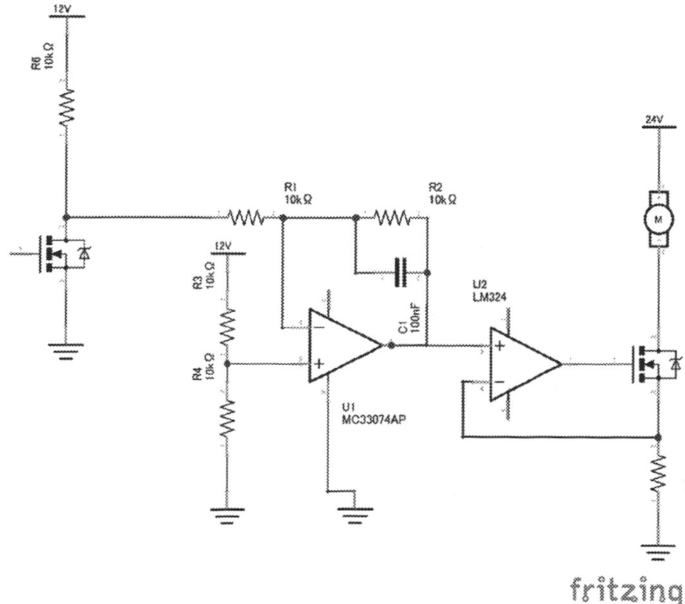

Fig. 2. Circuit diagram of control circuit for one module

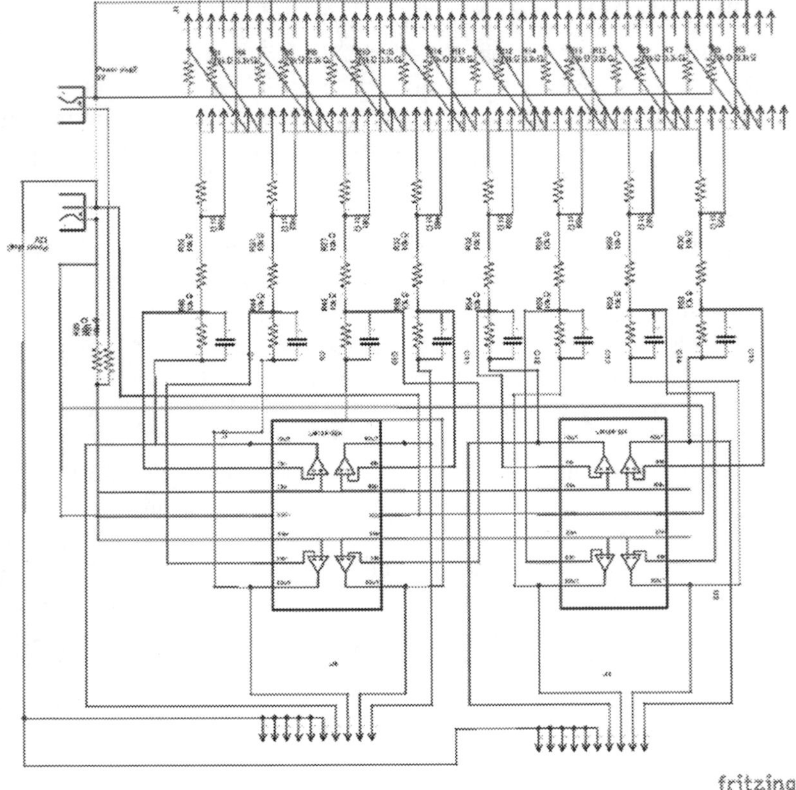

Fig. 3. Circuit diagram of the new entire control circuit

Table 1. Motor and pulley specification [6]

Maxon motor 118746 RE25 specification	
Nominal voltage [V]	24
Starting current [A]	3.1
Maximum continuous torque [mNm]	28
Stall torque [mNm]	136
Pulley radius [mm]	10

3 Stem Evaluation and Rod Tracking Task

3.1 Positional Error Measurement in the Workspace

The purpose of this experiment is to evaluate the constructed environment by measuring the positional error in the working space of SPIDAR-HS.

First, define the coordinate system of SPIDAR-HS. The center of SPIDAR-HS is the origin, and the right side and the front side of the horizontal direction and the upper of the vertical direction seen from user are the positive direction of the x, y, and z respectively. We set up a workspace cube with $-0.5 \leq x, y, z \leq 0.5$. The workspace is sufficiently large to execute the Rod Tracking Task. In measuring, the end-effector was moved from the origin to the vertex of the workspace cube (−0.5, −0.5, 0.5) at a uniform speed in 10 s. Because we adopted the two-point control, we recorded the position coordinates of two virtual end-effectors, point 1 and point 2, in Unity. We have done this 10 times.

We also investigated the accuracy of the encoders to determine the cause of the positional error. All strings were pulled out to the exit of each opposing upper or lower motor and measured the encoder count value. Individual pull out length were approximately 2.32 m. This was done three times for each encoder.

3.2 Presentation Force Measurement of SPIDAR-HS

The purpose of this experiment is to measure the presentation force in the workspace and to evaluate the constructed environment.

All 8 strings of SPIDAR-HS are connected at a single point on the origin. We measured the actual force intensity while varying the instruction value of the downward force from 0 to 5 N in Unity. This measurement was repeated three times. The above measurements were performed for the previous and the new circuits. A force gauge RZ-2 (Aikoh Engineering Co., Ltd. [8]) was used for the measurement.

3.3 Visual Rod Tracking Task

The purpose of this experiment is to clarify the effect of the proficiency in the visual Rod Tracking Task in the VR environment and its cause by using electromyography.

Table 2. Encoder specification [7]

HEDS 5540 specification	
Count/rotation (resolution)	500
Maximum frequency [kHz]	100
Maximum allowable speed [rpm]	12000

Figure 4 shows a schematic of the experiment. We also changed the end-effector for easy calibration to simplify the experiment. Figure 5 shows the new end-effector. In this experiment, force is not presented, and SPIDAR-HS only measures the position and orientation of the end-effector. Table 3 shows the details of the visual the Rod Tracking Task.

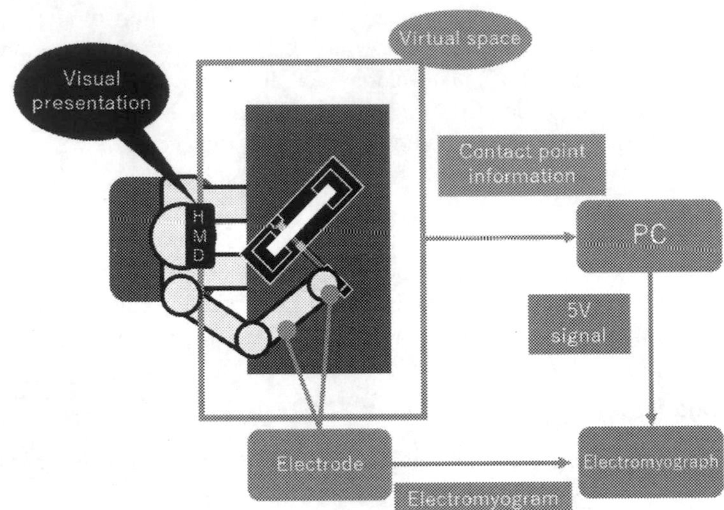

Fig. 4. Detail of visual Rod Tracking Task

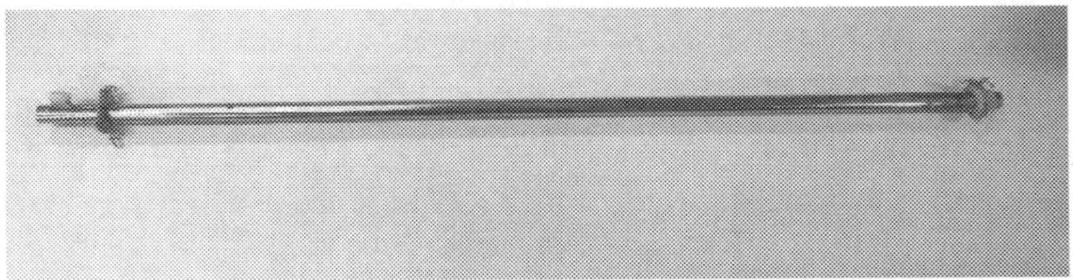

Fig. 5. End-effector used visual Rod Tracking Task

Table 3. Detailed conditions of visual Rod Tracking Task

Experiment conditions	
Rod Length [mm]	501
Rod diameter [mm]	10
Path size [mm]	200 × 300 × 20
Path width [mm]	20
Minimum require force[a] [N]	2.54
The muscles for measuring myoelectric potential	Abductor digiti minimi muscle, Flexor carpi ulnaris muscle

[a]The minimum required force is the minimum force required to move the end-effector while keeping the speed constant.

4 Results and Discussions

4.1 Position Error Measurement in Workspace

Two points in the two-point control system is called point 1 and point 2 respectively. The positional relative errors of point 1 and point 2 were calculated from the positional coordinates measured in Sect. 3.1. Table 4 shows the comparison of the relative errors of point 1 between the new and the previous system. And Table 5 shows the comparison of the relative errors of point 2. The relative errors in the new system are improved compared with in the previous one.

From the encoder count values measured in the Sect. 3.1, the relative error of each encoder count value were calculated from the average count value of all the encoders. Table 6 shows the Relative error of each encoder.

Table 4. Relative error of point 1 of two-point control system

	x component	y component	z component
Relative error (new) [%]	10.8	7.94	4.97
Relative error (previous) [%]	−16.2	−6.50	2.65

Table 5. Relative error of point 2 of two-point control system

	x component	y component	z component
Relative error (new) [%]	1.33	3.38	3.12
Relative error (previous) [%]	5.36	11.8	12.0

Table 6. Relative error of each encoder

Encoder number	Relative error [%]
0	−0.624
1	1.56
2	−2.00
3	0.913
4	−1.12
5	−0.408
6	1.65
7	0.0250

The error of the encoders in Table 6 is not as large as shown in Table 4 and Table 5. We think the improvement in positional accuracy is due to the fact that the minimum tension is increased, and the strings are wounded around the pulleys more tightly. This cause decrease in winding-thickening. So think about the thicker winding of the pulley. This is a problem that when the string winds around the pulley, it increases the radius of the pulley. In software configuration, the pulley radius is set to

10 mm. The position was calculated using this setting. But the actual pulley radius with the most string wounded around was 18.8 mm in the measurement. Assuming that the work is done at the center of the work space and the pulley radius is simply 14.4 mm, the length of the string wounded around for each revolution of the pulley in that area is $2\pi \cdot 14.4 = 90.4$ mm. If the pulley radius is 10 mm, the relative error is -30.5%. In the future, it is necessary to make a calculation that takes this winding-thickening into account.

4.2 The Measurement of the Force Intensity

Figure 6 shows the difference of force intensity between the new circuit and the previous circuit.

The horizontal axis is the specified force intensity and the vertical axis is the mean value of actual force intensity.

In the case of the new circuit, it is clear from Fig. 6 that the force intensity was enhanced compared to the previous one.

The results of the Sect. 3.3 will be described in another report in detail.

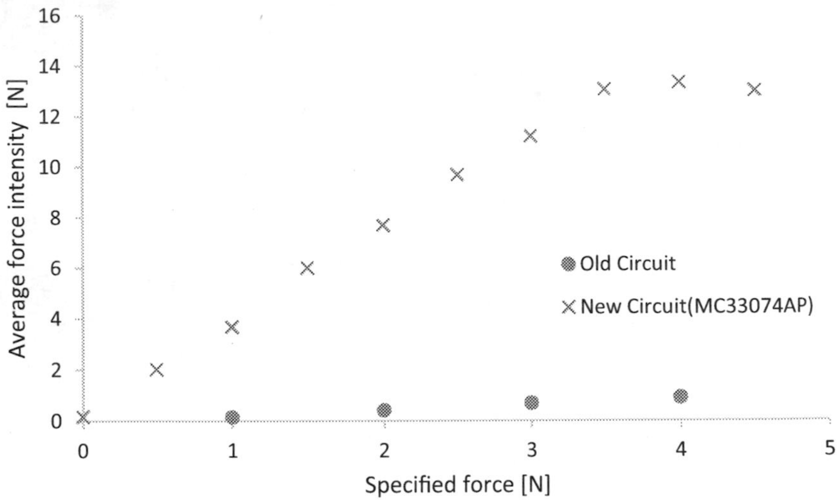

Fig. 6. The difference of force intensity between the new circuit and the previous circuit

5 Conclusions

The purpose of this paper is to establish an experimental environment for quantitative evaluation of SoE. To achieve this goal, we constructed a VR environment and conducted experiments with physiological behavioral measurements during the task.

In the previous VR environment, there is a limitation of the force sense intensity due to the control circuit. Therefore, in the previous environment of the rod tracking task, the system was not able to present sufficient force intensity, which caused the vibration of the end-effector. To solve this problem, we have improved the control circuit and the motors became to be controlled by 24 V instead of 5 V. As a result, stronger force became to be presented, but the minimum tension and the minimum required force were increased, and the end-effector became harder to move. It was found that the end-effector still vibrates in the Rod Tracking Task. The reason for this is that the combination of coefficients of the Spring-Damper Model has become more severe because the force sense has become stronger.

The positional accuracy of the end-effector was also examined. The effect of the encoder on the positional accuracy is not significant, and the main reason for this was thought to be the winding-thickening of the pulley. The winding-thickening produces a large error. To solve this problem, use thin threads, install an even winding device, or make a software-based correction. It is thought that the improvement of the positional accuracy in this study is due to the fact that the string was tightly wrapped around the pulley due to the increase of the minimum tension and the winding-thickening of the string was suppressed.

In this study, we strengthened the force and investigated the side effects of this on SPIDAR-HS. In addition, we were able to elucidate the main causes of positional accuracy.

Acknowledgments. We would like to thank all the research participants. This work was supported by JSPS KAKENHI Grant Number JP17H01782.

References

1. Tachi, S., Sato, M., Hirose, M.: Virtual Reality," The Virtual Reality Society of Japan, 1 st edn. (2011) (in Japanese)
2. Gallagher, S.: Philosophical conceptions of the self: implications for cognitive science. Trends Cogn. Sci. **4**(1), 14–21 (2000)
3. Limerick, H., Coyle, D., Moore, J.W.: The experience of agency in human computer interactions: a review. Front. Hum. Neurosci. **8**, 643 (2014)
4. Tsukikawa, R., et al.: Construction of experimental system SPIDAR-HS for designing VR guidelines based on physiological behavior measurement. In: Chen, J.Y.C., Fragomeni, G. (eds.) VAMR 2018. LNCS, vol. 10909, pp. 245–256. Springer, Cham (2018). https://doi.org/10.1007/978-3-319-91581-4_18
5. Suzuki, H., Tsukikawa, R., Kobayashi, D., Sato, M., Yamaguchi, T., Harada, T.: Implementation of two-point control system in SPIDAR-HS for the rod tracking task in virtual reality environment. In: Yamamoto, S., Mori, H. (eds.) HCII 2019. LNCS, vol. 11570, pp. 47–57. Springer, Cham (2019). https://doi.org/10.1007/978-3-030-22649-7_5
6. Maxon motor, "RE 25". https://www.maxonmotorusa.com/maxon/view/product/motor/dcmotor/re/re25/118746. Accessed 01 Mar 2019
7. Maxon motor, "HEDS 5540". https://www.maxonmotorusa.com/maxon/view/product/sensor/encoder/Optische-Encoder/ENCODERHEDS5540/110511
8. Aikoh Engineering Co., Ltd.: Sigital force gauge RZ Series. http://www.aikoh.co.jp/en/forcegauge/rz/. Accessed 12 Jun 2020

Social Computing and Social Media

The Tributes and Perils of Social Media Use Practices in Ethiopian Socio-political Landscape

Elefelious Getachew Belay[1(✉)], Getachew Hailemariam Mengesha[2], and Moges Ayele Asale[3]

[1] School of Information Technology and Scientific Computing, Addis Ababa Institute of Technology, Addis Ababa University, Addis Ababa, Ethiopia
elefelious.getachew@aau.edu.et
[2] School of Information Science, Addis Ababa University, Addis Ababa, Ethiopia
getachew.hailemariam@aau.edu.et
[3] School of Psychology, Addis Ababa University, Addis Ababa, Ethiopia
moges.ayeleA@aau.edu.et

Abstract. This paper explores various factors linked to social media use and its effect on the socio political realm of Ethiopia, which currently is a nation under political reform. Social media tools such as Facebook, Twitter and Telegram were extensively used by opposition movements that toppled the dictatorial regime that ruled the nation for over 27 years. While this is the positive side, currently ethnic extremist groups and different political factions are using social media to instigate violence and genocide that displaced millions of people across the nine Ethiopian regional states. In view of quelling the sporadically erupting ethnic violence, the Ethiopian government frequently disrupts Internet connections. This has caused untold distrust across hundreds of the diplomatic entities the nation hosts and among the business and scientific communities. The intermittently occurring political unrest as a result of fake news and hate speeches circulating through social media, has led the general public to develop negative impressions with regard to social media and Internet technologies. Further, the misuse of the technology is also urging the government to rethink its free use and is resorting to device protective measures. This study intends to explore the fortunes and caveats brought about by social media in the Ethiopian socio political life and propose mechanisms that would foster appropriate use of the technology. Data for the study was generated via focused group discussion and document review. The study revealed technical, legal, and ethical mechanisms surrounding social media use in the context of Ethiopia.

Keywords: Social media · Fake news · Hate speech · Social media use

C. Stephanidis et al. (Eds.): HCII 2020, LNCS 12427, pp. 199–209, 2020.
https://doi.org/10.1007/978-3-030-60152-2_16

1 Introduction

1.1 A Need to Study Social Media Use

The unprecedented social media engagement of the youth, political parties, and activists has led to a growing concern that social media use flare violence and hatred among the Ethiopian societies, which is currently divided along ethnic lines and political landscape. Particularly, the youth account about 70% of the Ethiopian population. Radicalizing this group along ethnic, political ideology, and religious lines tend to destabilize the nation. This study is motivated to partly address this concern of the general public, researchers, and policy makers and intends to explore the mechanism that would nurture appropriate use of the technology to foster public dialogues along the political, economic, and socio-cultural dimensions. Prior studies have detected scenarios of hate and dangerous speeches. For example, analysis of sampled Ethiopian Facebook users' data revealed that about 75% of hate speech and 58% of offensive speech have ethnic targets [1]. They further noted that about 92% dangerous statements of the sampled data were made by individuals who hide their identity. This signals the potential danger that may arise if social media spaces, like Facebook remain unmonitored. The study dwells on a contentious issue that requires striking a delicate balance between freedom of expression and national interest.

Ethiopia has the lowest Internet penetration and nearly 50% of the population has mobile connections [2]. Active Mobile Social Media Users are also less than five percent with a total number of 21.14 million internet users in Ethiopia. It has been stated that the total number of mobile connections in Ethiopia increased by 7.2 million (which is about 18%) between January 2019 and January 2020. Annual digital growth of active social media users is above 20% since 2017. Even if the country has limited internet penetration, it was paradoxical to see that before the advent of current political change in the country, Ethiopia was entangled with "Authoritarian Dilemma" which implies that the authorities of the country would like to expand the digital technology in the country yet they exercise very strict surveillance [3]. In relation to this, different agencies of the government played a great role in blocking more than 100 websites that were blamed for transmitting critical comments about the activities of the government [4].

Though this being the case in Ethiopia, nevertheless, it has been stated that these days social media sites enable users to freely share and interact with others at a very limited cost. By using these platforms people communicate their views and ideas and also they share news with one another [5]. Social Media offers a communication platform without gatekeepers, censoring, or monitoring for quality, so not only individual users can produce contents but also social groups, organizations, and parties [6]. These situations may result in the publication or sharing of information that is not validated and possibly false [7] and which might negatively affect users' thoughts and perceptions. This is to say that as there are no gatekeepers or censorship or monitoring contents on social media there is a viable and fertile ground for production and dissemination of fake news, which is a serious issue in recent days and became the focus of the people all over the world [8]. News that is disseminated through social media goes fast and reaches a large number of audience and due to various psychological mechanisms individuals fail to ascertain whether the news is fake or accurate. For instance, individuals are bombarded with huge amounts of information most of the time, and they simply accept the news as good if other people consider also it, as they process that information at a shallow level or peripherally [9–12]. In addition to the psychological mechanisms mentioned, other mechanisms include being repeatedly exposed to a piece of information or news or familiarity contribute for acceptance of that news as accurate [13] and also fake news often touches people's emotions and not their cognitive realm and thus individuals do not worry much in search of evidences that justify the validity of the information that they have got [14].

1.2 Social Media and Politics

Social media has been used in Ethiopia to promote digital activism especially during the times of elections [1]. This is mainly because the conventional media came under the strict control of the state and people have been forced to use social media as a viable alternative to express their anger and oppositions [15]. In Ethiopia, social media, specifically Facebook was served as a more preferable and favored source of information more than conventional sources of information such as radio, TV and newspapers. Besides, people use social media particularly Facebook to satisfy their various needs such as the need for entertainment, socialization, discussion or debate, and education. Moreover, most of the users of Facebook use it to express their high or low emotions at times when they experience these emotions [16]. Leaving aside the various advantage of social media in politics, it has also several negative sides. One serious dark side of these platforms is that people may share hate speeches that may polarize social groups and create social unrest among social groups as individuals come to use offensive and hateful words in 'us' versus 'them' way of exchanging views and ideas [5]. In nations like Ethiopia with weak governance and legal systems and strong social norm. Social medias are being used to disseminate hate and dangerous speech, attacking individuals or groups based on their ethnicity, and sharing antagonistic views against a given group (e.g. a religious group) [1]. Of course, the magnitude with which hate and dangerous speech communicated via social media in Ethiopia was limited to 0.4% and 0.3% respectively [1]. Yet, dangerous speech was committed more intentionally and targeted a given particular individual or group and almost exclusively (92%) carried out by those individuals who remain anonymous in their communication and essentially the contents of dangerous messages almost entirely focusing on ethnicity [1]. Ethnicity was at the core as the major target for other forms of inflammatory speech like hate speech (75%) and offensive speech (58%) [1]. Among social groups we often observe social segments holding their own beliefs, opinions or views, which might be reflected in their religious beliefs, political views, and the like that create the division that is often referred to as we and they. There could be polarization between these groups and one study showed that frequency of social media use particularly frequency of Facebook use was strongly associated with perception of polarization [17]. In this connection it has been reported that about 46.2% the research participants viewed that contents disseminated on Facebook may somehow or surely instigate ethnic hatred among various ethnic groups of Ethiopia [18]. In countries like Ethiopia hate speech may revolve around ethnic and religious spheres whereas in western countries such as US, UK and Canada it has been found that hate speech revolve around people's behavior and their physical characteristics largely. Of course this does not mean in these countries we do not have hate speech revolving around racial, and other minority features. For instance, in a study it was reported that in the US hate speech targeting black people accounted for 11% and the figure for white people was found to be 5% [5].

We may ask why people on social media engage in producing, storing, disseminating or using big data related to hate speech or inflammatory content that may evoke social unrest, violence or conflict among social groups. In this connection the psychologist Albert Bandura has provided us with a notion which he termed moral disengagement. According to this view, people exercise control over their thinking and behavior which is mediated through various constructs that may make people rationalize and morally disengage themselves from committing certain unethical actions on social media. These variables that were studied in Bandura's work include blaming others and not oneself on some morally challenging issues (attribution of blame), considering the targets of hate speech as not belonging to the same group (dehumanization), not assuming accountability on what one performs by disregarding or distorting the consequences, or diffusion of responsibility, or displacement of responsibility and thereby make oneself not responsible for performing harmful acts, finally through advantageous comparison, euphemistic language, and moral justification one may misinterpret the consequences of harmful acts as morally acceptable. According to Albert Bandura the issues presented here illustrate principal mechanisms through which individuals can disengage themselves from harmful actions [18]. Empirical study based on this Bandura's proposition derived support for the notion though further study is needed to uncover other situational and personality variables that contribute for engagement in unethical behaviors in relation to big data usage on social media [19].

2 Theoretical Foundation

In recognition of the multifaceted issues surrounding (social) media use, the theoretical foundation of this study is drawn from various policy documents and empirical research publications. Key constructs have been extracted from UNESCO [20], Gagliardone et al. [25], FDRE [21], and FDRE [22] to put together the theoretical framework indicated in Fig. 1.

The framework presented in Fig. 1 includes eight constructs from media and information literacy (such as locating and accessing information, assessing content and credibility, media engagement, communication skill, critical content analysis, content generation, content extraction and organization, content synthesis and understanding). The other composite constructs included are media policy, information access policy and ICTs policy. The inclusion of these constructs believed to regulate media use scenarios without compromising freedom of expression and national stability.

The media and information literacy dimension of the framework lays due emphasis on the need to enhance the capacity of citizens for accessing and locating relevant content, securitizing the credibility of the content before deciding to consume and react to it. Developing sound communication and media engagement skills are central to engage in meaningful dialogue and effective communication. Because of the convergence of the mainstream mass media and social media, quite often mainstream journalists are observed citing social media posts as evidence for their news. On the other hand, people tend to be curious when they engage with the mainstream media as they feel responsible for what they are uttering. This is not widely observed in social media use. After the censorship requirements have been lifted, mainstream media outlets are now supposed to adhere to self-censorship schemes they ought to put in place. However, with their engagement with the social media as well as with the mainstream media, some public figures seem not to abide by communication protocols, which undoubtedly reveals a deficiency in media literacy. As social media is an interactive environment, users are required to be competent in terms of generating contents they would share with other audiences, which presupposes skill in content extraction, analysis, and synthesis. UNESCO 2013 as well as the Ethiopian constitution derivative by laws recognize freedom of expression as a basic human right and at the same time reflect concerns to the right and dignity of other parties and the stability of the nation. These issues are properly pronounced in proclamation 560/2008 and proclamation 1158/2020. The theoretical framework included ICTs as a core enabling component of social media. Notwithstanding its enabling capabilities, the technology has also a deterrence feature that blocks unwanted contents from reaching the audience. Policy surrounding appropriate use of ICT helps to safeguard the right of individuals from receiving unwanted contents without violating individuals' rights of expression.

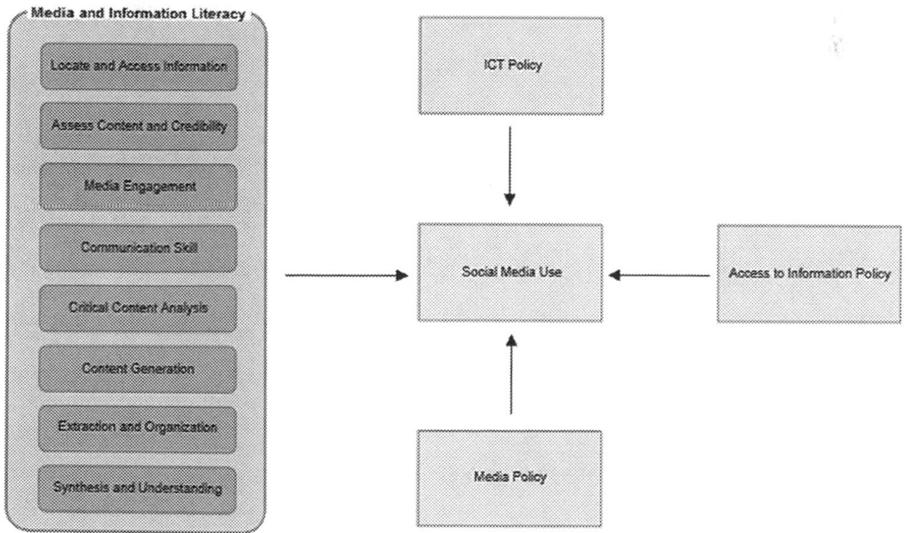

Fig. 1. Theoretical framework

3 Methodology

The study intends to gain a deeper understanding of the social media use practices along with the positive and negative effects in the socio economic and political life in Ethiopia. In doing so, the study endeavors to identify mechanisms that would promote sound use of media and that would help to instill a culture of dialogue and tolerance. To this end, focused group discussion and document analysis have been used to generate the study data. Domain-experts involved a half day focus group discussion was conducted on March 14, 2020 in the form of moderated workshop and oral presentations. Eight domain experts drawn from different disciplines and practitioners: journalist, lecturers from school of journalism, data scientist, social psychologist and IT professionals were involved in the workshop session. Table 1 presents the profiles of the focus group discussion participants. The entire session was recorded and later the recording was transcribed to generate the study data. Simple open coding was carried out and key concepts were identified and major themes emerged. In addition, policy documents such as media proclamation mass media law of 590/2008, the international UNESCO media policy, hate speeches and anti-terrorist proclamations were reviewed and augmented the identified themes.

Table 1. Participants profile.

Description	Educational level	Number
Journalist	MA	1
Psychologists	PhD/MA	2
Data scientist	MSc.	1
Information technology	PhD	2
Journalism lecturers	MA. and Assist Prof	2

4 Results

The findings of the study are mapped along the dimensions of the theoretical framework such as, media use scenarios, right to access information, media engagement and communication skills, and media policy, critical content analysis and scrutiny skills, and absence of fact checking platform and technology solutions.

4.1 Media Use

The Ethiopian young generation as in the case in the other part of the world, has strong attachment to digital technology tools and platforms. According to the world social media status report published in May 2020, Facebook is the most predominantly used social media platform. Table 2 shows a set of social media widely used in Ethiopia[1].

[1] https://gs.statcounter.com/social-media-stats/all/ethiopia

Table 2. Social Media Stats in Ethiopia – May 2020.

Social media site	Percentage
Facebook	53.99
Pinterest	19.11
Twitter	7.04
Youtube	17.73
Instagram	1.23

The extensive use of social media including SMS messages significantly contributed to the 2005 national election, which was the first Ethiopian election that had followed a fair democratic process. Following the vote rigging dispute occurred between the government and opposition party that culminated with brutal suppression measures, the conventional mass media fell under strict control of the government. The general public totally shifted to emerging social media such as Facebook, which later tend to be considered a credible source of information. Later, opposition party members and digital activists extensively used social media as a weapon to promote fierce protests against the government. After the 2005 election, the EPDRF government blocked about 260 critical websites, which are unblocked by the current reformist government. Although, this is a step forward in terms of liberalizing the media, the nation is now wrestling with the "dark" side of the social media as its improper use by radical political factions that disseminates dangerous messages is igniting devastating violence at various occasions. During the FGD it has been transpired that some armed political factions that currently operate in some pocket areas in the western part of the country, use Facebook to post lists of individuals and local government bodies that they would execute.

Polarization of the Federal Government and Regional State mainstream media is raised by the discussants of the FGD as a serious issue that needs timely attention. This incidence is causing further polarization in the social media as well. Without compromising the autonomy of the regional states and freedom of expression, reducing the antagonistic sentiments of these two government organs is critical for building a strong nation state.

After the EPDRF regime was toppled down, government organizations opened Facebook pages and began to interact with the public. Participants of the FGD noted a lack of principle and protocol among the government organs in the course of information dissemination. One of the participants made this statement, for instance if we mention the most recent incidence, "Who is the right person to disclose the first incidence of the corona virus reported to the public on March 13, 2020. Is it the Mayor of Addis Ababa or the Ministry of health?". The Mayor's swift announcement of this incident through his social media page is a good indicator of violation of one's mandate and inappropriate use of media.

4.2 Right to Access Information

The media proclamation 560/2008 clearly stipulates the right of individuals to access information held by public bodies and freedom of expression. This proclamation has not been enacted until the EPRDF regime is ousted. The current reformist government relaxed the restrictions imposed on the social and conventional mainstream media. This has undoubtedly boosted freedom of speech while clouding the issue of national interest and national security. Although the right to access public information is ensured, the long held bureaucratic culture has become a serious hurdle on information seekers. In addition, when censorship requirements are lifted, mechanisms for validating the quality of information have totally been removed. Absence of credible media sources like open public data sources and independent fact checkers or fact checking mechanisms has caused serious setbacks in media use and regulations. In a nation where a strict command and control system is enacted for a longer time, the current relative freedom has been perceived wrongly. The culture of assuming full responsibility for one's speech and the idea of self-regulation and self-censorship has not yet nurtured and developed. In relation to the misconception surrounding freedom of speech, participant of the FGD made the following remark.

"The notion of "Freedom of speech" is not clearly defined and delimited. This has led to untold dilemma between national interest and freedom of speech".

4.3 Media and Information Policy

With the intent to streamline media related affairs, the Federal Government has established an agency labeled "Ethiopian Broadcasting Authority" that is directly accountable to the parliament. The agency is responsible to regulate the media sector through licensing, standardization, and through monitoring the performance of private, governmental media as well as the social media. In addition, Media Proclamation 560/2008 is also expected as a potential policy instrument. However, the agency has not yet developed the required capacity to discharge its responsibility. Further, although the cited Proclamation 560/2008 is in place, it is already outdated. The proclamation's terms and conditions were formulated before the social media flourished. The regime change made has already eroded the fundamental assumptions behind the proclamation. Therefore, currently the absence of clear national media and communication policy coupled with lack of cross-border media legislation has put the nation at risk of frequent violence. Key government officials and leading opposition party leaders as well as key public figures were widely observed abusing the media. This shows absence of information policy at institutional level. Although some initiatives are undergoing, so far public and private media organizations lack clear institutional information policy. With regard to paucity of policy, one of the FGD participants stated the associated issues as follows "lack of regulatory instruments has led some individuals to mix activism and journalism, blatantly spreading hate messages through mainstream and mass media. Furthermore there is a confusion surrounding media regulation vs media monitoring and control". Scenarios of misinformation and disinformation are becoming a norm both in mainstream and social media.

4.4 Media Engagement and Communication Skills

Grand problems have been noted in relation media engagement and communication skills at all levels. This problem is deep rooted as educated individuals across the board found to have deficiency in media engagement and communication skills. This calls for total system overhaul right away starting from the early educational system. In this regard the participant of the FGD stated that "to address this hidden skill deficiency, it is mandatory to introduce media literacy courses as a subject in the school system". Even, some professional journalists that work.

In terms of ease of access and currency, social media is leading the conventional mainstream media in Ethiopian setting. Quite often the news being broadcasted by the mainstream media is found to be the one already posted in the social media. Although the mainstream and the social media are at the disposal of government officials, they fail to strategically engage the media to provide to the public swiftly the information that matters most. Opposition groups take advantage of this problem of transparency, attempt to feel the gap through hearsay and speculative information not supported by concrete evidence. With a bid to counter fake information that occasionally ignites violence, the government pursues a desperate total Internet shut down that flare further contentions.

4.5 Technology Platforms and Solutions

The social media platforms such as Facebook and Twitter are the most widely used by the government officials as well as by the general public. These platforms are owned by overseas companies. They abide by the legislation in their home country. Unless, these companies abide by their own company code of conduct, as they are for profit organizations they give precedence to their business benefit. The absence of cross-border legislations related to the governance of the Internet, harming emerging nations that have not yet developed the capacity to monitor adverse incidence in the Internet based platforms.

During the FGD participants disclosed that "The Ethiopian Broadcasting Authority has been striving to capture contents of leading mainstream media and store it for a limited period. Due to lack of repository and classification technology solutions, the authority is incapable of monitoring media contents as intended".

5 Discussion

This study has uncovered various issues related to the positive and dark side of social media in the Ethiopian socio-political landscape. The outcome of this study is in par with similar studies conducted in advanced countries. Social media implicitly shaped the Brexit referendum and the election of Donald Trump [23, 24]. For instance, "Euro Leave" supporters were more active in tweeting than their counterpart "Remain" twitters. Similarly, other researchers have come up with identical findings "Twitter bots coupled with the fragmentation of social media and the role of sentiment, could contribute to the vote outcomes" [25]. Social bots (computer algorithms) are becoming prominent in shaping people's opinion through creating a bit of information and a sense of consensus in the society in favor of a given candidate [23, 24]. These algorithms are believed to marginally contribute to the outcome of Brexit and the 2016 U.S. Presidential election. The outcome of this study is a good signal to policy makers to rethink how proper use bots is ensured. Extensive use of Facebook to call for widespread protests against the dictatorial minority EPDRF government and the eventual fall of the regime is fully in line with the influence of Twitter in Brexist and 2016 U.S. Presidential election.

6 Conclusion

This study reports the preliminary outcome of a three years thematic research project that spans 2019 to 2021. The research project intends to explore the impact of social media from legal, political, psychosocial, and technology dimensions. The theoretical framework presented in this paper is supposed to capture key issues surrounding mechanisms that fosters appropriate use of social media. The outcome achieved so far attests significant contribution of social media for transforming Ethiopian socio economic and political life. As media is a two-edged sword, gray areas witnessed along media use scenarios and absence of sound regulatory mechanisms is nullifying the positive outcomes realized by the social media.

Acknowledgments. This Thematic Research Work is funded by Addis Ababa University Research and Technology Transfer Office.

References

1. Gagliardone, I., Pohjonen, M., Beyene, Z., Zerai, A., Aynekulu, G., Bekalu, M., Bright, J., Moges, M., Seifu, M., Stremlau, N., Taflan, P.: Mechachal: online debates and elections in Ethiopia-from hate speech to engagement in social media. Available at SSRN 2831369 (2016)
2. DATAREPORTAL, https://datareportal.com/reports/digital-2020-ethiopia. Accessed 25 May 2020
3. Grinberg, D.: Chilling developments: digital access, surveillance, and the authoritarian dilemma in Ethiopia. Surveill. Soc. **15**(3/4), 432–438 (2017)
4. Freedom House, "Freedom on the Net 2016: Ethiopia." https://freedomhouse.org/sites/default/files/FOTN%202016%20Ethiopia.pdf. Accessed 28 May 2020
5. Mondal, M., Silva, L.A. Benevenuto, F.: A measurement study of hate speech in social media. In: Proceedings of the 28th ACM Conference on Hypertext and Social Media, pp. 85–94 (2017)
6. Jong, W., Dückers, M.L.: Self-correcting mechanisms and echo-effects in social media: an analysis of the "gunman in the newsroom" crisis. Comput. Hum. Behav. **59**, 334–341 (2016)
7. Berghel, H.: Lies, damn lies, and fake news. Computer **50**(2), 80–85 (2017)
8. Vargo, C.J., Guo, L., Amazeen, M.A.: The agenda-setting power of fake news: a big data analysis of the online media landscape from 2014 to 2016. New media Soc. **20**(5), 2028–2049 (2018)

9. Lang, A.: The limited capacity model of mediated message processing. J. Commun. **50**(1), 46–70 (2000)
10. Nyhan, B., Reifler, J.: Displacing misinformation about events: an experimental test of causal corrections. J. Exp. Polit. Sci. **2**(1), 81–93 (2015)
11. Metzger, M.J., Flanagin, A.J., Medders, R.B.: Social and heuristic approaches to credibility evaluation online. J. Commun. **60**(3), 413–439 (2010)
12. Sundar, S.S.: The MAIN model: a heuristic approach to understanding technology effects on credibility, pp. 73–100. MacArthur Foundation Digital Media and Learning Initiative (2008)
13. Pennycook, G., Cannon, T.D., Rand, D.G.: Prior exposure increases perceived accuracy of fake news. J. Exp. Psychol. Gen. **147**(12), 1865 (2018)
14. Meinert, J., Mirbabaie, M., Dungs, S., Aker, A.: Is it really fake?–Towards an understanding of fake news in social media communication. In: International Conference on Social Computing and Social Media, pp. 484–497 (2018)
15. Dugo, H.: The powers and limits of new media appropriation in authoritarian contexts: a comparative case study of oromo protests in Ethiopia. Africology J. Pan African Stud. **10**(10), 48–69 (2017)
16. Kumlachew, S.S.: Challenges and opportunities of Facebook as a media platform in Ethiopia. J. Media Commun. Stud. **16**, 99–110 (2014)
17. Valdez, A.C., Burbach, L., Ziefle, M.: Political opinions of us and them and the influence of digital media usage. In: International Conference on Social Computing and Social Media pp. 189–202 (2018)
18. Bandura, A.: Moral disengagement in the perpetration of inhumanities. Personal. Soc. Psychol. Rev. **3**(3), 193–209 (2019)
19. Beckmann, M., Scheiner, C.W., Zeyen, A.: Moral disengagement in social media generated big data. In: Meiselwitz, G. (ed.) SCSM 2018. LNCS, vol. 10913, pp. 417–430. Springer, Cham (2018). https://doi.org/10.1007/978-3-319-91521-0_30
20. UNESCO.: Media and Information Literacy Policy and Strategy Guidelines. 7, place de Fontenoy, 75352 Paris 07 SP, France Communication and Information Sector (2013)
21. FDRE.: Proclamation No. 590/2008. Freedom of the Mass Media and Access to Information. Proclamation Page 4322, (2008)
22. FDRE.: Proclamation No 1185/2020. Hate Speech and Disinformation Prevention and Suppression Proclamation Page 12339, (2020)
23. Hänska, M., Bauchowitz, S.: Tweeting for Brexit: how social media influenced the referendum (2017)
24. Gorodnichenko, Y., Pham, T., Talavera, O.: Social media, sentiment and public opinions: evidence from# Brexit and# USElection (No. w24631). National Bureau of Economic Research (2018)
25. Gagliardone, I., Pohjonen, M.: Engaging in polarized society: social media and political discourse in Ethiopia. In: Digital activism in the Social Media Era, pp. 25–44 (2016)

Towards Curtailing Infodemic in the Era of COVID-19: A Contextualized Solution for Ethiopia

Elefelious Getachew Belay[1(✉)], Melkamu Beyene[2], Teshome Alemu[2], Amanuel Negash[1], Tibebe Beshah Tesema[2], Aminu Mohammed[2], Mengistu Yilma[3], Berhan Tassew[3], and Solomon Mekonnen[2]

[1] School of Information Technology and Scientific Computing, Addis Ababa Institute of Technology, Addis Ababa University, Addis Ababa, Ethiopia
{elefelious.getachew, amanuel.negash}@aau.edu.et

[2] School of Information Science, Addis Ababa University, Addis Ababa, Ethiopia
{melkamu.beyene, teshome.alemu, tibebe.beshah, aminu.mohammed, solomon.mekonnen}@aau.edu.et

[3] College of Health Science, Addis Ababa University, Addis Ababa, Ethiopia
{mengistu.yilma, berhan.tassew}@aau.edu.et

Abstract. This paper focuses on infodemic response using semi-automated application, seeking to curtail the misinformation of COVID-19 related news and support reliable information dissemination in Ethiopia. We analyze the emerging news trend about COVID-19 in selected social media sites (Facebook and Twitter) and Language (Amharic/English) using the information extraction tool that we developed. This web-crawling tool extracts posts and tweets that have larger audience, high engagement and reaction in Ethiopian popular social media pages and profiles. Expert fact-checkers (group of three to five experts) are then used to verify the veracity and correctness of each of the selected posts/tweets. Posts and tweets are selected based on the keyword patterns emerged from the analysis. The system will present a dashboard to the experts with the required information to label the news as misinformation and educative (opted two broad categories) decided at this stage. The verified news and information will be pushed to various social-media sites, conventional media and to our COVID-19 related information dissemination website. This will provide counter-information with better evidence and proactively flag misinformation and disinformation, and furthermore convey accurate and timely information as educative. This can be achieved through three phase (problem identification, solution design and evaluation) design science approach by emphasizing the connection between knowledge and practice.

Keywords: Fake news · Social media use · Infodemic · Misinformation

C. Stephanidis et al. (Eds.): HCII 2020, LNCS 12427, pp. 210–221, 2020.
https://doi.org/10.1007/978-3-030-60152-2_17

1 Introduction

With the spread of COVID-19 pandemic all around the world, people have been in a constant state of stress, fear, panic and disillusion in the last couple of months. It has been given a lot of information and misinformation by individuals, experts, social media and mainstream media outlets and furthermore. These misinformation ranges from simple alteration of facts to a large conspiracy theory. Generally, there has been a massive infodemic that will undermined and disrupted national and global efforts to fight against the pandemic [1, 2].

Organization like WHO [2] primarily understands the magnitude of the repercussion that disinformation ultimately brought, and provides continuous information for governments and individuals on various media outlets. However, individuals consumed an overwhelming volume of information and unverified news from different sources, which includes COVID-19 hashtags in social media. This un-verified news and media headlines has inscribed fear and prejudice in the society.

As many scientists and healthcare professionals all over the world working towards understanding the characteristics of the virus, the transmission factors, the origin, and the vaccine. On the other side, individuals and various segment of the society overloaded by a large number of unverified news and information that are being produced every day and every hours. This phenomena of COVID-19 misinformation is being officially declared as 'infodemic' [2–4]. The WHO DG also bluntly mentioned "We are not just fighting epidemic, we are fighting an Infodemic". Infodemic is defined an excessive amount of information concerning a problem such that the solution is made more difficult. The information spreading through the social media channels also influences the people behavior and impact the response set forth by the government [1]. This become an immense threat to social, political and economic sphere of any country. The situation even be worst in developing economy like Ethiopia that has no fact-checking organization and weaker information flow policy.

Social media has become a major source of information for urban areas of Ethiopia following the increased penetration rate of mobile subscription and Internet [5]. Even though the country has less number of users compared to other counties, the social media information has been diffused faster through word of mouth communication. This has been shown in the last couple of years especially in the political arena of the country [6]. Social media sites (such as Facebook), microblogging services (such as twitter) and video hosting service (such as YouTube) vividly upsurge in providing various political, economic and social agendas for the society [5, 6]. Generally, social media has served as a more preferable and favored source information than the conventional media, it is then apparent that most of the COVID-19 discussions are based on those information emerged from this platforms.

Statements and/or words in news media and social discourse have a huge role and influence in shaping people's beliefs and opinions [7]. Each social media posts are accompanied by popularity ratings, and when someone post gets many likes, shares, or comments, then it is more likely to receive more attention by others [8]. This will then facilitate more visibility of the news and the author (or originator of the news). This popularity on social media leads to a self-fulfilling cycle, which in turn enable spreading of unverified news and information.

In recent years, fake news has captured much of the global and national attention, commonly in political context, but also in health, and stock values [7, 9]. Identifying and addressing fake news on social media poses several new and challenging research problems [10]. Certainly, fake news is not a new phenomenon but social media makes fake news more powerful and challenge the traditional journalism norm [10]. Among the many effort towards alleviating this issue; one is introducing fact-checker, and there has been a number of fact-checking organization established and tripled since 2014 [7]. However, this organizations are resides in the west and focused on their contextual issues, countries like Ethiopia differ in language and social norm.

In line with the fake news and regards to COVID-19, it is also important to pinpoint and surface the most important (critical and timely) educative news and information. This will help to drown the bad information and also act as a counter-information that provide better evidence and from the reliable source.

Thus, it is paramount to systematically address the disinformation and misinformation effect of COVID-19 and how to counteract false claims and provide persuasive information.

2 Related Works

With the proliferation of mobile device and Internet, social media has got much attention and became critical part of our information ecosystem [11]. It provides immense opportunities for various level of users, organizations, government. People across the whole world shared a large amount of information regardless of the time and location. This will set a stage to produce and disseminate a misinformation, conspiracy theory and myths to the society [12]. Information has a central and critical role at the time of national and global crisis. People should get the right information at the right time, like to be aware of the situation and respond accordingly [2, 13].

Misinformation and COVID-19: Misinformation is not new in the media land-scape, it has been there since the introduction of the earliest writing systems [14]. There are various terms that are interchangeably used (that includes misinformation, disinformation, Fake news, rumor and troll post) to define information that lacks veracity and accuracy. The term misinformation is linked to the motivation of falsehood, on the other hand disinformation assumes that inaccuracy of the information from deliberate intension, and rumor is for unconfirmed information [13]. A more detailed account of the conceptual difference among those terms are discussed in [13]. There are a number of research conducted related to COVID-19 misinformation in social media, for instance Kouzy et al. [15] studied the COVID-19 misinformation on Twitter and among the 673 tweets 24.8% included misinformation and 17.4% included unverifiable information regarding the COVID-19 Pandemic. The study also shows the rate of misinformation was higher among informal individual and group accounts and unverified accounts. It revealed prominent public figures continue to play an outsized role in spreading misinformation about COVID-19. Brennen et al. [16] also studied about the types, sources, and various claims of COVID-19 misinformation, and reported small percentage of the individual pieces of misinformation come from prominent politician, celebrities, and other public figures. Pulido et al. [17] work also aims on the type of tweets that circulated on Twitter around the COVID-19 outbreak for two days, in order to analyze how false and true information was shared.

There are also works related to the sources of misinformation to guarantee the truthiness of content sharing in social media. Nguyen et al. [18] work on multiple sources spreading misinformation at the same time. Efforts like [19] are worked on single source spreading misinformation. Comin and Costa [22] run multiple experiments to evaluate and compare degree, betweenness, closeness, and eigenvector centrality in identifying the sources of the misinformation. Shah and Zaman [23] presented a new centrality measure, named rumor centrality, and revealed that it outperforms all the previously considered centrality measures.

There are various strategies and platforms to fight against the misinformation and fake news, these includes drown the inaccurate information with accurate one, impose a suitable laws to regulate and monitor information dissemination, counter-information with evidence, flagging misinformation by the news provider and fact checker, media literacy, extract and prioritizing information from credible source, and promote free advertising for direct source [12, 20].

Fact-checking is the act of analyzing the given information to verify its veracity, correctness, and accuracy. Sometimes it also involves augmenting comments, nuances and link to reputable sources [7]. There are many fact-checkering website particularly in political context, these includes Politifact[1], FactCheck[2]. Organizations like International Fact-Checking Network (IFCN)[3] worked together with more than 100 fact-checkers around the globe (70 countries) in disseminating and translating facts (articles published around 40 languages) related to COVID-19 or coronavirus.

In fact-checking humans have a central role and their efforts has not yet been entirely automatized [10, 21]. Research on various issues of fact-checking and data journalism has been undergone that includes rating (classifying statements as false and true), approach (entailment from knowledge bases [10], linguistic approach [7, 24]), and others. In countries like Ethiopia that has unique and many languages with no established fact-checker institutes, the problem of addressing misinformation and its impact will be exacerbated. Thus, it is paramount to work on such issue at such critical time.

3 Methodology

Our goal in the general context of the research is to develop a contextualized COVID-19 ETH (Ethiopia) infodemic management framework that leverage reliable information dissemination. This component of the work involves a use of a design science

[1] http://www.politifact.org.

[2] http://www.factcheck.org.

[3] http://www.poynter.org/ifcn-covid-19-misinformation/.

(DS) research paradigm that contrive an artifact (a tool) for counter-information and persuasiveness.

Accordingly we have adopted Offerman's [25] three phase approach as indicated in the Fig. 1 below. The three phases are Problem identification, solution design and evaluation.

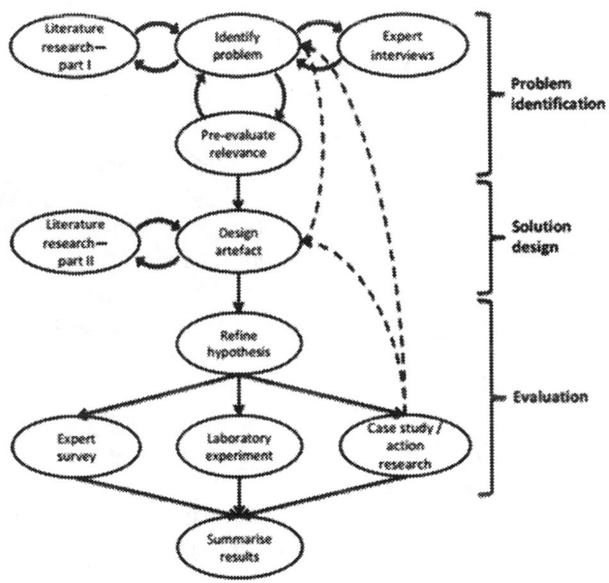

Fig. 1. Research process (Adapted from Philipp Offermann et al., 2019)

Problem Identification: In this phase of the research process, we identified and attest the problem has practical relevance for the context at hand and beyond. The relevance has been pointed out in various literatures that we reviewed, and in a recent WHO ad-hoc consultation workshop and expert discussion. Furthermore, we also reviewed and evaluated the local contextual factors and criteria of the problem relevance with Experts from media and health care professionals.

Solution Design: This phase has includes two basic tasks: 1) tool development and 2) supporting with literature. During the development various component of the artefact iteratively and meticulously experimented and compared with the relevant related work to ensure its rigor.

Evaluation: At this stage, our evaluation has been made using expert evaluations. We roll out the system and measure the applicability and use of the artifact and complied the results.

4 Experiment and Discussions

4.1 Description of the General Architecture of the Proposed Solution

This particular research has various experimental components that validate the rigor of the research and the designed artifact. Figure 2 shows components of the work. Various components and values have been analyzed and experimented to decide the best and sound values.

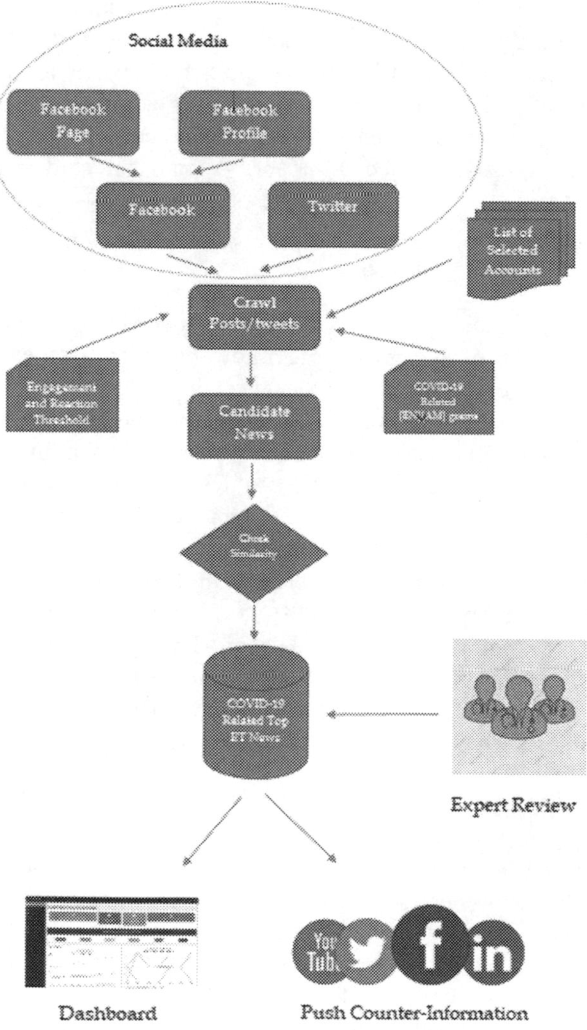

Fig. 2. General architecture of the proposed system

The research starts by tracing the origin and/or surrogate of the disinformation. In doing so, we tried to experiment two approach: the first one, we perform a desk research to select pages that have large number of Ethiopian followers (the selected pages are categorized as community, entertainment and media pages), and the second one, we used a snowballing techniques to identify the most influential and engaging pages. These results were stored in a file. From these selected page we extract posts/tweets every 24 h to get a news that received high (threshold were set by experiment) engagement and reaction. Then, the list of candidate news has emerged and recorded. We then tried to identify posts/tweets that have certain levels of similarity by implementing cosine similarly algorithm to circumvent redundancy. Once these news with high engagement and reaction stored, the system will present the posts/tweets to a panel of three to five health and media experts for verification (fact-checking). We opt this due to the limited availability of fact-checker in local languages. After review, the posts/tweets tagged as fake-news and educative, and were presented to our dashboard and also pushed to popular social and conventional media for information diffusion.

4.2 Tracing the Impact and Origin

Exploring for misinformation requires a systematic approach to get and to understand where the origin of the misinformation, and how much engagement and reaction it received. We focused on two approach that we believe helpful and plausible for the context under research. The first approach is examining and analyzing the two social media sites based on the number of audience using desk research. In doing so, we have identified several categories of pages and profiles that includes brands, celebrities, communities, entertainments, medias, places, societies and sports[4]. Among, the different categories community, entertainments and media are found to be an appealing outlets to misinformation. Others like celebrity and brands are less likely or unlikely to post such kind information. We identify and extract pages that have larger audience/fans both from Facebook and Twitter. To extract profile page, we used another techniques to address which we called snowballing technique, by chaining referrals, initially, we select the most prominent activist that have larger audience and recruit future subjects from among their acquaintances.

4.3 Extracting Posts/Tweets

Posts and tweets can be extracted from web in many ways, but the best one is using API's. Many social media sites like twitter, Facebook provide API's to pull their data in a structured manner. However, recently Facebook has released a new data policy that limit access even to the public post. So, in this case we need to use web scrapping, which involve search a page, inspect the structure, code the logic and store the result.

In our case, to analyze and extract the different posts and tweets across different origins of the news, we develop a web-crawling tool designed to effectively crawl and

[4] http://www.socialbakers.com.

extract post and tweets from the pages and profiles. For the Facebook pages, we have reviewed and tested different open source web scrapping tools, such as Facebook-Scraper[5], Facebook-Profile-Scraper[6], Ultimate-Facebook-Scraper[7] and Facebook-Profile-Scraper[8]. These third party tools are designed based on Beautiful Soup, a known Python library for extracting data from HTML/XML file. We have customized our program to scrap and extract the post from Facebook page and profile, based on the existing open source tools. For twitter we used the twitter APIs to capture the data. Twitter APIs is rich and flexible to pull twitter in many ways we want. For our purpose, we have requested the API to extract all tweets in Amharic and English language released within the last 24 h.

Every 24 h the system automatically run and check for posts/tweets that have high engagement and reaction in a form of number of likes and shares. Thresholds has been set, and every day posts/tweets will be checked whether if it surpass that threshold number or not. If it exceed the threshold then it will be retrieved and stored in our database. This number has been configurable depending on the conditions.

We extract posts that has keywords of the following. Table 1 displays the keywords that are used to extract posts/tweets.

Table 1. Keywords used in extracting posts/tweets

Uni-grams	Bi-grams	Tri-grams
Coronavirus ኮረናቫይረስ	Covid 19 ኮቪድ 19	Test Positive Coronavirus የኮረናቫይረስ በሽታ የተገኘበት
COVID ኮቪድ	Coronavirus Case የኮረናቫይረስ ታማሚ	Confirm Coronavirus case የተረጋገጠ የኮረናቫይረስ በሽታ
Case በሽታ	Confirmed Case. የተረጋገጠ በሽታ	Positive Covid 19 ኮቪድ 19 የተገኘባቸው
Test ምርመራ	Case Coronavirus የኮረናቫይረስ በሽታ	Test Postive Covid ኮቪድ በሽታ የተገኘበት
Pandemic ወረርሽኝ	Coronavirus Pandemic የኮረናቫይረስ ወረርሽኝ	Covid 19 Pandemic ኮቪድ 19 ወረርሽኝ
Health ጤና	Spread Coronavirus የኮረናቫይረስ ሥርጭት	Spread Covid 19 ኮቪድ 19 ሥርጭት

4.4 Identifying Similar Posts/Tweets

There are various reasons why we need to identify similar posts and tweets, these includes people may copy and paste similar contents, and also paraphrase the same news and information. As research also shows [13] false rumors (misinformation) tend to come back multiple times after the initial publication, facts do not. Thus, we applied

[5] https://pypi.org/project/facebook-scraper/.

[6] https://github.com/mhluska/facebook-profile-scraper.

[7] https://github.com/harismuneer/Ultimate-Facebook-Scraper.

[8] https://python.gotrained.com/scraping-facebook-posts-comments/.

cosine similarly algorithm to check the similitude of the any two post or tweets by certain margin.

4.5 Expert Fact-Checking

Twitter and Facebook has started displaying fact-checking labels, for instance labels on tweets that falsely claims 5G connectivity to the spread Covid-19 [26]. Even though there are efforts, however there is no automated system to date to check facts automatically, it is always essential to have and use humans to check facts [10, 21]. We used media and health care professional for fact-checking, to avoid biases of a single fact-checker, we used a panel of three to five experts to vote and rate the veracity and correctness of each of the selected posts for review. The following figure (Fig. 3) shows a screen shoot of some selected tweets for fact checking.

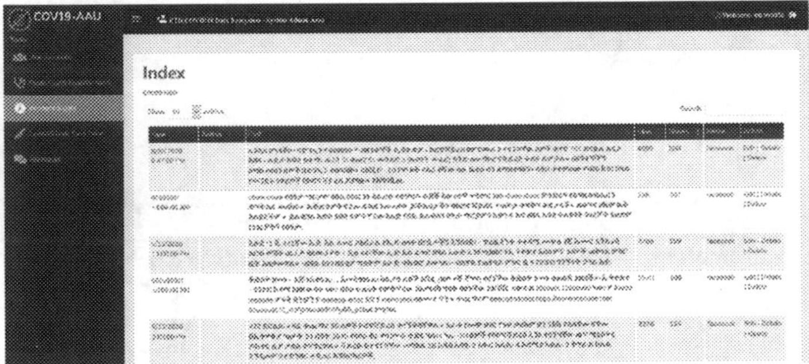

Fig. 3. Expert review dashboard

4.6 Presenting the Information

The results of all the process and the fact-checking were presented in different ways. These information are categorized into two group educative and misinformation. Educative posts/tweets was much engaging due to its educative nature. Misinformation are extracts of tweets and posts with a purpose of misleading readers. It will be presented in our dashboard and also push out in various media outlets for counter information that has been circulated around. The following figure shows educative and misinformation posts and tweets (Figs. 4 and 5).

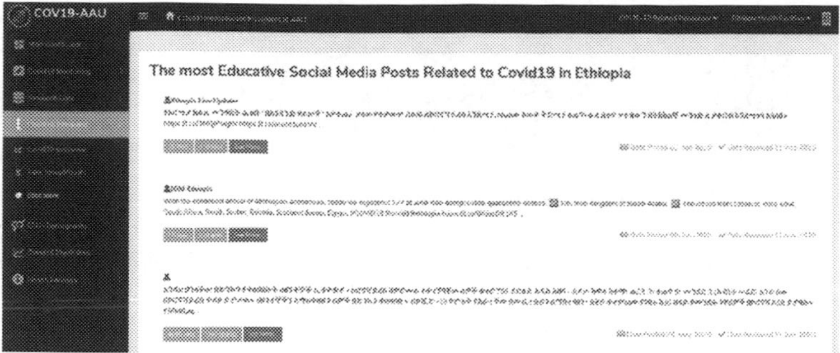

Fig. 4. Educative posts pushed to the dashboard

Fig. 5. Misinformation pushed to the dashboard

5 Conclusion

Social media can disseminate misinformation that can ultimately do harm the society. At the same time educative contents are also prevalent on such social media. Thus, it is important to put forwarded a comprehensive gatekeepers to respond to the situation and identify those extracts helping to educate the wide public.

In this paper we presented our semi-automated misinformation and disinformation surfacing and flagging system focused on Facebook and Twitter. Based on the result of the current study, we could be able to achieve acceptable result that can help the society to be aware of different rumors, myths and fake-news. Furthermore, it leverages diffusion of critical and essential information to the public.

Our future work will combine the results of this findings and further analysis of the context of use to provide a robust generic architecture that serves various context. We will also include other Ethiopian Languages and social media sites like Telegram.

Acknowledgments. This Research work is funded by Addis Ababa University Research and Technology Transfer office.

References

1. Cinelli, M., Quattrociocchi, W., Galeazzi, A., Valensise, C.M., Brugnoli, E., Schmidt, A.L., Zola, P., Zollo, F., Scala, A.: The covid-19 social media infodemic. arXiv preprint arXiv: 2003.05004 (2020)
2. WHO. https://www.who.int/emergencies/diseases/novel-coronavirus-2019. Accessed 15 May 15
3. Zarocostas, J.: How to fight an infodemic. Lancet **395**(10225), 676 (2020)
4. Singh, S.: How to fight an infodemic: COVID19 outbreak. Tathapi with ISSN 2320-0693 is an UGC CARE J. **19**(13), 399–401 (2020)
5. DATAREPORTAL. https://datareportal.com/reports/digital-2020-ethiopia. Accessed 25 May 2020
6. Dugo, H.: The powers and limits of new media appropriation in authoritarian contexts: a comparative case study of oromo protests in ethiopia. Africology J. Pan African Stud. **10** (10), 48–69 (2017)
7. Rashkin, H., Choi, E., Jang, J.Y., Volkova, S., Choi, Y.: Truth of varying shades: analyzing language in fake news and political fact-checking. In: Proceedings of the 2017 Conference on Empirical Methods in Natural Language Processing, pp. 2931–2937 (2017)
8. Thorson, E.: Changing patterns of news consumption and participation. Inf. Commun. Soc. **11**(4), 473–489 (2008)
9. Lazer, D.M., Baum, M.A., Benkler, Y., Berinsky, A.J., Greenhill, K.M., Menczer, F., Metzger, M.J., Nyhan, B., Pennycook, G., Rothschild, D., Schudson, M.: The science of fake news. Science **359**(6380), 1094–1096 (2018)
10. Shu, K., Sliva, A., Wang, S., Tang, J., Liu, H.: Fake news detection on social media: a data mining perspective. ACM SIGKDD Explor. Newslett. **19**(1), 22–36 (2017)
11. Stieglitz, S., Mirbabaie, M., Ross, B., Neuberger, C.: Social media analytics–challenges in topic discovery, data collection, and data preparation. Int. J. Inf. Manage. **39**, 156–168 (2018)
12. Brainard, J., Hunter, P.R.: Misinformation making a disease outbreak worse: outcomes compared for influenza, monkeypox, and norovirus. Simulation **96**(4), 365–374 (2019)
13. Shin, J., Jian, L., Driscoll, K., Bar, F.: The diffusion of misinformation on social media: Temporal pattern, message, and source. Comput. Hum. Behav. **83**, 278–287 (2018)
14. Tandoc Jr., E.C., Lim, Z.W., Ling, R.: Defining “fake news” a typology of scholarly definitions. Digit. J. **6**(2), 137–153 (2018)
15. Kouzy, R., Abi Jaoude, J., Kraitem, A., El Alam, M.B., Karam, B., Adib, E., Zarka, J., Traboulsi, C., Akl, E.W., Baddour, K.: Coronavirus goes viral: quantifying the COVID-19 misinformation epidemic on Twitter. Cureus **12**(3), e7255. (2020)
16. Brennen, J.S., Simon, F.M., Howard, P.N., Nielsen, R.K.: Types, sources, and claims of Covid-19 misinformation. Reuters Inst. **7** (2020)
17. Pulido, C.M., Villarejo-Carballido, B., Redondo-Sama, G., Gómez, A.: COVID-19 infodemic: more retweets for science-based information on coronavirus than for false information. Int. Soc. (2020). https://doi.org/10.1177/0268580920914755
18. Nguyen, D.T., Nguyen, N.P., Thai, M.T.: Sources of misinformation in Online Social Networks: who to suspect? In: MILCOM 2012–2012 IEEE Military Communications Conference, pp. 1–6 (2012)
19. Lappas, T., Terzi, E., Gunopulos, D., Mannila, H.: Finding effectors in social networks. In: Proceedings of the 16th ACM SIGKDD International Conference on Knowledge Discovery and Data Mining, pp. 1059–1068 (2010)

20. Brainard, J., Hunter, P.R., Hall, I.R.: An agent based model about the effects of fake news on a norovirus outbreak. Revue d'Épidémiologie et de Santé Publique (2020)
21. Goasdoué, F., Karanasos, K., Katsis, Y., Leblay, J., Manolescu, I., Zampetakis, S.: Fact checking and analyzing the web. In: Proceedings of the 2013 ACM SIGMOD International Conference on Management of Data, pp. 997–1000 (2013)
22. Comin, C.H., Costa, L.F.: Identifying the starting point of a spreading process in complex networks. Phys. Rev. E **84**(5), 056105 (2011)
23. Shah, D., Zaman, T.: Rumors in a network: who's the culprit? IEEE Trans. Inf. Theor. **57**(8), 5163–5181 (2011)
24. Popat, K., Mukherjee, S., Strötgen, J., Weikum, G.: CredEye: a credibility lens for analyzing and explaining misinformation. In: Companion Proceedings of the Web Conference 2018, pp. 155–158 (2018)
25. Offermann, P., Levina, O., Schönherr, M., Bub, U.: Outline of a design science research process. In: Proceedings of the 4th International Conference on Design Science Research in Information Systems and Technology, pp. 1–11 (2009)
26. TWITTER. https://blog.twitter.com/en_us/topics/product/2020/updating-our-approach-to-misleading-information.html. Accessed 10 June 2020

The Importance of Assessment and Evaluation in Higher Education Information Technology Projects

Dawn Brown(✉) and Nathan Johnson(✉)

Western Carolina University, Cullowhee, USA
{dawnbrown, nathan.johnson}@wcu.edu

Abstract. University planning requires a set of processes for continuous quality improvement but is often hampered due to the lack of project management across higher education. This paper discusses the benefits of carrying out continuous client feedback and communication processes throughout the lifecycle of information technology projects in higher education to assess and evaluate the condition of a project. The discussion includes the impact of customer participation on IT project performance as part of project quality management. Research shows that quality communication between stakeholders and project team members results in fewer misunderstandings and more precise definitions; however, there is a literature gap when it comes to assessing IT projects within higher education. Confirming expectations to improve customer perceptions of the project process and performance has a positive effect on client satisfaction. The paper emphasizes the importance of transparency on the part of IT in higher education project work, particularly in public universities, by involving higher education staff in all phases of the project, from project prioritization and selection to analysis, design, and implementation of the solution. Best practices for assessment and improving project outcomes are offered, as well as managing project quality and information flow. Future research directions are also offered.

Keywords: Stakeholder feedback · Information technology · Project management · Project success · Quality management

1 Introduction

Information technology (IT) projects in higher education require planning and a set of processes for quality improvement. The deliberate methodology and application of project management (PM) is one way to ensure proper planning is taking place, particularly when it comes to planning. Still, PM is a relatively new concept in higher education. The purpose of this paper is to analyze proposed methods of assessing and evaluating IT projects in higher education through continuous communication with stakeholders.

In the following pages, we first examine the pertinent literature regarding IT projects in higher education. Next we examine the importance of involving users throughout the lifecycle of the project and how it helps to improve their perception of

C. Stephanidis et al. (Eds.): HCII 2020, LNCS 12427, pp. 222–233, 2020.
https://doi.org/10.1007/978-3-030-60152-2_18

the IT project process while increasing customer satisfaction. Based on this analysis, we then consider best practices for defining project success with an emphasis on customer satisfaction. Finally, we provide recommendations for higher education IT projects and future research in this area.

2 Literature Review

Although there are now abundant resources for identifying project success factors and research results for best practices, IT implementation projects are still often not successful. A recent Gartner study (Booth 2000) revealed that across industries, at least 40% of IT projects are either terminated or fail to meet the requirements, and fewer than 40% of IT solutions purchased from vendors meet their goals. Some sources report failure rates as high as 70%, while others report as few as one in eight IT projects that are considered successful, with more than half experiencing cost and schedule overruns and still not delivering what was promised. According to the 2014 CHAOS Report (The Standish Group 2014), "only 35% of IT projects were completed on time, on budget, and met user requirements". While this number is higher than the original 1994 CHAOS Report (The Standish Group 1994) which reported 16.2%, approximately two-thirds of IT projects have significant problems, including 19% that failed outright (down from 31.1% in 1994). According to the CHAOS Report, the number one factor in project success is user involvement, while lack of user input is reported as the number one factor contributing to project failure.

IT projects are particularly challenging, especially within higher education. Universities desire to engage in integrated planning for their IT projects so they can continuously improve their institutions. This type of improvement requires successful PM; but to those in the academy, it often looks like a discipline that applies only to the corporate world. Commonly, faculty and staff see themselves as educators only, not business or project managers. In their study of PM within higher education, Austin et al. (2013) found that "historically, faculty are more focused on research and teaching and find project governance would require a refocus from their main passions and priorities of research and teaching, yet this conflicts with the structure of managing projects. PM may be viewed as "too corporate" of a way to make decisions, yet this is changing within higher education due to the need to be more effective" (Austin et al. 2013, p. 75).

2.1 Approaching IT Projects in Higher Education

Many times, university faculty and staff members while great at teaching, research, and administration, struggle to find ways for improving a process through a successful IT project.

According to the Project Management Institute (2013), projects are unique activities that must happen within a given schedule, cost, and scope that create change within an organization. In other words, projects are planned, look to complete a goal, have a set or resources, and have a defined start and finish. When approaching projects, expectations should be set using Smart Goals, in other words goals that are **S**pecific, **M**easurable, **A**ttainable. **R**ealistic, and **T**imely.

Many project team members have an issue as early as the initiation phase of a project when they fail to set Smart Goals from the beginning. For example, a goal to "improve student retention" is not specific enough. It does not have a benchmark to measure anything, and it fails to set realistic and timely goals. As it is written, this goal is not attainable. A better example would be to "improve student retention rates by 5% for undergraduate students in the Arts and Science college between 2016 and 2020."

Another common problem in higher education IT projects is the absence of valid current and historical data to use as a baseline when starting a project for accreditation, class management, or academic planning. Like any project, it is essential to use the correct data and prepare it for integration in an enterprise system like a university. This process can be time-consuming, but it is crucial to begin on the right foot with input from the stakeholders.

A project should follow a predefined set of processes with a series of deliverables over time, which means that we need to find the right support and resources for each step of the way. Educause (2014) recommends that "higher learning institutions clearly determine what services are required for PM, making sure they concentrate on the needs identified such as the right methodology, the right training, or a right coaching or mentoring program." On the other hand, it is also essential to reexamine success frequently, celebrate milestones, provide incentive and motivation, assess lessons learned, and update the project plan.

Croxall (2011) says that before starting any project, the university needs to ensure they understand what problem the project will fix, and if there are better alternatives or ways to expend scarce resources. Croxall (2011) further states that it can be difficult to do this in the university setting because projects are typically organic in nature and not something that is brought to them for evaluation. Consequently, it makes asking the questions about whether the project is pertinent, useful, efficient, viable, and sustainable even more important.

As with any project, communication is key. Communication with project stakeholders is extremely important and can come into sharp focus when working with university stakeholders. Clark (2013) posits that while managing several digital projects that involve "siloed" stakeholders, listening to and incorporating the broadest range of voices from across the university campus before prioritizing the work is important to success. Similarly, Educause (2014) endorses the thorough generation of a critical mass of support before on embarking on any one project.

Austin et al. (2013) expound on the unique challenges that the higher education environment brings, particularly where collaboration is concerned. Often, there is a divide between faculty and staff, departments, colleges, etc., and issues of shared governance can become sticking points. Ensuring the project team understand how the various stakeholders, their work, and their department, college, or division will be affected by a project is an extremely important factor in the success of a project or initiative (Austin et al. 2013).

2.2 Difficulty Defining IT Project Success in Higher Education

Dahlberg and Kivijarvi (2016) point out that there are many practitioner-oriented definitions of IT project success. In addition, academic research has investigated the elusive nature of "success" in projects (e.g.: Ika 2009; Kloppenborg et al. 2014; Shenhar and Dvir 1996). According to Kaplan and Harris-Salamone (2009), university IT projects present their own challenges and wide-ranging definitions of success in IT projects. As a result, there needs to be more understanding of various stakeholder views of success when undertaking a project. In addition, more qualitative studies of project failure are called for. Often, the complexity of both large-scale IT projects and the higher education environment make implementation more difficult because it is not just a technical process, but also a collaboration process between stakeholders with (often) very different goals, constituencies, political realities, and desired outcomes that just so happen to work within the same organization. On the one hand, success may be defined as just getting an application or system running, getting clients to use it, and getting at least reluctant acceptance. Unfortunately, that reluctance may turn to full on non-acceptance over time. Often, IT work involves simply offering "small wins" to stakeholders.

Additionally, Kaplan and Harris-Salamone (2009) state that what works for one group of users, such as faculty, may not work for another group, such as staff, and those users who benefit from the system may not be the same users who perform the actual work. For these reasons, there is often disagreement about what constitutes success or failure. There is a need for more teaching and research, especially qualitative research, to analyze why and how an institutional IT project failed, how an institution tries to turn itself around after failure, and what to do the next time differently.

2.3 Communication, Workflow, and Quality in Higher Education IT Projects

Communication breakdowns on projects can result in major setbacks and difficulties for project team members to continue working together (Daim et al. 2012). The difficulty of communicating across different groups in higher education makes it harder to identify IT project requirements and understand workflow. These communication and workflow issues add to project complexity. Higher education requires collaboration, as does system implementation, yet there is difficulty in translating among specialties, stakeholders, faculty, and staff. In addition to these communication challenges, is the difficulty of identifying requirements for the various groups involved. For instance, the individuals gathering requirements may not include all of the needed stakeholders, or the included individuals may not know how to communicate their requirements effectively.

Some IT projects may be selected because they are mandated due to university system or security compliance, legal requirements, needed improvements in student outcomes, or because developers like the people who want the project (Alexander 1999). When there is difficulty with fully understanding the workflow, there may be many workarounds and workflow changes. Sometimes this may be due to the inability of those doing the work to articulate what they do or need; sometimes it may be due to

IT not understanding the environment or process, or not in agreement with what needs to be done.

Importance of Quality. Quality is an essential element of PM along with three main components: time, cost and scope. Quality is a critical factor that may affect the project completion schedule or budget if quality management is not applied or the level of quality is not agreed upon by the customer (Goswami 2015). The purpose of quality management is first to understand the how the client perceives quality, and then put a plan in place to meet or exceed those expectations. According to Liberatore and Pollack-Johnson (2013, p. 518), "the value of a delivered project from a customer perspective can be measured by the level of quality associated with it." McKinsey & Company conducted a study on large-scale IT projects with the University of Oxford (2012) to determine the causes of project failure and improve their PM processes. The study found that on an average, "large IT projects exceeded their budgets by 45% and their schedules by 7% while delivering 56% less value than predicted" (University of Oxford 2012). The primary cause of failure was that the users and stakeholders did not participate in the daily or weekly project activities.

Stakeholders often perceive that a project is successful because it made their workflow easier (Parkes 2002); however quality issues need to be considered, especially in light of the importance of administrative and data reporting for public institutions. Administrative and quality indicators related to workflow need to be incorporated into policies and procedures, which further adds to the project complexity.

Quality Concepts in Agile Project Management. Agile PM is a well-proven methodology for IT projects with highly responsive, synergistic, and co-located teams (Chow and Cao 2008) that can provide opportunities for shareholders and users to participate actively. While agile methods may accept requirement changes late in the project life cycle, users are given occasions to provide input and feedback throughout the project lifecycle. While recent studies have not proven that agile methods perform better than traditional methods in large, distributed IT projects, many improvements can be observed on the quality, customer satisfaction, and the user's perception of the product (Serrador and Pinto 2015). Also, building good communication, team collaboration, and improved employee satisfaction seems to be a positive result of following the agile methodology. (Papadopoulos 2015, pp. 455–456.)

The lack of involvement of the client or stakeholder in a project may result in an inferior product or service delivered. This situation seems to occur quite frequently in higher education IT projects due to lack of knowledge in PM or by having the wrong methodology in place. In Agile methodology, stakeholders are involved at the beginning of the project and can provide feedback to iterate more frequent changes. It is essential to understand what the customer wants and keep them updated while at the same time managing their expectations (Holtsnider and Jaffe 2012, p. 264).

Service Quality Model. Service quality plays an important role when a customer is evaluating a new application or a process. Using several criteria in two specific areas to evaluate a new service or application makes it easier to assess the pieces and find the weak spot. Understanding how the services are evaluated enables IT organizations to influence these evaluations and push them in the desired direction. The two areas are the technical quality (the what) and the functional quality (the how) as demonstrated in Fig. 1.

The technical quality corresponds to the outcome of the provided service, for example, the speed of the broadband internet. The functional quality corresponds to the delivery of the outcome.

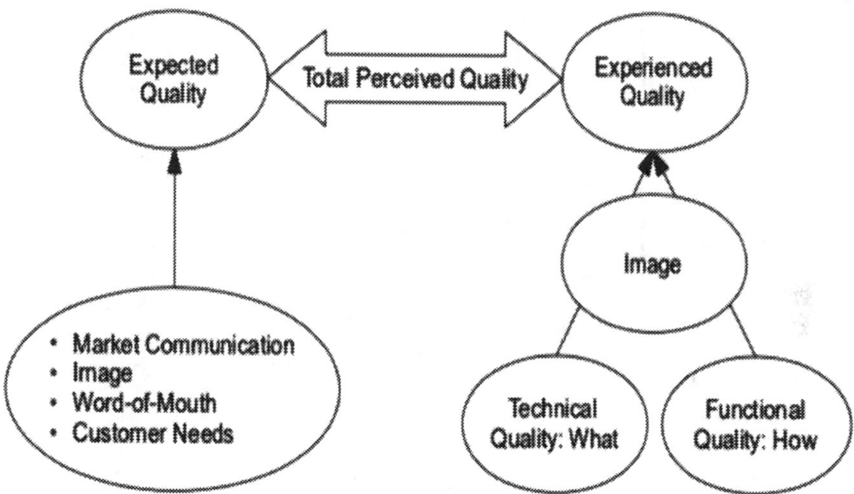

Fig. 1. Service quality model. (Grönroos 2007, p. 77).

Service quality can also be thought of as a measure of the gap between customer expectation and experience of the actual delivered service. According to Parasuraman et al. (1985) and Zeithaml et al. (1988), there could be four different types of gaps in the overall service process. Figure 2 demonstrates the various gaps in the service process.

- The first gap happens between consumer expectations and IT's perception of those expectations. This gap may occur when IT does not know what the customer wants due to a lack of communication with the customer about their needs.
- The second gap occurs partially because of the first gap. The design and standards may not align with customer expectations because IT did not collect the correct technical or functional requirements initially.

Measuring Service Quality. Measuring service quality for IT projects in higher education can be challenging. However, some tools have been developed and have been tested in many service-based companies to measure or critically analyze the service quality using quantitative or qualitative data. With quantitative tools, the service quality is investigated based on mathematical, statistical, or computational techniques. With qualitative tools, the service quality is investigated based on interviews, discussions, and observations. Talking with and listening to the customer is the best method to understand their real needs, expectations and experiences. Therefore, direct contact is preferred. This method is even more critical if a failure has happened, and the customer needs to be satisfied or the service still fulfilled. (Grönroos 2007, p. 125).

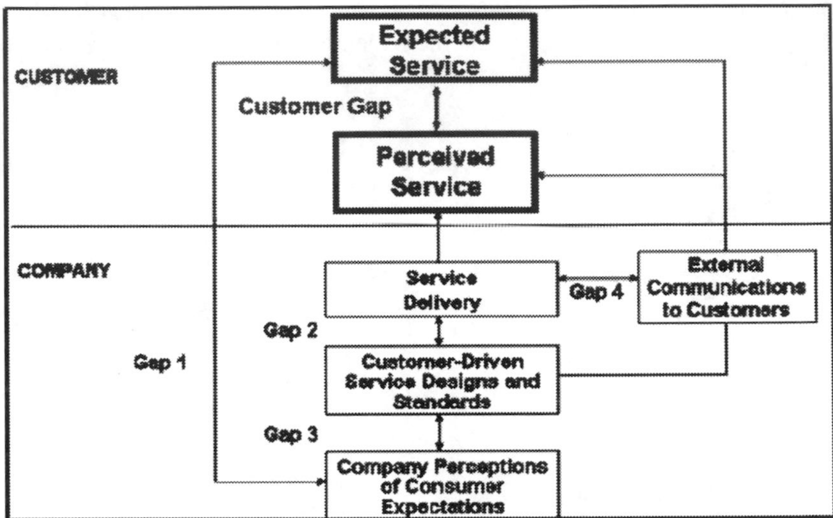

Fig. 2. Gap model of service quality. (Parasuraman et al. 1985).

Many service-based organizations establish a post-completion interview with project stakeholders and sponsors to determine the return on investment (ROI) and to confirm that the solution is effective. Surveys may consist of five to ten questions covering communication, processes, delivery, or overall value which the stakeholder may rate on a scale of 1 to 5 (5 = best). For example:

1. Scope, completion criteria, and goals were clearly defined at the beginning of the project.
2. The schedule provided the product/service in a timeframe that met the business need/requirement.
3. There were open, honest and timely communications.
4. When issues were encountered, they were brought to light quickly and discussed openly.
5. The project team demonstrated a professional and can-do attitude.
6. In regards to overall value, the IT project met its goals.

Based on that feedback, IT can compare and critically analyze the results and plan for continuous improvement in the future. Parasuraman et al. (1988) designed a questionnaire to measure service quality, the SERVQUAL, which can identify areas for intervention in the event of quality deficits. Later, another model was created, SERVPERF (Cronin and Taylor 1992) which is similar, but only measures the customers' expectations and perceptions of the service.

A summary of the literature review findings for the IT PM in higher education discussion can be found in Table 1.

Table 1. Literature summary: IT project management in higher education

IT project management in higher education	Reference
How to approach projects in higher education	Educause (2014), Croxall (2011), Clark (2013), Austin et al. (2013)
Importance of quality in project management	Goswami (2015), Liberatore and Johnson (2013, p. 518), University of Oxford (2012)
Quality concepts in agile project management	Papadopoulos (2015), Holtsnider and Jaffe (2012)
Service quality model	Grönroos (2007)
Service quality gaps theory	Parasuraman et al. (1985), Zeithaml et al. (1988)
Measuring service quality	Grönroos (2007), Parasuraman et al. (1988) Cronin and Taylor (1992)
Managing information flow in project quality management	Zeng, Lou, and Tam (2007)

3 Discussion

3.1 IT Project Improvement in Higher Education

There are several factors to consider when thinking about improving IT project success. The first factor is providing incentives and removing deterrents. End-users may perceive that they have no time to learn a new system, or that what they are being asked to do increases their workload. Faculty, for example, would be more engaged if they experienced applications that helped them directly rather than providing disincentives to adopt a new system. A second factor is identifying and mitigating risks. It is essential to determine any risks and assess them early and often during the project. These risks and possible ways to reduce or mitigate them should become part of new or existing policies and procedures about the new system and incorporated into training. A third factor is allowing resources and time for training, exposure, and learning. Sufficient training and learning time should be part of the implementation of any new system and needs to be on-going after implementation. Project success may be limited and user acceptance low when training is limited, especially for groups who have less exposure or access to computers. Finally, consider learning from the past and from others. It is essential to analyze successes, failures, and how failing situations can be turned around. After action and lessons learned sessions are invaluable for identifying successes, failures, wins and losses that can be integrated into future project efforts. Qualitative studies with more focus on higher education and incorporating information from change management may provide additional insight. Feedback from measurable evidence such as the feedback from a stakeholder satisfaction surveys can also provide great insight into project failure after the fact.

3.2 Best Practice Recommendations

Identifying best practices for higher education IT projects should cover system analysis, design, development, implementation, change management, support, how to identify all stakeholders and ensure a shared vision among them, workflow and process redesign, and providing benefits. Although much is already known about these areas from higher education research, as well as from research in other domains, it can be hard to translate general principles into practice in an actual university environment because the context can be very different across institutions. Therefore, we need more research studies that specifically investigate the effects of context on the implementation of IT systems. Additional research, knowledge bases, and examples are essential for identifying general principles as well as how they work in practice in higher education. Such information would help us gain more understanding on how to bring together regulations, workflow, policies, and IT practices in ways that make them easier to apply in higher education settings. See Table 2 for a synopsis of best practice for managing higher education IT projects.

3.3 Managing Information Flow in Project Quality Management

Information flow plays a vital role in improving the quality of projects. Project managers should collect information from various internal and external sources, extract the most relevant information for the project team and stakeholders, and store the information in a central repository during the project lifecycle. In many IT organizations, the information flow is not well defined due to lack of knowledge or control of the information, which may impact the quality of a project. Many IT organization have adopted ISO 9001 quality systems and obtained certifications in ISO 9001. An effective ISO 9001 implementation can benefit IT organizations by improving management control, efficiency, productivity, and customer service (Zeng, Lou, and Tam 2007, p. 30).

Information Requirements and Flows. The primary purpose of information management is to provide relevant information that project team members and stakeholders can easily access. Relevant information is critical for providing transparency and knowledge for any IT project and has a direct impact on the quality of the delivered service. Quality information management is concerned with effective communication and covers its acquisition, generation, preparation, organization and dissemination, evaluation and management of information resources. The information collected from different sources must be analyzed, processed critically, and used effectively to improve the overall quality management. Uncertainty and lack of information significantly increase the complexity of research on information flows.

Barriers to Information Flow. It is essential to identify information barriers in business processes and PM for managing the information flow in quality management. There are three primary type of barriers to information flow in quality management:

1. Organizational barriers: barriers caused by the organizational structure in an IT project; include multi-level structure barriers and horizontal communication barriers.

Table 2. Best practices for IT project management in higher education

IT project process	Best practice example
System analysis	Develop models (e.g., activity diagrams) to clarify requirements; activity diagrams can be used to improve system functionality
System design	Testing with user involvement is essential to successfully design a system that users will adopt
System development	Selecting development process that best suits the project and project team; choosing a waterfall or agile methodology can impact the success of the project
Implementation	Implementation should be completed as rapidly as possible with enough support; assistance in the form of written manuals, "how to" guides, and online tutorials can address the varied learning styles of individual users
Change management	A change management strategy should include processes for soliciting feedback at all stages of the change process
Support	End users should have the ability to reach out to respective groups for required technical and functional support by submitting a ticket in a ticketing tool
Identifying stakeholders and ensuring shared vision	Share information with stakeholders purposefully and consistently; actively involve them in planning and implementation; make sure they know the project plan and their role
Workflow and process redesign	Workflow redesigns completed and tested before a new system is introduced help prevent blame directed at the new system
Providing benefits	Identify problems and opportunities for process improvement so the organization can reap associated benefits

2. Behavioral barriers: barriers caused by behavioral characteristics of project team members and stakeholders, which may result in liability or lack of motivation.
3. Technical barriers: barriers caused by the technical characteristics of information in IT projects, such as the last of a central repository or a collaborative tool for project communication.

Some IT organizations are facing severe challenges with information barriers to information flow. However, lately, many companies have revealed that adequately managed information flow in the organization has improved relationships with customers, as well as the quality of products and services.

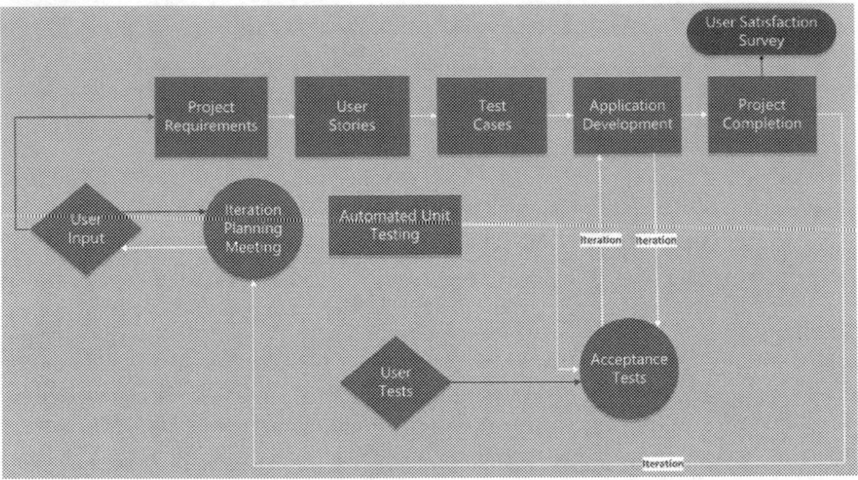

Fig. 3. Agile IT project framework for higher education

4 Conclusions and Suggested Future Research

This paper reviewed best practices for assessment and evaluation of IT projects through continuous stakeholder interaction and quality management. Although many IT organizations have quality management processes in place and knowledge, most IT organizations in higher education are still struggling with maintaining the quality level of project deliverables and communicating clearly and transparently with the stakeholders. IT organizations struggle to find a suitable PM methodology and framework which will not increase operational costs, while still meeting the needs of the university.

Figure 3 represents a theoretical framework for agile PM in higher education based on the need to keep operational costs low while addressing university needs. This framework gives a basic understanding of the different processes involved in PM and quality activities to ensure that the project will be delivered on time and within budget while meeting an agreed-upon quality level.

Suggested future research includes validating this framework by testing it with current higher education IT projects. It is hypothesized that by using this framework, projects will be much more likely to succeed because it considers customer input, user stories, customer testing, and acceptance.

References

Alexander, S.: An evaluation of innovative projects involving communication and information technology in higher education. High. Educ. Res. Dev. **18**(2), 173–183 (1999)

Austin, C., Browne, W., Haas, B., Kenyatta, E., Zulueta, S.: Application of project management in higher education. J. Econ. Dev. Manage. IT Financ. Mark. **5**(2), 75–99 (2013)

Booth, R.: Project failures costly. TechRepublic (2000). https://www.techrepublic.com/article/it-project-failures-costly-techrepublic-gartner-study-finds/

Chow, T., Cao, D.B.: A survey study of critical success factors in agile software projects. J. Syst. Softw. **81**(6), 961–971 (2008)

Clark, B.: Tips for managing digital projects in higher education. University Business Magazine, vol. 56, November 2013

Cronin Jr., J., Taylor, S.: Measuring service quality - a reexamination and extension. J. Mark. 56, 55–68 (1992). 10.2307/1252296

Croxall, B.: 12 Basic Principles of Project Management. The Chronicle of Higher Education (2011). https://www.chronicle.com/blogs/profhacker/12-basic-principles-of-project-management/31421

Dahlberg, T., Kivijärvi, H.: Towards an integrative, multilevel theory for managing the direct and indirect impacts of IT project success factors. In: 2016 49th Hawaii International Conference on System Sciences (HICSS), pp. 4971–4980. IEEE (2016)

Daim, T.U., et al.: Exploring the communication breakdown in global virtual teams. Int. J. Proj. Manage. **30**(2), 199–212 (2012)

Goswami, S.: Role of quality management system in project completion. Pipeline Gas J. **242**(5), 189–201 (2015)

Grönroos, C.: Service Management and Marketing: Customer Management in Service Competition, 3rd edn. Wiley, Hoboken (2007)

Holtsnider, B., Jaffe, D.: Manager's Handbook: Getting Your New Job Done, 3rd edn. Morgan Kaufmann Publishers Inc., San Francisco (2012)

Ika, L.A.: Project success as a topic in project management journals. Proj. Manage. J. **40**(4), 6–19 (2009)

Kaplan, B., Harris-Salamone, K.: Health IT success and failure: recommendations from literature and an AMIA workshop. J. Am. Med. Inf. Assoc. **16**(3), 291–299 (2009)

Kloppenborg, T.J., Tesch, D., Manolis, C.: Project success and executive sponsor behaviors: empirical life cycle stage investigations. Proj. Manage. J. **45**(1), 9–20 (2014)

Liberatore, M., Pollack-Johnson, B.: Improving project management decision making by modeling quality, time, and cost continuously. IEEE Trans. Eng. Manage. **60**, 518–528 (2013)

Papadopoulos, G.: moving from traditional to agile software development methodologies on large, distributed projects. Procedia Soc. Behav. Sci. **175**(12), 455–463 (2015)

Parasuraman, A., Zeithaml, V., Berry, L.: A conceptual model of service quality and its implication for future research (SERVQUAL). J. Mark. **49**, 41–50 (1985). https://doi.org/10.2307/1251430

Parkes, A.: Critical success factors in workflow implementation. In: Proceedings of the 6th Pacific Asia Conference on Information Systems, Jasmin, Tokyo, pp. 363–380, December 2002

Project Management Institute: A Guide to the Project Management Body of Knowledge (PMBOK Guide). Project Management Institute, Newtown Square (2013)

Serrador, P., Pinto, J.K.: Does agile work?—a quantitative analysis of agile project success. Int. J. Proj. Manage. **33**(5), 1040–1051 (2015)

Shenhar, A.J., Dvir, D.: Toward a typological theory of project management. Res. Policy **25**(4), 607–632 (1996)

The Standish Group (1994, 2014). Chaos Report. Project Smart. https://www.projectsmart.co.uk/white-papers/chaos-report.pdf. Accessed 15 Nov 2019

Zeithaml, V., Parasuraman, A., Berry, L.: SERVQUAL: a multiple- Item Scale for measuring consumer perceptions of service quality. J. Retail. **64**(1), 140 (1988)

Zeng, S.X., Lou, G.X., Tam, W.Y.: Managing information flows for quality improvement of projects. Meas. Bus. Excellence **11**(3), 30–40 (2007). https://doi.org/10.1108/13683040710820737

Recommendation or Advertisement? The Influence of Advertising-Disclosure Language with Pictorial Types on Influencer Credibility and Consumers' Brand Attitudes

Xinyi Deng(✉), Mengjun Li, and Ayoung Suh

School of Creative Media, City University of Hong Kong, Kowloon Tong, Hong Kong SAR

Xinyideng6-c@my.cityu.edu.hk, mengjunli3@um.cityu.edu.hk, ahysuh@cityu.edu.hk

Abstract. The impacts of advertising-disclosure language on influencer marketing have increasingly attracted researchers' attention. However, little research has focused on the role of pictorial types on social media influencers and brands. Thus, this study explores how different combinations of advertising-disclosure conditions and pictorial types can affect social media influencer credibility and consumers' brand attitudes. By using an experimental 3×2 between-subject design, we collected data from 264 followers of beauty influencers on Weibo. Results revealed that a post that disclosed sponsorship significantly decreased influencer credibility and consumers' brand attitudes compared to a post without any disclosure. Results also showed that a post with no-sponsorship disclosure significantly increased influencer credibility and consumers' brand attitudes than a post without any disclosure or a post disclosing sponsorship. With regard to pictorial types, the influencer-with-product type led to higher influencer credibility and more positive brand attitudes in consumers than the product-only type in all three disclosure conditions. This study highlights the role of pictorial types in strengthening the power of advertising-disclosure language, which can further affect influencer credibility and consumers' brand attitudes. This study contributes to the visual complexity literature for influencer marketing.

Keywords: Influencer marketing · Advertising-disclosure language · Pictorial types · Influencer credibility · Consumers' brand attitudes

1 Introduction

With the development of social media commerce, social media influencers have become powerful vehicles for brand endorsement [1, 2]. When compared to traditional advertising, consumers are more likely to value and trust the opinions of social media influencers [3]. These opinions are perceived to be more genuine and reliable than traditional advertisements [3, 4] because consumers trust social media influencers as they would trust their friends [5]. Furthermore, social media influencers' opinions effectively influence consumers' brand attitudes and guide purchasing decisions [6].

C. Stephanidis et al. (Eds.): HCII 2020, LNCS 12427, pp. 234–248, 2020.
https://doi.org/10.1007/978-3-030-60152-2_19

Due to their high popularity and large numbers of followers, social media influencers are paid and/or given products for free in order to promote or recommend these products on social media [7]. Given the commercial nature of sponsored posts, the disclosure of sponsorship is critical to prevent consumers from feeling deceived and misled [1]. A clear advertising disclosure is useful for consumers to promptly identify sponsored posts in an advertising-intensive environment on social media. However, some marketers and influencers are still reluctant to disclose sponsorship because they do not want to reveal their commercial motivations [1], which may decrease their credibility [6]. A decrease in an influencer's credibility can subsequently decrease consumers' brand attitudes [8, 9]. According to the information richness theory, visual presentation is richer than text because it conveys more information to consumers [10] and has the potential to affect influencer credibility and brand attitudes. However, prior research has predominantly focused on the effects of advertising-disclosure language on influencer credibility and brand attitudes [6]. Little attention has been paid to the ways in which influencers present and promote products visually. The extant literature lacks a systematic understanding of how different pictorial types (i.e., product-only and influencer-with-product) under the same disclosure conditions can have different effects on both influencer credibility and consumers' attitudes towards a brand.

To fill this research gap, the present study addresses how the combination of different disclosure conditions and pictorial types might affect influencer credibility and consumers' brand attitudes. Drawing on the model of De Veriman and Hudders [6], we posit that influencer-with-product pictures will strengthen the power of advertising disclosure language. Figure 1 presents the conceptual model we propose.

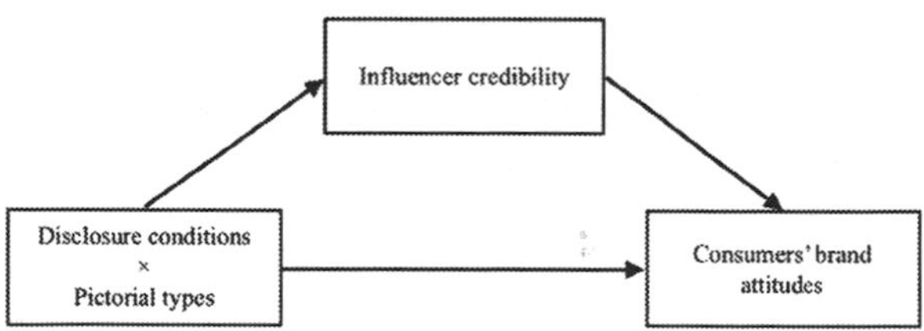

Fig. 1. Conceptual model

2 Literature Review

2.1 Social Media Influencers in Advertising

Social media influencers are online personalities with large followings on social media platforms (e.g., Facebook, Instagram, and Weibo), having an influence on their followers [5]. Social media influencers are ordinary people with relevant expertise in specific areas (e.g., travel, food, and beauty) rather than celebrities or popular public figures made famous by traditional media [3, 5].

Recently, endorsement by social media influencers has become a popular and profitable advertising approach for brands to reach both target and potential consumers [1]. The use of social media influencers to endorse products is considered more authentic and genuine than traditional advertising [5, 6]. In addition, many consumers are turning to social media influencers for their opinions and comments before buying a product [11–13]. Influencer marketing on social media has therefore grown exponentially. Social media influencers are generally paid and/or receive free products from brands [6] to promote or recommend these products and/or brands through their social media posts. Studies have found that endorsement from social media influencers is an effective tool for driving consumers' brand awareness, shaping consumers' attitudes toward the brand, and stimulating purchase intentions of products [2, 6, 14].

Given that paid endorsements can masquerade as unpaid endorsements if sponsorship is not disclosed in the post [1], influencers are advised to clearly disclose their relationship with sponsors by using hashtags (e.g., #Ad, and #Notsponsored) in their posts. This allows consumers to immediately identify whether they are viewing advertisements. In the long run, a clear identification of brand sponsorship is beneficial to retain more followers because consumers trust influencers; retention facilitates the development of long-term relationships between consumers, influencers, and brands [1, 15].

2.2 Influencer Credibility

Influencer credibility is defined as the degree to which consumers believe that an influencer is professional and attractive in a specific field and regard the information provided by the influencer as trustworthy [16]. Social media influencers are perceived as credible product endorsers because they are considered as experts who provide user-oriented recommendations or product comments based on real user experience [3]. Moreover, social media influencers act as opinion leaders [12], and their credibility plays an important role in affecting their ability to persuade their followers [5]. Prior studies have found that influencer credibility promotes positive brand attitudes, brand awareness, and purchase intentions in consumers [5, 6, 17, 18]. However, recent research has shown that a lack of transparency in the relationship between influencers and sponsors decreases influencer credibility [6]. When influencers are considered profit-oriented and do not share authentic opinions, they are perceived as unreliable information sources [1].

2.3 Consumers' Brand Attitudes

Consumers' brand attitudes refer to comprehensive judgments made by consumers as to whether or not a brand can meet their needs and expectations [19]. Previous research has suggested that consumers' brand attitudes play a significant role in marketing [16, 20]. Specifically, positive brand attitudes can cause consumers to consistently prefer brands and promote purchasing intentions of products [16]. Recently, researchers have applied consumers' brand attitudes to influencer marketing on social media [1, 6, 15, 21]. Kim and Kim [21], as well as Bakshy [15], have argued that using social media influencers can effectively enhance consumers' brand attitudes. Evans et al. [1] found that the advertising of disclosure language on social media and consumers' advertising recognition can promote the formation of consumers' brand attitudes. Furthermore, De

Veirman and Hudders [6] suggested that social media influencer credibility is positively related to consumers' brand attitudes.

3 Hypotheses Development

3.1 The Effects of Not Disclosing Sponsorship

A social media post with no-sponsorship disclosure reflects that the influencer's product recommendation is both genuine and unmotivated by commercial incentives [6, 22]. By disclosing a post without sponsorship relationships, consumers typically perceive the recommendation as authentic and sincere because the influencer is thought to be altruistically sharing an opinion [6]. Consequently, no-sponsorship disclosure can dispel consumers' uncertainty about the post as an advertisement [6]. However, a no-disclosure post may also increase suspicions that an influencer has ulterior motives, thus leading consumers to become sceptical that a post may be an advertisement [6, 23]. Consumers can immediately identify posts as advertisements if influencers use explicit disclosure language [1, 6]. Therefore, a post with no-sponsorship disclosure can cause customers to perceive the influencer as unbiased; this perception may increase the influencer's credibility and create more positive brand attitudes when compared to a no-disclosure post and an explicit-advertising post. Accordingly, we hypothesize the following:

H1a ~ b: A no-sponsorship disclosure in a social media post will lead to a) higher influencer credibility and b) more positive consumers' brand attitudes when compared to a no-disclosure post.

H2a ~ b: A no-sponsorship disclosure in a social media post will lead to a) higher influencer credibility and b) more positive consumers' brand attitudes when compared to an explicit disclosure post.

3.2 The Effects of Explicit Advertising Disclosure

A post with an explicit advertising disclosure has been found to reduce influencer credibility and can lead to negative brand attitudes when compared to a no-disclosure post [6]. Specifically, posts with explicit advertising disclosure have been found to effectively enhance consumers' awareness and recognition of such posts as advertising [1, 6, 24]. A high advertising recognition can further trigger scepticism as to whether or not the influencer has expressed a fair opinion about the product promoted [6, 10, 25]. Meanwhile, a post with an advertising disclosure is likely to negatively bias consumer impressions of an influencer's credibility, causing the recommendations to be perceived as commercially motivated [5, 6]. In addition, previous research has empirically validated that adverting disclosures can negatively affect influencer credibility and consumers' brand attitudes [5, 6]. Evans et al. [1] also found that a post with advertising disclosure resulted in more negative attitudinal and behavioural outcomes toward the brand when compared to a no-disclosure post. Given that a no-disclosure post is less transparent about the commercial relationship between an influencer and the sponsoring brand than a post with explicit advertising, it is possible for consumers to recognize a no-disclosure post as a genuine recommendation [8]. Therefore, we expect consumers to have higher perceptions of influencer credibility and increased positive brand attitudes than they would with an explicit disclosure post. Accordingly, we hypothesize the following:

H3a ~ b: A no-disclosure post in a social media post will lead to a) higher influencer credibility and b) more positive consumers' brand attitude when compared to an explicit advertising disclosure post.

3.3 The Effects of Influencer Credibility

Influencer credibility has been identified as an important way to promote products and brands on social media [6, 26]. Nowadays, an increasing number of brands are turning to social media to invest in social media influencers [5] because influencers are perceived as more authentic and credible than advertisers [26]. In addition, influencer posts are more effective at endorsing products than traditional advertisements because influencers have higher credibility [26]. Researchers have found that influencer credibility has a positive effect on consumers' brand attitudes [6, 9]. Chu and Kamal [27] have explained that this is because consumers unthinkingly accept the information provided by a trustworthy influencer, thereby affecting consumers' brand attitudes. Accordingly, we hypothesize the following:

H4: Influencer credibility is positively related to consumers' brand attitudes.

3.4 The Effects of Pictorial Types

In influencer marketing, many influencers recommend or endorse products by using captions and pictures of the product in their posts [1, 26]. The caption provides basic information about a product (e.g., the brand's name and its price), while the picture effectively conveys the product's visual details [28]. By directly viewing product pictures, an individual's interest is more likely to be aroused [14]. It is also a time-saving and effortless approach for an individual to learn more about products [28].

As pictures play a significant role in product recommendation and advertising [1, 6], the concept of visual complexity has increasingly attracted scholars' attention. Various studies have explored how the visual complexity of product pictures can affect influencer marketing [14, 28]. Visual complexity refers to the number of elements presented in a picture and the level of detail delivered by these elements [14]. Previous studies have found that visual complexity has a positive influence on customers' perceptions, judgments, attitudes, and behaviours toward the elements (e.g., products, brands, and endorsers) shown in the picture [14, 28]. For example, Wu et al. [28] reported that a complex design of product pictures can increase consumers' positive evaluations of a product's quality because of the richness of information provided. Kusumasondjaja and Tjiptono [14] found that visual complexity leads to better consumer responses and relates positively to consumers' purchase intention.

Considering that visual complexity brings positive outcomes, we expect that a complex design of product pictures will lead to higher influencer credibility and an increase of positive brand attitudes. In our study, we focus specifically on beauty products. We consider influencer-with-product pictures to be those with a high level of visual complexity, and product-only pictures to be those with a low level of visual complexity. When compared to product-only pictures, influencer-with-product images display more details about beauty products, such as the colors and effects of an eye-shadow on influencers. Influencers personally try products, which can imply an honest recommendation and then increase their credibility. This consequently can lead to more positive brand attitudes in consumers. Therefore, influencer-with-product pictures may result in higher influencer credibility than product-only pictures. In addition, previous studies have shown that a complex design of product pictures can lead consumers to positively evaluate products [28] and to an intent to purchase [14], showing the potential of complex design to enhance consumers' attitudes toward the brand. Accordingly, we hypothesize the following:

H5a ~ b: A post with an influencer-with-product picture will lead to a) higher influencer credibility and b) more positive brand attitudes when compared to a post with a product-only picture.

4 Method

4.1 Participants and Experimental Procedures

This study conducted an experiment based on a 3 × 2 between-subject design (disclosure conditions: no disclosure vs. no sponsorship vs. explicit disclosure × pictorial types: influencer-with-product vs. product-only). Appendix A shows the six experimental conditions. A total of 264 Chinese followers of beauty influencers on Weibo were recruited from an online survey company Sojump (https://www.wjx.cn/). First, we randomly assigned participants to one of the six experimental conditions (see Fig. 2). After reading a short introductory text, participants were asked to view a post released by a beauty influencer. Next, participants were asked to fill out a questionnaire regarding their demographic information and all constructs in our study (i.e., influencer credibility, consumers' brand attitudes, and manipulation checks of disclosure conditions and pictorial types).

	Influencer-with product	**Product-only**
No disclosure	Love my new lipstick and eye shadow, perfect gifts for Xmas! @Mayday makeup	Love my new lipstick and eye shadow, perfect gifts for Xmas! @Mayday makeup
Explicit disclosure	Love my new lipstick and eye shadow, perfect gifts for Xmas! @Mayday makeup #Ad	Love my new lipstick and eye shadow, perfect gifts for Xmas! @Mayday makeup #Ad
No sponsorship	Love my new lipstick and eye shadow, perfect gifts for Xmas! @Mayday makeup #Nonsponsored	Love my new lipstick and eye shadow, perfect gifts for Xmas! @Mayday makeup #Nonsponsored

Fig. 2. Six versions of the beauty post on social media.

4.2 Stimulus Material

Following the design guidelines of Evans et al. [1], we created six different stimuli (see Appendix A). The post in our study was designed with Adobe Photoshop software to resemble an authentic Weibo post. It included a picture caption and a photograph of either a product or an influencer with a product. We also used different colours to differentiate the types of sponsorship disclosure and other content in the picture caption. A hashtag (i.e., #) was added in front of the types of sponsorship disclosure. Specifically, advertising-disclosure conditions were manipulated by using three types of sponsorship disclosures in the picture caption – that is, explicit disclosure (i.e., #Ad), no sponsorship (i.e., #Nonsponsored), and no disclosure. Pictorial types were manipulated by two types of pictures, the product-only picture and the influencer-with-product picture.

4.3 Measures

All measurement items were adapted from previous research. The scales for influencer credibility were adapted from Wu and Shaffer [28]. The measures of consumers' brand attitudes were adapted from Spears and Singh [29]. Manipulation checks of disclosure conditions were adapted from De Veriman and Hudders [6]. Manipulation checks of pictorial types were adapted from Jin and Muqaddam [30]. All the measurement items are shown in Appendix B.

5 Data Analysis

Among the 264 participants, 74.2% were female and 25.8% were male. In terms of Weibo usage experience, most participants had used the platform for at least one year. In our study, we used SPSS 26.0 to test hypotheses. Table 1 summarized respondents' demographic characteristics. The next section shows the results of our study.

5.1 Manipulation Checks

As shown in Table 2, most participants correctly classified disclosure conditions and pictorial types. Results indicated that 82.8% of participants in the explicit disclosure condition, 73.8% of respondents in the no-sponsorship condition, and 77.2% of participants in the no-disclosure condition correctly recognized the disclosure conditions. Moreover, 90.8% of participants in the influencer-with-product condition and 97.5% of participants in the product-only condition correctly classified pictorial types.

5.2 Hypotheses Testing

To test hypotheses H1, H2, and H3, we used one-way ANOVAs to determine the effects of disclosure conditions on influencer credibility and consumers' brand attitudes. First, the results indicated a significant difference in influencer credibility among three disclosure conditions, $F(2,261) = 18.05$, $p < .001$, $\eta^2 = 0.12$. A no-sponsorship

Table 1. Demographic characteristics of the respondents

Item	Category	Frequency	Ratio
Age	19 or below	8	3.0%
	20–29	208	78.8%
	30–39	43	16.3%
	40 and above	5	1.9%
Gender	Female	196	74.2%
	Male	68	25.8%
Education	Below bachelor	16	6.0%
	Bachelor	232	87.9%
	Master	15	5.7%
	PhD	1	0.3%
Usage experience	≤ 1 year	2	0.8%
	> 1 year and ≤ 3 years	83	31.4%
	> 3 year and ≤ 5 years	93	35.2%
	5 years or above	86	32.6%

Table 2. Manipulation checks

Item	Category	Correct	Wrong
Disclosure type classification	Explicit advertising disclosure	82.8%	17.2%
	No sponsorship	73.8%	26.2%
	No disclosure	77.2%	22.8%
Pictorial types classification	Influencer-with-product	90.8%	9.2%
	Product-only	97.5%	2.5%

disclosure in a Weibo post (M = 5.24) leads to higher influencer credibility when compared to a no-disclosure post (M = 4.85) and a Weibo post with explicit advertising disclosure (M = 4.55). Thus, H1a, H2a, and H3a were all supported. Second, the results showed a significant difference in consumers' brand attitudes among three disclosure conditions, $F(2,261) = 42.21$, $p < .001$, $\eta^2 = 0.25$. A no-sponsorship disclosure in a Weibo post (M = 5.44) resulted in a positive increase in consumers' brand attitudes when compared to a no-disclosure post (M = 4.94) and a Weibo post with explicit advertising disclosure (M = 4.13). Therefore, H1b, H2b, and H3b were all supported (see Table 3).

We then checked for correlations between influencer credibility and consumers' brand attitudes. Our results demonstrated that influencer credibility is positively correlated with consumers' brand attitudes (r = 0.52, $p < .001$), supporting H4.

Finally, a one-way ANOVA was conducted to test H5, that is, the effects of disclosure conditions with pictorial types on influencer credibility and consumers' brand attitudes. We found that there was a significant difference in influencer credibility ($F\ (5,258) = 14.18$, $p <$,001, $\eta^2 = 0.22$) and consumers' brand attitudes ($F(5,528) = 42.37$, $p < .001$,

η^2= 0.45) among three experimental conditions. The results revealed that among three disclosure conditions, a post with an influencer-with-product picture leads to higher influencer credibility and a positive increase in consumers' brand attitudes when compared to a post with a product-only picture. Thus, H5 was supported (see Table 4).

Table 3. Descriptive statistics for disclosure conditions

Item	Disclosure conditions	M	SD
Influencer credibility	No sponsorship	5.24	0.77
	No disclosure	4.85	0.75
	Explicit advertising	4.55	0.78
Consumer' brand attitudes	No sponsorship	5.44	0.83
	No disclosure	4.94	0.97
	Explicit advertising	4.13	1.09

Note: M = Mean; SD = Standard Deviation.

Table 4. Descriptive statistics for pictorial types

Item	Category	Frequency	M	SD	*F*	Partial η^2	Sig.
Consumers' brand attitudes	Influencer-with-product (NSD)	48	5.53	0.81	42.37	0.45	0.000
	Product-only (NSD)	43	5.34	0.84			
	Influencer-with-product (ND)	43	5.41	0.65			
	Product-only (ND)	42	4.46	1.01			
	Influencer-with-product (EAD)	43	4.87	0.94			
	Product-only (EAD)	45	3.42	0.65			
	Total	264	4.84	1.10			
Influencer credibility	Influencer-with-product (NSD)	48	5.32	0.75	14.18	0.22	0.000
	Product-only (NSD)	43	5.15	0.79			
	Influencer-with-product (ND)	43	5.25	0.54			
	Product-only (ND)	42	4.45	0.73			
	Influencer-with-product (EAD)	43	4.72	0.94			
	Product-only (EAD)	45	4.45	0.73			
	Total	264	4.88	0.82			

Note: NSD = No-sponsorship disclosure; ND = No disclosure; EAD = Explicit advertising disclosure; M = Mean; SD = Standard Deviation; *F* = Variance Ratio; η^2 = Partial Eta Squared; Sig. = Significance level.

6 Discussion

This study explores how advertising-disclosure language with pictorial types affects influencer credibility and consumers' brand attitudes. To do so, we examined (1) the effects of advertising-disclosure language; (2) the effects of advertising-disclosure language with pictorial types on influencer credibility and consumers' brand attitudes, respectively; and (3) the interplay between influencer credibility and consumers' brand attitudes. Our study aims to highlight the role of visual complexity in enhancing influencer credibility and consumers' brand attitudes.

6.1 Theoretical Implications

This study provides several theoretical contributions to the literature. First, the extant visual complexity literature primarily examines how visual complexity operates in webpages or banner advertisements, whereas our study applies visual complexity to the context of social media. When accessing social media, visual information (such as videos, pictures, and graphics) is omnipresent. Our study focuses on how the visual design complexity of product pictures can affect influencer marketing.

Second, our study displays the two-way interactions between sponsorship disclosure and visual complexity in enhancing influencer credibility and consumers' brand attitudes. Prior literature predominantly examines the effects of sponsorship disclosure and seemingly neglects the complex design of product pictures. Our study contributes to the literature by providing a new perspective for introducing the role of visual complexity. Our results suggest that the visual design complexity of product pictures can strengthen the power of sponsorship disclosure and promote positive outcomes for both social media influencers and brands.

6.2 Practical Implications

This study also offers practical implications. First, our work extends the current understanding of how different combinations of advertising-disclosure types (i.e., text-based) and pictorial types (i.e., image-based) can jointly increase influencers' credibility and consumers' positive brand attitudes. The results imply that to better enhance influencer credibility and consumers' brand attitudes, influencers should use an influencer-with-product picture when promoting or recommending products. For the social media commerce industry, influencers can benefit from our study by learning better strategies to endorse brands, thereby causing influencer marketing to become even more effective.

Second, disclosing that a post is an advertisement can lead to lower influencer credibility and a decrease in positive brand attitudes. We suggest that influencers should clearly reveal sponsorship, as disclosure is helpful for building a long-term, trustworthy relationship between influencers and followers. Moreover, when influencers recommend products without sponsorship, a clear no-sponsorship disclosure should be included in the post as this can help consumers obtain certainty that no third-party was involved.

6.3 Limitation and Future Research

This study has several limitations. First, Weibo was examined as the target social media platform in our study. Because different social media platforms serve different functions, determining whether our results can be applied to other social media contexts would require further investigation. To address this issue, researchers may consider checking the generalizability of our results through multiple social media platforms (e.g., Twitter, Facebook, and Instagram). Second, because most participants in the study were young women, it is unclear whether our results are applicable to other Weibo users. Furthermore, as our research was conducted in China, we recommend the undertaking of a cross-cultural study to determine whether the results can be replicated using other samples. Finally, as we recruited only the followers of beauty influencers, future studies should also examine our model in different contexts, such as fashion, food, and travel.

7 Conclusions

This study proposed a conceptual model to understand the effects of advertising-disclosure language with pictorial types on affecting influencer credibility and consumers' brand attitudes. Prior research has shown that advertising-disclosure language can affect influencer credibility and consumers' brand attitudes, ignoring the effects of visual complexity in product pictures. By examining the effects of different pictorial types (i.e., product-only and influencer-with-product) under the same disclosure conditions on influencers' credibility and consumers' brand attitudes, we found that a post with an influencer-with-product picture leads to higher influencer credibility and a positive increase in consumers' brand attitudes when compared to a post with a product-only picture. This study confirms prior visual complexity literature and indicates that complex designs of product pictures can enhance the power of sponsorship disclosure more than simple designs, which can in turn lead to higher influencer credibility and consumers' brand attitudes.

Appendix A

Disclosure Conditions × Pictorial Types (blurred for legal and copyright reasons)

No disclosure × product only

No disclosure × influencer-with-product

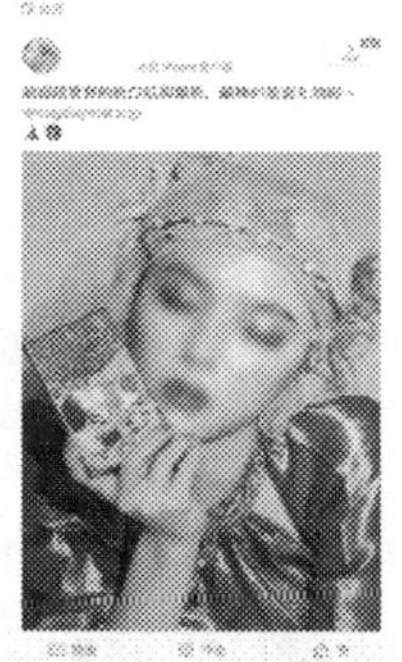

No sponsorship × product only

No sponsorship× influencer-with-product

Explicit disclosure × product only

Explicit disclosure × influencer-with-product

Note: Both pictures were adapted from the Weibo Account: @仇仇qiuqiu.

Appendix B

Measurement Items

Constructs	Items	Sources
Influencer credibility	Do you think this this beauty influencer is…	[28]
	1. Unattractive/attractive	
	2. Insincere/sincere	
	3. Not a beauty expert/a beauty expert	
	4. Unreliable/reliable	
	5. Unbiased/biased	
	6. Untrustworthy/trustworthy	
	7. Unconvincing/convincing	
Consumers' brand attitudes	Do you think the presented brand in the post is… 1. Unappealing/appealing 2. Unfavorable/favorable 3. Unlikable/likable 4. Unpleasant/pleasant	[29]
Manipulation checks of advertising disclosure conditions	Which of the following wording did you notice in the post? 1. #Ad 2. #Nonsponsored 3. None of the above	[6]
Manipulation checks of pictorial types	The picture within the post includes… 1. Products only 2. The beauty influencer and beauty products	[30]

References

1. Evans, N., Phua, J., Lim, J., Jun, H.: Disclosing instagram influencer advertising: the effects of disclosure language on advertising recognition, attitudes, and behavioral intent. J. Interactive Ad. **17**(2), 138–149 (2017)
2. Freberg, K., Graham, K., McGaughey, K., Freberg, L.A.: Who are the social media influencers? A study of public perceptions of personality. Public Relations Rev. **37**(1), 90–92 (2011)
3. Schouten, A., Janssen, L., Verspaget, M.: Celebrity vs. influencer endorsements in advertising: the role of identification, credibility, and product-endorser fit. Int. J. Ad. **39**(2), 258–281 (2020)
4. Chapple, C., Cownie, F.: An investigation into viewers' trust in and response towards disclosed paid-for-endorsements by YouTube lifestyle vloggers. J. Promot. Commun. **5**(2), 110–136 (2017)
5. Lou, C., Yuan, S.: Influencer marketing: how message value and credibility affect consumer trust of branded content on social media. J. Interactive Ad. **19**(1), 58–73 (2019)

6. De Veirman, M., Hudders, L.: Disclosing sponsored Instagram posts: the role of material connection with the brand and message-sidedness when disclosing covert advertising. Int. J. Ad. **39**(1), 94–130 (2020)
7. Woods, S.: #Sponsored: The Emergence of Influencer Marketing. PhD. University of Tennessee (2016).
8. Hwang, Y., Jeong, S.: "This is a sponsored blog post, but all opinions are my own": the effects of sponsorship disclosure on responses to sponsored blog posts. Comput. Hum. Behav. **62**, 528–535 (2016)
9. Reichelt, J., Sievert, J., Jacob, F.: How credibility affects eWOM reading: The influences of expertise, trustworthiness, and similarity on utilitarian and social functions. J. Market. Commun. **20**(1–2), 65–81 (2015)
10. Trevino, L., Lengel, R., Daft, R.: Media symbolism, media richness, and media choice in organizations: a symbolic interactionist perspective. Commun. Res. **14**(5), 553–574 (1987)
11. Gerdeman, D.: Lipstick tips: how influencers are making over beauty marketing. https://hbswk.hbs.edu/item/lipstick-tips-how-influencers-are-making-over-beauty-marketing. Accessed 26 May 2020
12. Casaló, L.V., Flavián, C., Ibáñez-Sánchez, S.: Antecedents of consumer intention to follow and recommend an Instagram account. Online Inf. Rev. **41**(7), 1046–1063 (2017)
13. Glucksman, M.: The rise of social media influencer marketing on lifestyle branding: a case study of Lucie Fink. Elon J. Undergraduate Res. Commun. **8**(2), 77–87 (2017)
14. Kusumasondjaja, S., Tjiptono, F.: Endorsement and visual complexity in food advertising on Instagram. Internet Research **4**(29), 659–687 (2019)
15. Bakshy, E., Hofman, J.M., Mason, W.A., Watts, D.J.: Everyone's an influencer: quantifying influence on twitter. In: Proceedings of the fourth ACM international conference on Web search and data mining. ACM, New York, pp. 65–74 (2011)
16. MacKenzie, S.B., Lutz, R.J.: An empirical examination of the structural antecedents of attitude toward the ad in an advertising pretesting context. J. Market. **53**(2), 14–65 (1989)
17. Pornpitakpan, C.: The persuasiveness of source credibility: a critical review of five decades' evidence. J. Appl. Soc. Psychol. **34**(2), 243–281 (2004)
18. Sallam, M.A.A., Wahid, N.A.: Endorser credibility effects on Yemeni male consumer's attitudes towards advertising, brand attitude and purchase intention: the mediating role of attitude toward brand. Int. Bus. Res. **5**(4), 55 (2012)
19. Simonin, B., Ruth, J.: Is a company known by the company it keeps? Assessing the spillover effects of brand alliances on consumer brand attitudes. J. Mark. Res. **35**(1), 30–42 (1998)
20. Kudeshia, C., Kumar, A.: Social eWOM: does it affect the brand attitude and purchase intention of brands? Manag. Res. Rev. **3**(40), 310–330 (2017)
21. Kim, D.Y., Kim, H.Y.: Influencer advertising on social media: the multiple inference model on influencer-product congruence and sponsorship disclosure. J. Bus. Res. (2020, in Press)
22. Stubb, C., Colliander, J.: "This is not sponsored content" – The effects of impartiality disclosure and e-commerce landing pages on consumer responses to social media influencer posts. Comput. Hum. Behav. **98**(1), 210–222 (2019)
23. Lee, M., Faber, R.J.: Effects of product placement in on-line games on brand memory: a perspective of the limited-capacity model of attention. J. Ad. **36**(4), 75–90 (2007)
24. Wojdynski, B.W., Evans, N.J.: Going native: effects of disclosure position and language on the recognition and evaluation of online native advertising. J. Ad. **45**(2), 157–168 (2016)
25. Tutaj, K., Van Reijmersdal, E.A.: Effects of online advertising format and persuasion knowledge on audience rections. J. Market. Commun. **18**(1), 5–18 (2012)
26. De Veirman, M., Cauberghe, V., Hudders, L.: Marketing through Instagram influencers: the impact of number of followers and product divergence on brand attitude. Int. J. Ad. **36**(5), 798–828 (2017)

27. Chu, S.C., Kamal, S.: The effect of perceived blogger credibility and argument quality on message elaboration and brand attitudes: an exploratory study. J. Interactive Ad. **8**(2), 26–37 (2008)
28. Wu, C., Shaffer, D.R.: Susceptibility to persuasive appeals as a function of source credibility and prior experience with the attitude object. J. Pers. Soc. Psychol. **52**(4), 677 (1987)
29. Spears, N., Singh, S.N.: Measuring attitude toward the brand and purchase intentions. J. Curr. Issu. Res. Ad. **26**(2), 53–66 (2004)
30. Jin, S.V., Muqaddam, A.: Product placement 20: "Do brands need influencers, or do influencers need brands?". Journal of Brand Management **26**(5), 522–537 (2019)

Characterizing Anxiety Disorders with Online Social and Interactional Networks

Sarmistha Dutta and Munmun De Choudhury(✉)

Georgia Institute of Technology, Atlanta, GA 30332, USA
{sdutta65,munmund}@gatech.edu

Abstract. Anxiety disorders are closely associated with an individual's interactions, manifested in the way an individual expresses themselves and interacts with others in their social environment. However, little is explored empirically about the association of social network structure and the interactions of an individual with aspects of mental health functioning, such as anxiety. In recent years, individuals have begun to appropriate social media to self-disclose about their mental illnesses, seek support, and derive therapeutic benefits. The study examines the online social network and interaction characteristics of Twitter users who self-disclose about their anxiety disorders. We analyze a sample of 200 Twitter users and their over 200,000 posts shared on the platform, who were expert-validated to have self-disclosed about suffering from an anxiety disorder. On their data, a variety of attributes of the users' online social networks, interactions, and social behaviors using natural language and network analysis approaches were modeled using state-of-the-art network science measures. A number of state-of-the-art supervised learning classification frameworks are built using these attributes, to identify whether an individual's anxiety disorder status could be automatically inferred. Results show that these social network, behavior, and interaction attributes, when incorporated in a support vector machine classifier, signal an individual's self-reported anxiety disorder status, in contrast to a demographically and activity matched control group, with 79% accuracy and 84% area under the receiver-operating characteristic curve. The work provides the first insights into the role that the social interactions and social network structure on online platforms play in characterizing an individual's mental health experience, such as anxiety. We discuss the implications of our work in instrumenting online social platforms in ways that yield positive affordances and outcomes for individuals vulnerable to mental illnesses.

Keywords: Anxiety disorder · Network structure · Online interactions · Social media · Twitter

1 Introduction

Anxiety disorders constitute one of the leading mental health concerns in the United States. The disorder has an estimated prevalence of 1.6% to 5.0% in the general population [1]. Sufferers of this condition are known to experience non-specific persistent fear and worry, and become overly concerned with everyday matters [2]. This

C. Stephanidis et al. (Eds.): HCII 2020, LNCS 12427, pp. 249–264, 2020.
https://doi.org/10.1007/978-3-030-60152-2_20

condition is known to incur high costs, including reduced productivity, diminished quality of life, and even heightened risk of suicide [3].

Anxiety disorders are closely associated with an individual's interactions, manifested in the way an individual expresses themselves and interacts with others in their social environment. In fact, they are associated with reduced social engagement and reduced perceived quality of social relations [4]. Conversely, support from friends, family, and loved ones has been shown to buffer the effects of anxiety [5, 6]: the presence of social ties may enhance coping strategies and increase a individual's sense of control over the situation. Researchers have employed theories in attempts to explain this role of social integration in the maintenance of improved mental health and reduced risk to conditions like anxiety disorders, including social causation, symbolic interactionist, social exchange, self-esteem, meta-cognitive, and stress-vulnerability theories [7].

However, little is explored empirically about the association of social network structure and the interactions of an individual with aspects of mental health functioning, such as anxiety. Empirical work examining the relationship of social interactions with anxiety has suffered from significant practical challenges [8]. A threat to validity in existing cross-sectional studies is the potential bias in the retrospective recall of social ties among anxiety-affected individuals. Moreover, while some studies have indeed found positive associations between the structural characteristics of social relations and the availability of instrumental and emotional support [9], how interpersonal dynamics are associated with anxiety experiences remains less understood. Additionally, prior work has largely recognized the value of strong social ties in mental wellbeing [8]; whether similar findings hold true for weak ties as well is relatively less known.

In recent years, individuals have begun to appropriate social media platforms like Twitter and online communities like Reddit to self-disclose about their mental illnesses [10], seek support [11], and derive therapeutic benefits [12]. Social media language in particular has also been established to be valuable in understanding and predicting different forms of mental illness like depression [13] and suicide [14].

Nevertheless, research so far has provided limited insights into the role that the social interactions and social networks on these online platforms play in characterizing an individual's mental health experience, such as anxiety. This is especially a notable gap, given that positive benefits of these platforms have been examined with considerable interest in the past. For instance, social affordances of these platforms have been argued to augment social relationships and support mental health [14–17]. In a way, these social functions have been touted to augment the benefits of face-to-face interactions in mental health due to the reach, accessibility, ubiquity, and pervasiveness of the platforms [11, 18].

Considering this gap in the literature and given that continued negative social exchanges, presence of unhealthy and unsupportive social relations, and negativity in ties can exacerbate anxiety disorders [19], we seek to answer the following research question in this paper: **Can online social network structure and interactions, and social behaviors signal an individual's risk of anxiety?**

To this end, we chose the social media platform, Twitter. Our study focuses on a sample of 200 Twitter users and their over 200 thousand posts shared on the platform, who were expert-validated to have self-disclosed about suffering from an anxiety disorder. On their data, we model, using state-of-the-art network science measures [20], a variety of attributes of their online social networks, interactions, and social behaviors using natural language and network analysis approaches. We find that several of these attributes, when incorporated in a supervised learning classifier, successfully help distinguish them from a control group. For instance, anxiety users demonstrate a strong tendency to engage with folks with whom they did not have any prior bidirectional interaction; they also connect to a diverse set of smaller, mutually disconnected sub-networks. Our findings provide novel empirical insights into the manner in which online social networks and interactions can be indicative of an individual's mental health status, specifically anxiety. We discuss the implications of our work instrumenting online social platforms in ways that yield positive affordances and outcomes for individuals vulnerable to mental illnesses.

2 Data and Methods

2.1 Identifying Anxiety Users

We started by collecting a sample of Twitter users who had self- disclosed about their diagnosis of anxiety on Twitter through a public post. To identify anxiety diagnoses, inspired from prior work [10], we utilized a set of carefully curated search queries: "diagnosed [me]* with anxiety" and "i got/was/am/have been diagnosed with anxiety".

Table 1. Descriptive statistics of acquired Twitter data.

Total number of anxiety users	200
Total number of tweets by anxiety users	209290
Median number of Tweets by anxiety users	695
Total size of the 1 hop network of the anxiety users	41,557
Total number of control users	200
Total number of tweets by control users	211318
Median number of Tweets by control users	658
Total size of the 1 hop network of the control users	43,345

We used a web-based Twitter crawler called GetOldTweetsAPI to obtain tweets with self-disclosures of anxiety between 2013 and 2017. Our initial search based on the queries listed above gave us 3856 tweets shared by 966 users. Thereafter, for each user, two human annotators familiar with social media content around mental health manually inspected the collected tweets to identify if they indicated a genuine self-disclosure of anxiety. Then we removed users who had less than 20 followers and followings as these networks were too small to compute meaningful social network metrics. We also removed accounts which had too few posts (less than 10) and with more than 10,000 followers or followings. Our final dataset at the conclusion of this data-gathering step contained 200 users. For each of these users, we collected their entire timeline data again using the Twitter API. Table 1 presents some descriptive statistics of this dataset.

2.2 Collecting Social Network and Interaction Data of Anxiety Users

Social Network Data
For each of the above 200 users we proceeded to collect their (Twitter) social network data. We define network data to consist of the list of other users who are following the anxiety users and the list of those who our anxiety users are following. Twitter allows unidirectional links i.e., if A follows B, B does not necessarily follow A. For our research question, we focus on bidirectional ties, which are better indicators of social connections than unidirectional ties. Therefore, for each anxiety user, we consider the intersection of the accounts that the user is following and the accounts that are following the user as their bidirectional ties. We refer to this set of users as "friends" of the user. We obtained 41,557 users who were the "friends" of our 200 anxiety users. For each of these friends, we further collected their friends in turn to get the two-hop social network for each anxiety user. This resulted in a total of 15,297,258 users in the two-hop network of the anxiety users.

Social Interaction Data
Next, we collected the timeline tweets for the 200 anxiety users and their network using the GetOldTweets API as before. For each tweet, we specifically collected the timestamp of the tweet, number of likes (or favorites), whether it is a retweet or quote and if yes, the original tweeter, whether it is a reply and if yes, to whom. Any tweet which is a reply or had @ <username> in it is considered as an interaction tweet. Then we used this stylistic convention of tweets to compile a list of search queries by appending an "@" symbol before the username of each of our anxiety users. This operation provided us, via the GetOldTweets API, with all tweets that were incoming interactions to an anxiety user. Thus, for each such user, alongside the Twitter users who interacted them and the number of such interaction tweets, we also compiled the textual content of these interaction tweets.

2.3 Gathering Control Data

Finally, we describe an approach to gathering "control data", that is a set of Twitter users without any self-disclosure of anxiety. For the purpose, we collected a matching set of 200 users who did not use any of the anxiety diagnosis and self-disclosure related phrases defined above. For this control dataset, using the methods described above, we collected data on their followers, followings, their two-hop network neighborhood, the tweets shared on their timeline, as well as their social interaction data. In Fig. 1 we present summary distributions of the social interactions and networks over the anxiety and control users.

2.4 Statistical Approach

Building an Anxiety Classifier Recall, our research question revolves around identifying specific characteristics of an individual's social network structure and interaction that would indicate whether or not they have an underlying anxiety disorder. To do so, we adopt a supervised learning based binary classification approach, utilizing the Twitter data of the anxiety users and the control users described above. We first present a set of attributes that can be used to characterize the social engagement, social network, social interactions, and social behavior based differences of these two classes. Then we use these attributes as features to build the classifier.

Social Engagement Attributes

We define four measures of engagement inspired from [13]: First, volume which is the normalized number of tweets per day of an (anxiety/control) user (given as the ratio between the total number of tweets of a user to the total number of days of activity of the same user); Second, proportion of reply posts (@- replies) from a user per day which shows her level of social interaction with other Twitter users; Third, the fraction of retweets, indicative of information propagation behaviors, and fourth, the fraction of quotes from a user per day, which signals how they participate in information sharing with their followers. In addition, we also use three more engagement measures differentiating engagement with friends (strong ties) compared to non-friends (weak ties). First, the fraction of replies to friends, second, fraction of retweets whose original

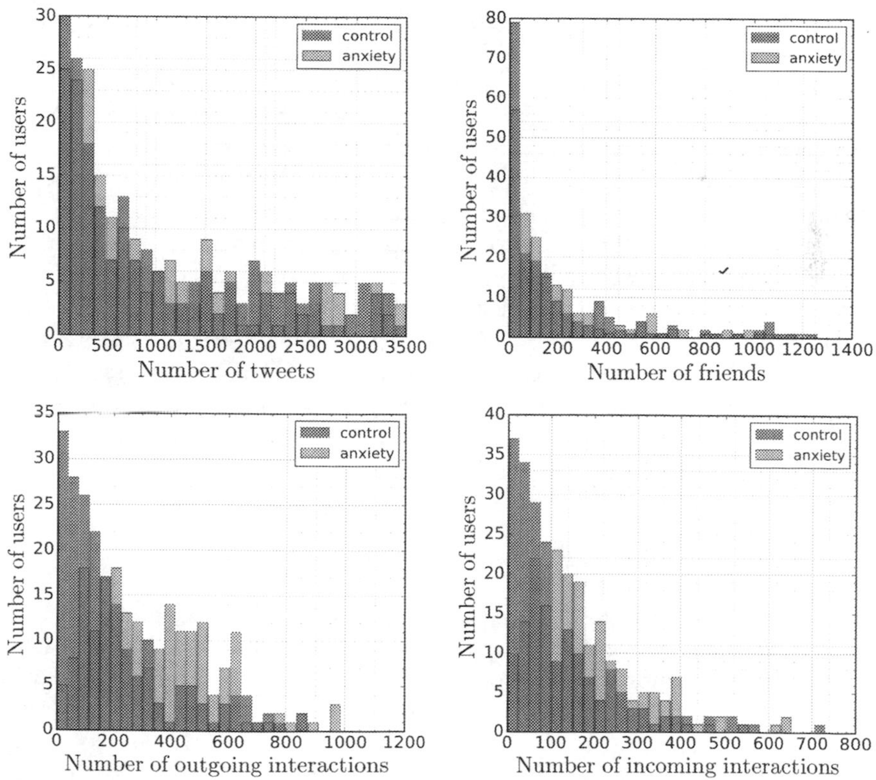

Fig. 1. Clockwise from top left: Distribution of tweets in the anxiety and control users; Distribution of friends (bidirectional ties) in the anxiety and control users; Distribution of outgoing interactions for anxiety and control users; Distribution of incoming interactions for anxiety and control users.

tweeters are friends of the user and third fraction of quotes whose original tweeters are friends of the user.

Attributes of Egocentric Social Graph Structure

Next we consider attributes of the egocentric social graph of an (anxiety/control) user. We define a user's ego-centric graph to be an undirected network of the set of nodes in their two-hop neighborhood (neighbors of the neighbors of the user in our dataset). Our egocentric attributes can be categorized into three types: drawing from the social network literature [13]:

- Node Properties: We define two attributes that characterize the nature of a user's egocentric social network. The first is the number of followers or inlinks of a user, the second is the count of their followees or outlinks. Additionally, we define a third feature, which is the count of their friends, which we define as the intersection of the followers and followees of the users (bidirectional links).
- Dyadic Properties: Here, we define two attributes. Our first attribute is a measure called reciprocity, which is measured as how many times a user *u* responds to another user *v* who had sent them @-reply messages. The second attribute is the prestige ratio, and is defined as the ratio of the number of @-replies that are targeted to *u*, to the number of @-replies targeted to a user *v*, where *v* is a user with whom *u* has bi-directional @-replies.
- Network Properties: In this category, we define seven attributes. The first is betweenness centrality – this attribute is for quantifying the control of a user on the communication between other users in a social network. The second attribute graph density is the ratio of the count of edges to the count of nodes in *u*'s egocentric (bidirectional) social network. The third attribute is the clustering coefficient of *u*'s ego network, which is a notion of local density. The fourth attribute, size of two-hop neighborhood is defined as the count of all of *u*'s neighbors, plus all of the neighbors of *u*'s neighbors. We define the fifth attribute embeddedness of *u* with respect to their neighborhood as the mean of the ratio between the set of common neighbors between *u* and any neighbor v, and the set of all neighbors of *u* and *v*. The sixth attribute in this category is the number of ego components in *u*'s ego network, defined as the count of the number of connected components that remain when the focal node *u* and its incident edges are removed [21]. The final attribute of this category is the normalized average size of the ego components of a user *u* in their ego network.

Egocentric Interaction Graph Attributes

Moving from social engagement and social network structure to social interactions, we further define a number of egocentric interaction graph attributes based the interactions of an anxiety or control user with others on Twitter (through @-replies). For the interaction graph, an edge between *u* and *v* implies that there has been at least one @-reply exchange each, from *u* to *v*, and from *v* to *u*. The interaction network is a subgraph of the egocentric social network graph as we consider interactions of a user with only friends of the user and the interactions of these friends with their own friends in turn, creating a two-hop interaction network of the anxiety/control users. We define the following specific attributes:

- Unsigned Network Properties: Here, we define seven attributes, similar to the network properties defined for the social network structure above.
- Signed Network Properties: While the general structure of a social network can be useful for a problem domain like ours, individuals often share rich relationships with their peers, which cannot necessarily be captured via simple pairwise unsigned links. To take into account both positive and negative pairwise interactions between individuals, we define attributes drawing upon the literature in modeling signed network properties [22]. For developing such a signed network, we assign polarity (positive or negative) to each edge of the interaction network. Specifically, we perform a sentiment analysis of all the interaction between a user *u* with a friend *v*, using an ensemble approach that combines the outcomes given by tools such as VADER [23], Stanford CoreNLP [24], and the NLTK library. Depending on whether there are more positive or negative interactions we define the net edge as positive or negative tie. From this signed network of every user, we define four attributes. The first measure is the fraction of negative ties. The second measure is the ratio of number of balanced triads to the total number of triads in the network. Balanced triad is a set of three connected users in which there is an odd count of positive edges (i.e., one or three positive edges) as shown in Fig. 2. The third attribute is the average degree of the nodes having negative ties with the user *u*. In order to understand how the users feel about their family and close relations, we defined the final attribute as the fraction of family category tweets which shows negative sentiment. To detect the tweets of family category we checked for the presence of 'family' category terms in the psycholinguistic lexicon Linguistic Inquiry and Word Count (LIWC) [25].

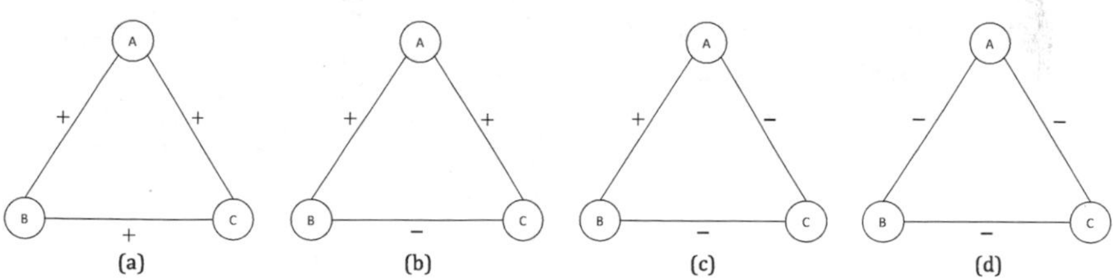

Fig. 2. Structural representation of balanced (a, c) and unbalanced signed triads (b, d), adapted from [22].

Social Behavioral Attributes

Finally, we consider four attributes of the emotional state of users in our dataset: positive affect (PA), negative affect (NA), anger, and sadness, drawing from prior work on psycholinguistics, mental health, and social media. These works have revealed that an individual's behaviors towards their friends and peers are often reflective of their underlying psychologies [25]. Daily measurements of these attributes per user were computed using LIWC. We also sought to measure social behaviors through a user's expression of linguistic style in their posts. For this, we again use LIWC, focusing on 22 specific linguistic style categories, drawing on prior research [25]: articles, auxiliary verbs, conjunctions, adverbs, personal pronouns, prepositions, functional words, assent, negation, certainty and quantifiers.

Classification Framework
Using the above-defined features, we use supervised learning to construct classifiers trained to predict anxiety in our two user classes. We compare several different parametric and non-parametric binary classifiers (Gaussian Process classifier, Decision Tree classifier, Random Forest classifier, Multi-Layer Perceptron classifier, Adaboost classifier, Gaussian Naive Bayes classifier, Logistic Regression classifier and Support Vector Machine or SVM classifier) to empirically determine the best suitable classification technique. In order to understand the importance of various feature types, for each classifier, we trained one model each using each category of attributes defined above. Additionally, we built a fifth classification model using dimensionality-reduced set of all features. The positive and negative examples for these classifiers came from the set of the 200 anxiety users and 200 control users respectively. We used k-fold (k = 5) cross validation alongside a 20% held out dataset to tune and then test our models.

3 Results

We begin by presenting the results of our best performing classifier, an SVM. The results are presented in Table 2. We find that the best performing model in the test set (s) yields an average accuracy of 79% (σ = 2.3 across the five cross validation folds) and high precision of 0.86, corresponding to the anxiety class. Note that a baseline chance model would yield accuracy of 50%, since the positive and negative classes are balanced. Good performance of this classifier is also evident from the receiver-operator characteristic (ROC) curves shown in in Fig. 3.

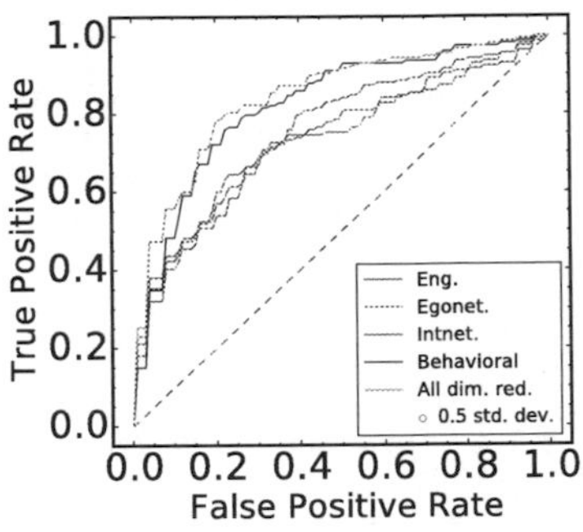

Fig. 3. ROC curve for the SVM based anxiety classification model.

We see that the model using the social behavioral attributes performs the best among all of the models. Results in prior literature suggest that use of linguistic styles such as pronouns and articles provide information about how individuals respond to psychological triggers [26]. Finally, the better performance of interaction network features than ego-network features shows that the friends with whom users interact with, bear more predictive cues to their psychological well-being (anxiety status) than just the structure of the social network in which they are embedded. We can conclude that social media network structure and interactions together with social behavioral attributes provide useful signals that can be utilized to classify if an individual is suffering from anxiety.

Table 2. Performance metrics of anxiety classification

Model	Precision	Recall	Accuracy	ROC-AUC
Engagement	0.75	0.62	0.68	0.73
Ego network	0.62	0.49	0.58	0.7
Interaction network	0.66	0.57	0.61	0.74
Social behaviors	0.74	0.73	0.73	0.8
All features dim. reduced	0.86	0.73	0.79	0.84

Based on the performance of the different classifiers above, we present some analyses of differences in the two classes of users in Table 3.

First, corresponding to the attributes relating to **social engagement**, we see that although is there no significant difference in the total volume (tweets, retweets, quotes and replies) of the posts in these two categories, the anxiety class shows marked higher volume of replies and low the volume of retweets. This suggests that people suffering from anxiety find it more comfortable interacting with other people instead of generating content. Further, the ratio of replies to non-friends compared to friends for the anxiety class is significantly higher than the control group. Non-friends are usually people who the users follow like celebrities, but they typically do not follow the users back as they might not know the users, whereas friends are the people who follow each other and know each other in some context. This suggests that users with anxiety may find it easier to interact where they have no preconceived image to maintain of themselves. People suffering from anxiety further show significantly less number of followers and followees but no significant difference in the number of friends. This indicates reduced desire to consume content beyond the known social sphere. This combined with less interactions in the form of replies with friends indicate that users prefer to consume the content produced by their friends in a passive manner, probably to know what they are doing, rather than directly communicating with them.

In the **egocentric social graph** attributes category, for dyadic properties, anxiety users also show reduced reciprocity to others' communications, indicating decreased desire for social interaction with friends. Since interaction with non-friends is not bidirectional, these do not contribute to building up the reciprocity of a user in our analysis.

For the **egocentric interaction graph** attributes, we observe that the network properties become pronounced between the two user classes. Anxiety users show higher number of ego components, higher betweenness centrality, lower clustering coefficient, higher size of two-hop neighborhood, and lower embeddedness. Lower embeddedness indicates that the anxiety users have lower number of mutual friends with their neighbors in their interaction graph. Next, anxiety users show significantly more number of negative ties, however the number of balanced triads shows no significant difference between the two groups. This implies although the anxiety users have stable networks (balanced triads), they have high negativity, expressed in the shared tweets, in their network. The anxiety users also show higher average degree of negative interaction ties suggesting that they have more negative ties of higher status in their social circles. Our next feature, the proportion of tweets expressing negative sentiment around family topics is also significantly higher for the anxiety users compared to the control group.

Table 3. Average values and differences of significant attributes for the anxiety and control groups (*** $p < 0.001$ following Bonferroni correction) for all the models.

Model	Control	Anxiety	Diff (%)	*p*-value
Engagement				
Replies	0.1998	0.2814	40.84	***
Followers	1402	482	−65.62	***
Replies to non-friends	0.426	0.506	18.77	***
Retweets	0.5093	0.3056	−39	***
Followees	1187	756	−36.31	***
Ego network				
Reciprocity	0.98	0.65	−33.67	***
Interaction network				
Embeddedness	0.028	0.017	−39.28	***
2-hop network size	398.4	425.8	6.87	***
clustering coefficient	0.0629	0.0381	−39.42	***
Fraction of neg. ties	0.14	0.19	35.71	***
Deg. of neg. ties	1.45	1.87	28.96	***
Family neg. tweets	0.041	0.053	29.26	***
#Ego components	15.2	20.2	32.89	***
Ratio of negative ties	0.075	0.095	26.67	***
Betweenness centrality	0.6365	0.7011	10.14	***
Social behaviors				
First person singular	0.0392	0.0766	95.4	***
Percept	0.0143	0.0218	52.44	***
Feel	0.0054	0.0062	14.81	***
Social	0.0511	0.0691	35.22	***
Health	0.0047	0.0063	31.91	***
Conjunction	0.0333	0.0504	51.35	***
Negative affect	0.0071	0.0102	43.66	***
Swear	0.0045	0.0067	48.88	***
Positive affect	0.0382	0.0473	23.82	***
Third person	0.0083	0.0091	9.63	***
Sadness	0.0037	0.0057	54.05	***

In the category of **social behavioral attributes**, we see significantly higher use of first person singular pronouns for the anxiety users. Increased self-attentional focus is known to indicate weak psychological functioning [13, 25]. We also observe higher use of social and health words which implies higher social and personal concern among the anxiety users. Additionally, for the anxiety users, we see greater emotional expressivity:

higher positive affect, negative affect, swear and sadness. Studies in the psychology literature [27] has observed that individuals suffering from anxiety disorders experience mood swings and these results confirm their expression in social media.

4 Discussion

4.1 Summary of Principal Results

This research offers many interesting and valuable findings that augment theoretical and empirical insights in prior work. First, we found that although weak ties are a hallmark of social media platforms like Twitter [28], individuals were indeed more affected by their interactions with the strong ties (or friends); when these interactions were negative, it resulted in greater risk to anxiety. Similarly, negative interactions centered around topics of close relationships like family were a significant contributor to anxiety. At the same time, the fact that anxiety users in our dataset were part of smaller, more disconnected diverse sub-networks goes on to show that, on the one hand, online social integration, like its offline counterpart, is limited in this population. But on the other hand, it questions to what extent vulnerable individuals are able to mobilize the social connectivity functionalities of social media to find access to larger communities of peers and support groups.

4.2 Comparison with Prior Work

As noted earlier, considerable work has been conducted to understand how various social environmental factors impact anxiety disorder risk or how anxiety disorders manifest in an individual's social interactions [4, 8]. However, whether these findings hold true for online social interactions as well has been less studied. Given the pervasive adoption of social media and their integration in daily lives, an understanding the relationship of online interactions with anxiety, is vital and our work has attempted to fill this gap.

Social Predictors of Anxiety

A rich body of literature exists in psychiatry and psychology to understand the causes and characteristics of anxiety disorders [6, 32]. Importantly, as noted above, social environmental factors can act as stressors to anxiety disorders. In fact, the link between social isolation and reduced psychological well-being is well established in sociology, dating back to Durkheim [30]. Smaller social networks, fewer close relationships, and lower perceived adequacy of social support have been linked to symptoms of anxiety [5, 8, 31].

Despite this attention, empirical work examining the impact of social interactions on anxiety experiences has faced significant challenges [8, 33, 34]. Our results showed empirical evidence in support of the existing theories, while mitigating some of the challenges of empirical work that relies heavily on retrospective recall and self-reports of social networks. Further, we evaluated whether online social network attributes like presence or absence of close-knit groups, high or low clustering in a person's egocentric social network, or positive or negative social ties can be indicative of anxiety, which in themselves are novel findings in this research area.

Mental Health and Social Media

It has been recognized that people share content about their emotional health on social media [35, 36], and there is a rich and growing body of work in social media research to develop quantitative and computational methods to understand various forms of mental and behavioral health and wellness states [37], like depression [13, 38–41], suicidal ideation [14, 42, 43], post traumatic stress disorder [44], schizophrenia [45, 46], social anxiety [47–50], and substance use [51, 52]; see Guntuku et al. [53] for a review.

However, we note that gaps exist in our understanding of the relationship of online social interactions and online social networks with the mental health of social media users. While existing works have largely established the value of social media language in helping characterize and even predict risk to various mental illnesses, the important role of social connectedness on these platforms on one's mental health experiences is relatively unexplored. As noted in a systematic review by Park et al. [55], others by Seabrook et al. [56] and by Dobrean and Pasarelu [57], social media use has been reported to be associated with lower levels of loneliness and greater belonging, social capital, and actual and perceived access to social support and is generally associated with higher levels of life satisfaction and self-esteem [58–60]. As a whole, the positive social components of social media (and broadly, Internet) use has been argued to serve a protective role against depression and anxiety [15, 16], although the converse – how negative online social interactions and specific network structures impact the risk of mental illnesses has remained an open question for investigation. Further, we note that anxiety disorders, as a mental illness [54], has received relatively less attention in the social media literature, with the exception of Tian et al. [61], who provided a qualitative thematic analysis of a random sample of 1000 anxiety-related postings on Sina Weibo. This work found anxiety-disclosures to be the biggest theme, but did not explore, in computational, scalable, or other capacity, the role of social connectedness in this illness.

Inspired by literature in psychology that have explored the impact of negative social ties and negative interactions on mental health of individuals (ref. above), we developed a variety of computational techniques to first characterize the social ties of anxiety sufferers as positive or negative and then assess their association with individual anxiety. Thereby, we have complemented the analyses in the above social media work —where the focus has largely been on identifying linguistic markers—to identifying what social attributes and social behaviors are associated with mental illnesses, particularly anxiety disorders.

4.3 Limitations and Future Work

There are some limitations to our work. We acknowledge an inherent population bias in our dataset. Not all demographic groups use Twitter and not all Twitter users self-disclose or share information indicative of their mental health (anxiety) status. Our observations are limited by individuals who choose to discuss about their anxiety on Twitter, maintain a social network with others on the platform, and engage in substantial interactions with them. Finally, in this work we did not obtain clinical validation on either the anxiety classification results or the empirical observations linking social interactions and anxiety. Future work can explore these opportunities through interdisciplinary collaborations as well as by gathering self-reported data. Last but certainly not the least, despite working with public Twitter data in this paper, future work, in the light of recent research [29], should also include discussions of people's privacy perceptions and ethical concerns around deriving such (sensitive) mental health assessments from social media data.

Nevertheless, our work offers several practical implications for instrumenting social media platforms to support the experiences of individuals with or at risk of anxiety disorders, which constitute promising directions for future research. Recall that we found that, although individuals suffering from anxiety disorders have more negative ties, the fraction of balanced triads in their network structure is not significantly different from the control group. Based on this finding, online platforms like Twitter can help users who self-disclose about their anxiety build and enrich their social networks. Platforms can suggest connections which are already observed to have positive interactions with other social ties of the user, or has negative interactions with the already existing negative ties of the user. Since individuals tend to reach out to weak ties in the broader online community during times of heightened anxiety, platforms can also enable provisions to recommend specific groups or lists where individuals can be less concerned about impression management and can engage in more disinhibiting discourse about their condition and experiences.

5 Conclusion

In this paper, we presented a large-scale data-based study examining the online social network and interaction characteristics of Twitter users who self-disclose about their anxiety disorders. We found that these attributes of network and interaction can be powerful in identifying, via a supervised learning based classifier, those at risk of this condition or those experiencing this condition, as self-reported on the Twitter platform. Our work provides one of the first results situating the relationship between online social interactions and networks, and anxiety disorders, and situated the important value of weak social ties prevalent on online platforms in understanding people's mental health experiences.

Acknowledgements. We thank Jennifer Ma for her involvement in writing some of the Twitter data collection scripts.

References

1. Grant, B.F., et al.: The epidemiology of social anxiety dis- order in the united states: results from the national epidemiologic survey on alcohol and related conditions. J. Clin. Psychiatry **66**(11), 1351 (2005)

2. Hettema, J.M., Neale, M.C., Kendler, K.S.: A review and meta-analysis of the genetic epidemiology of anxiety disorders. Am. J. Psychiatry **158**(10), 1568–1578 (2001)
3. Weisberg, R.B.: Overview of generalized anxiety disorder: epidemiology, presentation, and course. J. Clin. Psychiatry, **70**(suppl 2), 4–9 (2009)
4. Rapee, R., Heimberg, R.: A cognitive-behavioral model of anxiety in social phobia. Behav. Res. Therapy **35**(8), 741–756 (1997)
5. La Greca, A.M., Harrison, H.M.: Adolescent peer relations, friendships, and romantic relationships: do they predict social anxiety and depression? J. Clin. Child Adolescent Psych. **34**(1), 49–61 (2005)
6. Cohen, S., Wills, T.A.: Stress, social support, and the buffering hypothesis. Psychol. Bull. **98** (2), 310 (1985)
7. Achat, H., Kawachi, I., Levine, S., Berkey, C., Coakley, E., Colditz, G.: Social networks, stress and health-related quality of life. Quality Life Res. **7**(8), 735–750 (1998)
8. Kawachi, I., Berkman, L.F.: Social ties and mental health. J. Urban health **78**(3), 458–467 (2001)
9. Corrigan, P.W., Phelan, S.M.: Social support and recovery in people with serious mental illnesses. Commun. Mental Health J. **40**(6), 513–523 (2004)
10. Coppersmith, G., Dredze, M., Harman, C.: Quantifying mental health signals in twitter. In: Proceedings of the Workshop on Computational Linguistics and Clinical Psychology (2014)
11. Burke, M., Marlow, C., Lento, T.: Social network activity and social well-being. In: Proceedings of the SIGCHI Conference on Human Factors in Computing Systems, pp. 1909–1912. ACM (2010)
12. Ernala, S.K., Rizvi, A.F., Birnbaum, M., Kane, J., De Choudhury, M.: Linguistic markers indicating therapeutic outcomes of social media disclosures of schizophrenia. PACM HCI **1** (1), 43 (2017)
13. De Choudhury, M., Gamon, M., Counts, S., Horvitz, E.: Predicting depression via social media. In ICWSM (2013)
14. De Choudhury, M., Kıcıman, E.: The language of social support in social media and its effect on suicidal ideation risk. In: ICWSM (2017)
15. McCord, B., Rodebaugh, T.L., Levinson, C.A.: Face- book: social uses and anxiety. Comput. Hum. Behav. **34**, 23–27 (2014)
16. Heather Cleland Woods and Holly Scott: # sleepyteens: social media use in adolescence is associated with poor sleep quality, anxiety, depression and low self-esteem. J. Adolescence **51**, 41–49 (2016)
17. Ellison, N.B., Steinfield, C., Lampe, C.: The benefits of facebook friends: social capital and college students use of online social network sites. J. Comput.-Mediat. Commun. **12**(4), 1143–1168 (2007)
18. Steinfield, C., Ellison, N.B., Lampe, C.: Social capital, self- esteem, and use of online social network sites: a longitudinal analysis. J. Appl. Develop. Psychol. **29**(6), 434–445 (2008)
19. Lee, R.M., Robbins, S.B.: The relationship between social connectedness and anxiety, self-esteem, and social identity (1998)
20. Wasserman, S., Faust, K.: Social Network Analysis Methods and Applications, vol. 8. Cambridge University Press, Cambridge (1994)
21. De Choudhury, M., Mason, W.A., Hofman, J.M., Watts, D.J.: Inferring relevant social networks from interpersonal communication. In: Proceedings WWW (2010)
22. Heider, F.: The Psychology of Interpersonal Relations. Psychology Press, Hove (2013)
23. Hutto, C.J., Gilbert, E.: Vader: a parsimonious rule-based model for sentiment analysis of social media text. In: ICWSM (2014)

24. Manning, C.D., Surdeanu, M., Bauer, J., Finkel, J.R., Bethard, S., McClosky, D.: The stanford corenlp natural language processing toolkit. In: ACL (System Demonstrations), pp. 55–60 (2014)
25. Pennebaker, J., Francis, M., Booth, R.: Linguistic Inquiry and Word Count: LIWC 2001. Lawrence Erlbaum Assoc, Mahway (2001)
26. Pennebaker, J., Chung, C.: Expressive writing, emotional upheavals, and health. In: Handbook of Health Psychololgy, pp. 263–284 (2007)
27. Bowen, R., Clark, M., Baetz, M.: Mood swings in patients with anxiety disorders compared with normal controls. J. Affect. Disord. **78**(3), 185–192 (2004)
28. Kwak, H., Lee, C., Park, H., Moon, S.: What is twitter, a social network or a news media? In: Proceedings of the 19th International Conference on World Wide Web, pp. 591–600. ACM (2010)
29. Fiesler, C., Proferes, N.: Participant perceptions of twitter research ethics. Soc. Media + Soc. **4**(1) (2018). 2056305118763366
30. Durkheim, E.: Suicide: A Study in Sociology (ja spaulding & g. simpson, trans.), pp. 32–59. Free Press, Glencoe (1951)
31. Hill, H.M., Levermore, M., Twaite, J., Jones, L.P.: Exposure to community violence and social support as predictors of anxiety and social and emotional behavior among african american children. J. Child Family Stud. **5**(4), 399–441 (1996)
32. Wells, A.: A metacognitive model and therapy for generalized anxiety disorder. Clin. Psych. Psychother. **6**(2), 86–95 (1999)
33. Hassan, E.: Recall bias can be a threat to retrospective and prospective research designs. Internet J. Epidemiol. **3**(2), 339–412 (2006)
34. Golder, S.A., Macy, M.W.: Diurnal and seasonal mood vary with work, sleep, and day length across diverse cultures. Science **333**(6051), 1878–1881 (2011)
35. Settanni, M., Marengo, D.: Sharing feelings online: studying emotional well-being via automated text analysis of facebook posts. Front. Psychol. **6**, 1045 (2015)
36. De Choudhury, M., De, S.: Mental health discourse on reddit: self-disclosure, social support, and anonymity. In: ICWSM (2014)
37. Coppersmith, G., Dredze, M., Harman, C., Hollingshead, K.: From Adhd to sad: analyzing the language of mental health on twitter through self-reported diagnoses. In: Proceedings of the 2nd Workshop on Computational Linguistics and Clinical Psychology: From Linguistic Signal to Clinical Reality, pp. 1–10 (2015)
38. Schwartz, H.A., et al.: Towards assessing changes in degree of depression through facebook. In: Proceedings of the Workshop on Computational Linguistics and Clinical Psychology: From Linguistic Signal to Clinical Reality, pp. 118–125 (2014)
39. Reece, A.G., Reagan, A.J., Lix, K.L., Dodds, P.S., Danforth, C.M., Langer, E.J.: Forecasting the onset and course of mental illness with twitter data. Sci. Rep. **7**(1), 13006 (2017)
40. Reece, A.G., Danforth, C.M.: Instagram photos reveal predictive markers of depression. EPJ Data Sci. **6**(1), 15 (2017)
41. Eichstaedt, J.C., et al.: Facebook language predicts depression in medical records. In: Proceedings of the National Academy of Sciences, **115**(44), 11203–11208 (2018)
42. De Choudhury, M., Kiciman, E., Dredze, M., Coppersmith, G., Kumar, M.: Discovering shifts to suicidal ideation from mental health content in social media. In: Proceedings of the 2016 CHI conference on human factors in computing systems, pp. 2098–2110. ACM (2016)
43. Coppersmith, G., Ngo, K., Leary, R., Wood, A.: Exploratory analysis of social media prior to a suicide attempt. In: Proceedings of the Third Workshop on Computational Linguistics and Clinical Psychology, pp. 106–117 (2016)
44. Coppersmith, G.A., Harman, C.T., Dredze, M.H.: Measuring post traumatic stress disorder in twitter. In: ICWSM (2014)

45. Mitchell, M., Hollingshead, K., Coppersmith, G.: Quantifying the language of schizophrenia in social media. In: Proceedings of the 2nd Workshop on Computational Linguistics and Clinical Psychology: From Linguistic Signal to Clinical Reality, pp. 11–20 (2015)
46. Birnbaum, M.L., Ernala, S.K., Rizvi, A.F., De Choudhury, M., Kane, J.M.: A collaborative approach to identifying social media markers of schizophrenia by employing machine learning and clinical appraisals. J. Med. Internet Res. **19**(8), e289 (2017)
47. Fernandez, K.C., Levinson, C.A., Rodebaugh, T.L.: Profiling: predicting social anxiety from facebook profiles. Soc. Psychol. Personal. Sci. **3**(6), 706–713 (2012)
48. Prizant-Passal, S., Shechner, T., Aderka, I.M.: Social anxiety and internet use–a meta-analysis: what do we know? what are we missing? Comput. Hum. Behav. **62**, 221–229 (2016)
49. Weidman, A.C., Levinson, C.A.: Im still socially anxious online: offline relationship impairment characterizing social anxiety manifests and is accurately perceived in online social networking profiles. Comput. Hum. Behav. **49**, 12–19 (2015)
50. Deters, F., Mehl, Matthias R., Eid, Michael: Social responses to facebook status updates: the role of extraversion and social anxiety. Comput. Hum. Behav. **61**, 1–13 (2016)
51. Tamersoy, A., De Choudhury, M., Chau, D.H.: Characterizing smoking and drinking abstinence from social media. In: Proceedings of the 26th ACM Conference on Hypertext & Social Media, pp. 139–148. ACM (2015)
52. Liu, J., Weitzman, E.R., Chunara, R.: Assessing behavioral stages from social media data. In: CSCW (2017)
53. Guntuku, S.C., Yaden, D.B., Kern, M.L., Ungar, L.H., Eichstaedt, J.C.: Detecting depression and mental illness on social media: an integrative review. Curr. Opin. Behav. Sci. **18**, 43–49 (2017)
54. Noyes, R.: Comorbidity in generalized anxiety disorder. Psychiatric Clin. North Am. **24**(1), 41–55 (2001)
55. Park, S., Kim, I., Lee, W.S., Yoo, J., Jeong, B., Cha, M.: Manifestation of depression and loneliness on social networks: a case study of young adults on facebook. In: Proceedings of the 18th ACM Conference on Computer Supported Cooperative Work & Social Computing, pp. 557–570. ACM (2015)
56. Seabrook, E.M., Kern, M.L., Rickard, N.S.: Social networking sites, depression, and anxiety: a systematic review. JMIR Mental Health **3**(4), e50 (2016)
57. Dobrean, A., Pasarelu, C.-R.: Impact of social media on social anxiety: a systematic review. In: New Developments in Anxiety Disorders. InTech (2016)
58. Valkenburg, P.M., Peter, J., Schouten, A.P.: Friend networking sites and their relationship to adolescents' well-being and social self-esteem. Cyber Psychol. Behav. **9**(5), 584–590 (2006)
59. Vogel, E.A., Rose, J.P., Roberts, L.R., Eckles, K.: Social comparison, social media, and self-esteem. Psychol. Popular Media Cult. **3**(4), 206 (2014)
60. Barry, C.T., Sidoti, C.L., Briggs, S.M., Reiter, S.R., Lindsey, R.A.: Adolescent social media use and mental health from adolescent and parent perspectives. J. Adolescence **61**, 1–11 (2017)
61. Tian, X., He, F., Batterham, P., Wang, Z., Guang, Yu.: An analysis of anxiety-related postings on sina weibo. Int. J. Environ. Res. Public Health **14**(7), 775 (2017)

Gender Digital Violence - Study, Design and Communication of an Awareness-Raising Campaign from University to University

Mauro Ferraresi(✉)

IULM International University of Languages and Media, 20141 Milan, Italy
mauro.ferraresi@iulm.it

Abstract. This article aims to be a reflection, an analysis and a return to the reader of the path of conception and implementation of a communication campaign against gender violence. The project has been developed at the Inter-University Research Centre "Gender Cultures" in Milan, and wants to contribute and to give substance to the Istanbul Convention, which in its article 13 states that "to promote or implement awareness campaigns or programs (…) to increase awareness and understanding by the general public of the various manifestations of all forms of violence (…) as well as the need to prevent them".

To this end, three Italian and Milanese universities have been involved and a specific methodology has been used that has highlighted, from a communicative point of view, some important innovations. A group of university students, previously trained, carried out the creative work for the awareness campaigns by dividing themselves into subgroups and structuring a campaign in which the language was completely internal to the subculture of reference. Not only because the social segment to which the communication producers belonged was in fact homogeneous in terms of lifestyles, beliefs and/or worldview to the social and cultural segment of the campaign's target, but also because they were inspired by tales of violence experienced by other students at the Milanese universities. Another element of solid novelty concerns the field research carried out with neuromarketing techniques on a representative and very small sample of the reference universe.

Keywords: Gender studies · Communication · Sociological research

1 Foreword

1.1 The Interuniversity Research Centre 'Gender Cultures'

On 26 November 2013 the first inter-university Centre for the study and dissemination of gender cultures was born in Milan, Italy, with a Statute signed by the Rectors, the Scientific Council, the President and the relevant members.

As stated in the Statute, "The Centre aims to give permanent impetus to studies, research and positive action relating to the theme of gender cultures and thus contribute to the growth and spread of respect for the dignity and skills of women" (art. 2, 1). In particular, the Centre promotes and coordinates research programs on the subject, in

C. Stephanidis et al. (Eds.): HCII 2020, LNCS 12427, pp. 265–272, 2020.
https://doi.org/10.1007/978-3-030-60152-2_21

collaboration with other academic, governmental and international institutes. The varied scientific background of the members of the Scientific Council and its members is a guarantee of transdisciplinary approach to gender issues in all social and cultural aspects.

The Interuniversity Research Centre 'Gender Cultures' involves six Universities in Milan (University of Milan-Bicocca, University of Milan, Politecnico di Milano, Università Commerciale Luigi Bocconi, Università IULM, Università Vita - Salute San Raffaele), and assumes in the premises of its Statute the intention of the affiliated Universities to give impetus, with the instruments of their competence, to studies, research and positive action on the theme of gender cultures, in the spirit of the European Resolution of 8 March 2011, European Directive 113 of 2004 and Cedaw (Convention on the elimination of all forms of discrimination against women) adopted by the UN and ratified by 185 States.

2 The Plan of the Research

As part of the tasks of the inter-university Centre, in the spring of 2019, the nominees of the respective Rectors as members of the Scientific Council: Valeria Bucchetti for the Politecnico di Milano, Mauro Ferraresi for IULM, International University for Languages and Media IULM, plus Sveva Magaraggia for the University of Milan-Bicocca, started a three-step research.

The three-step research plan included participant observation and focus groups; design the system of communication artifacts and, finally, conduct sociological research implemented with neuromarketing techniques on the final four chosen advertising creative solutions.

2.1 Participant Observation and Focus Groups

Before starting the research, Sveva Magaraggia [5] made a list of suggestions on how to build social campaigns against gender-based violence. She reported that the literature suggests that gender violence should be represented as a process and not as an act [10], and to do this it is useful both to tell the cycle of violence present in the report and to give a precise image of the social context in which the violence occurred [4, 6].

The second knot concerns the familiarity of the campaigns that will have to be created. The images and texts of the advertisements must be familiar to the chosen target group, i.e. the students of the six universities involved [19]. This meant the need to know at least part of the students' language to represent a setting that was immediately recognizable to them. The underlying hypothesis was that familiarity with the language and images of the campaign would lead to an identification and a predisposition to implement those behaviors suggested to prevent male violence. In addition, the suggested behavior that advertising campaigns should induce should be presented as conduct that everyone could adopt, and not as a heroic or dangerous act [1, 19].

The third knot concerns the repertoire of possible discursive registers that males may exhibit in the reproduction of sexist discourse. Speeches can in fact also vary on the registers of irony and sarcasm, but it was felt that maintaining these registers even within the awareness campaign, in order to maintain that familiarity of language mentioned above, would have meant somehow diminishing their importance [13, 16].

The last aspect concerned the importance of intervening on men, and how strategic it is to call them into question both as authors and as allies. Including men in campaigns for the prevention of gender violence is not only a change in communication strategy, but it corresponds to a theoretical paradigm shift, to a recent transformation of perspective in the study of gender violence:

Sveva Magaraggia explains: "With this knowledge we started the first phase of the project, which aimed to know the moments in which students from different universities in Milan lived, witnessed or came to know about episodes of violence and know the language they use to talk about these issues. The methods adopted were two: a participatory observation in the places of sociality of students and five focus groups with 6–8 students and female students from different universities in Milan. Students talked mainly about verbal violence and both benevolent and hostile sexism [1, 4, 16] identified places in universities where they witnessed violence or sexism, most of the discussions were about online violence. The focus groups ended with discussions about their reactions and strategies to respond to violence" [5].

2.2 Design the System of Communication Artifacts

As for the second step, Valeria Bucchetti states: "The communication campaign took its own concrete form during a workshop dedicated to the design of communicative artifacts, which would allow to implement the communicative action useful to respond to the intentions of the issuer" [5].

The design role was assigned to six groups of students in the third year of the Communication Design degree course at the Politecnico di Milano. The students, assisted by Bucchetti, chose the poster format for the campaigns and tried to obtain, as a result of the communication process, to highlight the problem of male verbal violence against women within our universities and to counter the phenomenon, first of all by raising awareness.

"Specifically, it was decided to articulate the project brief taking into account the different university contexts, encouraging a reflection on the behaviours in formal areas such as classrooms, archives, libraries and informal university contexts, on what is experienced in areas of refreshment or leisure, in areas of passage (corridors, courtyards,…), until referring to the virtual contexts of social - Instagram, Facebook, Whatsapp etc., dwelling on the forms of verbal violence acted both in the presence and in the absence of the person to whom it is addressed, and on the forms of indirect verbal violence to which we are witnesses" [5].

The elements and directives of research that emerged in the first step (participant observation and focus groups) have constituted an essential data base to start reflections and have suggested interpretative keys around which to grasp and develop a creative thread. In this sense, particularly relevant was the frequency of the stories witnessed by the students, in which messages received from young people unknown to them, which were transformed into sexual advances, emerged.

There are more widespread forms of communication to deal with the issue of violence. Bucchetti in particular tried to avoid those negative implications that Semprini

[17] defines as "good distance"; that is, the tendency to adopt narrative strategies that make the object of discourse perceived as far away in space and time from the spectator. In other words, those messages that propose a stylized vision of male violence against women should be avoided because they risk transmitting an unrealistic account of the phenomenon, which therefore becomes difficult to identify in everyday reality. In the same way, the choice to omit points of reference that relate the episodes of violence told with the daily life of the public gives rise to a decontextualized narrative distant from the physical and cultural reality experienced by the individual recipient who thus risks not being involved and not being questioned by the campaign and its detached language. For all these reasons, the six groups have created as many creative ideas that have tried to use an immediate and direct language, using reference points drawn directly from the students' lived reality [2, 3]. The six proposals were then judged and selected and reduced to four.

2.3 Four Advertising Creative Solutions

Following the production of the communication artifacts and the implementation of six creative solutions, an ex ante analysis of the proposed communication was necessary [11]. An evaluation group was initially set up, composed of the research directors, which took into consideration the six creative proposals in order to reduce them to four on which it was necessary to carry out research before their dissemination.

Sociological research can be exploratory, descriptive or causal and is divided into quantitative or qualitative [9], but in the particular sub-sector of advertising research we are faced with a further complex panorama.

In advertising, researches are mainly divided into pre-test and post-test depending on whether they are carried out before or after the advertising campaign. These are qualitative researches that have to identify and select which of the different creative solutions can best meet the objectives of the campaign [8].

Recent guidelines of qualitative research have led to the use of sensory marketing or neuromarketing, a technique that can break down the bias of normal qualitative and non-qualitative market research, according to which respondents often produce answers in line with the self ideal and not necessarily sincere answers [14]; the answers are often decontextualized and suffer from the post hoc rationalisation according to which in retrospect the respondent tends to explain and justify what he has done and his choices, even the most inexplicable or unjustifiable [15, 20].

2.4 The Methodology of the Third Step

For these reasons it was decided to carry out research in the neuromarketing laboratories of IULM University during September 2019 to explore the impact of creative proposals against gender discrimination.

The project was structured following three distinct objectives. 1) Analysis of the emotional and cognitive impact of subjects in relation to the stimuli of the awareness campaign against gender-based violence. 2) Analysis of exploratory patterns and visual behaviour in relation to the stimuli. 3) Comparative evaluation of the rational analysis.

The stimuli concerned the vision, for about thirty seconds, of the four proposals selected for this last phase. The four proposals were: "Even no" ("Anche no"), "Auto Correct" ("Auto Correct"). "Deviations" ("Deviazioni"), "Out of place" ("Fuori Luogo") (Fig. 1).

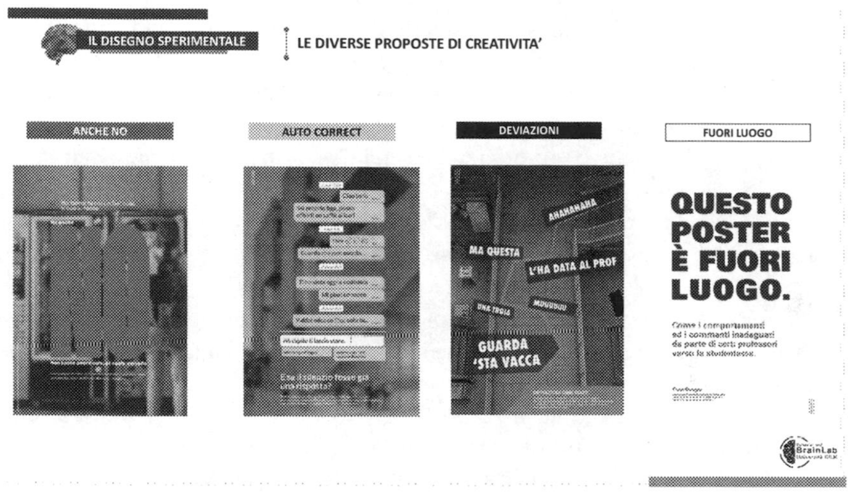

Fig. 1. The final four ad campaigns chosen

The sample consisted of eight people, 50% men and 50% women, university students. The research method of the Brain Lab at IULM University involves the use of different tools such as electroencephalographic analysis to assess which prefrontal areas are activated, physiological activation analysis that analyzes the microevaluations of sweating to assess the level of emotional activation of the sample, eye tracker system analysis that monitors the visual attention of the sample, and cognitive-rational analysis. To carry out the latter, at the end of the experiment the participants filled in a questionnaire on the aspects of liking and disliking, providing any suggestions and rationally explaining which was their favorite campaign. Furthermore, the questionnaire asked for an evaluation of the contents, the clarity of the message and, finally, whether the focus of the campaign had been centered.

The neurophysiological indices investigated and provided by the instrumentation that, one after the other, the sample under analysis wore are able to provide feedback on the value of the stimulus, i.e. the degree of interest or disinterest depending on the prefrontal areas of the brain involved.

In this way it is possible to calculate the attention index and the memorization index whose higher values indicate, respectively, a higher probability of attention to the stimulus and a higher probability of mnemonic processing [7].

2.5 The Final Results of the Sociological Research Implemented with Neuromarketing Techniques

All four campaigns have shown, through heatmap and scanpaths, that every element of creativity is explored in depth. This indicates that in the thirty seconds available the simulated space and the planar topology of the print ad are perfectly clicked and clearly legible. The invitation to decode is clear and the eye runs fast throughout the flatness of the print advertising [18].

On the whole, therefore, none of the campaigns presents any criticality in terms of visual exploration, and this is a valid consideration for both women and men, i.e. for both sub-samples. From the point of view of graphic construction, images and figures, each campaign therefore achieves a high degree of readability.

While in general the visual exploration tends to be slightly more in-depth in males, the opposite is true for the "Auto Correct" campaign: the exploration of females is more in-depth.

For all four campaigns the mode of exploration is similar: it starts from the top and moves downwards to deepen the areas where the aims of the campaign are explained.

In addition, the prefrontal asymmetry index reveals that all four campaigns' creativity has been appreciated, but in particular the "Auto Correct" creativity reaches a liking value of 0.75 (compared to 1 as maximum), while the other liking indexes do not exceed 0.34 which is only reached by the "Out of Place" campaign.

The attention rating is also higher for "Auto Correct", as expected due to the high text component.

Similarly, the memorization index of "Auto Correct" is twice as high as those campaigns that end in the second post ex aequo: "Also No" and "Out of Place".

Finally, the good results obtained through the neurophysiological indicators are also supported by a more than appreciable level of emotional activation: 0.90 for "Auto Correct"; 0.57 for "Also No"; 0.22 for "Deviations" and 0.59 for "Out of Place".

"Auto Correct" achieves excellent results in terms of valence, ability to elicit a state of attention, ability to activate mnemonic processes and ability to elicit emotional activation. Finally, the degree of concordance between males and females is positive.

In general, the performance of all four proposed campaigns is good, although some receive lower results in an absolute sense and in some cases there are more marked differences between males and females. The female sample prefers "Auto Correct" as much as the male sample, but the attention index and the memorization and emotional activation index are also very good for the "Out of place" campaign.

Finally, rational evaluation was required. At the end of the experiment, in fact, everyone was asked to answer a short questionnaire to compare the results of neuromarketing with the results of rationalization. Some surprises here happen because "No" and "Deviations" are the ones perceived as more effective in dealing with the topic, and "deviations" is defined as the clearest and most explanatory.

"Out of Place" is the only one on which critical issues are highlighted, since it requires, for a complete understanding, to be put really out of place (and therefore can generate the pun with "inappropriate"). The campaigns were designed to be distributed within the Milanese universities, as in fact happened in winter 2019\2020, but certainly "Fuori Luogo" should have been posted in places that are not their own, not on the notice boards but on the door of the rectorate, in front of a window, on the library, etc. In this way the text of the message would have been strengthened thanks to the context in which the message should have been placed. The context would play the role that Genette defined as paratext [12].

Males and females basically agree in these rational assessments.

As far as the exclusively creative side is concerned, always in the rationalization phase, a preferred campaign does not emerge. Each one receives in equal measure appreciations with the obvious criticalities, again, expressed for "Fuori Luogo".

We report some male verbatim referring to the campaign that, in general, was the best, or "Auto Correct".

"It's the one that most impressed me, aroused empathy towards the victims and disgust towards the authors. so it had a double action: not only is it wrong towards the victims (if a friend of mine was the victim it would bother me a lot) but it's also disgusting a guy who behaves like that (already I wouldn't do what the author did but with this advertisement I make myself count how disgusting I would be to do it)".

"He takes a theme often seen ironically but sees it critically."

And feminine.

"I personally feel her closer to my experience, which is why it touches me in a special way. The fact that it is reported involves those who observe it because it is something that is part of our everyday life".

3 Conclusions

In general terms, males are more critical of creativity than females. They mainly criticize aspects such as the graphic design and the lack of incisiveness of the message. Males are also the ones who provide the most suggestions.

The rational evaluation does not show a real "winning" creativity, although a slight preference goes to "Deviations". One can only deduce a lack of appreciation for "Out-of-Place" creativity.

When the subjects are asked to choose between the various versions, Deviations seems to agree almost completely with the female sample, receiving no preference from the males; creativity is also not chosen, instead, only by male subjects.

"Auto Correct" creativity gets good rational evaluations and discriminates less between males and females.

If from the rational evaluation only a "losing" creativity emerges, i.e. "Out of Place", it is the neurophysiological indicators and eye tracking outputs that help us to identify the "winning" campaign.

On "Auto Correct" creativity converge excellent neurophysiological data, excellent performance in terms of visual exploration and a good rational preference level for males and females.

In general terms, however, the creativity is well constructed, easily explorable and able to arouse a positive value in both males and females.

References

1. Banyard, V.L.: Toward the Next Generation of Bystander Prevention of Sexual and Relationship Violence. SC. Springer, Cham (2015). https://doi.org/10.1007/978-3-319-23171-6
2. Baule, G., Bucchetti, V. (eds.): Anticorpi comunicativi. Progettare per la comunicazione di genere. FrancoAngeli, Milano (2012)
3. Baule, G., Caratti, E. (eds.): Design è traduzione Il paradigma traduttivo, per la cultura del progetto. FrancoAngeli, Milano (2016)
4. Benwell, B.: New Sexism? Journal. Stud. **8**(4), 539–549 (2007)
5. Bucchetti, V., Ferraresi, M., Magaraggia S.: Violenza digitale di genere. Ricerca, progettazione, comunicazione di una campagna di sensibilizzazione dall'università per l'università, Sicurezza e scienze Sociali, VI, 2 (2019)
6. Capecchi, S.: La, comunicazione di genere. Carocci, Roma (2018)
7. Cerf, M., Garcia-Garcia, M.: Consumer Neuroscience. MIT Press, Cambridge Mass (2017)
8. Cheng, H.: The Handbook of International Advertising Research. Blackwell, London (2014)
9. De Luca, A.: Le, ricerche di mercato. Guida pratica e teorica. FrancoAngeli, Milano (2007)
10. Deriu, M.: Disonorare la violenza. La violenza maschile tra reputazione e discredito, Società degli individui (LA), 56, 80–94 (2016)
11. Ferraresi, M.: Pubblicità teorie e tecniche. Carocci, Roma (2017)
12. Genette, G.: Palimpsestes. La littérature au second degré. Seuils, Paris (1982)
13. Gough, B.: Men and the discursive reproduction of sexism: repertoires of difference and equality. Fem. Psychol. **8**, 25–49 (1998)
14. Morin, C.: Neuromarketing: the new science of consumer behavior. Society **48**(2), 131–135 (2011)
15. Noci, G.: Biomarketing. Egea, Milano (2018)
16. Riley, S.: Constructions of equality and discrimination in professional men's talk. Br. J. Soc. Psychol. **41**, 443–461 (2002)
17. Semprini, A.: Come analizzare la pubblicità, le immagini, I media. FrancoAngeli, Milano (2003)
18. La, T.F.: doppia spazialità in pittura: spazio simulato e topologia planare. In: Corrain, L., Valenti, M. (eds.) Leggere l'opera d'arte. Esculapio, Bologna (1991)
19. Wooten, S.C., Mitchell, R.W.: The Crisis of Campus Sexual Violence. Critical Perspectives on Prevention and Response. Routledge, New York (2015)
20. Zaltman, G.: How Customers Think: Essential Insight into the Mind of the Market. Harvard Business School Press, Boston (2003)

A Practice-Based Artificial Society Approach to Exploring the Evolution of Trust

Michael Heidt[1](✉) and Andreas Bischof[2]

[1] GeDIS, University of Kassel, Pfannkuchstraße 1, 34121 Kassel, Germany
mrbheidt@gmail.com

[2] Chair Media Informatics, Chemnitz University of Technology, Straße der Nationen 62, 09111 Chemnitz, Germany
andreas.bischof@informatik.tu-chemnitz.de

Abstract. Blockchain systems promise to fundamentally transform the way users interact with systems of governance, finance, and administration. Since relevant changes are expected to appear on a global scale, considerable methodological challenges exist regarding the sound description and adequate design-theoretic conceptualisation of respective social transformations. Artificial societies are a proven method for exploring the dynamics of large scale social systems which are otherwise hard to observe. They allow for precise modelling of social processes, facilitating an experimental attitude towards large scale systems which otherwise remain inaccessible to researchers. Within this paper, we describe how artificial societies can be employed as part of a practice-based research strategy aimed at elucidating a large scale transformation of this kind: the role of trust within social practices facilitated by Blockchain technologies.

Keywords: Blockchain · Trust · Practice-based research · Critical technical practice · Artificial societies · Digital hermeneutics · Generative art · Complexity

1 Introduction

Certain digital technologies operate with the promise of large scale social transformation: Artificial intelligence promises to upend the distinction between conscious users and mindless machine, robotics aspires to radically displace practices of manual labour, autonomous vehicles attempt to establish novel ways of organising transportation, distributed ledger technology promises novel ways of doing finance and contracting. At the same time our methods for technology design and analysis remain limited to relatively small, circumscribed areas of social reality.

Since the future development of society-scale interactions cannot be predicted in advance, designing for respective systems is exceptionally challenging.

C. Stephanidis et al. (Eds.): HCII 2020, LNCS 12427, pp. 273–284, 2020.
https://doi.org/10.1007/978-3-030-60152-2_22

The aforementioned problematic is exacerbated when considerations such as *Value Sensitive Design* [6] or *Ethically Aligned Design* [26] become part of the design process. While immediately designing for large scale transformations might not be feasible, the aforementioned paradigms do call for a research process which strives to give an account of possible large scale transformations effected by the technology in question.

Regarding Blockchain related transformations, a key phenomenon is that of *trust*. Trust can be described as the "glue" of social relations and practices, allowing individuals to relate to a stable set of expectations in the face of uncertainty [29]. Crucially, trust is involved when establishing contracts, since the contracting parties are required to trust one another to honour the agreement made.

Futhermore, Blockchain systems introduce a socially consequential innovation by allowing for the operation of *smart contracts* [7,46]. This novel form of contract facilitates the transferance of digital property in a manner ensuring that transfers are "safe and secure, everyone knows that the transfer has taken place, and nobody can challenge the legitimacy of the transfer" [1]. The respective mode of operation endows smart-contracts with a self-enforcing capability, automatically and securely transferring assets once certain formalized criteria are met. Consequently, systems built using smart contracts are described as being *trustless* [10,45].

Resultingly, for Blockchain systems to succeed, the position of trust within systems of practice has to change considerably. At the same time, it is hard to reason about or accurately predict the aforementioned global shift regarding the position of trust within social practices. In the following, we inquire into the potentials of Artificial Societies (AS) as a research tool for relating practice-based projects to the aforementioned large scale transformations.

2 Methodology

The research task of inquiring into patterns of trust evolution poses a unique set of methodological challenges [20]. Respective patterns of social behaviour could only be observed on a large scale and would only be fully expressed within a future state of society. Presently, they can be observed only in a nascent state if they can be studied at all.

At the same time, design decisions targetting blockchain based technologies have to be made in the here and now. Due to the incipient nature of the technology and respective systems of practice, any design decision will be based on expectations regarding future developments. These expectations might be implicit in nature and typically will remain only partly articulated. Oftentimes, implicit expectations are based on a mere forward projection of present social dynamics amended by inferred practices arising from specific situations of technology use. These kind of predictions, however, run the danger of missing fundamental large scale transformations of social phenomena connected to technological development.

In order to account for the described inherent future-directedness of blockchain artefacts, the applied methodology is based on a combination of methods of practice-based research and speculative design. This is in line with studies such as that of Elsden et al. [12] who explicitly call for combining design-led, speculative, and artistic methods in order to account for the specific set of challenges posed by Blockchain systems.

2.1 Practice-Based Research

Practice-based research is a methodology for achieving knowledge through the careful conduct of a constructive designerly or artistic process [39]. Following the principle of "Knowing through Making" [31], concepts are formed and explored through continuous engagement with constructed artefacts, paired with ongoing processes of reflection and observation.

In order to grasp the specificity of practice-based research as a method, we employ a schematic juxtaposition of research styles developed by digital media scholar Ashley Holmes (Table 1) [24]. Differentiating technology research and creative production projects, it provides a detailed set of criteria for identifying and comparing both.

Table 1. Characteristics of technology research and creative-production projects - reproduced from [24].

Technology research projects	Creative-production research projects
Artefact is produced	Artefact is produced
Artefact is new or improved	Artefact is of high quality and original in a cultural, social, political or/and aesthetic, *etc.*, context
Artefact is the solution to a known problem	Artefact is a response to issues, concerns, and interests
Artefact demonstrates a solution to problem	Artefact manifests these issues, concerns, and interests
The problem is recognised as such by others	These issues, concerns, and interests reflect cultural, social, political or/and aesthetic, *etc.*, preoccupations
Artefact (solution) is useful	Artefact generates apprehension
Knowledge reified in artefact can be described	Artefact is central to the process of apprehension
This knowledge is widely applicable and widely transferable	The creative-production process is self-conscious, reasoned and reflective
Knowledge reified in the artefact is more important than the artefact.	Knowledge may be a by-product of the process rather than its primary objective

2.2 Speculative Design

Speculative Design [11] and Research through Design Fictions [4,5,43] are methodological tools to extend the reach of design activities into the realm of the future. We argue for including speculative elements within the practice-based process in order to account for the disruptive and future oriented nature of many Blockchain systems. Including an element of speculation allows for a more adequate relationship to an anticipated, albeit not yet existing, situation of use. Artificial societies aid in the speculative design process by generating simulations of possible futures to which discussion processes can refer.

However, it has to be stressed that a conceptually sound combination of practice-based and speculative methods need not come as natural as it sometimes is made to appear. In order for practice-based methods to be fruitful, forming clear expectations is key. Without a clearly articulated set of expectations there can be no surprise and thus no productive perturbations to the conceptual apparatus formed during the research process [21,38,39].

Speculative activities, however, often appear to be concerned with building their own worlds. In order to solve the apparent antagonism between speculative methods and practice-based research, based on expectation and surprise, it might prove productive to adopt a more differentiated concept of speculation. Central to a speculative endeavour is the autonomy of an idea: clearly articulating a concept while following it through all its implications. The surprise might occur when we realise an idea to lead us to consequences we did not previously anticipate. Disciplined speculation thus has to differentiate itself from wishful thinking. Artificial societies might provide helpful impulses for a research endeavour situating itself at the intersection of speculation and practice orientation.

2.3 Artificial Societies Within the Context of Practice-Based Research

Artificial societies (AS) constitute agent based computational models that facilitate the simulation of social dynamics over time [13,17]. They have been employed in order to study market behaviour [14], opinion dynamics [18], or the formation of political identities [30]. Crucially, they allow for social systems to be observed and manipulated on a global scale as part of an overarching research process.

Within the overarching research strategy, artificial societies can serve the purpose of complementing practice-based and speculative methods with formalised components. They create a stream of "synthetic" material which can be used as input to discussion processes. However, it has to be stressed that these simulations always remain extremely limited in scope. They do not create meaning or interpretation on their own.

Crucially, different tools appeal to different target audiences. While some project participants might be amenable to narratives or case studies others might appreciate the formal clarity of an artificial society simulation [19,21].

3 Theory

3.1 Practice Theory

In order to productively conceptualize the impact of Blockchain technology on trust, we draw on a strand of social theory known as *practice theory*. Articulated by thinkers such as Reckwitz [34] and Schatzki [37], respective theories analyse the characteristics of social phenomena as interplay of social practices [8]. Practices themselves constitute habitual configurations of joint bodily actions and mental states which in tandem lead to stable patterns of social organization. Trust in turn is theorized as one of these mental attitudes necessary for instantiation of respective practices.

3.2 Complexity and Complication

A further conceptual differentiation is helpful in describing the specifics of artificial society simulations with respect to social processes encountered in the wild: the distinction between complexity and complication. The distinction will be briefly fleshed out in the following paragraphs, while a more nuanced discussion of the methodological implications for HCI design processes is available by virtue of [22].

Science and technology studies scholar Bruno Latour proposes distinguishing between *complexity and complication* in order to conceptually differentiate social phenomena according to the way they can be analytically decomposed: Complicated phenomena can be analysed in the form of a countable set of variables, while complex phenomena either consist of non-countable sets of variables or resist description through variables altogether [27,28].

The difference between complexity and complication is developed in "On Interobjectivity" [27] as well as the cross-media publication co-authored by Emilie Hermant [28]. A precursory usage of the term can be found within the text "Redefining the social link: from baboons to humans" authored by anthropologist Shirley Strum and by Latour [44].

Latour introduces complexity as a specific need for coordination: A complex situation forces us to take into account a large number of variables simultaneously [27, 233]: "Complex" will signify the simultaneous presence in all interactions of a great number of variables, which cannot be treated discretely." [27, p. 233].

Complexity is hence seen as characteristic of forms of social integration based on continuous interaction and negotiation. *Complication* on the other hand, refers to a process decomposable into a limited set of discrete variables, which, in principle, can be treated successively: [28, p. 30] "By contrast, we'll call "complicated" all those relation[s] which, at any given point, consider only a very small number of variables that can be listed and counted." It is not significant whether relevant operations are indeed performed sequentially or in parallel, but that they can, in principle, be performed in sequence, without interfering with one another. Examples of complicated situations can be found in bureaucratic processes such as scheduling or in scripted repetitive interactions such as buying stamps from a postal clerk [28, pp. 30–31][27, pp. 233–234].

Distinguishing complexity and complication in this manner, allows for a characterisation of processes on the level of computing as well as social complexity, employing a theoretical vocabulary already received within the field of HCI [3,9,15,16].

It has to be stressed, how Latour's approach contains an implicit valuation of complexity and complication. While he points out how evaluation of contemporary social structures entailed a development from complexity to complication, complex phenomena are repeatedly referred to as rich, while complicated procedures are associated with bureaucratic phenomena.

In the context of the present inquiry the complex/complication distinction is consequential both in relationship to the object of study, "trust", and to the description of artificial societies as a research tool. With respect to trust, the analysis is usually faced with a set of complex social phenomena, while trustless systems promise to solve the problematic through complication.

Furthermore, framing processes in the terms of complexity and complication allows us to analyse the role artificial societies play within the ongoing negotiations characteristic of practice-based research styles: Complicated structures themselves, artificial societies introduce an element of complication within the discursive network of the practice-based process. The act of expressing a certain social dynamic in the medium of artificial societies already implies the assertion it can sensibly be related to complicated processes. Consequently, special analytical attention has to be directed towards the question how complex and complicated structures interact.

4 Artificial Societies as Agents of Practice-Based Research

In order to inquire into the role of trust within emerging systems of Blockchain related practices, adequate artificial society simulations were employed as tool within a practice-based research process. The approach is based on the system developed by Holtz [25], which treats individuals as bearer of practices. Practices themselves are first-order citizens within the respective simulation.

Material, Meaning, Competence

Following Holtz [25], Shove and Pantzar [40], and Røpke [35], three levels of practice are modelled within the simulation: *material*, *meaning*, and *competence*.

On the basis of practice theoretic modelling, requisite conditions for the success of Blockchain based practices can be formulated. In order for Blockchain systems to become a viable source for instantiation of practices:

- the *material* requirements of respective practices have to be met,
- individuals have to combine actions with respective *meaning* of Blockchain linked actions,

- individuals need to possess adequate levels of *competence* with respect to the practice in question.

Adaptation and Habituation

Adaptation practices gradually effect matching characteristics of meaning and material. Following Holtz [25], three main processes can be modelled:

- *Adapt meaning to material*: Given a specific material (such as smartphone, car, or Blockchain technology) individuals adopt cognitive routines in line with the material used. As an example they might start regarding cars as practical, or learn about the worldview underlying distributed ledger technology.
- *Adapt competence to material*: Given a change in material (such as a new technology), individuals develop competence in handling it.
- *Habituation*: During the course of repeated performance of practices, these become habitualised.

Situating Artificial Societies within the Practice-Based Process

Simulated processes such as habituation and materialization allow for characteristic patterns of practices to emerge as objects of observation and discussion (see Fig. 1). Novel technologies such as blockchain systems enter into the simulation as new materials, which initially have not been subject to processes of habitualisation. They thus effect a perturbation of the simulation's state, which in turn can give rise to novel practices which gradually adapt to material conditions.

It has to be stressed that the present research process employs the tool of artificial societies in ways not expressly discussed within the modelling communities they originated from. Indeed, it is quite possible that some of the assertions in this paper run counter to the intentions of the authors it draws on. Holtz discusses observations applicable to evolution of social practices in general while subsequently drawing conclusions pertaining to the field of technological innovation (such as electric and hybrid cars).

The approach followed in the research project at hand is more limited in that it talks about a single technological artefact, the Blockchain distributed database. Artificial society models are used as an input for practice-based research processes. Reflection on the arbitrary nature of assigning signifiers to simulation models is part of the research strategy itself. "As such" models created do not possess the agency to talk about the Blockchain. Their relationship is deliberately created by project participants by assigning signifiers to formal objects within the model. However, once these symbolic links have been established, simulated scenarios are followed according to the logic of the formal model. Participants have to come to terms with the model's inner consistency, its history in the form of a sequence of states. Crucially, simulations provide a tangible complicated explanation for the evolution of practices of trust. They thus can serve as helpful boundary objects [41], mediating between otherwise unconnected interpretive communities.

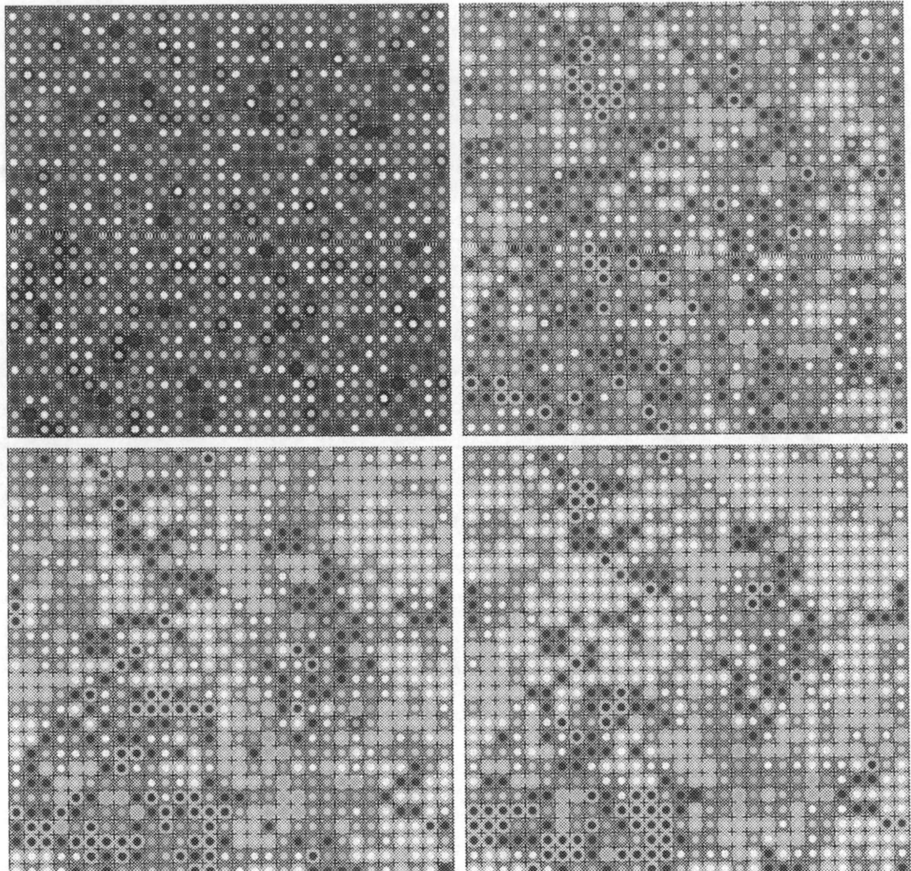

Fig. 1. Simulation run with four materials. The figure depicts the initial state in the upper left corner, the state at t = 200 is shown in the upper right corner, the lower left corner shows the state at t = 5000 while the lower right corner shows the state at t = 10000. Each cell within the grid shows a material component by virtue of a coloured circle. Colours refer to individual materials. Coherence is visualised by virtue of background brightness: A minimum brightness background refers to a coherence level of 0 while a maximum brightness background refers to a coherence of 1. For a detailed explanation of underlying modelling dynamics, refer to [25].

5 Related Work

While the fields of artificial societies and sociology entertain an intensive academic dialogue [36], the problematic of practice formation has received comparatively little attention within AS communities. Hofmann implements a simulation of social practices based on practice theory [23]. Following Tuomela [47], practices are conceived as being grounded through phenomena of collective acceptance.

Markey-Towler inquires into the possibility of Blockchain systems partially displacing state backed forms of governance [32]. He introduces a formalism intended to elucidate the basic mechanisms at work and describes the conditions necessary for social evolution into the direction of Blockchain-backed governance.

Methods from the field of generative architecture can be read as combination of the paradigm of information processing with the problematic of creating large (or at least meso-level) systems of practice [42].

Within the field of urbanism, Michael Batty employs evolutionary simulation as a method for design of the inherently complex system of the city [2]: The approach calls for growing cities within "digital laboratories" in order to arrive at tenable structures. Respective growth processes are regulated by continuous feedback and guided through specification of constraints.

6 Future Work

The system is designed as a practice-based research tool. It exposes large scale systems of practice to processes of analysis and discussion. It could however be repurposed as a tool for deliberation in the contexts of society-in-the-loop scenarios [33].

A Possible Future for Trust

The inherent future directedness of speculative methods can be read as obligation to formulate an alternative to the project of the "trustless system": The project we might collectively undertake could be one of redistribution and rearticulation of vulnerability. Blockchain need not be a technology to undo the evolution of trust. After all, its operation is simply that of a distributed database, the task of salvaging meaning from its records is entrusted to the complex social space surrounding its algorithms, links, and procedures. Artificial societies, their capacity to create patterns and engender interpretations might prompt us to realise the multiplicity inhering within the technology of the distributed ledger.

7 Conclusion

Artificial societies can be fruitfully employed as part of a practice-based research strategy. However, due to the synthetic character of generated scenarios, their use has to be carefully balanced against other forms of data generation and analysis. Furthermore, artificial societies present an interesting venue for substantiating practices of speculative design within practice-based research projects. In turn, generation of social practices constitutes an especially promising mode of doing artificial societies within a practice-based setting. At the same time, any use of artificial societies within practice-based contexts has to be discursively contained by a reflection on the limits of complicated systems of simulation. It is their differentiation against complex activities of interpretation which renders them productive.

Acknowledgements. This work was supported in part by the Andrea von Braun Foundation, Munich, under the grant "Blockchain – A Practice-Based Inquiry Into a Future Agent of Social Transformation".

References

1. Andreessen, M.: Why Bitcoin Matters. New York Times, 21 Jan 2014
2. Batty, M.: A digital breeder for designing cities. Archit. Des. **79**(4), 46–49 (2009)
3. Berger, A., Heidt, M., Eibl, M.: Conduplicated symmetries: renegotiating the material basis of prototype research. In: Chakrabarti, A. (ed.) ICoRD 2015 – Research into Design Across Boundaries Volume 1. SIST, vol. 34, pp. 71–78. Springer, New Delhi (2015). https://doi.org/10.1007/978-81-322-2232-3_7
4. Bleecker, J.: Design Fiction: A short essay on design, science, fact and fiction. Near Future Laboratory **29**, (2009)
5. Blythe, M.: Research through design fiction: narrative in real and imaginary abstracts. In: Proceedings of the SIGCHI Conference on Human Factors in Computing Systems, CHI 2014. ACM, New York, pp. 703–712 (2014)
6. Borning, A., Muller, M.: Next steps for value sensitive design. In: Proceedings of the 2012 ACM Annual Conference on Human Factors in Computing Systems - CHI 2012, p. 1125 (2012)
7. Buterin, V.: A next-generation smart contract and decentralized application platform. White Paper (2014)
8. Cetina, K.K., Schatzki, T.R., von Savigny, E.: The Practice Turn in Contemporary Theory. Routledge, London (2005)
9. Cordella, A., and Shaikh, M.: Actor-network theory and after: what's new for is research.Iin: European Conference on Information Systems 2003, Naples, Italy, June 2003
10. Dannen, C.: Introducing Ethereum and Solidity. Springer, Berlin (2017). https://doi.org/10.1007/978-1-4842-2535-6
11. Dunne, A., Raby, F.: Speculative Everything: Design, Fiction, and Social Dreaming. MIT Press, Cambridge (2013)
12. Elsden, C., Manohar, A., Briggs, J., Harding, M., Speed, C., Vines, J.: Making sense of blockchain applications: a typology for HCI. In: Proceedings of the 2018 CHI Conference on Human Factors in Computing Systems, CHI 2018, pp. 458:1–458:14. ACM, New York (2018)
13. Epstein, J.M., Axtell, R.: Growing Artificial Societies: Social Science from the Bottom Up. Brookings Institution Press, Washington, D.C. (1996)
14. Filatova, T., Parker, D., Van der Veen, A.: Agent-based urban land markets: agent's pricing behavior, land prices and urban land use change. J. Artif. Soc. Soc. Simul. **12**(1), 3 (2009)
15. Fuchsberger, V. Generational divides in terms of actor-network theory: potential crises and the potential of crises. In: Online Proceedings of the 7th Media in Transition Conference. MIT, Cambridge (2011)
16. Fuchsberger, V., Murer, M., Tscheligi, M.: Human-computer non-interaction: the activity of non-use. In Proceedings of the 2014 Companion Publication on Designing Interactive Systems, DIS Companion 2014, pp. 57–60. ACM, New York (2014)
17. Gilbert, N., Conte, R.: Artificial Societies: the Computer Simulation of Social Life. Routledge, London (2006)
18. Hegselmann, R., Krause, U.: Opinion dynamics and bounded confidence models, analysis, and simulation. J. Artif. Soc. Soc. Simul. **5**, 3 (2002)

19. Heidt, M.: Examining interdisciplinary prototyping in the context of cultural communication. In: Marcus, A. (ed.) DUXU 2013. LNCS, vol. 8013, pp. 54–61. Springer, Heidelberg (2013). https://doi.org/10.1007/978-3-642-39241-2_7
20. Heidt, M., Berger, A., Bischof, A.: Blockchain and trust: a practice-based inquiry. In: Nah, F.F.-H., Siau, K. (eds.) HCII 2019. LNCS, vol. 11588, pp. 148–158. Springer, Cham (2019). https://doi.org/10.1007/978-3-030-22335-9_10
21. Heidt, M., Kanellopoulos, K., Pfeiffer, L., Rosenthal, P.: Diverse ecologies – interdisciplinary development for cultural education. In: Kotzé, P., Marsden, G., Lindgaard, G., Wesson, J., Winckler, M. (eds.) INTERACT 2013. LNCS, vol. 8120, pp. 539–546. Springer, Heidelberg (2013). https://doi.org/10.1007/978-3-642-40498-6_43
22. Heidt, M., Wuttke, M., Ohler, P., Rosenthal, P.: Scaffolding a methodology for situating cognitive technology within everyday contexts. In: Marcus, A. (ed.) DUXU 2016. LNCS, Part I, vol. 9746, pp. 281–292. Springer, Cham (2016). https://doi.org/10.1007/978-3-319-40409-7_27
23. Hofmann, S.: Simulation einer sozialen Praxis. In: Dynamik sozialer Praktiken: Simulation gemeinsamer Unternehmungen von Frauengruppen. VS Verlag für Sozialwissenschaften, Wiesbaden, pp. 71–140 (2009)
24. Holmes, A.: Reconciling Experimentum and Experientia: Ontology for Reflective Practice Research in New Media
25. Holtz, G.: Generating Social Practices. J. Artif. Soci. Soc. Simul. **17**(1), 17 (2014)
26. Initiative, I.G.: Ethically Aligned Design. Technical report. The IEEE Global Initiative for Ethical Considerations in Artificial Intelligence and Autonomous Systems (2016)
27. Latour, B.: On interobjectivity. Mind Cult. Act. **3**(4), 228–245 (1996)
28. Latour, B., Hermant, E., Shannon, S.: Paris Ville Invisible. La Découverte Paris (1998)
29. Luhmann, N.: Trust and Power. Wiley, Hoboken (2018)
30. Lustick, I.S.: Agent-based modelling of collective identity: Testing constructivist theory. J. Artif. Soci. Soc. Simul. **3**(1), 1 (2000)
31. Mäkelä, M.: Knowing through making: the role of the artefact in practice-led research. Knowl. Technol. Policy **20**(3), 157–163 (2007)
32. Markey-Towler, B.: Anarchy, Blockchain and Utopia: A Theory of Political-Socioeconomic Systems Organised using Blockchain
33. Rahwan, I.: Society-in-the-loop: programming the algorithmic social contract. Ethics Inf. Technol. **20**(1), 5–14 (2018)
34. Reckwitz, A.: Toward a theory of social practices a development in culturalist theorizing. Eur. J. Soc. Theory **5**(2), 243–263 (2002)
35. Røpke, I.: Theories of practice—new inspiration for ecological economic studies on consumption. Ecol. Econ. **68**(10), 2490–2497 (2009)
36. Sawyer, R.K.: Artificial societies: multiagent systems and the micro-macro link in sociological theory. Sociol. Methods Res. **31**(3), 325–363 (2003)
37. Schatzki, T.R.: The Site of the Social: A Philosophical Account of the Constitution of Social Life and Change. Penn State Press, University Park (2002)
38. Scrivener, S. Reflection in and on action and practice in creative-production doctoral projects in art and design. Working Papers in art and design, vol. 1 (2000)
39. Scrivener, S., Chapman, P.: The practical implications of applying a theory of practice based research: A case study. Working papers in art and design, vol. 3 (2004)
40. Shove, E., Pantzar, M.: Consumers, producers and practices understanding the invention and reinvention of nordic walking. J. Consum. Cult. **5**(1), 43–64 (2016)

41. Star, S.L., Griesemer, J.R.: Institutional ecology, 'translations' and boundary objects: amateurs and professionals in berkeley's museum of vertebrate zoology, 1907–1939. Soc. Stud. Sci. **19**(3), 387–420 (1989)
42. Steenson, M.W.: Architectures of Information: Christopher Alexander, Cedric Price, and Nicholas Negroponte and MIT's Architecture Machine Group. Ph.D. thesis, Princeton University (2014)
43. Sterling, B.: Design fiction. Interactions **16**(3), 20–24 (2009)
44. Strum, S.S., Latour, B.: Redefining the social link: from baboons to humans. Soci. Sci. Inf. **26**(4), 783–802 (1987)
45. Swan, M.: Blockchain: Blueprint for a New Economy, 1st edn. O'Reilly Media, Beijing, Sebastopol (2015)
46. Szabo, N.: Formalizing and securing relationships on public networks. First Monday **2**, 9 (1997)
47. Tuomela, R.: The Philosophy of Social Practices: A Collective Acceptance View. Cambridge University Press, Cambridge (2002)

"If I'm Close with Them, It Wouldn't Be Weird": Social Distance and Animoji Use

Susan C. Herring(✉), Ashley R. Dainas, Holly Lopez Long, and Ying Tang

Indiana University, Bloomington, IN 47405, USA
{herring, ardainas, hdlopezl, ytl1}@iu.edu

Abstract. Each new type of graphical icon (graphicon) in CMC has been more complex and multimodal than its predecessor. For this reason, and because of their novelty, Konrad, Herring, and Choi (2020) claim that new graphicon types are initially restricted to use in intimate relationships. We explore this proposition qualitatively by interviewing student users of Animoji – dynamic, large-scale emoji on the Apple iPhone – about who they send Animoji messages to and why. The results of a think-aloud card sort task in which participants (N = 33) matched seven Animoji with seven relationship categories at different social distances and sent a message to each one were triangulated with responses to open-ended questions before and after the task. Participants sent Animoji to close friends, significant others, and siblings, and to a lesser extent, parents and other family members. They rejected the idea of sending Animoji to more distant relationships such as a teaching assistant, a mentor, and new friends. Different Animoji were considered more or less suitable for each relationship, as well as for recipients of different genders. The reasons given by the interviewees for sending Animoji to each relationship category centered around themes of politeness, (in)formality, familiarity, and self-presentation.

Keywords: Graphicons · Intimacy · Politeness · Relationships

1 Introduction[1]

The increasingly multimodal nature of computer-mediated communication (CMC) systems affects users' experiences of social connectedness. In particular, graphical enhancements such as emoticons, emoji, stickers, and GIFs (*graphicons*; [1]) facilitate playfulness and emotion expression in CMC, both of which are positively associated with intimacy [2, 3], or the feeling of closeness with another. Conversely, degree of intimacy as well as degree of multimodality can affect which relationships one uses graphicons with. As intimacy between interlocutors increases, they become more willing to use and accept graphicons, whereas in less intimate relationships, recipients have been known to react negatively, especially to graphicons that are more

[1] This work was partially funded by the Office of the Vice Provost of Research at Indiana University Bloomington through the Grant-in-Aid Program. Special thanks are due to Madeline Adams for assisting with the interviews and Dana Skold for modeling Fig. **1**.

C. Stephanidis et al. (Eds.): HCII 2020, LNCS 12427, pp. 285–304, 2020.
https://doi.org/10.1007/978-3-030-60152-2_23

dynamic and visually complex [4, 5]. Thus, senders may avoid using these icons with less intimate interlocutors, for fear of causing offense.

Another factor that influences to whom one sends graphicons is the age of the graphicon type. Konrad et al. [4] posit that a new graphicon type is initially restricted to very close relationships; sharing it with more socially distant relationships is only acceptable after it has achieved widespread adoption. This trend is evident in the evolution of emoticons and emoji, both of which can now be used with a much wider variety of addressees than when they were first introduced. Konrad et al. [4] interviewed and surveyed Facebook Messenger users in 2014 about their use of emoji and stickers. While the participants reported sometimes sending emoji to non-intimate relationships, stickers, which had been introduced more recently, were sent almost exclusively to intimate relationships. The reasons given again involved multimodality: The relatively larger size and greater complexity of stickers made them "intense," and their intensity was less likely to be misconstrued as inappropriate in intimate relationships.

Konrad et al.'s [4] proposal generates predictions about the use of Animoji, which are new and highly multimodal. Introduced in November 2017 with the Apple iPhone X, Animoji are large-format emoji that represent animals (Dog, Cat, etc.), imaginary creatures (e.g., Unicorn), and anthropomorphized objects (Poop, Robot, etc.). Of particular interest in this study, users can record short videos of themselves animating an Animoji with their facial expressions and voice that can be sent to other iPhone users on the iMessage app. Animoji are larger than the stickers described in Konrad et al.'s [4] study and, in addition to their dynamic nature, are visually complex and emotionally expressive. Thus, their use should be similarly restricted to close personal relationships at this relatively early stage of their existence.

We evaluate this proposition through a qualitative, exploratory user study involving 33 university students in the United States. Animoji messages are typically sent in private one-to-one communication, and collecting examples of authentic, private use is difficult. To get around this, we elicited information about the students' Animoji use in three ways. First, we asked them directly who they typically send Animoji to. They then performed a think-aloud card sort task in which they matched Animoji with relationship categories at different social distances and then recorded and sent an Animoji message to each relationship. After, they were asked about which pairings felt most natural or unnatural and why. Thus, the research questions that guided our analysis are:

RQ1: To whom do iPhone users send Animoji?
RQ2: Which Animoji do the users match with each relationship in the card sort task, and why?
RQ3: Which card sort pairings seem most natural or unnatural, and which would be most and least likely to occur in actuality?
RQ4: How, if at all, do the answers to RQs 1–3 vary according to the gender of the interlocutors?

The pairings were quantified using descriptive statistics, and the reasons provided by the participants were analyzed using thematic content analysis [6].

The interviewees reported sending Animoji primarily to close peer relationships (significant other, close friend, sibling) and secondarily to parents and other family. In the card sort task, they paired "weird" and "edgy" Animoji (e.g., Poop) with intimate relationships and paired "normal" (e.g., Dog, Monkey) and "polite" Animoji (e.g., Cat, Robot) with more distant relationships where risk avoidance was a concern, such as with a new friend or a teaching assistant. Additionally, some Animoji were more often described as representing the self, especially when matched with intimate relationships, whereas Animoji described in relation to the addressee were matched more often with socially distant relationships, seemingly as a way to show consideration or pay face to the addressee. However, weird and edgy Animoji were matched more with male addressees, and inoffensive Animoji were matched more with female addressees, independent of social distance. Overall, the participants said that it felt most natural and that they were most likely to send Animoji to social intimates.

These findings support Konrad et al.'s [4] claim that new graphicon types are initially restricted to use in intimate relationships. Further, the findings suggest that some Animoji could eventually be used to manage more distant social relationships.

2 Relevant Literature

2.1 Social Distance and Graphicon Use

Social distance is defined by Oxford Languages as "the perceived or desired degree of remoteness between a member of one social group and the members of another, as evidenced in the level of intimacy tolerated between them." The use of graphicons in CMC is an important strategy to establish, maintain, and manage relationships at different social distances. For example, emoji are sent as a form of low-cost phatic communication to indicate that the recipient is "on one's mind" [7]. Similarly, people use stickers in mobile messaging as icebreakers in private and group chat rooms, and also to manage existing relationships by expressing themselves as they wish to appear [8].

Further, graphicon use in online communication enhances feelings of intimacy, which is associated with familiarity, trust, self-disclosure, and personal/private contexts [9, 10]. In an experimental study, Janssen et al. [3] found that increasing the quantity of emoticons used in CMC led to higher levels of perceived intimacy. In another experimental study, Wang [11] analyzed perceptions of intimacy when people receive mobile messages with or without stickers; the combination of text and stickers in a message produced a higher level of perceived intimacy than a text-only response. Emoji, GIFs, and stickers are often sent for fun or to amuse the recipient [4, 12], and playfulness is a strong correlate of interpersonal closeness [2]. Thus when people aim to create proximity with the message recipient, visual elements are preferred over strictly verbal means of communication [13, 14].

The nature of the relationship can also affect which, if any, graphicons are used. For example, one user interviewed by Zhou, Hentschel and Kumar [15] said she actively searches for stickers about NBA stars and sends them to her husband, because she knows that he is a super fan. If the sender does not know the addressee well, though, it might not be appropriate to send graphicons. Xu et al. [5] posited that "the more intimate the receiver is with the sender … the more likely that emoticons … will be utilized and accepted by both parties." Conversely, less intimate recipients may experience irritation when they see emoticons, especially "those animated ones." Thus, the sender may avoid using frequent emoticons with receivers with whom there is a low level of intimacy. Similarly, Konrad et al.'s [4] interviewees reported sending stickers only to intimate relationships, in part because the stickers' "intensity" was less likely to be misconstrued as inappropriate in intimate relationships. Shared or lack of shared knowledge of a graphicon type is also a consideration. In an interview study of GIF users [16], interviewees expressed concern that the older generation would not understand GIFs, including what the reference was, how to "read" it, and whether it was a legitimate form of communication at all.

Finally, graphicon use may depend on the status differential between participants. In an experimental study, when the addressee occupied a higher position in the social hierarchy, people who normally used many emoji tended to use fewer emoji in their text messages, and people who used almost no emoji tended to increase their emoji frequency to match the emoji frequency of the higher status person [17]. It follows that if the higher status person in a relationship uses no graphicons, their interlocutor might hesitate to use any with them.

2.2 Animoji Use

Animoji are a novel, dynamic set of graphical filters that allow users to modify their digital self-presentation and that afford play with identity in mediated communication [18]. Memoji are Animoji in the shape of human heads with customizable facial features. Animoji and Memoji can be used to send short video clips in text messages, to videochat via FaceTime, and to generate sticker sets. This study is concerned with Animoji video clips sent in text messages.

As yet, Animoji have been discussed in few scholarly studies. Paasonen [19] mentions Animoji on the iPhone X in passing as an example of "foregrounding the functionalities of affective interaction" to downplay the "creepiness" of facial recognition technology. The affective nature of Animoji is also highlighted in an early interview study by de Costa and Prata [20]. Users of Animoji, Memoji, and AR Emoji in Brazil said that they mainly use these graphicons with close and trusted people, such as family and friends, as they can help create affective memories and develop friendships. Because of their expressiveness, dynamic movement, and customizability, they are also used to chat with and entertain younger siblings [20].

In an analysis of Animoji videos clips that were shared publicly on YouTube and Twitter, Herring et al. [18] found that users tended to modify their normal speaking voice and enact the Animoji characters in creative and playful ways. Men used the Dog, Monkey, Robot, and Poop Animoji more often, and women used the Unicorn, Pig, Chicken, and Cat more often, consistent with gendered stereotypes (e.g., "men are dogs, women are cats") and color preferences (e.g., brown tones vs. pastel colors). Herring et al. [21] delved deeper into gender differences in Animoji use in an interview study. Female interviewees self-reported using Animoji more often than male interviewees, and they were more likely to say they used certain ones because they were "cute"; however, the men were earlier adopters of the technology and displayed greater comfort experimenting with it. While both genders expressed a preference for Memoji over non-human Animoji, men were more adventurous in creating Memoji and had generated more versions of them, consistent with the Technology Acceptance Model [22]. Most interviewees were positive about Animoji and reported using them mostly for fun or to be amusing or entertaining.

Of the above studies of Animoji use, only de Costa and Prata [20] considered the nature of the relationship between Animoji sender and receiver, although they mentioned it only briefly. Herring et al.'s [18] observations were based on video clips posted publicly for the purpose of entertaining others; such videos may not reflect Animoji use in private communication and in authentic Animoji-mediated interactions. Finally, although Herring et al. [21] interviewed women and men about their actual Animoji use, their study did not take social distance into account. The present study seeks to expand on this early research and address its lacunae by conducting an in-depth analysis of Animoji and social relationships.

3 Method

3.1 Participant Recruitment and Demographics

To investigate how and with whom individuals use Animoji, in-depth semi-structured interviews were conducted with self-identified Animoji users between November 2019 and March 2020. Participants were recruited via email and flyers from the population of a Midwestern university town and screened by an online questionnaire to confirm that they had an Apple iPhone X, XR, or 11 (hereafter referred to as an iPhone). The screener also asked questions about their device preferences, demographics (age, gender, first language, education level), and frequency of Animoji use.

A total of 33 participants, including both undergraduate and graduate students, were interviewed face-to-face in a laboratory setting. All participants had some experience with Animoji. The majority of participants (58%, n = 19) identified as female; 39% (n = 13) identified as male; and 3.0% (n = 1) identified as non-binary. Since there was just one non-binary individual, only data from the self-identified males and females are included in the analysis of Animoji gender patterns. Participant ages ranged from 18 to 29 (M = 21.8, SD = 3.48; Females: M = 21.9, SD = 3.72; Males: M = 21.2, SD = 3.11). The majority of participants were monolingual American English speakers (55%, n = 18). Of those who did not speak English natively or reported English and another language as their native languages, 15% spoke Mandarin Chinese, 21% spoke a South Asian language, and 9% spoke other languages.

3.2 Interview Procedures

Interviews lasted approximately 60 min. After signing an informed consent form, each participant was seated diagonally across from the interviewer at a small table in front of a camcorder on a tripod and a laptop computer running a Zoom video conference, each of which recorded the participant's face and speech. The interviews were semi-structured, and questions were designed to elicit narratives about how participants made choices about the individuals with whom they used Animoji and their reasoning for particular Animoji-related decisions.[2] This included asking participants directly: "Who do you usually send Animoji to?"

Following this, participants were asked to complete a task aimed at understanding how they selected particular Animoji characters for different types of addressees. A second task was also conducted by asking participants to alternate between using Animoji and not using Animoji during FaceTime chat with a member of the research team. In this study, we focus on the first task, which is described in detail below.

Animoji and Social Relationship Card Sort Task. Participants were given two decks of cards. The first deck depicted seven Animoji characters: Dog, Cat, Monkey, Rabbit, Dragon, Robot, and Poop. The first four Animoji were selected based on Herring et al.'s [18] finding that Dog, Cat, Monkey, and Rabbit were popular Animoji used in video clips posted to social media platforms. The other three Animoji were selected to round out the representation of Animoji characters that were available at the time of the study: two anthropomorphized objects (Robot, Poop) and a fantasy character (Dragon).

The second deck described seven relationship categories: significant other (SO), close friend (CF), sibling (S), older relative or mentor (OR/M),[3] new female friend (NFF), new male friend (NMF), and teaching assistant (TA). These categories were selected to represent common relationships that our study participants, who were university students, were expected to have, and to whom they could potentially send Animoji messages. The relationships represent varying degrees of social distance. SO, CF, and S relationships are intimate, whereas a TA and new friends are typically not intimate, and an older relative or mentor typically falls in between. The relationships can be categorized more precisely according to the presence or absence of five features associated with intimacy [e.g., 22]: sexual intimacy, blood relationship, frequently shared social activities, long acquaintanceship, and low power differential, as shown in Table 1.[4] These five features together differentiate among all seven relationship categories except for NFF and NMF, which differ only by gender.[5] The sums of the features show SO and CF tied in terms of intimacy, but the individual features indicate that they are intimate in different ways (e.g., sexual vs. non-sexual).

The participants were asked to shuffle the decks, review the cards in both decks, and choose the Animoji that they would send to a person in each relationship category. Even if they did not have a particular relationship (such as a SO), they were asked to

[2] We also asked participants to describe their experiences and perspectives on Animoji use for self-representation; those findings are reported in [21].

[3] Mentor was included as an alternative to Older Relative to provide options for people who did not have or who did not speak with older relatives.

[4] This classification follows the approach of semantic feature analysis in linguistics; see, e.g., [24]. A value of 1 indicates that a feature is present, −1 indicates it is absent, and 0 indicates that the feature is sometimes present and sometimes absent.

[5] Women have been found to use Animoji more than men do [21]. Accordingly, NFF is ordered before NMF in Table 1.

Table 1. Feature profiles of seven social relationships.

	SO	CF	S	OR/M	NFF	NMF	TA
Sexual/Romantic	1	-1	-1	-1	-1	-1	-1
Frequently Shared Activities	1	1	-1	-1	-1	-1	-1
Blood Relative	-1	-1	1	1	-1	-1	-1
Long Acquaintance	0	1	1	1	-1	-1	-1
Low Power Differential	0	1	0	-1	1	1	-1
Sum:	1	1	0	-1	-3	-3	-5

imagine that they did. As participants were selecting among different matches, they were instructed to think aloud and describe their reasoning behind each choice. Figure 1 shows an example of how one person completed the Animoji-relationship matching task.[6]

After the participant finalized the matches, the researcher left the room while the participant recorded a 15–30 second, asynchronous video clip for each Animoji-relationship pair and sent it to the researcher's phone, as if they were sending it to the relationship for whom it was intended. They were told to use their imaginations and to say anything they wanted in the Animoji clips. The participants were instructed to make a screen recording of their phone during this part of the task.

After the participants finished sending the seven Animoji, the interviewer reentered the room. The participants were then asked three debrief questions about their experiences sending the Animoji to the relationships represented in the card sort task:

1. How natural did you feel when you recorded those Animoji messages?
2. Did you feel more natural with some more than others? Which ones?
3. Which of the hypothetical people you imagined sending the Animoji to are you most likely to send Animoji to in reality? Which people are you least likely to send Animoji to in reality? Why is that?

3.3 Analysis

Pairings produced during the card sorting task were aggregated and sorted. The reasons given by the participants for their matches were analyzed using a grounded theory approach [25]. Each reason was coded for themes in an iterative process, where each was open coded, followed by axial and selective coding. The codes were assigned

[6] Three pairings were excluded from analysis because the participants declined to match an Animoji. The other matches made by those participants are included in the analysis, however, resulting in 228 observations out of 231 total possible pairs.

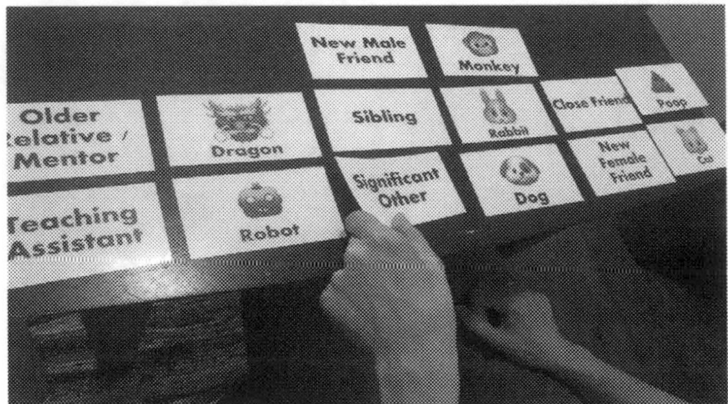

Fig. 1. An example of matches made in the card sort task.

jointly by at least two of the authors, and 100% agreement was reached. The findings were also sorted by gender. In light of the qualitative, exploratory design of the study and the small sample size, results are presented using descriptive statistics. Thus, no claims are made regarding the significance or generalizability of the findings.

4 Results

4.1 Open-Ended Question: Who Do You Send Animoji to?

When asked about the people to whom they send Animoji, the study participants typically provided more than one answer. Of the 33 participants, most mentioned close friends (58%) and/or their significant other (55%). Many participants also reported sending Animoji to their siblings (30%), and some reported sending them to their parents or grandparents (24%) and other family members (12%) such as cousins, aunts, and uncles. No one reported sending Animoji to any other relationships. However, other possible Animoji recipients were probed in the task that followed.

4.2 Card Sort Task

Relationship Type and Animoji Selection. The matches made in the Animoji-relationship card sort task are summarized in Table 2. The order of the first four columns reflects the relative frequency of Animoji use reported in Sect. 4.1; the order of the last three columns is from Table 1. The Animoji rows are ordered in the table to show the main associations between Animoji and relationship type on the diagonal as much as possible. (Boldface indicates the high value for the column; italics indicates the high value for the row).

Table 2 shows that certain Animoji were paired more with some relationships than others. The Poop Animoji was restricted to intimate relationships – significant other (SO), close friend (CF), sibling (S) – while at the other extreme, Robot was paired most often with the teaching assistant (TA). Similarly, Cat was selected most for a new

Table 2. Animoji selected for relationships.

	% CF (n=32)	% SO (n=32)	% Sib (n=32)	% OR/M (n=31)	% NFF (n=31)	% NMF (n=31)	% TA (n=32)
Robot (n=31)	6	9	6	10	3	16	***47***
Dragon (n=32)	19	3	6	**23**	3	***35***	13
Cat (n=31)	9	6	9	16	*29*	13	16
Rabbit (n=30)	3	13	16	**23**	*32*	3	6
Monkey (n=33)	16	22	16	16	*23*	10	3
Dog (n=32)	13	*19*	13	13	10	*19*	16
Poop (n=32)	***34***	**28**	*34*	0	0	3	0
Total	100	100	100	100	100	100	100

female friend (NFF) and Dragon for a new male friend (NMF), although Dragon was also the most popular choice for an older relative or mentor (OR/M). Monkey was paired often with both NFF and SO, whereas Dog was paired with all the relationships, especially SO and NMF. Monkey was also paired with all relationships except TA. Reasons for the preferred Animoji-relationship pairings are described in the next two sections.

Reasons for Sending to Relationships. We conducted a thematic content analysis of common reasons given by participants to explain why they paired an Animoji with a particular relationship type. A reason needed to be mentioned for a given relationship type at least five times to be included as a category in this analysis. Six themes (with variants in parentheses) were identified: *weird* (novel, surprising); *edgy* (risky, "could send them anything"); *funny* (fun, silly, goofy, playful); *cute*; *normal* (unthreatening); and *polite* (wise, respectful, or professional). The themes can be situated along a continuum from most risky (*weird*, *edgy*) to least risky (*normal*, *polite*), or, in the terms of politeness theory [26], from most to least threatening to the addressee's face.

The distributions of the reasons given for each relationship are shown in Table 3. The table is arranged to reveal the main associations between relationship type and reason on the diagonal as much as possible. (Boldface indicates the high value for the column; italics indicates the high value for the row.) Because participants sometimes mentioned more than one theme in the reason they gave for a given relationship, the totals in Table 3 add up to more than the total pairings in Table 2.

Certain reasons were more likely to be mentioned for certain relationships. For example, *weird* Animoji were paired especially with CF, *edgy* Animoji with CF and SO, and *funny* Animoji with CF, S, and SO – all intimate relationships. Conversely, *cute* was the main reason for matching an Animoji with a NFF, and *polite* the main reason for choosing an Animoji to send to a TA or OR/M, while Animoji considered

Table 3. Reasons sorted by relationships.

	% Weird (n=17)	% Edgy (n=22)	% Funny (n=35)	% Cute (n=33)	% Normal (n=44)	% Polite (n=20)
CF (n=29)	**35**	*41*	**29**	9	2	0
SO (n=28)	18	*45*	17	27	0	0
S (n=16)	*24*	14	20	6	0	0
NFF (n=21)	0	0	9	***30***	18	0
NMF (n=18)	12	0	14	3	*23*	0
OR/M (n=30)	0	0	6	18	25	*55*
TA (n=29)	12	0	6	6	**32**	*45*
Total	100	100	100	100	100	100

normal were paired with a TA, OR/M, and new friends of either gender – all less intimate relationships. Managing face is especially important in less intimate relationships [27], and thus participants paired *normal* Animoji with more distant relationships and *polite* Animoji with relationships where there is a power differential. OR/M and TA appear next to each other in Table 3 for this reason.

The six themes are illustrated in the following quotes from study participants. (Participant ID and gender are indicated in square brackets.)

1. Weird (to CF): "I feel like if I am close with them, it [Poop] wouldn't be weird." [p30, M]
2. Edgy (to SO): "Because I don't think poop is something that everyone will appreciate. I feel like, being an SO, you have built up to the relationship. Even though you're sending the poop Animoji it's like acceptable. I don't know if it will be appreciated but people won't like get mad." [p32, F]
3. Funny (to S) "If I am thinking of my sister, I would definitely put the monkey. Just because the only thing that I think of is funny business. Monkeys are always like … just messing around. That's just how siblings are." [p28, F]
4. Cute (to NFF): "[The rabbit] is not as exotic as the dragon or robot, and it's kinda cute. Girls like these little animals." [p17, F]
5. Normal (to NMF): "Personally I would only send a new male friend like just a plain smiling face, that's all… I don't want to send them something too cute or to make them feel like I'm interested or something. I don't want to send something that may make them feel like I crossed the line … So probably just a monkey." [p10, F]
6. Polite (to OR/M): "A dragon is like an old wise person. Older relatives and mentors are like that. I'm saying they are like the dragon by sending the dragon." [p29, M]
7. Polite (to TA): "[The robot] might be one of the more professional Animojis. You should be professional with a TA." [p21, F].

Reasons for Animoji Selection. We further sorted the reasons given for the pairings by individual Animoji, and this also revealed patterned regularities. That is, characteristics were identified for each Animoji that are somewhat independent of the relationship it was intended for, as shown in Table 4. The column order is the same as in Table 3; the rows are ordered to reveal the main associations between Animoji and reasons on the diagonal. (Boldface indicates the high value for the column; italics indicates the high value for the row).

The clearest association in Table 4 is for Rabbit, which was most often described as *cute*. Monkey was described as *funny* and *cute*, Dog as *normal* and *cute*, and Cat and Robot as *polite* and *normal*, although Robot was also sometimes characterized as *funny* and *weird*, depending on the relationship with which it was matched. Dragon also split depending on intended addressee, between *weird* (for CF and NMF) and *polite* (for OR/M). Finally, Poop was described variously as *edgy*, *funny*, and *weird*.[7]

Most of the top associations in Table 4 are illustrated in examples 1–7 above. The following participant quotes illustrate two others.

8. Dragon (weird): "Because it's a close friend so I can send her weird stuff. Dragon is kind of like weird stuff. ... It's like an old man, but also intimidating. So uh, I don't know, just uh weird stuff to close people." [p10, F]
9. Dog (normal): "Just because it is very neutral. Maybe it can be a connecting point like 'Oh I have a dog, you have a dog'. I'd say it's pretty neutral to the TA, not unprofessional like the poop or the robot." [p28, F]

Summary of Card Sort Task Results. The overall findings from the card sort task are summarized in Fig. 2. The column order follows Table 2, and the rows are ranked in order of descending pairing frequency (e.g., CF was most often paired with Poop, which was described as *edgy*, followed by Dragon, which was described as *weird*, etc.).

As Fig. 2 shows, certain Animoji were preferentially associated with certain relationships, although the reasons for choosing them often vary depending on the relationship. For example, an interviewee would orient toward Dog's cuteness when sending it to a SO but focus on its normalness when sending to a NMF or a TA. Dragon could be *weird* when sent to a CF but *polite* (wise) when sent to an older relative or mentor. In these split uses, a different aspect of the Animoji was highlighted as the reason for pairing it with each relationship (e.g., the Dragon's unusual appearance vs. what it symbolizes).

4.3 Self vs. Addressee Orientation

An unanticipated finding that emerged from the reasons given for the Animoji-relationship pairings was that the participants tended to adopt one of two perspectives: The Animoji in some way represented the participant's self, or the Animoji was representative of, or associated with, the addressee. The addressee orientation is illustrated in examples 3, 4, 6, and 7 above, while self-orientation is illustrated in 10–11 below.

[7] Each Animoji was also sometimes chosen last in the card sort by process of elimination. This occurred roughly the same number of times for each Animoji.

Table 4. Reasons sorted by Animoji.

	% Weird (n=17)	% Edgy (n=22)	% Funny (n=36)	% Cute (n=34)	% Normal (n=44)	% Polite (n=19)
Robot (n=21)	12	9	14	0	16	*26*
Cat (n=15)	0	0	0	9	18	*21*
Dog (n=25)	0	5	0	26	*32*	5
Rabbit (n=29)	0	9	6	*44*	18	11
Monkey (n=24)	0	5	*33*	21	9	0
Poop (n=30)	29	*59*	**33**	0	0	0
Dragon (n=28)	*59*	14	14	0	7	**37**
Total	100	100	100	100	100	100

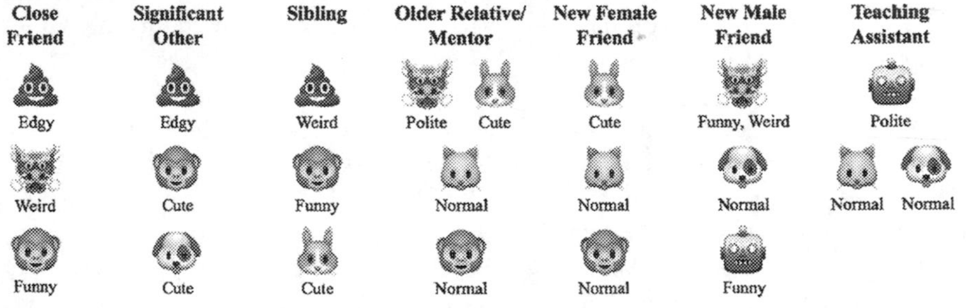

Fig. 2. Summary of card sort task results.

10. "When I put myself as a rabbit, I see myself as like- somehow, rabbit makes me feel like I am younger, cute, innocuous … safe to send to older people." [p14, M]
11. "Dragons kinda have a special meaning in China… Like all the emperors have dragons on their shirts. It's a figure of, like, power… [Interviewer: so you use it with a male friend to assert dominance?] Yes." [p32, F]

Both orientations were expressed for each relationship type, although the addressee orientation was more common, as shown in Fig. 3. As the figure shows, more reasons given for intimate relationships (i.e, SO and CF) were oriented toward the self, while reasons given for the less intimate relationships (NMF, NFF, and OR/M), with the exception of S and TA, were more focused on the addressee.

Reasons for Animoji choice also differed in self vs. addressee orientation (Fig. 4). Most Animoji choices were oriented toward the hypothetical addressee, especially Cat, Robot, and Dragon; these were also described as *polite*. In contrast, the reasons given for the Rabbit were mostly about the sender, as illustrated in example 10.

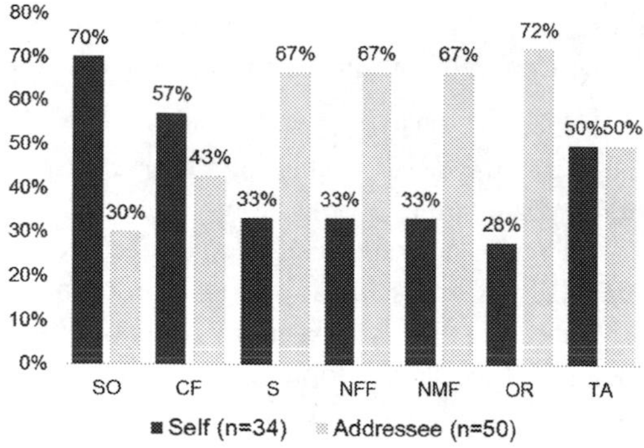

Fig. 3. Self vs. addressee orientation and relationships.

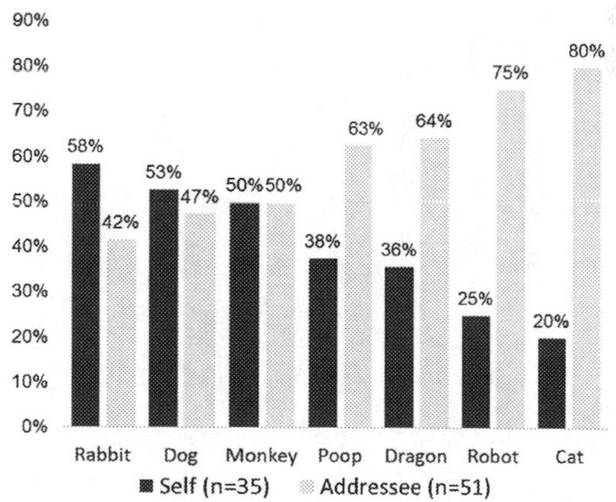

Fig. 4. Self vs. addressee orientation and Animoji.

Taken together, the patterns in Figs. 3 and 4 suggest that more socially distant recipients are accorded greater consideration by Animoji senders, who imagine what their addressees are like and what they might appreciate, and who pay them face through the use of *polite* Animoji. In contrast, representing oneself (or how one wishes to be seen) is more acceptable with intimate recipients, e.g., through *cute* Animoji.

4.4 Naturalness

After the participants had completed the card sort task and sent Animoji clips to the hypothetical addressees, they were asked debrief questions about how natural or unnatural it felt to record the Animoji messages, and how likely or unlikely they were to send Animoji to those relationship categories in reality. The answers to these questions overlapped considerably, since all referred to the same pairings. We combined them into two categories (relationships, Animoji) for the purpose of the analyses in this section.

Relationships. Similar to the responses to the open-ended question in Sect. 4.1, participants overwhelmingly said that they would be most likely to send Animoji to close relationships – SO, CF, and/or S. A smaller percentage specified they would send to an OR who was not a mentor. No one said they would send an Animoji to a mentor. Conversely, participants were least likely to send an Animoji to a TA – although NMF and NFF were also mentioned by some people as unlikely. See Fig. 5.

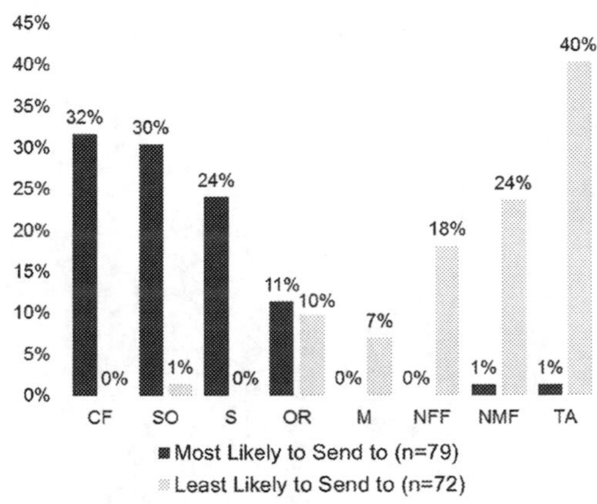

Fig. 5. Most and least likely relationships to send to.

Animoji. We also asked which Animoji felt most natural to send in reality. Most participants focused on relationships, but some mentioned Animoji in their answers. Dog and Monkey were the most natural, while Robot, followed by Dragon and Poop, were the least natural Animoji to send, as shown in Fig. 6. Miscellaneous responses such as "Animoji with a lot of movement" and "all of them" were coded as Other.

The most likely Animoji to be used are those described as *normal*, and the least likely Animoji are described as *weird* and *edgy* (Dragon and Poop). However, Robot, which was often described as professional (*polite*), was the least likely of all the Animoji to be used. In part, this was because it was usually paired with the TA, and sending an Animoji to a TA was felt by all participants to be unnatural and inappropriate.

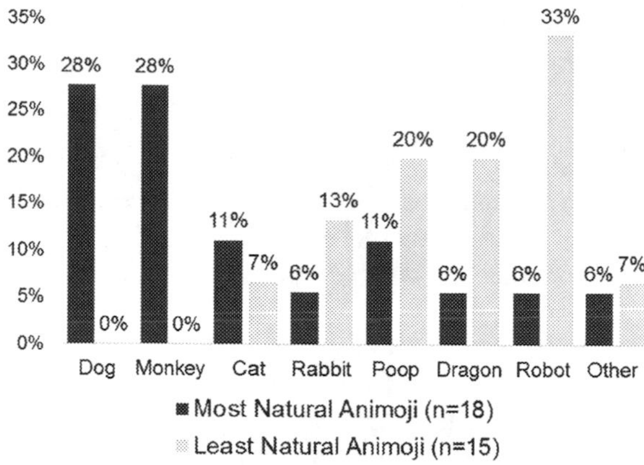

Fig. 6. Most and least natural Animoji to send.

Reasons for Saying Certain Relationships/Animoji Are Natural/Likely. Next, we categorized the reasons given for the naturalness and likeliness responses. Nine categories emerged from this analysis out of a total of 71 reasons.

The most common reasons involved politeness. Animoji were viewed as very *informal* (28%), so a relationship that called for formality was not right for Animoji. One participant explained that Animoji were too new to be used in a formal context:

> "I think Animojis are an extremely informal way of communication, so definitely I would not personally send them to someone that I would have not known very well. … It's not a traditional form of texting. It's something new, and I just feel it's just targeted to people who are young as a fun way [to communicate]." [p31, F]

There was also concern about both appropriateness (6%) and the possibility of giving offense (1%) to someone you do not know well.

Four of the categories that emerged are qualities associated with intimacy. Study participants said they would send Animoji to relationships where there was familiarity (13%) and closeness (13%) between the two, where the interviewee had a personal relationship with the receiver (7%) and was comfortable (13%) with them. One participant said that she would only send Animoji to someone she was familiar with, "because I know them pretty well, so I feel very comfortable just being silly or goofy." She added, "I feel like Animojis are especially about recording yourself, like are more personal, you have to know someone for a while in order to send them." [p27, F]

Related to this were cases where the participant already sent Animoji to that relationship type, having made a habit (13%) of it, which made it easier to pretend to send Animoji to them in our study. One participant said he was most likely to send Animoji to his brother and to his best friend "because I send them Animoji all the time" [p. 4, M].

Finally, age (8%) was a concern as regards older relatives, who were sometimes believed to be unable to understand either Animoji or technology [cf. 16]. A female participant said, "For an older relative, they probably don't know what it is. They wouldn't particularly, maybe even get it" [p27, F]. Conversely, several people commented that they would be likely to send Animoji to a younger sibling, e.g., to tease or amuse them [cf. 20].

4.5 Gender Variation

Although a full analysis of gender-based variation is beyond the scope of the present paper, several gender patterns must be noted briefly here.

First, the women more than the men reported sending Animoji to their significant other (63% vs. 38%) and close friends (63% vs. 46%), whereas fewer women than men sent Animoji to siblings (26% vs. 38%) and (grand)parents (21% vs. 31%).

Moreover, women and men matched different Animoji with their SOs and CFs in the card sort task. Women overwhelmingly chose Poop for their SO, while men mostly chose Dog and Monkey. With CFs, women mostly chose Dragon followed by Dog, and men mostly chose Poop. Assuming a default cross-sex bias in SO relationships and a same-sex bias in CF relationships,[8] it appears that participants of both genders preferentially send *risky* or *edgy* Animoji to men and *normal* and *polite* Animoji to women.

A similar pattern is evident in the Animoji that were paired with new female and male friends. Both genders paired NMF with Dragon (*weird*, *funny*) followed by Dog (*normal*) and Robot (*funny*). For NFF, both genders mostly chose Rabbit (*cute*), Cat (*normal*), and Monkey (*normal*). Again, more normal Animoji were chosen for the hypothetical female recipient.

Women and men generally agreed on the reasons for choosing particular Animoji. The exception is *polite*: Robot was described as polite (professional) by 36% of women and 0% of men, while Dragon was described as polite (wise) by 60% of men and 29% of women. In contrast, men often described Robot as *normal*, and most women described Dragon as *weird*. This may reflect differences in attitudes towards the entities that the Animoji represent (e.g., robots may be more ‘normal’ for men than for women).

No gender differences were found in self vs. addressee orientation or in perceptions of naturalness in the card sort task. In the reasons given for why certain pairings felt more or less natural, however, more men mentioned familiarity, and somewhat more women mentioned closeness, personal connection, and age-related considerations.

5 Discussion

5.1 Research Questions Revisited

Our first research question asked: “To whom do iPhone users send Animoji?” The students we interviewed send Animoji predominantly to close peer relationships: close friends, significant others, and siblings. They also reported sending Animoji to parents and other extended family members, albeit to a lesser degree.

Our second research question asked: “In the card sort task, which Animoji do the users match to each relationship, and why?” Based on the task results, *normal* Animoji

[8] At least two female and two male interviewees were in same-sex relationships, however.

such as Cat, Monkey, and Dog can be sent to anyone and were often paired with more distant relationships. *Polite* Animoji (such as Dragon and Robot) were paired with relationship categories with a larger power differential and where risk avoidance was a concern. *Weird* (i.e., Dragon and Poop) and *edgy* (i.e., Poop) Animoji were only appropriate when sent to close relationships. The principle underlying the pairings seems to be: Use safe Animoji with risky relationships, and risky Animoji with safe relationships. At the same time, although with certain Animoji there were strong specialized associations, there was considerable spread in our data, especially in Table 2. Intended addressee played a crucial role in how Animoji choices were rationalized, such that, for example, Dragon was *weird* when sent to a close friend but *polite* when addressing an older relative. The rationales for these choices focused variously on the recipient and on properties of the Animoji itself, such as its appearance.[9]

Third, we asked: "Which card sort pairings seem most natural or unnatural, and which would be most and least likely to occur in actuality?" Interviewees reported that they found sending Animoji to close personal and familial relationships most natural, and that they were most likely to send Animoji messages to those relationships. These are the same relationship categories reported in response to RQ1. The interviewees all rejected the idea of sending an Animoji to a TA and expressed discomfort at the idea of sending Animoji to new friends of either gender (although this may have been partially caused by the difficulty of thinking of a new friend to send to), as well as to an older relative or mentor. The most *normal* Animoji seemed the most natural to use, while *weird* and *edgy* Animoji were the least natural, although Poop was included in both categories. Robot stood out as being mostly *polite* (professional), while also being the least natural to use. The main driver of whether or not interviewees found pairings natural or unnatural was the (in)formality of the relationship, followed by whether they were familiar or comfortable with the person they imagined sending it to, as well as if they already had a habit of sending Animoji to them. Age was also mentioned, in that sending Animoji to younger people seemed natural and likely, but sending them to members of older generations could cause misunderstanding and thus was unlikely to occur.

Last, we asked: "How, if at all, do the answers to RQs 1–3 vary according to the gender of the interlocutors?" With regard to RQ1, female interviewees sent Animoji more often to their SO and close friends, whereas male interviewees reported sending Animoji more often to a sibling or parent. With regard to RQ2, participants preferentially sent *weird* and *edgy* Animoji to male recipients and inoffensive (i.e., *cute* and *normal*) Animoji to female recipients. This suggests that women were perceived to be more sensitive to face threats [25]. Our female and male participants also differed in which relationship they most often described in terms of politeness (TA or OR/M) and which Animoji they considered most *polite* (Robot or Dragon). As for RQ3, no gender

[9] Animoji may also have different associations in different cultures. Unfortunately, the small size of our sample and the diversity of backgrounds of the non-native English speakers made meaningful comparison of the native and non-native participants impossible.

differences were found in whether interviewees found Animoji-relationship pairings natural or unnatural.

5.2 Additional Findings

Initially, it was unclear whether to posit SO or CF as more intimate (Table 1). In fact, most of our results point to CF as more intimate in terms of Animoji use. However, interviewees were more self-focused (and self-revealing) when addressing a SO than a CF (Fig. 3), which suggests that SO is more intimate. Gender confounds these findings, in that women sent more and "riskier" Animoji to their SOs than men did, while men sent more of both to CFs. This suggests that for the women but not the men, the SO relationship was the most intimate. It could also be that women and men understand intimacy with a SO differently, as suggested by [29].

Another unexpected finding was the self vs. addressee distinction in reasons for Animoji-relationship pairings. The participants represented themselves with an Animoji more often when messaging a SO or CF. Overall, however, they more often oriented to their addressees. From their comments, it appears that our interviewees felt they could express their "true" natures (via Animoji) only to their closest intimates, while in less intimate relationships, which the card sort task required them to consider, they had to pay face to their addressees. They did this by appealing to their addressees' interests and sensibilities. With sibling relationships, particularly with younger siblings, this took the form of sending Animoji to entertain or reference private family jokes.

The Animoji that were used more to represent the self were Rabbit and Dog. Rabbit in particular was used to frame the sender as cute, small, and innocent in relation to the recipient, e.g., a grandparent or a SO. Much of the reasoning around representing the self with an Animoji focused on making the sender appear normal or non-threatening. At the same time, the rationale varied according to the relationship context and the Animoji. For example, the (female) self was represented as powerful and dominant over a NMF when sending the Dragon (ex. 11). Varied rationales were also given for Animoji that were used more often to represent the addressee, i.e., Cat, Robot/Dragon, and Poop, such as that the addressee might like or be amused by what is represented by the Animoji, or that the Animoji resembles the addressee in some way.

6 Conclusions

CMC platforms are commonly used as a means of maintaining relationships with family and friends. These environments foster a sense of belonging, shared space and time, and perceived proximity [28]. In this qualitative user study, we triangulated methods, including interviews and a think-aloud protocol, to answer the question: To whom do Animoji users send Animoji? The main contributions of this research are as follows.

The study is the first to report on why users do or do not send Animoji to particular relationships. As such, it contributes to the small but growing body of research on social distance and graphicon use. Moreover, it identified politeness, (in)formality, familiarity, and self-presentation as reasons for Animoji choices. It also uncovered differences in which Animoji are preferentially sent by and to women and men. These findings complement previous findings of gender differences in Animoji use [18, 21].

Last, the Animoji results provide strong support for the claim that new, multimodal graphicons tend initially to be restricted to use in intimate relationships, where they are less likely to be misconstrued or seen as inappropriate [4, 5]. Animoji pattern similarly to GIFs and stickers in this regard, and if their use catches on, they can be expected to follow a similar evolutionary trajectory. Our results suggest that *normal* Animoji such as Dog, Monkey, and Cat will be the first to extend to more distant social relationships.

6.1 Limitations and Future Directions

We acknowledge several limitations to this exploratory, qualitative user study. First, our small sample size (n = 33) combined with the study design meant that it was not possible to analyze our data using statistical methods. Future studies could build on this study by examining larger numbers of Animoji users with statistical analysis in mind. The artificiality of the card sort task is a limitation, as well. Future studies should aim to analyze actual Animoji messages sent by iPhone users outside of a lab environment. Also, the messages sent in the card sort task were not analyzed in this paper. Future research should examine how Animoji message content and vocal performance [1] vary depending on the social distance between sender and addressee. A methodological limitation is that OR and Mentor were combined in our analyses, yet they patterned differently in terms of naturalness, suggesting that they should be treated as separate categories in future research. Finally, our study interviewed Animoji senders. The perceptions and behaviors of Animoji recipients should also be investigated.

References

1. Herring, S.C., Dainas, A.R.: Nice picture comment! Graphicons in Facebook comment threads. In: Proceedings of the 50th Hawai'i International Conference on System Sciences (HICSS-50). Los Alamitos, CA, IEEE (2017)
2. Baxter, L.A.: Forms and functions of intimate play in personal relationships. Hum. Commun. Res. **18**, 336–363 (1992)
3. Janssen, J.H., Ijsselsteijn, W.A., Westerink, J.H.: How affective technologies can influence intimate interactions and improve social connectedness. Int. J. Hum Comput Stud. **72**(1), 33–43 (2014)
4. Konrad, A., Herring, S.C., Choi, D.: Sticker and emoji use in Facebook messenger: implications for graphicon change. J. Comput. Mediat. Commun. **25**(3), 217–235 (2020)
5. Xu, L., Yi, C., Xu, Y.: Emotional expression online: the impact of task, relationship and personality perception on emoticon usage in Instant Messenger. In: PACIS 2007 Proceedings, p. 79 (2007)
6. Braun, V., Clarke, V.: Using thematic analysis in psychology. Qual. Res. Psychol. **3**(2), 77–101 (2006)
7. Kelly, R., Watts, L.: Characterising the inventive appropriation of emoji as relationally meaningful in mediated close personal relationships. In: Experiences of Technology Appropriation: Unanticipated Users, Usage, Circumstances, and Design, Oslo, Norway (2015)

8. Lee, J.Y., Hong, N., Kim, S., Oh, J., Lee, J.: Smiley face: why we use emoticon stickers in mobile messaging. In: Proceedings of the 18th International Conference on Human-Computer Interaction with Mobile Devices and Services Adjunct, pp. 760–766. ACM (2016)
9. Prager, K.J.: Intimacy in personal relationships. In: Hendrick, C., Hendrick, S.S. (eds.) Close Relationships: A Sourcebook, pp. 229–242. Sage, New York (2000)
10. Reis, H.T., Shaver, P.: Intimacy as an Interpersonal Process. Handbook of Personal Relationships. Erlbaum, Hillsdale (1988)
11. Wang, S.S.: More than words? The effect of line character sticker use on intimacy in the mobile communication environment. Soc. Sci. Comput. Rev. **34**(4), 456–478 (2016)
12. Pazil, N.H.A.: Face, voice and intimacy in long-distance close friendships. Int. J. Asian Soc. Sci. **8**(11), 938–947 (2018)
13. Amit, E., Wakslak, C., Trope, Y.: The use of visual and verbal means of communication across psychological distance. Pers. Soc. Psychol. Bull. **39**(1), 43–56 (2013)
14. Torrez, B., Wakslak, C., Amit, E.: Dynamic distance: use of visual and verbal means of communication as social signals. J. Exp. Soc. Psychol. **85**, 103849 (2019)
15. Zhou, R., Hentschel, J., Kumar, N.: Goodbye text, hello emoji: mobile communication on WeChat in China. In: Proceedings of the 2017 CHI Conference on Human Factors in Computing Systems, NY, pp. 748–759. ACM (2017)
16. Jiang, J.A., Fiesler, C., Brubaker, J.R.: "The perfect one". Understanding communication practices and challenges with animated GIFs. In: Proceedings of the ACM on Human-Computer Interaction 2 (CSCW), NY, pp. 1–20. ACM (2018)
17. Kroll, T., Braun, L.M., Stieglitz, S.: Accommodated emoji usage: influence of hierarchy on the adaption of pictogram usage in instant messaging. In: ACIS 2018 Proceedings, p. 82. (2018)
18. Herring, S.C., Dainas, A.R., Lopez Long, H., Tang, Y.: Animoji performances: "Cuz I can be a sexy poop." Language@Internet 18, Article 1 (2020)
19. Paasonen, S.: Affect, data, manipulation and price in social media. Distinktion J. Soc. Theory **19**(2), 214–229 (2018)
20. de Costa, C.L., Prata, W.: Animoji, Memoji and AR Emoji: how the new emojis can contribute in the communication in chats of social networks. In: Proceedings of the 9th CIDI and 9th CONGIC, Belo Horizonte, Brazil, pp. 112–121 (2019)
21. Herring, S.C., Dainas, A.R., Lopez Long, H., Tang, Y.: Animoji adoption and use: gender associations with an emergent technology. In: Proceedings of Emoji2020, CA. AAAI (2020)
22. Broos, A.: Gender and information and communication technologies (ICT) anxiety: male self-assurance and female hesitation. CyberPsychol. Behav. **8**(1), 21–31 (2005)
23. Zhang, H., Wang, D., Yang, Y.: Explicit and implicit measures of intimate relationships and their association. Acta Psychologica Sinica **38**(06), 910–915 (2016)
24. Lipka, L.: An Outline of English Lexicology. Lexical Structure, Word Semantics, and Word-Formation, 2nd edn. Niemeyer, Tübingen (1990)
25. Glaser, B., Strauss, A.L.: The Discovery of Grounded Theory: Strategies for Qualitative Research. Aldine, Chicago (1967)
26. Brown, P., Levinson, S.: Politeness: Some Universals in Language Usage. Cambridge University Press, New York (1987)
27. Wolfson, N.: The bulge: a theory of speech behavior and social distance. In: Fine, J. (ed.) Second Language Discourse: A Textbook of Current Research, pp. 17–38. Ablex, Norwood (1988)
28. Lomanowska, A.M., Guitton, M.J.: Online intimacy and well-being in the digital age. Internet Interv. **4**(2), 138–144 (2016)
29. Orosan, P.G., Schilling, K.M.: Gender differences in college students' definitions and perceptions of intimacy. Women Ther. **12**(1–2), 201–212 (1992)

"OH MY GOD! BUY IT!" a Multimodal Discourse Analysis of the Discursive Strategies Used by Chinese Ecommerce Live-Streamer Austin Li

Haiyan Huang(✉), Jan Blommaert(✉), and Ellen Van Praet(✉)

Ghent University, Groot Brittanniëlaan 45, Ghent, Belgium
{Haiyan.Huang, Jan.Blommaert, Ellen.VanPraet}@UGent.be

Abstract. Ecommerce livestreaming, also known as live commerce or social commerce, has taken off over the past two years in East Asia and is showing the tendency of going global. Intrigued by the phenomenal success of ecommerce livestream, we concentrate on analyzing the most prominent and illustrative example of Chinese ecommerce live-streamer Austin Li. Through this individual case study, we aim to investigate discursive strategies employed in ecommerce livestreaming and reveal resources specific to this new media genre. Guided by multimodal discourse analysis, our research first accommodates the socio-economic context of Li's success to warrant social situatedness in interpreting data. After that we move into analyzing his discourse employed in livestreaming. Research findings suggest that in attention economy, Li strategically utilizes his male gender as a resource in trying on lipsticks for female customers. His discourse in multiple modes serves to build consumer trust and propagate products. An in-depth analysis of his discursive strategies indicates that, ecommerce livestreaming as a new form of advertising not only shares commonalities with traditional advertisement discourse but also embodies affordances that are specific to livestreaming platforms. To be more specific, livestreaming is featured with delimitation of time, real-time interactivity, and video-aided communication. These affordances enable Li to adopt more interactive and personalized persuasive discourse than conventional advertisements.

Keywords: Ecommerce livestreaming · Live commerce · Social commerce · Social media influencer · Multimodal discourse analysis (MDA)

1 Ecommerce Livestreaming

Ecommerce livestreaming, also known as live commerce or social commerce, has taken off over the past two years. In 2019, revenue generated by ecommerce livestreaming reached 433.8 billion Chinese Yuan ($61.4 billion) and the figure is projected to double in 2020 (iiMedia Research 2020a). There are now multiple platforms on which ecommerce livestreaming is available, mainly include Taobao— Alibaba's premier C2C ecommerce marketplace in China, Mugujie, and Kuaishou. Take Taobao, which is the most well-developed and occupies the largest market share, for example,

C. Stephanidis et al. (Eds.): HCII 2020, LNCS 12427, pp. 305–327, 2020.
https://doi.org/10.1007/978-3-030-60152-2_24

ecommerce livestreaming generated over 100 billion Chinese Yuan ($14 billion) in its transactions in 2018, and the revenue exceeded 2000 billion Chinese Yuan ($28 billion) in 2019 (CBNData 2020). A survey conducted by iiMedia Research (2020b) indicates that in China, there are now roughly 20,000 ecommerce live-streamers and more multi-channel networks (MCNs) are joining to train professional hosts (Topklout 2020). Such live ecommerce hype is going global too. Alibaba has brought the practice to Russia and other Asian countries, such as Thailand, Japan and South Korea. In the United States and Europe, Amazon, Wayfair, Monki, and luxury brand Luis Vuitton have noticed the power of live commerce and have started to use this new media genre to attract potential customer.

Current live ecommerce comprises two forms. The first is livestreaming on ecommerce platforms, such as Taobao (Alibaba's premier C2C ecommerce marketplace in China), and Mogujie on which live-streamers produce contents to propogate products. The second type occurs on content platforms such as Douyin and Kuaishou which collaborate with third party ecommerce platforms. Regardless of which form it takes, ecommerce livestreaming involves five parties, namely, producers (brands), platforms, MCNs, live-streamers and audiences. For professional live-streamers, such as Viya Huang, Austin Li and Xinba, they work with MCNs which negotiates with producers and then choose which products to be broadcast. Live-streamers then promote the products through livestreaming. When broadcasting, streamers have two devices at hand. The main equipment is the webcam in front of them and displays the livestreaming. A second mobile device is deployed nearby and allows them to read the barrages sent by audiences. For audience, they are able to watch sessions on their mobile devices, such as smart phones or tablets, and send their opinions to the platform and platforms. The messages (barrages) sent are visible on screen both to live-streamers and to other peer audiences. On the top of livestreaming page is the personal accounts of streamers, as shown in Image 2 below. Audience can follow them by clicking on the button. On the left bottom is a package of links of products promoted. Audience can buy the product just by clicking the link that will direct them to the shopping page, as shown in Image 3 below. Audiences can also choose to hide all these sub-sections just by tapping the center of their screens.

Fig. 1. Image 1: an overview of Taobao livestreaming on users' screen

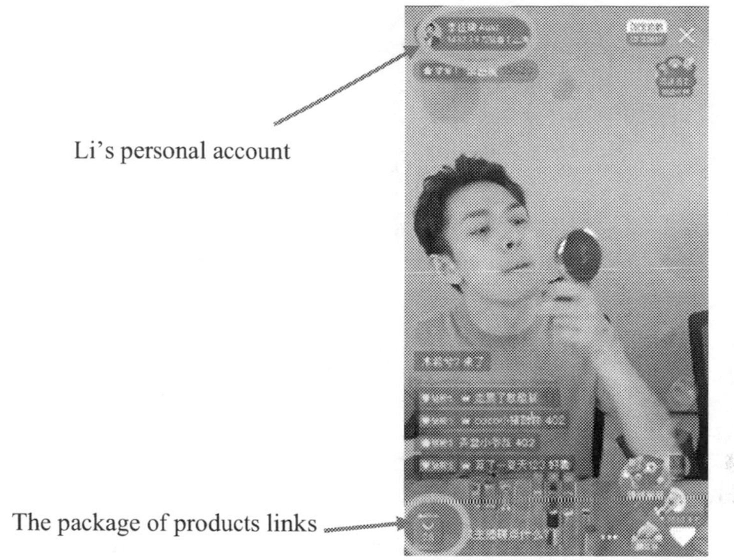

Fig. 2. Image 2: an overview of livestreaming session screen

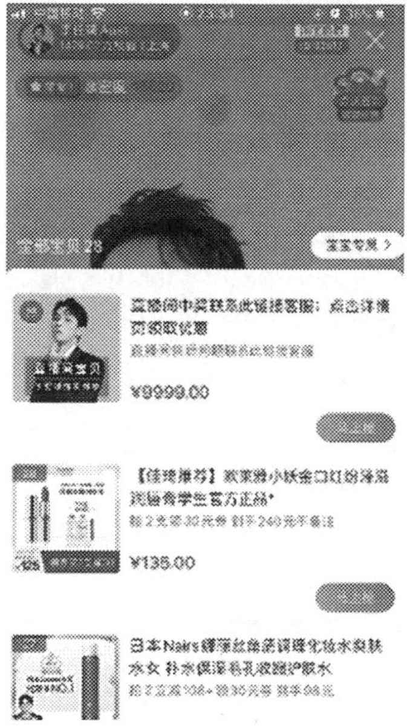

Fig. 3. Image 3: an overview of the screen after clicking product links

Amazed with the phenomenal success of live commerce and intrigued by this latest developed media genre and, we aim to investigate discursive strategies employed in ecommerce livestreaming and reveal resources specific to this new media genre. To fulfill our goals, we take Multimodal Discourse Analysis (MDA) as the research approach to look at an illustrative case—Austin Li, a Chinese ecommerce live-streamer who is known as "Lipstick King". Specifically, we aim to answer the following research questions:

- How does Li position himself as a male live-streamer who promotes lipsticks?
- How the characteristics of ecommerce livestreaming platforms impact Li's discoursc?
- What discursive strategies does Li employ in his livestreaming?
- What are the discursive differences between Li's livestreaming and traditional advertisements

2 Methodology

2.1 Multimodal Discourse Analysis

We rely on multimodal critical discourse analysis (MDA) as the research framework. According to Blommaert (2005), discourse comprises "a general mode of semiosis, i.e., meaningful symbolic behavior" (p. 2). Traditionally, discourse analysis heavily attends to linguistic resources (Fairclough 2001; Wodak 1989; Kress and Van Leeuwen 1996). However, the advent and wide application of digital technologies have diversified the modes of discourse, and expanded the subject of discourse analysis from texts to images, gazes, facial expressions and spatial positioning of people (or objects) inside the image, composition and among others (Kress 2010; Jewitt 2009). Inspired by theories of systematic functional linguistic and social semiotic (Halliday 1978 and 1994), traditional discourse researchers believe that non-linguistic modes also plays fundamental role in rendering social meanings (Kress and van Leeuwen 2001; van Leeuwen 2005) and thus appeal for MDA in addressing digital texts in which two or more semiotic systems are present. The aim of MDA is to understand the how multiple modes interact with one another to make complete meanings. Advocates of MDA hold that this method provides "more comprehensive and inclusive inquiries, analyses, and representations that can be socially, culturally and politically transformative" (Literat et al. 2018, p. 5). MDA has been applied in film, music, advertisements and new media studies where audio, visual and linguistic representations are intertwined (e.g., Gill, 2015; Baykal 2016). Jewitt (2009) summarizes that there three different approaches when conducting multimodal analysis. The first one is social semiotic approach which emphasizes the role of choices in rendering meanings. The second one is MDA and focuses on meaning interpretations. The subtle difference between social semiotic approach and MDA is that the former attaches more importance of "sign-maker" (ibid, p.36) than the latter. The last approach comprises multimodal interactional analysis which "addresses a dimension of the social semiotic that conventional multimodal analysis does not seem to commonly address and focuses on how multimodal texts are interfaced with and mediated by people" (ibid, p. 33). In this study, we choose MDA as the research framework.

2.2 Data Collection and Transcription

Our study focuses on the most prominent and illustrative example of Chinese ecommerce live- streamer Austin Li as the research subject. Such a method of drawing on individual case study enables scholars to have in-depth understanding of social behavior (Meyer 2001) and proves constructive in researching exploratory topics (Frey et al. 1999). In microcelebrities studies, researchers frequently relies on this method, such as Gamson's (2011) analysis of Marlene Dietrich, Marwick's (2013) analysis of Adam Jackson, Jerslev's (2016) analysis of Zoella and Li's (2019) analysis on Papi Jiang.

Our data comprises Li's livestreaming sessions online. Considering Li became famous by broadcasting lipsticks, we limit our data to his livestreaming merely about lipsticks. In collecting data, we found that neither Taobao nor Li's team releases have fully recorded or publicized his livestreamings about lipsticks. As an alternative, we choose to use data on Bilibili website since it possesses more abundant video clips about Li's livestreaming than other video platforms, such as Youtube, Tencent or Aiqiyi. The video clips on Bilibili are recorded by audiences of the time when they were watching the livestreaming sessions. We then input key words "Austin Li Lipstick" (李佳琦口红) in the searching bar and search results are rather miscellaneous, including clips of varying length and different people in camera. We then further set more specific rules and only choose those that are recordings of Li's livestreaming sessions and that are more than 5 min at least to ensure context of the livestreaming is fully demonstrated. In this way, we have more than 400 min of data that fulfills our criteria.

One challenge of MDA is data transcription (Kress 2010; Recktenwald 2017). In multimodal texts, multiple modes may impact data interpretation and researchers thus find it difficult to decide what should(not) be transcribed. Kress (2010) suggests scholars to understand their research purposes before transcribing data. As our study aims to analyze discursive strategies Li employs in livestreaming, we incorporate elements that contribute to persuasiveness in the data transcribed. Following the suggestion by Recktenwald (2017 and 2018), we adopts the multi-column transcription scheme as it allows us to present oral text and images in a concise and synchronized manner. In transcribing, we mainly follow the transcription codes followed by Dressler and Kreuz (2000).

3 Data Analysis and Findings

Before delving into the discursive strategies Li employs in broadcasting lipsticks, it is critical to be aware of the socio-economic context in which Li's success materializes. Such context consists of two levels—that of the society and of the ecommerce livestreaming platforms. Only when we situate the analysis within these backdrop shall we be able to draw accurate data interpretations and be aware of the connections of this media genre and discursive patterns that emerge.

We are in the age of participatory web in which numerous individual netizens contribute users generated contents (Walther and Jang 2012). Such abundant information online on one hand diversifies our daily life while on the hand other distracts us. Those who enjoy the most attention accumulate online fame and establish high online status. Attention thus is a rather valuable asset for microcelebrities. When analyzing Li's case, we repeatedly reflect on his success against this background and here propose that trying on lipsticks as a male streamer—a counterintuitive conduct, is capitalized on to attract audience's attention.

Besides attention economy, another distinct economic feature we shall point out here is network economy. The business mode of live ecommerce gives full play to networked economy. Ecommerce live-streamers have connections to millions of followers and audiences who are presented with literally numerous options in buying their desired items. Scholars (e.g., Sharma et al. 2019) suggest that organizations have realized the potential of social capital, i.e., drawing on social networks to promote sales, and understood the importance of trust in network economy. When it comes to online transactions, trust comes at greater importance as consumers are presented with countless choices and conflicting information (Azam et al. 2013; Yoon and Occena 2015; Tikhomirova and Shuai 2019). In Li's livestreamings, he employs a myriad of strategies to establish and further enhance his trustworthiness and credibility among his audiences, which serves as a very important antecedent to his successful livestreaming career.

3.1 Gender as Resource

Austin Li is a professional ecommerce live-streamer on Taobao. He started to broadcast on Taobao from 2017 and ever since, he has attracted millions of followers on his livestreaming channels. On November 11th 2018, which is known as Double Eleven Shopping Festival, he set the record of selling over 150,000 lipsticks within five minutes. According to his own accounts, Li now possesses over 50 million followers across his social media, including Taobao livestreaming platform, Douyin, and Weibo (iFeng Finance 2020).

At first glance, Li's phenomenal success comes unexpected given that he—a male broadcaster, promoting and trying on lipsticks, is against social norms. In social media 2.0 era when individuals are presented with all types of contents and smart devices, attention has become a valuable asset (Tufekci 2013; Romaniuk and Nguyen 2017). Researchers through empirical studies found that attention impacts product preference and brand consideration (Chandon et al. 2009; Janiszewski et al. 2013). Marwick (2015) suggests that microcelebrities grasped a range of techniques to appeal for attention, including traditional attention-seeking strategies employed by consumer brands, and behaviors that do not fit "the norms of mass culture" (p.138). In Li's case under scrutiny, Li, instead of seeing his gender as an disadvantage, strategically uses it a resource to conduct socially abnormal action—male putting on lipsticks. His counterintuitive behavior helps attract wide-scale attention from users and makes his first step towards a successful ecommerce livestreamer.

3.2 Constructing Trust

Presently the most widely accepted definition of trust is "the willingness of a party to be vulnerable to the actions of another party based on the expectation that the other will perform a particular action important to trustor, irrespective of the ability to monitor or control the other party" (Mayer et al. 1995, p. 712). Companies enjoying high level of credibility find their customers are more receptive to incoming promotional information and products (Fuoli and Hart 2018). Such trust is of particular important resource but also difficult to attain for e-retailers. Potential consumers of online transactions often are overwhelmed and intimidated by false information and shoddy products provided on ecommerce websites (Yoon 2002; Hajli et al. 2017).

Gibson and Manuel (2003) proposed that the key means to construct trust is through communication. This view is echoed by sociolinguists who hold the view that organizations that aspire to construct consumer trust, extensively capitalize on linguistic and other semiotic resources to promote a sense of belonging and a congenial relationship with their customers (Sinclair 2004). Prior research suggest that trust can be discursively constructed in three dimensions—competence, benevolence and integrity (Gibson and Manuel 2003). Competence concerns with individuals or organizations' expertise and professionalism in their fields. Benevolence refers to the extent to which altruism is shown and integrity is related to the qualities of being candid, honest and sincere. In analyzing the data we collected, we noticed that Li strategically constructs audiences' trust by demonstrating his quality of integrity.

Integrity. The salient characteristics of Li's livestreaming is that he directly criticizes the products he tries on. In this way, he manages to create the impression that he is a candid and honest person and that he aligns himself with audiences, rather than flattering products manufacturers. This strategy lends him credibility in the highly networked economy in China. The following excerpt is an example.

1 Extract

Spoken Texts	Livestreaming Images
1 Li: *What is this?*	
2 Fu: *This (..) The money for this is the sum of ten these here (pointing at lipsticks).*	
3 Li: *Hermes bag? A lipstick bag?*	
4 Fu: *Yes*	
5 Li: *{rolling his eyes}{sighing}*	 
6 Fu: *Let's take a look at the bag first!*	
7 Li: *I won't reimburse you the money for this bag. I didn't asked for a Hermes lipstick bag. Such a thing as lipstick bag, erm, really. [People with too much money, would like to buy it.]*	
8 Fu: *[They want to see it. They want to see it.]*	
9 Li: *Ok, Fine.*	
10 Fu: Don't against their, their, how to say it-	
11 Li: *>No one wants to see Hermes lipstick bag. They just want to see Hermes lipsticks.<*	
12 Fu: *They all want to see Hermes lipstick bag.*	
13 Li: *THEY DO NOT.{rolling eyes}*	
14 Fu: *They do.*	
15 Li: *They do not.*	
16 Fu: *No one wants to see it?*	
17 Li: *No.*	
18 Fu: *Am I the only one here?*	
19 Li: *{taking the package from Fu and shaking it} {frowning} This is too light!*	
...	

{Li opened the bag, threw it away immediately, and remained silent}	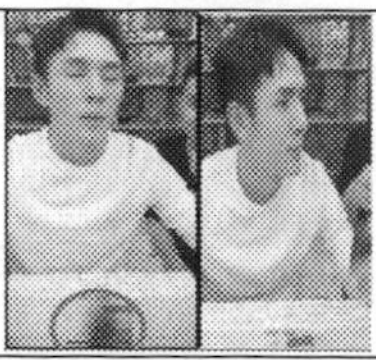
20 Fu: *Hey, don't be like this! It is not good. Your reaction is too real. Just making some ehm... noise is more than enough.*	
21 Li: *This is the kind of bag, (..), we used to make by ourselves when we were little.*	
22 Fu: *Hermes is famous for its handmade bags! Be quiet!*	
23 Li: *Famous for handmake. Of courses, they can have handmake bags. The leather for this is probably useless material, like the leftover of making other bags {frowning}.*	
24 Fu: *No, Hermes only uses the best leather in the world*	
...	
25 Li: *How much is it?*	
26 Fu: *More than 4000 yuan.*	
...	
27 Li: *This is really (..) {rolling eyes}. With 4000 yuan can do a lot of things. Really, I, I, I can never get the point of this bag for my whole life. Maybe I am not that fashionable, not that fancy. Oh, my gosh. It is really easy to make money these days {raising eyebrows, shaking head}.*	

Two themes in this argument stand out. First, through linguistic forms and bodily language, Li constantly displayed his disinterest and disappointments with the bag despite the fact it is made by the world famous bag company—Hermes. Conventionally, Hermes is well-known as a luxury brand that is particularly specialized in making quality and classic bags by handmade. These positive qualities associated with Hermes bag, however were all rejected by Li here. Rather, he explicitly stated that audiences are not interested in this lipstick bag and the bag is "too light", which renders the impression of low-quality here. He further showed his contempt by guessing the material for the tiny lipstick bag is made of "leftover" material of other Hermes bags. These dismissals form a sharp contrast with what Fu said afterwards in unit 24—"Hermes only uses the best leather in the world". His criticisms contribute to establishing his personality as honest. Besides these linguistic resources used, Li also resorted to a set of nonverbal signs, such as rolling eyes, sighing, raising eyebrows, and shaking head, to indicate his disagreement with Fu and disbelief on the price of Hermes lipstick bag. These bodily movements underscore his true feelings and opinions that are expressed through linguistic forms. In this way, his disinterest is strengthened and highlighted through a combination of linguistic and bodily semiosis. Second, throughout the conversation, he clearly aligns himself with audiences. Previous studies have found that such alignment strategies directly enhances trust and play an indirect role in fostering perceived credibility. Corporations and individuals strategically draw on stances resources to construct identities, legitimize behaviors and attain support (Fuoli 2018; Hart 2014; Bondi 2016; Fuoli and Paradis 2014). As discussed before, live ecommerce involves multiple parties, including product producers, ecommerce platforms, live-streamers, MCNs, and audiences. Live-streamers make profits by charging commissions from the items they sell. It is clear that the more they sell, the more financial benefits that they will enjoy. Live-streamers thus are motivated to take a stance with capitalist companies, from the point of financial benefits. However, the conversation here proves this prediction is not necessarily true. Rather than identifying himself with Hermes, Li firmly took the stance with his audiences. For instance, when Fu asked if anyone was interested in seeing the bag, Li directly spoke on behalf of the audience that they did not want to see the bag (units 13, 15 and17). In the end of the excerpt, he suggested that 4,000 yuan for a Hermes bag was overpriced and warned people that with that amount of money "can do a lot of things", which may dissuade them from buying the overpriced bag. The remark "Oh my gosh! It is really easy to make money these days!" (unit 27) obviously targeted at Hermes, an example of capitalist corporation, and accused the unfairness of how easy it is for capitalists to earn money from the mass. In this way, he aggrieved the greediness of the capitalism and sympathized the public who are exploited. This stance-taking act here is used as a resources to enhance consumer trust, which plays a fundamental role in maximizing persuasiveness.

3.3 Employing Discursive Strategies to Promote Products

Live ecommerce is a type of advertisement that is delivered on highly digitalized livestreaming platforms. Discursive strategies used for persuading audiences thus may overlap with those found in traditional advertisement discourse while at the same time, embodies characteristics that are specific to livestreaming.

Exclamations. Using exclamations, such as "God", "Oh my god", "Oh, my gosh", "Jesus Christ", are frequently used in advertisement to reveal strong emotions and indicated amazement and surprise (Syakur and Sukri 2018). As we see in almost Li's all livestreaming, he very frequently, if not always, uses exclamation expressions mentioned above. His catchphrases, "OH MY GOD", "OH MY GOSH" have become his personal identifiers and it is reported that his team has already trademarked the exclamation expression of "OH MY GOD". Below is one typical instance of extensive usages of exclamations.

2 Extract

Spoken Texts	Livestreaming Images
1 *Let me tell you, what can I say?*	
2 *Only... Not oh my god, what to say? Oh my mom! I have to say Oh my mother!*	
3 *This color is just <extremely beautiful>. Pretties, go to the mall tomorrow and buy one. Is it available online, Song Kang?{looking at his team member}*	
...	
4 *OH MY MOM!*	
5 *This color, my mom , oh no, oh my mother!*	
6 *I (..), oh my mom.*	
7 *It is REALLY TOO PRETTY. Lipstick 330*	
...	
{Putting on the lipstick}	
8 *WOW {screaming}It is so beautiful. Really.*	
9 *{adjusting the lipstick color} OH MY GOSH (...) It is really really pretty.*	
...	
10 *{approaching the camera to show his lips}OH MY GOSH! MY MOM! MY MOTHER! OH MY MOTHER! WOW!*	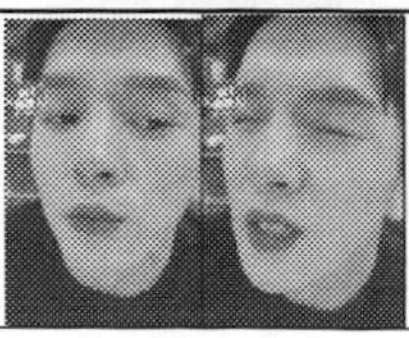

This review occurred in a livestreaming when Li was trying on Guerlain lipstick 330. After putting it on, he used strong exclamation words and expressions, "WOW", "OH, MY GOD", "OH MY GOSH", "OH MY MOM", "OH MY MOTHER" repeatedly to show how surprising and astound he is with the lipstick. Moreover, instead of merely saying "OH MY GOSH", "OH MY GOD"—his catchphrases, he explicitly remarked that English exclamations are not enough and thus suggested to use Chinese exclamations, "OH MY MOM" (wodiniangya), "OH MY MOTHER"(-wodemaya), which are oral expressions frequently used to show strong emotions, such as surprise, happiness, astonishment, etc. Li here expressed the need to use his first language—Chinese so that he would be able to show his emotional strength. Switching exclamations from English to Chinese plays an more effective role in communicating his emotion and views to audiences, who may find themselves emotionally mobilized by Li's discourse. His rising voice volumes (unit 8) and exaggerated facial expressions (units 3, 6, 8 and 10) here serve to strengthen the emotional power of his linguistic utterance.

Imperatives. Using imperative represents a typical grammatical features of advertisement discourse (Kaur et al. 2013; Labrador et al. 2014; Zjakic et al. 2017). Previous studies suggest that advertisers often use imperatives to persuade consumers into taking actions and to establish close relationships with them (Kaur et al. 2013). The frequent imperatives Li uses are listed as followings.

#3 Extracts

Spoken Texts	Livestreaming Images
1 *Buy it! Buy it! Buy it!*	
2 *Please buy it as soon as possible*	
3 *All girls, everybody, please go buy it right away!*	
4 *Put it in your shopping cart first, hurry up!*	
5 *You have to buy it!*	
6 *Bags, don't buy random ones.*	

In describing how decent red the lipstick color is, he chose to compare it to "a graduate from Tsinghua University an Peking University", both of which are the top-notch higher education institutions in China. Graduates from these two universities are considered as intelligent, cultured, and high-level. Via this metaphor, Li not only vividly expressed how fancy and cultured the lipstick is but also demonstrated his creativity—linking a lipstick to top level higher education institutions. Such novelty and creativity can trigger audiences' interest and attract their attention. Previous studies suggest that novel metaphors are more effective for persuasive purposes (Ringrow

2016; Sobrino 2017). Below are other examples in which his resorts to different but creative metaphors.

#5 Extract

Spoken Texts	Livestreaming Image
1 *OH MY MOTHER!*	
2 *With this color...Your lips are followers, fully blooming peonies.*	
3 *Wow, I think this color (..). Really.*	
4 *Girls who have rich husbands, have to buy it {approached the camera to show his lips}.*	
5 *If you go out with this lipstick on, your husband, people definitely will think your husband is rich. This is rich ladies' color.*	

In this extract, Li compares lips with the flower peony, which is considered the national flower of China. Therefore the metaphor here connotes two-level significance. First, peonies by themselves are fresh, delicate and colorful. This metaphor thus maps out these positives onto the product advertised. This metaphorical function, i.e., presenting "what is being advertised in terms of other entities that the characteristics which the advertisers want to associate to the product" (Hidalgo-Downing and Kraljevic-Mujic 2017, p. 324), has long been existing in advertisement discourse. Second, considered as a national flower of China, peony is closely associated with such qualities as beautiful, elegant and generous. Because of these social and cultural connotations associated with peonies, the metaphor used here indicates the semiotic meaning of having this lipstick; that is, it is a signifier of social status. This semiosis is corroborated by Li's explanations—"Girls who have rich husbands have to buy it. If you go out with this lipstick on, your husband, people definitely will think your husband is rich" (units 4 & 5). In this way, having this lipstick in some way satisfies people's fantasy of having a rich life.

#6 Extract

Spoken Texts	Livestreaming Images
1 *It is not sticky and the color applies evenly. It feels like putting on a layer of smashed potato on lips, like a layer of flour on lips.*	
2 *It is < just so comfortable>.*	
3 *The feeling is like that something very softly touches your lips. The feeling is, having melting chocolate on lips in summer.*	
4 *It is< really so comfortable>.*	

Li used several similes and metaphors in this extract to explain the feeling of using this lipstick. Very interestingly, he exclusively referred to food as vehicles—"smashed potato" (unit 1), "flour" (unit 1) and "melting chocolate" (unit 3). This pattern is in tandem with the trend that people in China, especially girls, who comprises the majority of Li's audiences (Sinolink Securities 2019), are generally obsessed with food of all types, as indicated by the booming of food-related documentaries, including A Bite of China, Once Upon A Bite, A Bite of Guangdong, Giving Cycles, and amongst others. In this way, making comparisons with food that is familiar and imaginable to the audiences, Li successfully described the feeling of having the lipstick on in a very simple but expressive way. Such concretization plays a fundamental role in live ecommerce streaming given that the audiences are not able to try on products by themselves. Their expectations are greatly shaped through live-streamers' linguistic strategies. Previous studies (e.g., Tehseem and Kalsoom 2015) indicate that audiences are more reception to promotions when they are presented with metaphorical claims.

Celebrity endorsement.

#7 Extract

Spoken Texts	Livestreaming Image
1 *The first one I am gonna try for you is from Tom Ford, color 69.*	
2 *I call it a color that Faye Wong would use.*	
3 *So I name it Faye Wong Color.*	

Referring to celebrities represents another distinctive strategy in Li's persuasive discourse, as indicated in this extract. "Faye Wong" (unit 2) that appears in this extract, is recognized as a one of the most popular Chinese Cantonpop diva in Chinese diasporas and represents such qualities as "independent, unconventional, and contentious" (Fung 2009, p. 252). Here, Li first associated this color with Faye Wong by assuming that Wong would like this color. He further nicknamed it "Faye Wong Color" to endorse this lipstick. By employing this strategy of celebrity endorsement, Li managed to attract audiences' attention and increase the product's credibility (Marwick 2013). Those audience who aspire to mimic fancy and up-class lifestyle of the famous people will be intrigued with the lipstick and highly motivated to purchase it. Besides fulfilling this materialistic fantasy, possessing this lipstick also encourages its users to think they share the same qualities as Wong do, just by using the lipstick that Li thinks she would like. Lipsticks thus are synonymous to fancy life and ideal personal qualities.

Personalization. Although addressing millions of indeterminable viewers, Li draws on a myriad of discursive resources to personalize his audiences. Fairclough (2001) named this discourse strategy as synthetic personalization—"a compensatory tendency to give the impression of treating each of the people 'handled' en masse as an individual" (p.52). This strategy plays a critical role in building relations between advertisers and viewers, and appeal to the needs to potential customers. In Li's livestreaming, he mainly adopts four types of strategies, including using direct address, presupposing represented groups, contextualization, and interactivity. Previous researchers (Kaur et al. 2013; Labrador et al. 2014; Tehseem and Kalsoom 2015) have found that direct address with audience with second pronoun you is a widespread linguistic feature across advertisement discourse. Designating represented groups and specifying contexts are also reported in advertisement (Hu and Luo 2016). Live commerce, or called by social commerce, values the social media property and thus necessitates interactivity with audiences. Such interactive feature is a representation of participatory web and personalize audiences with different needs by giving them voice (Cao 2019). Below are four illustrative examples to explain each strategy mentioned above.

Direct address.

#8 Extract

Spoken Texts	Livestreaming Image
1 *The first one, 316. This is a must buy. This color, no matter how much you hate MAC, you have to buy this one.*	
2 *No matter HOW MUCH you hate MAC the brand, if you have to buy a lipstick, then there it is*	
3 *It is really good. A really good one. You have to spend 170 yuan to buy it no matter what.*	
4 *You have to save 170 yuan for MAC. MAC deserves to ask for 170 yuan from you for this lipstick.*	

In this extract, Li directly addressed to his audience with you for seven times although he had no clue who are the recipients. Using the second pronoun allows advertisers, Li in this case, to communicate with potential customers in a direct and personal manner, which contributes to a close relationship with audience and fostering trust. When directly addressing audiences, Li also looked at the screen to read barrages (image in unit 1) and this gesture can be considered as having eye contacts with audiences through the texts they sent. From the perspective of audiences, by being individually addressed to instead of an unknown member among millions of viewers, they tend to have the feeling that they are highly valued (Kaur et al. 2013).

Represented groups.
#9 Extract

Spoken Texts	Livestreaming Image
1 *If you a girl that dances, you have to buy it. If you a girl who wears black clothes every day, you have to buy it.*	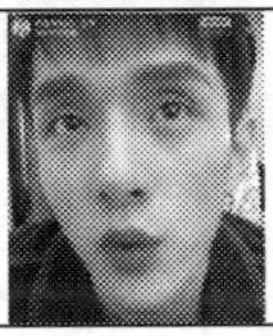
2 *Why? Because it elevates your charisma to full score!*	

Li here identified two groups of audiences—girls who "dance" (unit 1) or "wear black clothes every day" (unit 1). Presupposing groups of different characteristics has two benefits. First, these personal cues encourages identifications from audiences. That is, those who identify themselves as girls who "dance" or "wear black clothes every day" might be mobilized into purchasing the lipstick (Aaker et al. 2000). The usage of direct address, i.e., second pronoun you, further promotes the impression that targeted groups are valued and understood well. Audience thus may respond to Li's understanding by buying the item. Second, his group specific suggestions save audiences' trouble in choosing suitable products. The example below further illustrates this point.

#10 Extract

Spoken Texts	Livestreaming Images
1 *923, girls with fair skin have to buy it. If you don't have fair skin, don't but it.*	
2 *Because it requires makeups, ok? To wear this color, you have to have makeups. You HAVE TO HAVE makeup.*	
3 *If your skin is more yellowish, then you really have to have foundation.*	
...	
4 *And the color doesn't suit heavy makeups.*	
5 *Girls with eyeshadow this wide, eyeline till here, and eyebrows in the shape of a hook.*	
6 *Just light makeup. It matches with neat makeup.*	
7 *With a bit eyebrow and eyelines, some nude eyeshadow, and a bit highlighter then you can use it.*	

Li in this extract explicitly pointed out the target customer—girls who have fair skin (unit 1) as fair ladies do. Then he further gave very specific recommendations (instructions) on how to use the lipstick, including the shape and color of "eyeshadow", "eyelines" and "highlighters" (unit 7), which greatly help girls find appropriate makeup set in a very effective and efficient way. Potential customers who are inexperienced in doing makeups or too busy to think over what type of makeup suits the lipstick find Li very helpful. In this sense, Li's role is similar to "more recommendation" that appears on ecommerce platforms (e.g., Amazong and Tmall) when consumers place an order.

Contextualization.

#11 Extract

Spoken Texts	Livestreaming Images
1 *Summer is coming. It will be windy. Really, if you buy this matte lipstick, you hair won't get stuck on your lips.*	
2 *Pretties, if you go on business trips, or go abroad, please buy this one.*	
3 *Isn't it beautiful? It is a color for daily use.*	

In these three short extracts, Li enumerated four different circumstances—"in summer" (unit 1), "business trips" (unit 2), going "abroad" (unit 2), and "daily" life (unit 3). Such contextualization—"framing a message in a context meaningful to the recipient with the use of contextual variables" (Maslowska et al. 2016), comprises an effective persuasion strategy in the era of internet and social media as people are more often than not overwhelmed by the ocean of information and thus find it challenging to make a decision when shopping. First, by concretely specifying the occasions for different lipsticks, Li, personalized potential customers by imagining their lifestyles and saved the troubles for them who are not adept at or do not have time to choose appropriate cosmetics for different occasions. In this sense, he is an human artificial intelligence. Second, these contextualized messages contain personal cues that audiences might identify with. That is, Li evoked the similarities between the message he uttered and its recipients, and thus enhanced audiences' self-referencing and increased their motivation for social commerce (Aaker et al. 2000; Petty et al. 2002). This personalization strategy, aided by Li's frequent usages of imperatives, can often create artificial needs and foster unnecessary consumptions (Koteyko and Nerlich 2007). In other words, audiences are taught "to desire and generate demand for and consuming mass-marketed goods and services" (Cohen 2008, p.8)

Interactivity.

#12 Extract

Spoken Texts	Livestreaming Images
1 Assistant: *They are all sending messages asking which color suits their mother.*	
2 Li: *You want to know which color for your mother?*	
3 *I will try on some colors for mother in a bit, alright? I will show you later. Right now, I am still choosing colors. (..). I see you your messages asking which color suits mothers, and want to buy for your mom. I will tell you later what color mothers like, ok? Give me some time, I will tell you which color your mom should buy.*	
4 *I will let you know after I try on all colors.*	

This conversation between Li and one of his team members obviously was triggered after seeing messages sent by audiences. Such interactivity, represents one of the most salient features of participatory web—WEB 2.0 (Baldauf et al. 2017). Different from being passive recipients of traditional advertisements, audiences of ecommerce livestreaming are able to communicate their needs and feedback by sending real time comments to interact with live-streamers, Li in this case, and the livestreaming platform. Li, by selecting and reading out barrages on screen, makes his audiences heard and gives power to audiences. In unit 3, Li kept starring on his second device to read barrages and keep himself updated with audience's demands. The combination use of verbal languages and eye contacts creates the impression for the audience that they are valued and heard. Interactivity in this way foregrounds the role of consumers and construct perceived equality (Shanahan et al. 2019). As a consequence, the consumerist nature of traditional selling process is obscured. Rather than merely emphasizing positives of advertised products, ecommerce livestreaming accentuates the needs of consumers through interactivity and makes shopping highly personalized. Potential consumers are thus having the feeling that they play a proactive roles in consumption process in this live (social) commerce, as opposed to conventional practice in which they are passive recipients and lured to consumer by capitalists. By attaching social media property to ecommerce, corporations manages to background the notion of consumerism and foregrounds the role of needs of consumers.

4 Discussion and Conclusion

In this paper, we analyzed discursive strategies demonstrated in ecommerce livestreaming by concentrating on one illustrative case—ecommerce live-streamer Austin Li. As a male broadcaster, he achieved phenomenal success by selling lipsticks to female users. Intrigued by his achievement, we first examined the socio-economic context for the hyping live commerce and then particularly focused our attention on salient discursive strategies Li adopts. Our analysis suggests that in social media 2.0 era, attention is a valuable asset for microcelebrities (Marwick 2013) and conducting behaviors that are incongruent to social norms represents one of the strategies that microcelebrities often resort to. In our case, Li as a male streamer accumulated his fame by promoting and trying on female cosmetic—lipsticks. Such behavior apparent is against Chinese social norm that discourages males using lipsticks, even just for the purpose of work. However, rather than bringing forth harm, using his gender as a resource to accumulate attention proves a successful strategy.

After considering Li's personal peculiarities against current backdrop of attention economy, we then moved to Li's discourse in livestreaming session. From the analysis in previous section we can see that Li's discourse serves two main purposes: building trust and persuade audiences into engaging in online shopping. First, building trust features great important in ecommerce where false information represents one of consumers' biggest concerns (Gefen et al. 2003). There are several ways to construct trust, including benevolence, competence and integrity (Gibbson and Manuel 2003). In Li's case, he constructs his credibility through showing his integrity and impartiality via explicitly criticizing world famous brand and aligning himself with audiences. His discursive strategy of giving honest reviews and taking stances with audiences makes himself a reliable key opinion leader, which is an important antecedent for him to disseminate information and promote commercial products. Second, when promoting items, Li draws on a range of multimodal resources to fulfill various purposes. In terms of linguistic forms, he uses exclamations, imperatives and rhetorical devices (i.e., metaphors and similes), direct address with second pronoun *you*, celebrity endorsement to attract audience's attention, create equality, satisfy audience's desire for fancy life, all of which effectively contribute to maximizing persuasiveness (Cockcroft and Cockcroft 2013). With regards to non-linguistic semiosis, we identify that gestures, facial expressions, sounds and volumes comprise important resources to complement the persuasive power of language. Those strategies identified here are no rare to conventional advertising discourses. Beside them, in this study we found that unctions and features particular to ecommerce livestreaming provides Li news ways to disseminate his persuasiveness. The salient affordances of ecommerce livestreaming platforms consist of their delimitation of time, real-time interactivity, and video-aided communication. Aided by these affordances, Li is able to spend more time on personalizing and interacting with his audiences. For instance, Li often enumerates represented groups and specific contexts for using certain products in his livestreaming sessions. By selecting, reading out and constantly looking at audience's barrages sent, Li interacts with and further gives power to his audiences.

Appendix

Transcription Conventions

Symbol Meaning

TEXT Emphasis or higher volume

. Falling final intonation

? Rising final intonation

: Elongated vowel sounds

{} Physical actions by the streamer

(..) Brief pauses

[] Overlap in Speech

… Abridged content

References

Aaker, J.L., Brumbaugh, A.M., Grier, S.A.: Nontarget markets and viewer distinctiveness: the impact of target marketing on advertising attitudes. J. Consum. Psychol. **9**(3), 127–140 (2000)

Azam, A., Qiang, F., Sharif, S.: Personality based psychological antecedents of consumers' trust in e-commerce. J. WEI Bus. Econ. **2**(1), 31–40 (2013)

Baldauf, H., Develotte, C., Ollagnier-Beldame, M.: The effects of social media on the dynamics of identity: discourse, interaction and digital traces. Alsic. Apprentissage des Langues et Systèmes d'Information et de Communication **20**(1), 1–19 (2017)

Baykal, N.: Multimodal construction of female looks: an analysis of mascara Advertisements. Dilbilim Araştırmaları Dergisi **27**(2), 39–59 (2016)

Blommaert, J.: Discourse: A Critical Introduction. Cambridge University Press, Cambridge (2005)

Bondi, M.: The future in reports: prediction, commitment and legitimization in corporate social responsibility. Pragmat. Soc. **7**(1), 57–81 (2016)

Cao, X.: Bullet screens (Danmu): texting, online streaming, and the spectacle of social inequality on Chinese social networks. Theory, Cult. Soc. 1–21 (2019)

CBNData: 2020 New Economy Report on Taobao Livestreaming. https://www.cbndata.com/report/2219/detail?isReading=report&page=1. Accessed 08 June 2020

Chandon, P., Hutchinson, J.W., Bradlow, E.T., Young, S.H.: Does in-store marketing work? Effects of the number and position of shelf facings on brand attention and evaluation at the point of purchase. J. Market. **73**(6), 1–17 (2009)

Cockcroft, R., Cockcroft, S.: Persuading people: An introduction to Rhetoric. Macmillan International Higher Education, New York (2013)

Cohen, N.: The valorization of surveillance: towards a political economy of Facebook. Democratic Communiqé **22**(1), 5–22 (2008)

Dressler, R.A., Kreuz, R.J.: Transcribing oral discourse: a survey and a model system. Discourse Process. **29**(1), 25–36 (2000)

Fairclough, N.: Language and Power. Pearson Education, Essex (2001)

Frey, L.R., Botan, C.H., Kreps, G.L.: Investigating Communication: An Introduction to Research Methods. Pearson, Boston (1999)

Fuoli, M.: Building a trustworthy corporate identity: a corpus-based analysis of stance in annual and corporate social responsibility reports. Appl. Linguist. **39**(6), 846–885 (2018)

Fuoli, M., Hart, C.: Trust-building strategies in corporate discourse: an experimental study. Discourse Soc. **29**(5), 514–552 (2018)

Fuoli, M., Paradis, C.: A model of trust-repair discourse. J. Pragmat. **74**, 52–69 (2014)

Fung, A.: Faye and the Fandom of a Chinese Diva. Popular Commun. **7**(4), 252–266 (2009)

Gamson, J.: The unwatched life is not worth living: the elevation of the ordinary in celebrity culture. PMLA **126**(4), 1061–1069 (2011)

Gefen, D., Karahanna, E., Straub, D.W.: Trust and TAM in online shopping: an integrated model. MIS Q. **27**(1), 51–90 (2003)

Gibson, C.B., Manuel, J.A.: Building trust: effective multicultural communication processes in virtual teams. In: Gibson, C.B., Cohen, S.G. (eds.) Virtual Teams That Work, Creating Condition for Virtual Team Effectiveness, pp. 59–89. Jossey, Bass (2003)

Gill, S.K.: A multimodal analysis of cover stories on mobile phones: an ideational perspective. Research report, Universiti of Malaya Kuala Lumpur (2015)

Halliday, M.A.K.: Language as a Social Semiotic. Edward Arnold, London (1978)

Halliday, M.A.K.: An Introduction to Functional Grammar. Longman, London (1994)

Hajli, N., Sims, J., Zadeh, A.H., Richard, M.O.: A social commerce investigation of the role of trust in a social networking site on purchase intentions. J. Bus. Res. **71**, 133–141 (2017)

Hart, C.: Discourse, Grammar and Ideology: Functional and Cognitive Perspectives. Bloomsbury Publishing, London (2014)

Hidalgo-Downing, L., Kraljevic-Mujic, B.: Metaphor and persuasion in commercial advertisement. In: Semino, E., Demjén, Z. (eds.) The Routledge Handbook of Metaphor and Language, pp. 323–336. Taylor & Francis, Oxon (2017)

Hu, C., Luo, M.: A multimodal discourse analysis of Tmall's double eleven advertisement. English Lang. Teach. **9**(8), 156–169 (2016)

iFeng Finance. Cover: Talk with Lipstick King Austin Li. https://finance.ifeng.com/c/7rVclUCi5Ee. Accessed 15 Sep 2020

iiMedia Research. Statistical Analysis and Trends of Ecommerce Livestreaming Industryin China (2020–2021). http://www.jhsbggw.com/zhibo/970.html. Accessed 10 June 2020

iiMedia Research. Statistics of Livestreaming: Taobao Livestreaming Platform Boasts 20,000 Hosts in 2019. https://www.iimedia.cn/c1061/69517.html. Accessed 10 June 2020

Janiszewski, C., Kuo, A., Tavassoli, N.T.: The influence of selective attention and inattention to products on subsequent choice. J. Consum. Res. **39**(6), 1258–1274 (2013)

Jerslev, A.: In the time of the microcelebrity: celebrification and the YouTuber Zoella. Int. J. Commun. **10**, 5233–5251 (2016)

Jewitt, C.: The Routledge Handbook of Multimodal Analysis. Routledge, London (2009)

Kaur, K., Arumugam, N., Yunus, N.M.: Beauty product advertisements: a critical discourse analysis. Asian Soc. Sci. **9**(3), 61–71 (2013)

Koteyko, N., Nerlich, B.: Multimodal discourse analysis of probiotic web advertising. Int. J. Lang. Soc. Cult. **23**, 20–31 (2007)

Kress, G.: Multimodality: A Social Semiotic Approach to Contemporary Communication. Routledge, London (2010)

Kress, G.R., Van Leeuwen, T.: Reading Images: The Grammar of Visual Design. Psychology Press, Hove (1996)

Kress, G., Van Leeuwen, T.: Multimodal Discourse: The Modes and Media of Contemporary Communication. Arnold, London (2001)

Labrador, B., Ramón, N., Alaiz-Moretón, H., Sanjurjo-González, H.: Rhetorical structure and persuasive language in the subgenre of online advertisements. Engl. Specif. Purp. **34**, 38–47 (2014)

Li, A.K.: Papi Jiang and microcelebrity in China: a multilevel analysis. Int. J. Commun. **13**, 3016–3034 (2019)

Literat, I., et al.: Toward multimodal inquiry: opportunities, challenges and implications of multimodality for research and scholarship. High. Educ. Res. Develop. **37**(3), 565–578 (2018)

Marwick, A.E.: Status Update: Celebrity, Publicity, and Branding in the Social Media Age. Yale University Press, New Haven (2013)

Marwick, A.E.: Instafame: luxury selfies in the attention economy. Public Cult. **27**(75), 137–160 (2015)

Maslowska, E., Smit, E.G., van den Putte, B.: It is all in the name: a study of consumers' responses to personalized communication. J. Interactive Ad. **16**(1), 74–85 (2016)

Mayer, R.C., Davis, J.H., Schoorman, F.D.: An integrative model of organizational trust. Acad. Manag. Rev. **20**(3), 709–734 (1995)

Meyer, C.B.: A case in case study methodology. Field Methods **13**(4), 329–352 (2001)

Petty, R.E., Priester, J.R, Briñol, P.: Mass media attitude change: implications of the elaboration likelihood model of persuasion. In: Bryant, J., Zillmann, D.: (eds.) LEA's Communication Series. Media Effects: Advances in Theory and Research, pp. 155–198. Lawrence Erlbaum Associates Publishers (2002)

Recktenwald, D.: Toward a transcription and analysis of live streaming on Twitch. J. Pragmat. **115**, 68–81 (2017)

Recktenwald, D.: The discourse of online live streaming on twitch: communication between conversation and commentary. Doctoral dissertation, The Hong Kong Polytechnic University (2018)

Ringrow, H.: The Language of Cosmetics Advertising. Palgrave Macmillan UK, London (2016). https://doi.org/10.1057/978-1-137-55798-8_6

Romaniuk, J., Nguyen, C.: Is consumer psychology research ready for today's attention economy? J. Mark. Manage. **33**(11–12), 909–916 (2017)

Shanahan, T., Tran, T.P., Taylor, E.C.: Getting to know you: Social media personalization as a means of enhancing brand loyalty and perceived quality. J. Retail. Consum. Serv. **47**, 57–65 (2019)

Sharma, S., Menard, P., Mutchler, L.A.: Who to trust? Applying trust to social commerce. J. Comput. Inf. Syst. **59**(1), 32–42 (2019)

Sinclair, J.: Trust the Text: Language, Corpus and Discourse. Routledge, London (2004)

Sinolink securities. analytical report on ecommerce livestreaming of microcelebrities. http://www.767stock.com/2019/12/12/49335.html. Accessed 12 06 2020

Sobrino, P.P.: Multimodal Metaphor and Metonymy in Advertising. John Benjamins Publishing Company, Amsterdam/Philadelphia (2017)

Syakur, A.A., Sukri, M.: Text of cigarette advertisement: a semiology study of Roland Barthes. Int. J. Linguist., Lit. Cult. **4**(3), 72–79 (2018)

Tehseem, T., Kalsoom, U.: Exploring the veiled ideology in cosmetics adverts: a feminist perspective. Eur. J. Res. Soc. Sci. **3**(2), 81–98 (2015)

Tikhomirova, A., Shuai, C.: Assessment of trust building mechanisms of e-commerce: a discourse analysis approach. Profess. Discourse Commun. **1**(4), 23–32 (2019)

Topklout: China's MCN Industry White Book (2020). https://www.ershicimi.com/p/5d5291c8ecfcd54aa1f4f5681f7ee74d. Accessed 10 June 2020

Tufekci, Z.: "Not this one" social movements, the attention economy, and microcelebrity networked activism. Am. Behav. Sci. **57**(7), 848–870 (2013)

Van Leeuwen, T.: Introducing Social Semiotics. Psychology Press (2005)

Walther, J.B., Jang, J.W.: Communication processes in participatory websites. J. Comput. Med. Commun. **18**(1), 2–15 (2012)

Wodak, R. (ed.): Language, Power and Ideology: Studies in Political Discourse. John Benjamins Publishing Company, Amsterdam/Philadelphia (1989)

Yoon, S.J.: The antecedents and consequences of trust in online-purchase decisions. J. Interact. Mark. **16**(2), 47–63 (2002)

Yoon, H.S., Occeña, L.G.: Influencing factors of trust in consumer-to-consumer electronic commerce with gender and age. Int. J. Inf. Manage. **35**(3), 352–363 (2015)

Zjakic, H., Han, C., Liu, X.: "Get fit!"–The use of imperatives in Australian English gym advertisements on Facebook. Discourse Context Media **16**, 12–21 (2017)

Methods of Efficiently Constructing Text-Dialogue-Agent System Using Existing Anime Character

Ryo Ishii[1(✉)], Ryuichiro Higashinaka[1], Koh Mitsuda[1], Taichi Katayama[1], Masahiro Mizukami[2], Junji Tomita[1], Hidetoshi Kawabata[3], Emi Yamaguchi[3], Noritake Adachi[3], and Yushi Aono[1]

[1] NTT Media Intelligence Laboratories, NTT Corporation, 1-1, Hikari-no-oka, Yokosuka-shi, Kanagawa, Japan
ryo.ishii.ct@hco.ntt.co.jp

[2] NTT Communication Science Laboratories, NTT Corporation, 2-4, Hikaridai, Seika-cho, "Keihanna Science City", Kyoto, Japan

[3] DWANGO Co., Ltd., Kabukiza Tower, 4-12-15 Ginza, Chuo-ku, Tokyo, Japan

Abstract. Many surely dream of being able to chat with his/her favorite anime characters from an early age. To make such a dream possible, we propose an efficient method for constructing a system that enables users to text chat with existing anime characters. We tackled two research problems to generate verbal and nonverbal behaviors for a text-chat agent system of an existing character. In the generation of verbal behavior, it is a major issue to be able to generate utterance text that reflects the personality of existing characters in response to any user questions. For this problem, we propose the use role play-based question-answering to efficiently collect high-quality paired data of user's questions and system's answers reflecting the personality of an anime character. We also propose a new utterance generation method that uses a neural translation model with the collected data. Rich and natural expressions of nonverbal behavior greatly enhance the appeal of agent systems. However, not all existing anime characters move as naturally and as diversely as humans. Therefore, we propose a method that can automatically generate whole-body motion from spoken text in order to make it so that anime characters have human-like and natural movements. In addition to these movements, we try to add a small amount of characteristic movement on a rule basis to reflect personality. We created a text-dialogue agent system of a popular existing anime character using our proposed generation methods. As a result of a subjective evaluation of the implemented system, our models for generating verbal and nonverbal behavior improved the impression of the agent's responsiveness and reflected the personality of the character. In addition, generating characteristic motions with a small amount of on the basis of heuristic rules was not effective, but rather the character generated by our generation model that reflects the average motion of persons had more personality. Therefore, our proposed methods for generating verbal and nonverbal behaviors and the construction method will greatly contribute to the realization of text-dialogue-agent systems of existing characters.

C. Stephanidis et al. (Eds.): HCII 2020, LNCS 12427, pp. 328–347, 2020.
https://doi.org/10.1007/978-3-030-60152-2_25

Keywords: Text-dialogue-agent system · Existing anime character · Efficiently constructing method · Utterance generation · Motion generation

1 Introduction

With the practical use of conversational agents and robots using smartphones, the necessity for dialogue techniques for free chatting with agents and robots is increasing. In particular, in recent years, research on constructing dialogue agent systems for entertainment and counseling has been actively conducted, and attention has been paid to the development of actual services. We are aiming to realize a dialogue agent system that allows for text chatting with conversational agents of existing anime characters that have natural movements. Many have one or more favorite anime characters when they are child, and many will continue to dream about having a variety of daily conversations with these characters, even as adults. However, to date, the construction of a dialogue system that reflects the personality of a real character has not been realized to our knowledge. Thus, we here propose a construction method for making such a dream possible in a realistic way.

We try to develop a dialogue-agent system of an existing anime character that operates on the basis of text chat as a first attempt to realize such a system. So that the system generates the natural behaviors of a character, two elemental technologies are needed: verbal behavior (utterance) generation for responding to any user utterances and nonverbal behavior (body motion) generation for the system's utterances. We tackled these two research problems to generate the utterances and body motions of an existing character with the system.

In the generation of verbal behavior, it is a major issue to be able to generate utterances in the form of text that reflect the personality of existing characters in response to any user questions. If the utterances do not properly reflect a character's personality, the user may feel discomfort or get bored quickly without feeling that they are talking to the actual character. This could also break the personality that people expect of characters. Generally, a lot of data must be collected that reflects the personality of a specific character. To generate perfectly correct utterances, appropriate utterances must be created for every utterance that the user enters. However, collecting speech data while keeping a personality consistent is costly. In addition, appropriate answers to a variety of user questions must be generated from a limited amount of data. Therefore, it is necessary to 1) to efficiently collect high-quality data that accurately reflects the individuality of a character and 2) to use the collected data to generate appropriate answers to various user questions.

For this problem 1), we propose the use of a data collection method called "role play-based question-answering" [3], in which users play the role of the characters and answer the question which user ask the characters, to efficiently collect responses that accurately reflect the personality of a particular character for many questions. For this problem 2), we propose a new utterance generation method that uses a neural translation model with the collected data.

Rich and natural expressions of body motion greatly enhance the appeal of agent systems. Therefore, generating agent body movements that are more human and enriched is an essential element in building an attractive agent system. However, not all existing anime characters move as naturally and as diversely as humans. Characters can also have unique movements, for example, a specific pose that appears with a character-specific utterance. Therefore, it is also important to reproduce unique movements to realize a system that reflects a character's personality. It is known that there is a strong relationship between the content of an utterance and body motion. The motion should be suited to the content. If a skilled creator created a motion that satisfies these two aspects for every utterance, the dialogue agent would probably be able to express motion that completely matches that of a character. However, in dialogue systems that generate a variety of utterances, this is not practical.

Therefore, in this context, we propose the use of a motion generation method that is comprised of two methods of introduce movements in dialog agents of existing characters that are more human-like and natural and introduce character-specific motions. First, we propose a method that can automatically generate whole-body motion from utterance text in order to make anime characters have human-like and natural movements. Second, in addition to these movements, we try to add a small amount of characteristic movement on a rule basis to reflect the personality.

As a target for applying these proposed utterance and motion generation methods, we construct a dialogue system with "Ayase Aragaki," a character of the light novel "Ore no Imoto ga Konna ni Kawaii Wake ga Nai" in Japanese, which means "My Little Sister Can't Be this Cute" in English. The novel is a popular light novel that has sold over five million copies in Japan and has been animated. Ayase is not the main character, but she has an interesting personality called "yandere." According to Wikipedia, this means that she is mentally unstable, and once her mental state is disturbed, she acts out as an outlet for her emotions and behaves extremely violently. For this reason, she is a very popular character.

We constructed an agent text-chat dialogue system that reflects her personality by using the proposed construction method. We evaluated the usefulness of the implemented system from the viewpoint of whether her responses are good and her personality could be reflected properly. As a result, both of our proposed methods for generating utterances and body motions were found to be useful for improving the users' impression of goodness and the reflectiveness of the anime character's personality in the responses of the system. This suggests that our proposed construction method will greatly contribute to realizing text-dialogue-agent systems of characters.

Fig. 1. Site screen for data collection using "role play-base question-answering"

2 Utterance Generation Method

2.1 Approach

Generally, when generating an utterance that reflects personality in such a way that quality is guaranteed, there is a problem in that the cost is high because large-scale utterance-pair data must be prepared and used manually in advance. In this research, we propose using a data collection method called "role play-based question-answering" [3] to efficiently construct high-quality utterance pairs reflecting the individuality of a character. This is a method in which multiple users participate online, ask questions to a specific character, or answer a question as is, to efficiently collect high-quality character-like utterance pairs. Specifically, the user has two roles. One is to ask a character a question (utterance). The user asks a character a question that they want to ask on a variety of topics. This question is notified to all users. The second role is to become a character and answer that question. This makes it possible to efficiently collect utterance pairs of questions and answers by sharing and using the questions of the user and answering each question. In addition, the role-playing experience itself is interesting, so there is no need to modify it in order to make it easier for users to participate [12]. By using this method, we thought that it would be possible to efficiently collect utterance-pair data that seems to reflect an anime character.

Our proposed utterance generation method uses a neural translation model, which is one of the latest machine learning methods for text generation that has been attracting attention in recent years and that extracts appropriate answers from collected data.

2.2 Data Collection Using Role-Playing Question Answering

NICONICO Douga[1], a video streaming service, offers a channel service for fans of various characters. Use of the channel is limited to registered subscribers.

[1] http://www.nicovideo.jp/.

In our research, a bulletin board that responds to questions in tandem with this channel service was built for the channel about "Ayase Aragaki". Figure 1 shows a screenshot of the site. Users can freely ask Ayase Aragaki questions from a prepared text form. A user who wants to respond to Ayase Aragaki can freely answer a question. At the same time as answering, labeling of emotions accompanying an utterance is performed. There were eight classifications for the labeled emotions: normal, angry, fear, fun, sad, shy, surprise, and yandere.

To increase the users' motivation to participate, the website showed the ranking of users by the number of posts. In addition, a "Like" was placed next to each answer. If a user's answer seems to be Ayase Aragaki (-like), "Like" is pressed. We devised it so that the evaluation of users can be reflected in the quality of answers. In October 2017, a website was opened, and the service was operated for about 90 days. A total of 333 users participated. The collected utterance pairs exceeded 10,000 in about 20 days, and finally, 15,112 utterance pairs were obtained by the end of the service. Users voluntarily participated in this response site and were not paid. Nevertheless, the fact that we were able to collect such a large amount of data suggests that data collection using the role play-based question-answering method is useful.

Table 1. Results of user evaluation of experience with role play-based question-answering service

Evaluation item	Mean score
(Q1) Did you use the website comfortably?	4.08
(Q2) Did you enjoy the role play-based question-answering?	4.53
(Q3) Do you want to experience this web service again?	4.56

A questionnaire evaluation was conducted for participating users in order to determine their satisfaction with using the question-answering site. A total of 36 users cooperated in the evaluation and responded to the items shown in Table 1 on the basis of a five-point Likert scale (1–5 points). Table 1 shows the average of the user evaluation values.

Looking at the results, a high rating of 4.08 was obtained for the item "Did you use the website comfortably?" This suggests that the experience was comfortable for users when using the service on our website. A high rating of 4.53 and 4.56 also were obtained for the items "Did you enjoy the role play-based question-answering?" and "Do you want to experience this web service again?" These suggest that the service was attractive to the users.

Next, to evaluate the quality of the collected utterance data of Ayase, a subjective evaluation was performed by the participating users. For about 50 utterance pairs, which were selected randomly from all collected data of 15,112 utterance pairs, participating users evaluated whether they were natural and properly reflected her personality. The mean scores of naturalness and personality were 3.61 and 3.74 on a five-point Likert scale (1 to 5 points). This indicates that the quality of the response-utterance data collected through role play-based question-answering was reasonably high. However, it was a surprising that it was difficult to obtain a rating of 4.0 or more, even if the response data was created by human users. In other words, it was suggested that utterance generation that reflects the individuality of a particular character was a difficult task even for humans.

2.3 Proposed Utterance Generation Technique

We thus propose an utterance generation method that uses the collected utterance-pair data. Since the amount of data collected was not large enough to train an utterance generation model using neural networks [17], we used the approach of extracting optimal responses from the obtained utterance data. In other words, we addressed the problem of selecting a response for the most relevant utterance pair against a user's utterance. In this study, a neural translation model was used to select an appropriate utterance pair. Using the results obtained from LUCENE[2], a popular open source search engine, was not enough just to match words to user utterances. We developed a new method for this. This method focuses on recent advances in cross-lingual question answering (CLQA) [8] and neural dialogue models [17]. In addition, we matched the semantic and intention levels of the questions so that the appropriate answer candidates were ranked higher.

1. Given question Q as input, LUCENE searches the top N question-response pairs $(Q'_1, A'_1), \ldots, (Q'_N, A'_N)$ from our dataset.
2. For Q and Q', question type determination and named entity extraction are performed, and the question type and named entity (using Sekine's extended named entity [16] as the system) are extracted. To what extent a named entity asked by Q is included in A' is calculated, and this is used as the question type match score (qtype_match_score).
3. Using the focus extraction module, focus (noun phrase indicating topic) is extracted from Q and Q'. If the focus of Q is included in Q', the focus score (center-word_score) is set to 1; otherwise, it is set to 0.
4. The translation model calculates the probability that A' is generated from Q, that is, $p(A'|Q)$. We also calculate $p(Q|A')$ as the reverse translation probability. Such reverse translation has been validated in CLQA [8]. The generation probability is normalized by the number of words on the output side. Since it is difficult to integrate a probability value with other scores due to differences in range, we rank the answer candidates on the basis of this probability and translate the translation score (translation_score; _translation_score). Specifically, when the rank of a certain answer candidate is r, the translation score is obtained as follows.

$$1.0 - (r - 1)/\text{max\_rank} \quad (1)$$

[2] https://lucene.apache.org/.

Here, max_rank is the maximum number of possible answer candidates. The translation model was learned by pre-training about 500,000 general question-response pairs and then performing fine-tuning with the utterance pairs obtained from the complete question-response. In the reverse model, the same processing was performed, exchanging the questions and responses. For the training, the OpenNMT toolkit was used with the default settings.

5. The similarity between Q and Q' is measured by the semantic similarity model. Word2vec [13] is used for this. First, for each Q and Q', a word vector is obtained, an average vector is created, and cosine similarity is calculated, and this is set as a similarity score (*semantic_similarity_score*).
6. The previous scores are added by weight, and a final score is obtained.

$$
\begin{aligned}
&\mathrm{score}(Q, (Q', A')) \\
&= w_1 * \mathrm{search\_score} \\
&+ w_2 * \mathrm{qtypes\_match\_score} \\
&+ w_3 * \mathrm{center\text{-}word\_score} \\
&+ w_4 * \mathrm{translation\_score} \\
&+ w_5 * \mathrm{rev\_translation\_score} \\
&+ w_6 * \mathrm{semantic\_similarity\_score}
\end{aligned} \tag{2}
$$

Here, *search_score* is a score obtained from the ranking of search results by LUCENE, and it is obtained from expression 1. $w_1, \ldots, w_6$ is a weight, which is 1.0 in this study.

7. On the basis of the above score, the answer candidates are ranked, and the top items are output.
The most appropriate answer sentence to a question is selectively generated with this combination of various types of language processing.

3 Motion Generation Method

3.1 Approach

It has been shown that giving appropriate body movements to agents and humanoid robots not only improves the natural appearance but also promotes conversation. For example, actions accompanying utterances have the effect of enhancing the persuasiveness of utterances, making it easier for the other party to understand the content of the utterances [9]. Therefore, generating agent body movements that are more human and enriched is an essential element in building an attractive agent system. As mentioned above, however, not all existing anime characters move as naturally and as diversely as humans. Additionally, characters can have unique movements, for example, a special pose that appears with a character-specific utterance. Therefore, it is also important to reproduce unique movements so that the system reflects a character's personality. Since there is a strong relationship between utterance content and body motion, motion should be suited to the content. If a skilled creator were to create a motion that satisfies these two aspects for every utterance, the dialogue agent would probably be able to express motion that perfectly expresses a character. However, for dialogue systems that generate a variety of utterances, this is not practical.

Therefore, for practical motion generation, we propose using a motion generation method comprised of two methods that introduce characteristic movements that are more human-like and natural and introduce character-specific motions. First, we propose a method that can automatically generate whole-body motion from utterance text so that anime characters can make human-like and natural movements. Second, in addition, we try to add a small amount of characteristic movement on a rule basis to reflect personality.

The proposed motion generation method makes the motion of animated characters more natural and human. In a text dialogue system, linguistic information obtained from system utterances may be used as input to generate motions. In past research on motion generation using linguistic information, we mainly worked on the generation of a small number of motions using word information, such as the presence or absence of nodding and limited hand gestures [5–7]. In this research, we tried to generate more comprehensive whole body movements by using various types of linguistic information. As a specific approach, we constructed a corpus containing data on speech linguistic information and motion information obtained during human dialogue, learn the co-occurrence relationship of these using machine learning, and generate motion using speech linguistic information as input. In the next section, the construction of the corpus data, the motion generation method, and its performance are described. Then, we introduce a way to add specific body motions that reflect an anime character's personality.

3.2 Collecting Data for Motion Generation

A linguistic and non-linguistic multi-modal corpus, including spoken language and accompanying body movement data, was constructed for two-party dialogues. The participants in the two-way dialogue were Japanese men and women in their 20 s–50 s, who had never met. There was a total of 24 participants (12 pairs). Participants sat facing each other. To collect a lot of data on various actions, such as chats, discussions, and nodding and hand gestures associated with utterances, we used dialogues in which animated content was explained. In these dialogues, each participant watched an episode (Tom & Jerry) with different content and explained the content to their conversation partner. The conversation partner was free to ask the presenter questions and to have a free conversation. For recording utterances, a directional pin microphone attached to each subject's chest was used. A video camera was used to record the overall dialogue situation and the participants. The video was recorded 30 Hz.

The total time of the chats, discussions, and explanations was set to 10 min each, and in this study, the data of the first 5 min were used. For each pair, one chat dialogue, one discussion dialogue, and two explanation sessions were conducted. Therefore, we collected 20 min of conversation data for each pair, and we collected a total of 240 min of conversation data for 12 pairs.

Next, we show the acquired linguistic and non-linguistic data.

- Utterance: After manually transcribing the utterances from the voice information, the content of the utterances was confirmed, and the sentences were divided. Furthermore, each sentence was divided into phrases by using a dependency analysis engine [4]. The number of divided segments was 11,877.
- Face direction: Using the face image processing tool OpenFace [15], three-dimensional face orientation information was taken from the front of the participants with a video camera. The angles of yaw, roll, and pitch were obtained. Each angle was classified as micro when the angle was 10° or less, small when it was 20° or less, medium when it was 30° or less, and large when it was 45° or more.
- Nodding: Sections in the video where nodding occurred were manually labeled. Continuous nods were treated as one nod event. In addition, the number of times was classified into five stages from 1 to 5 (or more). In addition, for the depth of the nod, OpenFace was used to calculate the difference between the start of the head posture pitch at the time of the nod and the angle of the deepest rotation. The angle was classified as micro when the angle was 10° or less, small when it was 20° or less, medium when it was 30° or less, and large when it was 45° or more.
- Hand gesture: Sections in the video where hand gesture occurred were manually labeled. A series of hand-gesture motions were classified into the following four states.
  - Prep: Raise your hand to make a gesture from the home position
  - Hold: Hand held in the air (waiting time until gesture start)
  - Stroke: Perform gesture
  - Return: Return hand to home position

  However, in this study, for simplicity, a series of actions from Prep to Return were treated as one gesture event. Furthermore, the types of hand gestures were classified into the following eight types based on the classification of hand gestures by McNeil [10].
  - Iconic: Gestures used to describe scene descriptions and actions.
  - Metaphoric: Like Iconic, this is a pictorial and graphic gesture, but the specified content is an abstract matter or concept. For example, the flow of time.
  - Beat: Adjusts the tone of speech and emphasizes speech. Shake your hand or wave your hand according to your utterance.
  - Deictic: A gesture that points directly to a direction, place, or thing, such as pointing.
  - Feedback: Gestures issued in synchronization with, consent to, or in response to another person's utterance. A gesture that accompanies an utterance in response to an utterance or gesture in front of another person. In addition, gestures of the same shape performed by imitating the gestures of the other party.

  - Compellation: Gesture to call the other person.
  - Hesitate: Gesture that appears at the time of hesitation.
  - Others: Gestures that are unclear but seem to have some meaning.
- Upper body posture: We observed postures when the participants were seated, and there was no significant change in the seated position. For this reason, front and back position of upper body was extracted on the basis of the three-dimensional position of the head. Specifically, the difference between the coordinate position in the front-back direction of the head position obtained using OpenFace and the position of the center position was obtained. From the position information, the angle of the posture change of the upper body was calculated as micro when it was 10° or less, small when it was 20° or less, medium when it was 30° or less, and large when it was 45° or more.

Table 2. List of generated label for each motion part

Motion part	Number of labels	List of labels
Number of nods	6	0, 1, 2, 3, 4, more than 5
Deepness of nodding	4	micro, small, medium, large
Direction of head (yaw)	9	front, right-micro, right-small, right-medium, right-large, left-micro, left-small, left-medium, left-large
Direction of head (roll)	9	front, right-micro, right-small, right-medium, right-large, left-micro, left-small, left-medium, left-large
Direction of head (pitch)	7	front, up-micro, up-small, up-medium, up-large, up-micro, up-small, up-medium, up-large
Hand gesture	9	none, iconic, metaphoric, beat, deictic, feedback, compellation, hesitation, others
Upper body posture	7	center, forward-small, forward-medium, forward-large, backward-small, backward-medium, backward-large

Table 2 shows the list of parameters of the obtained corpus data. In addition, ELAN [18] was used for manual annotation, and all the above data were integrated with a time resolution 30 Hz.

3.3 Proposed Motion Generation Method

Using the constructed corpus data, we input a word, its part of speech, a thesaurus, a word position, and the utterance action of one entire utterance as input, and we created a model that generates one action class for each clause for each of the eight actions shown in the table by using the decision tree algorithm C4.5. That is, eight option labels were generated for each clause. Specifically, the language features used were as follows.

Table 3. Performance of generation model and chance level. The each score shows F-measure.

Motion part	Chance level	Proposed method
Number of nods	0.226	0.428
Deepness of nods	0.304	0.475
Face direction (yaw)	0.232	0.329
Face direction (roll)	0.297	0.397
Face direction (pitch)	0.261	0.378
Hand gesture	0.156	0.303
Upper body posture	0.183	0.311

- Number of characters: number of characters in a clause
- Position: the position of the phrase from the beginning and end of the sentence
- Words: Word information (Bag-of-words) in phrases extracted by the morphological analysis tool Jtag [1]
- Part-of-speech: part-of-speech information of words in a clause extracted by Jtag [1]
- Thesaurus: Thesaurus information of words in a phrase based on Japanese vocabulary
- Utterance act: Utterance act estimation method using word n-gram and thesaurus information [2,11]. Utterance act extracted for each sentence (33 types).

The evaluation was performed by cross-validation done 24 times, in which the data of 23 of the 24 participants were used for learning, and the remaining data of one participant was used for evaluation. We evaluated how much actual human motion could be generated from only the data of others. Table 3 shows the average of F-value as a performance evaluation result. The chance level indicates the performance when all classes with the highest number of correct answers are output. Table 3 shows that the accuracy was significantly higher than the chance level for all generation targets (results of paired t-test: $p < .05$). The results show that the proposed method, which uses words, their parts of speech and thesaurus information, word positions, and actions performed during the entirety of speaking, obtained from the spoken language, is effective in generating whole-body motions as shown in Table 3.

3.4 Additional Original Motion Reflecting Character's Personality

In addition to the motion generation proposed in the previous section, motions unique to Ayase were extracted from the motions in the animation, and the four original motions shown in Table 4 were added. These actions were selected in collaboration with Ayase's creators who have experience in creating animation. For these original actions, words and sentences that trigger the actions were set, and when these words and sentences appear in a system utterance, the original actions take precedence over the output results of the action generation model. Although the number of movements was as small as four, we could not find any more distinctive movements to note, so we thought that the number of movements was sufficient.

Fig. 2. Example of scene in which original motion of raising and lowering arm and protruding face is performed in accordance with text display of "Pervert!" included in system utterance.

Table 4. Additional original movements and example utterance text to trigger

Additional original movements	Example utterance text to trigger
Performs roundhouse kick	*I'm gonna take you down*
Crosses her arms	*Ewww*
Raises her arms and sticks face out	*Pervert*
Points to her opponent with right hand	*I'll report you*

4 Construction of Dialogue System Reflecting Anime Character's Personality

Using the proposed methods for utterance and motion generation, we constructed a dialogue system that can respond to user utterances with utterances and motion. Figure 2 shows a diagram of the system configuration.

The user enters input text from the chat UI. When the dialog manager receives it, the user text is first sent to the utterance generator. After acquiring the system utterance from the utterance generation unit, the system utterance is transmitted to the motion generation unit, and the motion information of each clause of the system utterance text is obtained.

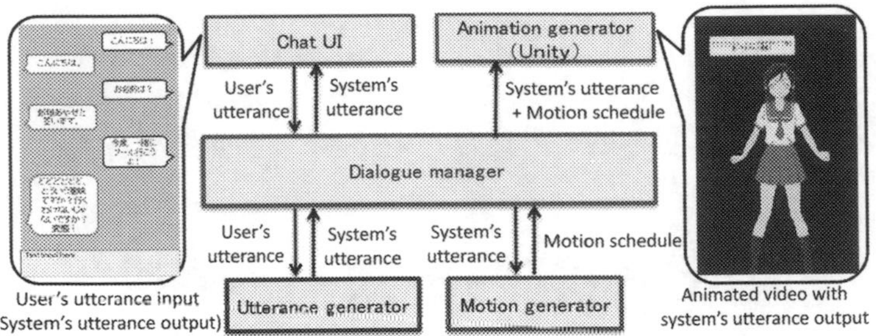

Fig. 3. System architecture of our system

In addition to this, while this is not necessary for text dialogues, it is also possible to obtain uttered speech with the speech synthesis unit. In this system, the speech obtained from the speech synthesis unit is used to generate the lip-sync motion.

The dialogue manager sends the system utterance text, motion schedules, and voices to the agent animation generator. In the agent animation part, the utterance text is displayed in a speech bubble above the character at equal time intervals from the first character, And the motion of the agent is generated in sync with the display of utterance characters according to a motion schedule. As a means of generating motion animation, a CG character was created in UNITY, and an animation corresponding to the motion list in Tables 2 and 4 list was generated in real time. At this time, the eight objects shown in Table 2 can operate independently, and the head motion is generated by mixing all parameters of the number of nods, deepness of head movement, and head directions (yaw, roll, pitch). When utterance text that is registered as a trigger for generating specific motion in Table 4. A specific motion is generated instead of a motion generated by our generation model. All motions of the agent are generated according to the timing of the utterance text display. An example of the presentation screen is shown on the right side of Fig. 3.

It is also possible to send a system utterance from the dialog management unit to the chat UI and to present the system utterance in the chat UI in addition to the user utterance shown on the left side of Fig. 3.

5 Subjective Evaluation

5.1 Evaluation Method

The effectiveness of the proposed method was evaluated in subject experiments using the constructed dialogue system. As an evaluation item, we evaluated the usefulness of the responses of the dialogue system by the proposed utterance- and motion generation methods. The purpose of this evaluation was to pay particular attention to character reproducibility (character-like) in addition to the goodness of the responses.

The following three conditions were set as experimental conditions for utterance generation.

- **U-AIML**: A rule-based method written in AIML, which is a general method used for utterance generation, was used. Specifically, we used a large-scale AIML database that has been constructed up to 300K utterance pairs. In Japanese, sentence-end expressions are some of the most important elements that indicate character, so these expressions were converted to expressions like those used by Ayase by using a sentence-end conversion method [14].
- **U-PROP**: A utterance is generated by using our proposed utterance generation method described in Sect. 2. The weight parameters w_1 to w_6 were experimentally set to 1.0 in Formula (1).
- **U-GOLD**: An utterance is generated by using collected data from the role play-based question-answering method in Sect. 2.2. When multiple answers were given to a question, one was selected at random.

By comparing the U-AIML and U-PROP conditions, the usefulness of the proposed utterance generation was compared with manual utterance generation and evaluated. We also compared the U-PROP and U-GOLD conditions to evaluate how much the proposed utterance generation method is useful to human-generated.

The following four conditions were set as the experimental conditions for motion generation.

- **M-BASE**: Generates basic character movements such as for lip sync and facial expressions. For generating facial expressions, we created animations for facial expressions corresponding to the eight emotions collected with the role play-based question-answering method in Sect. 2.2. Facial expressions under the U-PROP and U-GOLD conditions were generated by using the collected data. For the U-AIML condition, humans annotated the emotion label for each utterance manually. The labels were used to generate facial expressions.
- **M-RAND**: In addition to lip sync and facial expressions, whole body movements were randomly generated.
- **M-PROP1**: Motion generation method was used on basis of human data.
- **M-PROP2**: In addition to using the motion generation method with human data, a small amount of motion unique to the character was added.

By comparing the M-BASE and M-RAND conditions, we evaluated the usefulness of motion generation for the whole body, and we compared the M-RAND and M-PROP1 conditions to evaluate the usefulness of the proposed motion generation method by learning human motion data. Also, by comparing the M-PROP1 and M-PROP2 conditions, we evaluated the usefulness of adding a small amount of unique character-specific motions in addition to the proposed generation by learning human motion data.

Twelve conditions combining these three conditions of utterance generation and four conditions of motion generation were set as experimental conditions.

Table 5. Items and questionnaire of subjective evaluation

Items	*Questionnaire*
Goodness	*Is the overall response good?*
Character-likeness	*Does the response comprehensively reflect the personality of "Ayase Aragaki"?*

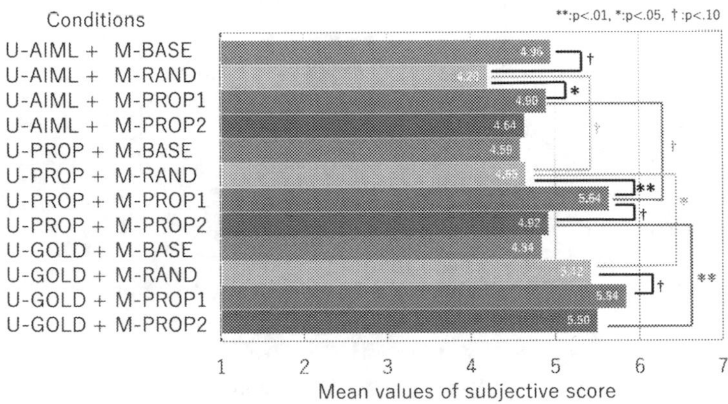

Fig. 4. Results of subjective evaluation of impression of "goodness" of overall response.

As an experimental method, the same user utterance was set for comparison under each condition, and the utterance and motion of the system in response to the user utterance were evaluated. Specifically, ten question utterances were randomly extracted from the collected question-answer data. The subjects observed the user's utterance text for 3 s and then watched a video showing the response of the system. At each viewing, the subjects evaluated the impression of the response of the dialogue system using a seven-point Likert scale (1 to 7 points). Specific evaluation items are shown in Table 5. Since video of 10 utterances was prepared for each of the twelve conditions, video viewing and evaluation were performed 120 times. Considering the order effect, the presentation order of the video presented for each subject was randomized.

5.2 Evaluation Result

An experiment was performed with seven subjects. The mean value of each subjective evaluation item for each subject under each experimental condition was calculated, and the mean values are shown in Figs. 4 and 5.

First, we performed a two-dimensional analysis of variance to evaluate the effect of the factors of the utterance and motion conditions on the rating value of the overall good responses. As a result, a simple main effect for both motion conditions was observed, and no interaction effect was observed (utterance condition: $F(2, 72) = 3.92, p < .05$, motion condition: $F(3, 72) = 3.82, p < .05$).

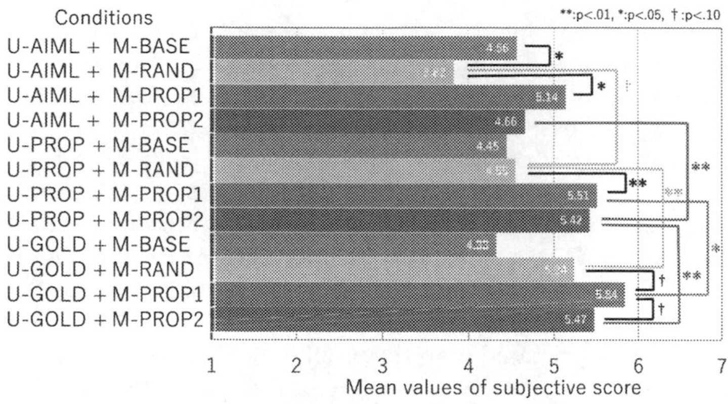

Fig. 5. Results of subjective evaluation about impression of "character-likeness"' of overall response.

Next, since a simple main effect for the factors of the utterance conditions was observed, multiple comparisons with the Bonferroni method were performed to verify which utterance conditions differed under each motion condition. Since there are many combinations of conditions and our main objective was to confirm to usefulness of the U-PROP condition, this paper mainly describes only the two differences between the U-AIML and U-PROP conditions and between the U-PROP and U-GOLD conditions.

First, a significant trend was observed between the U-AIML and U-PROP conditions only under the M-PROP1 condition ($p < .10$). Therefore, it was suggested that the proposed utterance generation method had a higher evaluation value than the general utterance generation using AIML when the proposed motion generation method (without original motion) was used. Under the M-RAND and M-PROP2 conditions, a significant difference was observed between the U-PROP and U-GOLD conditions ($p < .05$, $p < .01$). Therefore, it was suggested that an utterance made by a human was higher in terms of value than that of the proposed utterance generation method when random motion generation and the proposed motion generation method (with original motion) were used.

Next, since a simple main effect of the factors of the motion conditions was observed, similarly, it was verified by multiple comparisons to determine which motion condition had a difference under each utterance condition. Since there are many combinations of conditions and our main objective was to confirm the usefulness of our proposed motion generation method, this paper mainly describes only the three differences between the M-BASE and M-RAND conditions, the M-RAND and M-PROP1 conditions, and the M-PROP1 and M-PROP2 conditions.

First, a significant trend was observed between the M-BASE and M-RAND conditions only under the U-AIML condition ($p < .10$). Therefore, it was suggested that when utterance generation using AIML was used, the randomness of motion generation was lower in terms of the evaluation value than when no motion was generated. In addition, under all utterance conditions of U-AIML, U-PROP, and U-GOLD, a significant difference or a significant tendency was observed between the M-RAND and M-PROP1 conditions ($p < .01$, $p < .01$, $p < .10$). Therefore, it is suggested that the evaluation value of the goodness of overall response was higher (or the trend was higher) with the proposed motion generation method (without handmade original motion) than with random motion generation, regardless of the utterance generation conditions. In addition, a significant trend was observed between the M-PROP1 and M-PROP2 conditions under the U-PROP condition ($p < .10$). Therefore, it is suggested that when an utterance is generated with the proposed utterance generation method, the evaluation value tended to be higher when the original motion was not added than when the original motion was added to the method.

Next, the same analysis was performed for the evaluation value of the character likeness of the overall response. As a result, a simple main effect of the utterance and motion conditions was observed, and no interaction effect was observed (utterance condition: $F(2,72) = 3.92$, $p < .05$, motion condition: $F(3,72) = 3.82, p < .05$).

Next, using multiple comparison with Bonferroni's method, we verified which utterance conditions differed under each motion condition. First, under the M-RAND, M-PROP1, and M-PROP2 conditions, significant differences or significant trends between the U-AIML and U-PROP conditions were observed ($p < .01$, $p < .01$, $p < .10$). Under M-RAND and M-PROP1, significant differences were found between the U-PROP and U-GOLD conditions ($p < .05$, $p < .05$). Therefore, it was suggested that when using the random motion generation and the proposed motion generation method (without original motion), the evaluation value of the human response utterances was higher with the random motion generation than with the proposed utterance generation method.

Next, it was verified that there was a difference between the motion conditions under each utterance condition. First, under only the U-AIML condition, a significant difference was observed between the M-BASE and M-RAND conditions ($p < .05$). Therefore, it was suggested that when using AIML-based utterance generation, generating a random action would have a lower evaluation value than when not generating a motion. Under all utterance conditions of U-AIML, U-PROP, and U-GOLD, a significant difference or a significant tendency was observed between the M-RAND and M-PROP1 conditions ($p < .10$, $p < .10$, $p < .01$). Therefore, it is suggested that the proposed motion generation method (without original motion) has a higher evaluation value (or tendency) than random motion generation, regardless of the utterance generation conditions. In addition, a significant trend was observed between the M-PROP1 and M-PROP2 conditions under the U-GOLD condition ($p < .10$).

6 Discussion

From the evaluation results, it was confirmed that utterance generation and motion generation affected the impression in both items of the goodness of response and character-likeness. In addition, the rating value of the dialogue system constructed using the proposed construction method (U-PROP+M-PROP1) was as high as 5.83 for goodness of response and 5.98 for character-likeness in rating-value range of 1–7.

Although the number of samples was too small for seven subjects, there was no difference under all motion conditions, but the proposed utterance generation had better responses and character-likeness than utterance generation using AIML. It was also suggested that the proposed utterance generation method is more useful for the good response reflecting the character's personality than the manual utterance generation. It was also suggested that the proposed motion generation (without the original motion) similarly improves the impression as compared with the random motion generation.

As a result, it was found that, for the proposed motion generation method, the evaluation value of the original motion added was lower than that of the original action not added. After the experiment, subjects conducted a hearing survey, and it was found that the transition between the original and other normal motions was not smooth and that the timing did not completely match the display of the utterance sentence. Since the created original motion has a larger overall motion than the normal motion, it is conceivable that the transition with other motions did not go smoothly. Therefore, when adding the original motion to the proposed motion generation method, it was found that it is important to consider design considerations so that a smooth transition between previous motions is made when an original motion is generated.

In addition, under the motion conditions (M-PROP1) where the original motion was not included in the proposed motion generation, the rating value was very high at just a little under 6 points under the U-PROP+M-PROP1 condition (5.83). This suggests the possibility of generating a response that gives the impression of being good and character-like without inserting an original movement. Our motion generation model can generate the average motion of many people since it was trained with the movements of 24 people. This suggests that body motion that reflects the average movement can improve the impression of being good and character-like without inserting an original movement. Even if an average movement is given to the anime character agent, it would be possible to sense his/her individuality. This is a very interesting result. Of course, depending on the design and settings of the anime character, this technique may not always be effective. This is because it may be better for awkward robots not to behave like a human. However, if it is appropriate for a character to move like a human, our proposed motion generation method can be an effective means to enhance the quality of responses and character-likeness. Detailed evaluation of effectiveness using more diverse characters is one of our future tasks.

Finally, our proposed utterance and motion generation methods have made it possible to realistically realize a dialogue system of an existing animation character, which has been difficult to date. We cannot say that our proposed construction method has achieved a perfect system, but we believe it is worthwhile to prove the effectiveness of a new method that can efficiently realize such a system.

We have plans to carry out additional experiments to handle more samples and to verify in detail whether there is a mutual effect between speech and motion conditions. We will also improve the construction method to create a dialogue-agent system using other existing anime characters.

7 Conclusion

In this paper, we proposed a construction method for efficiently constructing a text chat system with animation for existing anime characters. We tackled two research problems to generate verbal and nonverbal behaviors. In the generation of verbal behavior, it is a major issue to be able to generate utterance text that reflects the personality of existing characters in response to any user questions. For this problem, we propose the use of the role-playing question-answering method to efficiently collect high-quality paired data of user questions and system answers that reflect the personality of an anime character. We also propose a new utterance generation method that uses a neural translation model with the collected data. Rich and natural expressions of nonverbal behavior greatly enhance the appeal of agent systems. However, not all existing anime characters move as naturally and as diversely as humans. Therefore, we propose a method that can automatically generate whole-body motion from spoken text so that anime characters can make human-like and natural movements. In addition to these movements, we try to add a small amount of characteristic movement on a rule basis to reflect personality. We created a text-dialogue agent system of a popular existing anime character by using our proposed generation models. As a result of a subjective evaluation of the implemented system, our models for generating verbal and nonverbal behavior improved the impression of the agent's responsiveness and reflected the personality of the character. In addition, the generation of characteristic motions with a small amount of characteristic movements based on heuristic rules is not effective, but rather the character generated by our generation model that reflects the average motion of persons had more personality. Therefore, our proposed generation models and construction method will greatly contribute to realizing text-dialogue-agent systems of existing characters.

References

1. Fuchi, T., Takagi, S.: Japanese morphological analyzer using word cooccurrence -JTAG. In: International Conference on Computational Linguistics, pp. 409–413 (1998)
2. Higashinaka, R., et al.: Towards an open-domain conversational system fully based on natural language processing. In: International Conference on Computational Linguistics, pp. 928–939 (2014)

3. Higashinaka, R., Sadamitsu, K., Saito, K., Kobayashi, N.: Question answering technology for pinpointing answers to a wide range of questions. NTT Tech. Rev. **11**(7) (2013)
4. Imamura, K.: Analysis of Japanese dependency analysis of semi-spoken words by series labeling. In: Proceedings of the Annual Meeting of the Association for Natural Language Processing, pp. 518–521 (2007)
5. Ishi, C.T., Haas, J., Wilbers, F.P., Ishiguro, H., Hagita, N.: Analysis of head motions and speech, and head motion control in an android. In: IEEE/RSJ International Conference on Intelligent Robots and Systems, pp. 548–553 (2007)
6. Ishi, C.T., Ishiguro, H., Hagita, N.: Head motion during dialogue speech and nod timing control in humanoid robots. In: ACM/IEEE International Conference on Human-Robot Interaction, pp. 293–300 (2010)
7. Kadono, Y., Takase, Y., Nakano, Y.I.: Generating iconic gestures based on graphic data analysis and clustering. In: The Eleventh ACM/IEEE International Conference on Human Robot Interaction, HRI 2016, Piscataway, NJ, USA, pp. 447–448. IEEE Press (2016)
8. Leuski, A., Patel, R., Traum, D., Kennedy, B.: Building effective question answering characters. In: Proceedings of the SIGDIAL, pp. 18–27 (2009)
9. Lohse, M., Rothuis, R., Gallego-Pérez, J., Karreman, D.E., Evers, V.: Robot gestures make difficult tasks easier: the impact of gestures on perceived workload and task performance. In: Proceedings of the SIGCHI Conference on Human Factors in Computing Systems, CHI 2014, pp. 1459–1466. ACM, New York (2014)
10. McNeill, D.: Hand and Mind: What Gestures Reveal About Thought. University of Chicago, Chicago Press (1996)
11. Meguro, T., Higashinaka, R., Minami, Y., Dohsaka, K.: Controlling listening-oriented dialogue using partially observable Markov decision processes. In: International Conference on Computational Linguistics, pp. 761–769 (2010)
12. Van Ments, M.: The Effective Use of Role Play: Practical Techniques for Improving Learning. Kogan Page Publishers, London (1999)
13. Mikolov, T., Sutskever, I., Chen, K., Corrado, G.S., Dean, J.: Distributed representations of words and phrases and their compositionality. In: Proceedings of the NIPS, pp. 3111–3119 (2013)
14. Miyazaki, C., Hirano, T., Higashinaka, R., Matsuo, Y.: Towards an entertaining natural language generation system: linguistic peculiarities of Japanese fictional characters. In: Proceedings of the SIGDIAL, pp. 319–328 (2016)
15. Schroff, F., Kalenichenko, D., Philbin, J.: FaceNet: a unified embedding for face recognition and clustering. CoRR, abs/1503.03832 (2015)
16. Sekine, S., Sudo, K., Nobata, C.: Extended named entity hierarchy. In: Proceedings of the LREC (2002)
17. Vinyals, O., Le, Q.: A neural conversational model. arXiv preprint arXiv:1506.05869 (2015)
18. Wittenburg, P., Brugman, H., Russel, A., Klassmann, A., Sloetjes, H : Elan a professional framework for multimodality research. In: International Conference on Language Resources and Evaluation (2006)

Utilization of Human-Robot Interaction for the Enhancement of Performer and Audience Engagement in Performing Art

Nihan Karatas[1(✉)], Hideo Sekino[2], and Takahiro Tanaka[1]

[1] Nagoya University, Nagoya 464-8601, Japan
karatas@mirai.nagoya-u.ac.jp, tanaka@coi.nagoya-u.ac.jp
[2] Stony Brook University, Stony Brook, New York, NY 11794, USA
hideo.sekino@stonybrook.edu

Abstract. Recently, computer-supported interactive technologies have played a significant role as complementary tools in creation of extraordinary artworks. These technologies have been used in art to explore their utility to compose new creative concepts, and to enrich the dimensions of artistic performances to strengthen the engagement between the performer and the audience. Since integrating the audience into the artistic performance has a significant role in enhancing an individual pleasure, a robotic medium holds great potential in bringing about new opportunities for artistic performances. In this preliminary study, we observed the effects of eye gazing behaviours of a minimal robot on audience engagement and connectedness in regard to an artistic performance. With this paper, the results from the data of a limited number of participants show that the audience tended to be distracted by the robot's existence, however, the gazing behavior of the robot maintain a feeling of connectivity between the robot and the audience.

Keywords: Human robot interaction · Performing arts · Engagement

1 Introduction

During an improvised interactive performance, the considerations of the artist not only about giving the best possible performance, but also interacting with the audience and the degree to which they are engaged. The interaction and engagement between the performer and the audience feeds the artist in a way that the artist can express their emotions vigorously.

Audience engagement with artwork is an essential part of the creative process where the audience passively join in making the performance together with the artist. The audience can interact with the artist cognitively through their emotional and expressive perceptions of the performance. In a sense, this kind of passive experiences allows for a greater appreciation the quality of performance. However, audience participation and engagement can be difficult when

C. Stephanidis et al. (Eds.): HCII 2020, LNCS 12427, pp. 348–358, 2020.
https://doi.org/10.1007/978-3-030-60152-2_26

the audience struggles to understand the structure of the performance. Interactive and multisensory human computer interaction (HCI) based technologies facilitate new approaches of audience participation and engagement in performing art that overcomes knowledge gaps and skill-related limitations. Research has been conducted to support audience participation through providing users with the different HCI manners [1–3]. In addition to maintaining audience participation within a performance, measuring their engagement and sustaining it is also important. Hassib et al. utilize brain-computer interfaces to obtain fine-grained information on audience engagement implicitly, and convey feedback to presenters for real time and post-hoc access [4]. Research showed that the audience engagement was improved when collecting audiences' bio-signals and applying content-related performing cues by considering these bio-signals when the engagement levels decreased [5]. However measuring and tracking the engagement level of the audience during certain performances via brain-computer interfaces may not be a practical approach. Herewith, it is important to investigate more natural and acceptable ways that are perceived as less distractive, in order to track and reinforce the audience engagement seamlessly.

Research in human-robot interaction (HRI) show that humans can engage with activities with the help of embodied robotic agents in a variety of domains. Embodied agents have been utilized in communication, education, health, transportation, entertainment, etc. due to our perception of them as more natural and familiar when compared to disembodied agents. Furthermore, they leverage more channels of communication such as proxemics [6], eye gaze interactions [7], gestures [8] engaging communications [9].

Humans engage in a wide variety of social interaction with the ability to reason with others' "intentional stance" and "social presence" . One of the critical precursors to expose social existence is shared attention which is defined as the ability to simultaneously allocate attention to a target of mutual interest. In human-human interaction, with the eye gaze following behavior people can convey information and understand their mental states and intentions [10].

A number of studies in HRI have shown that with eye gazing behaviors, robots can gain the ability to give information to their interlocutors [11,12]. In situated human-machine interaction, the robot's gaze could be used as a cue to facilitate the user's comprehension of the robot's instructions [13]. In this research, our aim is to observe audience engagement through simple eye gaze behavioral patterns (Fig. 1, 2). Specifically, our purpose in this study was to understand if it is possible to provide a better engagement of the audience on the artist and the performance with the shared attention behaviors of a robot when the dynamics of the performance change.

2 Method

In this study, we focused on observing the audience's engagement during a music performance, and the effect of a robot's gazing behaviours on their perception of the performance. As a music performance, we chose a traditional Japanese music piece called "Choushi" that was performed by a shakuhachi that is a Japanese, end-blown bamboo-flute. Choushi is defined as "pitch, tone, rhythm, manner, or condition" that also means "tuning life". Choushi is performed as a prelude to a performance. Even though the song itself is extremely simple, it contains all the elements which makes it a difficult song because of the complexity of its simplicity. The song starts with soft and low tones, and the latter half reaches a high-pitched tone and is where we expected the robot could increase the engagement between the performance and the audience through the gazing behaviours. The performer played two-minute long Choushi piece with his shakuhachi with their closed eyes in order to limit the human-human engagement during the performance to create room for the robot to emerge as a medium to maintain the interaction and engagement between the performer and the audience.

Fig. 1. The image of BB-9E from the Star Wars movie, was used in this study.

Because we only focused on utilizing the gaze behaviours of an embodied robotic agent, in order to not to reduce high expectations about the robot's adaptive capabilities and utilize only the most fundamental functions of the robot [14], we decided to use a minimaly designed robot. We chose BB-9E, an App-Enabled Droid, the off-the-shelf spherical robot developed by Sphero (Fig. 1). The robot consists of two parts: the head and the body. The head has a pair of LED lights to represent the eye of the robot. A magnet keeps the head attached to the body while it moves which gives the sense of the robot's moving and gazing directions.

2.1 Experimental Conditions

In order to evaluate the engagement of the audience with the performance with the existence of a robot and their perceptions of the robot's gazing behaviour, we conducted three experimental conditions:

No Robot condition (NR): The performer sits in front of the audience and performed his art with his eyes closed.
Robot with Random Gazing condition (RRG): The robot was placed between the audience and the performer but slightly closer to the performer in order to explicitly gaze at the audience. In this condition, the robot mainly gazing the

audience with the range of −30 and 30 degrees, while randomly gazing over at the performer for three second durations.
Robot with Joint Attention condition (RJA): The robot was again placed in the same position as the RRG condition. However, the robot was gazing the audience for the first half of the performance. When the music reached its high-pitch, the robot started joint attention behaviours. During this time, the robot would gaze back and forth at the performer and the audience for three second intervals. This behaviour was repeated until the song returned to its soft and low tones for the last ten seconds, then the performance ended.

Each condition took about two minutes. Upon arrival, the participants were given a brief explanation about the experiment. Informed consent was obtained from all participants, then they were asked to fill out a demographic questionnaire before the experiment started. After each experimental session, the participants were given questionnaires about their impressions on the performance and the robot' behaviours. After the NR condition, the participants were given a questionnaire about evaluating their engagement with the performance (Table 1). After the RRG and RJA conditions, they were given the same questionnaire and another questionnaire to evaluate their perception of the robot and its behaviors (Table 2). Each session was also recorded with a video camera to see the performer, the robot and the audience.

(a) Image depicts the interaction between the performer and the robot. The robot gazes at the performer.

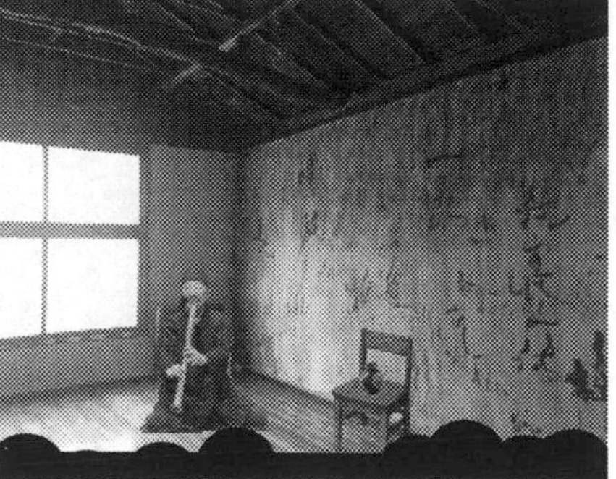

(b) Image depicts the interaction between the auidence and the robot. The robot gazes at the audience.

Fig. 2. The robot as an interaction medium between the performer and the audience. The images are recreation of the actual experimental environment.

3 Results

The experiment contained 3 participants (1 female, 2 males) ranging from 29–44 years old (M = 37, SD = 7.54). We conducted a within-subjects study. One participant was exposed to the experiment alone while other two participants exposed to the experiment together. Our sample size was very low in order to validate our hypothesis in this study. Nonetheless, here, we present our data, interpretation and inferences we obtained from these three participants.

3.1 Engagement with the Performance

The engagement questionnaire results showed no statistically significant differences between group means as determined by one-way ANOVA (ENQ1: $F(2, 6) = 1.33$, $p = .33$, ENQ2: $F(2, 6) = 1.4$, $p = .31$, ENQ3: $F(2, 6) = 1$, $p = .42$, ENQ4: $F(2, 6) = 3$, $p = .12$, ENQ5: $F(2, 6) = 1.3$, $p = .33$; Fig. 3). Even though we cannot validate our results, it can be said that the three participants were engaged best in the NR condition, then in the RRG condition. It can be said that the robot was a distractive element, and the joint attention behaviours of the robot broke the connection between the performance and the audience unlike our expectation.

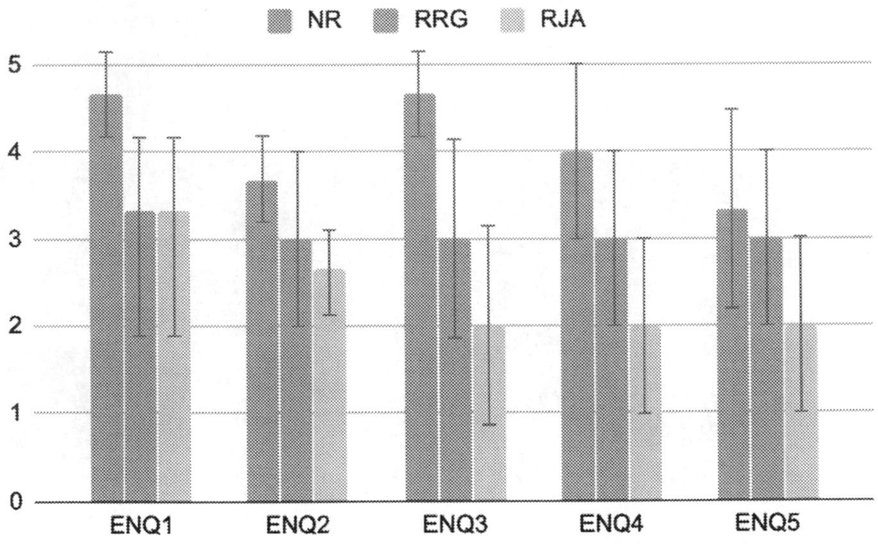

Fig. 3. One-way ANOAVA results of engagement questionnaire.

3.2 Perception of the Robot Behaviours

The robot perception questionnaire results also showed no statistically significant differences as determined by a paired t test (RBQ1: $t(2) = -0.8$, $p = 0.23$, RBQ2: $t(2) = -0.25$, $p = 0.41$, RBQ3: $t(2) = 1$, $p = 0.21$, RBQ4: $t(2) = 1.38$, $p = 0.14$, RBQ5: $t(2) = -1$, $p = 0.21$, RBQ6: $t(2) = -1$, $p = 0.21$, RBQ7: $t(2) = -1.7$, $p = 0.11$, RBQ8: $t(2) = -1$, $p = 0.21$; Fig. 4). With the first three questions, we wanted to understand if the robot was perceived as a

Table 1. The items of the engagement questionnaire.

Code	Engagement questions
ENQ1	How do you rate your engagement with the performance?
ENQ2	To what extent did you feel that you were connected with the human performer?
ENQ3	To what extent did you feel that you were connected with the music?
ENQ4	To what extent did you feel that you were a part of the performance?
ENQ5	To what extent did you think that you were an important part of the performance?

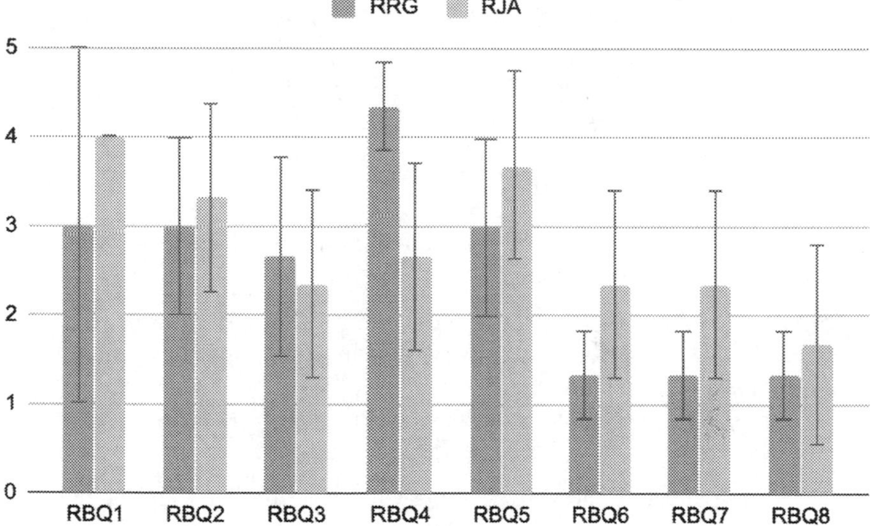

Fig. 4. One-way ANOAVA results of robot behaviour questionnaire.

performer or an audience. Because of the place of the robot where was closer to the performer and with its gazing the audience behaviours in both conditions, in some sense, might be perceived as a performer. Especially, in the RJA condition, the reactive behaviours of the robot (RBQ5) might give the audience the feeling of the performance occured between the human and the robot. The last three questions were evaluating the achievement of the joint attention. The results showed that in the RJA condition, the audience tended to feel more engaged, connected and they tended to feel that the robot shared their feelings more than in the RRG condition.

Table 2. The items of the robot behaviour questionnaire.

Code	Robot behaviour questions
RBQ1	To what extent do you think that the robot was a part of the performance?
RBQ2	To what extent do you think that the robot was a performer?
RBQ3	To what extent do you think that the robot was one member of the audience?
RBQ4	To what extent do you think that the robot was acting?
RBQ5	To what extent do you think that the robot was reacting to the performance?
RBQ6	To what extent do you think that the robot was trying to engage with you?
RBQ7	To what extent do you think that the robot wanted to be connected with you?
RBQ8	To what extent do you think that the robot shared your feelings?

3.3 Individual Assessments

We had a brief interview with the each participant to obtain their impressions on the each experimental session.

Participant 1: *NR Condition:* The participant stated that she felt a better engagement in this condition because it was a one-to-one performance. She was the only audience and it made her to feel special. From the video recordings, it was also shown that the participant was looking at the performer without any change in her position during this session.

RRG Condition: In this condition, the participant stated that she thought that she was not an important part of the performance anymore because the robot was included. From this statement, we can infer that she had a tendency to perceive the robot as an audience in this condition. The video recordings showed that she was towarding the performer and her posture did not change during this session as well.

RJA Condition: The participant stated that the robot was more reactive to the performance in this condition. She also stated that the performance was in between the human and the robot this time and she felt that she was excluded. From the video recordings, it was shown that her posture towarded the performer and did not show any change during the session.

Participant 2:

NR Condition: The participant stated that he was focused on the music. The video recording showed that he was looking at the performer and did not change his posture.

RRG Condition: In this condition, he stated that because he already listened to the music in the previous condition, this time, he only focused on the robot. He cared less about the music in this conditions, and observed the robot more. The video recording showed that he was explicitly looking at the robot during this session.

RJA Condition: He felt that the robot was reacting to the performance and it felt pressure with the high pitch of the music. He also stated that the robot was watching the audience and reacting, and the robot cared the audience more than in the RRG condition. The video recordings also showed that he was constantly looking at the robot in this session as well.

Participant 3:

NR Condition: The participant said that he could devote to the performance completely. In the video recordings, he was shown with no change in this posture and looking at the performer during the session.

RRG Condition: He stated that the robot's eye was shiny, so that he couldn't concentrate much on the performance. He was distracted by the light, and couldn't engage with the music when the robot was included. In the video recordings, we saw that the he was changing his hands positions and holding his head time to time which pointed that he was distracted.

RJA Condition: In this condition, he said that the robot was less disturbing, because he could understand the robot's movement behaviours from the RRG condition. In the video recordings, again it was shown that his body posture was moved constantly regarding to his distraction.

4 Discussion

In this study, we observed the effects of eye gazing behaviors of a minimal spherical robot on the engagement between an artist and their audience. Since the interaction between the performer and the audience was in the minimum level where the performer used his instrument with his eyes closed, we used robot as an interaction medium that used its gazing behaviours to maintain the engagement between the performer and the audience. Moreover, we focused on the audiences perception of the robot such as to what degree the eye gazing patterns make the robot to be perceived as a performer or an audience.

Based on the preliminary results that we obtained from three participants showed that the engagement between the performance and the audience was not improved by the robot's existence. It can be inferred that the interactivity level of the robot had a negative effect on the engagement. When the robot exhibited interactive joint attention gazing behaviours, the participants were disconnected from the performance. However, according to the results of RBQ6, RBQ7 and RBQ8 (Fig. 4), the robot's interactive gaze behaviours were perceived by some audience members as though the robot was sharing their feelings. However, the reason for the low ratings for these questionnaire items (RBQ6, RBQ7 and RBQ8) might be related to participants' lack of understanding of the robot's behaviour. Another consideration of the robot behaviour might be the appropriateness of the movements of the robot for this specific artistic performance. We chose a shakuhachi performance with the Choushi piece with the aim of keeping the simplicity due to the ease of the evaluation. However, the instrument, the music, the robot and its behaviours might create a different perception in the audience' mind where their expectation were not met with the harmony between the musical performance and the robot's behaviour, consequently the distraction occurred due to the presence of the robot. In our future study, we will consider this issue and conduct our experiment by using the robot within a more suitable and matching appearance and movements with the performance.

In this study, we also aim to introduce the people from art backgrounds with the opportunities of HRI, and enhance the opportunities to increase the pleasure of an art performance for both the artist and the audience. Despite widespread enthusiasm for social robots, users are still lacking of the knowledge of what to expect from their interactions from robots; therefore, not meeting these expectations can cause dissatisfaction. Thus, fresh perspectives and new approaches to create new interactions with robots to address these challenges is important.

Artistic performances stimulate the creation of new and intuitive interactions. Herein, the improvised performances present opportunities to situate and research Human-Robot Interactions (HRI) within a creative atmosphere. We believe that artists, such as Yi [15] and PARAMOUND [16], show that collaborating with embodied agents in artistic environments pave the way for accepting interactions with robots in different grounds. With this current study, we contribute to the investigation of new applications of human-robot interactions. The future aim of this study involves developing methods for remote participation of human artists to perform where they cannot physically appear.

5 Conclusion

In this paper, our main focus was to observe how the audience's engagement affected by the presence of a minimal spherical robot during a short traditional Japanese shakuhachi performance, and how their perception of the performance and their feeling towards the robot affected by the robot's gazing behaviours. The questionnaire and interview results that were obtained from three participants showed that the audience were distracted by the presence of the robot. On the other hand, the joint attention behaviours of the robot promoted connectedness and sharing feeling between the robot and the audience. In the future, we will continue to gather more participants in order to validate our preliminary results. In our future study, we will also consider to enhance the engagement between the audience and the performance through solidifing the role of the robot (performer or audience) depending on the dynamics of the performance. Moreover, with the potential positive effects of increased engagement of the audience by utilizing a robot, our future study may also contribute to musical therapy methods.

References

1. Striner, A., McNally, B.: Transitioning between audience and performer: co-designing interactive music performances with children (2017). arXiv preprint arXiv:1702.06236. https://doi.org/10.1145/3027063.3053171
2. Jessop, E.N., Torpey, P.A., Bloomberg, B.: Music and technology in death and the powers. In: NIME, pp. 349–354, May 2011
3. Freeman, J. (2005). Large Audience Participation, Technology, and Orchestral Performance. In: ICMC
4. Hassib, M., Schneegass, S., Eiglsperger, P., Henze, N., Schmidt, A., Alt, F.: EngageMeter: a system for implicit audience engagement sensing using electroencephalography. In: Proceedings of the 2017 CHI Conference on Human Factors in Computing Systems, pp. 5114–5119, May 2017
5. Yan, S., Ding, G., Li, H., Sun, N., Wu, Y., Guan, Z., Huang, T.: Enhancing audience engagement in performing arts through an adaptive virtual environment with a brain-computer interface. In: Proceedings of the 21st International Conference on Intelligent User Interfaces, pp. 306–316, March 2016. https://doi.org/10.1145/2856767.2856768
6. Takayama, L., Pantofaru, C.: Influences on proxemic behaviors in human-robot interaction. In: 2009 IEEE/RSJ International Conference on Intelligent Robots and Systems, pp. 5495–5502. IEEE, October 2009. https://doi.org/10.1109/IROS.2009.5354145
7. Mutlu, B., Kanda, T., Forlizzi, J., Hodgins, J., Ishiguro, H.: Conversational gaze mechanisms for humanlike robots. ACM Trans. Interact. Intell. Syst. (TiiS) **1**(2), 1–33 (2012). https://doi.org/10.1145/2070719.2070725
8. Breazeal, C., Kidd, C.D., Thomaz, A.L., Hoffman, G., Berlin, M.: Effects of nonverbal communication on efficiency and robustness in human-robot teamwork. In: 2005 IEEE/RSJ International Conference on Intelligent Robots and Systems, pp. 708–713. IEEE, August 2005. https://doi.org/10.1109/IROS.2005.1545011
9. Kidd, C.D., Breazeal, C.: Effect of a robot on user perceptions. In: 2004 IEEE/RSJ International Conference on Intelligent Robots and Systems (IROS) (IEEE Cat. No. 04CH37566), vol. 4, pp. 3559–3564). IEEE, September 2004. https://doi.org/10.1109/IROS.2004.1389967
10. Griffin, R., Baron-Cohen, S.: The intentional stance: Developmental and neurocognitive perspectives. Daniel Dennett, pp. 83–116 (2002)
11. Mutlu, B., Shiwa, T., Kanda, T., Ishiguro, H., Hagita, N.: Footing in human-robot conversations: how robots might shape participant roles using gaze cues. In: Proceedings of the 4th ACM/IEEE International Conference on Human Robot Interaction, pp. 61–68, March 2009. https://doi.org/10.1145/1514095.1514109
12. Karatas, N., Tamura, S., Fushiki, M., Okada, M.: The effects of driving agent gaze following behaviors on human-autonomous car interaction. In: Ge, S.S., et al. (eds.) ICSR 2018. LNCS (LNAI), vol. 11357, pp. 541–550. Springer, Cham (2018). https://doi.org/10.1007/978-3-030-05204-1_53
13. Admoni, H., Scassellati, B.: Social eye gaze in human-robot interaction: a review. J. Human-Robot Interact. **6**(1), 25–63 (2017). https://doi.org/10.5898/JHRI.6.1.Admoni
14. Komatsu, T., Kurosawa, R., Yamada, S.: How does the difference between users' expectations and perceptions about a robotic agent affect their behavior? Int. J. Soc. Robot. **4**(2), 109–116 (2012)

15. Have We Lost Our Anthropological Imagination? https://www.epicpeople.org/anthropological-imagination/. Accessed 02 October 2017
16. Swiss startup makes good on Broadway. https://www.therobotreport.com/swiss-startup-makes-good-on-broadway/. Accessed 29 July 2016

Personality Trait Classification Based on Co-occurrence Pattern Modeling with Convolutional Neural Network

Ryo Kimura and Shogo Okada(✉)

Japan Advanced Institute of Science and Technology, Nomi, Japan
okada-s@jaist.ac.jp

Abstract. In the modeling of impressions, a key factor for success is to extract nonverbal features that can be used to infer the target variable. To extract the effective features for capturing the relationship between the target subject which has the personality trait and other group members, Okada et al. propose a co-occurrence event-mining framework to explicitly extract the inter-modal and inter-personal features from multimodal interaction data. The framework is an unsupervised feature extraction algorithm by considering the relationship between nonverbal patterns of the target subject and group members. Though the label data of personality trait is useful to improve the accuracy, the valuable label data is not used for feature extraction. In this paper, we enhance the inter-modal and inter person feature extraction algorithm by using a deep neural network. We proposed a representation learning algorithm for capturing inter-modal and inter-person relationships by integrating using a convolutional neural network (CNN). In the experiment, we evaluate the effectiveness of the representation learning approach using the ELEA (Emerging Leadership Analysis) corpus, which includes 27 group interactions and is publicly available. We show that the proposed algorithm with CNN slightly improves the personality trait classification accuracy of the previous algorithms. In addition, we analyze which slice of multimodal time-series data is key descriptors to predict the personality trait using the proposed algorithm with CNN.

Keywords: Impression · Multimodal interaction · Personality traits · Convolutional neural network

1 Introduction

In recent comprehensive research on the computational multimodal analysis, the modeling of impressions is the focus of attention. The target variables vary widely, such as public speaking skills [1,2], persuasiveness [3], communication skill in job interviews [4], and leadership [5]. A key factor for success is to extract nonverbal features that can be used to infer the target variable. To extract the effective features, previous works have defined static features from audio and

C. Stephanidis et al. (Eds.): HCII 2020, LNCS 12427, pp. 359–370, 2020.
https://doi.org/10.1007/978-3-030-60152-2_27

visual data based on knowledge of social science. Speaking activity and prosodic features as audio cues, body activity, head activity, hand activity, gaze and facial expression as visual cues, are used for inference of personality traits. However such statistic features ignore the dynamics of nonverbal events observed in the whole meeting and the relationship between the target subject which has the personality trait and other group members. To solve the problem, Okada et al. [6,7] propose a co-occurrence event-mining framework to explicitly extract the inter-modal and inter-personal features from multimodal interaction data. The key approach of the framework is to discover co-occurrence patterns between modalities. The accuracy of the model trained with the inter- modal and inter person features outperforms that of models trained with the traditional statistic feature set. From another viewpoint, the framework is an unsupervised feature extraction algorithm by considering the relationship between nonverbal patterns of the target subject and group members. Though the label data of personality trait is useful to improve the accuracy, the valuable label data is not used for feature extraction. In this paper, we enhance the inter-modal and inter person feature extraction algorithm by using a deep neural network. We proposed a representation learning algorithm for capturing inter-modal and inter-person relationships by integrating using a convolutional neural network (CNN). In the experiment, we evaluate the effectiveness of the representation learning approach using the ELEA (Emerging Leadership Analysis) corpus, which includes 27 group interactions and publicity available. We show that the proposed algorithm with CNN improves the personality trait classification accuracy of the previous algorithm proposed in [6,7].

2 Related Work

Our research is related to personality-trait modeling and interaction mining. This study focuses on impression modeling in conversations. For multiparty interactions, different works included different variables: social roles [8,9], dominance [10], personality traits [11,12] and leadership [5]. As a common approach of these works, audio, and visual features are calculated using the mean, median, min, max, and X percentile of various statistics (count and length) from each pattern observed throughout an entire meeting or for a part of a meeting [5,11,13]. Although this approach can often fuse the total statistics of patterns observed within a specified duration, it cannot capture co-occurrence between multimodal patterns for each time period. For example, extracting co-occurrence events between an utterance and a body-motion pattern as a feature is useful if the utterance accompanying the body gesture makes a stronger impression on the listener than that utterance without the gesture. Our mining algorithm explicitly extracts such co-occurrence features. Several other studies have focused on extracting the correlations between modalities. Song et al. [14] proposed a multimodal technique that models explicit correlations among modalities via canonical correlation analyses (CCAs) [15]. The algorithm was evaluated using a recognition task for disagreement/agreement with a speaker in political debates

[16]. Chatterjee et al. [17] proposed an ensemble approach that combines a classifier based on inter-modality conditional independence with a classifier based on dimension reduction via a multiview CCA. Feature co-occurrence is often adopted in computer vision [18–23].

Preliminary works [6,24,25] have been performed using co-occurrence pattern mining similar to the proposed approach. Okada et al. [24] used a co-occurrence pattern-mining algorithm, which is a modified version of the algorithm in [26], to extract features to infer the performance level of storytelling in group interaction. The main difference with respect to our work is that the research focuses on the modeling of group performance and not the individual performance and that nonverbal features are extracted manually. The main limitation of these research works [24,25] is that only binary event (on/off) features are used for mining. Okada et al. [7] enhanced the co-occurrence pattern mining algorithm and also applied the algorithm for dyadic-interaction dataset. The enhanced algorithm proposed in [7] improves the classification accuracy of personality traits. In this paper, we enhance the inter-modal and inter person feature extraction algorithm by using a deep neural network. We proposed a representation learning algorithm for capturing inter-modal and inter-person relationships by integrating using a convolutional neural network (CNN). The main contribution of this paper is to analyze when the effective co-occurrence features are observed in the whole meeting using the co-occurrence pattern learning algorithm with CNN.

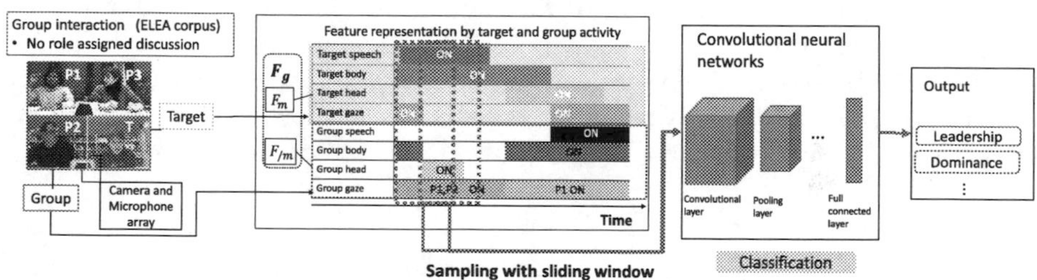

Fig. 1. Overview of proposed framework

3 Inter-person and Inter-modal Representation Learning

Figure 1 shows an overview of the proposed framework. We proposed Inter-person and Inter-modal Representation Learning by using convolutional neural networks, which are mainly used in the field of image recognition. The convolutional neural network extracts feature with a sliding window method applied to time-series data. Therefore, convolutional neural networks are expected to capture time-series dependency than feature extraction by co-occurrence patterns. In addition, there is a possibility that data features can be accurately captured by performing supervised learning using label data in feature extraction. However, the input of the convolutional neural network does not support multimodal features. Therefore, it is necessary to convert data with multimodal features into a form that can be used as input.

3.1 Multimodal Feature Representation

We propose a feature representation method for capturing the co-occurrence of the nonverbal patterns observed for each participant. We define co-occurrence patterns as multimodal events that overlap in time. Each event has a time length and corresponds to a segment denoted by "ON" in Fig. 1. We define an event as a segment in which the feature is active. Multimodal features are represented as follows. The feature representation for group interaction is described. We propose a feature representation for comparing nonverbal patterns that are observed for each participant in a group. The representation captures how a participant acts when other members execute any nonverbal activity by simultaneously observing the nonverbal activities of both the individual participant and the other group members. Let F_{group} be the feature set for a group interaction:

$$\boldsymbol{F}_{group} = \{F_m, F_g\}. \tag{1}$$

F_m is the feature representation for one specific person in a group, and F_g is the feature representation for a group composed of the other members without m. An example of $\{F_m,\ F_g\}$ is shown in Fig. 1. The co-occurrence pattern mining requires conversion of the time-series signal data into a sequence of events $(f_{m,i})$ with a finite time length as a preprocessing step. Multimodal behavior is inherently observed as time-series signals in a session. The binarization or discretization of continuous time-series data is described in Sect. 5. The modified audio-visual features f in F_{group} are also described at the bottom of Fig. 2, respectively.

3.2 Convolutional Neural Network (CNN) to Learn Co-Occurrence Pattern

The data type of multimodal features extracted from group meetings is multi-dimensional time series, so data is represented as two-dimensional matrix data with the vertical axis representing the number of features and the horizontal axis representing time. We segment the time-series data into slices with almost 1 min (58 s) and the two-dimensional data with $D \times 58$ s is defined as a training or test sample to input into CNN. The converted data is binary data in which the time-segment where the feature is recognized as "1" and the segment where the feature is not recognized is "0". The method for feature extraction is described in Sect. 5. The converted data is used as input to the convolutional neural network. The label data includes five labels for leadership ability evaluated on a seven-point scale. In this paper, we perform experiments of the classification task by replacing this label data with binary data that is above or below the average of participants.

4 Dataset and Features

4.1 ELEA: Group-Interaction Dataset

We used a subset of the ELEA corpus [5] for this study. The subset consists of audio-visual (AV) recordings of 27 meetings in which the participants performed a winter survival task with no roles assigned. A total of 102 participants were included (six meetings with three participants and 21 meetings with four participants). Each meeting lasted approximately 15 min. The synchronization of audio and video was performed manually by aligning the streams according to the clapping activity. Additional details on the ELEA AV corpus can be found in [27].

ID	Features	Symbol	Description
F_1	Speaking	ST	Speech segments of the target person
	Status	$SO1$	One person other than target speaks
	(ST)	$SO2$	More than two people speak.
		$Ssil$	Silent segment
F_2	Pitch	PUp, PDo	Sign of difference between utterance t and utterance $t-1$
	(PI)	PCL, PCM, PCH	Cluster index (low medium and high level) after clustering
		$PCNL$, $PCNM$, $PCNH$	Cluster index after clustering of normalized value
F_3	Energy	EUp, EDo	Sign of difference between utterance t and utterance
	(EN)	ECL, ECM, ECH	Cluster index after clustering
		$ECNL$, $ECNM$, $ECNH$	Cluster index after clustering of normalized value
F_4	Head	HMT	Motion segments of target person
	Motion	$HMO1$	One person other than target moves
	(H)	$HMO2$	More than two people move
		$HMsil$	Still motion segment
F_5	Body	BMT	Motion segments of target person
	Motion	$BMO1$	One person other than target moves
	(B)	$BMO2$	More than two people move
		$BMsil$	Still motion segment
F_6	MEI	MUp, MDo	Sign of difference between segments
	(MT)	MCL, MCM, MCH	Cluster index after clustering
		$MCNL$, $MCNM$, $MCNH$	Cluster index after clustering of normalized MEI
F_7	Gaze	GT	Target person looks at person
	(G)	$GTSp$	Target person looks at speaker
		$GOT1$	One person looks at the target
		$GOT2$	More than two people look at the target
		MGT	Mutual gaze between target and another person
		MGO	Mutual gaze between two people other than target

Fig. 2. Multimodal feature set [6] (The feature set used in this study is aligned to that used in [6] for comparing the accuracy of the proposed framework with that of [6]. This table is adopted by the article [6]).

The ELEA corpus also includes scores for traits of individuals with respect to dominance and leadership. After the meeting task, the participants completed a Perceived Interaction Score, which captures perceptions from participants during the interaction, in which they scored every participant in the group based on four items related to the following concepts: "Perceived Leadership (Leadership)", "Perceived Dominance (Dominance)", "Perceived Competence (Competence)" and "Perceived Liking (Liking)". Afterward, the "Dominance Ranking (Ranked Dominance)". Leadership captures whether a person directs the group and imposes his or her opinion. Dominance captures whether a person dominates or is in a position of power. Participants were asked to rank the group, assigning 1 to the most dominant participant and 3 or 4 to the less dominant participants. Additional details can be found in [5].

5 Multimodal Features

Multimodal features are extracted automatically from audio and visual cues in this study in same manner with [6,11]. The feature sets of the ELEA used in this study are summarized in Fig. 2. The detail of multimodal features used in this study is described in [6,7,11].

5.1 Audio Features

Speaking Status. Binary segmentation is performed to capture the speaking status (ST) of each participant. This binary segmentation is provided by the microphone array, and all speaking activity cues are based on the speaker segmentations obtained using the Microcone, which is used for the audio recordings and speaker diarization in [5,11] and [4]. We define a set of segments in which the speech status is "on" as the speaking-turn set ST.

Prosodic Features. Prosodic features are extracted for each individual member. Based on the binary speaker segmentation, we obtain the speech signal for each participant. Overlapping speech segments are discarded, only the segments in which the participant is the sole speaker are considered for further processing. Three prosodic speech features (energy, pitch) are determined based on the signal. We calculate the sign of the difference between the statistics of utterance j and utterance $j+1$ using statistical t-test. Energy and pitch signals are converted into three categorical data (low level, middle level and high level) using k-means clustering.

5.2 Visual Features

Visual Activity Features. The first approach is based on head and body tracking and optical flow, which provides the binary head and body activity status and the amount of activity as well. As done for speech states, binary segmentation is done and an activity state set is extracted for head and body motion.

Motion Template Based Features. As a second approach, we have used Motion Energy Images (MEI) [28] as descriptors of body activity. We used the length of the meeting segment to normalize the images. Motion Energy Images (MEI) are obtained by integrating each difference image from whole video clip. Significant changes of MEI have the possibility to capture behaviors related to personality traits. The features are extracted in same manner with prosodic features

Visual Focus of Attention Features. Visual focus of attention (VFOA) features were extracted and shared by the authors in [27], where a probabilistic framework was used to estimate the head location and pose jointly based on a state space formulation. We define a set of segments GT where the target participant looks at the other participants through the meeting. We also define a set of segments $GTSp$ where the target participant looks at the speaker. Looking at speaker is an important signal of the listener's interest and politeness [29]. We further define two features $GOT1, GOT2$ as group attention features $G_{/m}$. $GOT1$ is a set of segments where one member looks at the member m. $GOT2$ is a set of segments where more than two members looks at the member m.

Next, we extracted mutual gazing features (although mutual gazing is defined as co-occurrence pattern with GT and $GOT1, 2$). We prepare two group features for mutual gazing. MGT is a set of segments where one member x looks at the member m and vice versa. MGO is a set of segments where two members y, z except the member m look at each other.

6 Experiments

6.1 Experimental Setting

For the ELEA group interaction, the inference tasks were classification and regression in [11], and then further studied as classification in [6]. In this paper, we decided to focus only on the binary classification task in the same manner with [6,7], because the objective of this experiment is to compare with algorithms proposed in [6,7,11]. We classify binary levels of the impression index. In the classification task, impression values are converted to binary values (high or low) by thresholding using the median value. For example, this method is performed to represent people scoring high/low in terms of leadership. The trained model is evaluated based on the classification accuracy of the test data. In the experiments presented below, we use leave-one-out cross-validation and report the average accuracy over all folds. We normalize the data such that each feature has a zero mean and one standard deviation.

6.2 Setting of Proposed Algorithm

The network structure of CNN is shown in Fig. 3. Rectified linear function (Relu) is used as the activation function in all middle layers and cross-entropy function is used as the loss function. The loss function is defined as $E = -\frac{1}{N}\Sigma q(k)log(p(k))$, where $p(k)$ denote probability of each label for sample x_k, which is output by the CNN and $q(k)$ denotes the ground-truth distribution for sample x_k. In the testing phase, the output probability per time-series slice of multimodal features is output from CNN. The classification result for the slice is correct when the class with the highest output probability is equal to the true label. N time-series slices are obtained from a meeting (or a participant) and the classification accuracy is calculated as $\frac{Num.\,of\,correctly\,classified\,samples}{N}$ for each participant. If the classification accuracy is more than 0.5, we define that the classification for a participant is correct.

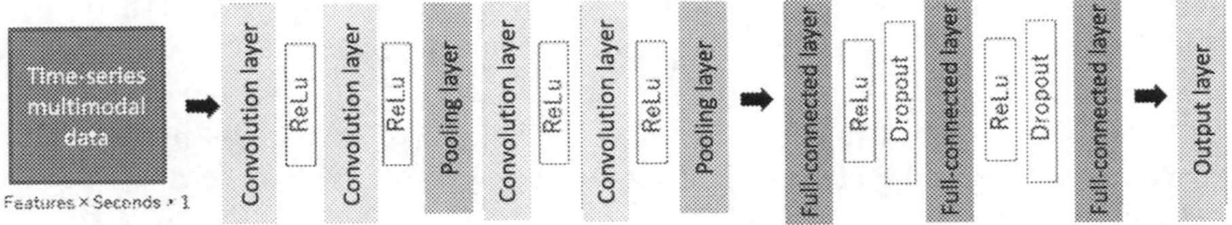

Fig. 3. Network structure of CNN

Table 1. Classification accuracy for leadership traits

	Perceived Leadership	Perceived Dominance	Perceived Competence	Perceived Liking	Ranked Dominance
Best in [7]	**73.53**	55.88	56.86	65.69	61.76
Best of [6]	72.55	61.76	**64.71**	53.92	64.71
Best of [11]	72.55	65.69	52.94	64.71	51.96
Co-occur CNN (All)	70.56	55.88	52.94	61.76	**66.67**
Co-occur CNN (Target)	64.71	**66.67**	50.98	**66.67**	**66.67**

6.3 Experimental Results

The Table 1 shows the classification accuracy for 5 leadership traits in ELEA corpus. We compared the accuracy of proposed models: Co-occur CNN (All) and Co-occur CNN (Speaker) with the best accuracy of proposed models of [6,7,11], which are reported in these articles. Co-occur CNN (All) is the CNN trained from time-series data which is composed of multimodal features observed from all members in a group. Co-occur CNN (Speaker) is the CNN trained from time-series data which is observed from only the target person who is the subject for the trait classification. From the Table 1, Co-occur CNN (Target) which is a proposed method obtained the best accuracy for "Perceived Dominance", "Perceived Liking" and "Ranked Dominance" with 66.67%. The proposed method improved the accuracy with 1–2 point. On the other hands, Best accuracy for

"Perceived Leadership" is obtained by [7] with 73.53% and that for "Perceived Competence"is obtained by [6] with 64.71%. These results show that applying CNN for co-occurrence pattern modeling is effective to improve the accuracy, though the improvement is limited. The reason why the improvement is limited is discussed as follows. The proposed framework regards time-series slices as independent training samples for input to CNN. Though some samples (time-series slices) are useful for improving the accuracy because the multimodal features in the slice can capture the personality traits of participants. Contraversely, samples are unnecessary as training samples if the observed slices (multimodal features) are noise data.

6.4 Time-Series Analysis of Classification Accuracy

The proposed algorithm with CNN enables us to analyze which slices of multimodal time-series data are key descriptors to predict the personality trait. In the proposed method, multimodal time-series slice is input to CNN and the slice is classified into binary classes. We can analyze when the key multimodal features are observed while the group meeting by comparing the classification accuracy per the time-series slice. Table 2 shows the classification accuracy per time-series slice. The horizontal axis denotes time-series and the vertical axis denotes the type of personality traits. The accuracy for the time segment ("T(X)" in Table 2) denotes the mean accuracy which is calculated by averaging accuracy over six slices (almost 6 min). The bold values denote the best and second-best accuracy for each trait. For "Perceived Leadership" and "Perceived Liking", the accuracy of T 4–6 is better than that of other segments. These results mean that the effective multimodal features are observed in the middle of the meeting, so the accuracy of a middle zone (T 4–6) tends to be higher than others. The accuracy of T 1 and T9 (62%, 61%) in "Ranked Dominance" and accuracy of T 8, 9 (59%, 55%) in "Perceived Competence" is better than that of other segments. These results mean that effective multimodal features are observed at the start of a meeting or at the end of the meeting. The effective features are observed in different timing per the types of traits.

Table 2. Classification accuracy per time-series slices (The accuracy for time segment ("T(X)" in Table 2) denotes the mean accuracy which is calculated by averaging accuracy over six slices (almost 6 min). The bold values denote the best and second-best accuracy for each trait.

Co-occur CNN (All)	T(1)	T(2)	T(3)	T(4)	T(5)	T(6)	T(7)	T(8)	T(9)
Leadership	66	65	65	**70**	**69**	65	64	59	59
Dominance	52	**57**	**57**	54	56	**57**	56	55	56
Competence	50	51	48	50	49	54	49	**59**	**55**
Liking	55	56	54	**57**	54	**60**	54	55	52
Ranked Dominance	**61**	60	59	60	60	54	59	56	**62**

7 Conclusion

In this paper, we enhance the inter-modal and inter person feature extraction algorithm by using a deep neural network. We proposed a representation learning algorithm for capturing inter-modal and inter-person relationships by integrating using a convolutional neural network (CNN). In the experiment, we evaluate the effectiveness of the representation learning approach using the ELEA (Emerging Leadership Analysis) corpus. The experimental results show the classification accuracy for 5 leadership traits in ELEA corpus. We compared the accuracy of proposed models: Co-occur CNN (All) and Co-occur CNN (Speaker) with the best accuracy of proposed models of [6,7,11], which are reported in these articles. Co-occur CNN (Target) which is a proposed method obtained the best accuracy for "Perceived Dominance", "Perceived Liking" and "Ranked Dominance" with 66.67%. The proposed method improved the accuracy with 1–2 point. Through time-series analysis, we found that the effective features are observed in different timing per the types of traits. The future work is to improve the accuracy by finding the effective features with using an attention mechanism.

References

1. Wörtwein, T., Chollet, M., Schauerte, B., Morency, L.P., Stiefelhagen, R., Scherer, S.: Multimodal public speaking performance assessment. In: Proceedings of the International Conference on Multimodal Interaction (ICMI), New York, NY, USA, pp. 43–50 (2015)
2. Ramanarayanan, V., Leong, C.W., Chen, L., Feng, G., Suendermann-Oeft, D.: Evaluating speech, face, emotion and body movement time-series features for automated multimodal presentation scoring. In: Proceedings of the International Conference on Multimodal Interaction (ICMI), pp. 23–30 (2015)
3. Park, S., Shim, H.S., Chatterjee, M., Sagae, K., Morency, L.P.: Computational analysis of persuasiveness in social multimedia: a novel dataset and multimodal prediction approach. In: Proceedings of the International Conference on Multimodal Interaction (ICMI), New York, NY, USA, pp. 50–57 (2014)
4. Nguyen, L., Frauendorfer, D., Mast, M., Gatica-Perez, D.: Hire me: computational inference of hirability in employment interviews based on nonverbal behavior. IEEE Trans. Multimed. **16**(4), 1018–1031 (2014)
5. Sanchez-Cortes, D., Aran, O., Mast, M.S., Gatica-Perez, D.: A nonverbal behavior approach to identify emergent leaders in small groups. IEEE Trans. Multimed. **14**(3), 816–832 (2012)
6. Okada, S., Aran, O., Gatica-Perez, D.: Personality trait classification via co-occurrent multiparty multimodal event discovery. In: Proceedings of the International Conference on Multimodal Interaction (ICMI), pp. 15–22 (2015)
7. Okada, S., Nguyen, L.S., Aran, O., Gatica-Perez, D.: Modeling dyadic and group impressions with intermodal and interperson features. ACM Trans. Multimed. Comput. Commun. Appl. **15**(1s), 1–30 (2019)
8. Vinciarelli, A.: Speakers role recognition in multiparty audio recordings using social network analysis and duration distribution modeling. IEEE Trans. Multimed. **9**(6), 1215–1226 (2007)

9. Zancanaro, M., Lepri, B., Pianesi, F.: Automatic detection of group functional roles in face to face interactions. In: Proceedings of the International Conference on Multimodal Interaction (ICMI), pp. 28–34 (2006)
10. Rienks, R., Heylen, D.: Dominance detection in meetings using easily obtainable features. In: Proceedings of the International Workshop on Machine Learning for Multimodal Interaction, pp. 76–86 (2005)
11. Aran, O., Gatica-Perez, D.: One of a kind: inferring personality impressions in meetings. In: Proceedings of the International Conference on Multimodal Interaction (ICMI), pp. 11–18 (2013)
12. Pianesi, F., Mana, N., Cappelletti, A., Lepri, B., Zancanaro, M.: Multimodal recognition of personality traits in social interactions. In: Proceedings of the International Conference on Multimodal Interaction (ICMI), pp. 53–60 (2008)
13. Nihei, F., Nakano, Y.I., Hayashi, Y., Hung, H.H., Okada, S.: Predicting influential statements in group discussions using speech and head motion information. In: Proceedings of the International Conference on Multimodal Interaction (ICMI), pp. 136–143 (2014)
14. Song, Y., Morency, L.P., Davis, R.: Multimodal human behavior analysis: learning correlation and interaction across modalities. In: Proceedings of the International Conference on Multimodal Interaction (ICMI), pp. 27–30 (2012)
15. Hotelling, H.: Relations between two sets of variates. Biometrika **28**(3/4), 321–377 (1936)
16. Vinciarelli, A., Dielmann, A., Favre, S., Salamin, H.: Canal9: a database of political debates for analysis of social interactions. In: Proceedings of the International Conference on Affective Computing and Intelligent Interaction (ACII), pp. 1–4 (2009)
17. Chatterjee, M., Park, S., Morency, L.P., Scherer, S.: Combining two perspectives on classifying multimodal data for recognizing speaker traits. In: Proceedings of the International Conference on Multimodal Interaction (ICMI), pp. 7–14 (2015)
18. Qian, X., Wang, H., Zhao, Y., Hou, X., Hong, R., Wang, M., Tang, Y.Y.: Image location inference by multisaliency enhancement. IEEE Trans. Multimed. **19**(4), 813–821 (2017)
19. Zhang, S., Tian, Q., Hua, G., Huang, Q., Gao, W.: Generating descriptive visual words and visual phrases for large-scale image applications. IEEE Trans. Image Process. **20**(9), 2664–2677 (2011)
20. Kumar, V., Namboodiri, A.M., Jawahar, C.V.: Visual phrases for exemplar face detection. In: Proceedings of the International Conference on Computer Vision (ICCV), pp. 1994–2002 (2015)
21. Wu, Z., Ke, Q., Isard, M., Sun, J.: Bundling features for large scale partial-duplicate web image search. In: Proceedings of the International Conference on Computer Vision and Pattern Recognition (CVPR), pp. 25–32 (2009)
22. Yang, X., Qian, X., Xue, Y.: Scalable mobile image retrieval by exploring contextual saliency. IEEE Trans. Image Process. **24**(6), 1709–1721 (2015)
23. Zhang, S., Yang, M., Wang, X., Lin, Y., Tian, Q.: Semantic-aware co-indexing for image retrieval. IEEE Trans. Pattern Anal. Mach. Intell. **37**(12), 2573–2587 (2015)
24. Okada, S., Hang, M., Nitta, K.: Predicting performance of collaborative storytelling using multimodal analysis. IEICE Trans. **99**(D(6)), 1462–1473 (2016)
25. Nakano, Y.I., Nihonyanagi, S., Takase, Y., Hayashi, Y., Okada, S.: Predicting participation styles using co-occurrence patterns of nonverbal behaviors in collaborative learning. In: Proceedings of the International Conference on Multimodal Interaction (ICMI), pp. 91–98 (2015)

26. Vahdatpour, A., Amini, N., Sarrafzadeh, M.: Toward unsupervised activity discovery using multi-dimensional motif detection in time series, pp. 1261–1266 (2009)
27. Sanchez-Cortes, D., Aran, O., Jayagopi, D.B., Mast, M.S., Gatica-Perez, D.: Emergent leaders through looking and speaking: from audio-visual data to multimodal recognition. J. Multimodal User Interfaces **7**(1–2), 39–53 (2013)
28. Davis, J.W., Bobick, A.F.: The representation and recognition of human movement using temporal templates. In: Proceedings of the International Conference on Computer Vision and Pattern Recognition (CVPR), pp. 928–934 (1997)
29. Turner, L.A., Perry, L.H., Sterk, H.M.: Constructing and Reconstructing Gender: The Links Among Communication, Language, and Gender. SUNY Press, Albany (1992)

Development of a Vision Training System Using an Eye Tracker by Analyzing Users' Eye Movements

Ryosuke Kita[1(✉)], Michiya Yamamoto[1], and Katsuya Kitade[2]

[1] Graduate School of Science and Technology, Kwansei Gakuin University, Sanda, Hyogo, Japan
r.kita@kwansei.ac.jp
[2] Joy Vision Training Center, Kobe, Hyogo, Japan

Abstract. Eye movements are important but, in Japan, not tested so. However, there are some reports that training eye movement can improve developmental and learning disorders. In the U.S. or EU countries, these tests and training are generally performed as medical procedures by optometrists. But these depend on their experience and knowhow. In this study, we develop an eye movement training system using a gaming eye tracker. First, we proposed a calibration method to realize highly precise eye tracking during training. Next, we developed a prototype of an eye movement training system. Then, we measured and analyzed the data of eye movements, and proposed some parameters with individual differences and high reliability for the test and the analysis of the effects of training. The results of the experiment demonstrated effectiveness of the system.

Keywords: Vision training · Eye movement · Eye tracker

1 Introduction

As a visual function, eye movement is important, as is visual acuity. In Japan, however, there has been no popular performance of tests of eye movements. On the other hand, there are some reports that the training of eye movements can be used to improve developmental and learning disorders [1]. For example, in reading a sentence, it is necessary to move the eyes accurately and quickly, but children whose eye movements are weak tend to skip words or read the same line over and over again. These symptoms may be improved by training eye movements.

In the U.S. and in EU countries, such tests and training are generally performed by optometrist and treated as medical procedures. In 2018, RightEye LLC released a computerized test and training device called EyeQ, introducing an eye tracking function using Tobii Dynavox I 15+, a part of the process of which is automated [2]. As for academic research, it has just begun to demonstrate the effects of such training of eye movements. For example, Parsons et al., showed that the percentage of secondary saccades decreases for a saccade 10○ away from one point to another point [3]. However, it is limited to some kinds of eye movement, and it is necessary to evaluate

C. Stephanidis et al. (Eds.): HCII 2020, LNCS 12427, pp. 371–382, 2020.
https://doi.org/10.1007/978-3-030-60152-2_28

the effects of eye movement training, in which various eye movement tasks are performed. To this end, it is necessary to analyze eye movements by extracting individual differences of visual functions using highly reliable data.

In this study, we develop a new eye movement training system using a gaming eye tracker, which can be purchased for a low price. First, we proposed a calibration method to realize highly precise eye tracking during the training. Next, we developed an eye movement training system. Then, we measured and analyzed the data of eye movements, and proposed some parameters with individual differences and high reliability for the test and analysis of the effects of training.

2 Related Studies

2.1 RightEye Oculomotor Test

Murray et al. developed a test of eye movement called the RightEye Oculomotor Test, which uses an eye tracker [4]. The tasks are composed of the pursuit of a target in circular, horizontal or vertical directions, looking at two distant points horizontally or vertically. They set eye movement parameters for each task. They analyzed the test-retest reliability for each parameter and showed that certain tasks can be tested with high reliability.

The approach is understandable, but the method cannot, as it is, be introduced into Japan. We focused on the importance of analysis of Japanese vertical text. The sample of EyeQ shows the large error in vertical eye movement as shown in Fig. 1, but such error can rarely be seen. We therefore needed to design the tasks for use by Japanese optometrists in the computerized system, and make use of the system available.

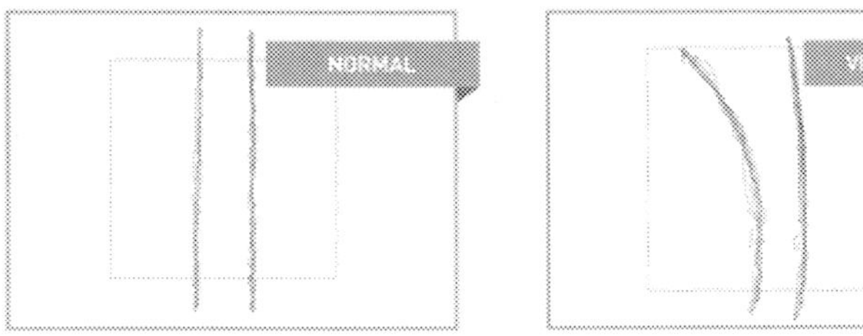

Fig. 1. Line of sight of vertical writing. (RightEye::What's Your EyeQ? (https://www.righteye.com/eyeq/eyeq-products))

2.2 Training of Eye Movements

The study of the effects of eye movements training has just begun. For example, Parsons et al. have developed a task that requires a subject to look at two points, 10° apart horizontally, as quickly and accurately as possible [2]. When this task was performed for 5 days, the ratio of primary saccades increased, as the ratio of secondary saccades decreased. They also showed a reduction in latency. This indicated that the ability to move the eye accurately to the target at once was developed by the training. It is an example of eye movement training effects, and it is required to evaluate the effects of the training of eye movement, in which various eye movement tasks are performed.

2.3 Approach

In this study, we propose a novel approach for Japanese eye movement tests and training as shown in Fig. 2. Currently, such tests and are is conducted by few experts and dependent on experiences and knowhow, as well as on subjective judgement. By computerizing the tests and trainings, users can use the system and measure accurately by employing a PC. Also, by defining some suitable parameters, we perform the test and an analysis of the effects of training by introducing some parameters with high reliability and individual differences.

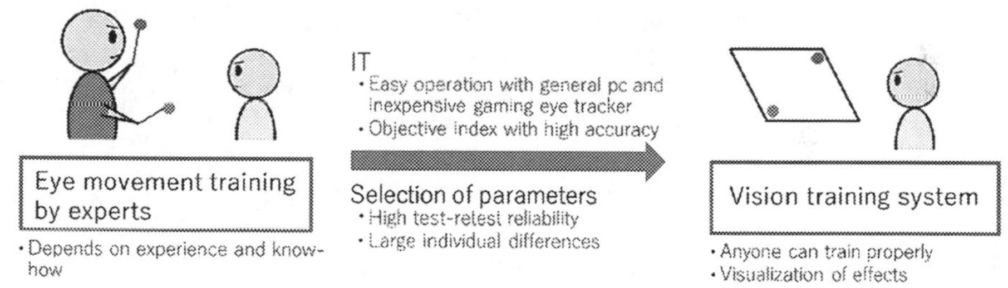

Fig. 2. Concept of the test and training system of eye movement.

3 High-Accuracy Gaze Measurement Using Gaming Eye Tracker

3.1 Problems in the Test and Training of Eye Movement

Generally, to perform gaze measurement with high accuracy and precision, it is necessary to look at several points, called calibration, to set various parameters to adjust to the arrangement of the system and individual differences. For this calibration, linear transformation is often performed using a second-order polynomial as described by Morimoto et al. [5]. In addition, as noted by Gomez et al., measurement errors increase with time [6], there are many tasks to be completed in a short period of time, and as a whole, a long time elapses. Therefore, we were aware of the need to propose a method that evaluates the measurement error between tasks and performs calibration each time.

3.2 Measurement Experiment

We used Tobii's Tobii Eye Tracker 4C and Tobii Gaming SDK as gaze measurement devices. This is a device for games, but we preliminarily performed experiments and confirmed that it has higher accuracy and stronger robustness than the Tobii Pro X2 for analysis released by the same company.

In this measuring experiment, we developed a program that displays markers at 35 points on the screen and displayed them for 3 s apiece. The positions are on a rectangle that includes the center of the screen and is 240 pixels wide and 135 pixels vertically, divided into eight equal parts vertically and horizontally on a screen of 1920 pixels by 1080 pixels. The marker is a circle with a diameter of 32 pixels, and the display angle is 1° in the measurement environment described later. When the button is pressed, markers are animated to cause the radius decreases with the sound and disappear after 3 s, making it easier to watch. When the animation is over, a marker appears at the next position. When the button is pressed again, the animation and the display are repeated at the next position. We used a PC (HP, ZBook 15) and a 23.8 in. display (EIZO, EV2451) for measurement, installed at a distance of 500 mm from the subject. The head was fixed with a chin rest (Fig. 3 left). We used this program to compel subjects to look at the markers in order. We told them that the timing for pressing the buttons was up to them. The time of the experiment was about 5 min. The subjects were 10 people, 8 men and 2 women, aged 21 to 24 years.

Figure 3 (right) shows a measurement example. The red point is the marker position and the blue point is the gaze point of the subject. Each point represents the median of the gaze measurement coordinates during 3 s of gazing, considering the effect of outliers. The average error for this particular subject was 1.05°. Across all subjects, the average error was 0.83° and the maximum value was 1.70°. In a few minutes' experiment time, there was not much change in the tendency of measurement error. When we analyzed the measurement results for all subjects, we identified a tendency toward deviations in specific directions from the actual gaze point. For example, in Fig. 3, **the gaze points** appears to be shifted to the lower right.

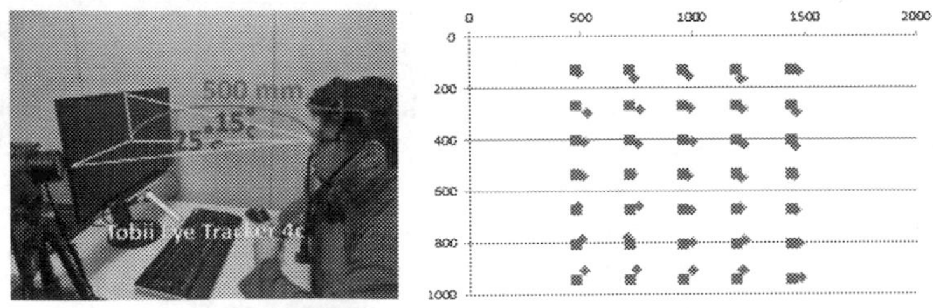

Fig. 3. Experimental environment (left) and measurement example (right).

3.3 4-Point Calibration Between the Tasks

We attempted to modify the transformation by using the median of the 35 measurement errors in the experiment. After the transformation, the average of the measurement errors decreased to 0.49, and 1.0° at maximum for all subjects. This is within the sum of the measurement error of 0.5° of Tobii's analytical gaze measuring device and the general microsaccade of 0. several degrees. We thereby determined this correction to be sufficient.

However, it is not realistic to look at 35 points between tasks and make corrections, so we wanted to perform calibrations of the same accuracy by simply looking at as few points as possible. Therefore, from Fig. 3, six calibration points were extracted (Fig. 4, left). When calibrating the measurement results for 10 persons using one point in (1), the average error was about 0.9°. Similarly, it was 0.7° at two points (1) and (2), 0.6° at three points (1) to (3), and about 0.5° at four points (1) to (4) (Fig. 4, right). At four points, it was almost equal to 35 points, and then even further increase of the points yielded no change. For this reason, **the optimal method of calibrating uses four points.**

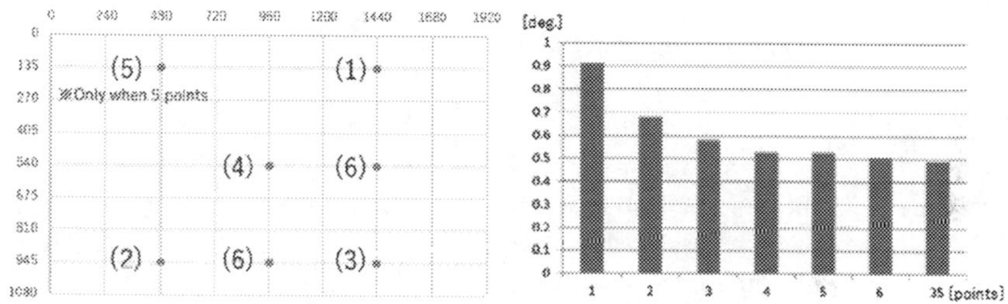

Fig. 4. Calibration points position (left) and effect of calibration points (right).

4 Development of the System

4.1 System Configuration

We developed a prototype of an eye movement training system using the configuration described in Sect. 3. We developed two eye movement tasks for the test and training system, with reference to paper based worksheets [1]. In addition, to measure the gaze point with high accuracy, we introduced 4-point calibration, as described in Sect. 3.3, **between** the tasks.

4.2 Saccade Test

The first test was a saccade test, which is the most basic test by experts, and uses sticks. We displayed markers instead of sticks and animated in eight directions, up, down, left and right, on a circle with a viewing angle of 30°. When a user looks at the marker, the marker is displayed on the opposite side. Then, the gaze point is measured while the marker is displayed (Fig. 5).

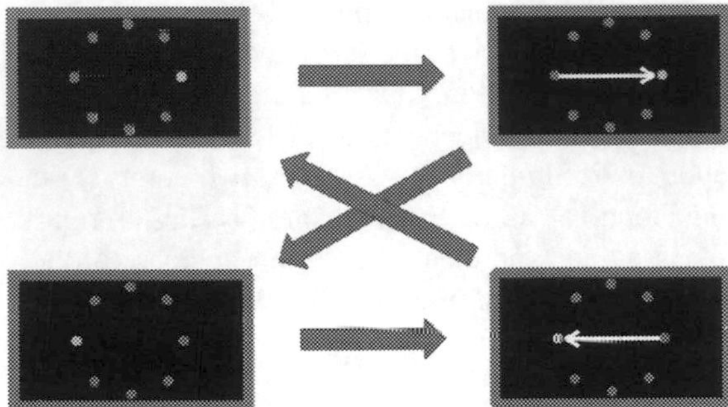

Fig. 5. Procedure of saccade test of two trials.

4.3 Pursuit Test

The second test is the pursuit test, in which there is a marker moving at a constant speed. When the button is pressed, a marker moves at a constant speed. Pursuant to the opinions of experts, the marker speed was set to 10° per second.

5 Feature Values for Eye Movement Training

5.1 Feature Values of Saccade Test

If eye movement functions are weak, certain symptoms manifest, such as eye movement drawing an arc even when subjects look at one marker and then another marker. To confirm this, we defined the saccade error as the distances between the gaze points and the straight line connecting the two marker points, as shown in Fig. 6. We used the values of median, standard deviation, and range (difference between the maximum and minimum errors) of the saccade error.

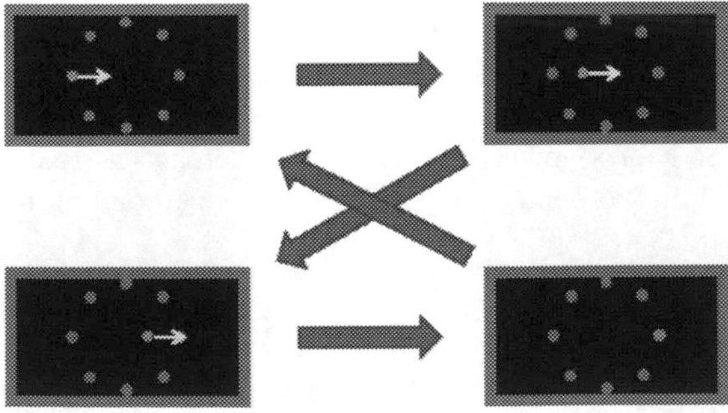

Fig. 6. Procedure of pursuit test of a trial.

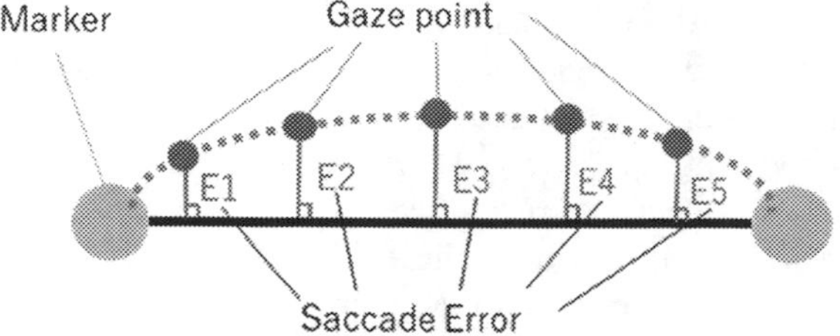

Fig. 7. The saccade error.

Some people cannot move their eyes quickly or at a constant speed. For this reason, we also used the speed of gaze movement as a parameter. For the feature values of this speed, we used the time between one marker position to another position. Because the effect of the error was large, due to the effect of blinking, we only used the median value for the analysis.

5.2 Feature Values of Pursuit Test

In the pursuit test, we also used the error from the target and speed as feature values. Further, Murray et al. set parameters for on target smooth pursuit (SP) where the error in a tracking test that moves on a circle. Its defined ration was within 2 degrees of the error. They showed that there was high reliability in the pursuit test, and we also used it for analysis (Fig. 8).

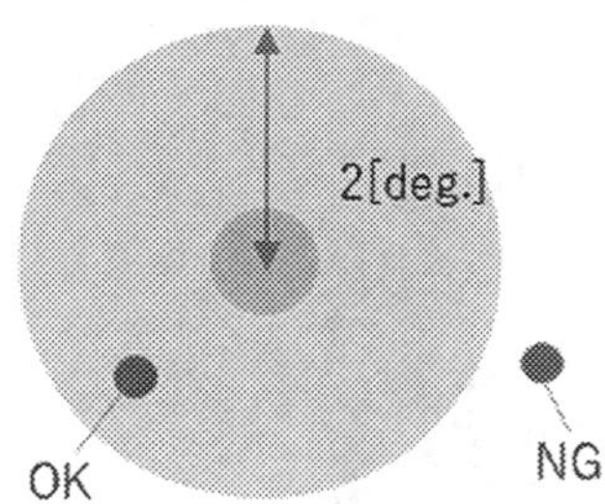

Fig. 8. On target smooth pursuit.

5.3 Procedure to Extract Feature Values

It is important to extract both individual characteristics and reliable data especially to perform the analysis of the effect of the training. We therefore decided to extract feature values according to the following procedure that considers test-retest reliability. There are four steps, as follows:

(1) We performed tests 10 times in each direction for the saccade and pursuit tests,
(2) We calculated the feature values, as described in Sect. 5.1 and Sect. 5.2, in each trial. Then, we measured the values in 10 trials,
(3) We calculated the median of the values across the10 trials to exclude outlier values,
(4) To extract individual characteristics, we calculated the feature values of each subject and selected the **feature values** with the largest individual differences (difference between maximum and minimum values), and
(5) To analyzc test-retest reliability, we evaluated the values using the intraclass correlation coefficient (ICC). Generally, it is said that 0.9 or more is excellent, 0.8 or more is good, 0.7 or more is normal, 0.6 or more is possible; if it is less than 0.6, the test must be reconsidered. This time, we used the value of ICC (3, k) in the analysis to evaluate reliability in the same measurement environment.

6 Measurement Experiment for Extracting Feature Values

6.1 Procedure

We conducted experiments using the saccade test (4.2), as well as the pursuit test (4.3) using the prototype system. In the experiment, the gaze points were measured as shown in Fig. 7. We used the data to calculate the feature values in Sect. 5.1 and Sect. 5.2 by using the data. Each task was performed 10 times for the extraction procedure described in Sect. 5.3 (1). Measurement errors were corrected between the tasks, using the calibration method in Sect. 3.3.

The subjects were 10 subjects, 8 men and 2 women aged 21 to 24 years. The same test was performed twice, no more than once a day (Fig. 9).

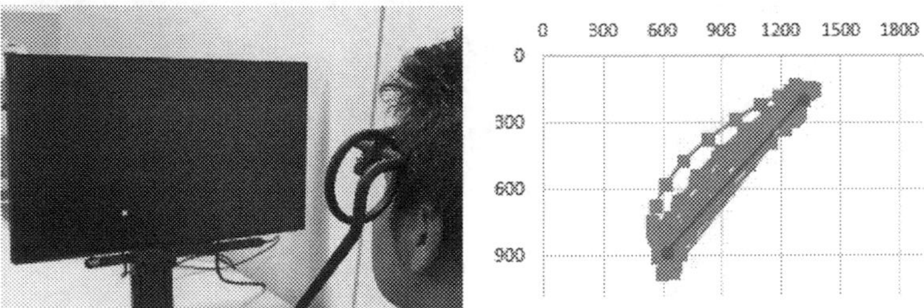

Fig. 9. Test situation and measurement example.

Considering the situation where the eye movements cannot be measured correctly, we calculated the 1-sigma section of the measurement error and used only the data within 1.5°. This is the sum of the marker diameter of 1° and the accuracy of the measuring device of 0.5°. As a result of this selection, we excluded 2 subjects' data and used a total of 8 subjects.

Table 1. Results of individual differences.

	Saccade Error[deg.]			Speed [deg./sec]
Direction[deg.]	Median	SD	Range	Median
0	0.44	0.19	0.48	27.0
45	1.40	0.57	1.69	16.3
90	0.64	0.27	0.48	67.7
135	0.89	0.38	1.16	27.7
180	0.61	0.11	0.33	40.7
225	0.70	0.83	2.50	43.9
270	0.17	0.21	0.68	45.1
315	1.32	0.64	2.23	30.3

6.2 Results of Saccade Test

For each subject, wc calculated the feature value as shown in Sect. 5.3 (2) and (3). Table 1 shows the results of the saccade test. In the table, the values of Sect. 5.3 (4) are shown. Range (error from the target) reflects a large difference in subject data in the oblique direction (Table 2), and can be described as an optimal test parameter.

Table 2. Test-retest reliability of range.

	First time		Second time		
Direction[deg.]	AV[deg.]	SD	AV	SD	ICC(3,k)
0	0.73	0.18	0.80	0.17	0.41
45	1.34	0.62	1.42	0.46	0.92
90	0.66	0.17	0.69	0.12	0.51
135	1.38	0.44	1.25	0.32	0.96
180	0.72	0.11	0.84	0.18	0.51
225	1.84	0.81	1.58	0.89	0.92
270	0.76	0.21	0.79	0.32	0.79
315	1.73	0.71	1.84	0.75	0.93

Then, we calculated the ICC (3, k) values for the range as described in Sect. 5.3 (5). In the oblique direction, the value exceeded 0.9. This showed high reliability, indicating that is optimal for a feature value of saccade test.

In the horizontal and vertical directions, the value was less than 0.6. In these directions, individual differences were within 0.5°, which was less than the sum of measurement error and 0.1° of general micro saccades. As such, there was considered to be little difference in eye movement characteristics.

In the orientation of 225°, where the individual difference is largest, the result of the subject with the minimum is shown on the left of Fig. 10, and the maximum is shown on the right of Fig. 10. When the error range is small, the gaze point is linear. By contrast, when the error range is large, the gaze point draws an arc. This is a typical symptom of a weak eye movement function.

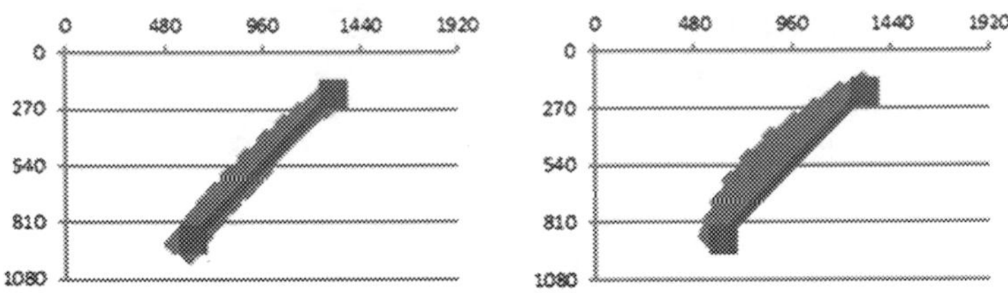

Fig. 10. Example of oblique measurement.

6.3 The Result of the Pursuit Test

We selected by the same procedure as in saccade test, and selected the SP as a feature value with a large individual difference. When we analyzed the test-retest reliability for this parameter, variation depended on the direction (Table 3). This is because all subject were university students, and there was minimal individual difference.

Table 3. Test-retest reliability of SP.

	First time		Second time		
Direction[deg.]	AV[%]	SD	AV	SD	ICC(3,k)
0	91.97	6.44	94.51	4.26	0.74
45	91.93	10.20	97.52	3.78	-0.21
90	83.85	15.61	89.64	15.60	0.85
135	93.11	8.85	96.54	3.48	0.29
180	92.69	9.74	96.16	5.58	0.91
225	91.14	10.01	90.81	6.20	0.47
270	87.71	13.80	96.01	1.97	0.22
315	99.07	8.77	84.78	17.00	0.71

7 Evaluation in Public Demonstration

In Sects. 5 and 6, there were few individual differences, because the subject was a college student. It was therefore insufficient to show whether the selected parameter was able to extract individual characteristics. We consequently evaluated whether it was possible to extract individual differences for more people in a public demonstration.

In the demonstration, we asked visitors to use the prototype of the system in the same way as in Sect. 6, and measured the data by using the system. The visitors consisted of a total of 44 subjects, aged 3 to 27 years.

Figure 11 shows the average of the range of the saccade test in a diagonal direction. In the figure, we plotted the values for each age and fit them by means of a quadratic. As shown in the figure, the value became smaller with age. This means that the extracted feature value was effective to show individual differences, because the ability of eye movement grows with age.

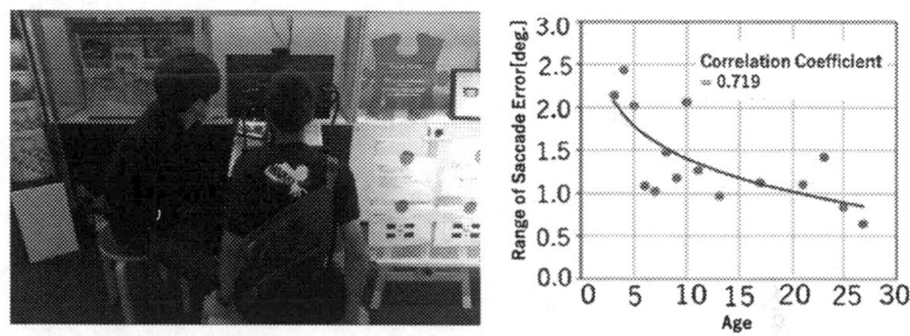

Fig. 11. Measurement example (left) and individual difference (right).

8 Conclusion

In this study, we developed a system to test eye movement, deploying a saccade test and a pursuit test, using a gaming eye tracker. By using a prototype of the system, we extracted feature values from the viewpoint of extracting individual characteristics and test-retest reliability. The effectiveness of the method was demonstrated in a public experiment. In the future, it will be possible to use more test data to make an automatic determination of the degree of the eye movement function of the subject.

Acknowledgement. Apart of this study was supported by JSPS Kakenhi 16H02860, 16H0 3225, etc.

References

1. Kitade, K.: Vision training that makes learning and exercise fun for children who are concerned about development, Natsume Inc. (2015). (in Japanese)
2. RightEye: What's Your EyeQ? https://www.righteye.com/eyeq/eyeqproducts. Accessed 1 Nov 2018
3. Parsons, B.D., Ivry, R.B.: Rapid alternating saccade training. In: Proceedings of the 2018 ACM Symposium on Eye Tracking Research & Applications (ETRA 2018), Article no. 30 (2018)
4. Murray, N., Kubitz, K., Roberts, C.M., Hunfalvay, M., Bolte, T., Tyagi, A.: Examination of the oculomotor behavior metrics within a suite of digitized eye tracking tests. Comput. Hum. Behav. **5**, 270–284 (2019)
5. Morimoto, C.H., Mimica, M.R.M., Gellersen, H.: Eye gaze tracking techniques for interactive applications. Comput. Vis. Image Underst. **98**(1), 4–24 (2005)

6. Gomez, A.R., Gellersen, H.: Smooth-I, smart re-calibration using smooth pursuit eye movements. In: Proceedings of the 2018 ACM Symposium on Eye Tracking Research & Applications (ETRA 2018), Article no. 10 (2018)
7. Imai, I., Shiomi, T.: A method for testing the reliability of evaluation in physiotherapy research. Physiother. Sci. **19**(3), 261–265 (2004). (in Japanese)

Digital Culture in YouTube Categories and Interfaces: User Experience and Social Interactions of the Most Popular Videos and Channels

Alberto Montero[1(✉)] and Jorge Mora-Fernandez[2,3,4(✉)]

[1] University of Castilla la Mancha, Cuenca, Spain
alberto.montero1@alu.uclm.es,
albertomonterogo@hotmail.com
[2] Arthur C. Clarke Center for Human Imagination,
University of California San Diego, San Diego, USA
j2morafernandez@ucsd.edu
[3] Research Group IAMNCEDC R+D+C+I/NAICECD I+D+C+i Interactive Arts and Media, Narrative Convergences and Edutainment in Digital Communications and Cultures R+D+C+I (Research+Development+Creation+Innovation), Universidad Nacional de Chimborazo, Riobamba, Ecuador
jorge.mora@unach.edu
[4] Laboratory of Digital Culture and Hypermedia Museology,
Universidad Complutense de Madrid, Madrid, Spain
multiculturalvideos@gmail.com

Abstract. YouTube is able to generate new forms of user interactions and an entire evolutionary cycle in audiovisual languages and digital culture through the interface. This platform offers the possibility for the user to assume the functions of prosumer of audiovisual contents, through generating their own digital identities. In this way they are able to influence and be part of the society of production, diffusion and audiovisual user interactions. The main aim of this research was to obtain a general idea and a first approximation of the tendencies of audiovisual popular user interactions and the digital culture on YouTube, which was undertaken by means of the quantitative analysis of YouTube video interfaces selected by YouTube categories. In order to extract this information, a quantitative analysis of the most viewed channels and videos was undertaken. These interface categories were those with more views, with more subscribers, with more likes and dislikes, with more comments, as well as the average time of some of the most viewed videos. This was done through a sample of 160 most viewed videos of the platform. The aim was to understand the common denominators and interface preferences that users manifest through their interactions with videos, channels and thematic categories in the YouTube digital culture interface. The conclusion from the study was that the YouTube categories that enjoy the most popularity are education and music, at the same time the interaction features of each category can also be defined on the YouTube platform.

C. Stephanidis et al. (Eds.): HCII 2020, LNCS 12427, pp. 383–401, 2020.
https://doi.org/10.1007/978-3-030-60152-2_29

Keywords: Digital humanities · User experience · YouTube categories · YouTube interfaces · Human interactions · Cultural studies · Social interactions

1 Introduction

YouTube, an online audiovisual platform has been developing its own interface ecosystem through sharing audiovisual content and information with the user and viewer community. Whilst YouTube started as a shared amateur video platform, which achieved its success due to the contributions of its users, its conflicts with the industry due to the improper use of copyrighted content, led to a steady transformation of its business logics. On the one hand there are a great number of users that produce user-generated content (UGC) through self-expression, and on the other hand professionals inside the formal media and music industries also use and shape the same platform [1]. However, in some cases professional generated content is superior in availability of number of videos but UGC is significantly more popular [2]. Due to this, YouTube had to find a way to maintain a stable relationship between users and industry. This was done by modifying the algorithm, which is as "a set of automated instructions that transform inputs into a desired output" [3], to allow viewing preferences to users who had contributed their content to enable growth of the platform and, at the same time, offered advantages through partner agreements to the traditional media. Not surprisingly, YouTube had a "migration from video sharing to commercial video streaming" [4], which shows that YouTube was invented by the industry [5].

Although YouTube initially offered the possibility of renegotiating the dominant media discourses more independently, the platform later adapted to the conventional conditions of the broadcast media, which led to closer collaboration between users and industry, e.g. YouTubers and brands. "Often generating as a consequence significant advertising and sponsorship revenue and increasingly the attention of mainstream media" [6]. Due to this, content producers act as independent social mediators showcasing their own audiovisual narratives, promoting the YouTube communities and engaging fan followers. Not surprisingly, YouTube uses a partner program in order to group YouTube partners who earn profits from their views and channel subscribers, promoting them on the interface and organizing YouTube events to increase the visibility of both. YouTube started monetizing their videos in late 2007 and reconfigured it again in 2012, allowing anyone to monetize videos. These changes granted positioning preferences on the platform to those users who had contributed their content to YouTube's popularity. "It was restricted in 2017 and again in 2018 in favor of larger, ad-friendly creators" [4]. Thus, by modifying its algorithm, YouTube rewards former partners but makes it difficult to promote new users who want to gain a foothold on the platform, making their positioning more difficult. Therefore, YouTube platform appears as a commercial broadcast media that protect its business model, changing its algorithm, modifying usage policies, establishing partners, and incorporating new technological innovations when needed. The transformation of the user experience was the direct consequence of modifying its interface algorithm and its platform recommendation system, including other interface changes such as automatic reproduction and the continuous stream of commercials.

1.1 The User's Experiences and Interface Cultural Evolution

Thanks to the arrival of Henry Jenkins theory of the culture of convergence, users begin to exchange the roles of sender and receiver, leading to multiple-way communication and giving rise to viral phenomena. Today's users act accordingly by uploading their posts to attract traffic, launching controversial messages, and analyzing their own statistics to see what works and what doesn't. They know how this dialogue works and try to reach the best practices and strategies to grow their audience, instead of simply serving content [7]. Thus, the user becomes a prosumer [8], being able to consume and produce contents as a result of the development of communication interfaces and the democratization of media. These new content producers have followed their own intuition, expanding the possibilities of interactive video, by generating open audiovisual dialogues through YouTube channels. Furthermore, they are responsible for building their identity based on the user profile and producing their own digital active life, selecting at all times the content they consume and share, and through which channel they download it [9]. In this context, the remix culture uses the creative forms of absorption, assimilation, and sharing through the video productions, making something original and valuable [10] while often seeking to make a profit. In social networks like YouTube, using words or images is acting by constructing subjectivities, contributing meanings that stabilize space and order time in a constant dialogue between subjective experiences [11]. Thus, the culture of appearances, spectacle and visibility arises, where the discourse is increasingly dedicated to offering the prosaic spectacle of its protagonists [12], leading to self-exhibitionism and the continued publication of commercial contents, or even a "chaos of useless information" [13].

Although YouTube establishes a classification by categories, which is selected when uploading audiovisual content, the platform does not, in the same way, allow a search for videos to be undertaken by categories only. Its filtering is done by upload date, type, duration, characteristics and sort by (YouTube.com), giving a "limited quantity of exploitable information" [14] to researchers of the audiovisual genres and categories. In this way, the YouTube search engine invites its users to a continuous flow of videos, whose automatic reproduction and advertisements avoid user interaction. Thus, YouTube stands out for being a video platform making available a huge volume of content but with search restrictions. However, by avoiding limitations on the logic of classification by the categories established by YouTube, and during the production of their content, users are capable of producing new interactive audiovisual genres thanks to the characteristics of the interface.

To understand the sciences of audiovisual images on YouTube in depth, it was important to dedicate hours of study and analysis to cultural productions and new modes of shared consumption through the interface. The key task of cultural studies is to understand the interactive relationships that users have with each other and the different cultural activities that can be carried out on media platforms. The theme of interactive audiovisual culture should be about what meanings are shared or questioned by certain users in certain places and conditions. In YouTube, there is an encounter between cultural activities through audiovisual genres that "can produce tension and friction, but also a process of creative and joint hybridization" [15]. The knowledge of the interactions in the YouTube interface through its quantitative analysis allows identification of the preferences of audiovisual user experience and the type of interactive relationships of the different categories that YouTube establishes.

1.2 The Emergence of New Narrative Genres

The interactive context and "the platformization of our societies" [16] with the appearance of audiovisual social networks such as YouTube, has led to the development of new interactive audiovisual genres. The medium itself was the one that demanded new interactive contents capable of audiovisual communication to fill the gaps offered by the new technology. As a consequence, the user experience changes completely, making users part of the social opinion and visual criticism through visualizations, subscriptions, likes, dislikes, and comments. In turn, all these social interactions promote the creation of communities and attracting content to the interactive audiovisual world. The need for different contents: comedy and humor, tutorials to share knowledge, the appearance of new advertising spaces, online videogame entertainment, shared betting pranks, activism videos, etc.; are some of the audiovisual and interactive needs that the information society demanded through new representation of the contemporary image. Therefore, in this interactive audiovisual context, a confluence of previous genres is produced, used in other media such as film and television, with the new interactive capabilities that the hyper-communication allows, giving rise to a broader audiovisual ecosystem open to new genres, incorporations and audiovisual typologies that change and develop the interactive audiovisual narratives.

Whilst the user structure the video in accordance with their own preferences, e.g. the pulse and the repetition, the unreality, the different graphic values included as the low resolution, the modification of the scale, the unusual casual relations, the intermediality and intertextuality, the comedy and the humor and the formal replication of contents [17], the videos can only be classified within one of the YouTube´s categories. From this we are able "to empirically analyze how these characteristics are assembled using various prototypical sequences and from a theoretical perspective study how inter-tipology works" [18]. Despite the fact that the videos can only be classified within one of YouTube´s categories, it is the prosumer who classified and give meaning to the productions, in turn establishing intertextual relations of the image and connections between communities based around thematic activities. All these visual meanings "require the interpretation of the viewer to discover the underlying theme, the apprehension of which will constitute the ultimate result of comprehensive activity" [19].

Although audiovisual genres usually come from literature, not all genres were previously found in other media. To understand the content classification established by YouTube, one must free the gaze from the culturally imposed one, "which tries to fit everything within the limits of that projection made up of the predominant representations and values" [20]. A YouTube category can be defined as an established content categorization by the platform to the recognition of common patterns of form and content, capable of establishing a stylistic system that organizes and classifies videos.

Based on the classification by YouTube categories, you can see videos of animals, cars, videogames, children's songs, activism, news, crafts, movies, entertainment, travel, etc. However, the borders between genres have been crossed to create hybrids as an extensive display of audiovisual hypertextuality, bringing new ways of thinking and structuring the digital narrative plots of new creations. Hence, the appearance of vlogs, tutorials, reviews, gameplays, unboxings, covers, challenges, fan video, video-reactions, etc. Tzvetan Todorov affirms that every literary genre comes from another: "a new genre is always the transformation of one or several old genres: by inversion, by displacement, by combination" [21]. If this statement is interpreted in interactive digital audiovisual contexts, it is possible to affirm that the new gender combinations start from previous genres and their multiple combinations. However, the interface features also give these genres the unique interactive properties of the medium, which requires an analytical perspective free of previous conventions.

Thanks to YouTube, audiovisual popular culture is empowered by showing alternative perspectives in their socio-economic and cultural contexts on the Internet. However, as Warren Buckland [22] exposes, following YouTube tags puts one on a cusp, precariously balanced and dangerously built over an abyss of thousands of similar or even the same videos, commented on and cross-referenced to yet more of the same and the similar, to immerse into an audiovisual homogenization of infinite repetition. For this reason, although YouTube shows the shared global spirit and can faithfully reflect the state of today's society and its ways of thinking, the lack of awareness of the repetitive user interactions through the stream of content and advertising, can lead to hypnotizing screen habits even without the need for users to share such ideals [23].

As Gilles Lipovetsky and Jean Serroy [24] explain, the hypermodern transformation "is characterized by affecting technologies and media, the economy and culture, consumption and aesthetics in a synchronous and global movement", which means that there is a temporal correspondence between the transformations of these factors. It is therefore necessary to study this type of interactive interface videos as a global memory based on a hypertextual network of superimposed connections, opening new paths of recombination of the image that remain for posterity, leading to new remixes and future reinterpretations. The same way that we revisit now and re-interpret the new narrative, techniques and social habits that photography and cinema brought to human cultural evolution, the new generations will study the meaningful changes that occurred during this digital culture re-evolution. It is our hope that this study serves as an initial insight of how contemporaries observed the influences of digital interfaces in the user experiences to become prosumers and viewers of the new audiovisual narratives and genres.

2 Materials and Methods

2.1 Methodology

Some of the YouTube studies are based on viewer interaction and the prediction of video popularity [25]. However, it is still interesting to focus on video categories and user's experiences in order to understand and contrast trends through the quantitative analysis of their audiovisual interactions. For example, some scholars have used the YouTube data API (Application Programming Interface), which allows a program to search for a video and retrieve its related information [26], to know how channels, uploads and views evolved over time [27]. On the contrary, other academics have exposed how YouTube imitates the rules of the old media [28].

This research was based on the study of quantitative data exposed on YouTube through video information and user interactions. As Davidson, Liebald, Lui, Nandy; Van Vleet [29] expose, there are two basic data sources; on the one hand, content data such as raw video streams and metadata (video title, description, etc.), and on the other hand, user activity data. In other words, some data are static and can be measured once: e.g. title, category, link, user update, length, etc. and other data are dynamic and can change over the time through user interaction on the YouTube interface [30]. Thus, the attractiveness and even the written opinion of every video are continuously changing over the lifecycle of a video based on the audience interactions. For these reasons, the quantitative analysis is considered the most appropriated tool to archive data streams of user experience and compare samples over the time.

2.2 Aims of the Research

The main objective of this work was to obtain a general idea of the trends in audiovisual user interactions, by means of the quantitative analysis of videos selected by YouTube categories. There were also several secondary objectives for this:

a) To know the themes, narrative and aesthetic content, which in general are most consumed by popular audiovisual culture on YouTube through identifying which are the most viewed categories on YouTube.
b) To recognize the most viewed videos in each of the 10 most viewed channels within the 16 YouTube categories.
c) To know the user interactions through the number of views, subscription, likes, dislikes, comments as well as the average duration of the videos for each category.
d) To establish inferences, when comparing various results on the data obtained, to get a global vision of the audiovisual user interactions of categories that takes place on YouTube.

2.3 Applied Methodology

The analysis of quantitative data of the most popular videos selected by categories was used to describe the user experience. The aim, from an analytical prism, was to identify the types of interactions on each YouTube category. As cultural trends change over time, the purpose of this research was to capture how these interactions were experienced as a whole, in the context of audiovisual culture on YouTube in a given period [31], in this case the data was collected in October 2017. To select a sample of 160 videos, from the most viewed 10 channels for each of the 16 YouTube categories were searched through www.socialblade.com. Once the 10 most popular channels by each category were known and ordered, the most viewed video was selected from each channel on www.youtube.com, 10 in total for each of the 16 YouTube categories. In this way, the data of the 10 most viewed videos from the 10 most viewed channels of the 16 YouTube categories were obtained, making a total of 160 videos to be analyzed using a quantitative analysis.

As the initial objective was to understand the trends in audiovisual consumption habits on YouTube based on quantitative data, the designed analysis model was in charge of collecting the data on the variables of: views, subscribers, likes, dislikes, number of comments and duration of the videos, in addition to other variables of qualifying interest such as the title of the video, the user name, the date of registration of the data, the date of uploading of the video and its link. These data were obtained by direct observation, to present and compare each variable together at the same time that the information was archived. This method was carried out to discover characteristic patterns of YouTube categories resulting from the total of interactions.

The data was collected in order to maintain a logic of analysis and exposure. The static variables were the title of the videos, date of upload, username and link, and the dynamic ones were the visualizations, subscribers, likes, dislikes, comments and the average duration of the videos for each category. The data collection included the following variables: the most viewed, with the most subscribers, with the most likes or dislikes, and with the most comments was carried out by means of a simple sum of each variable by category. The average duration of videos for each category was made by taking the total sum of the duration in minutes and seconds of the 10 most viewed videos, of each most viewed channel by category, divided by 10.

3 Results of the User's Interface Interactions by Categories

When expounding the results of this study through the analyzed quantitative data, an attempt was made to make it as clear as possible through Table 1. The order of the categories were organized according to the number of views in each category. Furthermore, graphs and descriptions show the results in a transparent way to relate and contrast data.

Table 1. Recorded data from YouTube categories. (Sources: www.youtube.com & www.socialblade.com). Own elaboration.

DATA OF THE MOST VIEWED VIDEOS INSIDE MOST VIEWED CHANNELS ORDERED BY AMOUNT OF VIEWS AND SELECTED BY CATEGORIES									
Order	Icon	Youtube Category	Views	Subscribers	Likes	Dislikes	Number of Comments	Average of Duration	TOTAL INTERACTIONS PER CATEGORY
1º		Education	24.530.224.462	52.740.000	4.458.000	3.049.000	154.396	28:00 m	24.590.625.858
2º		Music	9.886.869.983	154.020.000	27.595.000	3.111.000	2.997.605	7:35 m	10.074.593.588
3º		Shows	3.356.941.173	21.654.000	2.886.000	1.453.000	43.953	18:55 m	3.382.978.126
4º		Entertainment	1.816.011.947	90.400.000	2.401.000	941.000	189.916	9:25 m	1.909.943.863
5º		Comedy	1.509.089.491	153.400.000	4.217.000	252.000	269.030	7:57 m	1.667.227.521
6º		Film and animation	1.440.626.002	59.000.000	1.553.000	587.000	91.424	22:00 m	1.501.857.426
7º		How to and style	948.314.286	71.300.000	2.095.000	2.218.000	1.059.995	6:03 m	1.024.987.281
8º		People and Blogs	878.409.536	59.112.000	1.489.000	330.000	226.496	7:37 m	939.567.032
9º		Sports	848.402.816	82.400.000	221.800	82.025	133.125	8:53 m	931.239.766
10º		Pets and Animals	682.596.990	19.032.000	1.603.000	177.000	181.401	3:27 m	703.590.391
11º		Gaming	583.257.764	165.400.000	3.756.000	181.000	384.814	14:55 m	752.979.578
12º		Science and technology	358.687.383	67.200.000	2.208.000	455.000	270.195	4:18 m	428.820.578
13º		Non profit and activism	339.648.373	24.124.000	2.794.000	290.300	150.355	20:23 m	367.007.028
14º		Autos and vehicles	318.480.386	19.534.000	814.000	87.000	78.860	15:30 m	338.994.246
15º		Travels	294.079.846	9.216.300	674.000	54.571	102.242	9:26 m	304.126.959
16º		News and politics	260.899.102	52.406.000	512.000	123.000	82.585	5:02 m	314.022.687
TOTAL INTERACTIONS PER VARIABLE AND AVERAGE OF DURATION OF ALL CATEGORIES			48.852.539.540	1.100.938.300	59.276.800	13.390.896	6.416.392	11:50 m	49.232.561.928

3.1 Order of Categories by Number of Views

From the selected video sample, the category with the highest number of views was "education" (24,530,224,462), which presented mostly children's videos of children's songs. According to the data obtained, the next category with the most views was "music" (9,886,869,983), which featured different themed music videos from varying countries of production. The third category in the sample with the most views was "show", with varied content that included excerpts and sketches from series or television shows (3,356,941,173). The fourth category with the most views in the sample was "entertainment" (1,816,011,947), in which a multitude of animation videos and intertextual content were found that included intertextual aspects through Disney princesses, Mickey Mouse, Donald, Pluto, Daisy, Spiderman toys, etc. Other categories such as "comedy", "film and animation", etc. closely followed (see Fig. 1).

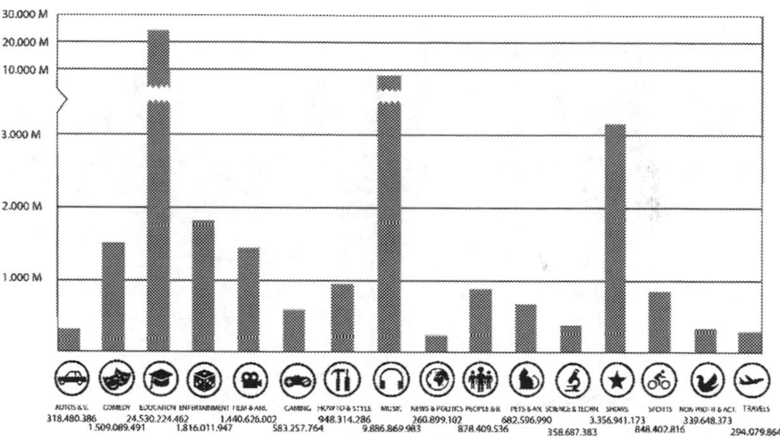

Fig. 1. Number of views of the most viewed videos of the most viewed channels selected by YouTube categories. (Sources: www.youtube.com & www.socialblade.com). Own elaboration.

3.2 Order of Categories by Number of Subscribers

Attending to the number of subscribers in the 16 YouTube categories, "gaming" (165,400,000) was in the first position, followed by "music" (154,020,000) and "comedy" (153,400,000). It was followed by the categories of "entertainment" (90,400,000), "sports" (82,400,000), "how to and style" (71,300,000), "science and technology" (67,200,000), "people and blogs" (59,112,000), "film and animation" (59,000,000), "education" (52,740,000) and "news and politics" (52,406,000). On the contrary, compared to far fewer followers, the channels that had the least subscribers were "non-profit and activism" (24,124,000), "show" (21,654,000), "cars and vehicles" (19,534,000), "animals" (19,032,000) and "travel" (9,216,300) (see Fig. 2).

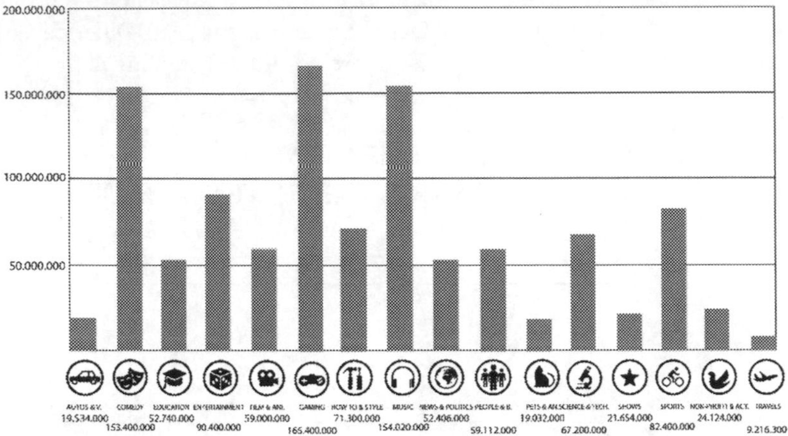

Fig. 2. Number of subscribers to the most viewed videos from the most viewed channels selected by YouTube categories. Sources: www.youtube.com & www.socialblade.com. Own elaboration.

3.3 Order of Categories by Number of Likes and Dislikes

Regarding the number of likes by categories, it seems that the clear winner was the "music" category, which had 27 million likes. The next categories far from the first were "education" (4,458,000), "comedy" (4,217,000), "gaming" (3,756,000), "show" (2,886,000), "non-profit and activism" (2,794,000), "entertainment" (2,401,000),

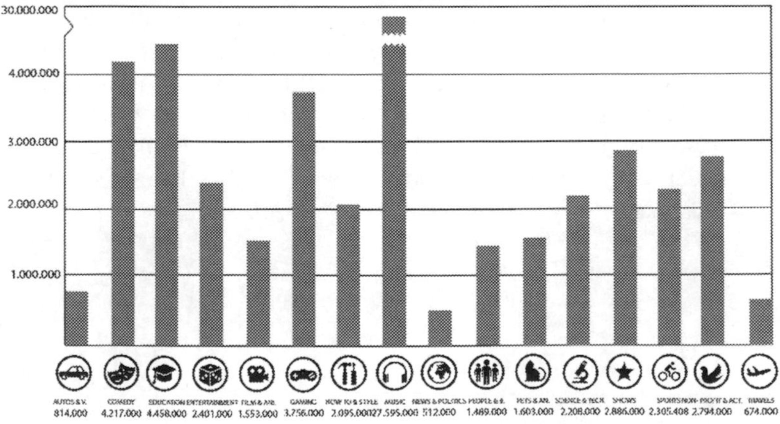

Fig. 3. Number of likes of the most viewed videos of the most viewed channels selected by YouTube categories. Sources: youtube.com & socialblade.com. Own elaboration.

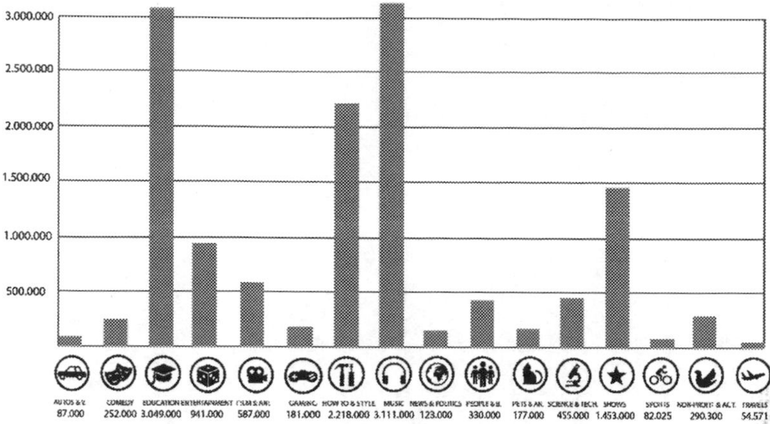

Fig. 4. Number of dislikes of the most viewed videos of the most viewed channels selected by YouTube categories. Sources: www.youtube.com & www.socialblade.com. Own elaboration.

"sports" (2,305,408), etc. (see Fig. 3). On the contrary, the most hated categories were also "music" (3,111,000), again in first position, "education" (3,049,000) and "how to and style" (2,218,000), showing the highest levels of dislikes (see Fig. 4).

3.4 Order of Categories by Number of Comments

The most commented category was "music" (2,997,605), with a lot of distance from the rest of the categories. In the second place were the videos "how to and style" (1,059,995). The next most commented categories were "gaming" (384,814), "science and technology" (270,195), "comedy" (269,030), "people and blogs" (226,496), etc. The data extracted that music videos were the most influential audiovisual genre in offering public opinion, followed by the interest in commenting on the tutorials or the didactic lessons of the videos "how to and style" (Fig. 5).

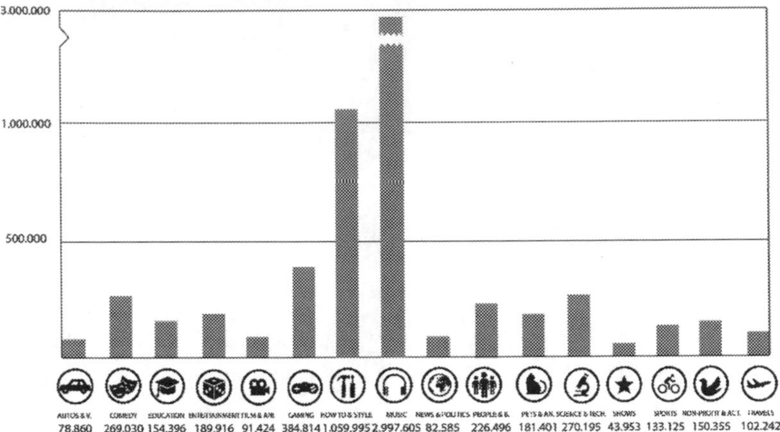

Fig. 5. Number of comments of the most viewed videos of the most viewed channels selected by YouTube categories. Sources: www.youtube.com & www.socialblade.com. Own elaboration.

3.5 Average Duration of the Most Viewed Videos by Categories

Based on the duration of the videos in each category, the average duration of the sum of time of the videos could be extracted. As a result, it was observed that the "education" category, which was mostly videos opening Kinder eggs and playing with toys, were of the longest duration (28 min). The next lengthiest videos were from "film and animation" (22:00 min), "nonprofit and activism" (20:23 min), "shows" (18:55 min), "cars and vehicles" (15:30 min) and "gaming" (14:55 min). Later the "travel" and "entertainment" categories were found with practically a tie between the two (9:26 min and 9:25 min). The videos in the following categories were "sports" (8:53 min), "comedy" (7:57 min), "people and blogs" (7:37 min), "music" (7:35 min), "how to and style" (6:03 min), "news and politics" (5:02 min), "science and technology" (4:18 min), and animals (3:27 min). These data allowed us to understand the average duration of the type of video according to its category (see Fig. 6). Taking into account the total number of categories, the average duration of the most viewed video on YouTube, regardless of the category, was 11:50 min (result of the sum of the means of all divided among the 16 categories).

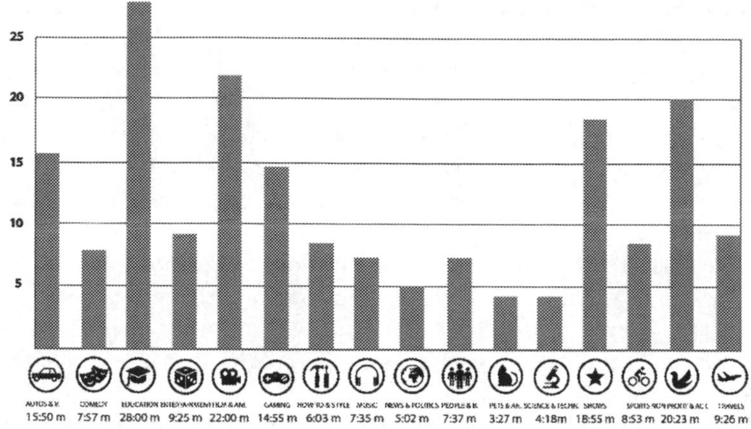

Fig. 6. Average duration of the most viewed videos from the most viewed channels selected by YouTube categories. Sources: www.youtube.com & www.socialblade.com. Own elaboration.

4 Data Discussion

In general terms, there is a great influence of YouTube on the current digital culture of user's interactions with online audiovisual interfaces. The massive totals of interactions that users have with each studied variable of the YouTube interface (views, subscribers, likes, dislikes and number of comments) make it evident that it is one of the most influential cultural interfaces. It counts with 48,052,539,540 interactions with the views category, 1,000,938,300 with subscriptions, 59,276,800 with likes, 13,390,896 with dislikes and 6,416,392 with comments, making a total of 49,232,561,928 interactions with all the variables.

If we compare these results with the previous research of Weilong Yang and Zhensong Qian [26], which was conducted in 2011 and based on YouTube categories, it was found in this current research that the category with more views was "education" instead of "entertainment." However, both research studies identified "music" as the second most viewed category, followed by other categories such as "comedy", "people and blogs" or "film and animation". Similar categories were also found in the top positions by number of views, such as "comedy", "entertainment" or "film and animation". On the contrary, "education" and "gaming" categories have gained more views compared to the previous study.

The fact that the "education" category has 24.5 million, 15 million more views than the second most view category "music" category (Table 1), denotes the great role that the social network YouTube plays in the context of educational development, and as an important digital tool for learning. However, it should be taken into account that when it comes to educational content on YouTube, this is mainly about educational entertainment content, namely edutainment, such as children's videos, which show modeling clay games and educational songs. The majority of videos in the "education" category are Kinder's egg opening and children's songs, including learning techniques for singing songs that use karaoke subtitles. In this way, children develop interactive visual self-taught skills from childhood while learning activities such as modeling clay or traditional children's songs. Thus, parents lead by example and provide the first learning experiences online, by teaching children to develop strategies to use search engines and select suitable and appropriate elements online [32]. Based on this data, the use that parents make of new technologies during the education of their children can be highlighted, using the videos on YouTube as a resource of learning. The use of videos also helps parents to teach and learn a wide variety of practical skills with visual examples, which can be repeated as many times as necessary through the YouTube interface video player. Whilst this category has the longest duration of videos, it has the least number of interactions in the variables of subscribers, likes, dislikes and number of comments, suggesting a low level of interactivity from the user (Table 1), both parents and children give more importance to the content than to the interface extra options.

Regarding the "music" category, the large number of subscribers, totaling around 154 million should be noted, 102 million more subscribers than "education." This shows the higher dynamic consumption of musical content and its ubiquity, in its cultural uses in different places, motivate greater loyalty to certain channels' subscriptions and musical artists. On the other hand, similar to music, is the comedy category, with 153 million subscribers. This denotes the subscribers' attraction to the latest videos of musical and comedy artists, which are some of the most shared contents with other social networks, such as Facebook, WhatsApp or Telegram. In short, the digital cultural habits of sharing the latest music and comedy videos are a very popular interaction habit that motivates larger numbers of subscribers in these categories.

Following the rationale of user interactions through the shared of entertainment contents, it was observed an increase in popularity of gameplays. The "gaming" category, with 165 million subscribers, is the one with the most number of subscribers. This category shows mostly videos of expert video gamers playing Minecraft, or other games such as Fortnite, accompanied by online competitions and commentaries, as if they were football matches, including multiplayer games and others popularly known as LP's (Let's Play). In addition to machinima productions, videos made with online video game animation engineering that help generate hypertextual and transmedia content [33], also support social interaction between channels and YouTube categories, as the video games based on films, and vice versa, demonstrate. Video gamers communities subscribe the most, their videos shows the broad and complex nature of usability, narratives, levels of interaction and immersion that video games allow. Those hidden tricks and new levels can only be solved through the collective experience and interactions that the videos illustrate. For those reasons, gamers are those who are most interested in obtaining the latest updates from those channels on video game advances through their emails, and for improving their quality to compete in tournaments. Compared to "music", "comedy" and gaming", the "pets and animals" and "travel" channels have less subscribers. There is less interest in viewing the last video of a cat flipping or the last trip of a vlogger; therefore, the update that the subscription provides is not as important for its subscribers. However, these channels also form part of the user experience and digital culture habits of sharing videos with related social groups, communities, friends and family since there a great number of subscribers 19 and 9 million, respectively in "pets and animals" and "travel" channels.

At the level of the variable of likes and dislikes, the category "music" has 27 million likes followed by "education" and "comedy" with 4 million likes each respectively. "Gaming" has 3.7 million and "non-profit and activism" with almost 3 million (Table 1). The great difference between "music" and the other categories exposes the degree of influence and interaction that music and its artists have on the tastes and hobbies of contemporary society, music market is the most influential themes representing likes interactions in the current digital culture. It should be appreciated the common denominator between categories, the digital cultural habit of emotionally sharing what one likes through the interface. It could be said that the common feeling of empathy, sharing a common human emotional space and experience, it happens through the user's interaction "likes" with the categories of music, education, humor and comedy, games, animal life and humanitarian campaigns. That emotional connection may have inspired YouTube interface designers to organize the categories based on the likes and similar contents.

In contrast to the above, but following the logic of the empathetic/non empathetic interactions through the "like" or "Do not like", "music" is also the category whose videos are more "disliked', or repudiated by haters. This is a term designated for the person who tends to criticize without argument on social networks. "Music" category accumulates 3,111,000 dislikes. In the case of the "education" category, although it is the most viewed, it has the second closest number of dislikes, 3,049,000, followed by "how to and style" and "show" (Fig. 4). This could be due to the fact that a large part of the interactive consumption that takes place on YouTube is by young people and adolescents who maintain a more critical position on educational contents, they may expect more updated semantic structures and production quality. These categories denote room for potential growth among these audiences. On the contrary, the least hated channels are "cars and vehicles" (87,000), "travel" (54,571) and "sports" (82,025), they are the contents least affected by public opinion, they focused on objects and actions more than in empathy.

There are a large number of people who participate in the YouTube community writing comments as subscribers (Fig. 5). The most commented category is "music", with almost 3 million, commenting mainly emotional over technical characteristics. However, the second most commented category is "how to and style" with 1 million comments, probably because Q&A interactions are most needed in how to descriptions. That is why many of the interactions are not only about the videos but also between the comments of some users and others, generating a richer debate in details than "I like" or "I don't like" in terms of what how to methods are more or less effective based on facts and styles. These comments are invaluable to both users and businesses since they provide data on user's engagement. They also allow prosumers to communicate with the audience and the audience among themselves, generating a sense of community or teams for or against some methods and implementation. For all this, some users are grateful that even negative comments provide them with improvement advice. In fact, in spite of the negative or positive nature of comments, YouTube congratulates users for having a highly commented video. However, these comments can also affect the self-esteem and acceptance of young and adolescents YouTube's users, and they may increase isolation if the proliferating comments are negative [34].

It can be seen that the YouTube categories with the lengthiest videos (Fig. 6), are "education", "shows", "gaming" and "film and animation" in contrast with the "how to and style" or "people and blogs," which are of a shorter duration. Average duration is an important factor both for the user to remain interested in the content and for the time they have to view the content, audiovisual digital consumption. In fact, there is an emotional, intellectual and social negotiation between users and time. To consider this further, and as previously explained, the lengthiest videos, 28 min on average, belong to the "education" category, which is also the most viewed category. The duration is probably tolerated because the investment of time is compensated by the acquisition of knowledge. The second lengthiest, with 22:00 min on average, is the "film and animation" category, which shows that feature and short films are still very present on the platform. The categories "non-profit and activism" follow with 20:23 min on average, where the duration is probably compensated by the socio-political and ethical message that these videos provide. "Show" category has 18:55 min in length on average, and where investment of time is likely justified by the nature of the videos. The videos in the "show" category are usually longer than others, this is because most of their most popular videos are television series or animations that follow the standard of television duration. "Cars and vehicles" videos are, on average, 15:50 min in length due to their documentary nature, followed by "gaming" with an average of 14:55 min, where entire plays are usually shown, being 15 min the medium-duration form of user gameplay experience. On the contrary, according to the data, the videos in the "pets and animals" (03:27 min) and "science and technology" (04:18 min) categories are the shortest. This fact is probably due to the more objective nature and narrative structure of their contents, based on concrete actions, animals or technological demonstrations. To contrast data, a more recent study, based on time average on YouTube categories, shows that the categories with more time average are "gaming" and "film and animation" [35]. This information confirms the emergence of the allowance of users' new digital cultural habits, uploading more traditionally paid entertainment in other paid platforms, united with the interest of YouTube to motivate its paid services since gaming and films are what audiovisual digital users are most willing to pay for. In brief, each video category establishes its own time standards according to its visual styles and narrative structures.

The contents on YouTube are very varied since prosumers can use the platform as they wish, influencing the information society in various ways. The contents that show high doses of innovation stand out because they are capable of establishing new audiovisual narrative genres. In addition, many of the most popular videos demonstrate the relationships between the industry and users in a particular way. For example, it is important to highlight that in the case of "science and technology", most of their most popular videos show short video tutorials with comedy content in which objects are broken to check their resistance. For this reason it may not be too striking that some of the most viewed videos in this category are videos that show the breakdown of the latest iPhone model. The interest in popular usability makes this type of video reach increasing views, offering huge profits to the creators of this content, since it serve the users and potential shoppers interest of checking quality control. Many of the most popular videos found in the YouTube categories expose the relationship and interactions between producers and consumers [36].

5 Conclusions

In conclusion, following the objectives of the study, carrying out a quantitative analysis of videos selected by YouTube categories, has allowed us to obtain a fairly approximate idea of the type of user interactions and their digital cultural habits that are recorded on the YouTube interface. To understand the digital culture dynamics between users and YouTube qualitative complementary analysis will be necessary to contrast the quantitative data. HCI and users' interaction on YouTube are a socio-cultural phenomena that requires a bigger scope of data to fully understand emerging digital cultural habits.

The user experience flow allowed through YouTube interface interactions will express how are the emerging media structures, the producer's mediation and the freedom to rich and meaningful interactive communications within a constantly updated online audiovisual network. Users' interaction can motivates media and interface transformations and can give the online media networks such as YouTube new properties, where the ability to synthesize information, interactivity and non-linear and arbitrary access gives rise to new variants of cultural habits that are still emerging.

The type of user participation in a new multidirectional communication system exposes the levels of interactivity, selective, transformative or constructive, being the more interactive the constructive level, the one that allows the user to become prosumer. Based on the quantitative characteristics of its interface, YouTube video visits interactions are selective because the interface allows to select from the content available to watch. The subscriptions interactions are transformative because the user can modify what date he receives according to several recombinant configurations. The likes and dislikes are selective interactions, focused on expressing simple emotions; and comments are constructive interactions, since each user can generate through textual content whatever they want [37]. From the perspective of the user experience all of these types of interactions generate certain freedom in the forms of HCI and flexibility in the change of communication roles that encourage interactions with YouTube interface.

The multidirectional and interactive media consumption and production encompasses all areas and habits of digital culture interactions. Even social relations and behaviors today represent a mediatized stage. An socio-cultural stage that goes from pure and simple abundance of social media accessibility and democratization of technologies, to the total conditioning of acts and time, having to be fully active and accessible through the cell phone, Facebook, Twitter, Instagram, etc. All passing through the network of globalized and systematized environments of commercial consumption on the internet [38]. The notion of "popular", built by the media and the digital humanities, follows the logic of the market. The popularly massive is what does not remain but in the instant, contrasting with the art and culture that last of ages, it does not accumulate as intergenerational experience, nor does it enrich itself with what has been acquired [39] but experienced in the here and now.

The YouTube ecosystem is constantly reconfiguring itself generating a hypertextual and reinterpretable audiovisual reality. Prosumers build their own mythologies according to categories, establishing their own meaning and contributing with new virtual constellations, new categories, formats and structures. New audiovisual interactive narratives can tell stories that cover all genres and styles, creating new hybrids that overcome universal meaning that evolves generationally. The new type of narrator in the big theater of cyber media interactions has great capacity for expression by being able to modify multiple narrative possibilities and characters, while being prepared to assimilate new content that will later be interpreted and reworked [40]. Future research analyzing the aesthetic and narrative expressions of the productions that take place on YouTube, from an analysis of content to forms and styles, will be the next necessary step for contrasting the quantitative results presented here and to determine emerging digital cultural habits. With all, the milestones for how to develop, create and motivate the evolution of contemporary audiovisual languages and digital cultural habits and consumption will be better understood. These interdisciplinary and comparative quantitative and qualitative studies are necessary to establish the solid theoretical and practical bases of digital humanities so the new generation of HCI users keeps humanizing the social media interfaces.

References

1. Burgess, J., Green, J.: YouTube: Online and Participatory Culture. Polity Press, Cambridge (2018)
2. Welbourne, D.J., Grant, W.J.: Science communication on YouTube: factors that affect channel and video popularity. Public Underst. Sci. 1–14 (2015). https://doi.org/10.1177/0963662515572068
3. Guillespie, T.: The relevance of algorithms. In: Guillespie, T., Boczkowski, P.J., Foot, K.A. (eds.) Media Technologies: Essays on Communication, Materiality, and Society, pp. 167–194. MIT Press, Cambridge (2014)
4. Lange, P.G.: Thanks for Watching. An Anthropological Study of Video Sharing on YouTube. University Press of Colorado, Louisville (2019)
5. Manovich, L.: Software Takes Command. Bloomsburt Academic, New York (2013)
6. Cunningham, S., Craig, D., Silver, J.: YouTube, multichannel networks and the accelerated evolution of new screen ecology. Converg. Int. J. Res. New Media Technol. 1–16 (2016). https://doi.org/10.1177/1354856516641620
7. Jenkins, H.: Convergence Culture. Where Old and New Media Collide. New York University Press, New York (2006)
8. Toffler, A.: The Third Wave. Bantam Books, New York (1980)
9. Brea, J.L.: Las tres eras de la imagen. Imagen-materia, film, e-image. Akal, Madrid (2010)
10. Prada, J.M.: Sampling – collage. EXIT, pp. 120–143 (2009)
11. Sibilia, P.: La intimidad como espectáculo. Fondo de Cultura Económica, Buenos Aires (2008)
12. Barthes, R.: Mitologías. Siglo XXI, Madrid (1999)
13. Keen, A.: The Cult of the Amateur. Doubleday, New York (2007)
14. Simonet, V.: Classifying YouTube channels: a practical system. In: WWW 2013 Companion, pp. 13–17 (2013)
15. Du Gay, P.: Production of Culture/Cultures of Production. The Open University, London (2006)
16. Van Dijck, J., Poell, T., De Waal, M.: The Platform Society. Public Values in a Connective World. Oxford University Press, New York (2018)

17. Vernallis, C.: Unruly Media. YouTube, Music Video and the New Digital Cinema. Oxford University Press, New York (2013)
18. Beaugrande, R.A., Dressler, W.U.: Introducción a la lingüística del texto. Ariel, Barcelona (1997)
19. Maldonado, M.: Texto y comunicación. Fundamentos, Madrid (2003)
20. Prada, J.M.: El ver y las imágenes en el tiempo de internet. Akal, Madrid (2018)
21. Todorov, T.: El origen de los géneros. Teoría de los géneros literarios, Arco, pp. 31–48 (1988)
22. Buckland, W.: Film Theory and Contemporary Hollywood Movies. Routledge, New York (2009)
23. Strinati, D.: An Introduction to Theories of Popular Culture. Reutledge, London (1995)
24. Lipovetsky, G., Serroy, J.: La cultura mundo. Anagrama, Barcelona (2010)
25. Cha, M., Kwak, H., Rodríguez, P., Ahn, Y.Y., Moon, S.: Analyzing the video popularity characteristics of large-scale user generated content system. IEEE/ACM Trans. Networking **17**(5), 1357–1370 (2009). https://doi.org/10.1145/1665838.1665839
26. Yang, W., Qian, Z.: Understanding the characteristics of category-specific YouTube videos (2011)
27. Bärlt, M.: YouTube channels, uploads and views: a statistical analysis of the past 10 year. Converg. Int. J. Res. New Media Technol. **24**(1), 16–32 (2018)
28. Kim, J.: The institutionalization of YouTube: from user-generated content to professionally generated content. Media Cult. Soc. **34**(1), 53–67 (2012). https://doi.org/10.1177/0163443711427199
29. Davidson, J., Liebald, B., Lui, J., Nandy, P., Van Vleet, T.: The YouTube video recommendation system. In: Proceedings of the 2010 ACM Conference on Recommender Systems, pp. 293–296 (2010). http://dx.doi.org/10.1145/1864708.1864770
30. Cheng, X., Dale, C., Liu, J.: Statistics and social network of YouTube videos. In: Proceedings of the 16th International Workshop on Quality of Service, pp. 229–238 (2008)
31. Hall, S.: Estudios culturales: Dos paradigmas. Causas y Azares, p. 1 (1994)
32. Brito, R., Dias, P.: La tecnología digital, aprendizaje y educación; prácticas y percepciones de niños menores de 8 años y sus padres. Revista Ensayos **31**(2), 23–40 (2016). http://dx.doi.org/10.18239/ensayos.v31i2.1001
33. Mora-Fernández, J.: Elementos narrativos que sirven para generar convergencias e inteligibilidad en narrativas transmediáticas o narrativas interactivas lineales. Revista Icono 14, **15**(1), 86–210 (2017). http://dx.doi.org/10.7195/ri14.v15i1.1032
34. Lange, P.G.: Searching for the 'You' in 'YouTube': An analysis of online response ability. In: Ethnographic Praxis in Industry Conference Proceedings, pp. 36–50 (2007). https://doi.org/10.1111/j.1559-8918.2007.tb00061.x
35. Turek, R.: What content dominates on YouTube? (2019). https://blog.pex.com/what-content-dominates-on-youtube-390811c0932d
36. Ardévol, E., Gómez-Cruz, E., Roig, A., San Cornelio, G.: Prácticas creativas y participación en los nuevos media. Quaderns del CAC 34 **13**(1), 27–28 (2010)
37. Moreno, I.: Musas y nuevas tecnologías. El relato hipermedia. Paidós comunicación, Barcelona, España (2002)
38. Baudrillard, J.: La sociedad de consumo. Sus mitos, sus estructuras. Siglo XXI, Madrid (2011)
39. García, N.: Culturas híbridas. Grijalbo, México D.F. (1990)
40. Murray, J.H.: Hamlet en la holocubierta. El futuro de la narrativa en el ciberespacio. Paidós, Barcelona (1999)

The Effect of Social Media Based Electronic Word of Mouth on Propensity to Buy Wearable Devices

David Ntumba and Adheesh Budree(✉)

Department of Information Systems, The University of Cape Town, Rondebosch, Cape Town, South Africa
ntmdav001@myuct.ac.za, adheesh.budree@uct.ac.za

Abstract. There has been an increase in both wearable devices and social media usage. This study sets to describe how social media influences the buying intentions of Wearable Devices (WD). In doing so, constructs from Information Acceptance Model (IACM) had been compared to the resulting themes from 9 one-on-one interview responses and a relationship between social media and wearable devices was formed.

Objectives: Primary objectives were to determine the relationship between social media and; (i) WD sales, (ii) WD features and application information (iii) ongoing interactions on WD's.

Secondary objectives were to determine (i) if the availability of a WD affected the volume of social media interactions and (ii) if the value associated with a WD was related to the social media posts of that device.

Design, Methodology, and Approach: This study took a qualitative approach to research and used phenomenology as a stance of exploring social media's influence on purchase intentions towards the wearable device. A positivist philosophical standpoint was applied to deduce themes from the ICAM by touching on interpretivism elements in analyzing the data collected from semi-structured interviews. The data was then categorized into themes, refined into sub-themes and then assessed according to the research propositions as well as the ICAM frameworks constructs.

Findings: Social media was viewed more of as an information outlet as opposed to a driver of purchase intention. From this, it was seen that word of mouth was the biggest driver in WD awareness. Information Quality, Attitudes towards information were the main influencers that social media has towards purchase intention from the ICAM model. This was evident in the construct scores which showed that there was a relationship between the adoption of social media information and purchase intention, that there is a relationship between the features of a WD and social media usage as well as showing that there is a relationship between the users' perception of the type of social media platform and the WD presented on that platform.

Research Implications: This study explored the influence of social media on the buying intention of wearable devices as well as provided channels for further exploratory research into the human perceptions of information provided on social media.

C. Stephanidis et al. (Eds.): HCII 2020, LNCS 12427, pp. 402–417, 2020.
https://doi.org/10.1007/978-3-030-60152-2_30

Keywords: Social media · E-Commerce · Influence · Electronic Word of Mouth · Wearable devices · Smartwatch · Information Acceptance Model

1 Introduction

Recent growth in social media and the usage thereof have shown a growing trend in online marketing and means of advertisements (Jashari and Rrustemi 2017) With the current online marketing trend and emergence of new technical devices in the forms of smartwatches and fitness bands, also known as wearable devices (WD), have been gained market traction seeing immense adoptions from markets for numerous reasons (Page 2015). Wearable devices have gathered interest as an emerging technology because of its visibility (Wu et al. 2016).

This study aims to describe the influence social media has on individuals' purchase intentions of wearable devices. To do this study has set out to achieve the following objectives:

- Determine the relationship between social media interactions and WD sales,
- Determine whether there is a cycle amounts social media interactions and WD features or applications,
- Determine whether there are specific social media interactions for a WD,
- Determine whether the WD availability affects the volume in social media interactions and
- Determine whether the value associated with the WD is related to social media posts.

2 Literature Review

Wearable devices are IoT artifacts and have various form factors, use specifications (Wairimu and Sun 2018). This study focuses on two WD form factors namely: smartwatches and smart bands which have distinct factors between them (Wu et al. 2016).

These devices are linked to social media applications which are also evident via the gamification of user data in mobile applications (von Entress-Fürsteneck et al. 2019). However, since social media and its influence through its interactions were at the core of this study, only social media platforms had been reviewed.

2.1 Social Media and Electronic Word of Mouth

Social media is a key element in this study and has been categorized in various forms and allowing for communications to be conducted online and the sharing of information between connected parties (Lee 2013).

In reviewing the literature on social media, influence and behavioral literature it was seen that. electronic Word of Mouth (eWOM) explains how information is shared between internet users (Erkan and Evans 2016). Furthermore, eWOM is the widespread of statements and/or communications made by potential, current, former customers about the product or company over the internet (Balakrishnan et al. 2014). Internet-based communications have allowed for a greater reach and audience when compared to traditional forms of word of mouth such as television and newspapers (Chuah et al. 2016). Therefore, social media was noted as an appropriate form of eWOM after the factors of eWOM and social media were discussed (Erkan and Evans 2016).

Social media has been closely related to web 2.0 platforms (Lee 2013). Below is an overall view of the social media components that are interconnected to form social media. The interconnection of their components has led to the phenomenon known as the socialization of information which can be conveyed to a larger global network (Lee 2013).

Figure 1 shows how the components of social interaction, communication media, and content interact with one another and merge to form social media. It further depicts how information can spread and be adopted by viewers. Therefore, the following section will cover the eWOM factors concerning social media as well as its various platforms.

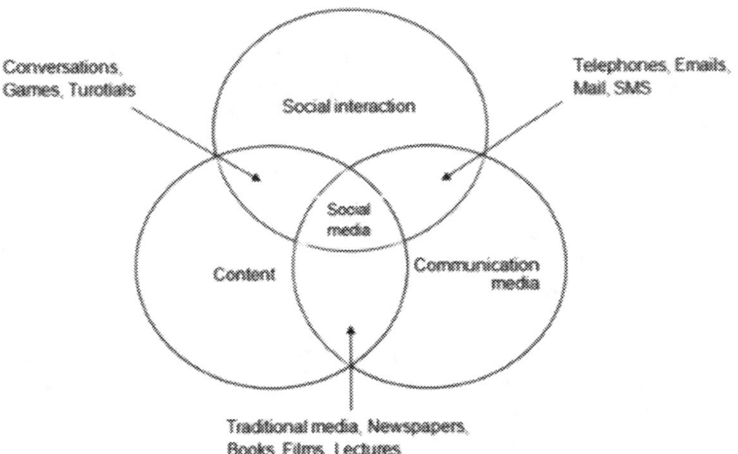

Fig. 1. Social media components (Lee 2013 p. 24)

2.2 Types of eWOM

Electronic Word of Mouth has been noted to be a combination of customer reviews and discussion forums of which involve e-commerce sites (Erkan and Evans 2016). This allows for members of those platforms to create a shared meaning of the brands or products discussed and portrayed and though those with purchase intention may not be able to physically review the device they rely highly on the information passed within eWOM platforms such as online discussions, forums, online customer review sites, blogs, social networking sites, and online brand/shopping sites (Teo et al. 2018).

2.3 Social Media Platforms

The following section explores the social media platforms contributing to eWOM as well as list the defining factors of each of the sites. Social media sites as defined by Lee (2013) may have multiple forms and this study focused on social media literature that covered factors such as influence and buying intention as they form the construct if the ICAM model which will be covered in the following sections. Literature shows that key social media sites are Facebook and Instagram and Twitter.

Facebook. Facebook, created 2004, has a substantial annual growth with registered members generating billions of views daily thus becoming an integral part of social media users' daily activities (Ellison et al. 2007). Interactions and content on Facebook are in both graphical and audio forms which include features such as friends list, walls, pokes, events and chat groups among others. The primary driving feature of Facebook is the friend's list because it allows users to publicly present their content to their connections which can be viewed, shared and further commented on (Nadkarni and Hofmann 2012).

Given the nature of eWOM platforms and Facebook's penetration and ability to generate large streams of content, this study has been directed towards a younger age group as a target population (Lee et al. 2018). Additional studies have also shown that there is a substantial Facebook sign up base within the age group of 12–17 years of age, who once members have high login rates (Pempek et al. 2009).

Firms have also used Facebook within their marketing strategy, using company Facebook pages firms can ensure brand loyalty from their customers who like and view the content on those pages (Lee et al. 2018). Therefore, the firms themselves can generate eWOM content of their products, though few users are noted to engage with them (Lee et al. 2018). Low engagement rates are since Facebook pages act as bulletin boards and users can view the content without necessarily providing any form of response (Pempek et al. 2009). This means that firms cannot rely on the number of shares and reactions on their posts as a true reflection of how many individuals have viewed.

Factors contributing to Facebook usage are due to the delivery of rich content coupled with the daily engagements from users that use their real identities thus removing anonymity and ensuring credibility (Lee et al. 2018).

Instagram. Instagram, founded in 2010, has a primary focus on sharing photographic content amongst users. The shared photos may be edited and beautified to generate a greater attraction and response to them through the online community known as "followers" who are also able to search for one another (Teo et al. 2018). However, Instagram allows for user profiles to be kept private meaning that content shared may only be viewed by authorized or approved individuals (Chen 2018). This provides a different perspective to the credibility of the posts on Instagram as the members viewing or posting the information are known to the sharer of the post to some degree.

Instagram's growth has been attributed to the improvements made in smartphones and tablet cameras to produce high-resolution images for sharing. Like Facebook, firms have also used Instagram for marketing purposes (Teo et al. 2018).

2.4 Internet of Things (IoT) Devices

IoT devices allow for interactions between people, technology and surrounding objects by connecting multiple devices on the internet primarily through cloud computing (Ray 2018) The connection of IoT devices allows for visual, audio and logical information to be collected and the actioning of jobs which includes transferring such data to other devices which is referred to as the devices "talking" to each other for users to make decisions (Al-Fuqaha et al. 2015). IoT also offers market opportunities for vendor manufactures. Although Al-Fuqaha et al. (2015) defines the market opportunities in the health care sector for mobile health, there is no definite mention that this is done using WD.

2.5 Wearable Devices

Wearable devices have been used in fitness tracking, this entails collecting and quantifying user data and offering it in a manner that is useful to the user by using graphics and metrics (Mekky 2014). User data is gathered through the combination of the accelerometer and gyroscope sensors built into the WDs (Kim et al. 2018). The data measured by WD comes in two forms namely; data from physical activities and physiological data. Physical activity data is based on what the user does such as a count of the user steps or calories. Alternatively, physiological data is the user's physiological data which is based on the changes in the users' body is which includes heart rate and temperature (Rupp et al. 2018).

Data from WDs may be used to provide the user of their health data and also drive the gamification of health applications (Lister et al. 2014). In the gamification of health applications, the use of rewards, levels, leader boards, and goal settings increases the attractiveness of respectively performing physical tasks, therefore, promoting users to become more active (Lister et al. 2014).

Wearable devices have the following characteristics:

- A portable device that does not hinder operational use.
- Allow for hands-free usage
- Make use of various sensory and measuring features (camera, GPS) to provide information about the users surrounding environment.
- Provide ambient communication to the user using notifications and alerts
- Constantly gather information about the user and surrounding environment, hence always active.

(Johnson 2014)

Using the above characteristics of WD the following section will explore the literature on smartwatches and smart bands whereby smartwatches literature focuses on purchase intention, usability, adoption, and retention. This forms a basis for the constructs which need to be considered when studying purchase intention. Smart bands have been seen in literature to be noted as fitness trackers and there has been a general health-themed trend in the literature concerning them. Also, in this section, the functions and or uses of both smart bands and smartwatches will be explored.

Smartwatch. Smartwatches, although taking the form of a traditional timepiece, have multiple functions and features that can be linked to smartphones and user accounts (Wu et al. 2016). Smartwatches can work independently of the user's smartphone device or maybe paired with such to all for cross-functionality (Gimpel et al. 2019).

Furthermore, new user experience has been created in the fashion industry trend in which users have preferences for various designs. This illustrates smartwatches, like mobile phones, to be attractive to users based on hedonic (the joy derived from using technology) or functional factors (Wairimu and Sun 2018).

Popular smartwatch vendors include Apple watches, Samsung Galaxy Gears, and Fitbits and are bought by a vast demographic group (Cho et al. 2019). However, the sales growth of smartwatches has seen a steady decline despite adoption studies report that users have a high intention to acquire smartwatches ((Chuah et al. 2016); (Wu et al. 2016)).

Drivers for Intent to Buy Smartwatches. The following are advantages of smartwatches which drive the intention to buy smartwatches:

- The smartwatches' form factor is already socially accepted, and the familiar wristwatch design allows users to feel comfortable wearing them.
- The interface of smartwatches provides a minimalistic approach to the information provided.
- Users can customize the information that is portrayed and, on the smartwatch, and this is further extended by the means of downloading applications.

(Johnson 2014)

Limitations to Intent to Buy Smartwatches. The following are disadvantages of smartwatches which speaks to the decline in smartwatch purchase intention seen in Wairimu and Sun (2018):

- Some designs although innovative are unable to measure all the data required.
- Users may be reluctant to using smartwatches based on the uncertainty and lack of control in what is being done with their data.
- Being an IoT device, users may need to be constantly connected to an internet source for their devices and or applications to process and display their data figures.

(Johnson 2014)

Uses and Adoption. Literature has also shown that customers benefited from the continuous use of smartwatches because they are a novelty technology. Meaning, the more they use it, the more they realize how it fits into their lives which is a gradual process (Nascimento et al. 2018) It has been noted that smartwatch purchases can derive from effective marketing from vendors, by someone else having the device hence WOM eWOM or because of them being an early adopter of smartwatches. This, however, does not guarantee usage of such (Nascimento et al. 2018). However, this study focuses on purchase intention and usage will only be considered when evaluating the eWOM from prior users of WD.

Other literature on smartwatches focused on factors surrounding purchase intention (Hsiao and Chen 2018), continuance intention (Nascimento et al. 2018), behavioral intention (Wu et al. 2016) and adoption with few studies also focusing on factors that influence purchase intention in the context of smartwatches (Hsiao and Chen 2018).

Smart Bands. Smart bands, also referred to as fitness trackers, like smartwatches are worn on the wrist and have similar features as the smartwatch. However, one distinct difference between smart bands and smartwatches is the screen size in which smart bands may have smaller or even no screen at all in the case of the Jawbone. Therefore, the Smart band form factor is that of a bracelet and the smartwatch form factor is that of a wristwatch (Lunney et al. 2016).

Smart band devices primarily track user information with step take, calorie count and work out intensity being the main measures. Research smart bands are still in early development however seeing as the main vendors in smart bands are Misfit, Jawbone, Germainly, and Fitbit with Fitbit also providing smartwatches this study has combined smartwatches and smart bands purchase intentions as WD purchase intention (Kaewkannate and Kim 2016).

3 Research Design

This study was descriptive as it examined the influence social media has on human behavior which, in addition to the (Erkans and Evans 2016) study of social media's influences buying intention through the use of eWOM, focuses on describing the phenomenon of how social media influences buying intention of WD.

A qualitative strategy had been applied within a cross-sectional timeframe whereby data was collected using 1-on-1 interviews and then responses were interpreted into themes that were then categorized and compared to the constructs of the ICAM model as a point to deduce relationships between them.

The target population forms a basis on analyzing the responses that are generated from the interviews and this study has focused on the South African 'youth' group. These individuals between the ages of 18–35 as age group (Bolton et al. 2013). Individuals of this age group, referred to as Generation Y, are highly skilled at using social media as well as ownership of mobile devices (Marketing Charts 2018). Given their high usage and ownership of both social media profiles and mobile devices, as well as being amongst the top percentile of social media users, generation Y stands as a harbinger of how future social media interactions will take place given their early exposure to the technology (Bolton et al. 2013).

In sampling the target population, it was ensured that interviewees own a social media account and are either looking to buy, own or have sold WD in connection to various social media interactions.

In gathering data ethical considerations had been made in such the interview had been performed only once consent had been given and participants were not obligated by any means to respond. Furthermore, the data recorded in the interview were solely for the use of the stated research objectives. Thus, data captured could not be represented in a manner that can identify the individual of which ensured anonymity.

3.1 Research Questions and Objectives

This study had focused on answering the following question:

How do social media influence and individuals' intention in buying a wearable device?

Following the research question was a list of sub research questions which supported the main research question as see below:

OB1. Can the acceptance of social media information be positively related to the consumers' buying intention?
OB2. How does social media introduce or increase the knowledge that users have of a WD?
OB3. Is there a relationship between social media interactions and the features or applications on WD?
OB4. What is the likelihood of users buying WD before and after being exposed to social media?
OB5. Is there a relationship between an interaction made on a given social media site and its following or credibility?

In answering the research question, primary, noted by 'OA', and secondary, noted by 'OB', objectives support the use of research tools, data collection and are listed below.

Primary Objectives

OA1. Determine the relationship between social media interactions and WD sales
OA2. Determine whether there is a cycle amounts social media interactions and WD features or applications
OA3. Determine whether there are specific social media interactions for a WD

Secondary Objectives

OB1. Determine whether WD availability affects the volume of social media interactions.
OB2. Determine whether the value associated with the WD is related to social media posts.

3.2 Research Model and Propositions

This study had made use of the theoretical model from a study focused on social media's influence on customers' purchase intentions through their eWOM interactions known as the Information Acceptance Model (ICAM) (Erkan and Evans 2016). Therefore Fig. 2 below illustrates the ICAM constructs of Information Quality, Information Credibility, Needs of Information, Attitude towards information, Information Usefulness, Information Adoption, and Purchase Intention.

Using the ICAM research model from Fig. 2, the following propositions are made to support the objectives:

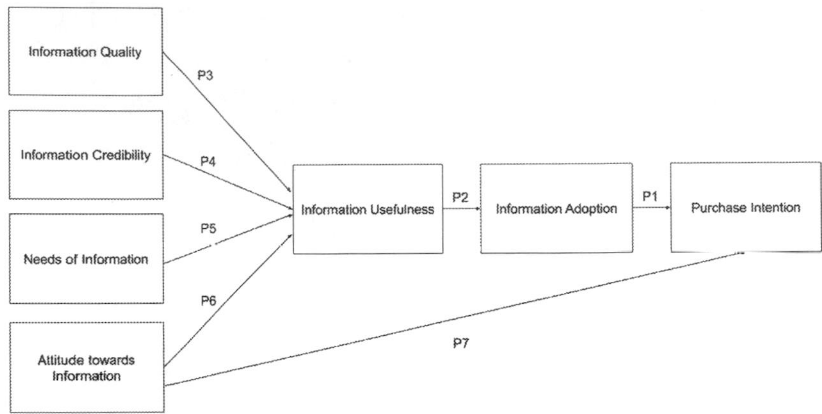

Fig. 2. Information Acceptance Model (Erkans and Evans 2016)

OA1. There is a relationship between the adoption of social media information and purchase intention

OA2. There is a relationship between the usefulness of social media information and the adoption of social media information

OA3. There is a relationship between WD features and social media usage

OA4. There is a relationship between the credibility of social media information and intention to buy WD

OA5. There is a relationship between the number of WD alternatives and the views of social media information

OA6. There is a relationship between the social media platform on the user's perception of the WD presented on that platform.

OA7. There is a relationship between the attitudes of towards social media information and purchase intention

4 Research Analysis and Findings

In undertaking the data analysis this study made use of the Memoing and Axial Coding methods to organize and classify the data that had come in (Bhattacherjee et al. 2012). In analyzing data, it was seen that responses and sentiments from owners of WDs had differed from potential customers.

4.1 Interview Response Themes

From the responses, respondents had elaborated on how various social media had influenced their interests in a WD, however, it was seen that social media was viewed to be merely a channel of information as instead of a driver of purchase intention.

This is supported by respondents expressing that they first have a set goal or reasoning into buying a WD and then proceed to rely on social media to confirm the choice that they had already made. Post validating the IACM model responses from the various interviews were used and the following table of themes was found:

From Table 1 the refined themes are as follows: Reliability, Experience, Word of Mouth, Monetary, Exclusion, Participation, Personal Preference, and Brand Loyalty had arisen.

Table 1. Interview themes

Theme	Brief description
Accuracy	The measure of Information found on a particular site versus the actual WD
Trusted/Credible	Relative to the influence of the site/site interaction has
Hands-on experience	Physically holding the WD
Feature	A technical feature between the WD and/or social media
Posts	An Interaction on social media
Speed	The time is taken for information to be gathered
Usability	How easily can the WD be used
Paid	Any monetary transaction needed to gain information
Value for money	The measure of a potential/existing sale versus the WD and/or its features
Not on social media	Alternative routes of gathering information
Encourage	Positive engagement of social media
Achievement	The completion of a task or goal set out by social media or WD
Competition	The ranking of achievements
Reasoning	The motive behind reviewing a WD
Alternative	In place of WD or social media
Awareness	Increase in audience
Comments	Posts made in response
Word of mouth	The spread of information through experience
Bias	Subjective posts
Personal Preference	Personal subjective views on social media and/or WD
Brand loyalty	Personal preference regarding a specific vendor(s)

In unpacking these themes it was seen that ‘Reliability’ encompassed the respondents view that WD information on social media was accurate, credible and trustworthy, had been retrievable promptly (relating to the ‘Speed’ theme) as well as being usable this was noted in the responses whereby it was said “Ï would first look around on Twitter to check on the device” also another respondent said, “I like to look at reviews to get an idea of the smartwatch”.

Word of Mouth. Word of Mouth covered ‘Posts’, ‘Alternative’, ‘Awareness’, ‘Comments’ and ‘Word of Mouth’ themes related to the social media information itself and respondents’ views on such. Phrases such as “I would spend more time on social media if there were posts on different devices” and “I have become more aware of devices on social media” support the Word of Mouth Theme.

Experiences. The experience was closely tied to how social media information allowed the user to know the WD without necessarily owning it. The grouped themes here were: ‘Hands-on Experience’ and ‘Feature’. Here respondents have mentioned that the use of Youtube reviews aided them in knowing the WD before deciding.

Monetary Value. Monetary themes such as ‘Paid’ and ‘Value for Money’ indicated that further factors needed to be considered in intention to buy a WD. One respondent mentioned, “I was going to get the higher Fitbit model, but it was too expensive”.

Exclusivity. Exclusion reflected cases whereby the respondent was not on a social media site and or their intention to buy a WD was not influenced by social media information. Responses such as “I will go to the store” and “I will check the company website” were used.

Participation. Participation was seen in the ‘Encourage’, ‘Achievement’ and ‘Competition’ themes which showed that engagement in social media information in terms of liking and reposting views affected the respondents’ intentions to buy the WD. Key phrases such as “I would post my steps on Facebook” and “I have joined the group” were noted in support of the Participation theme.

Personal Preference. Personal Preference covered ‘Reasoning’ and ‘Personal Preference’ themes and related to how the user felt about a WD and the social media information from their backing and ideals. Here respondents noted, “I would like to research before making a decision”.

Brand Loyalty. Brand Loyalty grouped ‘Bias’ and ‘Brand Loyalty’ themes together as respondents have shown that the choice of the WD had been impacted by the smartphone that they have been using as well as the brand that they have become accustomed to. Here the phrases “I only look at Samsung Products” as well as “This Smartwatch pairs well with Samsung Phones” illustrated how brand loyalty and bias were linked.

4.2 Summary of Themes

‘Word of mouth’ (WOM) was the biggest driver in WD awareness. This WOM would be based primarily on the smaller social groups be it the forums or the direct WhatsApp groups themselves. This relates to the IACM model whereby factors such as information usefulness and attitude towards usage are called. However, it was also noted through the responses that although social media is an influence to WD sales it does not represent the final sales transaction itself, rather it acts as a ‘self-advertising’ means of contact whereby the users can acquire as much information of the WD as they require, which again relates to the IACM model through needs for information. This is further supported by Fig. 3 below where word of mouth had the highest occurrence of themes.

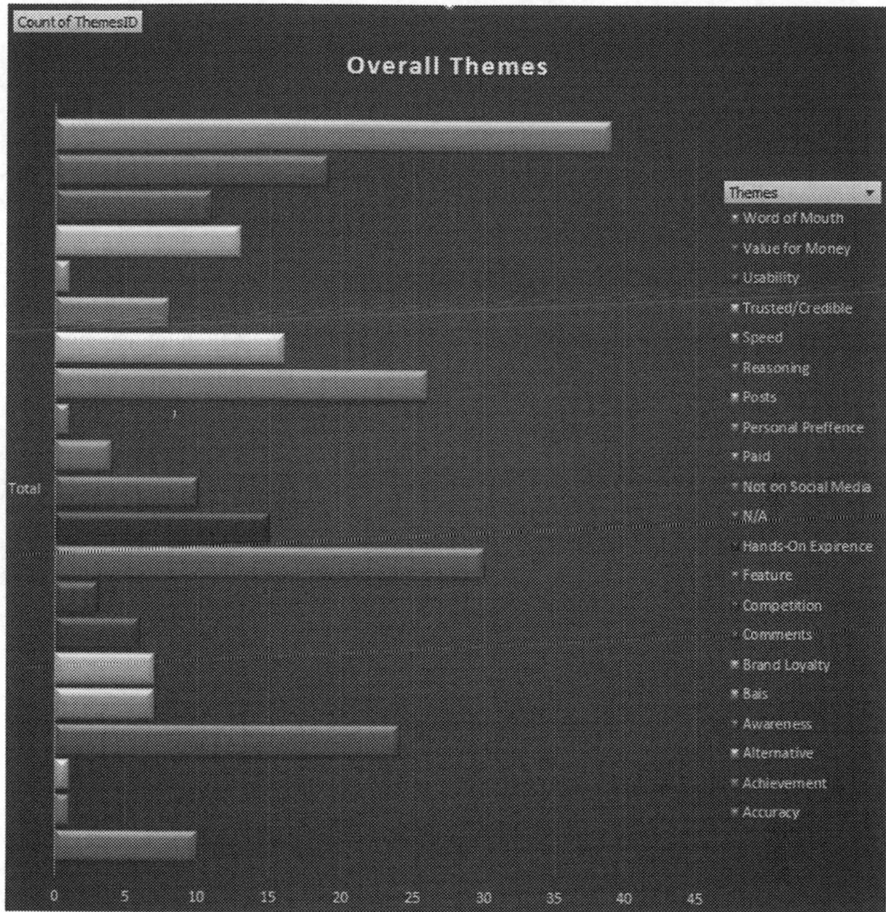

Fig. 3. Theme graph

Concerning the word of mouth, it was seen that themes relating to device features, personal preference, and awareness are closely coupled. This is because the respondents have sought information on social media after knowing or seeing the product in prior posts, therefore, social media aids in retaining or intensifying the users' awareness will and/or desire to buy the WD as well as aids in comparing various device features.

Another phenomenon noticed through the responses was that of brand loyalty. Therefore, a user had a smartphone of a brand then their searches on social media will be driven/filtered by that brand. Therefore, the influence of social interactions will be driven by the members closer to the brand. This brings about the issue of the source which drives the influence on the Social media platform. According to the responses the individual or company plays a significant role in attracting then to a specific WD, these may range to experts in the field, reviewers critiques or even the company themselves this then assures the users that the information on the Social media site is of quality and credible source.

Although many comments are seen to be biased, responses have shown that Social media can only provide direction as to which the users ‘should’ buy but cannot directly affect the buying decision of the users due to external factors such as price, ergonomics as well as availability.

Key social media platforms where noted as Facebook, Instagram and Youtube with Instagram noted to be the main drivers in allowing users to have an almost tangible experience with the WD.

Lastly, it was noted that social media had aided the users in viewing and discussing the usefulness of the devices by a measure of their features. Should not all the features be met, this will form a loop between the users searching through alternatives, discussing them and concluding that WD does not match their budget, brand or features required.

4.3 Summary of Findings

The primary and refined themes were linked to the ICAM constructs and summary of the primary and refined themes per construct is seen in the table below.

Table 2. Count of constructs

Construct	Number of primary themes	Number of secondary themes
Information quality	2	2
Source credibility	1	1
Needs for information	4	2
Attitude towards Information	4	2
Perceived usefulness	1	1
Information adoption	6	2
Purchase intention	4	2

Table 2 showed that the Information adoption construct has the highest count of themes. This supported the Erkans and Evans study where it was seen that information would more readily adopt due to viewers already deeming the source from which it originates as useful whereby “Since people usually receive the eWOM information from their friends and acquaintances in social media, they may already think that the information will be useful” (Erkans and Evans 2016 p. 52).

Information quality and information usefulness were considered to have a positive impact on information usefulness (Erkans and Evans 2016 p. 52). However, Unlike the (Erkans and Evans 2016) study information quality and information usefulness do not have a direct impact on information usefulness rather it is seen that the needs for information has more overlapping themes. This supports the (Teo et al. 2018) study that individuals require information on the items they intend to purchase.

5 Conclusion

Initially, this study had aimed to describe how social media influences individuals to purchase intentions towards wearable devices as outlined by the (Erkans and Evans 2016) literature. In the review of literature, it was seen that there has been an intent to acquire WD's (Chuah et al. 2016; Kaewkannate and Kim 2016; Wu et al. 2016; Hsiao and Chen 2018). However, there was little reference as to what role social media plays in doing so. From that gap in the literature, this study had applied the Information Acceptance Model and had deduced that social media influence is derived from its ability to produce large amounts of information.

Furthermore, from the responses, it was seen that individuals had other external influences as well as internal subjective norms for intending to buy a wearable device. These norms and influence were categorized within an array of resulting themes such as 'Value for Money' and 'Reasoning'. However, it was seen that of those influenced by social media, a significant amount has been deduced to intending to buy a wearable device due to being exposed to the electronic word of mouth found on those social media sites. Emerging themes such as 'Personal Preference' and 'Brand Loyalty' were seen to affect the type of exposure that individuals would see on social media sites whereas, on the other hand, other individuals had taken to spending more time on social media to find the best possible WD to buy.

By monitoring relevant social media platforms, it was seen that the interactions on social media had influenced the perceptions of users towards a WD and therefore their intention to buy it. These interactions were also seen to be more effective when related to social media groups which supported the (Erkans and Evans 2016) literature in that information from friends on social media sites is useful and is easily adopted.

Therefore, the ICAM constructs of the relationship between social media information adoption and purchase intention, the relationship between WD features and social media usage as well as the relationship between the social media platform and the users' perception of that platform stood out as high contributors to social media influence.

This study's value was derived from providing an understanding of how vendors, organizations and individuals can market their WD products as well as how consumers who intend on buying the devices are influenced by social media. Moreover, this study served as the basis for further means of examining findings by applying various contexts and methodologies for social media influence.

Overall it was deduced from the objectives and prepositions that this study had managed to determine a link between Social media and WD buying intention though the over-exposure and advertising of such, this does not necessarily represent a definite sale.

References

Al-Fuqaha, A., Guizani, M., Mohammadi, M., Aledhari, M., Ayyash, M.: Internet of Things: a survey on enabling technologies, protocols, and applications. IEEE Commun. Surv. Tutorials **17**(4), 2347–2376 (2015)

Balakrishnan, B.K., Dahnil, M.I., Yi, W.J.: The impact of social media marketing medium toward purchase intention and brand loyalty among generation Y. Procedia Soc. Behav. Sci. **148**, 177–185 (2014)

Bhattacherjee, A.: Social Science Research, 2nd edn. Anol Bhattacherjee, Tampa, Florida (2012). https://open.umn.edu/opentextbooks/BookDetail.aspx?bookId=79

Bolton, R.N., et al.: Understanding generation Y and their use of social media: a review and research agenda. J. Serv. Manage. **24**(3), 245–267 (2013)

Chen, H.: College-aged young consumers' perceptions of social media marketing: the story of instagram. J. Curr. Issucs Res. Advert. **39**(1), 22–36 (2018)

Cho, W., Lee, K.Y., Yang, S.: What makes you feel attached to smartwatches? The stimulus–organism–response (S–O–R) perspectives. Inf. Technol. People **32**(2), 319–343 (2019)

Chuah, S.H., Rauschnabel, P.A., Krey, N., Nguyen, B., Ramayah, T., Lade, S.: Wearable technologies: the role of usefulness and visibility in smartwatch adoption. Comput. Hum. Behav. **65**, 276–284 (2016)

Ellison, N.B., Steinfield, C., Lampe, C.: The benefits of Facebook "friends:" social capital and college students' use of online social network sites. J. Comput. Med. Commun. **12**(4), 1143–1168 (2007)

Erkan, I., Evans, C.: The influence of eWOM in social media on consumers' purchase intentions: an extended approach to information adoption. Comput. Hum. Behav. **61**, 47–55 (2016)

Gimpel, H., Nüske, N., Rückel, T., Urbach, N., von Entreß-Fürsteneck, M.: Self-tracking and gamification: analyzing the interplay of motivations, usage and motivation fulfillment (2019)

Hsiao, K., Chen, C.: What drives smartwatch purchase intention? Perspectives from hardware, software, design, and value. Telematics Inform. **35**(1), 103–113 (2018)

Jashari, F., Rrustemi, V.: The impact of social media on consumer behavior–case study Kosovo. J. Knowl. Manage. Econ. Inf. Technol. **7**(1), 1–21 (2017)

Johnson, K.M.: Literature review: an investigation into the usefulness of the smart watch interface for university students and the types of data they would require (2014)

Kaewkannate, K., Kim, S.: A comparison of wearable fitness devices. BMC Public Health **16**(1), 433 (2016). https://doi.org/10.1186/s12889-016-3059-0

Kim, S., Lee, S., Han, J.: StretchArms: promoting stretching exercise with a smartwatch. Int. J. Hum.-Comput. Inter. **34**(3), 218–225 (2018)

Lee, D., Hosanagar, K., Nair, H.S.: Advertising content and consumer engagement on social media: evidence from Facebook. Manage. Sci. **64**(11), 5105–5131 (2018)

Lee, E.: Impacts of social media on consumer behavior: decision making process. Turun ammattikorkeakoulu (2013). https://www.openaire.eu/search/publication?articleId=od_1319::728109eb8b479459b8eb5be8470655f4

Lister, C., West, J.H., Cannon, B., Sax, T., Brodegard, D.: Just a fad? Gamification in health and fitness apps. JMIR Serious Games **2**(2), e9 (2014)

Lunney, A., Cunningham, N.R., Eastin, M.S.: Wearable fitness technology: a structural investigation into acceptance and perceived fitness outcomes. Comput. Hum. Behav. **65**, 114–120 (2016)

Marketing Charts. (2018). Tech update: mobile & social media usage, by generation. https://www.marketingcharts.com/demographics-and-audiences-83363

Mekky, S.: Wearable computing and the hype of tracking personal activity. In: Paper Presented at the Royal Institute of Technology's Student Interaction Design Research Conference (SIDeR) Stockholm, Sweden, pp. 1–4 (2014)

Nadkarni, A., Hofmann, S.G.: Why do people use Facebook? Personality Individ. Differ. **52**(3), 243–249 (2012)

Nascimento, B., Oliveira, T., Tam, C.: Wearable technology: what explains continuance intention in smartwatches? J. Retail. Consum. Serv. **43**, 157–169 (2018)

Page, T.: Barriers to the adoption of wearable technology. I-Manager's J. Inf. Technol. **4**(3), 1 (2015)

Pempek, T.A., Yermolayeva, Y.A., Calvert, S.L.: College students' social networking experiences on Facebook. J. Appl. Dev. Psychol. **30**(3), 227–238 (2009)

Ray, P.P.: A survey on Internet of Things architectures. J. King Saud Univ. Comput. Inf. Sci. **30**(3), 291–319 (2018)

Rupp, M.A., Michaelis, J.R., McConnell, D.S., Smither, J.A.: The role of individual differences on perceptions of wearable fitness device trust, usability, and motivational impact. Appl. Ergon. **70**, 77–87 (2018)

Teo, L.X., Leng, H.K., Phua, Y.X.P.: Marketing on Instagram: social influence and image quality on perception of quality and purchase intention. Int. J. Sports Market. Sponsorship **20**, 321–332 (2018)

von Entress-Fürsteneck, M., Gimpel, H., Nüske, N., Rückel, T., Urbach, N.: Self-tracking and gamification: analyzing the interplay of motivations, usage and motivation fulfillment (2019)

Wairimu, J., Sun, J.: Is smartwatch really for me? An expectation-confirmation perspective (2018)

Wu, L., Wu, L., Chang, S.: Exploring consumers' intention to accept smartwatch. Comput. Hum. Behav. **64**, 383–392 (2016)

Effective Online Advertising Strategy

Marc Oliver Opresnik(✉)

Technische Hochschule Lübeck, Public Corporation, Mönkhofer Weg 239, 23562 Lübeck, Germany
Marc.Oliver.Opresnik@TH-Luebeck.de

Abstract. The Internet has changed the way people, organizations and institutions communicate. Accordingly, advertising planning is undergoing a dramatic change from traditional ATL communication tools such as newspapers and magazines to non-traditional BTL tools such as mobile and Internet marketing.

Marketers can use online advertising to build their brands or to attract visitors to their Web sites. Online advertising can be described as advertising that appears while customers are surfing the Web, including banner and ticker ads, interstitials, skyscrapers, and other forms.

An effective advertising strategy for online advertising aims at targeting the right advertisement message to the right person at the right time.

Keywords: Social media marketing · Marketing planning · Marketing management · Web 2.0 · Marketing 4.0 · Integrated marketing communication · Social computing · Social media

1 The Evolution of Online Marketing

The Internet has changed the way people, organizations and institutions communicate. Accordingly, media planning is undergoing a dramatic change from traditional ATL communication tools such as newspapers and magazines to non-traditional BTL tools such as mobile and Internet marketing. Figure 1 displays that mobile and desktop Internet adspend in Germany is already accounting for nearly 40% of the adspend market in 2019 [1].

A major strength of direct e-mail is its ability to qualify leads. Appropriate software allows the firm to track who is reading and responding along with the types of responses. This enables the firm to segment the audience accordingly, targeting future communications based on recipients' self-reported priorities.

A checklist for launching a successful e-mail marketing campaign includes the following aspects [2]:

Solid planning. Companies are required to have clear and measurable objectives, and they must carefully plan their campaign.

Excellent content. Standards are higher with e-mail, so firms have to make sure they are offering genuine value to the sub-scriber.

C. Stephanidis et al. (Eds.): HCII 2020, LNCS 12427, pp. 418–424, 2020.
https://doi.org/10.1007/978-3-030-60152-2_31

Appropriate and real 'from' field. This is the first thing recipients look at when they are deciding whether to open an e-mail.

Strong 'subject' field. The next place recipients look before deciding whether to open an e-mail is the subject field. Therefore, it needs to be compelling.

Right frequency and timing. Organizations must not overwhelm their audience. They are not supposed to send e-mail Friday through Monday or outside of normal business hours.

Appropriate use of graphics. Businesses should not get carried away. If graphics add real value and aren't too big, they could be used.

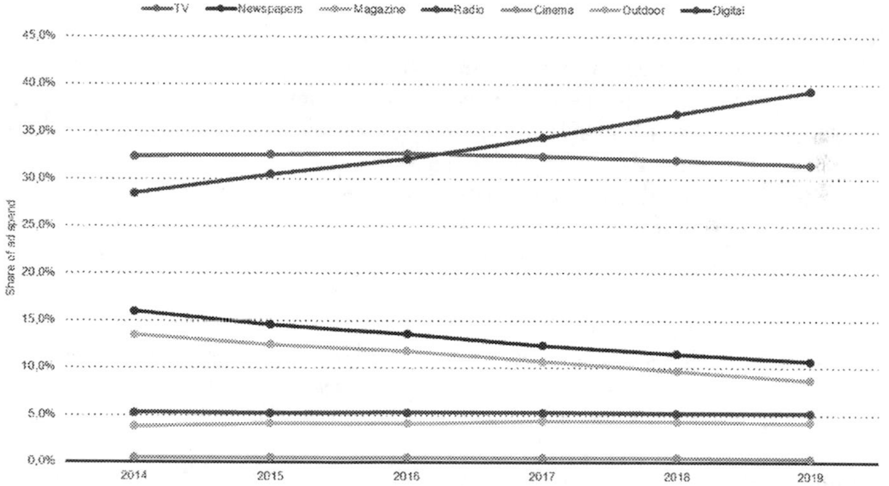

Fig. 1. Development of the media mix in advertisement 2014–2018 in Germany and forecast until 2019 (compared to the previous year) (Source: Statista, 2019)

Lead with company's strength. Companies should not bury the best content or offer. They need to ensure it is at the top or at the e-mail equivalent of 'above the fold'.

Shorter is better. Nobody reads a lot these days, and they read less in e-mail than anywhere else.

Personalize. Marketers should use just three or four elements of personalization, and response rates can potentially improve by 60%. They should try to go beyond just the first name and learn about the subscribers.

Link to company's web site. This is where the richness of content and interactivity can really reside. Marketers should tease readers with the e-mail so they will link to the Web site. Advertising can also be incorporated, serving the same role as the initial e-mail: to create a desire in the audience for more information. The Web site catch page is crucial to this tactic and is often where many people falter when integrating traditional advertising with online promotions.

Measure and improve. The ability to measure basics such as open and click-through rates is one of the main advantages of e-mail marketing, but companies should not stop there. They should also track sales or other conversions and learn from what works and make necessary adjustments.

Web 2.0 websites allow you to do more than just retrieve information, as this was mainly the case with Web 1.0. Web 2.0 transforms broadcast media monologues (one-to-many = Web 1.0) into social media dialogues (many-to-many). The term Web 2.0 was first used in 2004 to describe a new way software developers and end-users started to utilize the internet to create content and applications that were no longer created and published by individuals, but instead continuously modified by all users in a participatory and collaborative fashion. The popularity of the term Web 2.0, along with the increasing use of blogs, wikis, and social networking technologies, has led many in academia and business to work with these 'new' phenomena. For marketers, Web 2.0 offers an opportunity to engage consumers. A growing number of marketers are using Web 2.0 tools to collaborate with consumers on product development, service enhancement and promotion. Companies can use Web 2.0 tools to im-prove collaboration with both its business partners and consumers. Among other things, company employees have created wikis, which are Web sites that allow users to add, delete and edit content, and to list answers to frequently asked questions about each product, and consumers have added significant contributions. Another Web 2.0 marketing feature is to make sure consumers can use the online community to network among themselves on content that they choose themselves. Besides generating content, the Web 2.0 Internet user tends to proactively bring in a whole new perspective on established processes and approaches, so that the users create innovative ideas for the future development of companies [3].

With the creation of the World Wide Web and Web browsers in 1990 s, the Internet was transformed from a mere communication platform into a certifiably revolutionary technology. For consumers, digital technologies have not only provided the means to search for and buy products while saving time and money, but also to socialize and be entertained. The emergence of social networking sites such as MySpace and Facebook has enabled consumers to spend time socializing, and the development of video streaming and music downloads means that they can be entertained as well. A major challenge for marketers is to tap in to the huge audiences using the net.

The Internet is a global channel of communication, but the advertising messages are often perceived in the local context by the potential customer. Herein lays the dilemma that often causes the results from internet promotion to be less than anticipated.

Traditional media have two capabilities – building brands and direct marketing. In general, most promotional forms are useful for one or the other. The internet however, has the characteristics of both broadcast mass media and direct response advertising.

In the conventional model of communications in the marketplace, there are clear distinctions between the sender, the message and the recipient, and control of the message is with the sender. In 'market space', control of the message is shared between sender and receiver because of the interactivity of the medium, its ability to carry a message back in reply to that sent, and the impact of the information technology on time, space and communication. The above stated impacts on the feedback loop are built into the Internet and on the aspects of interference. In general, interference is more likely to be from internet clutter and less from external sources.

The web represents a change away from a push strategy in international promotion, where a producer focuses on compelling an intermediate to represent the products or services or a distributor to stock its goods, towards a pull strategy in which the producer communicates directly with the customer. In this transition process, promotional costs and other transaction costs are reduced. The differentiating feature of the Internet from other promotional vehicles is that of interactivity. This results in the special feature that Internet combines the attributes of both selling and advertising. Interactivity facilitates a completely innovative approach to reaching potential customers. Unlike television, for example, where the consumer passively observes, with the web there is an active intent to go onto the Internet and more attention to content as a result. In the Internet, the potential customer has a high involvement approach to advertising. A continual stream of decisions is demanded from the user. Each click represents a decision and therefore the web is a very high involvement medium. In addition, unlike traditional media, the web is a medium by which the user can click through and obtain more information or purchase the product. Web advertisements can and are often targeted to a user profile that in turn affects the way the message will be received. Increasingly, the ads displayed on the web are specific to user interests and appear as these interests are revealed while the user navigates the web [3].

2 Effective Online Advertising Strategy

Marketers can use online advertising to build their brands or to attract visitors to their Web sites. Online advertising can be described as advertising that appears while customers are surfing the Web, including banner and ticker ads, interstitials, skyscrapers, and other forms [4].

An effective advertising strategy for online advertising aims at targeting the right advertisement message to the right person at the right time [5].

Who to Advertise to?

Is online advertising for everyone? Knowledgeable marketers will state that advertisement design depends on the type of product or service be-ing sold and the desired target segment. In this respect, it is instrumental to divide the desired target segment according to first-time visitors to the company's Web site, registered users, and general information seekers. There is bound to be some overlap across these segments. However, this form of segmentation can provide useful insights while designing online advertising. Based on the user segment, the Web site can be programmed to respond appropriately. For example, every first-time visitor to a Web site can be made to see the same advertisement. Visitors identified as information seekers may be shown useful content instead of products and services directly, and registered users may see a customized advertisement message based on their profiles. Technologically, it is feasible to identify the type of user by studying their browsing behaviour through clickstream data and by using 'cookie' files [3].

How to Advertise?
After identifying the user or the Web site visitor, the next step is deter-mining how to advertise or what format to use for advertising. There are several different formats of Internet advertisements: banner ads (which move across the screen), skyscrapers (tall, skinny ads at the side of a Web page) and interstitials (ads that pop up between changes on a Web site).

Content sponsorships are another form of Internet promotion. Many companies achieve name exposure on the Internet by sponsoring special content on various Web sites, such as news or financial information. These sponsorships are best placed in carefully targeted sites where they can offer relevant information or service to the audience.

The type of advertisement chosen should be directed toward not only 'pushing' the message across but also 'pulling' the customer to click deep-er into the Web site by designing ads that contribute to the overall Web site experience. For example, a Web site with too many pop-up ads on the first page runs the risk of driving the user away [3].

What to Advertise?
People use the Internet to seek information as well as products and ser-vices. Marketers can be creative and design advertisements that could just give out helpful information to the user. For example, a user browsing for a digital camera may be offered useful tips and pointers on how to get the best results from digital photography. Non-commercial advertising like this may not have a short-term financial gain but may contribute to superior browsing experience leading to customer loyalty and repeat visits from the user. If customer profile or history of purchase is known, it is possible to predict future purchase behaviour and companies can pro-gram buying information in the Web site code. The next time the company's Web site detects a particular user returning to the Web site, there will be an advertisement ready with an appropriate and tailored content. If deployed properly, this approach can help marketers cross-selling products through combinations of online advertisement messaging [3].

When to Advertise?
The first three dimensions of the advertising strategy discussed so far would be rendered ineffective if the timing is not right. In the case of offline media, one can proactively call up the customer or send him/her a direct mailer at a specific time with a customized advertising message. However, these rules do not apply online. In the case of the Internet, users may decide to go online and visit the Web site during work, in the middle of the night, or whenever they want to. Therefore, timing in the Internet context would refer to the time from the instance a user is detected online.

The question is when to activate the advertisement. As soon as the user comes online, after he/she has browsed for a while, or at the time of the first purchase? Studies conducted with Internet ad timings have indicated that generally response (click-through) to pop-ups is greater when the ad appears immediately after the user enters the site. However, the results could vary greatly depending on the user segment and the user's information-seeking purposes.

Amazon.com employs a subtle form of advertisement in real time. Basically, while performing a search for a particular book, the search also throws up a list on the side or bottom of the page of relevant books that may complement the book the user was originally considering purchasing. Amazon.com was first to use 'collaborative filtering' technology, which sifts through each customer's past purchases and the purchasing patterns of customers with similar profiles to come up with personalized site content. Furthermore, the site's 'Your Recommendations' feature prepares personalized product recommendations, and its 'New for You' feature links customers through to their own personalized home pages. In perfecting the art of online selling, Amazon.com has become one of the best-known names on the Web [3].

Where to Advertise?
It is crucial to make Internet ads visible at vantage points to maximize their hit-rate with the intended target segment. Unlike other forms of media, where one can pick a well-defined spot within a finite set of possibilities, cyberspace offers an infinite number of possibilities across thousands of portals, search engines, and online publishers, as well as multiple possibilities within the vendor's Web site. Finding the perfect spot may seem like finding a needle in a haystack.

There are two ways to tackle this. The first is the easy way out. Follow intuition and place advertisements at obvious locations, such as frequently visited portals and search engines. However, this is not a costeffective solution. A more refined approach involves analyzing the browsing pattern of an Internet user on a company's Web site using the Web site's log files. Analysis of the log files can help model the browsing behaviour of a random visitor to the Web site. Based on this information, Internet ad displays may be placed at appropriate locations. Marketing managers can also leverage this model to sell complementary products to potential users. For example, a department store such as Marks & Spencer may advertise cosmetics on the page where a user is buying fragrances online. An electronics store like Best Buy may advertise the latest CD releases on the page listing different audio systems.

However, this form of analysis is limited to advertising within the company's Web site. A more advanced research approach involves modelling browsing behaviour at multiple Web sites using clickstream data. Information analyzed in this manner renders a total view of a customer's online habits before purchase consideration. Such information is invaluable to marketers who would be interested in knowing when and where they're most likely to find their potential customers and, based on that information, how they should place the Internet advertisements to pull the relevant customers to their site [3].

3 Conclusion

As digital communication becomes an increasingly dominant way for people exchange and share information, an effective online advertising strategy becomes an essential tool for any company and organization. The process and key questions of creating an appropriate strategy will help organizations and companies clarify what they want to achieve, understand how to engage their target market online and outline the key activities they need to take to market their business [6].

References

1. Statista.com. https://de.statista.com/statistik/daten/studie/380715/umfrage/werbeausgaben-in-deutschland-prognose-nach-segmenten/. Accessed 22 Nov 2019
2. Linkon, N.: Using e-mail marketing to build business, TACTICS, November, p. 16 (2004)
3. Hollensen, S., Opresnik, M.: Marketing: Principles and Practice, 2nd edition. Lübeck (2018)
4. Kotler, P., Armstrong, G., Opresnik, M.: Marketing: An Introduction, 14th ed. Berkshire, Omaha (2019)
5. Kumar, V., Shah, D.: Pushing and pulling on the internet. Market. Res. **16**(1), 28–33 (2004)
6. Kotler, P., Hollensen, S., Opresnik, M.: Social Media Marketing – A Practitioner Guide, 3rd ed. Lübeck (2019)

Gender- and Diversity-Oriented Design of Social Media for Participation in Public Transport

Cathleen Schöne(✉), Tobias Steinert, and Heidi Krömker

Technische Universität Ilmenau, PF 10 05 65, 98684 Ilmenau, Germany
{cathleen.schoene, tobias.steinert, heidi.kroemker}@tu-ilmenau.de

Abstract. The aim of the paper is to define design recommendations for gender- and diversity-sensitive social media appearances to strengthen participation in the context of local public transport and mobility planning. Social media are an effective instrument for actively and sustainably involving citizens and other interest groups in planning and evaluation processes [1]. A major advantage is its reach across gender, cultural, language and national boundaries, which means that a high diversity of users can be covered. For this very reason, social media have become one of the most important tools when it comes to implementing and promoting participation culture effectively and efficiently. Social media can not only be used to inform interested parties, but also to actively and above all interactively involve them. In order for this to happen, the needs and requirements of users must be taken into account when designing social media appearances. In addition to flexible user interface design and content design, this includes enabling and providing elements for an interactive exchange of information, opinions and experiences between providers and users as well as among users. Although this type of participation is already widespread (e.g. as marketing for companies and products, but also for political opinion-forming), it has not yet fully entered all areas. Especially in the transport sector, the social media appearances provide information, but often these are not designed to be gender- and diversity-transcending and therefore not understandable and usable for all target groups. Every day, countless and diverse people are moved and connected with each other over long distances via public transport, but this user diversity is not addressed and reflected on the social media appearances of public transport companies.

Keywords: Public transport · Social media · Gender · Diversity · Participation · Design elements

1 Introduction and Research Question

In the context of this paper, design recommendations are to be defined which can be used to determine the extent to which social media appearances enable as many users as possible to participate. The paper focuses on the transport sector, more precisely on the social media appearances of public transport companies. This focus results from the

C. Stephanidis et al. (Eds.): HCII 2020, LNCS 12427, pp. 425–443, 2020.
https://doi.org/10.1007/978-3-030-60152-2_32

participation of the authors in the TInnGO project, in which a pan-European knowledge platform (an observatory) for gender- and diversity-specific intelligent transport innovation will be developed. The TInnGO-platform will be used for data collection, analysis and dissemination of gender mainstreaming tools and will provide space for open innovation. Therefore, partners from 13 EU countries form 10 Hubs. The individual hubs deal with a variety of different current gender and diversity specific challenges in the transport ecosystem as well as the mobility needs of women. The German Hub specialized on gender- and diversity-sensitive participation culture. This of course also includes possible participation in the mobility planning process via social media (Fig. 1).

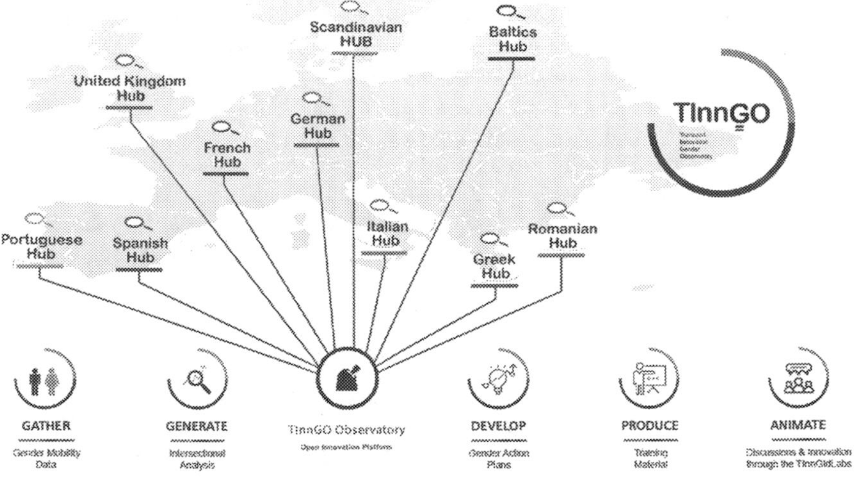

Fig. 1. The TInnGO-project: www.tinngo.eu

The concepts of gender and diversity are used to perceive persons in all their variations and situations in life. The resulting multitude and diversity of user groups has different needs and requirements with regard to the use of social media. For people with physical disabilities such as blindness or visual disorders in general, a different context of use applies than for people without disabilities. In order to meet these needs and requirements, there are already various standards and guidelines that recommend elements for gender- and diversity-oriented design of web content. However, the extent to which these have been complied with and are made available on the respective social media platforms varies. It has to be said that not only the platform's own elements and their utilisation can contribute to a gender- and diversity-oriented appearance, but also the providers themselves by means of a target group-oriented design and formulation of the posts. As participation in the mobility planning process via social media is to be investigated, a selection of social media appearances of international (EU) and national

(Germany) public transport companies will be examined in terms of the previously extracted recommendations and evaluated to what extent they are gender- and diversity-oriented, so that access, understanding and exchange of the content is guaranteed for the greatest possible number and diversity of users.

2 Related Work

On the basis of the foregoing explanations, it can be seen that in the context of this paper four fields of impact – Gender and Diversity, Participation, Social Media and specifications for UI and content – converge. In the following, these are briefly explained so that the interaction of all four can be understood in relation to the objective.

2.1 Gender and Diversity

Since the 1970s at the latest, concepts such as Gender and Diversity have increasingly appeared in the various fields of science. Over the years, their understanding has been consolidated and they have gained importance, especially in today's world. But it is only relatively recently that they have begun to play a role in the mobility planning process. Conditions and challenges of the sector are determined as well as gender- and diversity-specific habits and needs of mobility users in order to include them in future strategies for planning the structure and services of local public transport. This happens not only on a national (e.g. Mobility in Germany Data [2], Region Hannover Local Transport Plan [3] etc.), but also on an international level (e.g. CIVITAS Policy Note: Smart choices for cities - Gender equality and mobility: mind the gap! [4] etc.).

The term Gender is more than just the mere allocation of biological sex, it is rather "[…] a matter of the socially and culturally shaped gender roles of women and men, which (in contrast to biological sex) are learned and thus changeable" [5]. Diversity, on the other hand, refers to social diversity and different life situations. In order to be able to anchor the entire scope of aspects of gender and diversity as part of the social media appearances of public transport companies to strengthen the participation of users, the specific concerns of women, but also of persons with special contexts of use, such as age, religion, ethnic origin and restrictions due to special needs, must be taken into account.

2.2 Participation

The term participation has been shaped above all by political science: "Political participation is understood as the voluntary actions of citizens with the aim of influencing political material and personnel decisions at various levels of the political system or to participate directly in such decisions [6, p. 262]".

"With the growing spread of new information and communication technologies and especially the Internet, research on participation has once again received a significant boost, […]. Using terms such as "e-participation" or "e-democracy", an attempt is being made to summarize and discuss the entire range of forms of political and social participation which have become possible or have undergone change and renewal within the framework of the Internet and Internet-related services [6, pp. 263–264]".

"The concept of participation, which emphasizes above all the interactive and communicative aspects, has gained in popularity and expansion with the implementation of the so-called Web 2.0. While the term Web 2.0 strongly emphasizes the character of technical facilitation, the concept of social media emphasizes the moment of productive exchange and community building. Platforms such as YouTube […] and social networks such as Facebook or Twitter exemplify the defining element of social media: the participation of female users in the creation of content and in networking [6, pp. 265–266]".

Thus, not only the term participation itself is quite diverse, but also its range of use and its users. In this context, Wimmer et al. define participation as follows: "[..] an appropriation of public communication that can take different forms across different levels of organizational complexity and activity, ranging from the discussion and interaction between citizens in online forums or social media through social and protest movements to more general debates in the […] public sphere [7, p. 3]".

2.3 Social Media

According to Wessel, social media "[…] are a key feature of the ways in which people engage and participate in social and political life. Ordinary people of all ages and genders are increasingly using social media – there are currently approximately 2.51 billion social media users worldwide [8, p. 9]".

In Germany, the following picture for the use of social media platforms emerges (Fig. 2):

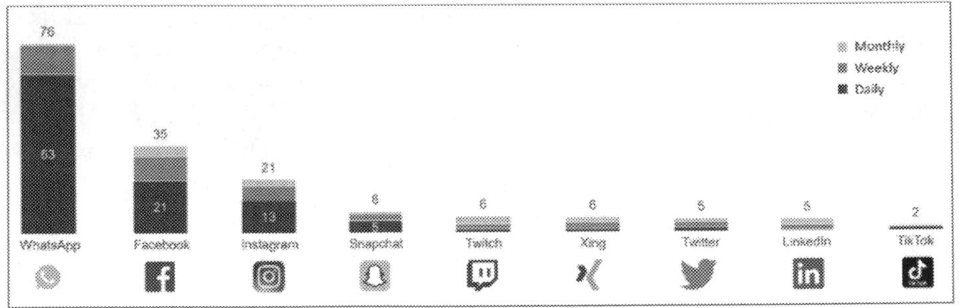

Fig. 2. Social media offers - user numbers Germany (translated in accordance with ARD/ZDF-Online Study 2019) [9]

The concept and the associated functions of the social media channels contribute to "replacing mere recipients of media with active participants in the public sphere […] [6, p. 267]".

The potential of social media appearances in terms of communication with users is certainly recognized by providers from different areas of activity (e.g. public transport companies), but it is rarely fully understood and exploited, especially with regard to the active (result-forming/influencing) participation of users.

To the advantages of social media (Facebook, Instagram, Twitter etc.) belong on the one hand the time and location independent and comparatively simple and technically low-level access and on the other hand the enormous reach, which makes it possible to address (potential) users across national, language, class and conventional boundaries. The mediated content can be provided in text, image, video and sound formats. The biggest advantage of social media, however, lies in their ability to connect and interact. In this way, they enable an exchange between provider and user and/or between users themselves. This includes not only the *Like*, *Comment* and *Share* functions, but also the possibility of concretely requesting opinions and entering into an active exchange with users. For example, surveys can be linked or votings can be carried out by directly using the platform functions.

Platform conditions such as character limits (Twitter) and format requirements can be seen as advantages and disadvantages in the same way. Although these make it easier for users to keep track of posts by contributing to a clearly structured and relatively short presentation of the content (e.g. *#Hashtag*), they can also limit or diminish their intention and content.

One disadvantage of social media can be seen, for example, in their fast pace of life. The flood of information and the associated activities of exchange lead to the fact that the topicality and issue of the contributions quickly loses importance and a large amount of content is completely passed by the user due to the short attention span depending on the medium.

To prevent this from happening, adjustments must be made. "The malleable design of social media's different platforms and devices enables people to use them in various creative ways. This creativity is located in spaces where literacy skills, social practices and cultural sensibilities of use interact with the user interface design of these tools and platforms [8, p. 62]".

2.4 Gender and Diversity Specifications for UI and Content

Regarding the consideration of gender- and diversity-specific user requirements in the design of user interfaces in the course of participation via social media as well as in the corresponding preparation of the associated content, various standards, guidelines and style guides are already available (Table 1 shows a small selection). Standards have the highest level of commitment, while style guides operationalize standards and guidelines for specific products.

Table 1. Selection of standards, guidelines and styleguides.

Title	Commitment	Topic
Web Content Accessibility Guidelines 2.1 [10]	Standard	Make content more accessible to a wider range of people with disabilities and therefore also often more usable to users in general
Google Material Guidelines [11]	Guideline	Accessibility in design to allow users of diverse abilities to navigate, understand, and use UI
UNESCO Guidelines on Gender-Neutral Language [12]	Guideline	Ambiguity, Stereotyping, Titles and forms of address
Recommendations for a gender-sensitive administrative language of Hannover Region [13]	Guideline	Instructions for a gender comprehensive formulation or the use of the gender star in the German speaking context
Guideline Gender-sensitive language of the Ludwig-Maximilians-University Munich [14]	Guideline	Symmetrical naming of men and women, avoidance of role clichés and stereotypes, gender-neutral expressions and abstractions, gender-specific individual forms and pair forms in the German speaking context
"Gender and Diversity in words and pictures" of the city of Freiburg [15]	Guideline	Gender-sensitive and anti-discriminatory Language actions; the most common forms in the German-speaking context (gender gap, gender star, gender-neutral formulations, etc.) Gender & Diversity in the visual language
Inclusion Europe - Information for all [16]	Guideline	European standards for making information easy to read and understand
Network Easy Language [17]	Guideline	Formulation of content in easy language for non-German speaking users or users with impairments
Guideline on Social Media platforms and accessibility [18]	Guideline	Explanations of a variety of elements and functions that allow barrier-free access to the individual platforms
Facebook [19, 20]	Guideline	How to use Facebook, Accessibility
Twitter [21]	Guideline	How to use Twitter, Accessibility
Instagram [19, 22]	Guideline	How to use Instagram, Accessibility
Apple Human Interface Guidelines [23]	Styleguide	Accessibility in design to allow users of diverse abilities to navigate, understand, and use UI by prioritizing simplicity and perceivability
…etc.		

If you are looking for style guides for Facebook, Twitter and Instagram, you will only be provided with templates that you can use to prepare a planned appearance or post [19–22]. In general, while it can be seen that the various platforms follow the standards and guidelines mentioned above with regard to the design elements of the UI and page layout, there are no explicitly accessible style guides. However, there are some references to available elements and functions that are related to accessibility and are therefore diversity-sensitive. With the help of these elements the user group is expanded and the possibility of participation is strengthened.

2.5 Conclusion

Existing research differentiates the target groups in terms of characteristics (e.g. elderly people, people with physical or cognitive impairments, children and ado-lescents, people with migration background). In concrete terms, this means that their circumstances, i.e. their usage context, is taken into account. As an example of the interplay of the above-mentioned fields of impact, mothers who, due to the effort involved in caring for their children, have no regulated leisure time and can therefore participate less frequently in local events (e.g. citizens' events) and thus benefit from the possibilities of using social media independent of time and place.

But despite the fact that participation, including that which is made possible via social media, plays a role in the design of software both at the level of usability engineering and at the political level in the sense of democratic participation processes, it hardly appears at all in connection with mobility planning. The use of a gender-sensitive or -neutral language and visual language in the context of social media appearance by transport companies is equally rare, although there are extensive recommendations in standards, guidelines and style guides. This means that even if the designs of the individual social media platforms are gender- and diversity-oriented, the same doesn't necessarily applies to the content/posts provided by individual appearances, e.g. public transport companies.

3 Method

The aim of the paper is to define design recommendations for gender- and diversity-oriented social media appearances to strengthen participation in the context of local public transport and mobility planning.

To this end, the methodological approach is based on an analytical study of standards, guidelines and style guides with regard to the consideration of gender- and diversity-specific design recommendations, as well as an empirical study that analyses the extent to which these design recommendations are implemented in the context of social media appearances by public transport companies.

3.1 Analysis of Standards, Guidelines and Style Guides with Regard to the Consideration of Gender- and Diversity-Specific Design Recommendations

The needs and requirements of the variety of user groups, that emerge, when considering gender and diversity-aspects, should be the basis when shaping social media appearances in a gender- and diversity-oriented manner in order to enable as many users as possible to participate. These user groups include, for example, children and young people, people with impairments (physical, mental, cognitive), people with a migration background (language, cultural barriers, etc.), seniors, people in different life situations (parents, childless, single, employed or not, etc.) and various other possible combinations.

In relation to the user groups, the standards, guidelines and style guides mentioned above consider various design and functional elements with regard to user requirements and needs. In the course of this, three element categories can be derived. The *UI-related gender- and diversity elements* are all covered by the platform and can be activated by the user if required. These elements enable basic access to the contributions provided and the information contained therein, as well as navigation on the platforms themselves. *Content-related gender and diversity elements* include all those that relate exclusively to the formulation of the content and can only be provided by the provider. All those elements that are inherent in the platforms and enable participation in the sense of interaction are considered *participation-related gender and diversity elements*. Table 2 lists standards, guidelines and style guides for the respective elements of the aforementioned categories, from which design recommendations can be derived. All elements to which these design recommendations are applied are listed under Subject, whereby the recommendations specify their characteristics and the user group to which they are primarily addressed.

Table 2. Extracted elements and functions for a gender- and diversity-oriented design

Standard/Guideline/Styleguide	Subject	Recommendations
UI-related gender- and diversity elements		
• WCAG [10] • Google Guideline on Accessibility [11] • Guideline on Social Media platforms and accessibility [18] • Facebook, Twitter, Instagram [19–22]	Scaling	Zoom possibilities especially for people with motor or visual impairments Responsive Design
• WCAG [10] • Google Guideline on Accessibility [11] • Guideline on Social Media platforms and accessibility [18] • Facebook, Twitter, Instagram [19–22]	Color and contrast	Contrast and colour regulation especially for people with visual impairments

(*continued*)

Table 2. (*continued*)

Standard/Guideline/Styleguide	Subject	Recommendations
• WCAG [10] • Google Guideline on Accessibility [11] • Guideline on Social Media platforms and accessibility [18] • Facebook, Twitter, Instagram [19–22]	Text, Font and Size	Possibility of enlargement and easy to read font especially for people with visual impairments
• WCAG [10] • Google Guideline on Accessibility [11] • Guideline on Social Media platforms and accessibility [18] • Facebook, Twitter, Instagram [19–22]	Reading	Offering an alternative to text and images, especially for blind and visually impaired people, but also for people who cannot read or can only read with restrictions Consider audio/sound quality Assistive Technology
• WCAG [10] • Google Guideline on Accessibility [11] • Guideline on Social Media platforms and accessibility [18] • Facebook, Twitter, Instagram [19–22]	Subtitle	Offering an alternative to sound, especially for deaf and hearing impaired people, but also as a support for non-native speakers Consider synchronicity subtitles and picture Assistive Technology
• WCAG [10] • Google Guideline on Accessibility [11] • Guideline on Social Media platforms and accessibility [18] • Facebook, Twitter, Instagram [19–22]	Images and Videos	Offering an alternative to text and sound, especially for hearing impaired people, but also as support for an easy overview/navigation/information for all users Consider picture quality and setting sizes as well as synchronicity of sound and picture, if necessary
• WCAG [10] • Google Guideline on Accessibility [11] • Guideline on Social Media platforms and accessibility [18] • Facebook, Twitter, Instagram [19–22]	Sound	for videos, useful for almost all user groups except for deaf or hearing impaired people and non-native speakers (except for music) consider sound quality, if necessary synchronicity of sound and picture
• Guideline on Social Media platforms and accessibility [18] • Facebook, Twitter, Instagram [19–22]	Multilingualism	Automated provision of translations or the possibility of adding translations independently for non-native speakers Consider translation quality

(*continued*)

Table 2. (*continued*)

Content-related gender- and diversity elements		
• UNESCO Guidelines on Gender-Neutral Language [12] • "Gender and Diversity in words and pictures" City of Freiburg [15] • Guideline Gender-sensitive language of the Ludwig-Maximilians-University Munich [14]	Gender-neutral Language	Addressing all users via the direct "you" address and neutral majority address without supposedly complicated adaptations to certain gender forms
• "Gender and Diversity in words and pictures" City of Freiburg [15] • "Recommendations for a gender-sensitive administrative language" Hannover Region [13] • Guideline Gender-sensitive language of the Ludwig-Maximilians-University Munich [14]	Gender-sensitive Language	Addressing in gendered form with adaptation of pronouns and spelling to specific gender forms Explicit reflection of the diversity of the people one wants to reach
• WCAG [10] • "Inclusion Europe - Information for all" [16] • Network Easy Language [17]	Easy Language	Simple sentence structure and reduced number of syllables, especially for people with cognitive impairments, but also for non-native speakers
• "Gender and Diversity in words and pictures" City of Freiburg [15]	Gender- and diversity-sensitive Visual Language	Images should also reflect the diversity of the people one wants to reach
Participation-related gender- and diversity elements		
• Guideline on Social Media platforms and accessibility [18] • Facebook, Twitter, Instagram [19–22]	Like	Participation in the form of expression of opinion in exchange with the provider and/or other users across all user groups
• Guideline on Social Media platforms and accessibility [18] • Facebook, Twitter, Instagram [19–22]	Share	Participation in the form of forwarding to/connecting with other users across all user groups
• Guideline on Social Media platforms and accessibility [18] • Facebook, Twitter, Instagram [19–22]	Comments	Participation in the form of expressions of opinion across all user groups
• Facebook, Twitter, Instagram [19–22]	Votings	Participation in the form of an opinion query by the provider Depending on the design using text, images or emoticons, more or less user groups can be reached
• Facebook, Twitter, Instagram [19–22]	Chats & Bots	Participation via direct contact between user and provider via Live Chat or indirectly via ChatBots

All the elements mentioned also have advantages or amenities for other user groups beyond their originally intended target group. This means that all user groups benefit from, for example, accessibility, regardless of whether they are subject to a restriction or not.

Since this paper deals with the gender- and diversity-oriented design of social media appearances, in order to promote participation, we have mapped the extracted elements with regard to their existence and provided configuration possibilities on social media platforms (Facebook, Twitter, Instagram) [18–22].

Table 3 shows how the previously extracted UI-, content- and participation-related elements are taken up by social media platforms. The elements are listed under *Subject*. Under *Platform*, the presence of the elements on one or all three platforms considered. *Role* can be providers and/or users, depending on their responsibility for the activities listed under *Specifications* in order to make the elements usable.

Table 3. How are UI-, Content- and Participation-related elements taken up by social media platforms.

Subject	Platform	Role	Specification
UI-related gender- and diversity elements			
Scaling	Facebook Twitter Instagram	User	Can be configured independently via the private account settings
Color & Contrast	Facebook Twitter Instagram	User	Can be configured independently via the private account settings
Text, Font and Size	Facebook Twitter Instagram	User	Can be configured independently via the private account settings
Reading	Facebook Twitter Instagram	Provider/User	Integrated VoiceOver for Apple users Assistive Technology: Screen reader
Subtitle	Facebook Twitter Instagram	Provider	Provider can add subtitles via the functions of the platforms
Images and Videos	Facebook Twitter Instagram	Provider/User	Independently formulated alternative texts can be read by Assistive Technology, Automatic Alternative Text (AAT) to voice (Facebook)
Sound	Facebook Twitter Instagram	Provider/User	Provider offers sound, user can de/activate it or define it as autoplay

(*continued*)

Table 3. (*continued*)

Subject	Platform	Role	Specification
UI-related gender- and diversity elements			
Multilingualism	Facebook	Provider/User	Provider offers translations, user must configurate via the private account settings
Content-related gender- and diversity elements			
Gender-neutral Language	Facebook Twitter Instagram	Provider	Is not operated by the platform, but can be performed by text/post formulation: *You* and/or generalizing plural terms
Gender-sensitive Language	Facebook Twitter Instagram	Provider	Is not operated by the platform side, but can be performed by text formulation: consideration of pronouns and naming of both male and female word forms
Easy Language	Facebook Twitter Instagram	Provider	Is not operated by the platform side, but can be performed by text formulation: simple sentence structure and reduced number of syllables
Gender- and diversity-sensitive Visual Language	Facebook Twitter Instagram	Provider	Is not operated by the platform side, but can be performed by image creation or selection
Participation-related gender- and diversity elements			
Like	Facebook Twitter Instagram	User/Provider	In the form of emoticons the user can communicate Likes – Likes can also be hidden (not public)
Share	Facebook Twitter Instagram	User/Provider	Can be restricted by the user himself or by the provider (not public etc.)
Comments	Facebook Twitter Instagram	User/Provider	Comments can be deactivated; Tweets and Comments on Tweets can be hidden
Votings	Facebook Twitter Instagram	User/Provider	Must be provided by the provider, who can limit access to it and specify the form of voting; User can vote using the emoticons, comments, selections, etc., depending on the provider's and platform's specifications Emoticons and little text are more inclusive than text and/or commentary formats

(*continued*)

Table 3. (*continued*)

Participation-related gender- and diversity elements			
Chats & Bots	Facebook Twitter Instagram	User/Provider	Is offered by the provider; can be used by the user Live chat with people who answer questions in real time and individually more inclusive than ChatBots, which answer automatically and are prone to errors

3.2 Comparative Analysis of National and International Social Media Appearances of Public Transport Companies

In the second step, the analysis of the social media appearances with regard to their gender- and diversity-oriented design is carried out based on the extracted recommendations. In total, the social media appearances of 16 public transport companies were analysed. A total of 5656 posts of a period of 3 months (15.11.2019–15.02.2020) were viewed. As the selected public transport companies only have a appearances on the social media platforms Facebook, Twitter and Instagram, the analysis is limited to these three platforms. All of the public transport companies are located in major cities, as there is a wider range of gender and diversity in terms of user groups. In the course of the evaluation, the public transport companies were made anonymous so that no "negative" conclusions can be associated with actual names (Fig. 3).

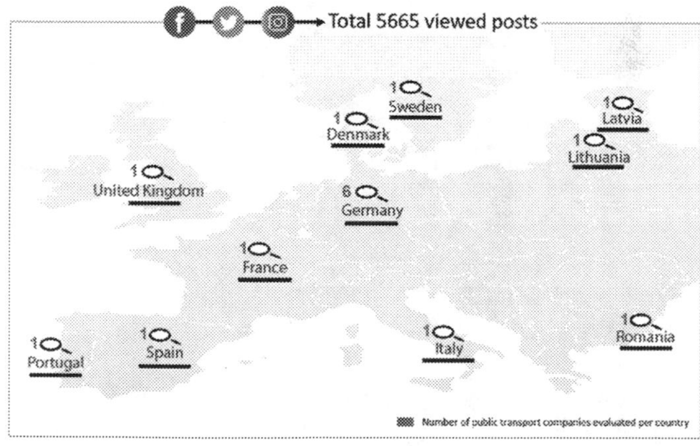

Fig. 3. Countries of the evaluated social media appearances of public transport companies

4 Evaluation and Outcome

According to the extracted recommendations for a gender- and diversity-sensitive design of social media appearances, the social media appearances of six selected German and ten selected international (EU) public transport companies were evaluated.

With regard to the *UI-related related gender- and diversity-sensitive social media elements*, only those were taken into account whose provision by the provider is necessary in order to be useful for the user. This includes reading, subtitle and multilinguism. Both the national and international evaluation shows that these recommendations are only considered in individual cases.

With regard to the *content-related gender- and diversity-sensitive social media elements*, the text-language recommendations could only be evaluated on a national level, as the authors do not have a command of the language features/characteristics of the ten foreign languages. The visual language, on the other hand, could also be evaluated at the international level.

Figure 4 shows that a gender- and diversity- sensitive language is rarely used in the context of the appearances of the six selected German public transport companies across the three platforms. Of a total of 2291 posts, only 12 contained gender-sensitive language. In addition, the interaction with Fig. 5 shows that Facebook and Instagram communicate more strongly via images, whereas Twitter is primarily used with text contributions. This is related to the fact that German public transport companies tend to use their Twitter appearances more as a kind of news ticker on the subject of delays and disruptions in the public transport network.

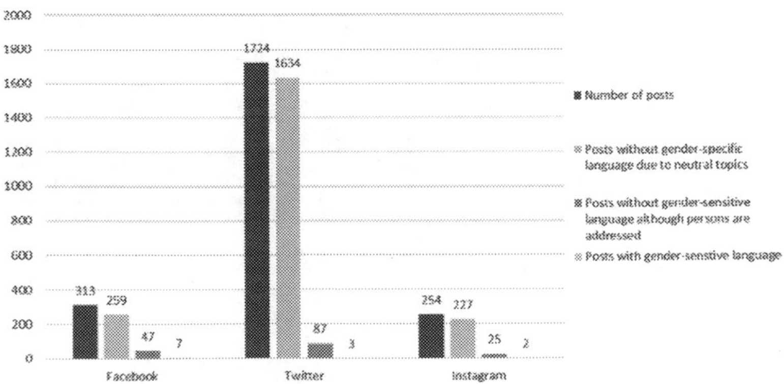

Fig. 4. Gender- und diversity-sensitive Language – evaluation based on the social media appearances of six selected German public transport companies.

When evaluating the visual language (Figs. 5 and 6), special attention was paid to the representation of different user groups, especially those of minorities (e.g. children and young people, elderly people, people with disabilities, etc.). The evaluation of the visual language of the German appearances clearly shows that for Facebook and Instagram images with neutral objects are used preferentially. Of a total of 952 image posts, 461 images showed only neutral things like objects or locations. On Twitter, the aforementioned text-heaviness even runs through the visual language, since here 280 of 385 images consist of sentences/text modules which are used like stickers. Across all platforms, however, it should be noted that minorities are represented only extremely rarely. Of a total of 952 image posts, only 12 represented minorities. In addition, women are represented significantly less than men (Figs. 5 and 6). For comparison, men are represented on 105 image posts, whereas women appear only on 66 image posts.

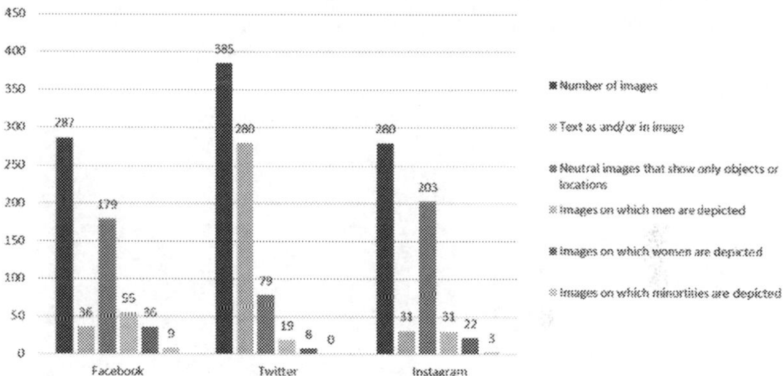

Fig. 5. Gender and diversity-sensitive visual language - evaluation based on the social media appearances of six selected German public transport companies.

The situation is similar with the evaluation of the international appearances, where a text-heavy visual language can be seen across all platforms. Of a total of 1567 image posts, 581 consist of text modules and 648 show objects or locations. In the international context, however, the gap is even wider with regard to the representation of minorities. Minorities are depicted in only 59 image posts. That is not even 5% of the total. Besides, the images that show women do not even reach 10%.

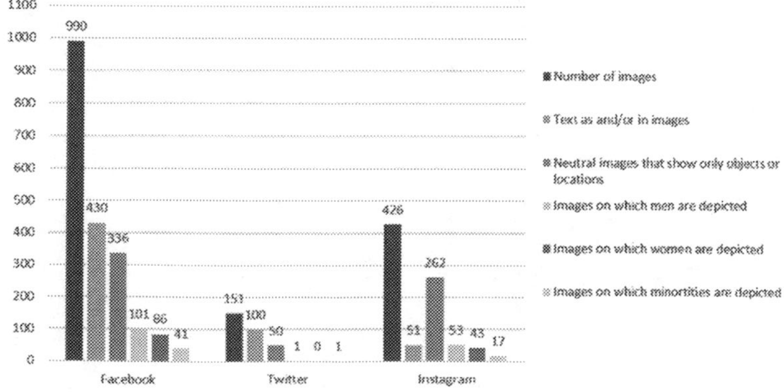

Fig. 6. Gender and diversity-sensitive visual language - evaluation based on the social media appearances of ten selected public transport companies within the EU.

With regard to the *participation-related gender- and diversity-sensitive social media elements*, the focus of the evaluation was placed on the actually traceable interaction between user and provider by counting user comments to which the public transport companies reacted. In the context of German appearances (Fig. 7), a certain level of interaction on Facebook can be identified, but the figures are mostly a result of the involvement of a single public transport company. The interaction on Twitter and Instagram, on the other hand, is extremely low, which can be attributed to the ticker character of Twitter. In total, the public transport companies only responded to 232 user comments out of a total number of 2291 posts. In the international context (Fig. 8), the inertia of interaction between users and public transport companies is a bit more pronounced across all platforms. In total, the public transport companies only responded to 377 user comments out of a total number of 3365 posts.

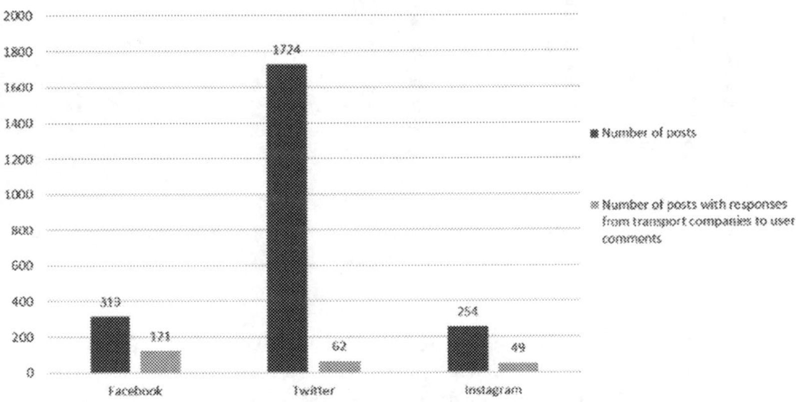

Fig. 7. Commentary interaction between user and provider - evaluation based on the social media appearances of six selected German public transport companies.

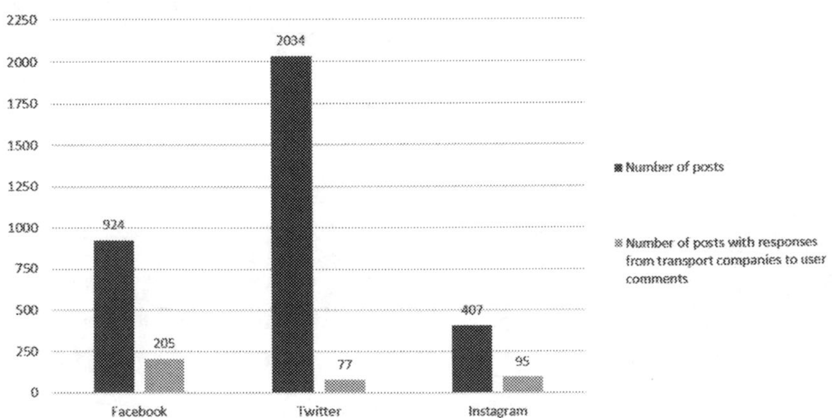

Fig. 8. Commentary interaction between user and provider - evaluation based on the social media appearances of ten selected public transport companies within the EU.

An aspect that is seriously neglected both at national and international level across all platforms is the involvement of users via voting or chats. No participation in decision-making processes was offered. For example, if voting was used, it was only in the form of humorous polls (Fig. 9).

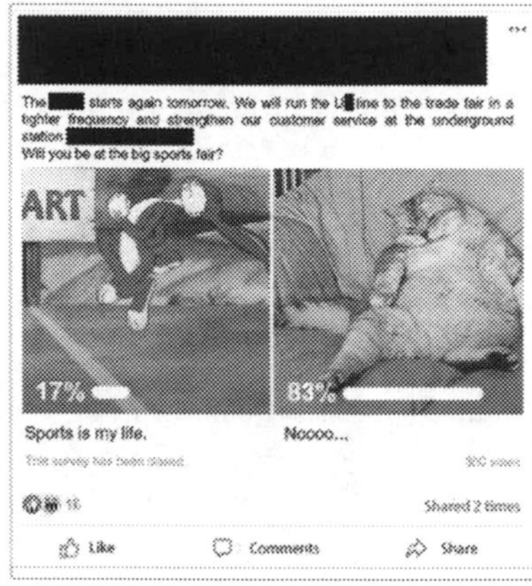

Fig. 9. Example for the provision of a humorous voting via Facebook – German Original and translated English version

Live chats were not offered, but at least on a national level, ChatBots could be observed occasionally, which, however, played a more service-oriented role.

In summary, it can therefore be stated that the possibilities for gender- and diversity-oriented design of social media appearances by public transport companies - both at national and international level - are hardly taken into account or used.

5 Discussion

The public transport companies were selected on the basis of their affiliation to a large city and the associated high diversity of users. For Germany, a nationwide analysis was carried out in order to analyse possible different participation cultures of the public transport companies. The analysis of the design guidelines has shown that gender and diversity are not perceived as influencing factors for design.

The analysis of the social media platforms of the transport companies showed that no gender and diversity appropriate design is implemented on the design levels text and image.

The possibility of influential participation is also hardly offered by the public transport companies. This can be observed both at the German and international level.

The results, which are based on a sample of over 5500 posts, provide the impetus for in-depth research with regard to the following questions:

- How can the gender- and diversity-oriented design recommendations from Table 2 be shifted more into the focus of design?
- How can the existing participation mechanisms be better used for the public transport companies in the future?
- How does compliance with the design rules contribute to an increase in the number of gender- and diversity-oriented users and to strengthening their participation?

The analysis did not provide any reliable starting points for the reasons for this lack of consideration. To what extent this is based on a lack of understanding of gender and diversity or of the possibilities of social media or even both is difficult to grasp.

References

1. Hutchins, A.L., Tindall, N.T.J. (Hrsg.): Public Relations and Participatory Culture: Fandom, Social Media and Community Engagement. Routledge New Directions in Public Relations and Communication Research. Routledge, London (2019)
2. Mobilität in Deutschland (Data on Mobility in Germany). http://www.mobilitaet-in-deutschland.de. Accessed 15 June 2020
3. Region Hannover Local Transport Plan. https://www.hannover.de/Leben-in-der-Region-Hannover/Mobilit%C3%A4t/Verkehrsplanung-entwicklung/Der-Nahverkehrsplan2. Accessed 15 June 2020
4. CIVITAS Policy Note: Smart choices for cities - Gender equality and mobility: mind the gap! https://civitas.eu/sites/default/files/civ_pol-an2_m_web.pdf. Accessed 15 June 2020
5. Gender Mainstreaming in Public Transport, Krause, J.: Gender Mainstreaming im Verkehrswesen - Einführung, Folie 2 "gender"/Folie 3 "gender mainstreaming". AG 1 Verkehrsplanung, Forschungsgesellschaft für Straßen- und Verkehrswesen e. V. (FGSV). https://www.fgsv.de/fileadmin/road_maps/GM_Einfuehrung.pdf. Accessed 15 June 2020
6. Rudolph, St.: Digitale Medien, Partizipation und Ungleichheit: eine Studie zum sozialen Gebrauch des Internets. Springer, Wiesbaden (2019). https://doi.org/10.1007/978-3-658-26943-2
7. Wimmer, J., Wallner, C., Winter, R., Oelsner, K.: Introduction to (Mis)understanding political participation: digital practices, new forms of participation and the renewal of democracy. In: Routledge Studies in European Communication Research and Education, 13. Routledge Taylor & Francis Group, London (2018)
8. Wessels, B.: Communicative Civic-ness: Social Media and Political Culture. Routledge, Taylor & Francis Group, London; New York (2018)
9. Results of the ARD/ZDF online study 2019. http://www.ard-zdf-onlinestudie.de/files/2019/Ergebnispraesentation_ARD_ZDF_Onlinestudie_PUBLIKATION_extern.pdf
10. Web Content Accessibility Guidelines (WCAG) 2.1. https://www.w3.org/TR/WCAG21/. Accessed 15 June 2020
11. Google Guideline on Accessibility. https://material.io/design/usability/accessibility.html. Accessed 15 June 2020
12. UNESCO Guidelines on Gender- Neutral Language. https://www.upm.es/sfs/Rectorado/Gerencia/Igualdad/Lenguaje/Gu%C3%ADa%20lenguaje%20no%20sexista%20en%20ingl%C3%A9s%20y%20franc%C3%A9s.pdf. Accessed 15 June 2020

13. Recommendations for gender-sensitive administrative language of Hannover Region. https://www.hannover.de/Leben-in-der-Region-Hannover/Verwaltungen-Kommunen/Die-Verwaltung-der-Landeshauptstadt-Hannover/Gleichstellungsbeauf%C2%ADtragte-der-Landeshauptstadt-Hannover/Aktuelles/Neue-Regelung-f%C3%BCr-geschlechtergerechte-Sprache. Accessed 15 June 2020
14. Guideline Gender-sensitive language of the Ludwig-Maximilians-University Munich. https://www.frauenbeauftragte.uni-muenchen.de/genderkompetenz/sprache/sprache_pdf.pdf. Accessed 15 June 2020
15. "Gender and Diversity in words and pictures" of the city of Freiburg. https://www.freiburg.de/pb/site/Freiburg/get/params_E2096115146/1114619/Leitfaden_GD2019.pdf. Accessed 15 June 2020
16. Inclusion Europe - Information for all. Guideline on Easy Language. https://www.inclusion-europe.eu/easy-to-read/. Accessed 15 June 2020
17. Network Easy Language. https://www.leichte-sprache.org/wp-content/uploads/2017/11/Regeln_Leichte_Sprache.pdf. Accessed 15 June 2020
18. Guideline on Social Media platforms and accessibility. https://www.netz-barrierefrei.de/wordpress/barrierefreies-web-2-0-ein-leitfaden-zu-social-media-und-behinderung/. Accessed 15 June 2020
19. Facebook and Instagram Accessibility. https://www.facebook.com/accessibility. Accessed 15 June 2020
20. Facebook Accessibility. https://de-de.facebook.com/help/273947702950567?helpref=hc_global_nav. Accessed 15 June 2020
21. How to use Twitter, https://help.twitter.com/en/using-twitter/picture-descriptions. Accessed 15 June 2020
22. How to use Instagram. https://help.instagram.com/. Accessed 15 June 2020
23. Apple Human Interface Guidelines. https://developer.apple.com/design/human-interface-guidelines/accessibility. Accessed 15 June 2020

Using Context to Help Predict Speaker's Emotions in Social Dialogue

Mei Si(✉)

Rensselaer Polytechnic Institute, Troy, NY 12180, USA
sim@rpi.edu

Abstract. Emotion plays a vital role in social interaction. Often, the speaker's attitude is as essential as, if not more important than, his/her words for communication purposes. In this paper, we present experiments for using conversational context to help text-based emotion detection. We used data from the Dialogue Emotion Recognition Challenge – EmotionX. BERT is used for encoding the input sentences. We explore four ways for encoding the input by varying whether to concatenate a dialogue history with the current sentence and whether to add the speaker's name as part of the input. Our results indicate that adding context can improve the results of emotion detection when the emotion categories do not overlap with each other.

Keywords: Emotion detection · BERT · Context

1 Introduction

1.1 Background and Motivation

Emotion plays a vital role in social interaction. Often, the speaker's attitude is as essential as, if not more important than, the content in his/her words for communication purposes.

In recent years, text-based sentiment analysis has gained increasing attention [1–4]. Much work has been conducted on sentiment analysis under the context of understanding the sentiment of customer reviews or posts on social media, i.e., the AI can automatically tell whether the user expressed a positive or negative view about an item or topic [4].

Social interaction, however, is much more complicated than customer reviews or social media posts, and people are not just expressing positive or negative attitudes towards each other. Instead, a rich set of social emotions such as anger, sadness, and happiness are expressed and detected by conversational partners. For helping AI systems understand social dialogue, it is therefore essential to enable the AI system to recognize emotions expressed by the speakers in social dialogues.

In most cases, emotion detection is conducted by using people's facial expressions, body postures, physical signals, or combing the verbal signals with these other signals. Text-based emotion detection has also been explored [5–7]. In this work, we want to focus on text-based emotion detection. We believe text-based emotion detection systems, even though may perform less effectively than a system that uses multimodal

C. Stephanidis et al. (Eds.): HCII 2020, LNCS 12427, pp. 444–453, 2020.
https://doi.org/10.1007/978-3-030-60152-2_33

data, can have a wider range of applications. Much online interaction is mainly in the form of text, e.g., online chat, tweets. And when multimodal data are presented, there are usually transcripts for the characters' lines.

Various labeled datasets exist for text-based emotion detection. In most of these datasets, the labeled text is not coming from natural social dialogues. For example, ISEAR – International Survey on Emotion Antecedents and Reactions – is a popular dataset [8]. It is composed of people's descriptions of emotion-provoking scenarios. The Alm's annotated fairy tale dataset consists of labeled sentences from children's fairy tales [9]. The SemEval-2007 dataset consists of news headlines annotated with their emotions [10].

The data set used in this paper comes from the Dialogue Emotion Recognition Challenge – EmotionX – at the SocialNLP Workshops, collocated at ACL 2018 and IJCAI 2019 [11, 12]. This challenge offers the EmotionLines dataset as the experimental materials and requires the participants to recognize the emotion of each utterance in dialogues. The EmotionLines dataset contains two types of conversations. One type is the Friend's TV show transcripts (Friends), and the other is people's chatting logs on Facebook messenger (EmotionPush). We used the Friends dataset for this work. We choose this dataset because its data are reprehensive of real-life dialogue. Therefore the results from our approach and experiments can be generable to other scenarios that involve realistic social interactions.

1.2 Friends Dataset

Table 1. Sample dialogue from friends

Character	Utterance	Emotion
Rachel	Oh okay, I'll fix that to. What's her e-mail address?	Neutral
Ross	Rachel!	Anger
Rachel	All right, I promise. I'll fix this. I swear. I'll-I'lll'll-I'll talk to her	Non-neutral
Ross	Okay!	Anger

The Friends dataset consist of dialogue from a famous American TV show Friends between 1994 and 2004. Amazon mTurk workers were hired to label each utterance with the emotion it expresses [11, 12]. Each utterance is labeled by five workers. The workers labeled the sentence with a single emotion selected from the list: neutral, joy, sadness, fear, anger, surprise, and disgust. The final label of the utterance is calculated by taking the absolute majority of the workers' labels. If there is no majority vote, the utterance is labeled as non-neutral. As reported in [12], the Fleiss' kappa was used for measuring the inner-annotator agreement. The Fleiss' kappa for the Friends dataset is 0.362, indicating a fair agreement [13]. The low Fleiss' kappa suggests identifying emotions from the utterances is a hard task even for humans.

Table 1 shows an example piece of dialogue from this dataset. For each line of the dialogue, the data contain the speaker, the utterance, and the emotion label. Sometimes the utterance is very short and only contains a single work, such as "OK". Because we didn't participate in the competition, we don't have the validation data from this dataset. We only have the training data, which contains 14323 lines. We randomly split this dataset and use 2/3 of it (7793) for training and 1/3 (4350) for testing.

Figure 1 shows the distribution of the emotion labels. The most popular label is neutral, followed by non-neutral, joy, surprise and anger. The task of this challenge is to identify anger, joy, neutral and sadness.

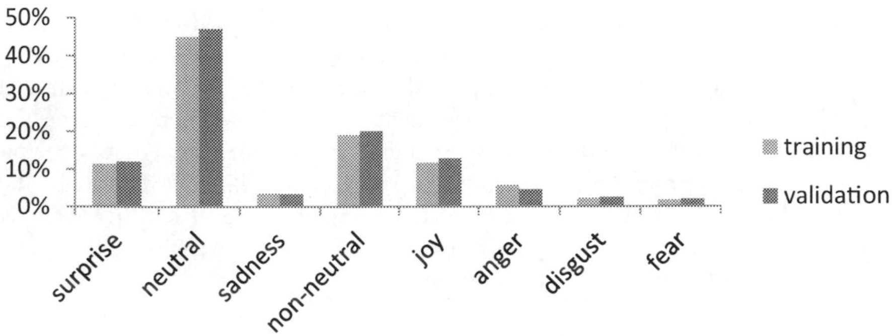

Fig. 1. Emotion label distribution.

2 Emotion Detection from Text

2.1 Related Work

To identify the emotion expressed in an utterance, various machine learning based approaches have been explored. These approaches can be roughly divided into two categories: feature-based approaches and deep learning based approaches. Feature-based approaches look for features, especially keywords with emotional or sentiment values [14, 15]. WordNet-Affect [17] and SentiWordNet [18] are often used for helping to identify the features. With these features, machine learning algorithms such as SVM [16] are applied for classifying emotions.

In recent years, deep learning based approaches have become increasingly popular for text classification tasks. Convolutional neural network (CNN), long short-term memory (LSTM) network, and Gated Recurrent Unit (GRU) have all been explored. According to a recent survey conducted in 2019 [5], the most common neural network architecture is bidirectional LSTM.

Transfer learning, where a pre-trained language model is fine-tuned for performing new tasks, i.e., emotion classification, is also a popular option. According to [5], BERT [19] is the most popular language model used in transfer learning. In fact, in the 2019 SocialNLP EmotionX competition, 5 out of 7 participating teams used BERT [12]. In this work, we also used BERT for encoding the input.

2.2 Considering Context

As we can see in Table 1, often the time in dialogue, people's sentences are very brief. This does not prevent people from understanding the sentiment and the meaning of these sentences, because people will naturally take the context of the sentence, i.e., the conversation before this point into account. If we only look at a short response itself, e.g., "ok", there is no way we can infer the speaker's emotions. Similarly, whom the speaker is also playing an essential role in predicting an utterances' emotion because there is usually a consistency in how the same person talks. Therefore, it is important to include previous conversations and the speaker's name or ID as context.

As people have paid more attention to the importance of context for text-based emotion detection, many works have taken some context as part of the input [5]. In the 2019 SocialNLP EmotionX competition, a couple of teams have used both previous conversation and speaker's name as context. Despite these efforts, the role of context has not been fully studied. In fact, the winning team from 2018 [20] did not include context as part of their input. They used a CNN-DCNN autoencoder based approach for emotion classification.

The goal of this work is to systematically vary the type of context used in emotion classification and determine their relative importance. The two types of context we study in this work are previous conversations and the speaker's name.

2.3 Experiment Setup and Implementation

In this work, we use Bidirectional Embedding Representations from Transformers (BERT) [19] as the encoder for input sentences. BERT is a pre-trained language model that was trained using a large general dataset. BERT can be fine-tuned to perform downstream tasks, and have shown many promising results.

Figure 2 shows how BERT takes input and produces output. More specifically, BERT takes two sentences – sentence1 and sentence 2 – as input. In its last layer, the vector at the <CLS> position is generally believed to contain the embedding of the input sentences. We used the transformer library from HuggingFace [21]. Following their examples, a linear layer is added after the last layer of BERT for making decisions.

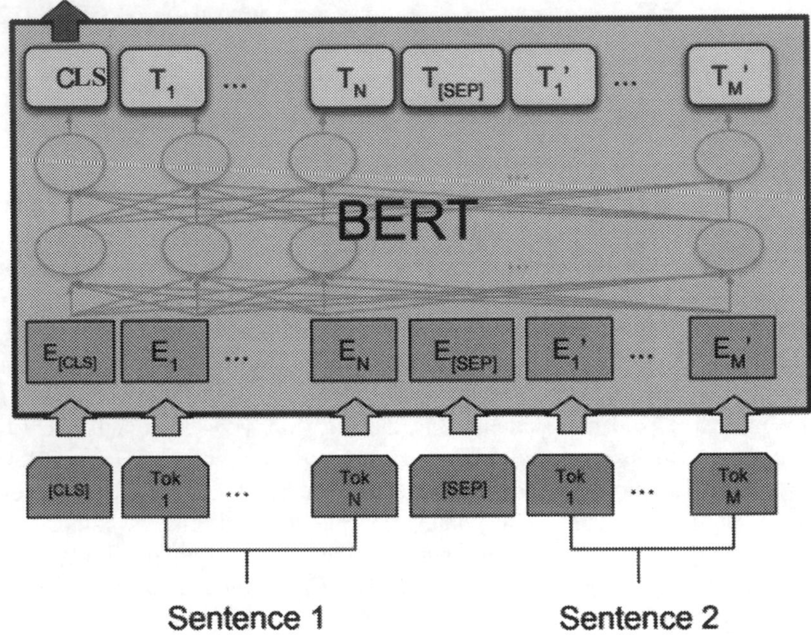

Fig. 2. BERT Input and output.

In this work, we used the "bert-large-uncased" model, which contains 24 layers. Our learning rate is 3e−5. The batch size for both training and testing is 16. When fine-tuning BERT, we first freeze BERT and only train the last layer for 1 epoch. We then unfreeze BERT for the rest of the training process. It has been suggested that it does not take many iterations to fine-tune BERT. This is also consistent with our observations. Therefore, after unfreezing BERT, we continue training the model for only 4 more epochs.

We conducted our experiments by varying two variables: whether to include the previous conversation as context and whether to include the speaker's name as context. Combining these two variables creates four conditions:

- **NPNN**: only use the current utterance as input. The current utterance is set as sentence 1 for being passed to BERT;
- **PNN**: use the current utterance plus two proceeding utterances as input. The three utterances are set as sentence 1 for being passed to BERT;
- **NPN**: use the current utterance and the speakers' names as input. The current utterance is set as sentence 1, and the speaker's name is set as sentence 2 for being passed to BERT;

- **PN**: use the current utterance plus two proceeding utterances and the speaker's name as input. The three utterances are set as sentence 1, and the speaker's name is set as sentence 2 for being passed to BERT;

We used BERT's BPE tokenizer [22] for tokenizing the utterances and did not do any pre-processing before passing the input to the tokenizer.

3 Results and Discussion

We conducted two experiments. In the first one, only the four emotions – anger, joy, neutral, and sadness – listed in the challenges are detected. When preparing the data, we filtered out entries whose emotion label is outside of this set. In the second one, we worked on detecting all the eight emotion labels that exist in this dataset: neutral, joy, sadness, fear, anger, surprise, disgust, and non-neutral.

3.1 Experiment 1

Table 2 and Fig. 3 show the results of using our model for detecting anger, joy, neutral and sadness.

Table 2. Results for detecting anger, joy, neutral and sadness

	Accuracy	Micro F1	Precision	Recall
NPNN	0.92	0.69	0.69	0.69
PNN	0.96	0.83	0.81	0.86
NPN	0.96	0.84	0.82	0.86
PN	0.96	0.83	0.82	0.83

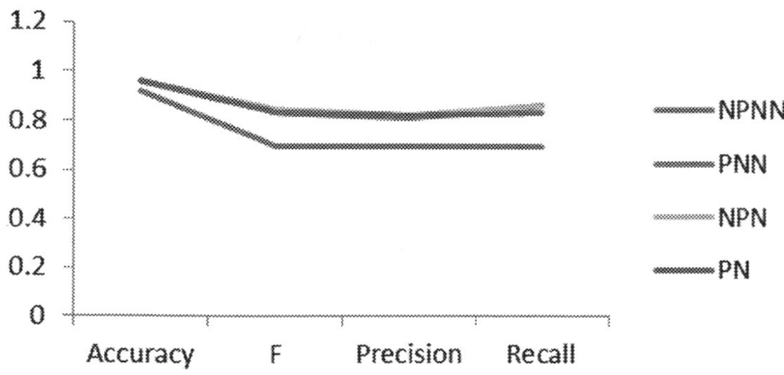

Fig. 3. Results for detecting anger, joy, neutral and sadness

It appears that adding any contextual information helps, and the amount of improvement in accuracy, f1, precision, and recall are very similar. An artifact caused by removing utterances whose labels are outside of the four emotions is sometimes the sentences we added as context are not immediately preceding the current utterance. However, this does not happen very often, nor seems to affect our results.

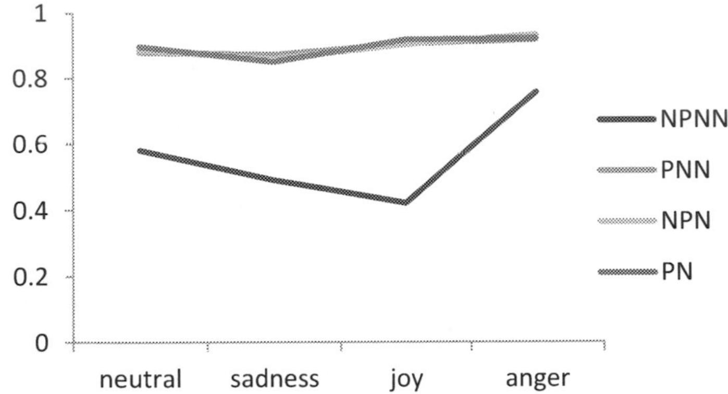

Fig. 4. ROC_AUC for neutral, sadness, joy and anger

We also computed the ROC AUC scores for each emotion as shown in Fig. 4. The ROC AUC score is an indicator of how differentiable an emotion is from other. Here we observe a similar pattern – adding context information proves the ROC AUC scores.

Because we don't have access to the actual validation data used in this competition, we cannot directly compare our results with other teams' results. By just comparing the f1 scores, our model performs better than the best model from last year's competition, which obtained an f1 of 81.5 [22]. The f1 scores for all other teams are lower than 80. In [22], both the past conversation and speaker's name are considered. Further, they also fine-tuned BERT. The difference is that in [22], the sliding window for including context is 2, which means only one proceeding sentence is added as context. [22] also pre-trained BERT with a larger Friends' dataset, which contains the complete scripts of all ten seasons. We didn't perform this pre-training.

The task for differentiating anger, joy, neutral, and sadness is relatively easy because these four emotions are very different from each other. In experiment 2, we tried a harder task by trying to identify all the eight emotion labels using the entire dataset. This task is more challenging because not only are there more emotion categories, but also these emotion categories overlap with each other more. In particular, if an utterance is labeled non-neutral, it means the utterance has received a diverse set of labels from people.

3.2 Experiment 2

Table 3 and Fig. 5 show the results of using our model for detecting neutral, joy, sadness, fear, anger, surprise, disgust, and non-neutral.

Table 3. Results for detecting eight emotion labels

	Accuracy	Micro F1	Precision	Recall
NPNN	0.90	0.61	0.58	0.67
PNN	0.89	0.59	0.57	0.64
NPN	0.90	0.60	0.57	0.65
PN	0.91	0.63	0.59	0.72

In this experiment, adding context information does not always help improving emotion detection. Only in the last condition, where both the proceeding sentence and the speaker's name are added, the model's performance is slightly better than using the current utterance alone as input.

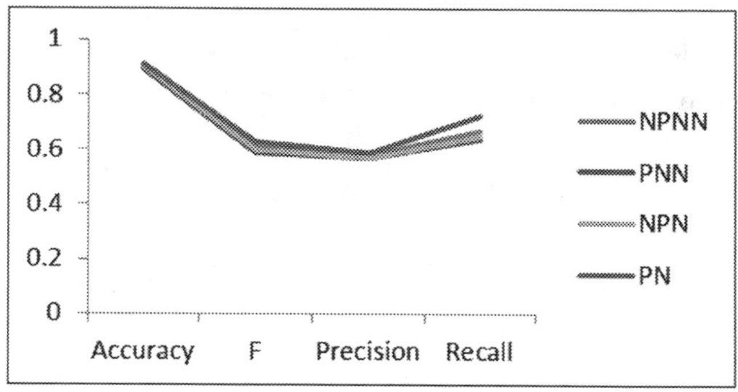

Fig. 5. Results for detecting eight emotion labels

We also computed the ROC AUC scores for each emotion, as shown in Fig. 6. It is not a surprise that the Non-neutral category always has the lowest ROC AUC scores in each condition. The pattern we can observe from Fig. 6 is consistent with that from Fig. 5 – adding context does not help the model to differentiate the emotion categories better.

4 Future Work

In this work, we demonstrated adding context information can help to fine-tune a pre-trained language model, i.e., BERT, in this work to classify the emotion expressed in an utterance's. Adding context information does not hurt the model's performance. It helps more when the emotion categories do not overlap with each other, i.e., the model does not need to differentiate non-neutral from other emotions. We suspect this is because when the emotion categories overlap with each other, even with additional relevant information, i.e., the conversational context of the utterance, the model still cannot extract useful features from the input to make the distinctions among very similar emotion categories.

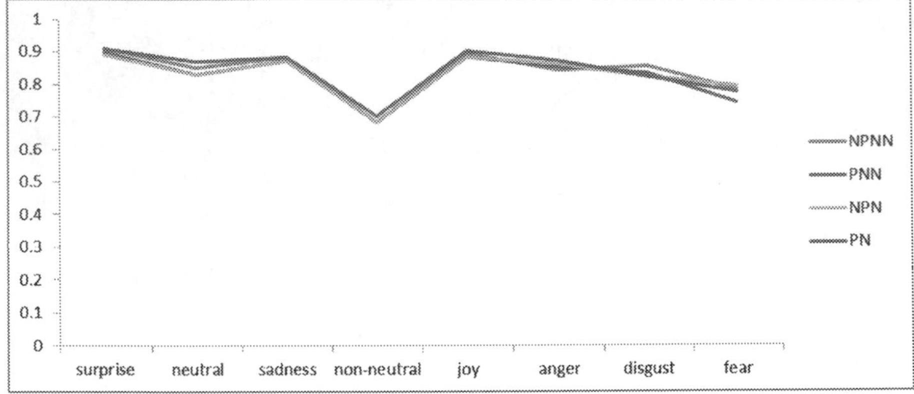

Fig. 6. ROC_AUC for surprise, neutral, sadness, non-neutral, joy, anger, disgust and fear

For our future work, we plan to combine the sentence embedding obtained from BERT with features that are either extracted manually or using an unsupervised learning method to improve the emotion detection results. We will also explore using alternative language models, such as XLnet [24], to encode the input sentences.

References

1. Wang, Y., et al.: Attention-based LSTM for aspect-level sentiment classification. In: Proceedings of the 2016 Conference on Empirical Methods in Natural Language Processing (2016)
2. Tang, D., Qin, B., Liu, T.: Aspect level sentiment classification with deep memory network. arXiv preprint arXiv:1605.08900 (2016)
3. Nakov, P., et al.: SemEval-2016 task 4: Sentiment analysis in Twitter. arXiv preprint arXiv: 1912.01973 (2019)
4. Hussein, D.M., Mohamed, E.-D.: A survey on sentiment analysis challenges. J. King Saud Univ. Eng. Sci. **30**(4), 330–338 (2018)
5. Chatterjee, A., et al.: SemEval-2019 task 3: EmoContext contextual emotion detection in text. In: Proceedings of the 13th International Workshop on Semantic Evaluation (2019)
6. Bandhakavi, A., et al.: Lexicon generation for emotion detection from text. IEEE Intell. Syst. **32**(1), 102–108 (2017)
7. Seyeditabari, A., Tabari, N., Zadrozny, W.: Emotion detection in text: a review. arXiv preprint arXiv:1806.00674 (2018)
8. Scherer, K.R., Wallbott, H.: International survey on emotion antecedents and reactions (isear) (1990). (2017)
9. Strapparava, C., Mihalcea, R.: Semeval-2007 task 14: affective text. In: Proceedings of the Fourth International Workshop on Semantic Evaluations (SemEval-2007) (2007)
10. Alm, C.O., Roth, D., Sproat, R.: Emotions from text: machine learning for text-based emotion prediction. In: Proceedings of the Conference on Human Language Technology and Empirical Methods in Natural language Processing. Association for Computational Linguistics (2005)

11. Hsu, C.-C., Ku, L.-W.: SocialNLP 2018 emotionX challenge overview: recognizing emotions in dialogues. In: Proceedings of the Sixth International Workshop on Natural Language Processing for Social Media (2018)
12. Shmueli, B., Ku, L.-W.: SocialNLP EmotionX 2019 challenge overview: predicting emotions in spoken dialogues and chats. arXiv preprint arXiv:1909.07734 (2019)
13. Sim, J., Wright, C.C.: The kappa statistic in reliability studies: use, interpretation, and sample size requirements. Phys. Ther. **85**(3), 257–268 (2005)
14. Balahur, A., Hermida, J.M., Montoyo, A.: Detecting implicit expressions of sentiment in text based on commonsense knowledge. In: Proceedings of the 2nd Workshop on Computational Approaches to Subjectivity and Sentiment Analysis. Association for Computational Linguistics (2011)
15. Sykora, M.D., et al.: Emotive ontology: extracting fine-grained emotions from terse, informal messages. In: Proceedings of the IADIS International Conference Intelligent Systems and Agents 2013, ISA 2013, Proceedings of the IADIS European Conference on Data Mining 2013, ECDM 2013 (2013)
16. Purver, M., Battersby, S.: Experimenting with distant supervision. In: Proceedings of the 13th Conference of the European Chapter of the Association for Computational Linguistics, pp. 482–491. Association for Computational Linguistics (2012)
17. Strapparava, C., Valitutti, A.: Wordnet affect: an affective extension of wordnet. In: Lrec, vol. 4. no. 1083–1086 (2004)
18. Esuli, A., Sebastiani, F.: SentiWordNet: a high-coverage lexical resource for opinion mining. Evaluation **17**(1), 26 (2007)
19. Devlin, J., et al.: BERT: pre-training of deep bidirectional transformers for language understanding. arXiv preprint arXiv:1810.04805 (2018)
20. Khosla, S.: EmotionX-AR: CNN-DCNN autoencoder based emotion classifier. In: Proceedings of the Sixth International Workshop on Natural Language Processing for Social Media (2018)
21. Wolf, T., et al.: Huggingface's transformers: state-of-the-art natural language processing. ArXiv, abs/1910.03771 (2019)
22. Gallé, M.: Investigating the effectiveness of BPE: the power of shorter sequences. In: Proceedings of the 2019 Conference on Empirical Methods in Natural Language Processing and the 9th International Joint Conference on Natural Language Processing (EMNLP-IJCNLP) (2019)
23. Huang, Y.-H., et al.: EmotionX-IDEA: emotion BERT–an affectional model for conversation. arXiv preprint arXiv:1908.06264 (2019)
24. Yang, Z., et al.: XLNET: generalized autoregressive pretraining for language understanding. In: Advances in Neural Information Processing Systems (2019)

A Novel Tool for Online Community Moderator Evaluation

Alicia J. W. Takaoka[1,2](✉)

[1] University of Hawai'i at Hilo, Hilo, USA
ajwilson@hawaii.edu
[2] University of Hawai'i at Mānoa, Honolulu, USA
https://cms.uhh.hawaii.edu/faculty/atakaoka/

Abstract. This study introduces a new instrument for leadership evaluation in online forums and other online communities which was developed using a grounded approach. Questions that emerged from the literature were then evaluated to create hypotheses that guided the development of an instrument for moderator evaluation. The Moderator Evaluation Contingency Scale (MECS) is modified from Fiedler's contingency model to determine if a moderator is more task- or relationship-oriented in his or her approach to moderation and interactions with other members of a community. The MECS was developed and tested on Reddit in 2013–2014 using random sampling for Forum selection, moderator selection, and interactions with users. A content analysis using the MECS to evaluate posts was found to be a viable measure of a moderator's ability to perform tasks like removing content as well as his or her ability to interact with users. Bots were analyzed using the MECS as well to determine bias. Next steps include making the instrument available for use by social media and niche community sites, administrators, and other moderators.

Keywords: Moderation · Online community · Online leadership · Fiedler's contingency model · Interaction instrument · Leadership evaluation

1 Introduction

Leadership is well-studied in academic literature. Studies across disciplines overlap and converge by observing the behavioral phenomena in different settings to evaluate qualities, like perception and emotion, of the ideal leader in a given situation [15]. This study updates Fiedler's contingency model of leadership to evaluate moderators and content moderation in online forums and communities. While Fiedler's model is not without its limitations [26], the model is versitile and adaptable for the both evaluating the needs of a community as well as evaluating existing work of a potential moderator. Researchers have been examining how to effectively lead teams in virtual settings (e.g. [5,11,25,28,30,31], and [35]), and leading in virtual classrooms [33]. Still most of the research is about leadership in the context of business and face-to-face interactions rather than

C. Stephanidis et al. (Eds.): HCII 2020, LNCS 12427, pp. 454–463, 2020.
https://doi.org/10.1007/978-3-030-60152-2_34

online interactions. Of the online leadership that is being studied, it relates to business. The body of work on online leadership is growing, but the research conducted on the behaviors of online leaders in forums and other types of social media is still minimal.

2 Background

This study evaluates the ability of measuring leadership styles based on moderator interactions, reminiscent of Vlachopoulos and Cowan's study of online moderation [34] before they become moderators. While this study does not assume a community informatics approach, it acknowledges the need for an administrator to evaluate a potential online moderation leader based on the needs of the group, forum, or community. This assumes that administrators will know the type of leadership style needed based on content and community composition [19].

Leadership is important in online spaces. Avolio and Kahai with others [2,3], and [4] have advocated the move to e-leadership for businesses in a current context [12] as well as re-evaluated the state of e-leadership studies ten years later [6]. Cosley et al. and others have stated that both oversight and moderation are necessary in online communities. Even if the behavior persists in the form of griefing, trolling, or becoming an archetype (e.g. [10,14,20], and [27]), moderation is a function of the site that is trusted to "key members who manage and maintain the community. These tasks include moderation, governance, welcoming new members, and building Frequently Asked Question (FAQ) lists" [10]. These functions are what separate some leaders from everyday users. There are several leadership styles that can be identified, but this paper focuses solely on Fiedler's contingency model of leadership.

2.1 Leadership Styles

According to Lewin, there are three primary leadership styles. They are authoritarian, participative [24], and laissez-faire [8]. All three styles operate effectively on Reddit in various subReddits dependent on different situational contexts. Authoritarian leadership is exhibited when the leader has very clear expectations and outlines how and when actions should be taken. Authoritative leadership is best used in situations where there is little time for group decision-making, as is the case when a moderator must choose to remove a post or, in some cases, ban a user. In this style of leadership, there is a clear division between the leader and the participants [29]. On Reddit, moderators have dual roles as both a user and as a moderator. When they participate in a conversation, however, the role is clearly visible. The moderator's name is green, and there is at least one insignia next to their name: **[M]**.

2.2 Fiedler's Contingency Model

Fiedler's contingency model of leadership was developed in the 1960s to determine if a person is more task-oriented (task motivated) or concerned with interpersonal relationships (relationship motivated) when making leadership decisions. Fiedler states, "a leader's performance is contingent on two interacting factors: (a) the individual's goals and needs or motivational structure and (b) the degree to which the situation provides the leader with control and influence over the outcome of the task" [18]. The instrument used to measure a leader's motivation is the least preferred coworker scale. A current leader is given this scale, which uses polar opposite adjectives, to describe his or her least preferred coworker. Leaders with lower scores who use mainly negative words to describe the least preferred coworker tend to be more task-oriented with leaders who have higher scores and use positive words to describe the least preferred coworker are more relationship-oriented. From this, the following question was developed: *Can Fiedler's leadership instrument be modified to evaluate online community moderators?*

2.3 Moderation

Moderation is important in online communities for many reasons. Administrators, designers, and community members have to regulate the content in the site in some way to ensure the quality of posts, the relevance of the topic, and that users interact in a way that is respectful or beneficial to one another. Grimes et al. explain that terms of use are usually obscured to the user by legal language, but they should be created by the users [22]. In this way, users can create a sense of ownership in the site and are aware of the policies. One way of making policies clearer to users are to have visible leaders that can be relied upon to explain why rules are in place and maintain a respectful environment.

Moderation is both a universal concept that transcends one group, yet a moderator is situated in a specific structure and culture. They are seen as leaders and figures of authority. Users look to moderators when other users break the rules of the community for punishment and resolution of a situation. In some online communities, users even look to moderators to facilitate discussions. This study examines whether content moderation is function of leadership styles. Specifically, this study looks at content moderation as a function of leadership determine if a moderator can switch between leadership styles depending on the situation.

2.4 Building Trust and Social Roles

The success of any online community is dependent on trust. Trust must be built between administrators and others in leadership as well as among the users who belong to the community. In some ways, this is done by users building an identity, creating a profile, and interacting with one another based on mutual friends or similar interests [9,16,23]. One way to build trust is through effective moderation of a site. Since a subReddit is a niche community with rules unique to that community, effective moderation is important for both new and existing users of those subReddits.

Moderation is a social role that a user can assume. It is a promotion in some ways, and a burden in others. Social role theory says that our behaviors, actions, thoughts, and wants are prescribed by a specific set of socially determined roles. Roles are the parts that we play, and these roles vary depending on the situation. One does not play the same role when she is a coworker as when they are wife, mother, or sister or husband, father, or brother. Each role is unique to a specific set of situations. The social construction of roles is maintained as we transition to interactions in online communities. Gleave, Welser, Lento, and Smith look at the foundation of a social role. The authors believe a role is rooted in and can only be understood from a context of structure and culture. The authors write:

> Our definition asserts that social roles begin from a structural foundation in simple commonalities in behavior...the role of father fills certain basic social needs and is therefore recognizable across cultures, both by outside observers and individuals living within a given social environment. Similarly, many social roles, especially those that are newly emerging, will have distinctive social structural foundations even if they have not yet developed the same level of recognition within and across cultural settings [21].

Moderation is both a concept that transcends borders and is unique to the specific community in which moderation occur. From this, the following question was developed: *Can an instrument be used to evaluate current and potential leaders according to the needs of the online community?*

2.5 Reddit

Reddit is an online community in which users post links to news and entertainment. A podcast derived from Reddit, called Endless Thread, can be heard on NPR. As Reddit is a collection of posts, it can be considered a bulletin board community, one where the main source of contribution and primary purpose is the posting of content, whether created or shared by links, to be discussed by users. As Anders writes, "Reddit is a giant bulletin board made up of 185,000 active forums with their own obsessions: science trivia, political arguments, video game critiques, jokes and photos—lots of photos, of which more than a sprinkling are of cats or naked women" [1]. Reddit has subReddits, created and managed by Redditors (any user belonging to the Reddit community), that boarder on illegal as they are extreme. National Socialism and Cocaine are two examples of extreme subReddits. Reddit gets over 3.4 billion page views a month. It was purchased by Advance Publications, which is also the parent company for Conde Nast, in 2006.

One of the appeals of Reddit is the anonymity afforded to users. Users can create as many usernames associated with a single account as they want, and they can change their names an infinite number of times. This is a freedom limited in most social media and niche communities.

3 Methods

This sequential study led to the development of an instrument that can be used by moderators to evaluate potential moderators in their online communities. The first phase was to determine the overall leadership style of Reddit and subreddits. Next was to determine the translatable dichotomous characteristics from the Least Preferred Coworker Scale for online moderation. Next was to identify moderators and their posts in order to test the instrument.

Ten Moderators were selected at random from the twenty main forums, and ten posts each moderator, or 100 total posts, specifically moderating another user were selected to evaluate the effectiveness of the dichotomous pairs. These posts include comments, acts of moderation, stating the rules of the subReddit, deleting posts or comments, and banning users. This data was scaled using Fiedler's Contingency Scale. Categorical analysis was used to evaluate the level of interaction between the moderator and Redditor receiving the moderation. The dichotomous pairs adapted for the MECS canbe seen in Fig. 1. This study was repeated with a larger data set. Twenty moderators were selected, and they moderated a total of 27 subreddits between them. Their posts, 270 in total, were selected across subreddits.

Unfriendly\ Friendly	Hostile\ Supportive	Unkind\ Kind	Guarded\ Open	Meant to Discuss?

Fig. 1. Categories adapted from Fiedler's Contingency Model- Least Preferred Coworker Scale.

The scale for analysis was reduced from 8 points to 5 points for categorical simplicity. Moderated posts were organized by moderator and post number in the order the content was retrieved from Reddit, as can be seen in Fig. 2. Each post was evaluated manually based on the moderated interaction, denoted by [M] only. Further interaction between the moderator and Redditor was only included in the content analysis if the interaction was relevant to the moderation. In other words, interactions that were irrelevant to the moderation- like conversations with other Redditors, images, and phatic communication- were not included in analysis. Response time was excluded but is available in the instrument, as seen in Fig. 4.

4 Results

Moderators employ a wide range of leadership styles. Figure 3 highlights this by displaying the total scores and averages of each moderator evaluated. Six moderators were task-oriented, thirteen were flexible or neutral, and only one

Post	Mod#	Timely\ Untimely	Unfriendly\ Friendly	Hostile\ Supportive	Unkind\ Kind	Guarded\ Open	Meant to Discuss?	Total	Notes
1	10		3	3	3	3	1	13	explanation
2	10		3	3	3	3	1	13	explanation
12	10							0	delete
4	10		1	1	1	1	1	5	
5	10		1	1	1	1	1	5	
6	10		2	2	2	2	1	9	
7	10		2	2	2	2	5	13	
8	10							0	post- gives rights to all redditors
3	10		2	2	2	2	1	9	
10	10							0	NA
11	10		3	3	3	3	1	13	
9	10		3	3	3	3	1	13	explanation
	10		20	20	20	20	13	93	10.33333333

Fig. 2. Example of the MECS criteria for evaluation in action.

moderator was relationship-oriented. Of the twenty moderators analyzed for this study, two were bots. As expected, their scores were neutral across the board. Their programmed responses were not meant to incite discussion. Both bots averaged a score of 15, according to the MECS. Scores ranged from 5.6 on the low end to 23 out of 25 as the highest. These scores accentuate different preferences for moderation as well as shape the role of the moderator in and the culture of a given subreddit. The given ranges for MECS overall scores and a moderator's leadership ability can be defined as:

- 5.00–11.99: Task-Oriented
- 12.00–18.99: Flexible to Neutral Moderator
- 19.00–25.00: Relationship-Oriented

Relationship-oriented moderators excel at reaching the user. They are willing to give users chances to modify behavior and are compassionate of users for errors like posting in the wrong place or breaking a rule once or twice. These moderators can acculturate new users to a community efficiently because they are less strict and more patient than task-oriented moderators. The median range is the flexible or neutral moderator. A flexible moderator is able to judge a situation and act accordingly. When the situation requires a moderator to acculturate or explain a situation to a user, the flexible moderator is able to do this. This is in contrast to [19] found and thus needs more exploration using MECS.

Moderators who have a lower score tend to be more abrupt in their interactions with users. They deal well with tasks but may lack in communicating meaningful information to the user. Moderators with higher scores tend to have better interactions with users. These moderators can successfully explain a situation or action with tact. It is worth noting that no person scored a 15. Only bots achieved a score of total neutrality by the crafted message that is sent to a user in the comments section. This message refers users to contacting a person if they have additional questions about the moderation.

The bots had neutral scores of 15. In all cases of a bot's presence, the responses were courteous, informative, provided instruction on either the removal

Mod#	Response Time	Unfriendly\ Friendly	Hostile\ Supportive	Unkind\ Kind	Guarded\ Open	Meant to Discuss?	Total	MECS Score
13		33	32	33	33	24	155	5.61
6		23	23	23	25	13	107	8.33
24		19	22	20	21	25	107	9.42
10		21	20	21	27	13	102	10.33
12		13	13	13	14	15	68	10.36
25		33	33	32	34	26	158	10.57
17		21	21	21	24	20	107	12.15
5		8	8	8	8	4	36	13
18		31	29	29	29	24	142	13
26		19	21	21	21	15	97	13.29
7		46	45	45	47	16	199	14.41
22		48	48	48	58	32	234	14.41
4		37	37	35	53	14	176	15
20		51	51	51	51	51	255	15
9		26	25	27	37	15	130	15.91
11		24	24	24	25	26	123	16.18
2		39	39	39	65	13	195	17.23
23		27	29	26	27	33	142	17.55
27		52	46	49	55	15	217	18
8		11	10	12	20	4	57	23

Fig. 3. MECS analysis of all moderators.

of content or comments about the post, and included disclosure of being a bot. In the disclosure, instructions for following up with a live person were also included. The same bot had different language written into its response for another type of violation. The response is less cordial and contains stronger language. The response includes the disclaimer to contact the moderators if the user has questions, but the tone is different, showing that users have pre-programmed responses containing different tones by using bold typeface to highlight matters of importance so that the bot still demonstrates the ability to be flexible depending on the severity of the moderation. However, a bot is still not as versatile in its moderation capabilities as a person. Each bot can only address a limited selection of violation types, and a person must continually check and update the code for accuracy.

5 Discussion

This study has found that it is possible to adapt an existing leadership mechanism for online communities and forums, and moderators can be evaluated from the third-person perspective based on the needs of a specific community. Moderators have different functions based on the needs of specific forums. This study has shown that relationship-oriented moderators are harder to find than task-oriented moderators, but more forums require task-oriented moderators than relationship-oriented moderators. However, more widespread testing of the instrument across subreddits and niche communities is required.

Moderators who fell into the flexible to neutral category with a score of 14 or higher on the MECS were able to positively interact with users during moderation. Moderators with a score below 14 tended to be more hostile or abrasive in their moderation style and interactions with their community of users. Although past interactions among moderators and the same users were unknown, the appearance of foul language indicated hostility on the MECS scale. It is possible to objectively review potential moderators based on their interactions with other users in a community before selecting them for a leadership role.

Post	Mod	Timely\ Untimely	Unfriendly\ Friendly	Hostile\ Supportive	Unkind\ Kind	Guarded\ Open	Meant to Discuss?	Total	Notes
1									
2									
3									
4									
5									
6									
7									
8									
9									
10									
11									
12									
13									
	[NAME]		0	0	0	0	0	00	

Fig. 4. Moderator evaluation contingency scale (MECS).

This study did not evaluate private messages between moderators and users on Reddit because there is no access to these interactions. It is possible that a moderator engages a user differently in private message than he or she does on a public forum. This is a limitation of this study.

6 Conclusion

The Moderator Equivalency Contingency Scale, or MECS, is a strong indicator for evaluating moderation behavior, and it is a reliable instrument for measuring a moderator's ability to interact with users on a site. It measured with accuracy the ability for a moderator to positively interact with a user regarding the removal of content. This instrument has the potential to streamline evaluation of moderators for researchers as well as site administrators.

Different sites have different needs, and this study has only evaluated publicly viewable content in forums. As Chen, Xu, and Whinston found, moderator interactions may affect the type, quality, and frequency of content created [7]. The MECS can be used to evaluate potential moderators' interactions with other users to determine if their interaction style matches the needs of the site or forum as interaction style may vary minimally. Site administrators can also use MECS evaluations to provide feedback as to why a potential moderator may not have gained a coveted position.

References

1. Anders, G.: Technology- what is reddit worth?-The Web's antibullying, meme-spawning, porn-flecked link forum may go down as one of the shrewdest buys in the media business. Forbes, 56 (2012)
2. Avolio, B.J., Kahai, S.S.: Adding the "E" to E-Leadership: how it may impact your leadership. Organ. Dyn. **31**(4), 325–338 (2003). https://doi.org/10.1016/S0090-2616(02)00133-X
3. Avolio, B.J., Kahai, S., Dodge, G.E.: E-leadership: implications for theory, research, and practice. Lead. Q. **11**(4), 615–668 (2000)
4. Avolio, B.J., Sosik, J.J., Kahai, S.S., Baker, B.: E-leadership: re-examining transformations in leadership source and transmission. Lead. Q. **25**(1), 105–131 (2014)
5. Balthazard, P.A., Waldman, D.A., Warren, J.E.: Predictors of the emergence of transformational leadership in virtual decision teams. Lead. Q. **20**(5), 651–663 (2009)
6. Cascio, W.F., Shurygailo, S.: E-leadership and virtual teams. Organ. Dyn. (2003)
7. Chen, J., Xu, H., Whinston, A.B.: Moderated online communities and quality of user-generated content. J. Manag. Inf. Syst. **28**(2), 237–268 (2011)
8. Cherry, K.: Lewin's leadership styles. Psychology (2011). http://psychology.about.com/od/leadership/a/leadstyles.htm
9. Cheshire, C.: Online trust, trustworthiness, or assurance? Daedalus **140**(4), 49–58 (2011)
10. Cosley, D., Frankowski, D., Kiesler, S., Terveen, L., Riedl, J.: How oversight improves member-maintained communities. In: Proceedings of the SIGCHI Conference on Human Factors in Computing Systems, pp. 11–20 (2005)
11. Crosby, B.C., Bryson, J.M.: Integrative leadership and the creation and maintenance of cross-sector collaborations. Lead. Q. **21**(2), 211–230 (2010)
12. Day, D.V., Fleenor, J.W., Atwater, L.E., Sturm, R.E., McKee, R.A.: Advances in leader and leadership development: a review of 25 years of research and theory. Lead. Q. **25**(1), 63–82 (2014)
13. Delort, J., Arunasalam, B., Paris, C.: Automatic moderation of online discussion sites. Int. J. Electron. Commer. **15**(3), 9–30 (2011)
14. Dibbell, J.: Mutilated furries, flying phalluses: put the blame on griefers, the sociopaths of the virtual world. Wired Mag. **16**(2), 16–20 (2008)
15. Dinh, J.E., Lord, R.G., Gardner, W.L., Meuser, J.D., Liden, R.C., Hu, J.: Leadership theory and research in the new millennium: current theoretical trends and changing perspectives. Lead. Q. **25**(1), 36–62 (2014)
16. Ellison, N.B., Steinfield, C., Lampe, C.: The benefits of Facebook "friends:" social capital and college students' use of online social network sites. J. Comput. Mediat. Commun. **12**(4), 1143–1168 (2007)
17. Fiedler, F.E.: Predicting the effects of leadership training and experience from the contingency model. J. Appl. Psychol. **56**(2), 114 (1972)
18. Fiedler, F.E., Mahar, L.: A field experiment validating contingency model leadership training. J. Appl. Psychol. **64**(3), 247 (1979)
19. Gairín-Sallán, J., Rodríguez-Gómez, D., Armengol-Asparó, C.: Who exactly is the moderator? A consideration of online knowledge management network moderation in educational organisations. Comput. Educ. **55**(1), 304–312 (2010)
20. Gazan, R.: Understanding the rogue user. In: Nahl, D., Bilal, D. (eds.) Information and Emotion: The Emergent Affective Paradigm in Information Behavior Research and Theory. Information Today, pp. 177–185 (2007)

21. Gleave, E., Welser, H.T., Lento, T.M., Smith, M.A.: A conceptual and operational definition of "social role" in online community. In: 42nd Hawaii International Conference on System Sciences, pp. 1–11 (2009). https://doi.org/10.1109/HICSS.2009.6
22. Grimes, J.M., Jaeger, P.T., Fleischmann, K.R.: Obfuscatocracy: a stakeholder analysis of governing documents for virtual worlds. First Monday **13**(9) (2008)
23. Huberman, B.A., Romero, D.M., Wu, F.: Social networks that matter: Twitter under the microscope. arXiv preprint arXiv:0812.1045 (2008)
24. Hussain, S.T., Lei, S., Akram, T., Haider, M.J., Hussain, S.H., Ali, M.: Kurt Lewin's change model: a critical review of the role of leadership and employee involvement in organizational change. J. Innov. Knowl. **3**(3), 123–127 (2018)
25. Huang, R., Kahai, S., Jestice, R.: The contingent effects of leadership on team collaboration in virtual teams. Comput. Hum. Behav. **26**(5), 1098–1110 (2010)
26. Kerr, S., Harlan, A.: Predicting the effects of leadership training and experience from the contingency model: some remaining problems. J. Appl. Psychol. **57**(2), 114–117 (1973)
27. Kirman, B., Lineham, C., Lawson, S.: Exploring mischief and mayhem in social computing or: how we learned to stop worrying and love the trolls. In: CHI 2012 Extended Abstracts on Human Factors in Computing Systems, pp. 121–130 (2012)
28. Lurey, J.S., Raisinghani, M.S.: An empirical study of best practices in virtual teams. Inf. Manag. **38**(8), 523–544 (2001)
29. Maslennikova, L.: Leader-centered versus follower-centered leadership styles. Lead. Adv. Online **14** (2007)
30. Mitchell, A.: Interventions for effectively leading in a virtual setting. Bus. Horiz. **55**(5), 431–439 (2012). https://doi.org/10.1016/j.bushor.2012.03.007
31. Purvanova, R.K., Bono, J.E.: Transformational leadership in context: face-to-face and virtual teams. Lead. Q. **20**(3), 343–357 (2009)
32. Reddit: Reddit Content Policy (2020). https://www.redditinc.com/policies/content-policy
33. Tseng, J.J., Tsai, Y.H., Chao, R.C.: Enhancing L2 interaction in avatar-based virtual worlds: student teachers' perceptions. Australas. J. Educ. Technol. **29**(3), 357–371 (2013)
34. Vlachopoulos, P., Cowan, J.: Reconceptualising moderation in asynchronous online discussions using grounded theory. Distance Educ. **31**(1), 23–36 (2010)
35. Zaccaro, S.J., Bader, P.: E-leadership and the challenges of leading e-teams: minimizing the bad and maximizing the good. Organ. Dyn. **31**(4), 377–387 (2003)

The Influence of Traits Associated with Autism Spectrum Disorder (ASD) on the Detection of Fake News

Jacqui Taylor-Jackson[1(✉)] and Sophie Matthews[2]

[1] School of Psychology, Western Sydney University, Sydney, NSW, Australia
J.Taylor-Jackson@westernsydney.edu.au

[2] Department of Psychology, Faculty of Science and Technology, Bournemouth University, Poole, Dorset, UK

Abstract. It has been suggested that neuro-diverse individuals may be particularly good at detecting online deception (Pick 2019). A small-scale exploratory study was conducted to investigate whether individuals with traits associated with Autism Spectrum Disorder (ASD) were more or less accurate in spotting different types of fake news. A non-clinical sample of university students completed an online identification task, where both fake and real articles items were manipulated in terms of their emotive content. When individuals with low and high scores on the Autism-Spectrum Quotient (Baron-Cohen et al. 2001) were compared, there were no significant main effects on detection accuracy. However, there were two significant interactions, indicating an interesting relationship between message emotiveness, ASD and fake news detection. The results contribute to an understanding of how psychological differences, in particular ASD, may affect online judgements and will contribute to a developing body of work relating positive skills of neuro-diverse individuals to the cybersecurity industry.

Keywords: Autism Spectrum Disorder · Cybersecurity · Neuro-diverse · Fake news

1 Introduction

1.1 Background

Fake news has been defined as, "false stories that appear to be news, spread on the internet or using other media, usually created to influence political views or as a joke" (Cambridge Advanced Learner's Dictionary 2018). Other definitions focus on the intentional fabrication of news to mimic real news content to mislead others (Lazer et al. 2018). The term has grown in popularity and use, particularly so during and after the 2016 American election (Allcott and Gentzkow 2017). Much of the previous research has focussed on factors that affect the believability of fake news. For example, Pennycook et al. (2018) found that when fake news headlines gathered from social media were displayed, just a single prior exposure to the item increased the perception that the headline was perceived as accurate (a 'familiarity effect'). Believability was

C. Stephanidis et al. (Eds.): HCII 2020, LNCS 12427, pp. 464–472, 2020.
https://doi.org/10.1007/978-3-030-60152-2_35

further increased when the headline was labelled as 'contested by fact checkers' or supported by a perceived reliable source.

1.2 Individual Psychological Characteristics and Fake News

The implications of spreading fake news can be great and far-reaching, yet there is relatively little research investigating factors relating to individual psychological characteristics and the detection of fake news. Pennycook and Rand (2019) investigated the 'cognitive psychological profile' of individuals and their ability to differentiate fake news from real news. They found that those individuals with a less analytical thinking style and those who had a tendency to believe information based on its conformity with their own ideology were poorer at differentiating fake news. Pennycook and Rand (2019) explain this by saying that people may believe in fake news because they 'fail to think'. This is supported by the finding that through analytical and open-minded thinking, people can help to protect themselves from disinformation (Bronstein et al. 2019).

1.3 Autism Spectrum Disorder

Individuals with traits associated with Autism Spectrum Disorder (ASD) understand information in different ways compared with those individuals without these traits and they may perceive emotion in different ways. ASD as outlined in the ICD-11 (2018), is a lifelong neuro-developmental disorder that is believed to affect 1% of the population, at a male to female ratio of approximately 2 to 1. ASD can cause an individual to experience difficulties in social and communication aspects of everyday life that are severe enough to result in impairments in other aspects of functioning (ICD-11 2018; Baron-Cohen et al. 2009; Carpenter et al. 2019). Individuals with ASD may display inflexible patterns of behaviour, deficits in Theory of Mind, (recognising the mental states of others), and/or deficits in their ability to empathise with others (Baron-Cohen et al. 1985; Smith 2009). As ASD is a spectrum condition, while many individuals with the condition may share symptoms, each individual with ASD will have their own unique set of symptoms and difficulties. However, despite research focusing on 'difficulties', other research has focused on the positive abilities of individuals displaying ASD traits. For example, Hayashi et al. (2008) found that individuals with ASD exhibit superior fluid intelligence (the ability to solve novel problems and reason independently from previous knowledge), when compared to a control group.

Brosnan et al. (2014) reported that individuals with ASD showed a more circumspect decision-making style, (rather than jumping to conclusions) and DeMartino et al. (2008) suggest that there is a level of reliance upon rational and logical decision making by individuals with ASD. In a study investigating the life experiences of individuals with high functioning ASD, Luke et al. (2012) found that these individuals find decision making to be exhausting and may often avoid decision making if they can. Bronsan, Chapman and Ashwin (2014) also propose that an association between decision-making and anxiety or mental 'freezing' can result in individuals with ASD delaying the decision-making process. Pennycook and Rand (2019) propose that susceptibility to fake news can be explained by lack of reasoning, therefore this might suggest that individuals with ASD traits would be less susceptible. It has been suggested by Pick (2019) that the unique set of aptitudes and skills of neuro-diverse individuals may make them particularly suited to a career in cybersecurity, where there is currently a 'cyber skills gap'. However, Dawson and Thomson (2018) discuss the lack of empirical assessment regarding the cognitive aptitudes, communication skills and team-working needed for individuals to work effectively in the cyber security profession. Recently though, researchers have attempted to remedy this lack of empirical research, and the findings of this study will contribute to this developing body of work.

1.4 Emotiveness and Fake News

Fake news is often created to include highly affective content, particularly through using images and other visual cues (Bakir and McStay 2018). Affective content is used to provoke interest, and to arouse emotion and empathy in the reader (Grabe et al. 2001), while at the same time often forsaking the accuracy of the information presented (Molek-Kozakowska 2013). DeMartino et al. (2008) found that individuals with autism gave less attention to emotional context when making decisions, compared to individuals without ASD traits. Therefore, due to differences in the way that emotions are interpreted, individuals high in autistic traits may be less affected by emotive content in fake news, compared to those with less or no traits. Horne and Adali (2017), suggest that when attempting to differentiate between fake and real news, the title or headline of an article can be a strong indicator in decision making, due purely to its wording and structure. Headlines using emotive language are often associated with topics such as scandal, crime or disaster (Chesney et al. 2017). Exaggeration, embellishment and use of emotive language is often a prominent feature in sensationalist news, especially fake news and clickbait headlines (Molek-Kozakowska 2013).

1.5 Hypotheses

Two hypotheses are proposed:

H1: Individuals with higher levels of autistic traits will correctly identify more articles as being real or fake, compared to individuals with low levels of autistic traits.
H2: For individuals with higher levels of autistic traits, there will be no difference in accuracy of detection for emotive and non-emotive articles.

2 Method

2.1 Design

The study used a 2 × 2× 2 mixed factorial design. Two between-subjects independent variables were: whether the article was emotive or not emotive, and whether the article was real or fake. The within-subjects independent variable categorised participants as having high or low levels of autistic traits (measured using the Autism-Spectrum Quotient by Baron-Cohen et al. 2001). This categorisation was based on a median split of the data, where participants scoring 15 or higher on the Autism-Spectrum Quotient were assigned to the high-level group, and those scoring less than 15 assigned to the low-level group.

2.2 Participants

Participants consisted of a volunteer sample of 35 students. The sample had a mean age of 20.69 years and consisted of 3 male and 32 female participants. A separate sample, used to pilot materials, consisted of 4 participants (1 male and 3 females), with a mean age of 22 years. All participants were rewarded with 0.5 course participation credits.

2.3 Materials

The Autism-Spectrum Quotient (Baron-Cohen et al. 2001) was designed to measure autistic traits in adults of normal intelligence, and to not be significantly affected by IQ. The Autism-Spectrum Quotient was chosen as it is a measure of autistic traits that is not used for clinical assessment or diagnosis of ASD. However, it is effective at measuring traits associated with the disorder and it has been used in many studies involving non-clinical populations. Baron-Cohen et al. provide evidence that the questionnaire has face and construct validity and that it is a valid measure of the five main points of interest that are considered in ASD; social, communication, imagination, attention to detail and attention switching.

Twenty article headlines with an accompanying image were developed: ten of these were sourced from real news reports, whilst the remaining were ‘fake news’, created for this study. The articles were piloted to ensure that they were of similar levels of difficulty for participants to determine whether they were real or fake. Both fake and real articles were categorised by one of the co-authors as either emotive or not emotive, with an equal number of fake and real articles in each category. In order to categorise these articles, the definition of ‘emotive’ as given by the Oxford Dictionary of English (Stevenson 2010) was used, where an emotive issue is one ‘expressing a person’s feelings rather than being neutrally descriptive’. As suggested by Chesney et al. (2017), use of emotive language was designed to evoke emotion or empathy from the reader.

2.4 Procedure

After participants gave their informed consent and their age and gender were collected, they were asked to rate on a scale of 10 how confident they believed themselves to be at detecting fake news, giving an initial confidence score. After this, each article was displayed for as long as a participant desired, and participants were asked to decide whether it was real or fake. Once all 20 articles had been displayed, participants completed the Autism-Spectrum Quotient. After completing the study, participants were thanked and debriefed and informed as to which articles were fake and the role of emotiveness.

3 Results

A 2 × 2 × 2 mixed ANOVA statistical test was conducted. The manipulation of emotiveness was not significant, $F(1,33) = .90$, $p = .35$, therefore, whether the article was emotive or not made no significant difference upon participant's correct identification of fake and real articles. The main effect of level of autistic traits was also not significant, $F(1,33) = .98$, $p = .33$, suggesting that having high or low levels of autistic traits does not affect an individual's ability to correctly identify fake and real news articles. The main effect of article type (fake or real) was significant, $F(1,33) = 5.89$, $p = .02$ and is illustrated in Fig. 3. This suggests that whether the article presented was real or fake had a significant effect on whether the article was identified correctly as real or fake.

There was a significant interaction between level of emotiveness and level of autistic traits, $F(1, 33) = 7.50$, $p = .01$. Figure 1 illustrates this interaction and shows that for individuals with a low level of autistic traits, the article emotiveness significantly affected detection rates; where the high emotive articles were correctly identified more than the non-emotive articles.

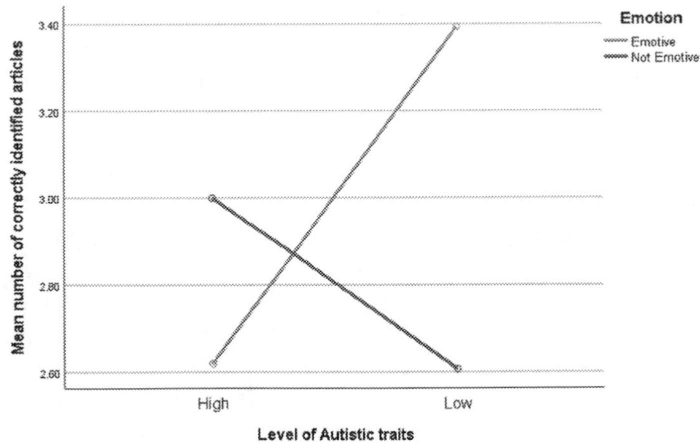

Fig. 1. The interaction between emotiveness of articles (emotive or non-emotive) and level of autistic traits (low or high).

The interaction between emotiveness and article type (fake or real) was also significant, $F(1,33) = 13.38$, $p = < .01$ (Fig. 2). This indicates that when articles were emotive, they were significantly more likely to be correctly identified when they were real, compared to when they were fake. The interaction between level of autistic traits and article type was not significant, $F(1,33) = .58$, $p = .63$, as illustrated in Fig. 3.

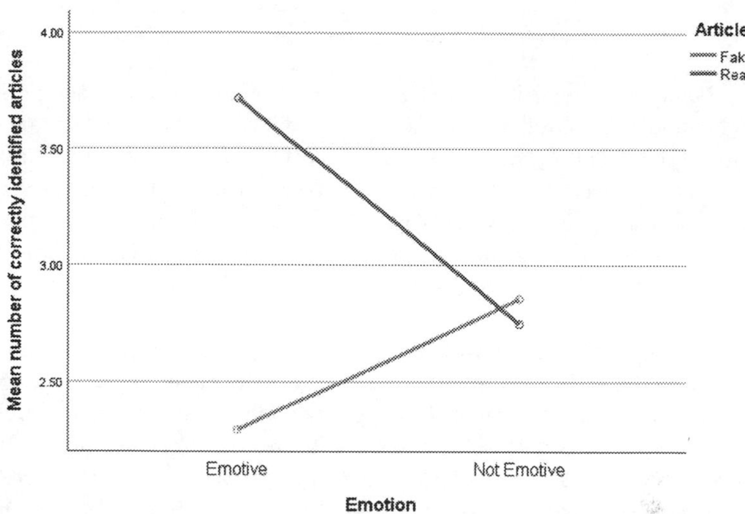

Fig. 2. The interaction between emotiveness of articles (emotive or non-emotive) and article type (real or fake).

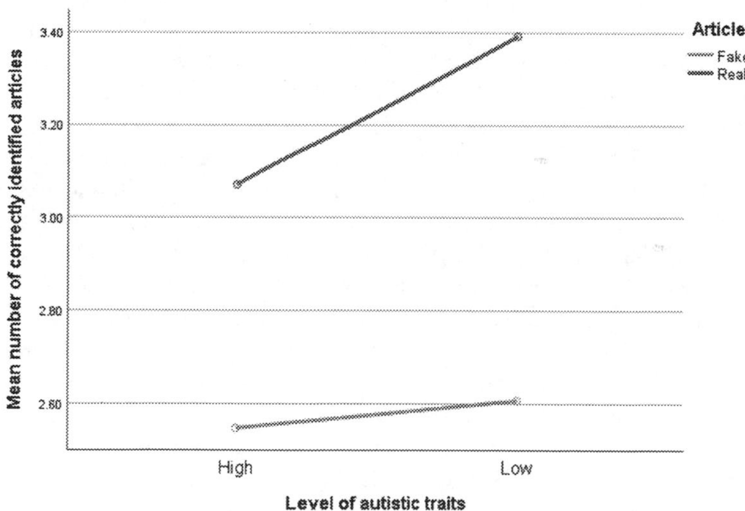

Fig. 3. The interaction between article type (real or fake) and level of autistic traits (low or high).

An independent-samples t-test compared individuals with high and low levels of autistic traits on their initial confidence scores. Although participants in the group with low levels of autistic traits rated their own fake news detection ability to be higher (M = 6.79, SD = 1.53), compared to those in the group with high levels of autistic traits (M = 6.33, SD = 1.85), this difference was not significant, t (33) = .76, p = .45, Cohen's d = 0.26.

4 Discussion

This study found no significant main effect for level of autistic traits, suggesting that having high or low levels of autistic traits does not significantly affect the correct identification of fake and real news articles. Therefore, Hypothesis 1 that individuals who display higher levels of autistic traits will correctly identify more articles as being real or fake, compared with individuals with lower levels of autistic traits is rejected and the null hypothesis is accepted. Both groups correctly identified a higher number of real articles than fake news articles and this was statistically significant. One potential reason for this may be due to the fact that the real news articles were sourced from real world news and therefore may have been seen previously by participants before they took part in the study. In future research, fake news articles should be sourced from real fake news articles and not manipulated to appear as fake news. Future research could explore the reasons why people believe news to be real or fake, and what features of fake news articles participants consider in order to identify them as fake. In contrast to the mainly quantitative studies so far conducted, this could be done through an in-depth qualitative analysis of reasons individuals give for believing an article to be true or fake. Accounts could then be compared for individuals with high and low levels of autistic traits.

The results showed no significant main effect of emotiveness on correct identification of fake news articles, however two significant interactions between emotiveness and autistic traits and between emotiveness and article type were found. As shown in Fig. 1, individuals with low levels of autistic traits correctly identified significantly more articles when they were emotive than when they were non-emotive; while for individuals with higher levels of ASD, emotiveness had no influence. Therefore Hypothesis 2, that individuals with higher levels of autistic traits will not be influenced by the level of emotiveness, is accepted. This provides support for De Martino et al. (2008) that individuals with ASD may be less affected by the affective or emotive content included in fake news and the wider research showing that those with ASD are less likely to take emotional context into account when making decisions. A significant interaction was also found between emotiveness and article type for all participants. Figure 2 shows that emotive content played a prominent role in participant's decision making. When articles were both emotive and real, correct detection almost doubled from that when articles were emotive and fake. While correct detection for the non-emotive articles was the same for fake and real articles. This provides support for the findings of Bakir and McStay (2018), who suggest that fake news is intentionally created to include highly affective content; when fake news articles are more emotive, participants are less accurate at correctly identifying them as fake, compared to emotive real articles. Future research is needed to further investigate how emotive or affective content influences decision making in individuals with ASD.

An independent samples t-test, analysing the initial confidence ratings given by participants regarding how good they believed themselves to be at detecting fake news, showed no significant differences between individuals with low or high levels of ASD. Considering the difficulties and reported dislike of decision-making highlighted by Luke et al. (2012) it may have been predicted that individuals with ASD would rate their own ability as worse than those without ASD. Although this pattern was shown it was not significant and perhaps further research using larger samples may allow this test stronger power. Future research is needed to explore ASD and confidence in decision-making. Also, further research could collect the time needed to make decisions regarding the judgment of fake news considering that Goldstein et al. (2001) and Brosnan et al. (2014) found that individuals with ASD took longer to make decisions.

A key limitation of this study is that non the sample of individuals with a high level of ASD were not clinically diagnosed as having ASD. Further research needs to sample populations of clinically diagnosed ASD individuals. Although the Autism-Spectrum Quotient is a valid measure of traits that are associated with ASD, it is not one used for clinical diagnosis. Another limitation that may have arisen from measuring only traits associated with ASD is that when performing a median split upon the data collected from participants, those who scored higher than 15 were allocated to the high level group. This score is lower than the level that Baron-Cohen et al. (2001) outline as a cut-off for a significant level of traits, with only two participants in the current sample scoring 32 or higher. ASD is a spectrum condition, one in which each individual diagnosed may have symptoms or traits that are shared with and experienced by others as well as having some which are more unique to themselves. Therefore, the results may not be generalisable to all individuals with ASD, due to the fact that each individual's experience with the condition is so unique. Also, the current sample consisted mainly of females, with only three males taking part in the study. Considering that ASD is more common in males than females, (Carpenter et al. 2019), future research needs to use an equal number of male and female participants. Further research could also explore gender differences in the detection of fake news.

Although the study found no significant support for Hypothesis 1 relating level of autistic traits to the correct detection of fake news, there was support for the second hypothesis that individuals with high levels of ASD were less biased by the level of emotiveness in news articles, compared to those with low levels of ASD. Future studies using clinical samples may be able to investigate whether fake news detection is different in participants with a formal diagnosis. Further research is also needed to investigate whether individuals with ASD possess other skills that may relate to the cybersecurity sector which differ from neuro-typical participants. As noted previously, Pick (2019) suggested that adults diagnosed with ASD may have the skill sets needed to fill the current skill gap in the cybersecurity sector. According to the Cyber Neurodiversity Group (2020), potential future empirical findings which support these suggestions could have an important influence in diversifying the sector and help the group in their current aims to utilise the skills of neuro-diverse individuals such as those with ASD within the cybersecurity industry.

References

Allcott, H., Gentzkow, M.: Social media and fake news in the 2016 election. J. Econ. Perspect. **31**(2), 211–236 (2017)

Bakir, V., McStay, A.: Fake news and the economy of emotions: problems, causes, solutions. Digit. Journal. **6**(2), 154–175 (2018)

Baron-Cohen, S., Leslie, A.M., Frith, U.: Does the autistic child have a "theory of mind"? Cognition **21**(1), 37–46 (1985)

Baron-Cohen, S., et al.: Prevalence of autism-spectrum conditions: UK school-based population studies. Br. J. Psychiatry **194**(6), 500–509 (2009)

Baron-Cohen, S., Wheelwright, S., Skinner, R., Martin, J., Clubley, E.: The autism-spectrum quotient (AQ): evidence from asperger syndrome/high-functioning autism, males and females, scientists and mathematicians. J. Autism Dev. Disord. **31**(1), 5–17 (2001)

Bronstein, M.V., Pennycook, G., Bear, A., Rand, D.G., Cannon, T.D.: Belief in fake news is associated with delusionality, dogmatism, religious fundamentalism, and reduced analytic thinking. J. Appl. Res. Mem. Cogn. **8**(1), 108–117 (2019)

Brosnan, M., Chapman, E., Ashwin, C.: Adolescents with autism spectrum disorder show a circumspect reasoning bias rather than 'jumping-to-conclusions'. J. Autism Dev. Disord. **44** (3), 513–520 (2014)

Cambridge Advanced Learner's Dictionary, 4th edn. https://dictionary.cambridge.org/dictionary/english/fake-news. Accessed 19 Jan 2020

Carpenter, B., Happé, F., Egerton, J. (eds.): Girls and Autism: Educational, Family and Personal Perspectives. Routledge, UK (2019)

Chesney, S., Liakata, M., Poesio, M., Purver, M.: Incongruent headlines: yet another way to mislead your readers. In: Proceedings of the 2017 EMNLP Workshop: Natural Language Processing meets Journalism, Copenhagen, Denmark, pp. 56–61. Association for Computational Linguistics (2017)

Dawson, J., Thomson, R.: The future cybersecurity workforce: going beyond technical skills for successful cyber performance. Front. Psychol. **9**, 1664–1078 (2018)

De Martino, B., Harrison, N.A., Knafo, S., Bird, G., Dolan, R.J.: Explaining enhanced logical consistency during decision making in autism. J. Neurosci. **28**, 10746–10750 (2008)

Goldstein, G., Johnson, C.R., Minshew, N.J.: Attentional processes in autism. J. Autism Dev. Disord. **31**(4), 433–440 (2001)

Grabe, M.E., Zhou, S., Barnett, B.: Explicating sensationalism in television news: content and the bells and whistles of form. J. Broadcast. Electron. Media **45**(4), 635–655 (2001)

Hayashi, M., Kato, M., Igarashi, K., Kashima, H.: Superior fluid intelligence in children with Asperger's disorder. Brain Cogn. **66**(3), 306–310 (2008)

Horne, B.D., Adali, S.: This just in: fake news packs a lot in title, uses simpler, repetitive content in text body, more similar to satire than real news. In: Proceedings of Eleventh International AAAI Conference on Web and Social Media, Montreal, Quebec, pp. 759–766. AAAI Press (2017)

ICD-11 Mortality and Morbidity Statistics Section 6A02. https://icd.who.int/browse11/l-m/en#/http://id.who.int/icd/entity/437815624. Accessed 20 Jan 2020

Lazer, D.M., et al.: The science of fake news. Science **359**(6380), 1094–1096 (2018)

Luke, L., Clare, I.C., Ring, H., Redley, M., Watson, P.: Decision-making difficulties experienced by adults with autism spectrum conditions. Autism **16**(6), 612–621 (2012)

Molek-Kozakowska, K.: Towards a pragma-linguistic framework for the study of sensationalism in news headlines. Discourse Commun. **7**(2), 173–197 (2013)

Pennycook, G., Rand, D.G.: Lazy, not biased: susceptibility to partisan fake news is better explained by lack of reasoning than by motivated reasoning. Cognition **188**, 39–50 (2019)

Pennycook, G., Cannon, T.D., Rand, D.G.: Prior exposure increases perceived accuracy of fake news. J. Exp. Psychol. Gen. **147**(12), 1865 (2018)

Pick, K.: Neurodiversity in Cybersecurity. Infosecurity Mag. **Q2**, 10–13 (2019)

Smith, A.: The empathy imbalance hypothesis of autism: a theoretical approach to cognitive and emotional empathy in autistic development. Psychol. Rec. **59**(3), 489–510 (2009)

Stevenson, A. (ed.): Oxford Dictionary of English, 3rd edn., p. 574. OUP, Oxford, UK (2010)

The Cyber Neurodiversity Group. https://www.neurocyber.uk/. Accessed 21 Jan 2020

Analysis of Imitating Behavior on Social Media

Ying Zhong(✉) and Haihan Zhang

University of Tsukuba, Tennodai 1-1, Tsukuba, Ibaraki, Japan
ying.zhong.2018d@mlab.info, zhang.haihan@image.iit.tsukuba.ac.jp

Abstract. In the age of web 2.0, the concept of "the User Generated Content (UGC)" is put forward. Now millions of images posted on the photo-sharing service such as Instagram. In this paper, we choose the popular photos "follow me to" on Instagram as an example for analysis. By utilizing the comparative analysis of the "follow me to" photos of the photographer and the imitators, we explored the pattern of image propagation on Instagram and the characteristic of the popular images. In this paper, we also established a research method by content analysis and image analysis based on the image itself to study social media.

Keywords: Social media · Imitating behavior

1 Introduction

From the 1990s, the Internet began to appear and now has become a part of people's working lives. Social Network Service (SNS) has become an important part of the development of the network in recent years. Then Milner [1] mentioned that web 2.0 began to emerge, and the concept of User Generated Content (UGC) began to come up so that people could start creating the adapted content by themselves. In addition, with the continuous development of technology, mobile phones can easily take and store photos and connect to the network. Photo sharing services began to appear, and the expression of photos is more simple than words.

Today's society is an eye-catching society, the expression of photos are simpler than words. With the continuous development of technology, social media continued to evolve to the era of web 2.0, but also the advent of photo-sharing services. Mobile applications like Instagram that combine image modification and community features also make it easy for users to modify, share, and view photos.

The advent of Instagram has changed users' shooting habits and community experiences. Hochman et al. [3] introduced Instagram's combination of image modification tools, location marking, and instant sharing in a simple mobile application. Instagram makes photos modified and a large number of pictures are uploaded to Instagram every day, showing how fast they are growing.

C. Stephanidis et al. (Eds.): HCII 2020, LNCS 12427, pp. 473–485, 2020.
https://doi.org/10.1007/978-3-030-60152-2_36

Instagram combines photo modification with community features to significantly reduce the difficulty of image production.

A large number of photos have a lot of imitation of other people's photos and other photos adapted. Photos were created, circulated, and altered by countless participants, transforming web users, and creating a shared cultural experience. Imitation requires social media as a platform. Many people upload their photos to Instagram and share them with friends. Some users will imitate the past oil painting, film, popular MV and other content or style to shoot [5]. Besides, there will also be a lot of people imitating photos of well-known photographers on Instagram to imitate shots and upload.

Most of the research on social media now is based on words. In this paper, we hope to establish a research method by content analysis and image analysis based on the image itself to study social media.

In this paper, we want to explore the following three questions in this paper: 1) How is the style of popular images on the social media set up? 2) How do imitated images on social media maintain and change? 3) Do imitated images have a specific image style?

The contribution to this paper is as follows. First, through the comparative analysis of the "follow me to" photos of the photographer and the imitators, we explored the pattern of image propagation on Instagram and the characteristic of the popular images. Second, this paper established a research approach by using image-based content analysis and image analysis to study social media.

The remainder of this paper is organized as follows. Section 2 describes the related work. Sections 3, describes the research method. Section 4 describes the result of our method. Section 5 gives the discussion. Section 6 concludes the paper and presents future work.

2 Related Works

2.1 Web 2.0 with Photo Sharing Service Development

Mirzoeff et al. [6] noted that "by 2012, 380 billion photos were taken in a year, almost all digitally." In 2014, the number of photographs taken was one trillion. That is, "In 2014, the world's photo gallery increased by about 25%. Most of these photos are created by the user."

Web 2.0. From the very beginning of BBS to Classmate.com in 1995 and the same period of SixDegrees.com Blogger, the community network era was revealed. Web 2.0 has created a user-centric, multi-directional, bottom-up communication mode. In the era of web 1.0, the most common actions of users are "search", "reading" and "download" information, the information of the website is mostly one-way flow, the content arrangement is determined by the website operator, the user can only receive, less sense of participation. Web 2.0 emphasizes "mass intelligence" where information comes from people in many different roles, and users are also providers of content when the information provided by those users comes together. One of the great features of Web 2.0 is the creation of a mass-written area, emphasizing the writer as the center, the past one-way, top-down organization, gradually transformed into a multi-directional, bottom-up assembly mode. The website may give the corresponding information service according to the user's preferences or needs, and link users with the same preferences into a community, so much emphasis on interaction and sharing of information flow to the site information volume increased significantly.

Image-Based SNS. Flickr[1] is a widely used photo-sharing platform on the photo-sharing service, uses many Web 2.0 ideas such as content sharing, group sharing, open program code, and so on.

Van Dijk et al. [2] mentioned Flickr as a combination of a social media platform and a database. On the one hand, it is often seen as a common view and experience space through photo exchange, and on the other hand, it is also considered to be a visual archive database of physical memories. It also provided the construction of infinite connections. Embedded in a linked culture, and the social network structure of this culture has gradually penetrated into people's daily life.

Cox et al. [8] focused on the Flickr group and developed a set of stereotyped Flickr community standards.

We can see that Flickr is still a photo-sharing platform. A large number of photographers are taken with a professional camera and may be able to modify the photo on their computer before uploading it to the platform for sharing.

2.2 The Study About Instagram

Highfield et al. [4] mentioned that Instagram was a small community on the iPhone platform since 2010, and was acquired by Facebook in 2012 and can be used directly with Facebook, until now, and has been around the world more than 500 million users.

Instagram[2], which began in October 2010 and was a small app on iPhone, has grown rapidly since then, and was acquired by Facebook in 2012 and can be integrated with Facebook, which can be used directly on Facebook's photo, becoming a very important social media. Instagram's official blog post now has more than 500 million users and an average of more than 70 million photos and videos uploaded to Instagram every day. Hochman et al. [3] mentioned that Instagram combines features such as image modification tools, spot marking, and real-time sharing in a simple mobile app.

Instagram has the following key features: (1) Filters and modifications; (2) Time and space tag; (3) Hashtag; (4) Community features.

Instagram has created a culture of individual records of the world, breaking what was originally provided by big companies (e.g.. google street view). It is similar to early visual art impressionism. Now that the two are coming together,

[1] Flickr: https://www.flickr.com.
[2] Instagram: https://www.instagram.com.

Google now allows users to provide geographic information to present on Google Earth [3].

Instagram can transform a media city by uploading photos of daily life, combining personal experiences with urban spaces. In Instagram, it is entirely built by user-created and user-organized content. Instagram can be forwarded to Facebook, or it can upload photos on a mobile platform. Unlike other platforms, Instagram is primarily creating, editing, uploading, and sharing from the mobile side. Instagram is a platform to observe complex issues and do further research [9].

Instagram combines photo modification and community features to greatly reduce the door to image production, which is often considered a natural state of life. In fact, many photos are created to be life-like and naturalized.

There are also a lot of people on Instagram who take selfies to think about others. Tifentale et al. [10] noted that selfies continue the traditional self-portrait to present themselves. Lev Manovich and his software research lab have a Selfiecity[3] research program that collected 3,200 selfies from five cities and studies from multiple perspectives in photography, art, digital humanities, and software research. The study looked at the selfie-taker's style of shooting based on sex, eye opening, expression, angle, and so on.

In addition to personal expressions like selfies, there are many things or styles that mimic past paintings, movies, popular MV, etc. Some brands or well-known filmmakers will also lead a style. Manovich et al. [5] mentioned that a magazine Kinfolk has an obvious style, and there are plenty of personal accounts on Instagram to go back to imitate his style.

Instagram photos are not just personal expressions, but also a glimpse into the architectural environment of a city. Manovich et al. [11] analyzed the digital footprint of a large number of people constructing cities by analyzing photos of landmarks or moods uploaded by visitors, or by locals, etc.

Social media content, such as Instagram photos, tags, and descriptions, is an important form of urban life and shares the experiences and self-expression of city life. The study found that photos tend to be concentrated in some areas, and in most areas, they are rarely uploaded. Analysis and research on social media provide scant possibilities for urban research and planning.

To sum up, Instagram often gives the impression of life, naturalization, but many of the photos are carefully designed and created life. People are carefully designed to show themselves, such as self-photographing. In addition, many people will imitate the past oil paintings, movies, popular MV or other well-known filmmakers, such as content or style. In addition to being a personal expression, Instagram photos can also paint a picture of a city's environment.

[3] Selfiecity: http://selfiecity.net.

2.3 Imitation on Social Media

Meindl et al. [11] explained the role of the media in promoting generalized imitation and proposed a generalized imitation model how one person's behavior can influence another person to engage in similar behavior.

On social media, there are also a lot of imitating behavior. Stefanone et al. [12] explained that social cognitive theory suggests a likely relationship between behavior modeled on increasingly user behavior modeled on social networking sites (SNSs).

One imitation on social media is meme. Milner et al. [13] referred to meme, which was first to describe the flow and flux of culture. Meme evolved in online media. Meme is an amateur man-made object that blends the participation of different participants in the community. Meme is a unit of popular culture that has transformed web users and created a shared cultural experience. It is a multi-pattern symbol component that is created, circulated, and altered by countless participants. Borzsei et al. [14] mentioned that the research of meme can better understand the digital culture. Meme, as a user-created and constantly changing online content, grew from the 1980s to the 2010s.

Milner et al. [1] mentioned the following characteristics of meme: First, meme needs to have some understanding of the network culture. Second, meme contains a wide range of public expressions, including subcultures. Third, although the public expresses many elements, the perspective sits narrowly.

Heylighen et al. [15] proposed that meme must go through four stages for successful replication, and proposed a selection criteria for a successful meme.

Meme is also a form of public discussion of public affairs. The public's use of meme to convey their views is an expression of citizens outside the political system. Meme is created, disseminated and changed outside of the traditional media gatekeepers [13].

3 Method

3.1 Date Set

This study chooses the hand-in-hand photo of "follow me to" as the research dataset for three reasons: First, this type was a photo posted on Instagram in 2011 by Murad Osmann and his wife, Natalie Nataly Zakxarova, on Instagram. Murad Osmann has 4.6 million followers, and there are plenty of "followmeto" hashtags on Instagram, suggesting that this category of photos has attracted a lot of imitation. Manikonda et al. [7] looked for 5659795 photos of 369828 users and got the hashtag, of which "love" ranked first, "follow" ranked second, "me" ranked fourth. It can be seen that the theme of this category is the most popular and representative of Instagram. Second, this type of photo is easy to imitate. There are imitations all over the world and less cultural differences. Third, Murad Osmann who is the first to upload a hand-in-hand photo of "follow me to" is a professional photographer. His photos also have obvious style characteristics. This type of imitation is a meme shot of the type of photo in a meme that is directly imitated.

3.2 Research Design

This study is based on the concept of relevant photography theory as a guide to style analysis. Figure 1 shows the research process. First, we crawled photos and related data through the Instagram API. Second, we use OpenCV[4] to capture messages such as the brightness of the picture, including total brightness, body brightness, background brightness, body position, and Project Oxford[5] to extract tones, including main, foreground, background, and picture labels. In conjunction with the content analysis method, the content of the picture foreground and background are determined manually. Finally, we analyzed the results and got conclusions.

Table 1 show the components we analyzed. The details are as follows.

- Brightness
  The inner orientation of light creates mood and atmosphere. The background dark will produce a mysterious "low" mood, the background bright produces a more happy "high" mood [16].
  - The overall average brightness
    This study uses OpenCV to take the r, g, b values of each pixel in the picture, and then calculate the brightness of each pixel. The brightness formula is brightness = 0.3 * r + 0.6 * g + 0.1 * b. The total brightness is the average of all pixel brightness. The brightness range is 0 to 255.
  - Body brightness
    We used Project Oxford to take the image's high and one-third wide-body thumbnail.
  - Contrast
    Contrast refers to the difference in brightness between the brightest white and darkest blacks, or how much of a gradient from black to white.
- Composition
  - Body position
    We used Project Oxford's thumbnail feature to cut out the main part of the photo. Then match the midline of the thumbnail, one-third left, one-third middle, or one-third right, to determine the left, middle, or right.
  - Scenic
    We divide photos into long shot, medium shot, and close shot by manual classification. long shot refers to the entire body that can hold the subject person. medium shot is the body shape that encompasses the subject from the knee or above the waist. close shot means to focus on small objects, such as a person's face.
- Content: foreground and background.

[4] OpenCV: https://opencv.org.
[5] Project Oxford: https://azure.microsoft.com/.

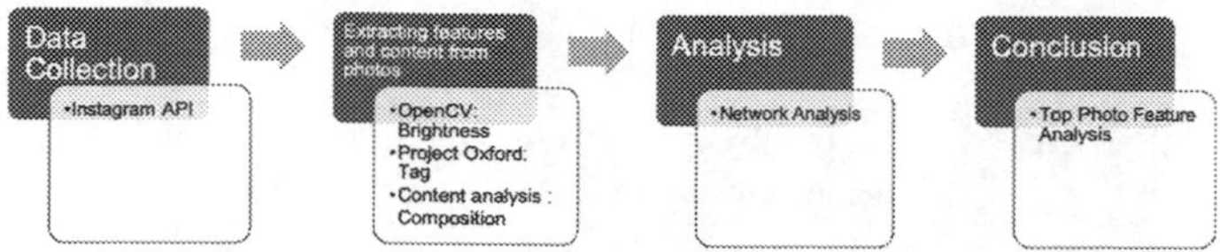

Fig. 1. Overview of the research process

Table 1. Style and content of photography

Style	Brightness	The overall average brightness	The average brightness of the photo
		Body brightness	Whether to highlight the subject
		Contrast	Difference in brightness
	Composition	Body position	Whether in the prominent position
		Scenic	Long shot, medium shot and close-up
Content	Content	Foreground	The body of the photo
		Background	The background of the photo

4 Result

4.1 Brightness

- The overall average brightness
  This study classifies brightness as high, medium, and low. The brightness is low (below 100), medium (100 to 150) or high (above 150). Of these, 132 photos were high in brightness, accounting for 70%, 38 photos with low brightness, accounting for 20%, and 19 photos with high brightness, accounting for 10%.
  We find that most of the brightness is distributed between 100 and 150, and **most of the brightness is distributed in the middle brightness**. It is also possible that the photographer has dimming to make the photo brightness more suitable for the public to see.
  The brightness range of the imitator's photo is 30 to 230, the brightness range is larger than that of the photographer, and **the imitator may not pay attention to the dimming**.
- Body brightness
  Of the photographer's photos, a total of 116 of the 189 photos have greater brightness than the overall average brightness, which shows that photography will make the subject brighter, **highlighting the subject**.
  Only 372 of the 808 photos of the imitator were brighter. This means that the imitator does not focus on highlighting the subject in terms of brightness.

Table 2. The content of the background

Person	Assisted by a character, such as a person holding a camera in the background
Landscape	A photo of the landscape, including natural scenery, urban, rural landscape, etc.
Scene	The background of the photo mainly represents the scene and the event's photos
Close-up	A close-up of a finc part of something, typically taken using a telephoto lens or a zoomed-in lens

- Contrast
  Of photographer's photos, 147 photos have high contrast, accounting for 78%, 38 photos with medium, accounting for 20%, and 4 photos with low, accounting for 2%. Photographer Murad Osmann's photo can be seen with **high contrast**.
  The imitator's 808 photos, 473 with high contrast, accounting for 59%, 276 photos with medium contrast, accounting for 34%, and 59 photos with low contrast, accounting for 7%. Compared to the original photographer, **the imitator's photo has a higher contrast ratio than the original photographer's**.

4.2 Composition

- Body position
  Of photographer Murad Osmann's 189 photos, 186 are in the middle, 2 on the right and 1 on the left. The right side of the visual senses heavier than the left, so you can see that the vast majority of photos are in the **main position, highlighting the main body**.
  Compared with the photographer, 798 of the 808 photos of the imitator were in the middle, 4 on the left and 6 on the right. **The situation is similar to that of a photographer**.
- Scenic
  f photographer Murad Osmann's 189 photos have 180 mid-view, accounting for 95%, 9 photos are long shot, accounting for 5%. We find that **most part of the photos is medium shot**.
  In terms of scenery, of the 808 copies, 721 were **medium shot**, accounting for 89%, 86 were long shot, accounting for 11%, and 1 was close-up.

4.3 Content

- Foreground
  Photographer Murad Osmann's 189 photos of the prospect are all **women**,

also known as the photographer's wife.

For the photos of imitators, based on the subject, the study divided the foreground into women, men, children, multiple people, and non-human.

In terms of foreground, of the 808 photos, the foreground content was **women** accounting for 81% (653), animal accounting for 16% (125), men accounting for 3% (26), and 2 photos were non-human and multi-person.

In terms of foreground content, photographer Murad Osmann's photo looks female. The foreground content of the imitators' photos is also dominated by women.

- Background

  This study divides the background content into characters, landscapes, scenes and close-ups. The classification rules are in the following Table 2.

  The background content has 152 photos for the **landscape**, accounting for 80%, 25 photos for the scene, accounting for 13%, 11 photos as persons, accounting for 6%, 1 photo as a close-up, accounting for 1%.

  For the photos of imitators, 775 of the 808 photos of the imitator were **landscapes**, accounting for 96%, 26 for scenes, accounting for 4%, and 3 persons and 1 close-up.

  Unlike photographer Murad Osmann's photographs, the imitators are mostly landscapes, with fewer scene classes in which the characters are dressed and related to the background, and the scene category is highly narrative.

Fig. 2. Text cloud of photographer's content label

Fig. 3. Text cloud of imitator's content label

4.4 Content Label Analysis

Frequency Analysis. Using Project Oxford's labeling feature, which takes a label above 0.8, photographer Murad Osmann's 189 photos have a total of 130 labels. Words that appear more than 10 times are outdoor, person, sky, water, building, grass, crowd, scene, tree (see Fig. 2). Of these, 155 are labeled with outdoors, and most of the photos can be seen taken **outdoors**. Most of the photos are sky, earth, water, architecture, lawns, trees and so on.

Project Oxford's labeling feature sits with a confidence rating of 0.8 or more, with a total of 223 volumes for 808 photos of the imitator. Words that appear more than 100 times are outdoor, person, sky, water, tree, grass (see Fig. 3). Of these, 658 are labeled with outdoors, and most of the photos can be seen taken outdoors. Most of the photos are the sky, earth, water, architecture, lawns, trees, and so on. A total of 105 labels in the photo are the same as photographer Murad Osmann. The high-frequency content of the copycat's photo is similar to that of photographer Murad Osmann.

Co-exist Networks. This study takes each label as a node, weighting the number of times the label appears on the image at the same time, constructing a label co-exist network in Gephi[6], photographer Murad Osmann and the imitator's co-exist network Fig. 4 and Fig. 5.

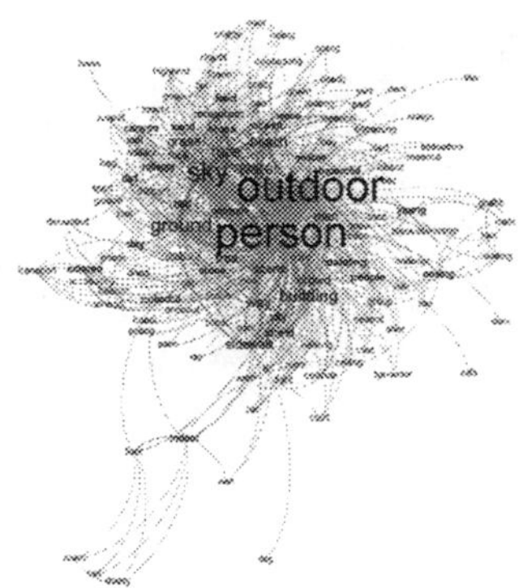

Fig. 4. Co-exist network of photographer's content label

[6] Gephi: https://gephi.org.

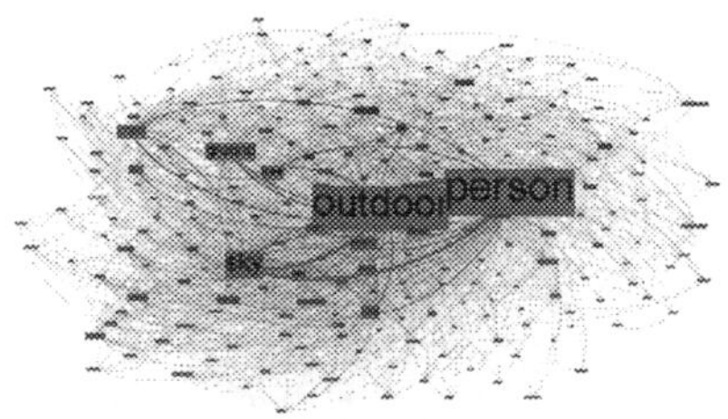

Fig. 5. Co-exist network of imitator's content label

Photographer Murad Osmann's network diagrams can be broadly divided into five categories. The first category is **outdoor photography**, the main words are person, outdoor, sky, ground, and scene. The second category is **natural photography**, the main words are beach, water, ocean, and mountain. The third category is **indoor photography**, the main words are indoor, floor, wall, hall, and library. The fourth and fifth categories have less content, the fourth category is **decorative photography**, the main words are colored, colorful, balloon, and dressed. The fifth category is **graffiti photography**, the main words are graffiti, text, skating, and posing.

The imitator's network diagram can be broadly divided into three categories. The first category is the **general outdoor photography**, the main words are outdoor, person, sky, and water. The second category is **natural photography**, the main words are ground, tree, grass, and dog. The third category is **street view**, the main words are sidewalk, building, street, and road.

Comparing the contents of the imitator and photographer Murad Osmann, the content categories are roughly similar, while photographer Murad Osmann has more indoor photography.

5 Discussion

We also observed the photographs of photographers and imitators. In addition to the above results, we found the following characteristics.

First, the photographer focuses more on clothing and the narrative of images. There are a large number of photographs of the woman's clothes associated with the background. For example, a ball game, tailoring, and so on. Such topics tend to highlight the intention of emotional expression. In the classification of content, the classification of a scene refers to whether the main person in the photo is associated with the background. Of the 189 photos taken by photographer Murad Osmann, 25 are scenes, accounting for 13%. For example, a woman dressed as Brazilian fans stand in front of football stadiums, dressed as an astronaut dressed in front of the capsule, and wore suits around the suit shop with someone helping to measure. On the contrary, the imitator does not pay attention to the narrative of the image.

Second, shooting location. Photographers have visited many countries and famous attractions. He has been to India (including the famous Ganges and

Taj Mahal), China, the United States (including the famous Hollywood and the Statue of Liberty), the United Arab Emirates, Singapore, France (including the famous Eiffel Tower), Brazil (including the Statue of Christ in Rio de Janeiro), Italy, Germany, and more. In contrast, a single imitator doesn't go to that many locations to shoot.

6 Conclusion

Through the comparative analysis of the "follow me to" photos of photographer Murad Osmann and the imitators, the popular images on social media have characteristics: 1) It is easy to imitate the photo's composition and view. Most of the photos are medium shot, people in the middle, the viewing angle and the flat perspective are mostly. The content of the shoot is often set against a background of similar scenes and locations that are relatively easy to find and imitate. 2) Paying attention to the model's clothing and image narrative. 3) Having enough funds to take pictures around the world. 4) Having arts beauty and skilled photography. The photos of imitators and photographer Murad Osmann have the same composition and contents.

The above from the degree of ease to achieve from simple to difficult, the first point is directly imitate the form of the photo can be, the second point needs to have emotional expression in the photo, the third point is the need to have a certain economic ability to go to different attractions to shoot scenes, the fourth point needs long-term training of photo processing and photographic literacy.

There also have some differences between the photographer and imitators: 1) Richer in shooting scale and contents. 2) Paying no attention to image narrative. 3) Having not enough funds to take pictures around the world. 4) Having not enough arts beauty and skilled photography. In conclusion, compared to hot photos, most of the imitators' don't have their own characteristics. But some imitators have some characteristics.

The contribution to this paper is as follows. First, through the comparative analysis of the "follow me to" photos of the photographer and the imitators, we explored the pattern of image propagation on Instagram and the characteristic of the popular images. Secondly, this paper established a research approach by content analysis and image analysis based on the image itself to study social media.

In this paper, there is one of the photographers on Instagram, Murad Osmann, took a case to explore image style. In the future, we will explore more case studies.

References

1. Milner, R.M.: The world made meme: discourse and identity in participatory media. University of Kansas (2012)
2. Van Dijck, J.: Flickr and the culture of connectivity: sharing views, experiences, memories. Memory Stud. **4**(4), 401–415 (2010)

3. Hochman, N., Manovich, L.: Zooming into an Instagram City: reading the local through social media. First Monday 18(7) 2013
4. Highfield, T., Leaver, T.: Instagrammatics and digital methods: studying visual social media, from selfies and GIFs to memes and emoji. Commun. Res. Pract. **2**(1), 47–62 (2016)
5. Manovich, L.: Notes on Instagrammism and contemporary cultural identity (2016). manovich.net
6. Mirzoeff, N.: How to See the World: An Introduction to Images, from Self-Portraits to Selfies, Maps to Movies, and More. Basic Books, New York (2016)
7. Manikonda, L., Hu, Y., Kambhampati, S.: Analyzing User Activities, Demographics. Social Network Structure and User-Generated Content on Instagram. Physics and Society, Social and Information Networks (2014)
8. Cox, A., Clough, P.: Developing metrics to characterize Flickr group. J. Am. Soc. Inform. Sci. Technol. **62**(3), 493–506 (2011)
9. MacDowall, L.J.: I'd double tap that!!': street art, graffti, and Instagram research. Media Cult. Soc. **40**, 3–22 (2017)
10. Tifentale, A.: Art of the masses: from Kodak Brownie to Instagram. Network. Knowl. **8** (2015)
11. Meindl, J.N., Ivy, J.W.: Mass shootings: the role of the media in promoting generalized imitation. Am. J. Public Health **107**(3), 368–370 (2017)
12. Stefanone, M.A., Lackaff, D., Rosen, D.: The relationship between traditional mass media and "social media": reality television as a model for social network site behavior. J. Broadcast. Electron. Media **54**(3), 508–525 (2010). https://doi.org/10.1080/08838151.2010.498851
13. Milner, R.M.: Pop polyvocality: internet memes, public participation, and the occupy wall street movement. Int. J. Commun. Syst. **7**, 2357–2390 (2013)
14. Börzsei, L.K.: Makes a meme instead: a concise history of internet memes. New Media Stud. Mag. **7** (2013)
15. Heylighen, F.: What makes a meme successful? Selection criteria for cultural evolution. In: Proceedings of 15th International Congress on Cybernetics, pp. 418–423 (1998)
16. Zettl, H.: Sight, Sound, Motion: Applied Media Aesthetics. Cengage Learning (2016)

HCI and Social Media in Business

A Comparison Study of Trust in M-commerce Between Qatari and Non-qatari Customers

Eiman Al-Khalaf(✉) and Pilsung Choe

Qatar University, P.O.Box 2713, Doha, Qatar
e_alkhalaf@qp.com.qa, pchoe@qu.edu.qa

Abstract. Over the past few years, mobile commerce (M-commerce), which is a new channel for conducting online businesses has revolutionized the global market. In Qatar, due to lack of trust, an only small percentage of Qatari populations has shown an interest in mobile shopping although the average annual individual's expenditure is impressive. Therefore, it is crucial to investigate antecedents of consumer' trust in the context of mobile commerce [1]. The objective of this paper is to explore whether trust perceptions towards mobile shopping among diverse segments of consumers living in Qatar vary significantly or not by proposing a conceptual framework for trust based on the Technology Acceptance Model. An online survey was conducted to empirically validate the newly developed trust factors between consumers with different nationalities, genders, and age ranges by performing permutations and partial lease squares multi-group analysis algorithms. Our research findings revealed that nationalities differences related to the factor perceived usability had positive indirect effect on Qataris trust, female consumers' trust could be enhanced by perceived usability, and young consumers could easily trust mobile commerce via the endorsement of social media influencers. The results also emphasized that perceived security was one of the most significant factors that affect drastically mobile commerce trust between all groups of consumers. We ended our work offering some valuable theoretical implications for scholars and interesting strategies for practitioners to help to create customer's trust especially during crises, such as COVID-19.

Keywords: M-commerce · M-trust · Localization · Social media · Luxury brands · Perceived privacy and security · PLS-MGA · Qatar · Multicultural

1 Introduction

M-commerce is defined as the capability of conducting online transactions facilitated through wireless devices; such as smartphones and tablets [1–4]. The major feature that leads more online users to carry out every day's business transactions by using their mobile devices is the ability of accessing the internet anytime and anywhere, which provides (1) convenience; (2) simplicity; (3) usefulness; as well as (4) ease of use to the users [1, 5–7].

However, comparing the enormous evolution of M-commerce with its expectations, several weaknesses in Qatar have been realized [4, 5, 7–9]. According to [10], mobile phone penetration in Qatar has reached 100%; at which, 75% of the population are

C. Stephanidis et al. (Eds.): HCII 2020, LNCS 12427, pp. 489–506, 2020.
https://doi.org/10.1007/978-3-030-60152-2_37

using smartphones; making it the highest penetration rate in the Middle East. Additionally, Qatar has all the favorable critical factors that are important to support a robust online commerce ecosystem; for example, (1) a population with great disposable income; (2) trustworthy information, and communication technology infrastructure; and (3) a very connected society [11]. Nevertheless, only 10.2% of the population in Qatar are conducting transactions over their smartphones [1, 12]. The key reason that causes the lack of acknowledging M-commerce in Qatar is the absence of trust affecting the intention to buy online [13–15]. Given that, M-commerce is comparatively a new business channel in Qatar, there is still a lack of understanding of how to make the context of mobile shopping more efficient. Furthermore, the market of M-commerce is comprising of different segments, which make marketers, and developers having difficulties in (1) anticipating consumers' usage of M-commerce apps and (2) finding reasons why customers trust perspectives are lagging behind expectations [1]. Therefore, it is important to recognize the social as well as cultural beliefs factors that could motivate consumers' trust in M-commerce [9].

While several studies have emphasized the significance of trust for the consumers to accept mobile shopping (e.g., [9, 16, 17]), it appears that in this region of the world, trust in M-commerce has never been explored and empirically tested. Accordingly, we believe that in a multicultural country such as Qatar, a thorough exploration and validation of motivation factors that promote consumer's trust, especially during crises when mobility is limited, such as COVID-19, is crucial and may provide insights for developers and concerned merchants to help them move to a higher level of M-commerce maturity. The objectives of this study are to broaden the knowledge of what influences the perception of trust among the Qatari population in the field of mobile commerce and understand to what extent the factors affect different groups of consumers' trust.

2 Literature Review and Question Statements

This section briefly discusses prior studies that dealt with trust in the context of M-commerce, and accordingly, the questions statements of this research have been developed.

2.1 Trust in M-commerce (M-trust)

The concept of reaffirming customer's trust in electronic commerce (E-commerce) has already been studied and eventually recognized [18–21]. However, regardless of how convenient mobile devices can be used to do online shopping anytime and anywhere, it is still challenging to maintain this trust when it comes to M-commerce [22, 23]. There are several potential reasons, such as small screens, keypads, and low resolution in some mobile devices. Other problems could be linked to wireless networks and users' vulnerability to risks while their data is being transmitted wirelessly [1, 23]. Thus, trust means the readiness of an individual to take a risk rather than merely taking the risks [19, 24].

In this research, trust is decomposed into two fields as stated by [9, 25]. The first field is labeled as hard trust; which is related to secure interactions and technology solutions [25]. The other area is called soft trust; which is mostly associated with the privacy of the information as well as the quality of services offered by vendors [9]. Accordingly, the development of favorable consumers trust towards M-commerce could be realized by blending several factors, like (1) usability of M-commerce; (2) quality of information; (3) privacy of user data; (4) security of mobile transactions; (5) credibility of the vendors; (6) quality of the products; and (6) culture effects [26].

Table 1 summarizes the core factors that have been examined in the recent studies that are related to consumer's trust and intention on M-commerce.

Table 1. Recent studies on consumer's trust and adoption towards M-commerce.

Core factors	Outcome	Study
– Perceived cost – Perceived entertainment – Perceived usefulness	Attitude towards using M-commerce	[27]
– Perceived ease of use – Perceived usefulness – Perceived mobility – Perceived compatibility – Social influences	Intention to use M-commerce	[28]
– Ease of use – Usefulness – Enjoyment – Mobility – Contextual offer – Online trust – Mobile trust – Offline trust	Intention to purchase via mobile devices	[16]
– Website design – Website reliability/fulfillment – Website security, privacy, and trust – Website customer service	Online purchase intention	[29]
– Design aesthetics – Ease of use – Usefulness – Customization	M-commerce trust	[19]
– Interactivity – Customization – Usefulness – Ease of use – Responsiveness – Brand name	Customer satisfaction and trust in M-commerce	[30]

(*continued*)

Table 1. (*continued*)

Core factors	Outcome	Study
– Interface quality – Information quality – Perceived privacy	User trust in M-commerce apps	[31]
– Familiarity – Compatibility – Perceived security	Trust in M-commerce	[32]
– Perceived privacy – Perceived security – Perceived ease of use – Quality of information – Disposition to trust – Reputation – Willingness to customize	Trust in M-commerce	[17]

2.2 Antecedents of Consumer Trust Towards M-commerce in Qatar

Based on the prior studies listed in Table 1 we propose that trust is a vital factor that can influence the acceptance of M-commerce. However, it appears that only a few studies have examined trust antecedents in the context of mobile commerce. Therefore, for this study, we aim to explore the determinants motivating online user trust in mobile commerce based on (1) core inhibitors behind the low penetration rate of mobile commerce amongst people living in Qatar and (2) existing literature.

According to [11], several hurdles are currently impeding the consumers for trusting mobile businesses and they are as discussed below;

1. For mobile commerce, payments are critical, and it is still the biggest obstacle for Qatari consumers. There are several reasons for this, and one major reason is security. In Qatar, many users are still having uncertainties regarding mobile payments and they perceive that their financial and personal information is not secured when used for mobile transactions [11]. Another important issue is that most of the issued cards are not having the capabilities to be used for online payments and though most of the countries are accepting debit cards to be used for online payments, in Qatar many banks do not permit mobile purchases to be made by using debit cards [1, 33]. Accordingly, for this study, security and privacy can be perceived as crucial factors that can enhance mobile user trust in the Qatari market.
2. The available limited mobile commerce apps in Qatar are having several usability issues, such as apps that are (1) not attractive; (2) using navigation tools that are not user friendly; (3) and do not meet the needs of the diverse segments of the population. Moreover, since users are less patient to deal with difficult to understand and use mobile applications, and they are in search of apps where they can get the information they seek quickly [1, 33] It can be assumed that well-designed mobile commerce apps could offer good recognition regarding the vendor. Hence, M-commerce usability and localization are essential factors to be considered in this research in order to improve trust among the cultural diversity of consumers live in Qatar.

The following trust influences have been extracted from the report of consumer's lifestyle in Qatar, that offers information on the unique attitudes, behaviors, and spending patterns of Qatari consumers [34].

3. While people generally take advice from family members and close friends regarding purchases nowadays, it is becoming increasingly popular to seek advice from social media influencers [35]. Given that, Qatar has been reported to have the highest global diffusion of social media via mobile devices and popular social media influencers have a positive impact on Qatari consumers when promoting products and services [36]. Correspondingly, examining whether there are direct or indirect significant relationships between the endorsement of social media influencers and trust in M-commerce apps amongst different segments of consumers is one of this study's goals.
4. Consumers in Qatar have embedded uncertainties towards M-commerce apps that provide unfamiliar brands to them as they believe that these products are not trustworthy particularly when it comes to privacy and mobile payments, which are the core concepts of the success of any online commerce [37]. This can be witnessed in the luxury buying behavior, which in Qatar is not narrow to a specific group of people but has prolonged to a middle as well as a lower levels of society over time [34]. Therefore, offering luxury brand products through M-commerce apps may be considered in this study to influence consumer's trust positively in the multi-cultural Qatari society.

2.3 Proposed Question Statements

After identifying the potential factors that may be incorporated into the technology acceptant model, this research will explore and assess how localization of mobile stores, an offering of luxury brands, and endorsement from social media influencers can promote trust towards different groups of consumers in Qatar. These factors have been termed motivation factors and they are as well independent. Perceived usability, privacy, and security factors, which are already existing in literature have been proposed to be cognitive factors and they are working as both dependent and independent elements. M-commerce trust has been considered to be an Affective response and it is the primary aim of this exploratory research. Figure 1 illustrates the relationships of the trust factors that are examined in this study.

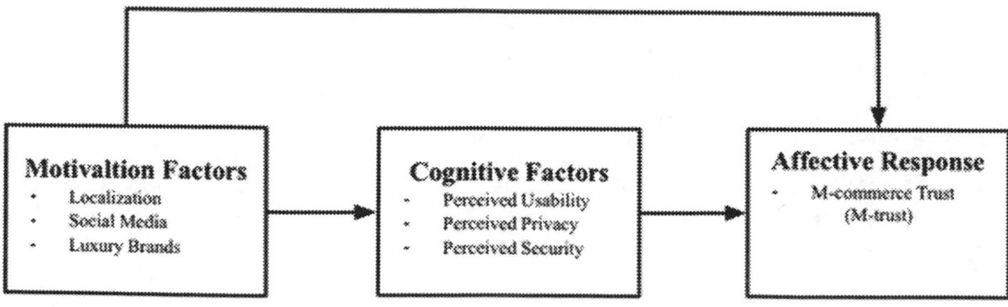

Fig. 1. Relationships of the examined trust factors.

The associated definition for each of the studied trust factors is presented in the following Table 2.

Table 2. Definitions of the proposed trust factors.

Motivational factors		
These are independent factors that can encourage consumers to improve their trust in M-commerce and they will be investigated in this research		
Factors	Definition	Author(s)
Localization	M-commerce webstores or apps appear local to users who belong to a particular cultural group by incorporating local languages and currencies	[38]
Social media	M-commerce webstores or apps are endorsed by influencers who can affect a large number of audience through the use of social media channels	[39]
Luxury brands	M-commerce webstores or apps are offering luxury brand products that users perceived as having uniqueness, authenticity, and quality	[40]
Cognitive factors		
These are dependent factors that have been confirmed previously as trust predictors.		
Factors	Definition	Author(s)
Perceived usability	The degree to which users believe M-commerce webstores or apps are being free of effort and improving their performance	[41]
Perceived privacy	The degree to which users believe M-commerce webstores or apps are protecting their personal information and providing them confidence to control their privacy details	[16]
Perceived security	The degree to which users believe M-commerce webstores or apps are providing secure and reliable measures as well as offering different payment methods	[42]
Affective response		
The feelings during the interaction with mobile commerce apps or webstores		
Factors	Definition	Author(s)
M-commerce Trust (M-trust)	The degree to which users believe M-commerce webstores or apps are reliable and trustworthy	[43]

The associated question statements that have been developed for this research study are listed next;

– QS1: Localization of M-commerce webstores or apps has a significant difference among different groups of consumers (nationality, gender, age, language, etc.…) on M-trust.

- QS2: Localization of M-commerce webstores or apps has a significant difference among different groups of consumers (nationality, gender, age, language, etc.…) on perceived usability.
- QS3: M-commerce apps or websites endorsed by social media influencers have a significant difference among different groups of consumers (nationality, gender, age, language, etc.…) on M-trust.
- QS4: M-commerce webstores or apps endorsed by social media influencers have a significant difference among different groups of consumers (nationality, gender, age, language, etc.…) on perceived usability.
- QS5: Offering luxury brands via M-commerce webstores or apps has a significant difference among different groups of consumers (nationality, gender, age, language, etc.…) on M-trust.
- QS6: Offering luxury brands via M-commerce webstores or apps has a significant difference among different groups of consumers (nationality, gender, age, language, etc.…) on perceived privacy.
- QS7: Perceived usability has a significant difference among different groups of consumers (nationality, gender, age, language, etc.…) on M-trust.
- QS8: Perceived usability has a significant difference among different groups of consumers (nationality, gender, age, language, etc.…) on perceived security.
- QS9: Perceived usability has a significant difference among different groups of consumers (nationality, gender, age, language, etc.…) on perceived privacy.
- QS10: Perceived privacy has a significant difference among different groups of consumers (nationality, gender, age, language, etc.…) on M-trust.
- QS11: Perceived privacy has a significant difference among different groups of consumers (nationality, gender, age, language, etc.…) on perceived security.
- QS12: Perceived security has a significant difference among different groups of consumers (nationality, gender, age, language, etc.…) on M-trust

3 Research Methodology

To examine the validity of the proposed question statements, this section briefly examines the research methodology of this study.

3.1 Instrument Design

An online survey questionnaire was conducted between diverse online users living in Qatar. For the data collection, a structured self-administered questionnaire was used, and it was offered in Arabic and English languages. The questionnaire comprised of three parts. Demographic characteristics were captured in part A; part B was related to the participant's experience with online shopping; and part C measured the postulated trust factors from both experienced participants and those who were interested to shop online in the future.

A set of 23 scale items was used for assessing the proposed trust's factors. Items associated with perceived usability and M-trust were adopted from [44], perceived privacy from [45], and perceived security from [42]. Motivation factors items were new; hence, they have been created by applying the inductive approach [46]. A five-point Likert scale ranging from 1 = Strongly Disagree to 5 = Strongly Agree, with 3 = Neither Agree nor Disagree as the neutral point was used to measure the questionnaire items.

3.2 Pilot Test

A pilot test with a sample of 30 potential participants was conducted. The implications from the test were that three questions, one associated with "M-trust" factor, and two to "Perceived Security", had to be amended and rephrased in Arabic and English.

3.3 Data Collection and Sampling

The survey was distributed via different platforms, such as universities, company's communication centers, and social media channels; in addition, the sampling method was based on self-selection. Two hundred and fifty completed surveys were received; in which, 235 of them from experienced users as well as those who were interested to do mobile shopping in the future, and the remaining 15 respondents were coming from percipients who were not interested in M-commerce. Only 228 from experienced and interested participants were providing valid answers to all survey questions, which means missing cases represented 2.8% of the whole sample; thus, deleted safely from the study [44]. Table 3 summarizes the demographic information of the valid 228 cases.

Table 3. Demographic characteristics [1].

Demographics	Items	Frequency (N = 228)	Percentage (%)
Gender	Male	114	50
	Female	114	50
Age range	<25	62	27
	25–34	87	38
	≥ 35	79	35
Nationality	Qatari	121	53
	Non-Qatari	107	47
Primary language	Arabic	173	76
	Non-Arabic	55	24
Occupational status	Employed	160	70
	Self-employed	7	3
	Unemployed	20	9
	Student	41	18

3.4 Data Analysis Method

As this is an exploratory study, partial least squares equation modeling (PLS-SEM) was applied. According to [44], this approach (1) runs more efficient with small sample size; (2) does not requires assumptions for the distribution of the primary data, and (3) can be used for multifaceted structural models like the case of the present research; wherein, the proposed model comprises of seven constructs and 23 items. Additionally, to determine and better recognize whether trust perceptions towards mobile shopping among consumers vary significantly or not, advanced multi-group analysis has been conducted by applying permutation and multi-group algorithms available in SmartPLS3.

3.5 Empirical Data Examination

Before applying PLS-SEM procedures, a preliminary assessment of the collected data was performed. First, the minimum sample size requirement has been confirmed by employing the ten times heuristic rules suggested by [47]; therefore, a minimum sample size of 60 valid cases was needed for this research. Second, assuming that participants of this study have given true opinion and keeping this assumption valid, all cases (N = 228) were considered true; thus, detected outliers could not be deleted [44, 48]. Lastly, IBM SPSS was used to perform descriptive statistics and verify the gathered data were not extremely non-normal by measuring the values of the skewness and kurtosis of the survey's items [49]. The maximum absolute values of skewness and kurtosis were (1.682) and (2.947), correspondingly and they were fulfilling the cut-off values recommended by [50].

4 Results

The obtained findings of this research are presented in this section. The results comprise both measurement model (the survey items) assessment and multigroup analyses.

4.1 Internal Reliability, Content Validity, and Discriminant Validity

Internal reliability and content validly were assessed by verifying that the outer loadings for all the items were exceeding 0.7. However, when developing new scales especially in social science studies, outer loadings between 0.4 and 0.7 are accepted and should be removed only when the value of the composite reliability or the average variance extracted (AVE) increases above the recommended threshold values [51]. Therefore, one item under the construct "Perceived Security" was deleted. The internal reliability as shown in Appendix 1 has been established since all the values were above the minimum required level of 0.7 [52]. The AVE values of all the constructs were higher than 0.5, which implies that the constructs' measures had strong content validity [49, 51]. The results in

Appendix 2 imply that all of the proposed factors satisfied the Fornell-Larcker [51]; consequently, the discriminant validity of this study was confirmed.

4.2 Multigroup Analyses

Three multi-group analyses have been performed for this study. The first analysis to compare consumers' nationalities that have been categorized as Qataris and Non-Qataris, the second analysis was to compare between male and female users, and the last analysis between young and post-millennials adult users. However, language-wise assessment has not been conducted as the sample size of consumers who were non-Arabic speakers was relatively small (n = 55), which was below 60 observations per group.

The following sections provide the results obtained from the multi-group analysis but before performing this procedure, measurement invariance as a pre-test has been conducted.

Measurement Invariance. Since the suggested factors were defined by their associated indicators in the measurement model, measurement invariance must be established before carrying out the multi-group analysis. This is because a lack of measurement invariance signifies that the constructs are measuring different things although the indicators between the groups are similar [53]. Furthermore, as addressed by [54] without verifying the measurement invariance of the outer model the analysis can produce misleading results.

Table 4 displays the permutation p-values that assess if the outer loading of each construct is invariant across the group.

Table 4. Results of measurement invariance test.

Construct	Permutation p-Values		
	Nationalities	Gender	Age range
Localization	0.176	0.734	0.559
Luxury brands	0.316	0.131	0.560
M-trust	0.665	0.334	0.682
Perceived privacy	0.807	0.281	0.234
Perceived security	0.298	0.169	0.171
Perceived usability	0.859	0.184	0.389
Social media	0.951	0.507	0.771

PLS - Multigroup Analysis. One common approach to assess the significance of the path differences as suggested by [55] is the PLS-MGA, which is a non-parametric significance test. This test considers the absolute group-specific path coefficients difference to be significant if the p-value is either less than 0.05 or more than 0.95. In addition, as stated by [53], p-values that are less than 0.1 can be considered marginally significant for exploratory study related to social science; therefore, p-values more than 0.9 and smaller than 0.1 were also assumed to be slightly significant for the multi-group analysis.

The next sections explain the obtained results by defining nationality, gender, and age range as a basis for multi-group comparisons.

Nationality: Qataris and Non-Qataris. Table 5 provides a summary of the results for all relationships in the model for the Qatari and Non-Qatari groups.

Table 5. Results of Qataris and non-Qataris groups.

Path	Group 1	Group 2	Group 1 vs. Group 2	
	$p^{(1)}$	$p^{(2)}$	$\lvert p^{(1)} - p^{(2)} \rvert$	p Value
Localization → M-trust	0.046	−0.026	0.072	0.693
Localization → Perceived Usability	0.448	0.453	0.005	0.483
Luxury Brands → M-trust	−0.161	0.022	0.183	*0.094
Luxury Brands → Perceived Privacy	0.230	0.135	0.094	0.799
Perceived Privacy → M-trust	0.044	−0.018	0.062	0.643
Perceived Privacy → Perceived Security	0.263	0.395	0.133	0.197
Perceived Security → M-trust	0.564	0.531	0.034	0.593
Perceived Usability → M-trust	−0.078	0.060	0.138	0.183
Perceived Usability → Perceived Privacy	0.427	0.456	0.029	0.388
Perceived Usability → Perceived Security	0.313	0.109	0.205	*0.929
Social Media → M-trust	0.166	0.115	0.051	0.641
Social Media → Perceived Usability	−0.142	0.022	0.163	0.123
Sample size (n)	121	107		

Notes; Group 1: Qatari, Group 2: Non-Qatari
p (1) and p (2) are path coefficients of Group 1 and Group 2, respectively
$*p > 0.9$ or $*p < 0.1$

As shown from the results table, only two relationships (Luxury Brands → M-trust and Perceived Usability → Perceived Security) differ significantly across the Qatari and Non-Qatari groups. The results reveal that there is a somehow significant negative effect of offering luxury brands products on M-trust for Qatari consumers ($\beta = -0.161$). However, Qataris consumers are preferring an easy to use mobile apps ($\beta = 0.313$) as the usability factor enhances the security perceptions towers M-commerce.

Gender: Male and Female. A summary of the results for all relationships in the model for male and female groups under the gender category is provided in Table 6. By examining the p-value column in the results table, it can be said that there is a significant difference between male and female groups when it comes to the usability factor and its impact on privacy. Women shoppers are perceiving usability features of mobile commerce will help them significantly ($\beta = 0.611$) in protecting their personal details and reducing their privacy concerns. Male consumers, on the other hand, are having a slightly significant negative impact ($\beta = -0.157$) on viewing social media influencers to be as facilitators in assisting them to make the mobile shopping experience easy.

Table 6. Results of male and female groups.

Path	Group 1	Group 2	Group 1 vs. Group 2	
	$p^{(1)}$	$p^{(2)}$	$\lvert p^{(1)} - p^{(2)} \rvert$	p Value
Localization → M-trust	−0.053	0.027	0.080	0.288
Localization → Perceived Usability	0.409	0.425	0.016	0.449
Luxury Brands → M-trust	−0.104	−0.037	0.066	0.308
Luxury Brands → Perceived Privacy	0.252	0.134	0.118	0.853
Perceived Privacy → M-trust	0.089	−0.013	0.102	0.737
Perceived Privacy → Perceived Security	0.285	0.319	0.033	0.427
Perceived Security → M-trust	0.534	0.523	0.011	0.533
Perceived Usability → M-trust	0.055	−0.004	0.059	0.658
Perceived Usability → Perceived Privacy	0.270	0.611	0.341	**0.000
Perceived Usability → Perceived Security	0.188	0.275	0.088	0.278
Social Media → M-trust	0.181	0.190	0.009	0.464
Social Media → Perceived Usability	−0.157	0.059	0.216	*0.075
Sample size (n)	114	114		

Notes; Group 1: Male, Group 2: Female
p (1) and p (2) are path coefficients of Group 1 and Group 2, respectively
**p < 0.05, *p < 0.1

Age Range: Millennials and Post-millennials. Table 7 provides a summary of the results for all relationships in the model for respondents who were below the age of 18 years (young consumers) and those whose age ranges between 18 and 34 years (adults).

Table 7. Results of millennials and post-millennials groups.

Path	Group 1	Group 2	Group 1 vs. Group 2	
	$p^{(1)}$	$p^{(2)}$	$\lvert p^{(1)} - p^{(2)} \rvert$	p Value
Localization → M-trust	0.086	−0.089	0.175	0.192
Localization → Perceived Usability	0.408	0.423	0.015	0.507
Luxury Brands → M-trust	−0.121	0.028	0.149	0.746
Luxury Brands → Perceived Privacy	0.231	0.066	0.165	0.144
Perceived Privacy → M-trust	0.152	0.124	0.029	0.452
Perceived Privacy → Perceived Security	0.069	0.474	0.405	**0.982
Perceived Security → M-trust	0.508	0.492	0.016	0.465
Perceived Usability → M-trust	−0.012	−0.084	0.072	0.353
Perceived Usability → Perceived Privacy	0.379	0.564	0.185	*0.904
Perceived Usability → Perceived Security	0.418	0.000	0.418	**0.015
Social Media → M-trust	0.233	0.005	0.228	*0.096
Social Media → Perceived Usability	−0.233	−0.061	0.172	0.827
Sample size (n)	62	87		

Notes; Group 1: <18 years, Group 2: 18–34 years
P (1) and p (2) are path coefficients of Group 1 and Group 2, respectively
**p < 0.05, *p > 0.9 or *p < 0.1

The results show the significant difference between adults and young consumers in Qatar. Older consumers are perceiving the usability of mobile commerce apps to have a favorable influence on protecting their privacy information (β = 564), which can lead to a significant positive impact on perceived security (β = 0.474). Furthermore, the results disclose that younger respondents are having a marginally significant effect (β = 0.233) to trust mobile commerce apps that are endorsed by social media influencers. Finally, perceived usability is playing a vital role in promoting security towards M-commerce for young consumers (β = 0.418) as revealed from the results Table 7 above.

5 Discussion and Conclusions

The process of building user's trust towards mobile commerce has been recognized in most of the present studies as complex and complicated [19]. However, measuring how introducing motivational factors, such as localization, an endorsement from social media influencers, and offering of luxury brand products can enhance M-commerce trust still has a lack of clarity [1]. Hence, our study contributes to filling this gap in literature empirically by examining the effect of the newly developed motivation factors on M-commerce trust together with the most cited trust predictors (e.g., perceived usability, perceived privacy, and perceived security).

From the multi-group analyses, it has been found that mobile users with different group identities have different perceptions on trust towards mobile commerce in the Qatari society.

The sub-group analysis categorized by nationality revealed that the effect of motivation factors set on M-trust for the sub-sample results are similar to the full sample except for luxury brands factor that shows a marginally significant negative impact on trust for Qatari consumers. These results indicate that for both Qatari and non-Qatari consumers, an endorsement from social media influencers is a significant predictor of their trust towards mobile commerce. In addition, results show that for Qataris, mobile commerce trust is developed through their perceived usability together with perceived security; thus, this implies that they build their mobile commerce trust under particular conditions. Mobile-friendly apps and websites are the key drivers for Qatari users to feel more secured, and too time-consuming shopping processes will lead indirectly to distrust. On the other hand, for non-Qatari consumers, the results disclose that the perceived usability factor is not a big concern for them in order to trust the M-commerce app and a plausible justification could be due to their familiarly with mobile commerce as they may have purchased through mobile stores for many years.

Furthermore, the results highlight that there are no significant differences with regards to gender in Qatar towards mobile commerce trust. However, women compared to men have been shown to be more concerned with the usability of the app or website. This means that in order for female consumers to feel that their private details are protected, the mobile store should be user friendly and easy to use. The finding agrees with [56] study that connected gender with perceived ease of use and perceived usefulness.

The results also show that there is a significant difference between adults and young consumers when it comes to the effect of social media influencers on trust towards mobile commerce. Young consumers are heavily impacted by social media in regard to mobile shopping and this is mainly because these consumers are making social media channels as part of themselves. In addition, these consumers have been observed to trust social media influencers because these influencers are more approachable, and they can leave a lasting impression.

Given that presently post-millennial consumers are evolving very fast into the most influential group for mobile shopping, mobile commerce marketers must realize that it is no longer enough to market their products and services by using old approaches. Additionally, considering the COVID-19 pandemic, the implementation of social distancing, and more people are currently staying indoors spending their time watching contents form social media influencers, the wisest way to add a new level of consumers trust is by incorporating influencers marking. As social media platforms are presenting various opportunities (e.g., sponsored Instagram posts, Snapchat stories) mobile commerce businesses should make sure that the endorsement from social media influencers appear to consumers both believable and genuine. Marketers should also find influencers with followers that are meeting their target consumers since this is one of the key influencer's marketing tactics that can allow mobile store vendors to be trusted and ultimately visited by consumers. By applying these marking strategies, practitioners can reach a high volume of young users to build trust and eventually purchase from their mobile commerce apps or websites.

Finally, this research has some limitations and offers some opportunities for future studies. First, this is an exploratory study that has been conducted before the COVID-19 outbreak, which pushes more consumers to do their shopping online. This means that people's perceptions might change; therefore, confirmatory factor analysis could be conducted in future research. Next, nationality and primary language could make such biases since most of the research's participants were Qatari who were speaking Arabic; therefore, the proposed trust antecedents can be extended further by examining them to other groups of people residing in Qatar. Last, our sample size was comparatively small, although it surpassed the minimum required limit for PLS; hence, larger sample sizes could be used in the future as they are preferable to produce more stable results [1, 57].

Appendix 1

Internal reliability and content validity [1].

Construct	Items	Outer loadings	Cronbach's Alpha	Composite reliability	Average variance extracted
Localization	LOC-1	0.772	0.798	0.868	0.622
	LOC-2	0.786			
	LOC-3	0.854			
	LOC-4	0.737			
Social media	SM-1	0.894	0.884	0.928	0.811
	SM-2	0.904			
	SM-3	0.904			
Luxury brands	LUX-1	0.902	0.905	0.940	0.840
	LUX-2	0.934			
	LUX-3	0.913			
Perceived usability	PU-1	0.778	0.731	0.847	0.649
	PU-2	0.805			
	PU-3	0.832			
Perceived privacy	PP-1	0.851	0.775	0.865	0.689
	PP-2	0.781			
	PP-3	0.857			
Perceived security	PS-1	0.609	0.708	0.836	0.635
	PS-3	0.884			
	PS-4	0.867			
M-trust	MT-1	0.787	0.706	0.835	0.629
	MT-2	0.757			
	MT-3	0.832			

Appendix 2

Discriminant validity [1].

Construct	(1)	(2)	(3)	(4)	(5)	(6)	(7)
(1) Localization	**0.788**						
(2) Social media	0.233	**0.901**					
(3) Luxury brands	0.238	0.413	**0.916**				
(4) M-trust	0.134	0.120	0.023	**0.793**			
(5) Perceived security	0.172	−0.006	0.055	0.529	**0.797**		
(6) Perceived usability	0.410	0.040	0.140	0.204	0.364	**0.805**	
(7) Perceived privacy	0.374	0.105	0.247	0.249	0.406	0.462	**0.830**

References

1. Al-Khalaf, E., Choe, P.: Increasing customer trust towards mobile commerce in a multicultural society: a case of qatar. J. Internet Commer. **19**(1), 32–61 (2020)
2. Hałabuda, P.: Introduction to M-Commerce: Benefits of M-Commerce (2015). http://whallalabs.com/introduction-to-m-commerce/. Accessed 09 Oct 2017
3. Clarke, I.: Emerging value propositions for m-commerce. J. Bus. Strateg. **18**(4), 464–494 (2001)
4. Esmaeili, M., Eydgahi, A.: Main factors influencing mobile commerce adoption. In: Proceedings of The 2016 IAJC-ISAM International Conference, pp. 1–11 (2016)
5. Anckar, B., D'Incau, D.: Value creation in mobile commerce: findings from a consumer survey. J. Inf. Technol. Theory Appl. **4**(1), 43–64 (2002)
6. Kalakota, R., Robinson, M.: M-Business: The Race to Mobility. McGraw-Hill Education, Bosten (2001)
7. Venkatesh, V., Ramesh, V., Massey, A.P.: Understanding usability in mobile commerce. Commun. ACM **46**(12), 53 (2003)
8. Eze, U.C., Ten, M.A.T.M.Y., Poong, Y.-S.: Mobile commerce usage in Malaysia: assessing key determinants. Int. Conf. Soc. Sci. Hum. **5**, 265–269 (2011)
9. Hillman, S., Neustaedter, C.: Trust and mobile commerce in North America. Comput. Human Behav. **70**, 10–21 (2017)
10. Metodieva, V.M.: Qatar's Smartphone Market Q4 2011 Consumers' Perspective: A Nielsen syndicated study (2012)
11. Qatar National E-Commerce Roadmap 2017 (2017)
12. Khan, H., Talib, F., Faisal, M.N.: An analysis of the barriers to the proliferation of M-commerce in Qatar A relationship modeling approach. J. Syst. Inf. Technol. **17**(1), 54–81 (2015)
13. AlGhamdi, R.: Diffusion of the Adoption of Online Retailing in Saudi Arabia. Griffith University (2012)
14. Jing, R., Yu, J., Jiang, Z.: Exploring influencing factors in E-commerce transaction behaviors. In: 2008 International Symposium on Electronic Commerce and Security, 3–5 August 2008, pp. 603–607 (2008)
15. Sohaib, O.: A study of individual consumer level culture in B2C e-commerce through a multi-perspective itrust model. University of Technology Sydney City Campus (2015)
16. Giovannini, C.J., Ferreira, J.B., da Silva, J.F., Ferreira, D.B.: The effects of trust transference, mobile attributes and enjoyment on mobile trust. BAR - Braz. Adm. Rev. **12** (1), 88–108 (2015)
17. Junqueira, E.T.: The antecedents of trust in mobile commerce: a quantitative analysis of what drives mobile trust, in the Brazilian market (2016)
18. Doney, P.M., Cannon, J.P.: An examination of the nature of trust in buyer-seller relationships. J. Mark. **61**(2), 35–51 (1997)
19. Li, Y.M., Yeh, Y.S.: Increasing trust in mobile commerce through design aesthetics. Comput. Hum. Behav. **26**(4), 673–684 (2010)
20. Jarvenpaa, S.L., Tractinsky, N., Vitale, M.: Consumer trust in an internet store. Inf. Technol. Manag. **1**, 45–71 (2000)
21. Selnes, F.: Antecedents and consequences of trust and satisfaction in buyer-seller relationships. Eur. J. Mark. **32**(3/4), 305–322 (1998)
22. Siau, K., Lim, E.-P., Shen, Z.: Mobile commerce: promises, challenges and research agenda. J. Database Manag. **12**(3), 4–13 (2001)

23. Siau, K., Shen, Z.: Building customer trust in mobile commerce. Commun. ACM **46**(4), 91–94 (2003)
24. Mayer, R.C., Davis, J.H., Schoorman, F.D.: An integrative model of organizational trust. Acad. Manag. **20**(3), 709–734 (1995)
25. Head, M., Hassanein, K.: Trust in e-commerce: evaluating the impact of third-party seals. Q. J. Electron. Commer. **3**(3), 307–325 (2002)
26. Siau, K., Sheng, H., Nah, F.: Development of a framework for trust in mobile commerce. In: SIGHCI 2003 Proceedings, pp. 85–89 (2003)
27. Zheng, H., Li, Y., Jiang, D.: Empircal study and model of user's acceptance for mobile commercial in China. IJCSI Int. J. Comput. Sci. Issues **9**(6), 278–283 (2012)
28. Batkovic, I., Batkovic, R.: Understanding consumer acceptance of mobile-retail - an empirical analysis of the revised technology acceptance mode. University of Gothenburg (2015)
29. Lee, W.O., Wong, L.S.: Determinants of mobile commerce customer loyalty in Malaysia. Procedia Soc. Behav. Sci. **224**, 60–67 (2016)
30. Suki, N.M.: A structural model of customer satisfaction and trust in vendors involved in mobile commerce. Int. J. Bus. Sci. Appl. Manag. **6**(2), 17–30 (2011)
31. Deepika, R., Karpagam, V.: Antecedents of smartphone user satisfaction, trust and loyalty towards mobile applications. Indian J. Sci. Technol. **9**, 1–8 (2016)
32. Alqatan, S., Noor, N.M.M., Man, M., Mohemad, R.: An empirical study on success factors to enhance customer trust for mobile commerce in small and medium-sized tourism enterprises (SMTES) in Jordan. J. Theor. Appl. Inf. Technol. **83**(3), 373–398 (2016)
33. Qatar National e-Commerce Roadmap 2015 (2015)
34. Consumer Lifestyles in Qatar (2015)
35. Oppenheim, M.: New data reveals people trust social media influencers almost as much as their own friends (2016). http://www.independent.co.uk/news/people/new-data-reveals-people-trust-social-media-influencers-almost-as-much-as-their-own-friends-a7026941.html. Accessed 03 Sept 2017
36. Iqbal, A.: Qatar & UAE have highest social media penetration globally (2017). http://www.qatar-tribune.com/news-details/id/48811. Accessed 09 Oct 2016
37. Vel, K.P., Captain, A., Al-Abbas, R., Al Hashemi, B.: Luxury buying in the United Arab Emirates. J. Bus. Behav. Sci. **23**(3), 145–160 (2011)
38. Singh, N., Park, J.E., Kalliny, M.: A framework to localize international business to business web sites. ACM SIGMIS Database DATABASE Adv. Inf. Syst. **44**(1), 56–77 (2012)
39. Freberg, K., Graham, K., McGaughey, K., Freberg, L.A.: Who are the social media influencers? A study of public perceptions of personality. Public Relat. Rev. **37**(1), 90–92 (2011)
40. Bauer, M., von Wallpach, S., Hemetsberger, A.: My little luxury' - a consumer-centred, experiential view. Mark. ZFP **33**(1), 57–67 (2011)
41. Chiu, J.L., Bool, N.C., Chiu, C.L.: Challenges and factors influencing initial trust and behavioral intention to use mobile banking services in the Philippines. Asia Pacific J. Innov. Entrep. **11**(2), 246–278 (2017)
42. Gustavsson, M., Johansson, A.-M.: Consumer Trust in E-commerce. Kristanstad University (2006)
43. Xin, H., Techatassanasoontorn, A.A., Tan, F.B.: Exploring the influence of trust on mobile payment adoption. In: PACIS 2013 Proceedings, vol. 143, no. 18, pp. 1–17 (2013)
44. Pittayachawan, S.: Fostering consumer trust and purchase intention in B2C e-commerce. RMIT University (2007)
45. Belanger, F., Hiller, J.S., Smith, W.J.: Trustworthiness in electronic commerce: the role of privacy, security, and site attributes. J. Strat. Inf. Syst. **11**, 245–270 (2002)

46. Hinkin, T.R.: A brief tutorial on the development of measures for use in survey questionnaires. Organ. Res. Methods **2**(1), 104–121 (1998)
47. Barclay, D., Higgins, C., Thompson, R.: The partial least squares (PLS) approach to causal modelling: personal computer adaptation and use as an illustration. Technol. Stud. **2**(2), 286–309 (1995)
48. Durkheim, E., Allcock, J.B.: Pragmatism and Sociology. Cambridge University Press, Cambridge (1983)
49. Hair, J.F.J., Hult, G.T.M., Ringle, C., Sarstedt, M.: A primer on partial least squares structural equation modeling (PLS-SEM), vol. 46, no. 1–2 (2014)
50. Fabrigar, L.R., Wegener, D.T., MacCallum, R.C., Strahan, E.J.: Evaluating the use of exploratory factor analysis in psychological research. Psychol. Methods **4**(3), 272–299 (1999)
51. Hulland, J.: Use of partial least squares (PLS) in strategic management research: a review of four recent studies. Strat. Manag. J. **20**(2), 195–204 (1999)
52. Nunnally, J.C., Bernstein, I.H.: Psychometric Theory, vol. 3. McGraw-Hill Education, New York (1994)
53. Garson, G.D.: Partial Least Squares: Regression & Structural Equation Models. Statistical Associates Publishers, Asheboro (2016)
54. Hult, G.T.M., et al.: Data equivalence in cross-cultural international business research: assessment and guidelines. J. Int. Bus. Stud. **39**(6), 1027–1044 (2008)
55. Ringle, C.M., Sinkovics, R.R., Henseler, J.: The use of partial least squares path modeling in international marketing. In: New Challenges to International Marketing, vol. 20, pp. 277–319. Emerald Group Publishing Limited (2009)
56. Venkatesh, V., Morris, M.G.: Why don't men ever stop to ask for directions? Gender, social influence, and their role in technology acceptance and usage behavior. MIS Q. **24**(1), 115–139 (2000)
57. Zhou, T., Lu, Y., Wang, B.: A comparative analysis of Chinese consumers' increased vs decreased online purchases. J. Electron. Commer. Organ. **9**, 38–40 (2011)

Factors Shaping Information Security Culture in an Internal IT Department

Peter Dornheim(✉) and Rüdiger Zarnekow

Technische Universität Berlin, Straße des 17. Juni 135, 10623 Berlin, Germany
peter.dornheim@campus.tu-berlin.de,
ruediger.zarnekow@tu-berlin.de

Abstract. Companies are exposed to the risk of falling victim to cyber attacks every day and therefore invest in optimizing their information security (IS) level. In addition to technological investments, the focus is now increasingly on employees, as their attitude and behavior has a significant influence on the IS level of a company. There is already extensive research on the influencing factors and the establishment of an IS culture in companies. However, little attention has been paid to the group of IT employees, although it has been proven that their attitude, behavior and judgment regarding cyber attacks and IS in general differ from non-IT employees. Even within an IT department, one can expect to see different degrees of these factors, as a distinction must be made between IS employees and employees with traditional IT functions (software development, server operation, etc.).

Based on 25 recent IS studies, a literature review has identified four factors that influence the IS attitude and behavior of employees in an internal IT department and thus have an impact on the IS culture. These four components are IT tools, IS skill, appreciation and sub-cultures.

The results show that more qualitative research with a focus on employees of an IT department is necessary to advance the findings in the area of IS culture. In addition, CIOs and CISOs benefit from the results, as they specify fields of action that can be tested and optimized in their own organizations.

Keywords: Information security culture · Information security behavior · IT employees · IT organization

1 Introduction

In 2020 the Allianz Risk Barometer identifies becoming the victim of a cyber attack and thus risking a total IT failure, losing sensitive data or having to pay high fines as the greatest business risk [1]. In the Global Risk Report of the World Economic Forum, a cyber attack also ranks second among the riskiest events that will have a large impact on business activities over the next 10 years [2].

CIOs and CISOs are aware of this and provide extensive resources to increase the information security (IS) level: the largest investments over the next three years are planned in the field of IS, and projects in the areas of data protection and IS are at the top of the agenda [3].

C. Stephanidis et al. (Eds.): HCII 2020, LNCS 12427, pp. 507–521, 2020.
https://doi.org/10.1007/978-3-030-60152-2_38

In addition to a technical and a process-related dimension, the IS level of a company also has an employee-related or behavior-oriented dimension. The importance of this dimension has been confirmed by countless studies in recent years. The declared objective is therefore often to establish a company-wide IS culture which demands and promotes good IS behavior. It was also shown that IS behavior differs between IT and non-IT employees [4].

However, it must be taken into account that differences also exist within the group of IT employees. On the one hand there are the traditional IT functions such as software development, IT project management and server & client operation. On the other hand, there are IS functions such as security operation center analysts and IT forensic scientists. The two functions differ in many ways: while performance, functionality, usability and availability are the main priorities for traditional IT employees, employees in the IS function focus on confidentiality, integrity and availability. The level of knowledge regarding current IS vulnerabilities, possible attack scenarios and the corresponding defense measures also differs. This fact means that even within an IT department, there is often a different understanding of IS, a heterogeneous IS risk assessment and thus a differentiated IS behavior, which in turn cannot contribute to a homogeneous IS culture. However, an IT department has the overall task within a company of implementing IS in projects, especially in day-to-day work, and of setting a good example in the company.

The present literature review therefore identifies factors that influence IS culture within an IT department. This allows approaches to be identified to promote a good IS culture in this area and eliminate differences between traditional IT functions and the employees working in an IS function.

This study thus contributes to the current state of knowledge by identifying on the basis of existing research factors that influence IS culture among an extremely important group of people. Since very few studies exist for this sub-area, the different sources already indicate the need for further research: Hooper and Blunt [4] describe the IS behavior of IT employees as not very transparent and indicate the need for qualitative research to clarify the importance of different factors influencing the IS culture of IT employees. Nel and Drevin [5] agree with this by suggesting that the factors of an IS culture should be evaluated in terms of their importance in the corporate context. Moreover, it is not clear what specific characteristics of personal attitudes, values and beliefs make for good IS behavior [6].

CIOs, CISOs and IS experts benefit from the results, as the factors described can be checked in their own organization and, if necessary, concrete improvement measures can be planned and implemented.

2 Background and Related Work

In the following, the research question underlying the literature review is explained, the terms contained therein are defined and a brief overview of existing and fundamental studies is given.

2.1 Research Question and Definition of Terms

This literature review focuses on the identification of factors that influence the IS culture of an internal IT department. The following research question is therefore posed and answered within the framework of this study:

- RQ1: What are the relevant factors influencing the IS culture of an internal IT department?

The term IS culture is described and formally defined in a number of ways, which have been analyzed and summarized by Da Veiga et al. [7] on the basis of 16 studies. The result is a consolidated definition, which is the basis for the present literature review:

> *"IS Culture is contextualized to the behavior of humans in an organizational context to protect information proceed by the organization through compliance with the IS policy and procedures and an understanding of how to implement requirements in a cautious and attentive manner as embedded through regular communication, awareness, training and education initiatives"* [7].

An internal IT department is the functional unit of a company which provides IT services in the form of hardware, software, projects, operation and consulting for the company with its own employees. In the context of this study it is assumed (as is very often the case in practice) that the internal IT department is also responsible for IS or the operational implementation of IS. Thus, the employees of the internal IT department can be differentiated into IT and IS employees.

2.2 Related Work

Due to the increasing importance of IS not only in the business environment but also in all areas of public and private life, the number of research projects is constantly growing. Different aspects of IS culture are analyzed and continuously developed and in part, contradictory statements are made. These studies often focus on end users or companies as a whole, e.g. to determine their level of IS compliance and identify potential for improvement. Sometimes a distinction is made between IT and non-IT employees, but there are very few studies that focus on the area of an internal IT department or on the behavior of IT employees and their internal differences.

However, Awawdeh and Tubaishat [8], for example, conclude that the first step towards improving the company-wide IS culture must come from the IT department – even if the argument that IS is the sole responsibility of the IT department seems questionable. The study by Lin and Wittmer [9] focuses on the concept of proactive security behavior (PASB) and examines whether, among other things, a decentralized IT governance structure has a positive effect on the PASB. However, this could not be confirmed by the study. Al-Mohannadi et al. [10] put the focus on an internal IT department: the study analyzes the differences in skills and awareness between IS and IT employees and comes to the conclusion that there are significant differences in both areas. It is also argued that communication and coordination between these two parties often does not work well and should be improved. Hooper and Blunt [4] deal with the factors that positively influence the IS behavior of IT employees. They come to the conclusion that these include self-efficacy (one's own conviction that one can successfully overcome difficult situations by one's own efforts) and the possible impact of an IS incident. They also show that the influencing factors differ between IT and non-IT employees.

In contrast to the few IT-related studies, there is a large amount of research on general IS culture frameworks. Thus, Nasir et al. [11] evaluated 79 studies that have dealt with the definition and components of IS culture between 2000 and 2017. Here it becomes transparent that the majority of all IS culture frameworks are based on Schein's organizational culture model. This represents organizational culture through the three levels of *basic assumptions*, *values* and *artifacts*. Based on existing studies and an online survey, Nel and Drevin [5] have also defined the 21 most important factors that must exist for a good IS culture. A further comprehensive analysis of relevant factors of an IS culture and initial comparison of the theoretical view and practical implementation was published by Da Veiga et al. [7]: 25 factors were identified as relevant in this study, and at the same time it was found that the theoretical view and the practical implementation differ significantly in many areas.

The field of IS behavior is also represented extensively in current research. A study by Hutchinson and Ophoff [12] has identified 26 theories describing the IS behavior of individuals in research over the last 10 years. The general deterrence theory (GDT), the theory of planned behavior (TPB) and the protection motivation theory (PMT) have proved to be the leading theories. The dominance and applicability of these theories is confirmed by further studies [13, 14]. There is also criticism of the above theories, however: Menard et al. [15] note, for example, that the factor of *motivation* is not considered in any of the theories, but has a significant influence on IS behavior. They therefore propose linking PMT to self-determination theory (SDT).

To unite the components of IS culture and IS behavior, Da Veiga and Eloff [16] have developed a model that shows how components influence IS behavior and therefore IS culture. The model includes Schein's organizational culture and can be extended to include further aspects of Tolah et al. [17], Nel and Drevin [5] and Van Niekerk and Von Solms [18], and consolidated as shown in Fig. 1.

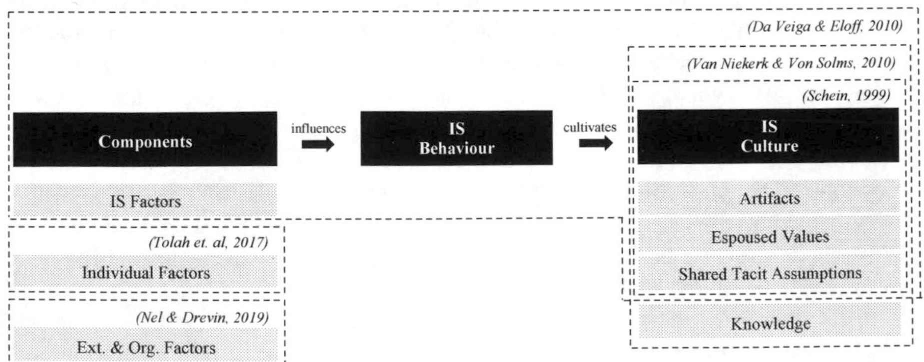

Fig. 1. Consolidated view of existing IS culture models

Since the relevant factors influencing a good IS culture have not yet been extensively researched in the context of an internal IT department or for its employees, but it is generally accepted how such a culture is defined, the model as shown in Fig. 2 is defined as the basis for identifying the influencing factors.

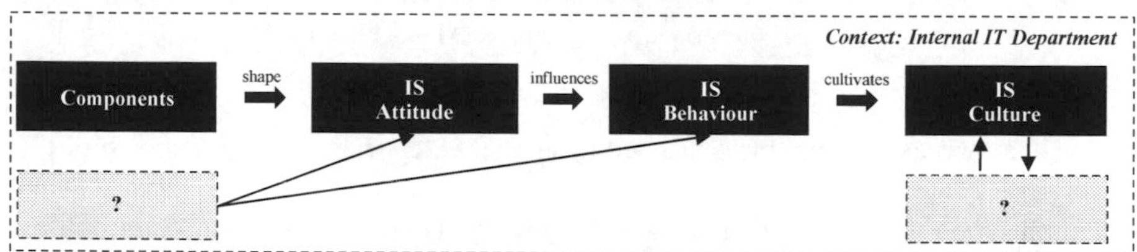

Fig. 2. Research model to identify factors influencing IS culture in an internal IT department

In addition to the existing models for IS culture, the model also includes the aspect of IS attitude and thus takes into account common behavioral theories.

The model does not claim to replace or question existing models. The literature research will be carried out on the basis of this model, however, with the aim of identifying the concrete characteristics of the components and elements of IS culture which are relevant for employees of an IT department and factors which influence their IS culture.

3 Research Method

The recommendations of Webster and Watson [19] were followed to identify and analyze the existing literature. Accordingly, the literature review process can be divided into three areas: selection of literature sources, selection of literature and structuring of literature.

3.1 Selection of Relevant Sources for Literature

To ensure a high-quality literature review, a 4-step process was applied in the selection of relevant literature sources as follows:

1. Selection of recognized top journals
2. Selection of recognized conferences
3. Reverse search, i.e. checking the sources used for the articles from 1 & 2
4. Forward search, i.e. checking articles that quote articles from 1 & 2

The selection of the top journals was determined with the help of Scimago Journal & Country Rank. Table 1 shows the four different subject areas and categories which were searched.

Table 1. Areas for selecting relevant sources

Subject Area	Subject Category	#Journals
Computer Science	Computer Science (miscellaneous)	254
Computer Science	Information Systems	267
Computer Science	Human Computer Interaction	88
Decision Science	Information Systems & Management	88

Since a journal is sometimes assigned to several categories, the search result of 697 hits included duplicates, which were reduced to 638 unique journals. In the following, all journals were selected which have an H-index of at least 75 and according to their title deal with the topics of management, human or behavior. In addition, three journals were selected which, although they have an H-index <75, according to their title deal exclusively with IS (Table 2).

Table 2. Selected journals for literature search

Journal	H Index	Country
MIS Quarterly: Management Information Systems	195	United States
Communications of the ACM	189	United States
Information and management	142	Netherlands
Computers in Human Behavior	137	United Kingdom
Journal of Management Information Systems	128	United States
Cyberpsychology, Behavior, and Social Networking	119	United States
Journal of the ACM	117	United States
International Journal of Human Computer Studies	109	United States
International Journal of Information Management	91	United Kingdom
ACM Transactions on Computer-Human Interaction	78	United States
Computers and Security	77	United Kingdom
ACM Transactions on Information and System Security	75	United States
Information and Computer Security	42	United Kingdom
International Journal of IS	35	Germany
Computer Fraud and Security	19	United Kingdom

In total, there are 15 journals from the USA, the Netherlands, England and Germany which are used for the literature search.

The selection of relevant conferences proved to be difficult: the search for relevant conference proceedings revealed that very few hits were achieved at the leading IS conferences such as the IEEE S&P, USENIX Sec and ACM CCS. This can be explained by the fact that mainly technically oriented papers are submitted and presented at these conferences. For this reason, no pre-selection of conferences was made and the search was carried out in all available conference proceedings.

3.2 Selection of Relevant Literature

A literature search was carried out in the aforementioned journals using WebOfScience and Elsevier. In order not to exclude relevant articles at the outset, the search was kept general: a search was carried out for all articles between the years 2015 and 2020, in which the string {*Information Security* OR *Cyber Security*} is used in the title. In order to include corresponding conference contributions, the databases WebOfScience, Elsevier and ACM were also searched. Here the string {*Information Security* OR *Cyber Security*) was searched for in the title and at the same time {*culture* OR *behavior* OR *skill* OR *knowledge* OR *employee* OR *human*} in the abstract.

In a multi-stage selection process, 25 studies were finally identified as the basis for the literature review as shown in Fig. 3.

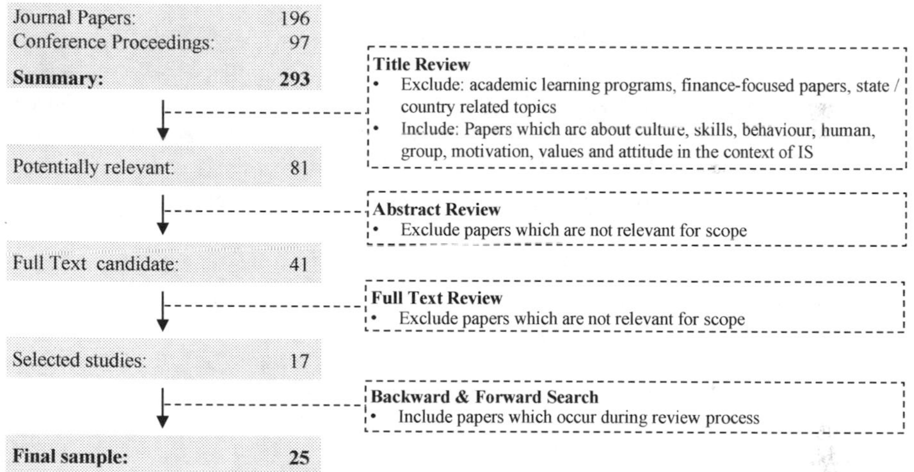

Fig. 3. Flowchart of literature review process

Concepts and models that were empirically proven and literature reviews based on empirical data were included. Cross-references and citations from other studies (forward and backward searches) were considered in the literature analysis and investigated with the help of Google Scholar.

3.3 Structuring of the Literature

For each item it was verified whether the results or parts of the results refer to the context of an IT department or its employees or whether they can be applied to it. The analysis process resulted in three categories into which the selected articles were classified:

- Research method (quantitative/qualitative)
- Target group (IT or non-IT employees)
- Relevant IS culture factor (in the area of the IT department)

In the category *research method*, the studies could be clearly assigned to the quantitative, qualitative or mixed-method characteristics. An equally clear classification was made in the category of *target group*. In the *IS culture factor* category, the studies were assigned several times depending on their content.

4 Analysis of the Results

In the following, the studies are classified into three categories. On the one hand, the classification indicates a focus of previous research, while at the same time future fields of research can be identified.

4.1 Research Method

As can be seen from Table 3, the majority of the studies analyzed are based on a quantitative method for determining the research results. Quantitative research has the advantages of greater objectivity and comparability due to the necessary standardization. At the same time, quantitative research assumes that social conditions can be explained objectively and evaluated statistically [20]. In the context of IS culture, this means that existing hypotheses and theories are tested by means of a deductive approach. However, since culture is shaped by behavior, values and basic assumptions, among other things, this research approach can certainly not provide fully comprehensive explanations. A qualitative approach is characterized by its openness and exploratory style. He tries to understand the perspective of the actors and on this basis to generate hypotheses and theories to explain social phenomena [20]. For an analysis of factors that influence an IS culture, this approach is therefore at least as important as a quantitative analysis.

Table 3. Applied research methods of considered studies

Type	Paper
Quantitative	[4, 6, 7, 10, 15, 21–34]
Qualitative	[8, 13, 35–37]
Mixed-method	[5]

The quantitative studies are largely based on online questionnaires, the answers to which were measured or evaluated using the Likert scale. Qualitative research work, which is mainly based on semi-structured expert interviews, is only sparsely available, as previously mentioned: Of the 25 studies considered, 19 were conducted on a quantitative level and only six were based on a qualitative or mixed-method approach.

The result suggests that qualitative research in the field of IS culture – especially if it is restricted to a specific area – is underrepresented. This view is also held by Bauer et al. [38], who specifically indicate the need for more qualitative research in this environment.

4.2 Target Group

Of the studies analyzed in the literature reviews, only five have a clear focus on IT employees. The remaining studies focus on non-IT users, mix the participants or do not specify them further (Table 4).

Table 4. Target groups of considered studies

Type	Paper
Business & IT User	[5, 21, 23–25, 27, 31, 32]
IT User only	[4, 8, 10, 34, 37]
Business User only	[15, 28, 33, 35]
Users not specified	[7, 22, 29, 30]
No Users involved	[6, 13, 26, 36]

It becomes clear that in the environment of an internal IT department, studies of IS culture or the factors interacting with it were only carried out selectively. It has been proven that the IS behavior of IT employees differs demonstrably from that of non-IT employees and that there are even differences within an IT department [4, 10].

A closer look at the IT staff – which includes both the traditional IT functions and the IS functions – thus seems to make sense.

4.3 Relevant IS Culture Factors

The main focus of this literature review is to identify relevant factors for an IT department or its employees which have the potential to (positively) influence the IS culture within the IT department. Derivation and identification of the factors is based on existing IS culture models [5, 11, 17]. On the other hand, an obvious connection to the IT department of a company is an important criterion. From the context of the respective articles, factors could be identified which are found in existing IS culture models and at the same time seem to have a special meaning in the context of an IT department or IT staff.

In the articles, different terms are used to describe the same factor (e.g. skill, competence, expertise). Such terms have been combined.

The factor *Subculture & Groups* subsumes group dynamic affiliation and processes such as the lemming effect and the impact of personal attributes and societal norms on IS attitudes, behavior and risk tolerance. According to Da Veiga and Martins [32], different IS subcultures develop in a company, which establish themselves on team, division or location level. Snyman et al. have partially described the effects and demonstrated that positive IS behavior is adopted in a group. However, it is not discussed whether negative IS behavior is also adapted by colleagues [30]. IS subcultures therefore also exist within an IT department, since even there people with different IS risk propensities, IS skills, IS attitudes and thus differentiated IS behavior come together [10].

Under the factor *IS skill* all terms are summarized, which refer to experience, expertise, theoretical knowledge but also self-efficacy regarding the handling of IS. This factor is described in al standard IS culture frameworks, but plays a special role in the IT department: often the IT department of a company is still seen as solely responsible for IS [8, 37]. In addition, situations arise in which, for example, the administration of an IS system – such as a firewall or antivirus – is assigned to employees who have no experience in this area [34]. At the same time, there is also the expectation that every employee of the IT department is able to provide information on current IS issues and also takes these into account in his or her daily work. IT employees who perform a traditional function such as client management are thus faced with the challenge of acquiring specialized IS knowledge in addition to their technical skills. This is advocated by Awawdeh and Tubaishat [8], who propose dedicated IS training with a clear focus on the respective traditional IT function such as helpdesk or database administrator.

Appreciation was identified as a further factor: this included the terms recognition, (intrinsic) motivation and the fulfilment of personal needs. Recognition for work done contributes to intrinsic motivation, which is an important driver of good IS behavior [15]. For an IT department, this factor is particularly interesting: when implementing IS measures, the motivation of employees in traditional IT functions differs from that of employees in the IS function [34].

The *Distortion* factor includes statements about the IS paradox as well as different types of cognitive bias and false assumptions or prejudices about IS.

This factor is relevant for representatives of the IT department, as an overestimation of the IS level has been identified there (optimistic bias). This leads to the fact that those affected assess themselves better than the average and thus, if necessary, do not carry out essential IS measures [37].

The *IT tools* factor includes the influence and impact of increased complexity, the use of heterogeneous systems and the expectation of or trust in technological solutions. All this influences the IS attitude and the IS behavior of all IT users in a company. The Technology Acceptance Model according to Davis [39], which makes statements about the acceptance and application of an IT tools, often serves as the basis for this. Hwang and Cha [31] argue that technology stress caused by complex IT tools contributes to poor IS behavior. For employees of an IT department, however, this factor must be interpreted more broadly: in accordance with their function, these employees are not only users of the IT tools, but are also responsible for its implementation and operation. The responsibility for a highly heterogeneous IT landscape, which on the one hand contains modern cloud solutions and on the other hand often IT tools from the last millennium, has an impact on IS attitude and behavior.

The factor of *Monitoring* covers all areas that can be used to monitor the compliance of the IS behavior. According to Ahmad et al. [28], awareness of active monitoring activities promotes IS behavior of employees. On the other hand, it is also shown that users prefer the highest possible degree of freedom and that this should be regarded as the basis for good IS behavior [15]. This factor is relevant for employees of an IT department, since in some cases monitoring of activities is already implemented (e.g. in the case of administration with highly privileged user accounts), but on the other hand this monitoring can also be interpreted as mistrust.

IT satisfaction includes attributes that can influence it positively or negatively and thus also have demonstrable effects on IS behavior. Montesdioca and Maçada [23], for example, explain that while users understand the need for technological IS measures, the application or implementation of these measures has a negative impact, as they often result in a loss of productivity, which in turn leads to dissatisfaction. This statement is confirmed by Amo and Cichocki [33] and supplemented by the fact that satisfaction with IT resources leads to good IS behavior and dissatisfaction has a negative effect on it. Just like the technical factor, IT satisfaction is also relevant for the internal IT department. On the one hand, IT staff are also users of the technology, and on the other hand all complaints and malfunctions are reported directly to them (Table 5).

Table 5. IS culture factors for internal IT departments

Factor	Paper
Sub-Culture & Groups	[5, 7, 22–25, 27, 28, 30, 32, 35, 36]
IS skills	[4, 5, 7, 8, 10, 21, 23, 25, 34, 36]
Appreciation	[5, 7, 13, 15, 34]
Distortion	[28, 30, 37]
IT tools	[7, 25, 26, 29, 31–33, 35, 36]
Monitoring	[21, 25, 37]
IT satisfaction	[26, 33]

Although seven factors were identified and studies have described more than one factor, it is clear that the factors *SubCulture & Groups, IS skill, IT tools and Appreciation* were named most frequently (at least in 20% of the relevant papers). Further and more detailed consideration of these factors in the context of an IT department is therefore an obvious requirement.

5 Conclusion

The results of this literature review show that in the past, mainly quantitative studies have been conducted to analyze IS culture. Of the 25 studies analyzed, only six could be assigned to a qualitative or mixed-method approach. With regard to the target group, it has been shown that only five focus on the internal IT department of a company. The other studies used employees of other company functions or mixed groups as a data basis.

If one considers the results of the analysis of the IS factors for an IT department, the four most relevant ones can be classified in the model already presented as shown in Fig. 4:

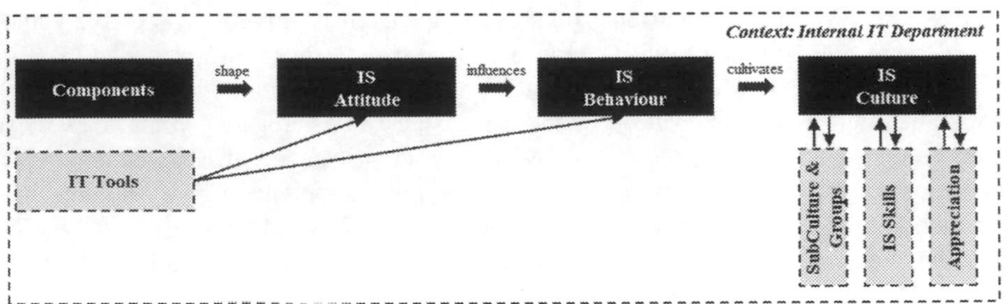

Fig. 4. Relevant factors influencing the IS culture in an IT department

The factor of IT tools influences the IS attitude and the IS behavior of employees of an IT department to a different extent than employees of other departments. If you are confronted with IS vulnerabilities, IS patches and cyber attacks on a daily basis and at the same time have to make sure that the required applications and infrastructure components are available despite all this, you will behave differently with regard to IS. This is especially the case when it is known, for example, that existing weaknesses in the IT architecture cannot be eliminated, but new IT applications have to pass a perceived excessive IS assessment. Such an imbalance does not occur with users from other departments and is specific to the IT department.

Multiple functions in the IT department (traditional IT functions and IS function) result in different manifestations of an IS culture. Not all employees have consistent prioritization of IS measures, and project teams and sub departments handle IS risks differently. Thus, representatives of the IS function often have to deal with resistance within the department and still have to make sure that the IS behavior of all employees in the IT department improves. It is important to exemplify positive IS behavior and to convince departmental IS cultures which are neglecting IS of the relevance of IS.

The IS skill within an IT department plays an essential role for the IS function as well as for traditional IT functions. From the perspective of the IS function, its representatives must be able to rely on the implementation of basic IS recommendations, even if a guideline or instruction for action does not exist for every detail. From the perspective of traditional IT functions, insufficient training can increase the fear of making glaring mistakes in system administration or software development, and thus encourage an IS incident. This fear can only be eliminated through targeted training and good cooperation.

The recognition of personal performance is as important for employees in the IT department as it is for all other employees. However, the continued presence of IS threats and the associated countermeasures taken by an IT department can result in employees in the IS function receiving special recognition from managers and colleagues. On the one hand, these employees are praised for their successful defense against cyber attacks, while on the other hand traditional IT functions may be neglected. This situation has a direct impact on the IS behavior and therefore on the IS culture of the IT department and should not be neglected. Representatives of both traditional and IS functions are mutually dependent on each other and can only improve the IS culture of the IT department and thus the IS level of the company together.

6 Limitations and Future Research

The results of the literature review should be evaluated with consideration for two limitations. Firstly, despite the greatest care in conducting the literature review, it cannot be ruled out that a relevant study was not discovered and thus not considered in the literature review. Secondly, the identification of IS culture factors and the classification of studies was not based on a strict mathematical foundation, but was carried out within the framework of the literature review process using qualitative assessments. It is likely that further factors can be identified and that individual terms or further study results can also be included.

However, the three areas of results show that qualitative research into IS culture is underrepresented in the environment of an internal IT department and the identified influencing factors need to be analyzed in more detail for a better understanding. It is to be expected that complementary insights will be gained through focused qualitative research, e.g. by conducting semi-structured interviews with employees of an IT department.

References

1. Allianz: Allianz Risk Barometer Report – Identifying the major business risks for 2020 (2020)
2. World Economic Forum: The Global Risks Report 2020 (2020)
3. International Data Group: Otto Beisheim School of Management – Wissenschaftliche Hochschule für Unternehmensführung, Bechtle AG (2020) CIO Agenda (2020)
4. Hooper, V., Blunt, C.: Factors influencing the information security behaviour of IT employees. Behav. Inf. Technol. **1–13**, (2019). https://doi.org/10.1080/0144929X.2019.1623322
5. Nel, F., Drevin, L.: Key elements of an information security culture in organisations. Inf. Comput. Secur. **27**, 146–164 (2019). https://doi.org/10.1108/ICS-12-2016-0095
6. Cram, W.A., D'Arcy, J., Proudfoot, J.G.: Seeing the forest and the trees: a meta-analysis of the antecedents to information security policy compliance. MISQ **43**, 525–554 (2019). https://doi.org/10.25300/MISQ/2019/15117
7. Da Veiga, A., Astakhova, L.V., Botha, A., Herselman, M.: Defining organisational information security culture—Perspectives from academia and industry. Comput. Secur. **92**, 101713 (2020). https://doi.org/10.1016/j.cose.2020.101713
8. Awawdeh, S.A., Tubaishat, A.: An information security awareness program to address common security concerns in IT unit. In: 2014 11th International Conference on Information Technology: New Generations, Las Vegas, NV, USA. IEEE, pp. 273–278 (2014)
9. Lin, C., Wittmer, J.L.S.: Proactive information security behavior and individual creativity: effects of group culture and decentralized IT governance. In: 2017 IEEE International Conference on Intelligence and Security Informatics (ISI). Univ Arizona, Artificial Intelligence Lab; Univ. Chinese Acad. Sci., pp 1–6 (2017)
10. Al-Mohannadi, H., Awan, I., Al Hamar, J., Al Hamar, Y., Shah, M., Musa, A.: Understanding awareness of cyber security threat among IT employees. In: 2018 6th International Conference on Future Internet of Things and Cloud Workshops (FiCloudW), Barcelona, pp. 188–192. IEEE (2018)

11. Nasir, A., Arshah, R.A., Hamid, M.R.A., Fahmy, S.: An analysis on the dimensions of information security culture concept: a review. J. Inf. Secur. Appl. **44**, 12–22 (2019). https://doi.org/10.1016/j.jisa.2018.11.003
12. Hutchinson, G., Ophoff, J.: A descriptive review and classification of organizational information security awareness research. In: Venter, H., Loock, M., Coetzee, M., Eloff, M., Eloff, J. (eds.) ISSA 2019. CCIS, vol. 1166, pp. 114–130. Springer, Cham (2020). https://doi.org/10.1007/978-3-030-43276-8_9
13. Gangire, Y., Da Veiga, A., Herselman, M.: A conceptual model of information security compliant behaviour based on the self-determination theory. In: 2019 Conference on Information Communications Technology and Society (ICTAS), Durban, South Africa, pp. 1–6. IEEE (2019)
14. Nasir, A., Abdullah Arshah, R., Rashid Ab Hamid, M.: The significance of main constructs of theory of planned behavior in recent information security policy compliance behavior study: a comparison among top three behavioral theories. IJET **7**, 737 (2018). https://doi.org/10.14419/ijet.v7i2.29.14008
15. Menard, P., Bott, G.J., Crossler, R.E.: User motivations in protecting information security: protection motivation theory versus self-determination theory. J. Manag. Inf. Syst. **34**, 1203–1230 (2017). https://doi.org/10.1080/07421222.2017.1394083
16. Da Veiga, A., Eloff, J.H.P.: A framework and assessment instrument for information security culture. Comput. Secur. **29**, 196–207 (2010). https://doi.org/10.1016/j.cose.2009.09.002
17. Tolah, A., Furnell, S.M., Papadaki, M.: A Comprehensive framework for cultivating and assessing information security culture, p. 13 (2017)
18. Van Niekerk, J.F., Von Solms, R.: Information security culture: a management perspective. Comput. Secur. **29**, 476–486 (2010). https://doi.org/10.1016/j.cose.2009.10.005
19. Webster, J., Watson, R.T.: Analyzing the past to prepare for the future: writing a literature review. MIS Q. **26**, xiii–xxiii (2002)
20. Yilmaz, K.: Comparison of quantitative and qualitative research traditions: epistemological, theoretical, and methodological differences. Eur. J. Educ. **48**, 311–325 (2013). https://doi.org/10.1111/ejed.12014
21. Ahmad, Z., Ong, T.S., Liew, T.H., Norhashim, M.: Security monitoring and information security assurance behaviour among employees: an empirical analysis. Inf. Comput. Secur. **27**, 165–188 (2019). https://doi.org/10.1108/ICS-10-2017-0073
22. Sommestad, T.: Work-related groups and information security policy compliance. Inf. Comput. Secur. **26**, 533–550 (2018). https://doi.org/10.1108/ICS-08-2017-0054
23. Halevi, T, et al.: Cultural and psychological factors in cyber-security. In: Proceedings of the 18th International Conference on Information Integration and Web-Based Applications and Services, New York, NY, USA, pp. 318–324. Association for Computing Machinery (2016)
24. Dang-Pham, D., Pittayachawan, S., Bruno, V.: Applying network analysis to investigate interpersonal influence of information security behaviours in the workplace. Inf. Manag. **54**, 625–637 (2017). https://doi.org/10.1016/j.im.2016.12.003
25. AlHogail, A.: Design and validation of information security culture framework. Comput. Hum. Behav. **49**, 567–575 (2015). https://doi.org/10.1016/j.chb.2015.03.054
26. Montesdioca, G.P.Z., Maçada, A.C.G.: Measuring user satisfaction with information security practices. Comput. Secur. **48**, 267–280 (2015). https://doi.org/10.1016/j.cose.2014.10.015
27. Gratian, M., Bandi, S., Cukier, M., Dykstra, J., Ginther, A.: Correlating human traits and cyber security behavior intentions. Comput. Secur. **73**, 345–358 (2018). https://doi.org/10.1016/j.cose.2017.11.015
28. McCormac, A., Zwaans, T., Parsons, K., Calic, D., Butavicius, M., Pattinson, M.: Individual differences and information security awareness. Comput. Hum. Behav. **69**, 151–156 (2017). https://doi.org/10.1016/j.chb.2016.11.065

29. Topa, I., Karyda, M.: Identifying factors that influence employees' security behavior for enhancing ISP compliance. In: Fischer-Hübner, S., Lambrinoudakis, C., Lopez, J. (eds.) TrustBus 2015. LNCS, vol. 9264, pp. 169–179. Springer, Cham (2015). https://doi.org/10.1007/978-3-319-22906-5_13
30. Snyman, D.P., Kruger, H., Kearney, W.D.: I shall, we shall, and all others will: paradoxical information security behaviour. Inf. Comput. Secur. **26**, 290–305 (2018). https://doi.org/10.1108/ICS-03-2018-0034
31. Hwang, I., Cha, O.: Examining technostress creators and role stress as potential threats to employees' information security compliance. Comput. Hum. Behav. **81**, 282–293 (2018). https://doi.org/10.1016/j.chb.2017.12.022
32. Da Veiga, A., Martins, N.: Defining and identifying dominant information security cultures and subcultures. Comput. Secur. **70**, 72–94 (2017). https://doi.org/10.1016/j.cose.2017.05.002
33. Amo, L.C., Cichocki, D.: Disgruntled yet deft with IT: employees who pose information security risk. In: Proceedings of the 2019 on Computers and People Research Conference, Nashville, TN, USA, pp. 122–124. ACM (2019)
34. Govender, S.G., Loock, M., Kritzinger, E.: Enhancing information security culture to reduce information security cost: a proposed framework. In: Castiglione, A., Pop, F., Ficco, M., Palmieri, F. (eds.) CSS 2018. LNCS, vol. 11161, pp. 281–290. Springer, Cham (2018). https://doi.org/10.1007/978-3-030-01689-0_22
35. Karjalainen, M., Siponen, M., Sarker, S.: Toward a stage theory of the development of employees' information security behavior. Comput. Secur. **93**, 101782 (2020). https://doi.org/10.1016/j.cose.2020.101782
36. Glaspie, H.W., Karwowski, W.: Human factors in information security culture: a literature review. In: Nicholson, D. (ed.) AHFE 2017. AISC, vol. 593, pp. 269–280. Springer, Cham (2018). https://doi.org/10.1007/978-3-319-60585-2_25
37. Tariq, M.A., Brynielsson, J., Artman, H.: The security awareness paradox: a case study. In: 2014 IEEE/ACM International Conference on Advances in Social Networks Analysis and Mining (ASONAM 2014), China, pp. 704–711. IEEE (2014)
38. Bauer, S., Bernroider, E.W.N., Chudzikowski, K.: Prevention is better than cure! Designing information security awareness programs to overcome users' non-compliance with information security policies in banks. Comput. Secur. **68**, 145–159 (2017). https://doi.org/10.1016/j.cose.2017.04.009
39. Davis, F.D.: Perceived usefulness, perceived ease of use, and user acceptance of information technology. MIS Q. **13**, 319 (1989). https://doi.org/10.2307/249008

Research on Payment UX Status During the Share Cycle Services Between Japan and China

Jiahao Jiang(✉), Keiko Kasamatsu, and Takeo Ainoya

Tokyo Metropolitan University, 6-6 Asahigaoka, Hino, Tokyo, Japan
grapekun1995@gmail.com

Abstract. Different from traditional services, online services can simply develop from the place of origin to any region of the world. However, users in different regions are mostly used to local service design. Therefore, when online services are developed in different regions, service designers should consider that users have different perceptions of user experience. In this study, the rapid development of shared bikes around the world is used as a reference object. By comparing the different payment systems of online Share Cycle services between China and Japan, investigate the reasons, and research on User Experience of the Payment System (include Pre-charge or Post Payment, Manual Payment or Automatic Payment etc.) by questionnaire survey. In the subsequent research, I hope to propose user experience and service design methods that can be applied to China and Japan.

Keywords: Share Cycle · Mobile payment · User experience

1 Introduction

1.1 Online Service

Conventional services are designed according to cultural differences at the regional, regional and individual levels across countries. The spread of Internet-based smartphones has enabled online services to operate almost completely ignoring national borders. It is often heard that the game started from somewhere, but in a world without the borderline of online, it can easily spread to the world. Different cultures have different perceptions of users from services to user experiences.

Designers of online services must always consider the user interface (UI) and user experience (UX) and imagine and infer patterns of various types of user's sense, recognition, and behavior. However, each user has different knowledge and familiarity, and different physiques, physical abilities, sensory abilities, and cognitive abilities. There are also differences in culture at the level, region, attributes, and individual levels across countries.

While various things are being digitized, designers need to fully understand the hobbies, preferences, and values of the recipient, and then consider products and services for different targets. In this study, we compare the current state of the Japan-China share cycle service and the analysis of its user experience. In the future, we will

C. Stephanidis et al. (Eds.): HCII 2020, LNCS 12427, pp. 522–534, 2020.
https://doi.org/10.1007/978-3-030-60152-2_39

also conduct questionnaire and interview surveys for users and observe culture. We clarified the relationship between psychological features arising from different cultures and differences in UX cognition.

1.2 Share Cycle Service

The urban development of modern society has broadened the market demand for Share Cycle. For a long time, no matter how the public transportation industry develops and perfects, it is impossible to solve the short-distance mobility problems of users from subway stations to destinations, from home to bus stations. Share Cycle solves the problem of "last mile" for users, and bicycles have become an excellent choice for users to travel short distances.

In addition, in New York, Tokyo, Beijing and other world-class metropolises and other first-tier cities, the huge traffic pressure cannot be eased, and the emergence of shared bicycles has reduced the congestion of subways and bus stations. When users have the same or even better choices, they will naturally generate a large number of Share Cycle users. In addition, with the popularization of industry and automobiles, the planet is facing environmental problems such as rising global temperatures and increasing carbon emissions. As an environmentally friendly and healthy way of green travel, Share Cycle is very popular among users, and it is changing the way people travel more subtly.

For the reason that, we use Share Cycle as a reference case for this study.

Share Cycle Service is referred to as Share Cycle System or Public Bicycle Scheme (PBS). It is a service that allows each user to share and use the last mile bicycle of the modern society temporarily or just for free.

The share cycle service is mainly divided into four parts according to functions and roles.

- Spot

Spot exert its action as parking Bicycles and storing the cycles to avoid confusion. In the share cycle system in countries such as Japan where public space management laws are strict, users rent bicycles from "spots" and return them to other spots belonging to the same system. There are also spots of some systems are a special bike rack that locks the bike and can only be released under their system computer control. But in less demanding areas of China and other Asian countries, you can park your bicycle in the city. However, in recent years, designated bicycle parking areas have been dictated by city management regulations enforced by each region. The user places the bicycle at the spot and returns it to lock the bicycle.

- Cycle

Most bikes have numbers written on them to distinguish them from users and be recognized by the system. And will be equipped with a smart lock that can only be unlocked through the system application. As well usual sharing bicycles are often designed in a size that can be used by people of all sizes.

- System Application

Share Cycle applications can be divided into web applications and smartphone applications. Users can register as Share Cycle users through the application or web pages. In different countries or regions, the Share Cycle service uses a local third-party online map navigation service to display the user's own location and the location of the bicycle or spot. After the user finds a bicycle, he can unlock it automatically/manually by scanning the QR code of the vehicle body or entering the password. When the user uses it, some applications will automatically record the usage line and usage time.

- Payment

Depending on the bike-sharing system, payment occurs before or at the end of the entire usage process. It is usually done by a third-party payment platform. It is common to bind a credit card or pay online before use. In some systems, payment can also be done through pre-sold IC cards and smart lock induction.

2 Research of Share Cycle Services in China and Japan

2.1 Share Cycle in China

According to data released by the Ministry of Transport, as of the end of August 2019, there are 19.5 million Internet rental bicycles in China, covering 360 cities across the country, with more than 300 million registered users and an average daily order of 47 million.

Survey data show that as of May 2018, China's Share Cycle service users were concentrated in ofobike and mobike, occupying more than 90% of the market share. By 2019, the proportion of users has changed, and new brands such as hellowbike and didibike appeared. In future research, services such as ofobike, mobike, hellowbike, didibike will be used as reference objects for user experience research.

36.8% of users use shared bicycles for commuting purposes, such as commuting to work and school; 36.7% of users use shared bicycles for time-saving purposes (excluding commuting), including cycling to commercial supermarkets; leisure demand accounts for 22.9%.

In addition, it is found in the investigation that the damage and quality of the bicycle will be an important factor affecting the user experience. This often requires the service operator to make a proper operation and maintenance plan.

2.2 Share Cycle in Japan

Currently, Japan's share cycle is still developing. Although the total number is uncertain, the number of installation locations is increasing every year, with 852 locations in 2016 and 1028 locations in 2018. As of 2018, share cycle business operates in 135 cities in Japan.

The survey study showed that nearly half of users use Share Cycle for personal behaviors such as shopping, eating, and pastimes, and other usage purposes such as commuting, business, and sightseeing can also be observed. Maybe because of the geographical relationship, there are not many users who use shared bicycles to go to schools in this area.

In the future research, we will try to collect a wide range of Japanese Share Cycle user samples to describe the user portraits and mental states of Japanese users more clearly.

3 Comparison of Share Cycle Payment UX

3.1 UX Map of Share Cycle Services Between Japan and China

When discussing mobile o2o services such as Share Cycle, we should note that all online services need to have a legal and easy-to-use mobile payment function as a prerequisite. Before discussing the differences in user experience of Share Cycle between China and Japan, we believe that the difference in payment systems is an important factor affecting the user experience of the entire service (Fig. 1).

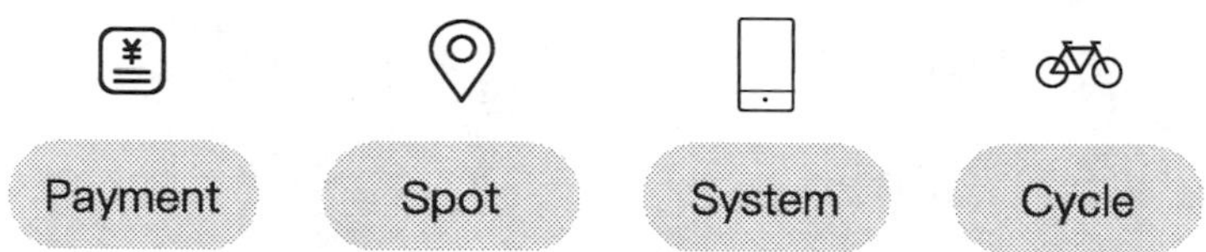

Fig. 1. 4 components that make up Share Cycle

Bicycle sharing in China is based on bicycles, which can be unlocked by scanning the QR code with Bluetooth and can be parked anywhere. Finally, payment is made through a bound third-party mobile payment. The Share Cycle in Japan is based on the spot, and the password is unlocked by entering the password prompted by the mobile phone application. After the end, the bicycle must be parked to a spot near the destination. Finally, the payment is made via a bound credit card.

After not considering the existence of spots due to regulatory issues and the difference in the operation of smart locks, we noticed that the biggest difference is the design of the two parties in payment.

In the user experience analysis of bicycle sharing services in China and Japan, we selected ofobike and DoCoMo ShareBike, which have the highest number of users in each country (Figs. 2 and 4).

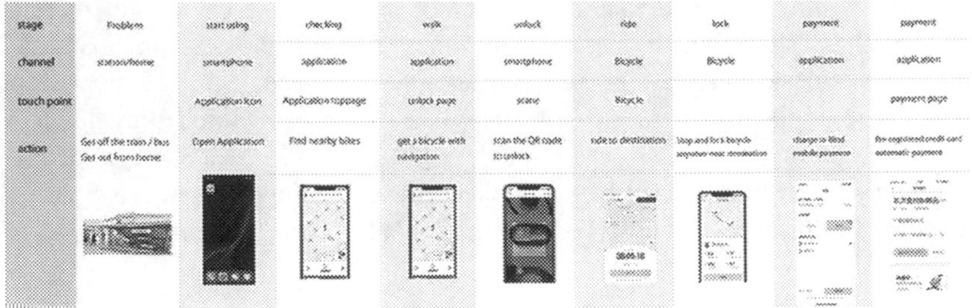

Fig. 2. ofobike user experience map

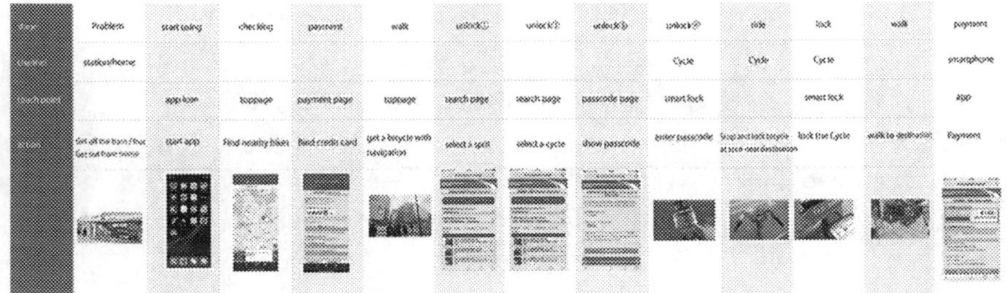

Fig. 3. DoCoMo sharebike user experience map

3.2 Different Payment Methods

According to the survey of the payment system mentioned above, due to the different development history of mobile payment on both sides, the penetration rate and usage rate of mobile payment are greatly different, which shows that the user's acceptance and awareness of mobile payment exist difference.

After investigating the payment methods of China-Japan Share Cycle Service, it can be seen that the payment of Share Cycle in China relies almost on third-party payment platforms represented by Alipay and WeChat Pay. Examples are mobike, ofobike, hellobike (Fig. 4).

In Japan, the shared bicycle payment system basically uses a credit card or a carrier payment provided by a mobile communication operator to make payments. It is rare to support mobile payments. Such as DoCoMo sharebike, PIPPA, Hello Cycling (Fig. 5).

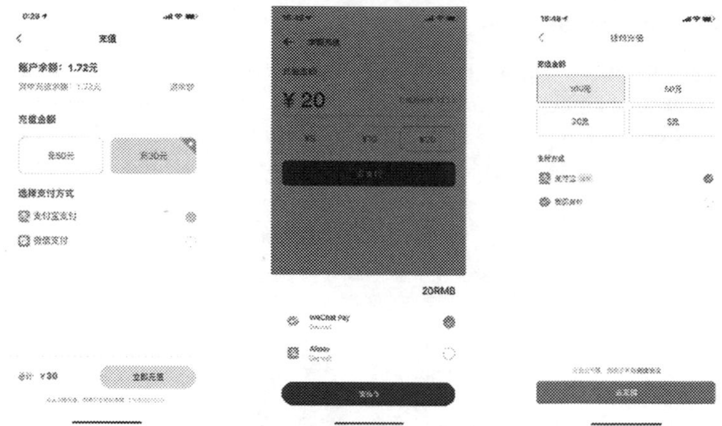

Fig. 4. China share cycle service payment interface (ofo, mobike, hellobike)

Fig. 5. Japan share cycle service payment interface (Pippa, DoCoMo Sharebike, Hello Cycling)

4 Investigation of Online Payment in China and Japan

In the comparison of the above UX Map, we see that there is a big difference in the setting of the payment system between the two parties. However, local online service designers must design the entire service usage process based on the most familiar usage habits of local users. It also leads to different perceptions of user experience. Chinese users seem to accept mobile payments more, while Japanese users seem to use more credit cards, e-money cards. Here we hope to find different reasons that lead to the design of bicycle sharing payment system. Therefore, we need to investigate the status of mobile payment development in the two countries and the factors that affect its development.

4.1 Mobile Payment in China

China has become the world's largest mobile payment market, and it is in a leading position in terms of mobile payment user size, transaction size, and penetration rate. As of the first half of 2018, the number of mobile payment users in China was about 890 million, and the penetration rate of mobile payment among mobile phone users reached 92.4% (Fig. 6).

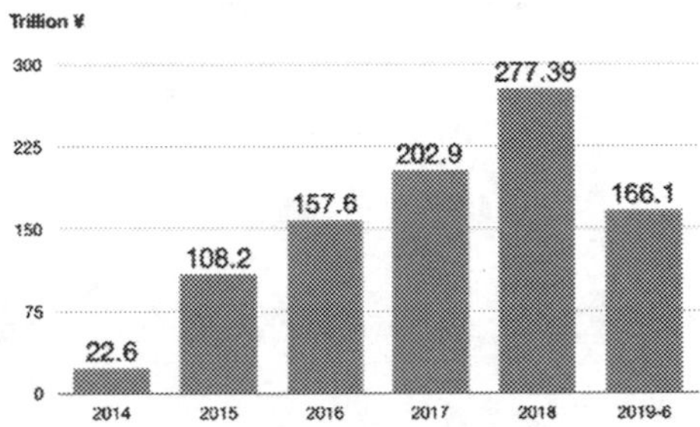

Fig. 6. 2014-2019H1 China Mobile Payment Transaction Scale 1)

By 2018, the number of mobile payment users in China has exceeded 500 million, and it is estimated that there will be more than 800 million users in 2019, which is almost half of China's population. Mobile payment is used in all aspects of life, such as online shopping, offline payment, transfer, payment and so on. Users are distributed across all ages, which shows that Chinese people are basically used to mobile payment services.

The 2019 Q3 third-party mobile payment user research report released by the market research group Ipsos surveyed and published by two thousand current users pointed out that the proportion of cash payments in daily life is only 14%, and those using credit or savings cards to pay It accounts for 25%, and the proportion of using mobile payment and Internet payment is as high as 61%. It can be seen that Chinese people rely heavily on mobile payments for their daily lives (Fig. 7).

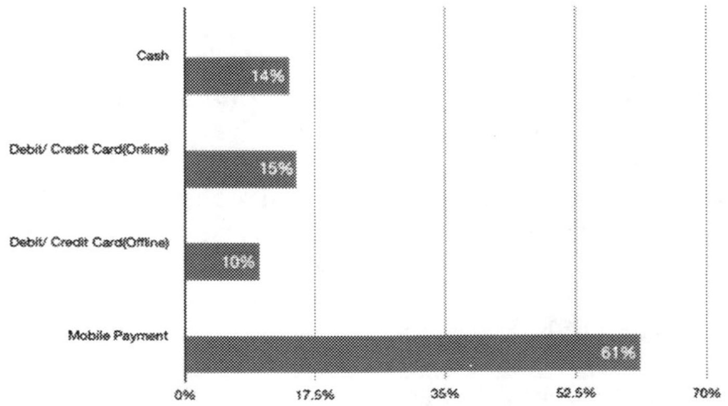

Fig. 7. Payment methods often used in Chinese daily life 4)

4.2 Mobile Payment in Japan

In the mobile payment situation in Japan, the size of the domestic mobile payment market in FY2017 expanded to 1 trillion yen. In the Japanese market, the introduction of Apple Pay and the full-scale rollout of Google Pay have led to a rapid expansion of mobile contactless payments, and the market has been expanding due to the spread and use of QR code payments.

In mobile payments in Japan, "Osaifu-Keitai", which appeared from NTT DoCoMo in 2004, impressed that it is an advanced initiative worldwide, and that payments can be made with mobile phones in Japan. However, since then, it has not spread rapidly, and there has been no great excitement until the appearance of Apple Pay, an overseas service (Fig. 8).

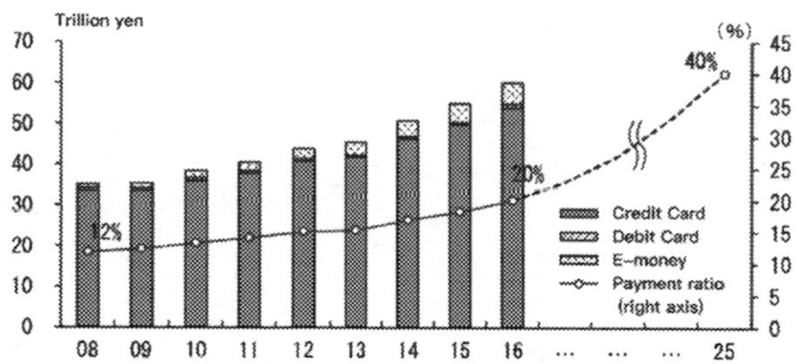

Fig. 8. Cashless spending and percentage of private final consumption expenditure 5)

Also, based on the results of the September 2019 Survey on Smartphone Payments conducted by the MMD Research Institute, we asked 37,040 men and women between the ages of 18 and 69 about their usual payment methods. "Next was "credit cards" followed by "card-type transportation electronic money", with mobile use only 16.4% (Fig. 9).

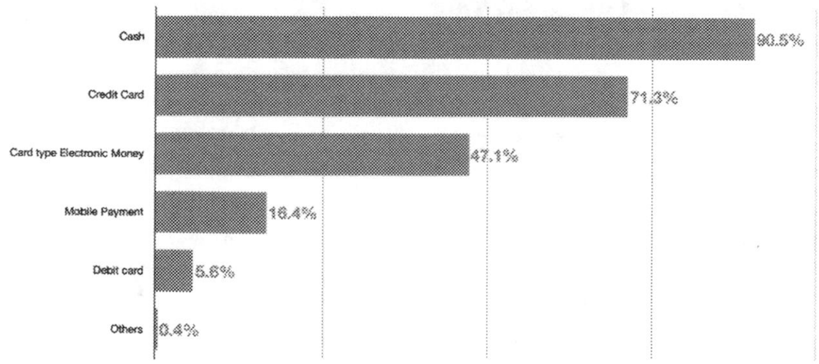

Fig. 9. Payment methods often used in Japanese daily life 6)

The different scales of mobile payment usage have led to different considerations for the design of payment functions in online services. As Chinese people are used to satisfying various living needs through mobile payment, developers have defaulted to using mobile payment when designing. And because mobile in Japan is relatively slow, mobile payment services are rare.

An investigation of the history of the development of physical payments in the two countries may find the answer. According to the "Status of Cashless Payments" report issued by the Bank of Japan, as of the end of 2017, the average Japanese had 2.15 credit cards, 3.48 debit cards, and 2.90 e-money cards. According to the "Overall Situation of the Payment System in the Fourth Quarter of 2017" announced by the People's Bank of China, the average Chinese person holds 4.45 debit cards, compared with 0.39 credit cards. This means that in the development of payment systems, Chinese people do not often use credit cards, while Japanese people are used to spending with cash and credit cards (Fig. 10).

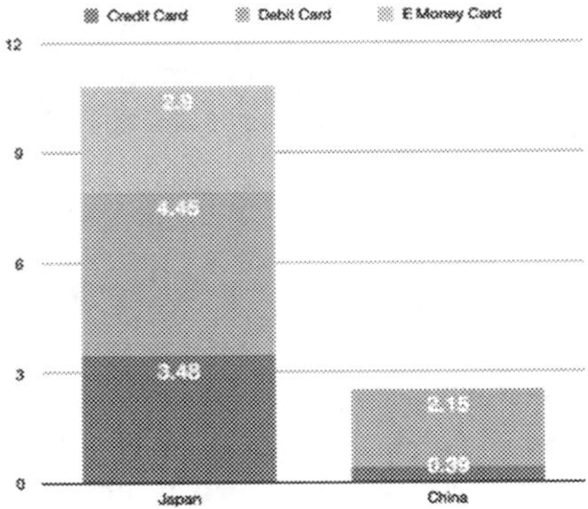

Fig. 10. Payment methods often used in Japanese daily life 5) 7)

5 Research of Payment Methods

We conducted a questionnaire survey to investigate user concerns about the security and convenience of mobile systems.

Through the survey, we collected 82 valid answers. There are 75 Chinese and 7 Japanese. 39 were male and 43 were female. The majority of them are in their 20 s, accounting for 68.2% of the total. Of those, 35 were students and 41 were office workers. Therefore, we can know that the respondents are basically familiar with and accustomed to the process of online services, and have a certain experience of using online services.

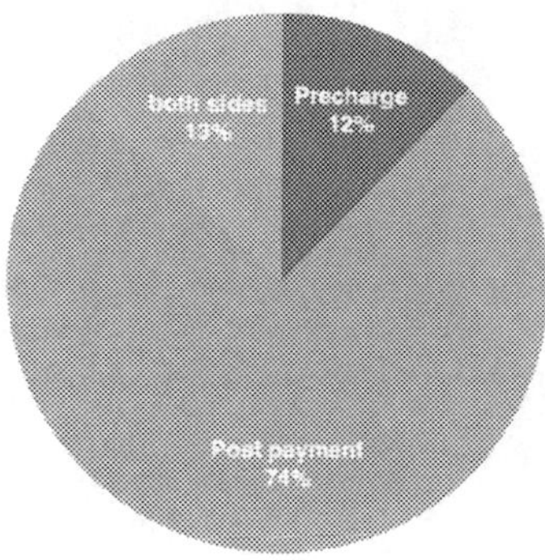

Fig. 11. Ratio of payment methods

In "What payment method do you usually use for services that require online payment?" 74% of the respondents chose post payment, and only 12% chose pre-charge as Fig. 11.

After distinguishing the nationalities of the respondents, we found a big difference as Fig. 12. 80% of Chinese respondents chose postpay, and only 8% chose precharge.

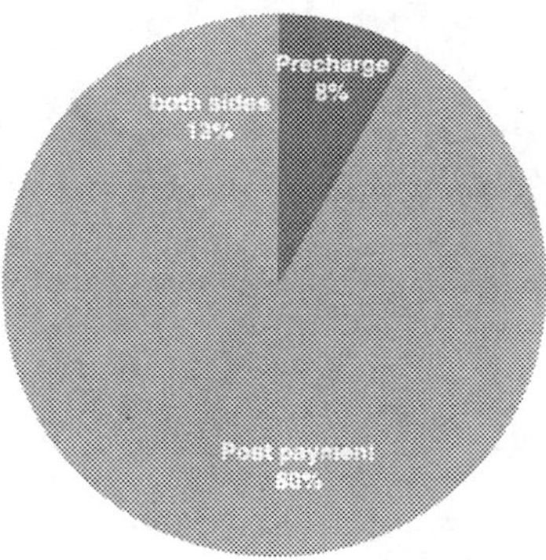

Fig. 12. Ratio of payment methods of Chinese

In addition, when it is mentioned that after using the online service, the system prefers automatic payment after the service ends or manual payment after confirming the charge details, 60% answered the manual payment and 40% answered the automatic payment (Fig. 13).

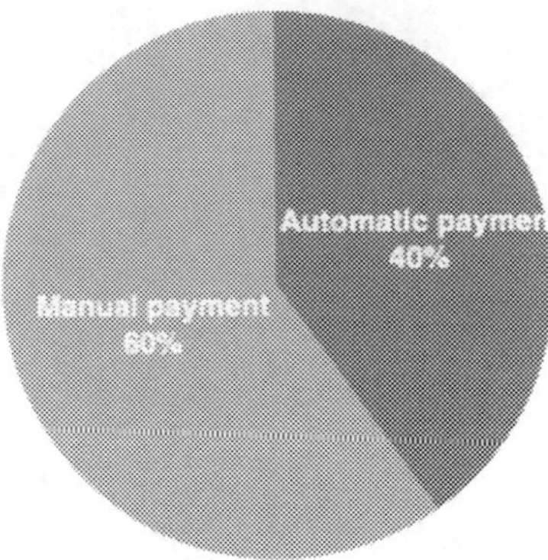

Fig. 13. Tendency to pay manually or automatically

Result shows that 57% of Chinese respondents are more accustomed to automatic payments as Fig. 14.

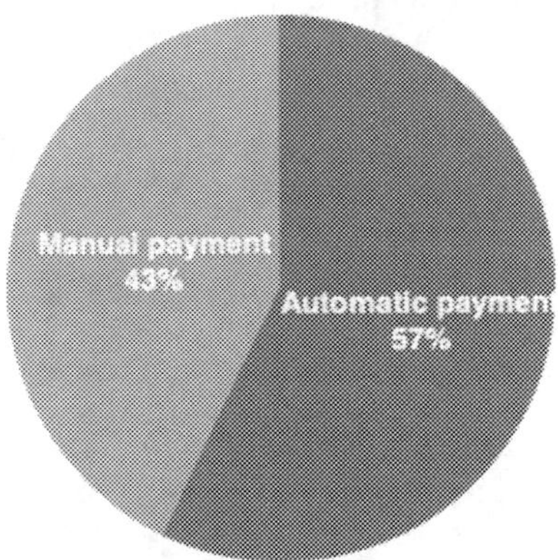

Fig. 14. Tendency to pay manually or automatically of Chinese

In addition, we also analyzed and evaluated the safety and convenience of users for automatic payment and manual payment. Most users who choose automatic payment think that convenience has been improved and security is relatively high. Users who choose manual payment think that the security of manual payment is reassuring, but the convenience is reduced. Shows as Fig. 15.

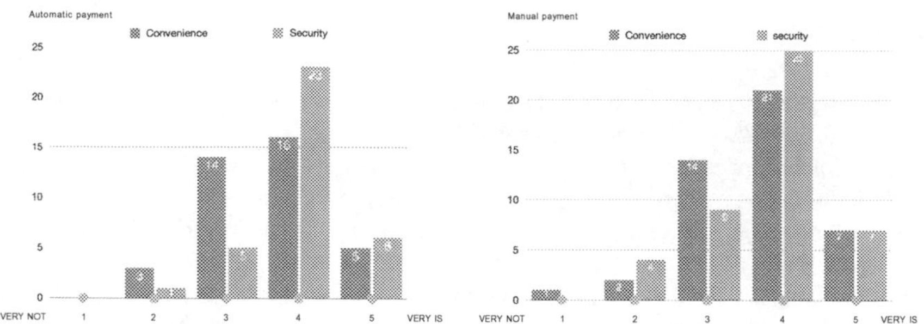

Fig. 15. Evaluation of security and convenience of automatic/manual payment

In this survey, due to the small number of Japanese questionnaires collected, the results are not included in the comparison of results. However, the small sample data seems to show that there is a large difference between the choice of Japanese and Chinese in the choice of pre-charge and post-payment, automatic payment and manual payment.

6 Consideration and Future Research

This study uses bicycle sharing as a reference object to make UX maps of bicycle sharing services in China and Japan, finds and compares the differences in user experience, and focuses on extracting payment systems for exploration. Through data surveys and literature review, we understand that the development and popularization of e-money in China and Japan's credit cards have led to differences in the development and current status of mobile payment services. This has affected the user's acceptance of credit cards and mobile payments. Then we conducted a questionnaire survey of target users who are familiar with online services. The results show that under the influence of the above reasons, China users are more likely to accept automatic payments, and Japanese users prefer manual payments. Users who choose manual payments pay more attention to security, and users who choose automatic payments pay more attention to convenience.

However, due to the small number of Japanese user samples collected, we are unable to effectively compare their results with those of Chinese users. But this sample also seems to show that Japanese users are more likely to choose pre-charges and manual payments. In future research, we will collect more samples and establish a more comprehensive survey method to achieve effective Chinese-Japanese user comparison.

Therefore, when comparing different user experiences, we should investigate the reasons for their occurrence and conduct user surveys and analysis on them. In future research, we will conduct more in-depth investigations on the different perceptions of the user experience of online services in order to propose UX design techniques that can smoothly change the perceptions of different user experiences.

References

1. iiMedia Research: The first half of 2018 China's shared bicycle industry monitoring report, 21 June 2018. https://www.iimedia.cn/c400/63243.html. Accessed 2020 Jan 30
2. Ministry of Land, Infrastructure, Transport and Tourism: Share cycle initiatives, November 2018. https://www.mlit.go.jp/common/001267508.pdf. Accessed 30 Jan 2020
3. Iimedia Data, https://data.iimedia.cn/page-category.jsp?nodeid=24733915. Accessed 23 Feb 2020
4. Ipsos Data. http://www.199it.com/archives/962869.html. Accessed 23 Feb 2020
5. Bank of Japan: Payment and Settlement Systems Report – Annex. https://www.boj.or.jp/research/brp/psr/data/psrb180928a.pdf. Accessed 20 Feb 2020
6. Mobile Marketing Data Labo. https://mmdlabo.jp/investigation/detail_1821.html. Accessed 20 Feb 2020
7. People's Bank of China: Overall operation of the payment system in the fourth quarter of 2017. http://www.pbc.gov.cn/goutongjiaoliu/113456/113469/3492259/2018030510583487991.pdf. Accessed 23 Feb 2020

A Combined AHP-TOPSIS Approach for Evaluating the Process of Innovation and Integration of Management Systems in the Logistic Sector

Genett Jiménez-Delgado[1], Gilberto Santos[2(✉)], María João Félix[2], Paulo Teixeira[3], and José Carlos Sá[4]

[1] Department of Industrial Engineering, Engineering Faculty, Institucion Universitaria ITSA, Soledad, Atlántico, Colombia
gjimenez@itsa.edu.co

[2] Design School, Polytechnic Institute Cávado Ave, Barcelos, Portugal
{gsantos,mfelix}@ipca.pt

[3] Technology School, Polytechnic Institute Cávado Ave, Barcelos, Portugal
pteixeira@ipca.pt

[4] School of Engineering (ISEP), Polytechnic of Porto, Porto, Portugal
cvs@isep.ipp.pt

Abstract. Logistics is increasingly essential for the economy and social development. Statistics show incremental contributions in GDP, job creation, and in the movement of goods worldwide, connecting suppliers with companies and distributors and stimulating different economic sectors of manufacturing and services. However, the challenges of the logistics sector include optimizing its operations, reducing costs, and generating an offer of services with added values for its clients and interest groups. In this sense, innovation and management systems can support companies in the sector in the search for competitive advantages and operational efficiency aligned with corporate strategy and the environment. Different models and methodologies have been developed to work in an integrated way on innovation processes and management systems, with a focus on continuous improvement where evaluation and development are crucial to identifying the level of performance, gaps, and improvement strategies. Besides, the use of multi-criteria decision methodologies was identified in the scientific literature as an objective approach and easy to implement in companies, but of little application in the evaluation of innovation and the integration of management systems for the logistics sector. This paper presents the application of a combined multi-criteria decision-making methodology to evaluate the innovation process in conjunction with the integration of management systems used in the logistics sector. First, the Analytic Hierarchy Process AHP is applied to determine the weights of each criterion and sub-criterion through qualitative pair comparisons. Then, The Technique for Order of Preferences by Similarity to Ideal Solution (TOPSIS) is used to evaluate three big companies of the logistic sector considering their ability to develop the process of innovation under criteria associated with Integrated Management Systems (Quality, Environmental, Health, and Safety). A case study is presented to validate the combined approach.

C. Stephanidis et al. (Eds.): HCII 2020, LNCS 12427, pp. 535–559, 2020.
https://doi.org/10.1007/978-3-030-60152-2_40

Keywords: Analytical Hierarchy Process · AHP · Multicriteria decision making · MCMD · TOPSIS · Integrated management systems · Quality · Safety · Environment · Logistic sector · Performance evaluation

1 Introduction

The logistics sector is currently considered as a source of competitive advantages, from the business level, where companies compete not only in products or brand strategies but also through the speed, flexibility, and optimization of their supply chains. At the level of countries, the logistic sector can represent around 10% to 12% of GDP. However, the logistics sector, which being involved with activities related to the flow and transformation of goods and products, from the procurement of raw materials to the delivery to the final customer for consumption, is called to plan to implement and control the efficient and effective in the flow of goods, services, and information, involving infrastructure and integrating the services provided through it, which allows optimizing the structure of logistics costs and physical distribution of products, improving the conditions of connectivity of production areas with the of consumption. The aforementioned raises the logistics chain companies the incorporation of innovation processes that allow them to respond to an increasingly complex, dynamic, uncertain, and ambiguous environment, generating value offer by optimizing their processes and aligned operating costs with the new technologies and management trends. In this sense, the innovation processes have also evolved with different models, systems and methodologies that currently can be carried out under comprehensive approaches, combining, in addition to the innovation's criteria, quality, environment and safety and health criteria that allow the companies to design products, processes and service models with environment-oriented. In this sense, the development of methodologies for the evaluation and improvement of the integration processes of innovation is essential to maintain the sustainability of the sector. In this regard, the low adoption of integration processes of innovation generates low levels of operational performance, a decrease in competitiveness, and the value of the company before its market. In the scientific literature, some studies work with different methodologies such as sector analysis, analysis of strengths and weaknesses as well as the use of multi-criteria techniques, of which the most widely used method is the Analytical Hierarchy Process (AHP). Despite these considerations, few studies delve into the integrated evaluation of the innovation process, taking into account other criteria such as quality, environment, and health and safety, and applying multi-criteria approaches. For the considerations mentioned above, this work intends to bridge this gap by extending the Multi-criteria decision-making approach in the innovation process and the integration of management systems in the logistics sector.

The paper is organized as follows: Sect. 2 presents the literature review of approaches of MCDM for applied in the innovation process and integration of management systems. In Sect. 3 its detail the methodological approach. Then, in Sect. 4 it presents the results and analysis of the combined methods. Finally, the conclusions and future works are analyzed in Sect. 5.

A case study is presented that considering four criteria, 20 sub-criteria, seven decision-makers, and a survey for the evaluation of the innovation process and the integration of man-aging systems that was applied in four big companies of the logistic sector. The outcomes evidence that the criteria "Idea generation" (38.15%), and "Selection and formulation of innovation projects" (28.4%), are the most relevant criteria in the evaluation of the integration of innovation process and management systems applied in logistic companies. Also, the ranking of the logistics companies with the best performance is presented later through the use of the TOPSIS technique, identifying gaps and improvement strategies.

2 Approaches of MCDM for Evaluation of Innovation Process and Integration of Management System in Logistic Sector: A Literature Review

According to Frascati Manual [1], innovation is defined as the transformation of an idea into a marketable product or service, a new or improved manufacturing or distribution procedure, or new business or marketing models. In this sense, different models, techniques, and methodologies have been developed to support organizations to generate added value at the economic and social level as well as significant improvements in increasingly complex and dynamic sectors such as logistics, where the implementation of innovation processes is a competitive necessity to optimize their operations and diversify its offer of services with an impact on the national economy.

Despite the above considerations, there are difficulties for companies in adopting organizational processes and structures for innovation, given the complexity and uncertainty of these processes, the high demand for human, technical, and financial resources and the risks in the framework of the innovation processes, due to the failure of a project affects the continuity of the company [2]. In this sense, the integration of innovation processes with different management standards can minimize the risks associated with the development of innovations and generate a culture of change and knowledge from within the organization, towards its stakeholders [3–5]. Concerning this integration, the scientific literature shows the interrelationship of management systems, particularly quality, safety, and health and environment, with innovation processes, aligning these processes with the organizations' strategy [2, 6, 7]. Standards such as UNE 166002 and ISO 56002 propose integration with other management standards such as ISO 9001, ISO 14001 or OHSAS 18000 in its structure and to a large extent of its content, establishing particular requirements for the management of R+D+I activities [8].

One of the essential aspects in the adoption of innovation processes integrating management standards in quality, environment and health and safety, is the focus towards continuous improvement, which offers companies, in addition to planning and implementation, the permanent verification of its processes to evaluate its performance and identify its opportunities for development [9]. Different evaluation approaches have been used to assess performance in the implementation of innovation processes and integrated management systems, such as the Balanced Scorecard, the evaluation of internal capacities, among others. However, due to the multiple factors that affect the innovation process and the different stakeholders involved, there are techniques, the evaluation becomes complex and even subjective, making it difficult to identify gaps and action plans. In this regard, multicriteria decision-making methods are proposed as an alternative for evaluation, already used in supplier selection processes, technology selection, and performance evaluation in sectors such as health, manufacturing, among others.

Concerning the goal of this study, we found several articles focused on the application of the multicriteria decision methods (MCDM) in the innovation process and integrated management systems, as seen in Table 1.

Table 1. Literature review in innovation and IMS based in MDCM approaches

Authors (Year)	Topic	Application	Objective	MDCM applied/Other
Falak, J., Kunjan, M., Nagaraju, D. and Narayanan, S. (2020) [10]	Integrated Management Systems	Continuous improvement	Evaluate the continuous improvement techniques using hybrid MCDM techniques under fuzzy environment	Fuzzy AHP, VIKOR
Ying-Hsun Hung, Seng-Cho T. Chou, Gwo-Hshiung Tzeng. (2011) [11]	Innovation	Knowledge management	Knowledge management adoption and assessment for SMEs by a novel MCDM approach	ANP, DEMATEL, VIKOR, TOPSIS
Ikrama, M., Sroufe, R., Qingyu, Z. (2020) [12]	Integrated management system	Barriers in IMS	Prioritizing and Overcoming Barriers to Integrated Management System (IMS) Implementation Using AHP and G-TOPSIS	AHP, G-TOPSIS
14. Ming Lang Tseng, M., and Chiu, A. (2012) [13]	Innovation	Green innovation	Integrated MCDM techniques that are grey theory, entropy weight and the analytical network process to evaluate the green innovation practices under uncertainty	ANP
Sadeghi, A. (2018) [14]	Innovation	High-tech SMEs	Success factors of high-tech SMEs in Iran: a fuzzy MCDM approach	ANP, TOPSIS

(*continued*)

Table 1. *(continued)*

Tseng, M., Lin, Y., Lim, M. and Teehankee, B. (2015) [15]	Innovation	Service innovation	Using a hybrid method to evaluate service innovation in the hotel industry	Fuzzy triangular numbers
De, D., Chowdhury, S., Dey, P., and Ghosh, S. (2020) [16]	Innovation	Innovation oriented lean and sustainability performance	Impact of Lean and Sustainability Oriented Innovation on Sustainability Performance of Small and Medium Sized Enterprises: a Data Envelopment Analysis-based framework	DEA
Jimenez-Delgado G., Balmaceda-Castro N., Hernández-Palma H., de la Hoz-Franco E., García-Guiliany J., Martinez-Ventura J. (2019) [17]	Integrated management systems	Performance evaluation in OHS	Integrated Approach of Multiple Correspondences Analysis (MCA) and Fuzzy AHP Method for Occupational Health and Safety Performance Evaluation in the Land Cargo Transportation	Fuzzy AHP and MCA method

The articles found oriented on the application of MCDM methods for the evaluation of continuous improvement techniques, performance in OHS, integration of the innovation and sustainability, service innovation, knowledge management, implementation of integrated management systems, and high-tech in companies. However, in the literature review, we don't found papers focused on the integral assessment of innovation processes, that consider the alignment with elements of integrated management systems (quality, health and safety, and environmental). Besides, the applications of the MCDM techniques in innovation/integrated management system in the logistics sector, in particular, are oriented of the supplier's selection.

The novelty of the present study is based on the combination of the AHP with the TOPSIS method to evaluate the process innovation and the integration of management systems in the logistic sector. The choice of AHP and TOPSIS methods is justified for the following reasons: AHP is an easy-to-use, scalable methodology and adapts to different complex problems by designing the hierarchical decision structure [18] and TOPSIS is a technique that quickly and objectively allows the classification of alternatives taking into account their proximity to the ideal scenario and their remoteness to the anti-ideal scenario [19]. The AHP-TOPSIS combined approach can provide complete classification results by using weights and objective data to calculate relative distances [20].

Therefore, this research contributes to the scientific literature and provides a combined approach for evaluating the innovation process and the integration of management systems in the logistic sector through the application of MCDM techniques and can support to managers in their activities with the aim to generate a culture of innovation in products, process, and business models, with a focus in the strategic alignment and incorporate elements of quality health and safety, and environment, as drivers of competitive advantages and added value for the companies of the logistic sector, customers and other stakeholders.

3 Proposed Methodology

The proposed approach aims to evaluate the process of innovation and the integration of management systems in companies of the logistic sector by the combined method using AHP and TOPSIS. In this regard, the methodology is comprised of five phases (refer to Fig. 1):

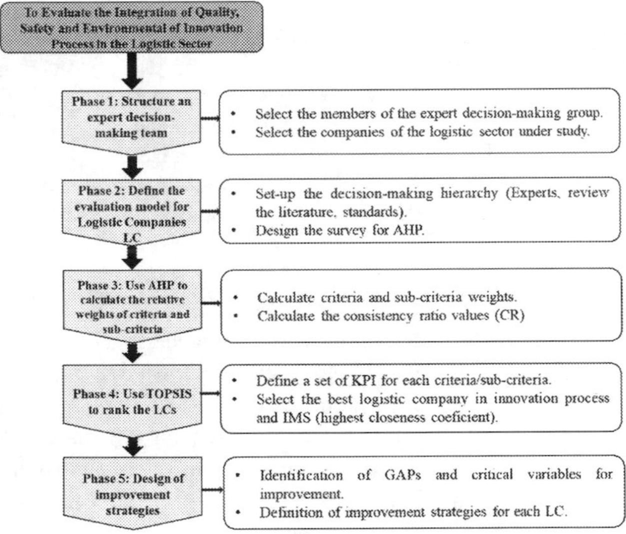

Fig. 1. The methodological approach for evaluating the process of innovation and the integration of management systems in companies of the logistic sector.

- **Phase 1 (Structure an expert decision-making team):** A decision-making group is chosen based on their experience in innovation process, integrated management systems and logistic. The experts will be invited to be part of the decision-making process through AHP method.
- **Phase 2 (Define the integrated innovation process evaluation model for logistic companies AHP):** The criteria and sub-criteria are established to set up a decision hierarchy considering the opinion of the expert decision-makers, the literature review, and the international standards in innovation, quality, environmental, and

health and safety management and its application in logistic sector. Then, the survey for the application of the AHP technique was designed.

- **Phase 3 (AHP application):** In this step, AHP is used to estimate the global and local weights of criteria and sub-criteria. In this phase, the experts were invited to perform pairwise comparisons, which are subsequently processed following the AHP method, as detailed in Sect. 3.1.
- **Phase 4 (TOPSIS application)**: In this step, TOPSIS (described in Sect. 3.2) is applied for ranking the logistic companies and identifying the best company in adoption of process innovation under integration of management systems.
- **Phase 5 (Design of improvement strategies)**: In this step, GAPs and critical variables were identified taking into account the outcomes of the use of TOPSIS technique to improve the performance in process of innovation and the integration of management systems in quality, health and safety and environment. Subsequently, were defined the improvement proposals for each company [21].

3.1 Analytic Hierarchy Process (AHP)

The Analytical Hierarchy Process (AHP) is one of the most used multicriteria techniques for determining weights for criteria and sub-criteria. This technique was proposed and developed by Thomas L. Saaty [22] as a methodology of easy application in different contexts and because it allows the combination with other methods offering greater robustness and reliability in the results [23, 24].

The AHP technique is applied based on the definition of the team of experts in the objective of multicriteria selection and the design of the hierarchical structure that breaks down a complex selection or evaluation problem at different levels, the highest level that represents the goal, the middle level that contains the criteria and sub-criteria of evaluation and the lowest level that shows the various alternatives of selection [25, 26]. In AHP method, paired comparisons are made using a scale to assess the importance of criteria and sub-criteria. For this study, a three-point scale was used to facilitate expert understanding and reduce inconsistencies in the paired comparison process [27, 28] as described below (refer to Table 2).

Table 2. Linguistic terms and reduced AHP scale

Reduced AHP scale	Definition
1	Equally important
3	More important
5	Much more important
1/3	Less important
1/5	Much less important

The steps of the AHP technique as follows:

- Step 1: Perform pairwise comparisons between criteria/sub-criteria by using the linguistic terms and the corresponding reduce AHP scale established in Table 1. To facilitate the collect of the comparisons by the experts, is recommended the design and application of a survey.
- Step 2: Incorporate the comparisons of the experts by using applying the geometric mean formula [29] as shown in Eq. 1. where n' represents the number of experts and a_{ij} represents the relative importance of the ith criterion/sub-criterion compared to the jth criteria/sub-criteria:

$$\left(\prod_{k=1}^{n'} a^{k}{}_{ij}\right)^{1/n}. \tag{1}$$

- Step 3: Arrange the judgments into an $n \times n$ pairwise comparison matrix A for criteria using Eq. 2. and matrix B for sub-criteria via Eq. 3. In Eq. 2 and Eq. 3, the diagonal values in the matrices A and B are equal to 1 since $i = j$. For the case of a decision-making team, *aij* and *bij* are obtained by applying the geometric mean of all the judgments related to the comparison:

$$A = \begin{bmatrix} 1 & a_{12} & \ldots & a_{1n} \\ a_{21} & 1 & \ldots & a_{2n} \\ \ldots & \ldots & \ldots & \ldots \\ a_{n1} & a_{n2} & \ldots & 1 \end{bmatrix}, \tag{2}$$

$$B = \begin{bmatrix} 1 & b_{12} & \ldots & b_{1n} \\ b_{21} & 1 & \ldots & b_{2n} \\ \ldots & \ldots & \ldots & \ldots \\ b_{n1} & b_{n2} & \ldots & 1 \end{bmatrix}, \tag{3}$$

- Step 4: Calculate the weights of criteria using Eq. 4 and the importance of sub-criteria via Eq. 5. In this regard, the relative importance degree of each sub-criterion i compared to each of the other sub-criteria in the same criterion c is denominated local weight (LW_i^c). Besides, it is must determine the relative importance of each criteria c concerning the hierarchy goal or global weight (GW_c):

$$LW_i^c = \frac{\left(\prod_{j=1}^{n} b_{ij}\right)^{1/n}}{\sum_{i=1}^{n}\left(\prod_{j=1}^{n} b_{ij}\right)^{1/n}}, i,j = 1,2\ldots,n, \tag{4}$$

$$FW_i^c = \frac{\left(\prod_{j=1}^{n} a_{ij}\right)^{1/n}}{\sum_{i=1}^{n}\left(\prod_{j=1}^{n} a_{ij}\right)^{1/n}}, i,j = 1,2\ldots,n, \tag{5}$$

- Step 5: Calculate the consistency ratio (CR) to evaluate the suitability in the judgments made by the experts through paired comparisons. First, the consistency index (CI) is calculated by applying Eq. 6 where λ_{max} is the eigenvalue and n is the size of the matrix. Subsequently, the consistency ratio (CR) is calculated using Eq. 7. If CR $\leq$ 10 percent is considered as consistent. Otherwise, the judgments are classified as inconsistent, and the decision-makers should then review the comparisons:

$$CI = \frac{\lambda_{max} - n}{n - 1}, \tag{6}$$

$$CR = \frac{CI}{RI}, \tag{7}$$

- Step 6: Calculate the relative importance degree of each sub-criteria i in relation to the hierarchy goal, which is denominated global weight GW_i, using Eq. 8:

$$GW^i = LW_i^c \times FW^c. \tag{8}$$

3.2 Technique for Order of Preference by Similarity to Ideal Solution (TOPSIS)

The Technique for Order Preference by Similarity to Ideal Solution (TOPSIS) developed by Hwang and Yoon [30] is a technique to evaluate the performance of alternatives through the similarity with the ideal solution [31]. Concerning this technique, TOPSIS chooses the best option; that is, say the alternative that simultaneously is closest to the positive-ideal solution (PIS) and farthest from the negative-ideal solution (NIS) through the calculation of the closeness coefficient [32]. The positive-ideal solution is composed of all best values attainable of criteria, and the negative-ideal solution consists of all the worst values attainable of criteria [31].

The procedure of TOPSIS method is given as follows:

- Step 1: Create a decision matrix X with "m" logistic companies and "n" sub-criteria via Eq. 9. X_{ij} is the value of the sub-criterion $S_j(j = 1, 2, 3, \ldots, n)$ in each logistic company $GD_i(i = 1, 2, \ldots, m)$.

$$X = \begin{matrix} LC_1 \\ LC_2 \\ LC_3 \\ \vdots \\ LC_m \end{matrix} \begin{bmatrix} S_1 & S_2 & \cdots & S_n \\ x_{11} & x_{12} & \cdots & x_{1n} \\ x_{21} & x_{22} & \cdots & x_{2n} \\ x_{31} & x_{32} & \cdots & x_{3n} \\ \vdots & \vdots & \cdots & \vdots \\ x_{y1} & x_{y2} & \cdots & x_{yn} \end{bmatrix} \tag{9}$$

- Step 2: Calculate the normalized decision matrix R using Eq. 10. Let n_{ij} be the norm used by TOPSIS via Eq. 11. Besides, r_{ij} is defined as the element of this matrix.

$$R = X \cdot n_{ij} \tag{10}$$

$$n_{ij} = \frac{x_{ij}}{\sqrt{\sum_{i=1}^{y} x_{ij}^{2}}} \tag{11}$$

- Step 3: Obtain the weighted normalized decision matrix V using Eq. 12. The sub-criteria weights (w_j) are calculated from the AHP method.
-

$$V = \left[w_j r_{ij}\right] = \left[v_{ij}\right] \tag{12}$$

- Step 4: Establish the ideal (C^{+}) and anti-ideal (C^{-}) scenarios according with Eq. 13 and Eq. 14 respectively:

$$C^{+} = \left\{\left(\substack{max\\i} c_{ij} | \mathrm{j} \in J\right), \left(\substack{min\\i} c_{ij} | \mathrm{j} \ \in J'\right) for\ i = 1, 2, \ldots, m\right\}\left\{c_1^{+}, c_2^{+}, \ldots, c_j^{+}, \ldots, c_n^{+}\right\} \tag{13}$$

$$C^{-} = \left\{\left(\substack{min\\i} c_{ij} | \mathrm{j} \in J\right), \left(\substack{max\\i} c_{ij} | \mathrm{j} \ \in J'\right) for\ i = 1, 2, \ldots, m\right\} = \left\{c_1^{-}, c, \ldots, c_j^{-}, \ldots, c_n^{-}\right\} \tag{14}$$

Considering that,

$$J = \{j = 1, 2, \ldots, n | j\ \ associated\ \ with\ \ the\ \ benefit\ \ sub-criterion\}$$

$$J\prime = \{j = 1, 2, \ldots, n | j\ \ associated\ \ with\ \ the\ \ cost\ \ sub-criterion\}$$

- Step 5: Calculated the separation measures of each logistic company to C^{+} and C^{-} using Euclidean separation via Eqs. 15–16.
- *Euclidean separation from ideal scenario*

$$d_i^{+} = \sqrt{\sum_{j=1}^{n} \left(c_{ij} - c_j^{+}\right)^2} i = 1, 2, \ldots, m \tag{15}$$

- *Euclidean separation from anti-ideal scenario*

$$d_i^{-} = \sqrt{\sum_{j=1}^{n} \left(c_{ij} - c_j^{-}\right)^2} i = 1, 2, \ldots, m \tag{16}$$

– Step 6: Calculate the relative closeness coefficient (CC_i) by using Eq. 17. If $CC_i = 1$, the performance of the logistic company performs according to PIS; therefore, larger values of CC_i denote satisfactory proficiency.

$$CC_i = \frac{d_i^-}{(d_i^+ + d_i^-)}, 0 < CC_i < 1, i = 1, 2, \ldots, m \tag{17}$$

– Step 7: Rank the logistic companies according with the preference order of CC_i (Very low performance: 0 $\leq$ CCi $\leq$ 0,25 || Low performance: 0,25 < CCi $\leq$ 0,5 || Medium performance: 0,5 < CCi $\leq$ 0,75 || High performance: 0,75 < CCi $\leq$ 1).

4 Application of the Combined Approach: Evaluating the Process of Innovation and Integration of Management Systems in the Logistic Sector

4.1 Structure an Expert Decision-Making Team

In this section, a case study is presented to validate the proposed methodology. The case study is illustrated in four big companies in the logistics sector located in Colombia. The first step was the setup of the decision-making team. In this respect, decision-making group was selected to validate the criteria and sub-criteria through the application of AHP method for to evaluate the innovation process and the integration of management systems (quality, health, and safety, and environment), according to their expertise in these topics with application in the logistic sector. In this regard, four types of experts were selected meaningful for the decision-making process: three leaders of the companies of the logistic sector that participated in the research, two experts consultors in innovation and integrated management systems, and two representatives of universities linked to the innovation processes and integrated management systems in different companies including logistic industry. The team of experts is described below:

- Experts 1, 2, and 3 are the professionals in Business Administration or Industrial Engineering with more than 10 years of experience in the logistics sector in roles of executive strategy, operations management, and innovation projects. These experts work in the companies participants in the study.
- Expert 4 is an Industrial Engineer, master in administrative engineering with more than 12 years of experience as a coach and consultant in both private and public organizations in innovation projects, management systems, and organizational change processes.
- Expert 5 is Industrial Engineer, Master in Innovation with 5 years of experience as a consultant in the innovation processes and entrepreneurship for different organizations public and private.
- Expert 6 is a Ph.D. in Mechanical Engineering with an emphasis in production technologies and Full-time professor of the Design Scholl in an international university, with more than 20 years of experience in integrated management systems,

in-novation projects and application of methodologies such as Lean Manufacturing in companies of manufacturing or services.

- Expert 7 is an Industrial Engineer, Specialist in Quality Engineering and Management, Master in Industrial Engineering and Full-time professor in higher education with 7 years of experience in the formulation and development of projects, and the application of multi-criteria models for evaluation performance in integrated management systems. The expert 7 acted as a facilitator to take over the judgment process.

4.2 Define the Multi-criteria Decision-Making Model to Evaluate the Innovation Process and the Integration of Management Systems in the Logistic Companies

Concerning to hierarchy of the decision-making model, the experts identified five criteria and 20 sub-criteria. In this particular case, four logistic companies were evaluated (LC1, LC2, LC3, and LC4). Both criteria and sub-criteria were determined considering the experts' experience, the analysis of the standards in integrated management systems such as ISO 9001, ISO 14001 and ISO 45001, the review of different models of innovation such as UNE 166002 [33, 34] and the pertinent scientific literature to support each criteria and sub-criteria. Subsequently, the experts validate these criteria and sub-criteria according to models, standards in innovation and integrated management systems, and the literature review presented in order to provide an MCDM model to responding to the goal of evaluation. Then, the multi-criteria hierarchy was then verified and discussed through different sessions with the expert decision-making team to establish the comprehension of the model and the hierarchy. Finally, the decision model is shown in Fig. 2.

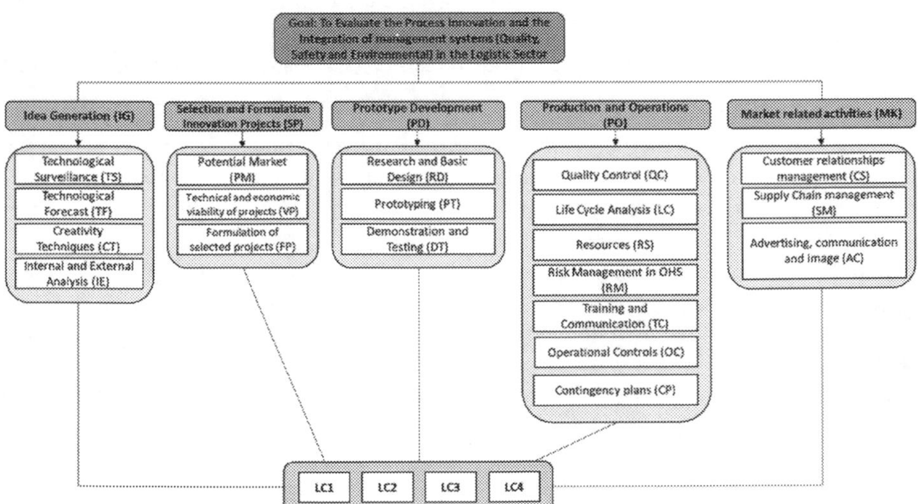

Fig. 2. Multi-criteria decision-making model to evaluate the innovation process and the integration of management systems in the logistic sector

The criteria and sub-criteria were labeled and described in Table 3.

Table 3. Description of criteria

Criterion (C)	Sub-criteria (SC)	Criterion description
C1. Idea Generation (IG)	SC1. Technological Surveillance (TS) SC2. Technological Forecast (TF) SC3. Creativity Techniques (CT) SC4. Internal and External Analysis (IE)	Idea generation is defined as the ability of the company to identify opportunities for innovation using tools such as technological surveillance, forecasting of technological trends, the use of creativity techniques, and analysis of the internal and external context of the company [8, 33, 34]
C2. Selection and Formulation of Innovation Projects (SP)	SC5. Potential market (PM) SC6. Technical and economic viability of projects (VP) SC7. Formulation of selected projects (FP)	This criterion refers to the basic aspects that the organization must develop for the selection and formulation of innovation projects based on the identification of opportunities for innovation. For the selection and formulation of R+D+I projects, it is important to identify the potential market, technical and economic feasibility, as well as the selection and structuring of R+D+I project portfolios [8, 33, 34]
C3. Prototype Development (PD)	SC8. Research and Basic Design (RD) SC9. Prototyping (PT) SC.10 Demonstration and testing (DT)	This criterion evaluates the company's ability to carry out or access basic research for the design of products/processes, as well as the use of prototyping techniques and the development of the validations and tests of products or processes before taking them to the production or implementation stage [8, 33–36].
C4. Production and Operations (PO)	SC11. Quality Control (QC) SC12. Life Cycle Analysis (LC) SC13. Resources (RS) SC14. Risk Management in OHS (RM) SC15. Training and Communication in R +D+R (TC) SC16. Operational Controls (OP) SC17. Contingency Plans (CP)	This criterion assesses the company's ability to develop the production and operations process according to factors of innovation and integrated management systems such as quality control [37–41, 48–52], use of methodologies of lean and cleaner production, and life cycle analysis of products [42–44], the provision of the technical, human and economic resources [8, 37–40], the risk management in OHS [17, 56, 57], the training and communication to generate skills and knowledge in innovation [8, 33, 35, 45, 46], the implementation of procedures or mechanisms of operational controls and contingency plans to reduce the negative impacts in the production process derivated of the operational risks [8, 35, 37, 54]

(*continued*)

Table 3. *(continued)*

C5. Market related activities (MK)	SC18. Customer relationships management (MR) SC19. Supply Chain Management (SM) SC.20. Advertising, communication, and image (AC)	This criterion evaluates the company's ability to integrate innovation in the market, taking into account aspects such as implementation time, price, among others, through the development of processes and integrating platforms for the management of the supply chain and the relationship with the client, the advertising management, the corporate communication, and the activities of branding of the innovations generated and launched on the market [35, 47, 53, 55]

4.3 Design of Data Collection Tool for AHP

In this step, a data collection instrument was designed for the paired comparisons process performed by the experts (refer to Fig. 3). In this respect, for each pairwise evaluation, the participants answered the following question: According to the goal/criteria, ¿how important is each element on the leftover the item on the right? The experts used Table 1 to represent their judgments until finalizing all the criteria and sub-criteria. Then, via Eq. 1–5, the weights of criteria and sub-criteria were determined.

According to your experience with respect to "Prototype Development" sub-criterion, ¿how important is each sub-criterion on the left concerning the sub-criterion on the right when evaluating the Integration of Quality, Safety and Environmental of Innovation Process in the Logistic Sector?

		1	2	3	4	5		
Research and Basic Design	is	O	O	O	O	O	Important than	Prototyping
Research and Basic Design	is	O	O	O	O	O	Important than	Demonstration and Testing
Prototyping	is	O	O	O	O	O	Important than	Demonstration and Testing

① Much less ③ Equally ⑤ Much more

② Less ④ More

Fig. 3. Data-collection instrument implemented for AHP judgments

4.4 Calculating the Relative Weights of Criteria and Sub-criteria Using AHP

In this phase, through the application of the AHP technique, the local and global weights of criteria and sub-criteria were determined considering. First, the judgments of the experts were collected by paired comparisons using the survey designed from the hierarchical model explained in Sect. 4.2. Then, the geometric means of the experts' judgments were calculated and incorporated into the comparison matrix for criteria and sub-criteria. Subsequently, the values of global and local weights were calculated. Tables 4–5, it is shown an example of the results of the AHP process, applying Eqs. 1–5 of the methodology detailed in Sect. 3.1. Finally, Table 6 presents the local and global weights of all the criteria and sub-criteria.

Table 4. AHP comparison matrix for criteria "patient" cluster

	C1 (IG)	C2 (SP)	C3 (PD)	C4 (PO)	C5 (MK)
C1 (IG)	1	2.76	1.85	3.47	3.73
C2 (SP)	0.36	1	3.47	2.76	3.73
C3 (PD)	0.54	0.29	1	1.87	2.19
C4 (PO)	0.29	0.36	0.53	1	1.87
C5 (MK)	0.27	0.27	0.46	0.53	1

Table 5. AHP comparison matrix for criteria "market related activities" cluster

	SC17 (CS)	SC19 (SM)	SC20 (AC)
SC18 (CS)	1	2.36	1.37
SC19 (SM)	0.42	1	0.97
SC20 (AC)	0.73	1.03	1

Table 6. Local and global weights of criteria and sub-criteria

Cluster	GW	LW
C1. Idea Generation (IG)	**0.382**	
SC1. Technological Surveillance (TS)	0.123	0.323
SC2. Technological Forecast (TF)	0.082	0.216
SC3. Creativity Techniques (CT)	0.077	0.203
SC4. Internal and External Analysis (IE)	0.097	0.256
C2. Selection and Formulation of Innovation Projects (SP)	**0.284**	
SC5. Potential Market (PM)	0.149	0.524
SC6. Technical and Economic Viability of Projects (VP)	0.096	0.338
SC7. Formulation of Selected Projects (FP)	0.039	0.137
C3. Prototype Development (PD)	**0.157**	
SC8. Research and Basic Design (RD)	0.066	0.423
SC9. Prototyping (PT)	0.048	0.310
SC10. Demonstration and Testing (DT)	0.041	0.265
C4. Production and Operations (PO)	**0.104**	
SC11. Quality Control (QC)	0.017	0.163
SC12. Life Cycle Analysis (LC)	0.011	0.111
SC13. Resources (RS)	0.028	0.270
SC14. Risk Management in OHS (RM)	0.012	0.122
SC15. Training and Communication (TC)	0.010	0.100
SC16. Operational Controls (OC)	0.015	0.144
SC17. Contingency Plans (CP)	0.009	0.087
C5. Market Related Activities (MK)	**0.073**	
SC18. Customer Relationships Management (CS)	0.034	0.470
SC19. Supply Chain Management (SM)	0.017	0.238
SC20. Advertising, Communication, and Image (AC)	0.021	0.291

Figure 4 shows the classification of the criteria given their importance, obtained from the application of the AHP method. According to these outcomes, “Idea Generation” (GW = 38.15%) is the most significant criterion in the innovation process and integration of management systems. In this regard, the generation of ideas allows companies to favor the innovation process to generate products that respond to market needs and are oriented towards the environment, that is, incorporating aspects of quality, environment, and health and safety. Within the “Idea Generation” criterion, the sub-criteria of “Technological Surveillance” (LW = 32.36%) and “Technological Forecast” (LW = 21.64%) are highlighted as strategies that can underpin in the identification of opportunities for innovation.

The second criterion in order of importance is “Selection and Formulation of Innovation Projects” (GW = 28.40%) as an essential stage to select R+D+I projects that are relevant to the stakeholders and the objectives of the organization. This criterion allows the companies to optimize the resources for the development of these projects, taking into account sub-criteria such as “Potential Market” (LW = 52.47%) and “Technical and Economic Viability of Projects” (LW = 33.81%). On the other hand, the criteria “Prototype Development” (GW = 15.74%), “Production and Operations” (GW = 10.42%), and “Market and Related Activities” (GW = 7.28%), presented a consolidated ponderation of 33.44%, which also shows the relevance of these factors where the integration of quality, environment and health and safety management systems contributes to the implementation and placing on the market of innovative products, processes, and services. These results demonstrate the importance of establishing integrated innovation processes in companies in the logistics sector that consider not only variables of R+D+I but also incorporate aspects of quality, environment, and health and safety to enhance competitive advantages and the sustainability of the sector.

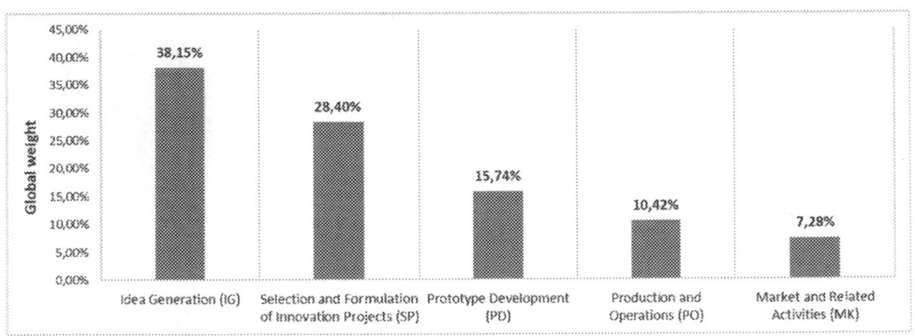

Fig. 4. Global weights of criteria in the performance evaluation in occupational health and safety in companies of electric sector

Finally, we calculated the consistency (refer to Table 7) to guarantee the reliability of the judgments obtained by the expert team. The results evidence that all criteria present adequate consistency values (CR $\leq$ 0.1). Therefore, the criteria and sub-criteria can then be applied to evaluate and rank the companies of the logistic sector in the innovation process and integration of management systems.

Table 7. Consistency values for AHP matrices

Cluster	Consistency ratio (CR)
Criteria	0.055
Idea Generation (IG)	0.002
Selection and Formulation of Innovation Projects (SP)	0.025
Prototype Development (PD)	0.019
Production and Operations (PO)	0.034
Market Related Activities (MK)	0.022

4.5 Ranking of Logistic Companies via TOPSIS Method and Identifying Improvement Opportunities

The final phases of this combined approach involve the application of the TOPSIS technique for ranking the logistic companies according to their performance in the innovation process and the integration of management systems. Subsequently, the weaknesses of each company that requires intervention were identified, and it defined improvement strategies. For the application of the TOPSIS methodology, an indicator was first determined for each sub-criterion, based on the UNE 166002 standard and the review of the relevant literature on innovation and integrated management systems, as observed in Table 8.

Table 8. Performance indicators for sub-criteria

Decision element	Indicator	Formula
SC1. Technological Surveillance (TS)	Application of technological surveillance	If complies (2), otherwise (1)
SC2. Technological Forecast (TF)	Application of technological forecast	If complies (2), otherwise (1)
SC3. Creativity Techniques (CT)	Use of creativity techniques	If complies (2), otherwise (1)
SC4. Internal and External Analysis (IE)	Application of internal and external analysis	If complies (2), otherwise (1)
SC5. Potential Market (PM)	Develop of market research	If complies (2), otherwise (1)
SC6. Technical and Economic Viability of Projects (VP)	Develop of technical and economic studies of viability for R+D+I projects	If complies (2), otherwise (1)
SC7. Formulation of Selected Projects (FP)	Application of techniques to select R+D+I projects	If complies (2), otherwise (1)
SC8. Research and Basic Design (RD)	Implementation of the process of research and basic design for innovation	If complies (2), otherwise (1)
SC9. Prototyping (PT)	Implementation of the process of development and prototyping for product/process	If complies (2), otherwise (1)

(*continued*)

Table 8. (*continued*)

Decision element	Indicator	Formula
SC10. Demonstration and Testing (DT)	Application of development test and demonstrations of prototypes	If complies (2), otherwise (1)
SC11. Quality Control (QC)	Certification ISO 9001	If complies (2), otherwise (1)
SC12. Life Cycle Analysis (LC)	Certification ISO 14001 or LCA	If complies (2), otherwise (1)
SC13. Resources (RS)	Definition of budget for R+D+I	If complies (2), otherwise (1)
SC14. Risk Management in OHS (RM)	Implement controls for risk in OHS	If complies (2), otherwise (1)
SC15. Training and Communication (TC)	Implementation of training programs in R+D+I	If complies (2), otherwise (1)
SC16. Operational Controls (OC)	Application of procedures and mechanisms for operational controls	If complies (2), otherwise (1)
SC17. Contingency Plans (CP)	Application of contingency plans	If complies (2), otherwise (1)
SC18. Customer Relationships Management (CS)	Index of Customer Satisfaction	% Customer satisfaction
SC19. Supply Chain Management (SM)	Use of ERP platforms in SCM	If complies (2), otherwise (1)
SC20. Advertising, Communication, and Image (AC)	Investment in Advertising, Communication, and Image (COP)	Advertising budget (COP)

The indicators data of each company were incorporated into decision matrix D and applying Eqs. 9–17 of Sect. 3.2, the normalized decision matrix R, the weighted normalized decision matrix, the separations from positive and negative solutions, and finally, the calculation of the closeness coefficients for each company were obtained. Table 9 shows the normalized scores for TOPSIS application. Besides, this table also presents the negative ideal solution (NIS) and the positive ideal solution (PIS).

Table 9. Normalized scores for TOPSIS application

	CL1	CL2	CL3	CL4	PIS	NIS
TS	1	2	2	2	2	1
TF	1	2	1	1	2	1
CT	1	2	2	1	2	1
IE	2	2	2	2	2	1
PM	1	2	2	2	2	1
VP	1	2	2	1	2	1

(*continued*)

Table 9. (*continued*)

	CL1	CL2	CL3	CL4	PIS	NIS
FP	1	2	1	1	2	1
RD	1	2	1	1	2	1
PT	1	2	2	1	2	1
DT	1	1	1	1	2	1
QC	2	2	2	2	2	1
LC	1	2	1	2	2	1
RS	1	2	2	1	2	1
RM	2	2	2	2	2	1
TC	1	2	1	1	2	1
OC	1	2	2	2	2	1
CP	1	2	2	1	2	1
CS	80	92	90	85	92	80
SM	1	2	2	1	2	1
AC	348	1200	850	520	1200	348

Table 10 shows the calculation of the closeness coefficient and the separation from NIS and PIS when evaluating the innovation process and the integration of management systems via the TOPSIS method. In this regard, the closeness coefficient represents the overall performance of each company.

Table 10. Results of evaluation of innovation process and integration of management systems in companies of logistic sector via TOPSIS method

Company	CC_i	d_i^+	d_i^-	Rank
LC2	0,804	0,021	0,086	1
LC3	0,606	0,048	0,074	2
LC4	0,478	0,065	0,060	3
LC1	0,228	0,085	0,025	4

The results evidenced that logistic company 1 obtained the first place with 80.4% ranking at the high-performance level concerning innovation processes integrating management systems. In comparison, Logistics company 1 achieved the lowest score (22.8%), and its located at a very low-performance level. On the other hand, company 3 was located at the medium level (60.6%), and company 4 was classified in the low-performance level (47.8%). The aforementioned implies the analysis of the weaknesses of each of the logistics companies to establish the appropriate improvement strategies.

Finally, the identified gaps between the ideal performance in innovation and integration of management systems were analyzed concerning each of the logistics companies evaluated (refer to Figs. 5–6). In this sense, we analyze the Euclidean distance to the positive ideal solution (PIS) and the negative ideal solution (NIS) to provide logistics companies with strategies that improve their processes of innovation and integration of management systems. For example, the indicator of *customer satisfaction* for company 1 is of 85%, the lowest of the companies evaluated (Euclidean distance = 0.00013), which raises deficiencies in the process of marketing and relationship with customers. Besides, is evidenced in other related indicators that must also be improved, such as the *development of market research* (Euclidean distance = 0.00171) and *technological surveillance* (Euclidean distance = 0.00117), for which alliances with innovation centers and technology transfer offices are proposed as a strategy to support this company in the development of market studies and the application of technological surveillance tools that allow the identification of innovation opportunities aligned with the needs of the market.

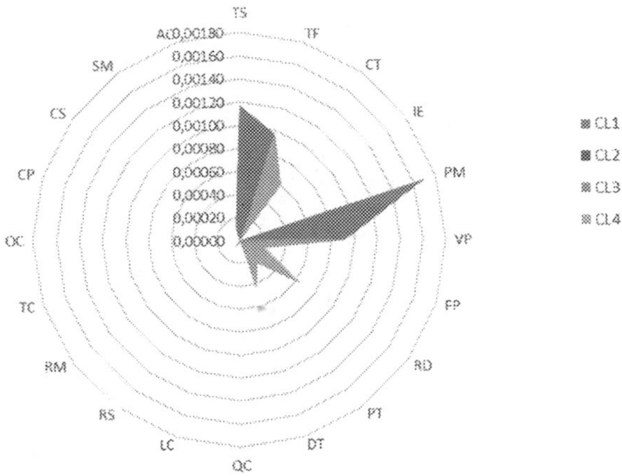

Fig. 5. Spider diagram for separation of the logistic companies from the PIS

For company 4, indicators of the *technical and economic viability of projects* (distance Euclidian = 0.00092) were identified as opportunities for improvement, due that affects the company's ability to make an adequate selection projects of R+D+I, therefore, strategies such as the application of methodologies for the technical and financial evaluation of projects, the acquisition of sector reports, and the training of personnel in the structuring of R+D+I projects are recommended. On the other hand, all the indicators associated with the development of prototypes, such as *research and basic design for innovation* (Euclidean distance = 0.00064), *prototype development* (Euclidean distance = 0.00024) and *test and demonstrations* (Euclidean distance = 0.00044) are located far from the PIS, influencing negatively in the process of developing product and process innovations. In this sense, strategies such as the acquisition of scientific and patent databases, the use of rapid techniques and software for prototyping, and support with technology transfer offices and business acceleration centers are recommended for the development of product demonstrations and tests and processes. The gaps for the rest of the indicators and all the companies can observe in Figs. 5–6.

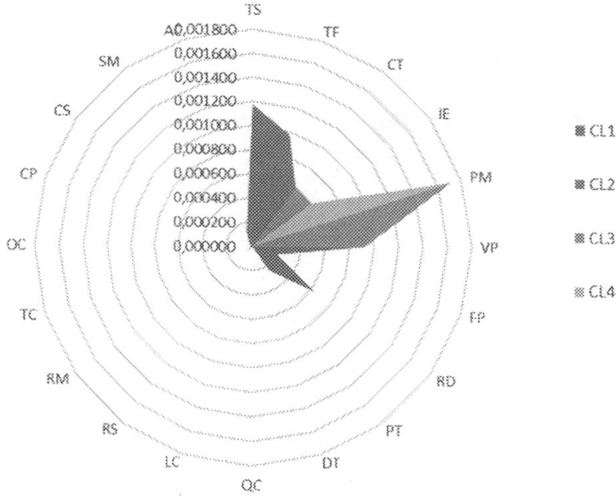

Fig. 6. Spider diagram for separation of the logistic companies from the NIS

5 Conclusions and Future Work

This study presents a combined AHP-TOPSIS approach to evaluate the innovation process and the integration of management systems in the logistics sector. The proposed methodology includes five phases: Design of the MCDM model, application of AHP technique to calculate the relevance of the criteria and sub-criteria, use of TOPSIS method to rank the logistic companies, and definition of improvement opportunities.

According to the outcomes of this research, it was obtained two essential conclusions. The first conclusion was the identification of the most relevant criteria in the evaluation of innovation process, and integration of management systems considering particular elements of the innovation models and factors in quality, health and safety, and environmental that allows logistic companies to generate added value in their products and processes. The second conclusion is the ranking of logistic companies according to several indicators to measure the level in the innovation process and integration of management systems that involving idea generation, selection, and formulation of R+D+I projects. prototype development, production and operations, and market-related activities.

In this sense, the results of the AHP method show that the criteria "Idea generation" and "Selection and formulation of R+D+I projects" were identified as the most relevant with GW contributions of 38.15% and 28.40%, respectively. On the other hand, company 2 was classified as the best performance in the innovation process and integration of management systems with a closeness coefficient of 80.4%. However, for companies 3, 4, and 1, we identified several critical points and establish opportunities for improvement that can be implemented by the companies with the aim of generating added value in products and processes through the integration of innovation-process and the management systems.

Finally, as future work, the proposed methodology will be extended in other industries. Besides, we intend to continue the development of the approach using different hybrid methods as FAHP, DEMATEL, VIKOR, and DEA to analyze other aspects such as interdependency and interrelations between criteria and sub-criteria, and for improving the performance of the method.

Acknowledgments. The authors would like to thank the support of the logistic companies participants and the Institucion Universitaria ITSA.

References

1. OECD: Frascati Manual 2015: Guidelines for Collecting and Reporting Data on Research and Experimental Development, The Measurement of Scientific, Technological and Innovation Activities. OECD Publishing, Paris (2015) https://doi.org/10.1787/9789264239012-en
2. Santos, G., Afonseca, J., Murmura, F., Félix, M.J., Lopes, N.: Critical success factors in the management of ideas as an essential component of innovation and business excellence. Int. J. Qual. Serv. Sci. **3**(3), 214–232 (2018)
3. Santos, G., et al.: Value creation through quality and innovation – a case study on Portugal. TQM J. **31**(6), 928–947 (2019)
4. Hauser, C., Siller, M., Schatzer, T., Walde, J., Gottfried Tappeiner, G.: Measuring regional innovation: a critical inspection of the ability of single indicators to shape technological change. Technol. Forecast. Soc. Chang. **129**, 43–55 (2018)
5. Zoo, H., de Vries, H.J., Lee, H.: Interplay of innovation and standardization: exploring the relevance in developing countries. Technol. Forecast. Soc. Chang. **118**, 334–348 (2017)
6. Santos, G., Mendes, F., Barbosa, J.: Certification and integration of management systems: the experience of Portuguese small and medium enterprises. J. Clean. Prod. **19**(17–18), 1965–1974 (2011)
7. Santos, D., Rebelo, M., Santos, G.: The integration of certified management systems. Case study – organizations located at the district of Braga, Portugal. Procedia Manuf. **13**, 964–971 (2017)
8. UNE 166002. R&D&I Management: R&D&I Management System Requirements (2014)
9. Barbosa, L.C.F., Oliveira, O.J., Santos, G.: Proposition for the alignment of the integrated management system (quality, environmental and safety) with the business strategy. Int. J. Qual. Res. **12**(4), 925–940 (2018)
10. Falak, J., Kunjan, M., Nagaraju, D., Narayanan, S.: Evaluation of continuous improvement techniques using hybrid MCDM technique under fuzzy environment. Mater. Today Proc. eedings **22**(2020), 1295–1305 (2020)
11. Hung, Y.-H., Chou, S.-C.T., Tzeng, G.-H.: Knowledge management adoption and assessment for SMEs by a novel MCDM approach. Decis. Support Syst. **51**(2), 270–291 (2011)
12. Ikrama, M., Sroufe, R., Qingyu, Z.: Prioritizing and overcoming barriers to integrated management system (IMS) Implementation using AHP and G-TOPSIS. J. Clean. Prod. **254** (1), 120121 (2020). https://doi.org/10.1016/j.jclepro.2020.120121
13. Tseng, M.L., Chiu, A.: Grey-entropy analytical network process for green innovation practices. Procedia Soc. Behav. Sci. 57(9), 10–21 (2012)
14. Sadeghi, A.: Success factors of high-tech SMEs in Iran: A fuzzy MCDM approach. J. High Technol. Manage. Res. **29**(1), 71–87 (2018)

15. Tseng, M., Lin, Y., Lim, M., Teehankee, B.: Using a hybrid method to evaluate service innovation in the hotel industry. Appl. Soft Comput. **28**, 411–421 (2015)
16. De, D., Chowdhury, S., Dey, P., Ghosh, S.: Impact of lean and sustainability oriented innovation on sustainability performance of small and medium sized enterprises: a data envelopment analysis-based framework. Int. J. Prod. Econ. **219**(2020), 416–430 (2020)
17. Jimenez-Delgado, G., Balmaceda-Castro, N., Hernández-Palma, H., de la Hoz-Franco, E., García-Guiliany, J., Martinez-Ventura, J.: An integrated approach of multiple correspondences analysis (MCA) and fuzzy AHP method for occupational health and safety performance evaluation in the land cargo transportation. In: Duffy, V.G. (ed.) HCII 2019. LNCS, vol. 11581, pp. 433–457. Springer, Cham (2019). https://doi.org/10.1007/978-3-030-22216-1_32
18. Saaty, T.L.: Decision making with the analytic hierarchy process. Int. J. Serv. Sci. **1**(1), 83–98 (2008)
19. Tavana, M., Hatami-Marbini, A.: A group AHP-TOPSIS framework for human spaceflight mission planning at NASA. Expert Syst. Appl. **38**, 13588–13603 (2011)
20. Ortiz, M., De Felice, F., Parra, K., Alemán, B., Yaruro, A., Petrillo, A.: An AHP-TOPSIS integrated model for selecting the most appropriate tomography equipment. Int. J. Inf. Technol. Decis. Making **15**(4), 861–885 (2016)
21. Hernandez, L., Jimenez, G.: Characterization of the current conditions of the ITSA data centers according to standards of the green data centers friendly to the environment. In: Silhavy, R., Senkerik, R., Kominkova Oplatkova, Z., Prokopova, Z., Silhavy, P. (eds.) CSOC 2017. AISC, vol. 574, pp. 329–340. Springer, Cham (2017). https://doi.org/10.1007/978-3-319-57264-2_34
22. Saaty, T.L.: The Analytical Hierarchy Process. McGraw-Hill, New York (1980)
23. Julián Mayor, J., Botero, S., González, J.: Modelo de decisión multicriterio difuso para la selección de contratistas en proyectos de infraestructura: caso Colombia. Obras y Proyectos **20**, 56–74 (2016)
24. Jato-Espino, D., Castillo-Lopez, E., Rodriguez-Hernandez, J., Canteras-Jordana, J.C.: A review of application of multi-criteria decision-making methods in construction. Autom. Constr. **45**, 151–162 (2014)
25. Cannavacciuolo, L., Iandoli, L., Ponsiglione, C., Zollo, G.: An analytical framework based on AHP and activity-based costing to assess the value of competencies in production processes. Int. J. Prod. Res. **50**(17), 4877–4888 (2012)
26. Ortiz-Barrios, M.A., Herrera-Fontalvo, Z., Rúa-Muñoz, J., Ojeda-Gutiérrez, S., De Felice, F., Petrillo, A.: An integrated approach to evaluate the risk of adverse events in hospital sector: From theory to practice. Manag. Decis. **56**(10), 2187–2224 (2018). https://doi.org/10.1108/MD-09-2017-0917
27. Ortiz-Barrios, M.A., Kucukaltan, B., Carvajal-Tinoco, D., Neira-Rodado, D., Jiménez, G.: Strategic hybrid approach for selecting suppliers of high-density polyethylene. J. Multi-Crit. Decis. Anal. **24**, 1–21 (2017). https://doi.org/10.1002/mcda.1617
28. Meesariganda, B.R., Ishizaka, A.: Mapping verbal AHP scale to numerical scale for cloud computing strategy selection. Appl. Soft Comput. **53**, 111–118 (2017)
29. Ishizaka, A., Balkenborg, D., Kaplan, T.: Influence of aggregation and measurement scale on ranking a compromise alternative in AHP. J. Oper. Res. Soc. **62**(4), 700–710 (2011). https://doi.org/10.1057/jors.2010.23
30. Hwang, C.L., Yoon, K.P.: Multiple attributes decision making methods and applications. Springer-Verlag, Berlin (1981). https://doi.org/10.1007/978-3-642-48318-9
31. Renato, A., Krohling, R., Pacheco, A.: A-TOPSIS – an approach based on TOPSIS for ranking evolutionary algorithms. Procedia Comput. Sci. **55**, 308–317 (2015)

32. Mao, N., Song, M., Pan, D., Deng, S.: Comparative studies on using RSM and TOPSIS methods to optimize residential air conditioning systems. Energy **144**, 98–109 (2018)
33. Santos, G., Mandado, E.: Technological and quality differences between North and Southern European Countries. In: Proceedings 19th QMOD-ICQSS International Conference on Quality and Service Sciences, 21–23 September. University of Roma Tre, Rome, Italy (2016). ISBN 978-91-7623-086-2
34. Santos, G., Engenharia, P.T.: Uma via verde para o desenvolvimento tecnológico e económico de Portugal. Vida Económica Publisher, Oporto (2014)
35. Félix, M., Gonçalves, S., Jimenez, G., Santos, G.: The contribution of design to the development of products and manufacturing processes in the portuguese industry. Procedia Manuf. **41**, 1055–1062 (2019)
36. Bravi, L., Murmura, F., Santos, G.: Manufacturing labs: where new digital technologies help improve life quality. Int. J. Qual. Res. **12**(4), 957–974 (2018)
37. Jimenez, G., Novoa, L., Ramos, L., Martinez, J., Alvarino, C.: Diagnosis of initial conditions for the implementation of the integrated management system in the companies of the land cargo transportation in the city of Barranquilla (Colombia). In: Stephanidis, C. (ed.) HCI 2018. CCIS, vol. 852, pp. 282–289. Springer, Cham (2018). https://doi.org/10.1007/978-3-319-92285-0_39
38. Jimenez, G., Zapata, E.: Metodología integrada para el control estratégico y la mejora continua, basada en el Balanced Scorecard y el Sistema de Gestión de Calidad: aplicación en una organización de servicios en Colombia. In: 51a Asamblea Anual del Consejo Latinoamericano de Escuelas de Administración CLADEA 2016, Medellín, Colombia, pp. 1–20 (2016)
39. Jimenez, G., Hernandez, L., Hernandez, H., Cabas, L., Ferreira, J.: Evaluation of quality management for strategic decision making in companies in the plastic sector of the Colombian Caribbean region using the TQM diagnostic report and data analysis. In: Stephanidis, C. (ed.) HCI 2018. CCIS, vol. 852, pp. 273–281. Springer, Cham (2018). https://doi.org/10.1007/978-3-319-92285-0_38
40. Jimenez, G.: Procedimientos para el mejoramiento de la calidad y la implantación de la Norma ISO 9001 aplicado al proceso de asesoramiento del Centro de Investigaciones y Desarrollo Empresarial y Regional en una Institucion de Educación Superior basados en la gestión por procesos. In: Congreso de Gestión de la Calidad y Protección Ambiental GECPA 2014, Habana, Cuba, pp. 1–22 (2014)
41. Sá, J.C., Amaral, A., Barreto, L., Carvalho, F., Santos, G.: Perception of the importance to implement ISO 9001 in organizations related to people linked to quality-an empirical study. Int. J. Qual. Res. **13**(4), 1055–1070 (2019)
42. Azevedo, J., et al.: Improvement of production line in the automotive industry through lean philosophy. Procedia Manuf. **41**, 1023–1030 (2019)
43. Jimenez, G., et al.: Improvement of productivity and quality in the value chain through lean manufacturing – a case study. Procedia Manuf. **41**, 882–889 (2019)
44. Jimenez, G., Santos, G., Félix, M., Hernández, H., Rondón, C.: Good practices and trends in reverse logistics in the plastic products manufacturing industry. Procedia Manuf. **41**, 367–374 (2019)
45. Santos, G., Doiro, M., Mandado, E., Silva, R.: Engineering learning objectives and computer assisted tools. Eur. J. Eng. Educ. **44**(4), 616–628 (2019)
46. Hernandez, L., Jimenez, G., Baloco, C., Jimenez, A., Hernandez, H.: Characterization of the use of the internet of things in the institutions of higher education of the city of Barranquilla and its metropolitan area. In: Stephanidis, C. (ed.) HCI 2018. CCIS, vol. 852, pp. 17–24. Springer, Cham (2018). https://doi.org/10.1007/978-3-319-92285-0_3

47. Silva, J., et al.: Factors affecting the big data adoption as a marketing tool in SMEs. In: Tan, Y., Shi, Y. (eds.) DMBD 2019. CCIS, vol. 1071, pp. 34–43. Springer, Singapore (2019). https://doi.org/10.1007/978-981-32-9563-6_4
48. Talapatra, S., Santos, G., Uddin, K., Carvalho, F.: Main benefits of integrated management systems through literature review. Int. J. Qual. Res. **13**(4), 1037–1054 (2019)
49. Santos, G., Rebelo, M., Barros, S., Silva, R., Pereira, M., Lopes, N.: Developments regarding the integration of the occupational safety and health with quality and environment management systems. In: Kavouras, I.G., Chalbot, M.C.G. (eds.) Occupational Safety and Health, pp. 113–146. Nova Science, New York (2014). ISBN 978-1-63117-698-2
50. Araújo, R., Santos, G., da Costa, J.B., Sá, J.C.: The quality management system as a driver of organizational culture: an empirical study in the Portuguese textile industry. Qual. Innov. Prosperity **23**(1), 1–24 (2019)
51. Santos, G., Sá, J.C., Oliveira, J., Ramos, D.G., Ferreira, C.: Quality and safety continuous improvement through lean tools. In: Lean Manufacturing: Implementation, Opportunities and Challenges, pp. 165–188 (2019)
52. Bravi, L., Murmura, F., Santos, G.: The ISO 9001: 2015 quality management system standard: companies' drivers, benefits and barriers to its implementation. Qual. Innov. Prosperity **23**(2), 64–82 (2019)
53. Castro, C., Pereira, T., Sá, J.C., Santos, G.: Logistics reorganization and management of the ambulatory pharmacy of a local health unit in Portugal **80**, 101801 (2020). https://doi.org/10.1016/j.evalprogplan.2020.101801
54. Santos, G., Bravi, L., Murmura, F.: Attitudes and behaviours of Italian 3D prosumer in the Era of Additive Manufacturing. Procedia Manuf. **13**, 980–986 (2017)
55. Santos, G., Murmura, F., Bravi, L.: Developing a model of vendor rating to manage quality in the supply chain. Int. J. Qual. Serv. Sci. **11**(1), 34–52 (2019)
56. Ferreira, N., Santos, G., Silva, R.: Risk level reduction in construction sites: towards a computer aided methodology – a case study. Appl. Comput. Inf. **15**(2), 136–143 (2019)
57. Gonçalves, I., Sá, J.C., Santos, G., Gonçalves, M.: Safety stream mapping—a new tool applied to the textile company as a case study. Stud. Syst. Decis. Control **202**, 71–79 (2019). https://doi.org/10.1007/978-3-030-14730-3_8

A User Interface for Personalized Web Service Selection in Business Processes

Dionisis Margaris[1], Dimitris Spiliotopoulos[2(✉)], Costas Vassilakis[2], and Gregory Karagiorgos[3]

[1] Department of Informatics and Telecommunications, University of Athens, Athens, Greece
margaris@di.uoa.gr

[2] Department of Informatics and Telecommunications, University of the Peloponnese, Tripoli, Greece
{dspiliot,costas}@uop.gr

[3] Department of Digital Systems, University of the Peloponnese, Sparti, Greece
greg@us.uop.gr

Abstract. Nowadays, due to the huge volume of information available on the web, the need for personalization is more than necessary. Choosing the right information for each user is as important as the way this information is presented to him or her. Currently, user-triggered recommendation requests for web services are implemented as an automatic recommendation based on parametric computation. This work reports on a specialized user interface for business processes, where writing code entails invocation of business process information. The paper presents the user interface design for Personalized Web Service Selection in Business Process scenario execution and the user evaluation by business process engineers.

Keywords: User interface · Personalization · Web services · Adaptation · Business processes · WS-BPEL

1 Introduction

Nowadays, web services (WSs) are typically offered by multiple providers and hence under different quality of service (QoS) parameters, such as availability, cost, response time and so forth [1]. In this context, it is extremely useful for users to be able to specify the QoS criteria that a WS should fulfill and accordingly the software designed to model business processes should select only WSs that meet the QoS criteria set by users. In this direction, WS-BPEL is the typical language that allows for building high-level business processes through orchestrating individual WSs. Contemporary research allows users not only to select which processes they believe that better support their needs, but also to ask for a WS recommendation, specifying their desirable QoS criteria [2–5].

For example, in a summer holiday planning business process, the user may request a specific luxury hotel service and a specific car rental service and may then ask for a recommendation concerning the airline ticket. The user may also specify that the price

C. Stephanidis et al. (Eds.): HCII 2020, LNCS 12427, pp. 560–573, 2020.
https://doi.org/10.1007/978-3-030-60152-2_41

of the flight must be relatively "low" or at least "medium", while at the same time the invoked WS's reliability must be "high". Many research works exist that support the aforementioned operation [6–9], however all of them invoke the WS which they compute that it better covers the user's need, rather than return a list of the recommended WSs to the user, and let him decide which one he really wants to invoke. As a result, in many cases, the invoked WS is not the one that a user would select on his own and, hence, from a personalization point of view, the WS-BPEL scenario execution adaptation typically fails.

In this work, we present a user interface (UI) for personalized WS selection in WS-BPEL scenarios, which accommodates the QoS criteria set by the users. The main objective is to create a UI that can efficiently present the recommended WSs to WS-BPEL users, so they can decide which WS better fulfills their needs and finally invoke the selected one.

The rest of the paper is structured as follows: Sect. 2 overviews related work, while Sect. 3 presents the necessary foundations for our work, from the area of business processes and WSs, for self-containment purposes. Section 4 discusses the UI design, rationale and the paper experiment, while Sect. 5 presents the results of the user evaluation. Finally, Sect. 6 concludes the paper and outlines the future work.

2 Related Work

Since the information systems in the contemporary WWW are very frequently nowadays implemented by composing WSs offered by individual providers, the adaptation process of WS scenarios is a research field that has attracted significant research efforts over the last years [10–13].

Chen et al. [14] present a QoS-aware WS composition method by multi-objective optimization to assist users make variable decisions. The problem of QoS-aware WS composition is formulated to a multi-objective optimization model where the individual optimization objective is either QoS performance or risk, in order to solve the aforementioned model, an efficient multi-objective evolutionary algorithm is developed. Rodriguez-Mier et al. [15] present a theoretical analysis of graph-based service composition utilizing its dependency with WS discovery. Their work defines a composition framework integrating fine-grained I/O service discovery that enables graph-based composition generation. The composition includes the semantically relevant set of services for input-output requests. Their proposed framework also presents an optimal composition search algorithm for extracting the optimal composition from the graph by minimizing the number and the size of WSs, as well as several graph optimizations for improved system scalability. In order to model WSs with uncertain effects, Wang et al. [16] propose an approach that introduces branch structures into composite solutions to cope with uncertainty in the service composition process. Liu et al. [17] present a WS composition method based on QoS dynamic prediction and global QoS constraints decomposition. Their approach includes 2 phases. The first phase happens before service composition, by decomposing global QoS constraints into local ones, thereby transforming the WS dynamic composition problem to a local optimization one. The second phase happens at runtime, using predicted QoS values to select the optimal WS for the current abstract service. Margaris et al. [4] perform a horizontal QoS-based adaptation, supporting both the parallel and sequential execution structures within the WS-BPEL scenario, while Margaris et al. [18] present an associated execution framework and employ collaborative filtering (CF) techniques to direct the adaptation. However, this approach uses very limited QoS-based criteria (optionally specifying lower and upper bounds for each QoS attribute), hence running the risk of inferior overall QoS of the formulated solutions when compared to the possibly attainable optimal composition QoS, especially for the cases where CF has to address known issues such as gray sheep and cold start. Margaris et al. [7] present a CF-based algorithm which also takes into account the WSs' QoS parameters, in order to personalize the business processes execution to the users' preferences. Furthermore, they introduce an offline clustering technique for supporting the scalable and efficient execution of the presented algorithm under the presence of large volumes of data in the WS repository. Cardellini et al. [19] present a software platform supporting QoS-driven adaptation of WS-oriented systems, named MOSES, which integrates within a unified framework different adaptation mechanism, achieving a greater flexibility in facing various operating environments and the possibly conflicting QoS requirements of several concurrent users.

Although all the aforementioned works successfully address the WS adaptation problem, they also complete the adaptation process, by invoking the WS that they have found that better satisfies the WS-BPEL designer rather than only recommend one or more WSs to the WS-BPEL designer and let him select the WS he prefers the most, as all the modern general purpose RSs do [20–26].

The complexity of the enterprise solutions that interconnect with several services, locally or in the cloud, affect the performance of systems and users [27–30]. To effectively implement business processes, graphical notation can be useful, especially when used for process driven WS-BPEL design [31]. Haihong et al. [32] propose an asynchronous, process-driven mechanism for easier combination of web services. Jose et al. [33] extend the process driven approach using ontologies to describe the relevant aspects and the user interface to display the type of match to the business process WS. Another approach is the use of static call graphs to contain the semantic knowledge, which is then used for the generation of WS-BPEL test cases that support the subprocesses between WS-BPEL files [34].

The aforementioned approaches to user interface design, try to automate the WS selection matching processes. However, none of those works attempt to simplify the user interaction. The necessity for involvement of the user in more complex tasks, casts a burden to the WS-BPEL designer, resulting in reduced usability of the programming environment. This is similar to complexity-induced cognitive load that is usually present in complex user interfaces [35, 36], multi-source information processing [37, 38] and multimodal interaction [39–42].

In this work, we approach the WS-BPEL design through a user centric view, catering to the preference and selection of WS candidates by the designer. The WS-BPEL designer may have preference and selection criteria that differ per business process and that may apply to ones that are matched to the same parameters.

The motivation is to examine how the WS-BPEL designers make selections from multiple choice and define how the user interface could support the interaction.

3 Prerequisites

The following subsections include a summarize concepts and underpinnings from the domains of WS QoS attributes and WS substitution relationships that are used in our work, for conciseness purposes.

3.1 Web Services QoS Attributes

WS QoS attributes describe measurements of the overall performance of a WS [43, 44]. For conciseness purposes and without loss of generality, in this paper we will consider only the attributes *cost* (c), *response time* (rt) and *reliability* (rel), adopting their definitions from Comerio et al. [45].

In the context of a WS-BPEL scenario execution, a user may want to define constraints (i.e. upper and lower bounds) on the QoS delivered by each WS executed in the context of a specific invocation.

More specifically, for each WS ws_x included in a WS-BPEL scenario, the user may provide the following two vectors:

- $MIN_x(min_{rt,x}, min_{c,x}, min_{rel,x})$ and
- $MAX_x(max_{rt,x}, max_{c,x}, max_{rel,x})$.

where the MAX_x vector includes the upper bounds of the QoS attributes for the WS implementation that will be selected to realize the functionality of s_x (e.g. booking an airticket) and the MIN_x vector includes the respective lower bounds.

Additionally, the user provides a weight vector $W(rt_w, c_w, rel_w)$, indicating the importance of each attribute in the context of the particular invocation of the WS-BPEL scenario, which is multiplied by the value of the corresponding QoS attribute, and these products are then aggregated in order to produce the WS-BPEL scenario's overall score. It has to be mentioned that, in contrary to the upper and lower bound vectors, which are applied to each WS invocation, the weight vector is applied to the whole WS-BPEL scenario (the whole composition).

As far as attribute values are concerned, clearly different attributes may have different measurement units (e.g. milliseconds, Euros and kilograms), hence in order to have meaningful results from combining all the attribute values of a WS-BPEL, we consider that all attributes are normalized in the range [0, 10], through the application of a standard normalization formula [46]:

$$norm(v) = 10 * \frac{v - \min(v)}{\max(v) - \min(v)} \tag{1}$$

where v is the value to be normalized, while *max(v)* and *min(v)* are the maximum and minimum values of the corresponding attribute in the database extension. Furthermore, we consider that larger attribute values correspond to higher QoS levels: for attributes where the inverse holds (such as latency and price), the following transformation must be done (altering the normalization formula):

$$norm(v) = 10 * \left(1 - \frac{v - \min(v)}{\max(v) - \min(v)}\right) \tag{2}$$

This approach is typically followed for the handling of QoS attribute values in the context of WS selection and composition [4, 11, 47, 48].

An example of the repository form is shown in Table 1, containing three indicative WSs.

Table 1. Example of repository contents.

Service	Response time	Cost	Reliability
S_1	6	6	6
S_2	8	8	4
S_3	2	2	2

It has to be noted that our framework is also able to handle (a) sequential and parallel (concurrent invocations) structures, (b) service selection affinity and (c) exception resolution in WS-BPEL scenarios. For more details on the aforementioned concepts, the interested reader is referred to [4].

3.2 Substitution Relationship Representation

When a WS-BPEL scenario is adopted to match the QoS designations provided by the user, the execution adaptation process selects WS service implementations to realize functionalities that appear in the context of the WS-BPEL scenario. Therefore, the adaptation software needs to be able to test whether within a particular execution, a WS service implementation *A* can realize some requested functionality *B*. In this work, in order to address this issue, we use the WS matchmaking relationships [49], where any WS that either (i) provides the *same functionality as F* or (ii) provides a *more specific functionality than F*. For more information on matchmaking relationships, the interested reader is referred to [18, 49].

An example of a WS taxonomy is depicted in Fig. 1, concerning an airline ticket booking service. In this figure, the orange shaded rectangles (upper level) represent (sub-)categories, while the white shaded rectangles representing WS implementations. The line between two rectangles denotes that the higher-level node includes the lower-level one in a superclass-subclass relation.

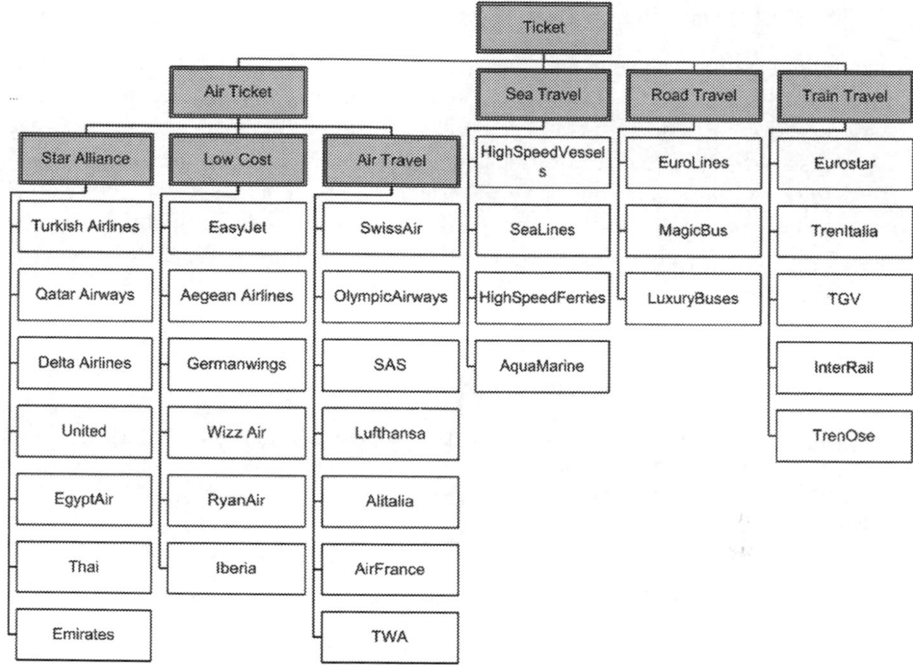

Fig. 1. Hierarchy of WSs (sub-)categories and implementations for the airline ticket booking service.

Each node corresponding to a WS implementation stores information regarding its QoS attributes values in the WS repository. An excerpt of this repository is depicted in Fig. 2.

Category: Ticket
  Category: Air Ticket
    Category: Star Alliance
      Implementation: Turkish Airlines (rt=6, cost=2, rel=9)
      Implementation: Qatar Airways (rt=5, cost=8, rel=3)
      Implementation: Delta Airlines (rt=2, cost=6, rel=6)

Fig. 2. Excerpt of the WS taxonomy repository.

3.3 WS-BPEL Scenario and Dataset Example

To exemplify the proposed approach, we will use the case of adapting the invocation of a simple WS-BPEL scenario (for simplicity only one sequential construct is used). This scenario describes a business process of booking a summer holiday vacation package (modeled as a WS-BPEL scenario, termed as *SHVP*), which includes the functionalities of (i) *asking a recommendation* for an airline ticket, (ii) booking a specific luxury hotel room (explicitly defined by the user) and (iii) renting a car (also explicitly defined by the user). Figure 3 depicts the pseudocode of the aforementioned WS-BPEL scenario.

```
SHVP
WEIGHTS(respTime=0.2, cost=0.3, reliability=0.5)
  SEQ
    (name=bookAirTravel, REC, Ticket / Air Ticket / Low Cost , min=(-,3,5), max=(-,9,-) )
    (name=bookHotel, INV, "Hilton")
    (name=bookCarRental, INV, "Rent a Car")
  END_SEQ
```

Fig. 3. Pseudocode of business process execution request example

In Fig. 3, the *WEIGHTS* entity corresponds to the weight vector that indicates the importance of each attribute in the context of the particular adaptation (as defined in Subsect. 3.1). Then, each construct follows, either indicating an invocation to a specific WS (term *INV*) or asking for a WS recommendation (term *REC*). In the latter case, the user must enter the full path to the WS (sub-)category in the repository (see Fig. 1), where the WS to be recommended must belong (in our example, the user is interested in booking a *Low Cost Air Ticket*), along with the lower and upper bounds for the QoS attributes follow (the '-' symbol is used when a user does not want to set an upper/lower bound for a specific attribute).

4 Experiment

The user interface design was evaluated by 9 WS-BPEL designers, each with more than 3 years of experience in BPEL design using various commercial or in-house software development platforms. The mean age of the participants was 34 years old, 7 male and 2 female. The participants took part in the study using a custom BPEL design IDE designed to provide the facilitators the choice to adjust the number of items to display as well as alternative visualization elements for standard BPEL.

The participants were asked to create simple BPEL code (consisting only of WS invocations) using the pseudocode supported by the user interface. They were presented with visual elements for WS candidates and parameter fine-tuning options. Each participant was asked to design two randomly assigned business processes, out of the five in total that were available as scenarios, that included at least two explicitly user-defined functionalities.

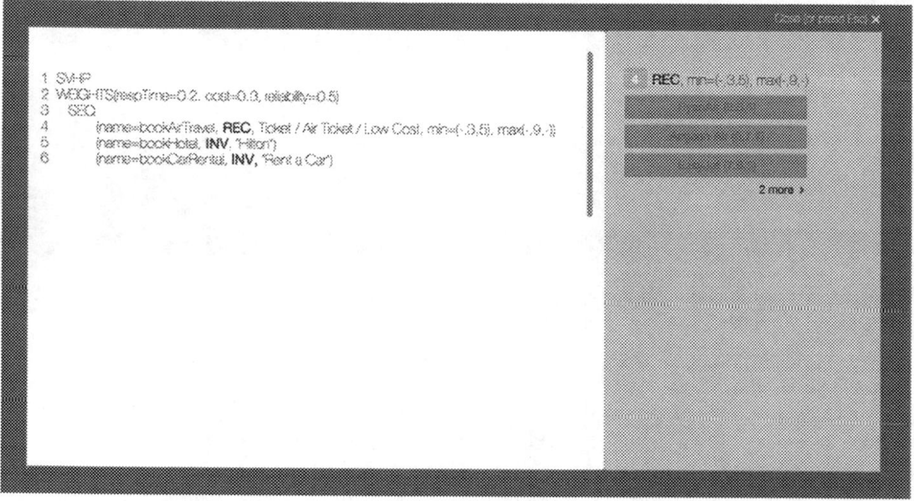

Fig. 4. BPEL design environment that supports recommendation of multiple functionality options and user selection.

The participants were asked to complete the BPEL design and report on their user experience. Before the study, they had the opportunity to engage in BPEL design using the standard interface and parameterization options albeit without the recommendation options.

Figure 4 depicts the user that typed a *REC* command and the parameters of his choice. The interface, once the line feed was pressed, triggered the recommendation algorithm and calculated the functionalities that satisfied the user criteria. The functionalities are shown on the right panel. The right panel is showing the line number, the *REC* information and the list of functionalities, ordered (descending) based on their total calculated score. Therefore, the first airline has a total score of 6.7, the second has a total score of 6.6 and so on.

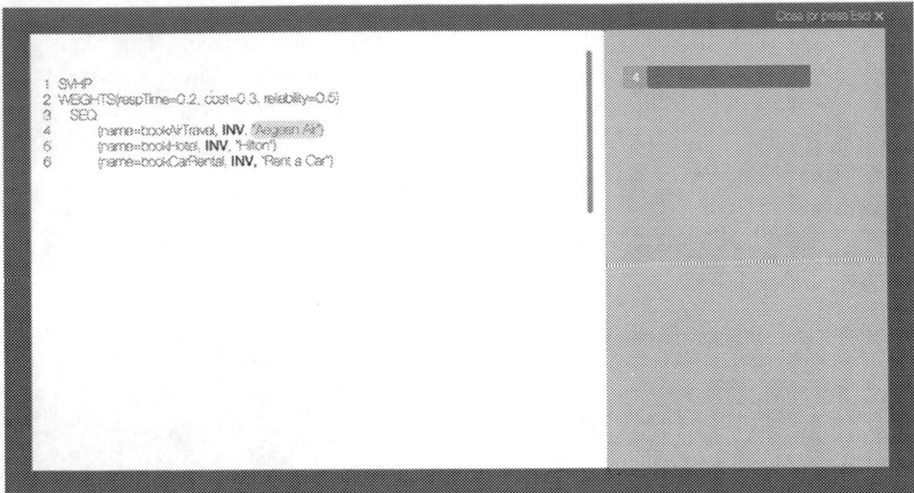

Fig. 5. BPEL design environment depicting recommendation and user selection for a WS-BPEL scenario.

The user decided that the most fulfilling functionality was the second one in the ordered list since it has a higher reliability score. Based on his personal preference for the higher reliability score, the user selected the second choice and the code was auto filled with that information, changing the *REC* to *INV* with the selected value (Fig. 5). The former *REC* is shown in green, allowing the user to click to re-select, if required. The right panel now shows the user current choice. The re-selection can be triggered from either the code or the right panel.

5 Evaluation

The participants were asked to design two business processes and report their rationale to the facilitators using the think-aloud method. It was the method of choice based on previous experience with professional developers and IDE environments. From the objective metrics, 83% of the times the users selected a functionality, their choice was not the first presented functionality (WS). That WS would have been automatically invoked in traditional approaches, requiring additional human effort and time to repair (and in most cases the result is irreversible). This shows that the proposed approach fills a gap in the core functionality of WS-BPEL design user interfaces and builds toward better user experience.

After the end of the session, the participants were asked to fill in a usability questionnaire and then invited to discuss suggestions for improvement. Figure 6 shows that the users reported very high acceptance and relatively high satisfaction. Regarding the latter, the follow up discussion was focused on the next steps of the user interface development and additional functionalities.

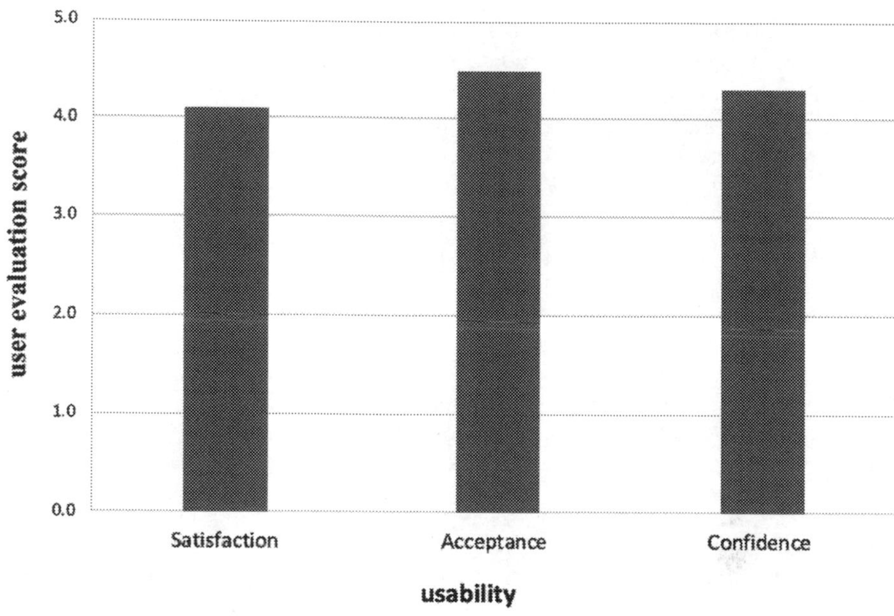

Fig. 6. BPEL design environment usability evaluation overview.

Most of the users reported that they would use this IDE in its current state for their WS-BPEL design. All users mentioned that they would definitely use the user interface if additional functionality for BPEL overview is available. The overview was defined as a visual summary of their choices and the capability to edit and fine tune specific choices for overall targets, such as maximum cost.

6 Conclusion and Future Work

This paper presented a specialized UI for WS-BPEL designers that allows personalized recommendation and selection of business process functionalities based on user generated criteria. The proposed design allows WS-BPEL designers to ask for recommendations and select functionalities out of ordered lists, based on the total or particular scores, per user choice. Allowing the users to have the final selection choice, mitigates the issues caused by automatic system selection that leads to adaptation failure, as shown in the experimental study. The user empowerment is further strengthened from the capability to edit back the selections at will, via the UI.

The UI and the recommendation-selection process were evaluated by WS-BPEL designers. The majority of the participants made selections that were not the highest scoring (based on the total score) in the ordered list. This reinforces the identified requirements and needs that user personalization is reflected in the user-entered parameters. Additionally, it clearly shows that scores are perceived by the users differently and so does their importance in the user selection.

The future work has been identified during the user evaluation. The users suggested that the same recommendation-selection capability be expanded on the business process level, providing an overview of the selections and the functionality to edit individual selections to achieve global scores, such as maximum reliability or minimized costs.

References

1. O'Sullivan, J., Edmond, D., ter Hofstede, A.: What's in a service? Distrib. Parallel Databases **12**, 117–133 (2002). https://doi.org/10.1023/A:1016547000822
2. Margaris, D., Vassilakis, C., Georgiadis, P.: An integrated framework for adapting WS-BPEL scenario execution using QoS and collaborative filtering techniques. Sci. Comput. Program. **98**, 707–734 (2015). https://doi.org/10.1016/j.scico.2014.10.007
3. Furusawa, Yu., Sugiki, Y., Hishiyama, R.: A web service recommendation system based on users' reputations. In: Kinny, D., Hsu, J.Y., Governatori, G., Ghose, A.K. (eds.) PRIMA 2011. LNCS (LNAI), vol. 7047, pp. 508–519. Springer, Heidelberg (2011). https://doi.org/10.1007/978-3-642-25044-6_41
4. Dionisis, M., Costas, V., Panagiotis, G.: An integrated framework for QoS-based adaptation and exception resolution in WS-BPEL scenarios. In: Proceedings of the 28th Annual ACM Symposium on Applied Computing - SAC 2013, p. 1900. ACM Press, New York (2013). https://doi.org/10.1145/2480362.2480714
5. Chen, X., Zheng, Z., Yu, Q., Lyu, M.R.: Web service recommendation via exploiting location and QoS information. IEEE Trans. Parallel Distrib. Syst. **25**, 1913–1924 (2014). https://doi.org/10.1109/TPDS.2013.308
6. Mukherjee, D., Jalote, P., Gowri Nanda, M.: Determining QoS of WS-BPEL compositions. In: Bouguettaya, A., Krueger, I., Margaria, T. (eds.) ICSOC 2008. LNCS, vol. 5364, pp. 378–393. Springer, Heidelberg (2008). https://doi.org/10.1007/978-3-540-89652-4_29
7. Margaris, D., Georgiadis, P., Vassilakis, C.: A collaborative filtering algorithm with clustering for personalized web service selection in business processes. In: 2015 IEEE 9th International Conference on Research Challenges in Information Science (RCIS), pp. 169–180 (2015). https://doi.org/10.1109/RCIS.2015.7128877
8. Dionisis, M., Costas, V., Panagiotis, G.: A hybrid framework for WS-BPEL scenario execution adaptation, using monitoring and feedback data. In: Proceedings of the 30th Annual ACM Symposium on Applied Computing – SAC 2015, pp. 1672–1679. ACM Press, New York (2015). https://doi.org/10.1145/2695664.2695687
9. Margaris, D., Vassilakis, C., Georgiadis, P.: Improving QoS delivered by WS-BPEL scenario adaptation through service execution parallelization. In: Proceedings of the 31st Annual ACM Symposium on Applied Computing, pp. 1590–1596. Association for Computing Machinery, New York (2016). https://doi.org/10.1145/2851613.2851805
10. Gupta, R., Kamal, R., Suman, U.: A QoS-supported approach using fault detection and tolerance for achieving reliability in dynamic orchestration of web services. Int. J. Inf. Technol. **10**(1), 71–81 (2017). https://doi.org/10.1007/s41870-017-0066-z
11. Halfaoui, A., Hadjila, F., Didi, F.: *QoS*-aware web services selection based on fuzzy dominance. In: Amine, A., Bellatreche, L., Elberrichi, Z., Neuhold, E.J., Wrembel, R. (eds.) CIIA 2015. IAICT, vol. 456, pp. 291–300. Springer, Cham (2015). https://doi.org/10.1007/978-3-319-19578-0_24

12. Comes, D., Baraki, H., Reichle, R., Zapf, M., Geihs, K.: Heuristic approaches for QoS-based service selection. In: Maglio, P.P., Weske, M., Yang, J., Fantinato, M. (eds.) ICSOC 2010. LNCS, vol. 6470, pp. 441–455. Springer, Heidelberg (2010). https://doi.org/10.1007/978-3-642-17358-5_30
13. Zeng, L., Benatallah, B., Ngu, A.H.H., Dumas, M., Kalagnanam, J., Chang, H.: QoS-aware middleware for Web services composition. IEEE Trans. Softw. Eng. **30**, 311–327 (2004). https://doi.org/10.1109/TSE.2004.11
14. Chen, F., Dou, R., Li, M., Wu, H.: A flexible QoS-aware Web service composition method by multi-objective optimization in cloud manufacturing. Comput. Ind. Eng. **99**, 423–431 (2016). https://doi.org/10.1016/j.cie.2015.12.018
15. Rodriguez-Mier, P., Pedrinaci, C., Lama, M., Mucientes, M.: An integrated semantic web service discovery and composition framework. IEEE Trans. Serv. Comput. **9**, 537–550 (2016). https://doi.org/10.1109/TSC.2015.2402679
16. Wang, P., Ding, Z., Jiang, C., Zhou, M., Zheng, Y.: Automatic web service composition based on uncertainty execution effects. IEEE Trans. Serv. Comput. **9**, 551–565 (2016). https://doi.org/10.1109/TSC.2015.2412943
17. Liu, Z.Z., Jia, Z.P., Xue, X., An, J.Y.: Reliable Web service composition based on QoS dynamic prediction. Soft. Comput. **19**(5), 1409–1425 (2014). https://doi.org/10.1007/s00500-014-1351-4
18. Margaris, D., Georgiadis, P., Vassilakis, C.: Adapting WS-BPEL scenario execution using collaborative filtering techniques. In: Proceedings - International Conference on Research Challenges in Information Science, pp. 174–184 (2013). https://doi.org/10.1109/RCIS.2013.6577691
19. Cardellini, V., Casalicchio, E., Grassi, V., Iannucci, S., Lo Presti, F., Mirandola, R.: MOSES: a platform for experimenting with QoS-driven self-adaptation policies for service oriented systems. In: de Lemos, R., Garlan, D., Ghezzi, C., Giese, H. (eds.) Software Engineering for Self-Adaptive Systems III. Assurances. LNCS, vol. 9640, pp. 409–433. Springer, Cham (2017). https://doi.org/10.1007/978-3-319-74183-3_14
20. Aivazoglou, M., et al.: A fine-grained social network recommender system. Soc. Netw. Anal. Min. **10**(1), 1–18 (2019). https://doi.org/10.1007/s13278-019-0621-7
21. Margaris, D., Vassilakis, C., Georgiadis, P.: Query personalization using social network information and collaborative filtering techniques. Futur. Gener. Comput. Syst. **78**, 440–450 (2018). https://doi.org/10.1016/j.future.2017.03.015
22. Margaris, D., Vassilakis, C., Spiliotopoulos, D.: What makes a review a reliable rating in recommender systems? Inf. Process. Manag. **57**, 102304 (2020). https://doi.org/10.1016/j.ipm.2020.102304
23. Margaris, D., Vassilakis, C.: Exploiting Internet of Things information to enhance venues' recommendation accuracy. SOCA **11**(4), 393–409 (2017). https://doi.org/10.1007/s11761-017-0216-y
24. Margaris, D., Vassilakis, C., Georgiadis, P.: Recommendation information diffusion in social networks considering user influence and semantics. Soc. Netw. Anal. Min. **6**(1), 1–22 (2016). https://doi.org/10.1007/s13278-016-0416-z
25. Margaris, D., Vassilakis, C., Spiliotopoulos, D.: Handling uncertainty in social media textual information for improving venue recommendation formulation quality in social networks. Soc. Netw. Anal. Min. **9**(1), 1–19 (2019). https://doi.org/10.1007/s13278-019-0610-x
26. Margaris, D., Kobusinska, A., Spiliotopoulos, D., Vassilakis, C.: An adaptive social network-aware collaborative filtering algorithm for improved rating prediction accuracy. IEEE Access **8**, 68301–68310 (2020). https://doi.org/10.1109/ACCESS.2020.2981567

27. Sturm, R., Pollard, C., Craig, J.: Application programming interfaces and connected systems. In: Application Performance Management (APM) in the Digital Enterprise, pp. 137–150. Elsevier (2017). https://doi.org/10.1016/B978-0-12-804018-8.00011-5
28. Risse, T., et al.: The ARCOMEM architecture for social- and semantic-driven web archiving. Futur. Internet. **6**, 688–716 (2014). https://doi.org/10.3390/fi6040688
29. Demidova, E., et al.: Analysing and enriching focused semantic web archives for parliament applications. Futur. Internet. **6**, 433–456 (2014). https://doi.org/10.3390/fi6030433
30. Bernaschina, C., Falzone, E., Fraternali, P., Gonzalez, S.L.H.: The virtual developer. ACM Trans. Softw. Eng. Methodol. **28**, 1–38 (2019). https://doi.org/10.1145/3340545
31. Stiehl, V., Danei, M., Elliott, J., Heiler, M., Kerwien, T.: Effectively and efficiently implementing complex business processes: a case study. In: Lübke, D., Pautasso, C. (eds.) Software Engineering for Self-Adaptive Systems III. Assurances. LNCS, vol. 9, pp. 33–57. Springer, Cham (2019). https://doi.org/10.1007/978-3-030-17666-2_3
32. Haihong, E., Lin, Y., Song, M., Xu, X., Zhang, C.: A visual web service composition system based on process tree. In: 2019 5th International Conference on Information Management (ICIM), pp. 274–278. IEEE (2019). https://doi.org/10.1109/INFOMAN.2019.8714694
33. Jose, H.S.A.S., Cappelli, C., Santoro, F.M., Azevedo, L.G.: Implementation of aspect-oriented business process models with web services. Bus. Inf. Syst. Eng. **17**(1), 1–24 (2020). https://doi.org/10.1007/s12599-020-00643-2
34. Bousanoh, W., Suwannasart, T.: Test case generation for WS-BPEL from a static call graph. J. Phys: Conf. Ser. **1195**, 12004 (2019). https://doi.org/10.1088/1742-6596/1195/1/012004
35. Pino, A., Kouroupetroglou, G., Kacorri, H., Sarantidou, A., Spiliotopoulos, D.: An open source/freeware assistive technology software inventory. In: Miesenberger, K., Klaus, J., Zagler, W., Karshmer, A. (eds.) ICCHP 2010. LNCS, vol. 6179, pp. 178–185. Springer, Heidelberg (2010). https://doi.org/10.1007/978-3-642-14097-6_29
36. Androutsopoulos, I., Spiliotopoulos, D., Stamatakis, K., Dimitromanolaki, A., Karkaletsis, V., Spyropoulos, C.D.: Symbolic authoring for multilingual natural language generation. In: Vlahavas, I.P., Spyropoulos, C.D. (eds.) SETN 2002. LNCS (LNAI), vol. 2308, pp. 131–142. Springer, Heidelberg (2002). https://doi.org/10.1007/3-540-46014-4_13
37. Schefbeck, G., Spiliotopoulos, D., Risse, T.: The Recent challenge in web archiving: archiving the social web. In: Proceedings of the International Council on Archives Congress, pp. 1–5 (2012)
38. Antonakaki, D., Spiliotopoulos, D., Samaras, C. V., Ioannidis, S., Fragopoulou, P.: Investigating the complete corpus of referendum and elections tweets. In: Proceedings of the 2016 IEEE/ACM International Conference on Advances in Social Networks Analysis and Mining, ASONAM 2016, pp. 100–105 (2016). https://doi.org/10.1109/ASONAM.2016.7752220
39. Kouroupetroglou, G., Spiliotopoulos, D.: Usability methodologies for real-life voice user interfaces. Int. J. Inf. Technol. Web Eng. **4**, 78–94 (2009). https://doi.org/10.4018/jitwe.2009100105
40. Xydas, G., Spiliotopoulos, D., Kouroupetroglou, G.: Modeling emphatic events from non-speech aware documents in speech based user interfaces. Proc. Hum. Comput. Interact. **2**, 806–810 (2003)
41. Spiliotopoulos, Dimitris., Xydas, Gerasimos, Kouroupetroglou, Georgios: Diction based prosody modeling in table-to-speech synthesis. In: Matoušek, Václav, Mautner, Pavel, Pavelka, Tomáš (eds.) TSD 2005. LNCS (LNAI), vol. 3658, pp. 294–301. Springer, Heidelberg (2005). https://doi.org/10.1007/11551874_38

42. Spiliotopoulos, D., Stavropoulou, P., Kouroupetroglou, G.: Acoustic rendering of data tables using earcons and prosody for document accessibility. In: Stephanidis, C. (ed.) UAHCI 2009. LNCS, vol. 5616, pp. 587–596. Springer, Heidelberg (2009). https://doi.org/10.1007/978-3-642-02713-0_62
43. Canfora, G., Di Penta, M., Esposito, R., Villani, M.L.: An approach for QoS-aware service composition based on genetic algorithms. In: Proceedings of the 2005 Conference on Genetic and evolutionary computation - GECCO 2005, p. 1069. ACM Press, New York (2005). https://doi.org/10.1145/1068009.1068189
44. Hammas, O., Ben Yahia, S., Ben Ahmed, S.: Adaptive web service composition insuring global QoS optimization. In: 2015 International Symposium on Networks, Computers and Communications (ISNCC), pp. 1–6. IEEE (2015). https://doi.org/10.1109/ISNCC.2015.7238593
45. Comerio, M., De Paoli, F., Grega, S., Maurino, A., Batini, C.: WSMoD. Int. J. Web Serv. Res. **4**, 33–60 (2007). https://doi.org/10.4018/jwsr.2007040102
46. Daqing He, Wu, D.: Toward a robust data fusion for document retrieval. In: 2008 International Conference on Natural Language Processing and Knowledge Engineering, pp. 1–8. IEEE (2008). https://doi.org/10.1109/NLPKE.2008.4906754
47. Liu, Y., Ngu, A.H., Zeng, L.Z.: QoS computation and policing in dynamic web service selection. In: Proceedings of the 13th International World Wide Web Conference on Alternate Track Papers & Posters - WWW Alternate 2004, p. 66. ACM Press, New York (2004). https://doi.org/10.1145/1013367.1013379
48. Margaris, D., Georgiadis, P., Vassilakis, C.: On replacement service selection in WS-BPEL scenario adaptation. In: Proceedings - 2015 IEEE 8th International Conference on Service-Oriented Computing and Applications, SOCA 2015, pp. 10–17 (2015). https://doi.org/10.1109/SOCA.2015.11
49. Bellur, U., Kulkarni, R.: Improved matchmaking algorithm for semantic web services based on bipartite graph matching. In: IEEE International Conference on Web Services (ICWS 2007), pp. 86–93. IEEE (2007). https://doi.org/10.1109/ICWS.2007.105

A Comparative Study of Data Augmentation Methods for Brand Logo Classifiers

Matheus Moraes Machado, Aléssio Miranda Júnior(✉), and Marcelo de Sousa Balbino

Centro Federal de Educação Tecnológica de Minas Gerais Timóteo, Timóteo, Brazil
matheus7mm@hotmail.com
{alessio,marcelobalbino}@cefetmg.br
http://dcctim.cefetmg.br/

Abstract. Social networks have become a widely used way to share information, including images. This information can be useful for companies, which need to know their customers' opinions and their interest in their brands. For these purposes, an image classifier has a great utility. Currently, deep learning using convolutional neural networks are heavily employed for image classification. However, they require a large number of training image samples, which are not always accessible. In order to solve this problem, we can use data augmentation, which is a regularization technique that is based on expand the original dataset to increase classification accuracy and avoid overfitting. This present work aims to compare the use of different data augmentation methods for brand logo classification. For tests, seven methods (flip, crop, rotation, gaussian filter, gaussian noise, scale, and shear) were selected based on previous studies. Two convolutional neural networks were used, AlexNet and SmallerVGGNet, this way we could see if some combinations lead to better results in different artificial network architectures. Ten different combinations of those methods show that the combination of flip, crop, rotation, and scale is a more effective combination for a brand logo classifier in the both convolutional neural networks used, improving accuracy by 22.09% using AlexNet and 12.57% using SmallerVGGNet. Other good results are found with the combination of all seven methods; flip, crop, rotation, plus gaussian noise; and the combination of flip, crop, and rotation. Further research is intended to use another regularization technique along with data augmentation to improve even more the accuracy and reduce overfitting.

Keywords: Data augmentation · Image classification · Brand logo · Convolutional neural network · Computer vision

1 Introduction

Since the advent of Twitter in 2006, social networks have rapidly become influential ways to share information [16]. Known as microblogs, a growing number

C. Stephanidis et al. (Eds.): HCII 2020, LNCS 12427, pp. 574–584, 2020.
https://doi.org/10.1007/978-3-030-60152-2_42

of users are interacting on them at any time and everywhere. They generate a massive amount of media content through creating, describing, and posting text, images and multimedia messages [4,15].

Companies disseminate products and supervise the development of their brands on microblog platforms. Messages showing feelings or sentiments about products posted in there offer companies a good opportunity to understand the real thoughts from customers [15].

Consumers normally provide positive or negative comments when they post brand related information on social networks. Such knowledge and insights have important marketing values for enterprises, which need to know about brand exposure and user acceptance. Even for individual users, such insights are extremely useful to help them make purchase decisions about brands and products. The rapidly increasing amount of live information in social media streams demands the development of effective brand tracking techniques for data gathering and media content analysis [1].

Figure 1 exemplifies a customer's opinion on a product of the American company Starbucks. It does not refer to the brand textually or even by using hashtag, but the posted image makes it possible to know it is referencing the company.

Object classification is one approach enterprises can use to find social networks content related to their brand. Object classification is the task of identifying a given object in an image as belonging to one of a predefined set of classes [10].

In the last few years, we have seen a great development in this field using Artificial Neural Networks (ANN) as Deep Nets. However, there are enormous challenges which must be overcome. A disadvantage of ANNs is that they typically need a very large amount of annotated training data, known as datasets, which restricts their use to situations where big data is available [6]. Yet for many applications only relatively small training data exist [7].

Adding modified images to the dataset prior to training in ANN yields better results in cases where extensive training data is unavailable. This method, called data augmentation (DA), can drastically increase the accuracy for ANNs [7].

In the literature, researchers use different approaches to generate new data for the dataset [5,7,8,10]. Choosing the right methods for data augmentation have long been used for deep learning and selecting appropriate strategies for this approach is even more important than choosing the ANN model [2].

1.1 Related Work

The most common techniques described in the literature apply affine transformations known as translation, zoom, flip, shear, mirror, and color perturbation. Their use in the literature, along with other techniques, have shown an increase in the efficiency of a classifier, improving its accuracy and avoiding over-fitting[1] [3].

[1] Over-fitting of a classification model occurs when too many and/or irrelevant model terms are included and it may lead to low robustness/repeatability when the classification model is applied to independent validation data [9].

Elon Musk
@elonmusk
My coffee tastes unusually good this morning

DUMB STARBUCKS
COFFEE
5:15 PM · Feb 18, 2016 · Twitter for iPhone

Fig. 1. A customer's opinion. Source: [14].

In [12], a study was done to explore the impact of various data augmentation methods on image classification tasks with a deep convolution neural network where Alexnet was employed as the pre-training network model and a subset of CIFAR10 and ImageNet (10 categories) were selected as the original dataset. The data augmentation methods used in this work were flip, crop, shift, color jitter, noise, rotation and their combinations. Experimental results showed that crop, rotation, and flip applied individually performed better; while noise had the worse results. Among the possible pair combinations, flip and crop was the most effective. However, other results showed the overall performance of triple combinations is superior to the pair combinations. Nonetheless, some triple combinations may produce performance degradation. The combination of flip, crop, and rotation had the best outcome.

Other recent work [6], applied some data augmentation strategies in the context of medical imaging. The results are reported on Table 1. This paper also had worse results using noise, while shear, scale, flip, rotate, and gaussian filter had the best outcomes.

Table 1. Results of data augmentation methods applied in the context of medical imaging [6].

DA method	Training accuracy	Testing accuracy
Noise	0.625	0.660
Gaussian filter	0.870	0.881
Jitter	0.832	0.813
Scale	0.887	0.874
Powers	0.661	0.737
Rotate	0.884	0.880
Shear	0.891	0.879
Flip	0.891	0.842

1.2 Proposed Methods

To verify which combinations of data augmentation methods have better results for brand logo recognition, were applied sixteen cases and a comparison of the accuracy of each case was performed. Based on the work [12] and [6], the following data recognition cases were applied to a dataset before the classifier training:

- Case 1: Original dataset;
- Case 2: Flip and gaussian filter;
- Case 3: Flip, crop, and rotation;
- Case 4: Flip, crop, and scale;
- Case 5: Flip, rotation, and shear;
- Case 6: Flip, crop, rotation, and gaussian filter;
- Case 7: Flip, crop, rotation, and gaussian noise;
- Case 8: Flip, crop, rotation, and scale;
- Case 9: Flip, crop, rotation, gaussian filter, gaussian noise, scale, and shear;
- Case 10: Dataset original, but with the same size of case 8.

These methods and combinations were chosen because they had better results in both studies. The gaussian filter was chosen because it had the worse result in both studies.

1.3 Convolutional Neural Network Architectures

Two convolutional neural networks were used, AlexNet and SmallerVGGNet, this way we could see if some combinations lead to better results in different artificial network architectures.

SmallerVGGNet was proposed for [11]. It is a more compact adaptation of the VGGNet network, introduced by Simonyan and Zisserman in their 2014 paper [13].

VGGNet-like architectures are characterized by [11]:

1. Using only 3×3 convolutional layers stacked on top of each other in increasing depth.
2. Reducing volume size by max pooling.
3. Fully-connected layers at the end of the network prior to a softmax classifier.

1.4 Data Augmentation Methods

As seen in Fig. 2, I use the data augmentation methods: Flip, crop, rotation, gaussian filter, gaussian noise, scale and shear. They are described below:

- Flip: I perform a horizontal and vertical flip for each original image. For this, I use the NumPy function fliplr (flip left to right) and flipud (flip up to down).
- Crop: A random part of the original image is cropped. This cut can be in their left, right, top or bottom side and cannot exceed one-third the size of the original image.
- Rotation: I perform a random rotation of the original image using the ImageDataGenerator function from Keras.
- Gaussian filter: A gaussian filter is applied to the original image using the GaussianBlur function from OpenCV; the gaussian kernel size used is 11×11 and a gaussian kernel standard deviation between 1 and 10 in X direction.
- Gaussian noise: A random gaussian noise is applied to the original image using random_noise function from Scikit-image.
- Scale: An affine transformation is applied to the original image with random scale factors between 0.7 and 1.3 for X and Y; for this, I use AffineTransform function from Scikit-image.
- Shear: An affine transformation is applied to the original image with shear between 0.1 and 0.4 using AffineTransform function from Scikit-image.

For those methods, except for flip, I generate five new images for each original image.

2 Experiments

The small dataset consists of 96 images of tree brands, from Coca-Cola, Lacoste, and Starbucks.

1. Coca-Cola (32 images).
2. Lacoste (32 images).
3. Starbucks (32 images).

The new datasets are generated for each case. The number of images of each dataset, and their sizes are shown in Table 2.

Table 2. Dataset of each case of study

Case of study	Images	Size in SmallerVGGNet	Size in AlexNet
1	96	20.74 MB	115.94 MB
2	768	165.89 MB	927.52 MB
3	1248	269.57 MB	1507.22 MB
4	1248	269.57 MB	1507.22 MB
5	1248	269.57 MB	1507.22 MB
6	1728	373.25 MB	2086.92 MB
7	1728	373.25 MB	2086.92 MB
8	1728	373.25 MB	2086.92 MB
9	3168	684.29 MB	3826.03 MB
10	1728	373.25 MB	2086.92 MB

The datasets for each case and the original dataset are trained in 400 epochs.

The validate of data augmentation techniques are make for 120 images, which are very different from the dataset trained. For each class it uses images of the logo only in its dataset for train. In the images for tests it uses the logo used in the real world (a logo used with a t-shirt or shoes, for example).

3 Results and Discussion

Table 3 and Table 4 show the results of the experiments, and Fig. 3 the graphs of the network training for case 8 and 10 using AlexNet (in the complete paper it will show the graphs for all cases). As can see in them, the better case of study (8) has less loss than using the original dataset in case 10, showing how using data augmentation can reduce over-fitting.

3.1 Quality

As observed on Table 3 and Table 4, case 8 has the best result of all cases for both CNNs. That means flip, clop, rotation, and scale is the best combination, it improves 22.09% of the accuracy using AlexNet, and 12.57% using SmallerVGGNet. Other good results are the combination of all methods (case 9), and the combinations flip, crop, and rotation (case 3).

Table 3. Results from SmallerVGGNet.

Case of study	Training Acc.	Loss	Testing Acc.	Loss
1	1	0.0018	0.7305	2.6099
2	1	6.50×10^{-4}	0.68	5.8117
3	1	3.51×10^{-6}	0.8181	2.1946
4	0.9984	0.0059	0.8076	2.4972
5	1	6.44×10^{-5}	0.8029	3.2131
6	1	7.06×10^{-4}	0.7495	5.0537
7	0.9994	0.0121	0.8095	2.6849
8	1	2.57×10^{-6}	0.8371	2.512
9	0.9991	0.007	0.781	4.3662
10	1	2.66×10^{-7}	0.7114	5.147

Table 4. Results from AlexNet

Case of study	Training Acc.	Loss	Testing Acc.	Loss
1	0.9688	1.2553	0.6362	9.4334
2	0.9948	0.5416	0.7381	2.1429
3	0.9968	1.0512	0.799	2.8605
4	0.9952	0.6865	0.7771	2.4486
5	1	0.2655	0.7238	1.4029
6	0.9931	0.4242	0.721	2.8487
7	1	0.3668	0.8371	1.0632
8	0.9954	0.3532	0.8571	1.1775
9	0.9965	0.1634	0.8124	1.0159
10	0.9959	0.7742	0.7229	5.6727

(a) Original Image
(b) Flip
(c) Crop
(d) Rotation
(e) Gaussian Filter
(f) Gaussian Noise
(g) Scale
(h) Shear

Fig. 2. Data augmentation methods used

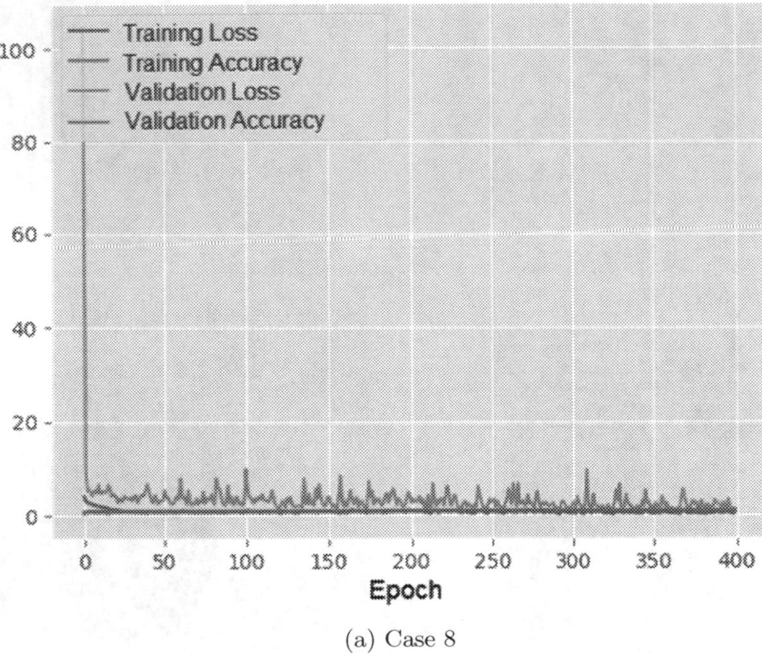

(a) Case 8

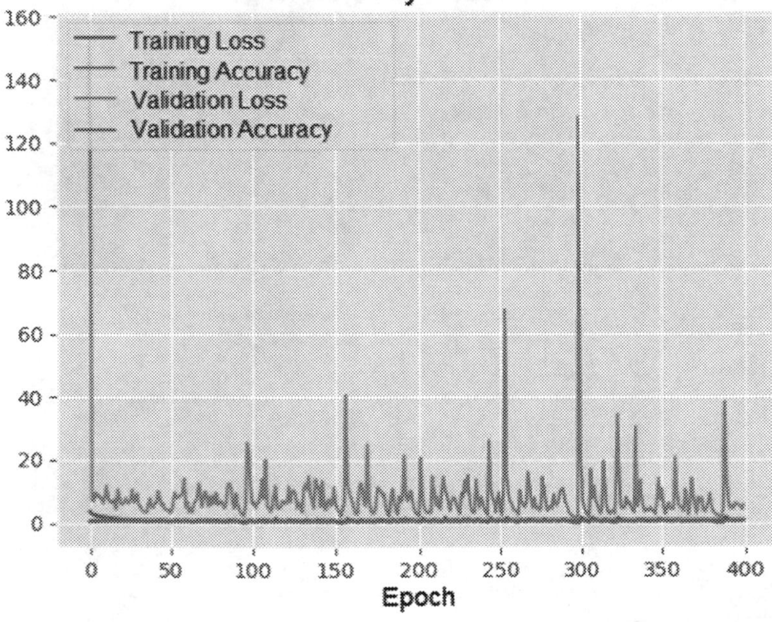

(b) Case 10

Fig. 3. Graphs from AlexNet (Cases 8 and 10)

4 Conclusion

This paper introduced some methods for data augmentation and its use for brand logo classification. It achieved some improvements, as seen in three of the most successful cases. However, using all the methods together will not necessarily produce superior results. For that, the best combination is required.

References

1. Gao, Y., Wang, F., Luan, H., Chua, T.S.: Brand data gathering from live social media streams. In: Proceedings of International Conference on Multimedia Retrieval, pp. 169:169–169:176. ICMR 2014. ACM, New York (2014). https://doi.org/10.1145/2578726.2578748
2. Goodfellow, I., Bengio, Y., Courville, A.: Deep Learning. MIT Press, Cambridge (2016). http://www.deeplearningbook.org
3. Han, D., Liu, Q., Fan, W.: A new image classification method using CNN transfer learning and web data augmentation. Expert Syst. Appl. **95**, 43–56 (2018). https://doi.org/10.1016/j.eswa.2017.11.028, http://www.sciencedirect.com/science/article/pii/S0957417417307844
4. Hossain, M.S., Alhamid, M.F., Muhammad, G.: Collaborative analysis model for trending images on social networks. Futur. Gener. Comput. Syst. **86**, 855–862 (2018). https://doi.org/10.1016/j.future.2017.01.030, http://www.sciencedirect.com/science/article/pii/S0167739X17301383
5. Hu, W., Hu, R., Xie, N., Ling, H., Maybank, S.: Image classification using multiscale information fusion based on saliency driven nonlinear diffusion filtering. IEEE Trans. Image Process. **23**(4), 1513–1526 (2014). https://doi.org/10.1109/TIP.2014.2303639
6. Hussain, Z., Gimenez, F., Yi, D., Rubin, D.: Differential data augmentation techniques for medical imaging classification tasks (2017). https://www.ncbi.nlm.nih.gov/pmc/articles/PMC5977656/
7. Ismail Fawaz, H., Forestier, G., Weber, J., Idoumghar, L., Muller, P.A.: Data augmentation using synthetic data for time series classification with deep residual networks. arXiv e-prints arXiv:1808.02455, August 2018
8. Molina, J.F., Gil, R., Bojacá, C., Díaz, G., Franco, H.: Color and size image dataset normalization protocol for natural image classification: a case study in tomato crop pathologies. In: Symposium of Signals, Images and Artificial Vision - 2013: STSIVA - 2013, pp. 1–5, September 2013. https://doi.org/10.1109/STSIVA.2013.6644938
9. Nansen, C., Geremias, L.D., Xue, Y., Huang, F., Parra, J.R.: Agricultural case studies of classification accuracy, spectral resolution, and model over-fitting. Appl. Spectrosc. **67**(11), 1332–1338 (2013). http://as.osa.org/abstract.cfm?URI=as-67-11-1332
10. Rizvi, S.T.H., Cabodi, G., Gusmao, P., Francini, G.: Gabor filter based image representation for object classification. In: 2016 International Conference on Control, Decision and Information Technologies (CoDIT), pp. 628–632, April 2016. https://doi.org/10.1109/CoDIT.2016.7593635
11. Rosebrock, A.: Keras and convolutional neural networks (CNNs) (2018). https://www.pyimagesearch.com/2018/04/16/keras-and-convolutional-neural-networks-cnns/

12. Shijie, J., Ping, W., Peiyi, J., Siping, H.: Research on data augmentation for image classification based on convolution neural networks. In: 2017 Chinese Automation Congress (CAC), pp. 4165–4170, October 2017. https://doi.org/10.1109/CAC.2017.8243510
13. Simonyan, K., Zisserman, A.: Very deep convolutional networks for large-scale image recognition. In: International Conference on Learning Representations (2015)
14. Twitter: Postagem do twitter (2018). https://twitter.com/elonmusk/status/700398127044362240
15. Wang, F., Qi, S., Gao, G., Zhao, S., Wang, X.: Logo information recognition in large-scale social media data. Multimedia Syst. **22**(1), 63–73 (2016). https://doi.org/10.1007/s00530-014-0393-x
16. Yan, Q., Wu, L., Zheng, L.: Social network based microblog user behavior analysis. Physica A **392**(7), 1712–1723 (2013). https://doi.org/10.1016/j.physa.2012.12.008, http://www.sciencedirect.com/science/article/pii/S0378437112010540

COVID-19 Pandemic – Role of Technology in Transforming Business to the New Normal

Fiona Fui-Hoon Nah(✉) and Keng Siau

Missouri University of Science and Technology, Rolla, MO, USA
{nahf, siauk}@mst.edu

Abstract. COVID-19 has disrupted our lives and the economy. In this paper, we outline approaches in which information technology can be used to implement business strategies to enhance resilience by coping with, adapting to, and recovering from adversity resulting from the COVID-19 pandemic. We discuss how information technology such as digital supply chain, data analytics, artificial intelligence, machine learning, robotics, digital commerce, and Internet of Things can be used to enhance resilience and continuity of business.

Keywords: Pandemic · Covid-19 · Coronavirus · Business transformation · Business continuity · Technology · Resilience · New Normal

1 Introduction

Technology is an enabler of resilience and agility [11, 15, 57]. According to McKinsey & Company, Asian countries have utilized technology quicker to respond to disruptions caused by the COVID-19 pandemic [15]. For example, technology has been utilized by some Asian countries to assist with contact tracing and testing for COVID-19 [14], sharing medical resources and information related to COVID-19 (e.g., https://www.theschwartzcenter.org/covid-19), and informing customers about the availability of personal protection equipment in the stores [13]. Technology has enabled the digitalization of service offerings and product delivery, and hence, plays an important role in the implementation of safety and precautionary measures to minimize the spread of COVID-19.

Not only has the COVID-19 pandemic caused health issues and death, but it has also negatively impacted businesses and the economy [103]. With precautionary measures in place (i.e., including lockdowns and sheltering in place), the impact of the pandemic is much greater on brick-and-mortar businesses than businesses that have digital commerce in place where the digital arms may still operate and continue with the business. Businesses without digital commerce are racing to create a virtual store image [37] for business continuity. Research has shown that information technology can increase the resilience and agility of organizations and enable them to perform better than their counterparts under turbulent or unstable environments [11, 16, 57], including in the case of a pandemic.

C. Stephanidis et al. (Eds.): HCII 2020, LNCS 12427, pp. 585–600, 2020.
https://doi.org/10.1007/978-3-030-60152-2_43

2 Theoretical Perspective on Organizational Resilience

Organizational resilience is key to business continuity in a pandemic, such as the COVID-19 pandemic. Businesses need resilience capacities to cope with, adapt to, and recover from a pandemic. Burnard and Bhamra [8] stressed the importance of cultivating resilience in organizations in order to create the capabilities to adapt to new or different circumstances during turbulent times. Hamel and Valikangas [30] indicated that in times of turbulence, "the only dependable advantage is a superior capacity for reinventing your business model before circumstances force you to" (p. 53).

The concept of resilience has been examined from multiple perspectives in the literature [6–8, 19, 30, 44, 58, 66, 97]. Resilience has been referred to as "the incremental capacity of an organization to anticipate and adjust to the environment" [58, p. 6]. It has also been referred to as "a firm's ability to effectively absorb, develop situation-specific responses to, and ultimately engage in transformative activities to capitalize on disruptive surprises that potentially threaten organization survival" [44, p. 244]. Vogus and Sutcliffe [97] defined resilience as "the maintenance of positive adjustment under challenging conditions such that the organization emerges from those conditions strengthened and more resourceful" (p. 3418), while Reinmoeller and van Baardwijk [66] defined resilience as "the capability to self-renew over time through innovation" (p. 61). Similarly, Hamel and Valikangas [30] referred to resilience as "a capacity to continuous reconstruction" (p. 55). Other researchers view resilience from the recovery perspective by viewing it as "bouncing back to a state of normalcy" [7, p. 431] and "developable capacity to rebound from adversity" [47, p. 28].

Hence, the literature has viewed resilience in the business and organizational context as comprising three successive stages: anticipation, coping, and adaptation [19]. Duchek [19] reviewed the literature on organizational resilience and conceptualized resilience as the capabilities to anticipate (i.e., observe, identify, prepare), cope (accept the problem, then develop and implement solutions), and adapt to the unexpected (i.e., reflect and learn, followed by organizational change). Duchek's [19] conceptualization is similar to Ponomarov and Holcomb's [61] conceptualization of resilience as comprising three phases – readiness, response, and recovery. These three capabilities will be explained next.

Anticipation is "the ability to detect critical developments within the firm or in its environment and to adapt proactively" [19, p. 225]. Based on the notion of anticipation developed by researchers in the context of resilience [8, 40, 87], Duchek [19] conceptualized the anticipation stage as comprising three specific capabilities: (i) the ability to *observe* internal and external developments; (ii) the ability to *identify* critical developments and potential threats; and (iii) the ability to *prepare* for unexpected events. Duchek [19] reiterated Somers' [87] original quote that highlighted anticipation capabilities as building "resilience that is not presently evident or realized" [87, p. 13].

The ability to cope with unexpected events is one of the most important capabilities of resilience. Coping is the "effective handling of unexpected events so as to resist destruction" [19, p. 227]. Duchek [19] identified two aspects of coping: (i) the ability to *accept* a problem; and (ii) the ability to *develop* and *implement* solutions. By accepting the problem, organizations can move on to deal with the adversity arising from the critical events in order to react quickly to them. Developing and implementing solutions can be viewed from the perspective of sensemaking [100], where "there must be continual feedback between understanding and action, which means that sense must continually be made and remade" [19, p. 228]. Collective sensemaking is key to success in coping with turbulence or pandemics.

The ability to adapt and change, which is another important capability of resilience, refers to making "adjustments following crises and is directed toward organizational advancement" [19, p. 230]. Two types of capabilities are involved in adaptation [19]: (1) *reflection* and *learning*, and (2) organizational *change* capabilities. The ability to reflect and learn drives the adaptive capacity of organizations and is a precursor to developing new norms, values, and practices in organizations [19] and to improving future preparedness [8].

Burnard and Bhamra [8] provided the following definition of organizational resilience that encompasses all three stages discussed earlier:

> "*Resilience is the emergent property of organisational systems that relates to the inherent and adaptive qualities and capabilities that enable an organisation's adaptive capacity during turbulent periods. The mechanisms of organisational resilience thereby strive to improve an organisation's situational awareness, reduce organisational vulnerabilities to systemic risk environments and restore efficacy following the events of a disruption.*" (p. 5587)

Citing Norris et al. [56], Burnard and Bhamra [8] also indicated that the resilience of an organizational system is determined by its dynamic capabilities and resources that enable the adaptive capacity of the system. Such capabilities and resources reside in the "individuals, systems, structures, infrastructure, procedures and parameters of the organization" [8, p. 5586]. Hence, the human, social, and psychological capital of an organization and its resources, which include the IT infrastructure and systems, play an important role in building and developing organizational resilience [46]. In the following sections, we discuss the transformation of business strategies enabled by human, social, and psychological capital of an organization, as well as the role that technology plays in business transformation to the new normal.

3 Transformation of Business Strategies to the New Normal

The COVID-19 pandemic is forcing businesses to adapt and rethink their business strategies and operations in order to stay in business. Information technology can predict a pandemic by detecting the first sign of an outbreak and providing warnings to businesses to increase their readiness for a potential pandemic. Given industry trends in data analytics, artificial intelligence (AI), machine learning, and Internet of Things (IoT) [65, 78], anticipating or predicting the occurrence of a pandemic is possible. BlueDot (https://bluedot.global/), which is a Canadian start-up that offers a software-as-a-service health monitoring platform, is among the first in the world to identify the emerging risk of the COVID-19 outbreak in Wuhan, China. It issued warnings on December 31, 2019, before notifications were issued by the US Centers for Disease Control and Prevention and the World Health Organization [55, 91]. Early warnings of potential pandemics could trigger businesses to revisit their business strategies and operations, and set up contingency plans to enhance business continuity. For example, with early warnings of a potential pandemic, simulations of market trends could be generated to predict upward (e.g., spikes) and downward trends in the demand of products and services in order to adjust procurement, production, distribution, sales, and manpower. Trends in online and in-store sales as well as changes in demands for deliveries and courier services could also be projected. Strategies by businesses include switching their focus to essential products (e.g., personal care and household cleaning products such as personal protection items/equipment and disinfecting products as well as dry goods and shelf-stable/frozen foods such as canned foods, rice, pasta, and frozen vegetables) and making adjustments to prolong the shelf life of products due to the potential pandemic and possible lockdowns. Businesses offering in-person services should also take into account travel restrictions and social distancing requirements and adjust to working in a virtual collaborative environment.

In the case of a pandemic, the value chain needs to be revisited and managed even more carefully. A value chain model comprises primary and secondary activities [62]. The primary activities include inbound logistics, operations, outbound logistics, marketing and sales, and service. These activities need to operate without disruptions. The secondary activities that support the primary activities include firm infrastructure, human resource management, technology development (e.g., integrated supply chain management), and procurement. During a pandemic, procurement could become an issue that could adversely affect the primary activities. Having a flexible and integrated supply chain is particularly critical for business continuity as it serves as a backbone for the primary activities. Other advanced information technology such as data analytics, AI, machine learning, robotics, digital commerce, and IoT are also of paramount importance.

Businesses that utilize the blue ocean strategy to innovate and transform their business can gain advantages over the competitors [4]. The blue ocean strategy, in contrast to the red ocean strategy, focuses on creating a new market space by capturing new demands in the market [41]. To utilize the blue ocean strategy, the human, social, and psychological capital in an organization plays a key role in thinking outside the box in order to address and capture the new demands. For example, people are unlikely or less likely to eat out during the pandemic. Due to social distancing restrictions and consumers' desires to minimize trips to grocery stores, the meal kit delivery service has increased in popularity and demand. Digital commerce in the meal kit delivery service industry is growing at a rapid rate. Due to market demands arising from restrictions associated with the pandemic, the revenue for the meal kit delivery service industry is estimated to increase by 20.6% in 2020 alone [34]. Even restaurants are offering meal kits to go [73] and many more are offering free delivery or are partnering with meal delivery services such as UberEats, Grubhub, and DoorDash to sustain their business during the pandemic [67]. Restaurant chains such as Papa John's, Domino's, Pizza Hut, Little Caesars, and Wingstop that offer delivery and takeout are hiring and staffing up as quickly as possible to keep up with the demand [59]. Chain restaurants are even entering the market to sell groceries to stay in business [60] while grocery stores are realizing that their business will be negatively affected without offering online ordering and delivery or pickup service [50]. Some Whole Foods, Kroger, and Giant Eagle grocery stores have even transformed into dark fulfillment centers that fill pickup and delivery orders only [20]. Digital commerce is playing a critical role in the transformation of the food supply chain.

Fitness is a $34 billion industry that has undertaken a radical transformation during the COVID-19 pandemic [21]. Technology-enhanced offerings of in-person fitness through live streaming are rapidly rising in demand [12, 21]. Most of these sessions are also recorded and made available to customers with subscriptions. ClassPass is a subscription service that originally focused on physical access to fitness centers but now includes live-streamed fitness classes and on-demand workouts through their website and mobile app [12]. Customers can join virtual classes hosted by studios across the nation, increasing workout offerings and removing many of the constraints imposed by brick-and-mortar studios by working out from home [12]. Similar to the fitness industry, traditional institutions in the education industry also need to innovate in a transformative way [23, 77]. Although some educational institutions, such as the Missouri University of Science and Technology (formerly University of Missouri–Rolla), or Missouri S&T for short, has a long history of offering in-person classes using the online synchronous delivery mode, it was not until the onset of the COVID-19 pandemic that many other educational institutions followed suit [25]. At Missouri S&T, distance students can reside anywhere in the world but attend and participate in the same classes as on-campus students in traditional classrooms. Healthcare is another area where the use of telemedicine is in an upward trend despite its limitations [35, 72, 76, 83]. Furthermore, the use of robots as assistants is becoming popular in healthcare institutions [1, 89, 96]. Hence, COVID-19 has radically transformed many industries including fitness, education, and healthcare.

4 Role of Technology in Transforming Business to the New Normal

The COVID-19 pandemic has not only greatly accelerated digital transformation, but it has also made businesses realize the importance of utilizing technologies to build resilience. Business resilience is a key determinant of survival in a pandemic. Resilience is determined by the dynamic capabilities and resources that give rise to adaptive capacity [8, 56]. Hence, the adaptive capacity of systems in an organization not only enhances resilience but also serves as a foundation for organizational resilience.

In this section, we examine how information technology can be used to help build resilience in organizations while physical distancing and associated lockdowns have dramatically changed people's behavior. We discuss the role of the following technologies: digital supply chain, data analytics, AI, machine learning, robotics, digital commerce, and IoT.

4.1 Digital Supply Chain

Even though digital supply chains have traditionally been viewed from the efficiency and cost perspectives [79, 84], they now take the center stage in enabling business resilience and transformation. According to the IBM website at https://www.ibm.com/topics/supply-chain-management, supply chain management is the "handling of the entire production flow of a good or service to maximize quality, delivery, the customer experience, and profitability." Supply chain resilience refers to the ability of the supply chains to "incorporate event readiness, provide an efficient and effective response, and be capable of recovering to their original state or even better post the disruptive event" [61, p. 124]. Supply chain resilience is more formally defined as the "adaptive capability of the supply chain to prepare for unexpected events, respond to disruptions, and recover from them by maintaining continuity of operations at the desired level of connectedness and control over structure and function" [61, p. 131]. Not only is adaptive capability critical to resilience, but Ponomarov and Holcomb [61] also highlighted the role of control, coherence, and connectedness in supply chain resilience. Control refers to the ability to direct and regulate activities and actions within the supply chain network. Coherence refers to the enhanced meaning and understanding resulting from the pandemic and putting in place processes and procedures to reduce uncertainty by creating order and structure. Connectedness refers to the systematic coordination across different parties of a network to achieve efficiency and effectiveness. Control, coherence, and connectedness are also characteristics of supply chain resilience.

In times of crisis such as the COVID-19 pandemic, companies recognize the need for a resilient supply chain that can predict changes and trends in the environment (e.g., supply and demand shocks) in order to adapt and respond quickly in managing the high degree of uncertainty. Additionally, customer demands change drastically in a pandemic, further justifying the need for resilient supply chains. The end-to-end visibility of supply chains, including information on tier two and tier three suppliers, is paramount to business resilience.

The supply chain is disrupted during the pandemic. Factories stop working and borders are closed. Cost optimization of the supply chain is not the only concern. Supply chain resilience is key to business survival and continuity. Adaptive supply chains will identify alternative sourcing and shift the focus to high-demand and essential goods. A shift from globalization to localization could be a step forward. Businesses, in the new or next normal, should enhance the resilience of their supply chain by expanding the number of suppliers for each item and giving higher priority to suppliers in proximity to minimize disruptions during a pandemic or crisis.

Price Waterhouse and Coopers & Lybrand (PwC) [63] provides a list of suggestions to businesses on minimizing operations and supply chain disruptions. Some of these suggestions include transporting inventory away from quarantine or high-risk areas, securing capacity and delivery status from tier-2 and tier-3 suppliers, adding overtime assembly capacity where possible, buying ahead to procure inventory and raw material in short supply, activating pre-approved parts or raw material substitutions, informing customers about delays, and adjusting customer allocations to optimize profits or to meet contractual terms [63].

In the context of the US food supply chain, if drastic changes can be predicted in advance (such as a decline or cancellations in orders from full-service restaurants), the supplies could be re-packaged and channeled elsewhere such as to retailers or meal kit delivery services. In the case where supply shortages can be predicted in advance (due to limited availability of workers or the closing of processing plants), customers can be notified in advance to make alternative plans or substitutes which will help minimize customer dissatisfaction. If acute shortages of supplies occur, trade-offs and adjustments in the form of reallocations will need to be made to distribution in such a way that they minimize disruptions to customers/buyers, especially to end customers in vulnerable populations. Adaptive capabilities that manage these uncertainties and risks from farmers to end-customer channels are critical to business continuity and the overall customer experience.

4.2 Data Analytics

Data analytics play a predominant role in all three aspects of business resilience – anticipation, coping, and recovery. Data analytics and business intelligence can enhance an organization's agility [11]. The data used for analysis may originate from any source, including digital supply chains, digital commerce, IoT, and location-based services [78]. Advanced analytics can increase the accuracy of the supply forecast by up to 60% [29].

The ability to utilize data analytics to anticipate and quickly spot new trends, requirements, and demands will provide businesses with opportunities and options. For example, personal protective equipment such as face masks are necessary items during the pandemic, and face masks can be marketed as fashion items. Some apparel companies such as Uniqlo are producing face masks that serve both health and fashion purposes. High-tech companies offer customized printing on these masks and hence, it is possible for the face to look complete in appearance with the mask.

Healthcare is an industry where data analytics can help manage, cope better with, and recover from the pandemic [81]. For example, QuadMed, a health and wellness provider, is collaborating with BSG Analytics, LLC to take a data-driven approach to minimize the impact of COVID-19 by calculating aggregate risks of different segments of its clients (e.g., based on health conditions and age groups) to better allocate resources and programs for its clients [48, 64]. QuadMed implemented a home recovery program that checks in with its clients daily and provides round-the-clock support to assist with recovery without unnecessary hospitalization [64]. Real-time analytics are carried out by healthcare professionals to monitor intensive care capacities, track safety and fatigue of their staff, and optimize the resources available to better manage resources and help make informed decisions quickly [33]. It is projected that analytics will be used increasingly to offer personalized medicine and healthcare service to patients [33].

4.3 Artificial Intelligence, Machine Learning, and Robotics

AI and machine learning are used by BlueDot to identify unusual healthcare trends or patterns such as the possibility of virus outbreaks or a global pandemic [55, 86, 91]. AI and machine learning have achieved success in medical areas, including identifying breast cancer and spotting tumors from x-ray films [85, 98]. One of the organs affected by COVID-19 is the lung. AI and machine learning can be trained to spot COVID-19 from chest x-rays and predict which patients will get more serious complications from COVID-19 as well as detect vital signs associated with COVID-19 [49, 86]. For example, Closedloop has developed a predictive model that identifies people by their

COVID-vulnerability index, called the C-19 index, that is used by healthcare systems and healthcare management organizations [86].

COVID-19 is highly contagious. Reports abound about doctors and nurses infected by COVID-19. Furthermore, a large number of people are being quarantined because of exposure to infected people. AI and machine learning can be used to understand how COVID-19 spreads and to enable organizations to scale and adapt [86]. They can be used to estimate the number of COVID-19 infections that had gone undetected and analyze how the virus mutates [86]. They can also be used to help businesses in the food supply chain industry by assessing satellite images of crops to identify potential issues and problems in order to better manage inventory, supply, and procurement [86].

Robots can be utilized in hospitals to reduce the risks of infection to doctors and nurses [1, 89, 96]. Robots have also been used to deliver meals to people quarantined in hotels to minimize human-to-human contact. Similarly, drones can be used for deliveries of food or essential items, particularly in remote or difficult to access areas or terrains [22]. Self-driving cars that utilize AI and machine learning are also increasingly used for deliveries [31].

In addition, robotics can play an important role in helping to maintain social distancing restrictions at work and in public areas. For example, a British online supermarket, Ocado, has no chain stores and instead, delivers groceries directly to customers' homes from its warehouses [94]. It utilizes robotics extensively in its warehouses, which helps to ease social distancing restrictions. Robots developed by Boston Dynamics are used in public areas in Singapore to remind people about social distancing [42, 95]. The robots are managed by remote operators and are able to sense their surroundings to avoid collisions with people or obstacles.

4.4 Digital Commerce

The world as we know it has come to a stop! Potential infection and social distancing have minimized face-to-face commerce. Restaurants were forced to provide only take-out service to tame the spread and some restaurants are required to stop providing dine-in services because of the resurgence of infections. Many cities are experiencing the second wave and third wave of COVID-19 infections. Businesses that have a digital and web presence are doing better than those that are purely brick-and-mortar. In the new or next normal, businesses will be eager to create and develop their digital commerce capability to enhance their resiliency against future crises. Until an effective vaccine is widely available, digital commerce will be an important element to keep many businesses afloat and the economy moving.

Social distancing and sheltering in place have resulted in a big shift from traditional commerce to digital commerce. People are making purchases online in order to avoid trips to physical stores. Sales on Amazon.com soar during the pandemic even though its costs also increase due to additional expenses incurred to deliver products to customers as quickly as possible while keeping its employees safe [28]. Almost every retailer is leveraging on digital commerce to stay in business during the pandemic. Even the service industry, such as casual dining, fast food, fitness, and car racing, are using digital means to conduct commerce. Casual dining and fast-food chain restaurants, such as Papa John's, Domino's, Wingstop, McDonald's, and Chipotle, are experiencing increased sales through online orders for pickup or delivery [24, 39, 43, 45]. These restaurants are successful because of their digital-enabled off-premise channels. Fitness centers have moved to conducting classes virtually [12]. Formula 1 hosted E-sports - Virtual Grand Prix series in place of its traditional annual series and achieved a record-breaking viewership of 30 million [26]. Hence, digital commerce is soaring during the COVID-19 pandemic. Businesses need to rethink their business models using the blue ocean strategy to create a new market space to gain competitive advantages.

4.5 Internet of Things (IoT)

The IoT is a system of interrelated computing devices, mechanical machines, and digital appliances with the ability to transfer data over a network without requiring human-to-human or human-to-computer interaction. Wearable IoT devices can be utilized to control the spread of COVID-19. For example, smart and low-cost IoT devices can be implemented to track the locations of people in quarantine, to assist in contact tracing, and to enforce social distancing restrictions. South Korea is using IoT for contact tracing. In Singapore, an IoT startup, Nodle.io, launched Coalition, which is a contact tracing application to contain the spread of COVID-19 [36].

The COVID-19 pandemic has revealed not only the shortage of ventilators but also a shortage of trained healthcare professionals to manage the large volume of ventilators in hospitals. Furthermore, frequent bedside visits increase the risk for healthcare professionals. A Singapore company, ABM's Tele-Ventilator, addresses these issues by using IoT to enable healthcare professionals to securely monitor and adjust ventilator settings through their online portal from any location [32, 70].

Ramco Systems (https://www.ramco.com/facial-recognition/) has introduced a touchless facial recognition-based time and attendance system (RamcoGEEK) that includes temperature recording and IoT based sliding doors, turnstiles, and kiosks [70, 99]. The IoT system can allow or restrict access to staff or visitors based on their temperature range. The IoT system can also monitor the movement of staff who has a high temperature within the workplace premises and issue alerts to relevant parties including management and human resources.

Pensees has released its new Intelligent AI-based Non-contact Body Temperature Monitoring System [9, 70] that is appropriate for use at high volume locations. The front-end devices of the system include a face recognition access control system with a body temperature monitoring module to provide fast screening. The system can also issue automatic alerts at public premises such as subways, airports, train stations, bus stations, schools, communities, and organizations. The automatic and non-contact features are advantageous when dealing with a highly contagious virus.

SenseGiz developed an IoT-based People Tracking and Historic Contact Tracing solution for companies to track their employees in real-time. Employees can be notified if someone who came in proximity to them is later diagnosed with a COVID-19 infection [70]. The system can also be used to enforce social distancing restrictions at the workplace.

In short, in order for businesses to adapt to and recover from the COVID-19 pandemic, IoT solutions can be deployed. To return to the new normal, it is important to have such mechanisms in place for business continuity. Furthermore, the interest in creating connected health related ecosystems using IoT is increasing due to the availability of smart health devices and individuals' interest in monitoring and managing their own health [3]. Individuals can be empowered to manage their health more proactively with Internet of Medical Things [3].

5 Issues, Challenges, and Implications

Given the ubiquitous application of data analytics, privacy and quality of data are two emerging issues and challenges, particularly in the healthcare context. A strong data strategy covering collection, assurance, preparation, and use is necessary to ensure high quality of data that can maximize the benefits and impact of healthcare data analytics [33]. Security and privacy of personal data also need to be addressed [10, 11, 27, 75, 80, 85]. The more complete the data, the greater the potential benefits that can be generated. To what degree would individuals be willing to give up some aspects of the privacy of their personal information for the good and benefits of the greater population? To identify and quarantine people who might be infected with COVID-19, contact tracing is necessary. Would people give up privacy for security and safety? This trade-off between individual privacy and community safety poses the greatest challenge.

To identify hotspots of COVID-19 transmission, aggregate and anonymized consumer data have been used [69]. Hence, it is possible to utilize anonymized data to enhance the security and safety of the general public. Hence, Privacy by Design principles [71], which take into account privacy measures from the outset of the design and engineering process, should be used to the greatest extent possible to maximize the anonymity of individuals' data as well as to maximize the safety and security of the community and the general public.

Maintaining the well-being, health, and eudaimonia of individuals during the pandemic is critical given the stress and inconvenience that the pandemic has brought to our lives [90]. Some people are indulged in virtual fitness/workout and online gaming, while others still pursue outdoor activities like camping, hiking, and sports while maintaining social distancing. Wearable devices such as fitbits and smartwatches could be used to track progress in fitness programs and possibly entertain the wearers/owners by cracking or sharing jokes periodically with them to help enhance their well-being. User experience is paramount [5, 18, 38, 68, 88, 101, 102]. In addition to maximizing the functionality and usability of applications, hedonic design elements can also be incorporated to enhance the overall experience of the users [2, 17, 51–54, 93].

6 Conclusions

In times of a pandemic, organizations that have the psychological capital to innovate using the blue ocean strategy and are supported by modern and advanced IT infrastructure and development will have an edge in adapting to and recovering from the pandemic. Resilience is key to business continuity during a pandemic and it can be enhanced by adaptive capabilities, including capabilities enabled by information technology as well as the human and psychological capital of the firm. Human capital in the form of competence, psychological capital factors of hope, optimism, and confidence, and adaptive capabilities enabled by information technology are three factors that will enhance one another to maximize business resilience.

In returning to a new normal, businesses need to build long-term resilience in the following capabilities: anticipation/readiness for preparedness, coping/response for adaptability, and adaptation and recovery for change. Companies want to leverage the power of their supply chain and their network of suppliers to avoid disruptions to their business. They also need to enhance and expand their digital channels such as in the form of digital commerce and/or service offerings through digital means, particularly in mobile commerce [74, 82, 92]. To capitalize on the power of AI and machine learning as well as data analytics, data from a variety of sources will need to be collected, cleansed, and combined. Robots can function as workers or assistants to help maintain social distancing restrictions. Data from IoT and other sources can be helpful for contact tracing, monitoring quarantines, tracking sources of outbreaks, and even reminding offenders about social distancing restrictions. Hence, businesses can leverage the capabilities of technologies to build resilience and enhance business continuity.

References

1. Ackerman, E., Guizzo, E., Shi, F.: Video Friday: how robots are helping to fight the Coronavirus outbreak. IEEE Spectr. 31 January 2020 https://spectrum.ieee.org/automaton/robotics/robotics-hardware/robots-helping-to-fight-coronavirus-outbreak
2. Adapa, A., Nah, F., Hall, R., Siau, K., Smith, S.: Factors influencing the adoption of smart wearable devices. Int. J. Hum. Comput. Interact. **34**, 399–409 (2018)
3. Adarsha, A.S., Reader, K., Erban, S.: User experience, IoMT, and healthcare. AIS Trans. Hum. Comput. Interact. **11**, 264–273 (2019)
4. Baskaran, L.: How successful businesses pivot during COVID19? Medium, 5 April 2020. https://medium.com/swlh/how-successful-businesses-pivot-or-persevere-during-covid19-a68dcee79915
5. Benbunan-Fich, R.: User satisfaction with wearables. AIS Trans. Hum. Comput. Interact. **12**, 1–27 (2020)
6. Bhamra, R., Dani, S., Burnard, K.: Resilience: the concept, a literature review and future directions. Int. J. Prod. Res. **49**, 5375–5393 (2011)
7. Boin, A., van Eeten, M.J.G.: The resilient organization. Pub. Manag. Rev. **15**, 429–445 (2013)
8. Burnard, K., Bhamra, R.: Organisational resilience: development of a conceptual framework for organisational responses. Int. J. Prod. Res. **49**, 5581–5599 (2011)
9. Business Wire: Fight virus with AI! Pensees releases its intelligent non-contact body temperature monitoring system. Yahoo! News, 27 April 2020. https://news.yahoo.com/fight-virus-ai-pensees-releases-042600154.html
10. Campbell, D.E.: A relational build-up model of consumer intention to self-disclose personal information in e-commerce B2C relationships. AIS Trans. Hum. Comput. Interact. **11**, 33–53 (2019)
11. Chen, X., Siau, K.: Business analytics/business intelligence and IT infrastructure: impact on organizational agility. J. Organ. End User Comput. (forthcoming)

12. Chilkoti, A.: As home workouts rise during Coronavirus, gyms sweat. Wall Street J. (2020). https://www.wsj.com/articles/as-home-workouts-rise-during-coronavirus-gyms-sweat-11588784616
13. Cho, M.-H.: South Korea launches mask inventory apps to address shortages from COVID-19. ZDNet, 11 March 2020. https://www.zdnet.com/article/south-korea-launches-mask-inventory-apps-to-address-shortages-from-covid-19/
14. Choudhury, S.R.: Singapore says it will make its contact tracing tech freely available to developers. CNBC Online, 25 March 2020. https://www.cnbc.com/2020/03/25/coronavirus-singapore-to-make-contact-tracing-tech-open-source.html
15. Choudhury, S.R.: How technology helped Asian countries cope in the coronavirus pandemic. CNBC Online, 13 May 2020. https://www.cnbc.com/2020/05/13/coronavirus-mckinsey-says-technology-helping-asia-cope-better.html
16. Chung, T., Liang, T., Peng, C., Chen, D., Sharma, P.: Knowledge creation and organizational performance: moderating and mediating processes from an organizational agility perspective. AIS Trans. Hum. Comput. Interact. **11**, 79–106 (2019)
17. Coursaris, C.K., Van Osch, W.: A cognitive-affective model of perceived user satisfaction (CAMPUS): the complementary effects and interdependence of usability and aesthetics in IS design. Inf. Manage. **53**, 252–264 (2016)
18. Djamasbi, S., Strong, D.: User experience-driven innovation in smart and connected worlds. AIS Trans. Hum. Comput. Interact. **11**, 215–231 (2019)
19. Duchek, S.: Organizational resilience: a capability-based conceptualization. Bus. Res. **13** (1), 215–246 (2019). https://doi.org/10.1007/s40685-019-0085-7
20. Duprey, R.: Amazon is turning some Whole Foods markets into ‘dark stores’. Nasdaq, 16 April 2020. https://www.nasdaq.com/articles/amazon-is-turning-some-whole-foods-markets-into-dark-stores-2020-04-16
21. Eschner, K.: COVID-19 has changed how people exercise, but that doesn’t mean gyms are going away. Fortune, 11 June 2020. https://fortune.com/2020/06/11/coronavirus-gyms-workouts-fitness-apps-reopening/
22. Edwards, D.: FedEx to test Wing Aviation drone for walgreens deliveries. Robotics & Automation, 7 September 2019. https://roboticsandautomationnews.com/2019/09/27/fedex-to-test-wing-aviation-drone-for-walgreens-grocery-deliveries/25970/
23. Erickson, J., Siau, K.: Education. Commun. ACM **46**, 134–140 (2003)
24. Fast Casual: Chipotle beats COVID-19 odds as Q1 digital sales soar. Fast Casual News, 21 April 2020. https://www.fastcasual.com/news/chipotle-beats-covid-19-odds-as-q1-digital-sales-soar/
25. Flaherty, C.: Zoom boom. Insider Higher Ed., 29 April 2020 https://www.insidehighered.com/news/2020/04/29/synchronous-instruction-hot-right-now-it-sustainable
26. Formula 1: Formula 1 Virtual Grand Prix series achieves record-breaking viewership. F1, 19 June 2020. https://www.formula1.com/en/latest/article.formula-1-virtual-grand-prix-series-achieves-record-breaking-viewership.7bv94UJPCtxW0L5mwTxBHk.html
27. Galanxhi, H., Nah, F.: Privacy issues in the era of ubiquitous commerce. Electron. Mark. **16**, 222–232 (2006)
28. Greene, J.: Amazon sales soar as coronavirus-worried consumers shop from home, but costs rise. The Washington Post 30 April 2020. https://www.washingtonpost.com/technology/2020/04/30/amazon-earnings-coronavirus/
29. Guillot, C.: COVID-19 spurs a leap to digital supply chains. Supply Chain Dive (2020). https://www.supplychaindive.com/news/covid-19-digital-transformation-supply-chains/579452/
30. Hamel, G., Valikangas, L.: The quest for resilience. Harvard Bus. Rev. **81**, 52–65 (2003)

31. Hawkins, A.J.: Cruise redeploys some of its self-driving cars to make food deliveries in San Francisco. The Verge, 29 April 2020. https://www.theverge.com/2020/4/29/21241122/cruise-self-driving-car-deliveries-food-banks-sf
32. Hospital and Healthcare Management: ABM respiratory care creates world's first IOT-enabled tele-ventilator for COVID-19 pandemic (2020). https://www.hhmglobal.com/industry-updates/press-releases/abm-respiratory-care-creates-worlds-first-iot-enabled-tele-ventilator-for-covid-19-pandemic
33. Huiskens, J.: What is the future of data and analytics in healthcare after COVID-19? Health IT Outcomes (2020). https://www.healthitoutcomes.com/doc/what-is-the-future-of-data-and-analytics-in-healthcare-after-covid-0001
34. IBIS World: Meal kit delivery services industry in the US – market research report, May 2020. https://www.ibisworld.com/united-states/market-research-reports/meal-kit-delivery-services-industry/
35. La Monica, P.R.: Teladoc soars on bet that virtual health is here to stay. CNN Business, 9 April 2020. https://www.cnn.com/2020/04/09/investing/teladoc-coronavirus-virtual-health/index.html
36. Ledger Insights: IoT blockchain platform Nodle launches Coalition COVID-19 contact tracing app (2020). https://www.ledgerinsights.com/covid-19-contact-tracing-iot-blockchain-nodle/
37. Katerattanakul, P., Siau, K.: Creating a virtual store image. Commun. ACM **46**, 226–232 (2003)
38. Katerattanakul, P., Siau, K.: Factors affecting the quality of personal websites. J. Am. Soc. Inf. Sci. Tech. **59**, 63–76 (2008)
39. Kelso, A.: Why Chipotle will come out of the Coronavirus crisis stronger than ever. Forbes, 23 April 2020. https://www.forbes.com/sites/aliciakelso/2020/04/23/why-chipotle-will-come-out-of-the-coronavirus-crisis-stronger-than-ever/#1dafd18368ca
40. Kendra, J.M., Wachtendorf, T.: Elements of resilience after the world trade center disaster: reconstituting New York City's emergency operations centre. Disasters **27**(1), 37–53 (2003)
41. Kim, W.C., Mauborgne, R.: Blue ocean strategy. Harvard Bus. Rev. **82**, 75–84 (2004)
42. Kooser, A.: Boston Dynamics Spot robot dog reminds park visitors to maintain distance. CNET, 8 May 2020. https://www.cnet.com/news/boston-dynamics-spot-robot-dog-reminds-park-visitors-to-maintain-distance/
43. Kumar, U.S., Russ, H.: Chipotle deliveries, online orders soar due to coronavirus, sending shares higher. Reuters, 21 April 2020. https://www.reuters.com/article/us-chipotle-results/chipotle-deliveries-online-orders-soar-due-to-coronavirus-sending-shares-higher-idUSKCN223348
44. Lengnick-Hall, C.A., Beck, T.E., Lengnick-Hall, M.L.: Developing a capacity for organizational resilience through strategic human resource management. Hum. Resour. Manag. Rev. **21**, 243–255 (2011)
45. Liddle, A.J.: An early look at the impact of coronavirus on restaurant sales. Nation's Restaurant News, 21 April 2020. https://www.nrn.com/fast-casual/early-look-impact-coronavirus-restaurant-sales
46. Luthans, F., Luthans, K.W., Luthans, B.C.: Positive psychological capital: beyond human and social capital. Bus. Horiz. **47**, 45–50 (2004)
47. Luthans, F., Vogelgesang, G.R., Lester, P.B.: Developing the psychological capital of resiliency. Hum. Resour. Develop. Rev. **5**, 25–44 (2006)
48. Matthews, K.: How data analytics are being applied to COVID-19 recovery strategies. Information Age (2020). https://www.information-age.com/how-data-analytics-are-being-applied-covid-19-recovery-strategies-123489564/

49. McDermid, B.: This AI model has predicted which patients will get the sickest from COVID-19. World Economic Forum (2020). https://www.weforum.org/agenda/2020/05/we-designed-an-experimental-ai-tool-to-predict-which-covid-19-patients-are-going-to-get-the-sickest
50. Moyer, J.W, Ruane, M.E.: Demand for online ordering leaves grocery stores scrambling, customers waiting. The Washington Post, 27 March 2020. https://www.washingtonpost.com/local/demand-for-online-ordering-leaves-grocery-stores-scrambling-customers-waiting/2020/03/27/8c246b48-6ed7-11ea-96a0-df4c5d9284af_story.html
51. Nah, F., Eschenbrenner, B., Claybaugh, C., Koob, P.: Gamification of enterprise systems. Systems **7**, 1–21 (2019)
52. Nah, F., Eschenbrenner, B., Zeng, Q., Telaprolu, V., Sepehr, S.: Flow in gaming: literature synthesis and framework development. Int. J. Inf. Syst. Manag. **1**, 83–124 (2014)
53. Nah, F.F.-H., Telaprolu, V.R., Rallapalli, S., Venkata, P.R.: Gamification of education using computer games. In: Yamamoto, S. (ed.) HIMI 2013. LNCS, vol. 8018, pp. 99–107. Springer, Heidelberg (2013). https://doi.org/10.1007/978-3-642-39226-9_12
54. Nah, F.F.-H., Zeng, Q., Telaprolu, V.R., Ayyappa, A.P., Eschenbrenner, B.: Gamification of education: a review of literature. In: Nah, F.F.-H. (ed.) HCIB 2014. LNCS, vol. 8527, pp. 401–409. Springer, Cham (2014). https://doi.org/10.1007/978-3-319-07293-7_39
55. Niller, E.: An AI epidemiologist sent the first warnings of the Wuhan virus. Wired, 25 January 2020. https://www.wired.com/story/ai-epidemiologist-wuhan-public-health-warnings/
56. Norris, F.H., Stevens, S.P., Pfefferbaum, B., Wyche, K.F., Pfefferbaum, R.I.: Community resilience as a metaphor, theory, set of capacities, and strategy for disaster readiness. Am. J. Comm. Psychol. **41**, 127–150 (2008)
57. Oh, L.-B., Teo, H.-H.: The impacts of information technology and managerial proactiveness in building net-enabled organizational resilience. In: Donnellan, B., Larsen, T.J., Levine, L., DeGross, J.I. (eds.) TDIT 2006. IIFIP, vol. 206, pp. 33–50. Springer, Boston, MA (2006). https://doi.org/10.1007/0-387-34410-1_3
58. Ortiz-de-Mandojana, N., Bansal, P.: The long-term benefits of organizational resilience through sustainable business practices. Strat. Manag. J. **37**, 1615–1631 (2016)
59. Patton, L.: Old restaurant chains get second look by quarantined America. Bloomberg, 23 March 2020. https://www.bloomberg.com/news/articles/2020-03-23/old-restaurant-chains-get-a-second-look-from-quarantined-america?srnd=premium
60. Pomranz, M.: Even chain restaurants are selling groceries to stay in business. Food & Wine, 8 April 2020. https://www.foodandwine.com/news/chain-restaurants-selling-groceries-cpk-panera-subway
61. Ponomarov, S.Y., Holcomb, M.C.: Understanding the concept of supply chain resilience. Int. J. Logist. Manage. **20**, 124–143 (2009)
62. Porter, M.E.: Competitive Advantage: Creating and Sustaining Superior Performance. Free Press, New York (1985)
63. PwC: COVID-19: Operations and supply chain disruption (2020). https://www.pwc.com/us/en/library/covid-19/supply-chain.html
64. QuadMed, LLC.: QuadMed leverages data analytics to lessen financial and population health impact during COVID-19 pandemic. Cision PR Newswire (2020). https://www.prnewswire.com/news-releases/quadmed-leverages-data-analytics-to-lessen-financial-and-population-health-impact-during-covid-19-pandemic-301049320.html
65. Ravindran, S., Nah, F.: Prescriptive analytics: a game changer for business. Cutter Bus. Tech. J. **30**, 11–17 (2017)
66. Reinmoeller, P., van Baardwijk, N.: The link between diversity and resilience. MIT Sloan Manage. Rev. **46**, 61–65 (2005)

67. Roberts, C.: DoorDash, Grubhub, Postmates, and Uber Eats: how food delivery services perform. Consumer Reports, 21 May 2020. https://www.consumerreports.org/food-delivery-services/food-delivery-services-apps-review/
68. Rochford, J.: Accessibility and IoT / Smart and Connected Communities. AIS Trans. Hum. Comput. Interact. **11**, 253–263 (2019)
69. Ross, C.: After 9/11, we gave up privacy for security. Will we make the same trade-off after Covid-19? STAT news, 8 April 2020. https://www.statnews.com/2020/04/08/coronavirus-will-we-give-up-privacy-for-security/
70. Roy, D.: Singapore demonstrates innovative IoT uses during COVID pandemic. Geospatial World, 5 June 2020. https://www.geospatialworld.net/blogs/singapore-demonstrates-innovative-iot-uses-during-covid-pandemic/
71. Schaar, P.: Privacy by design. Identity Inf. Soc. **3**, 267–274 (2010)
72. Schwalb, P., Klecun, E.: The role of contradictions and norms in the design and use of telemedicine: healthcare professionals' perspective. AIS Trans. Hum. Comput. Interact. **11**, 117–135 (2019)
73. Shaw, I. These popular St. Louis restaurants offer meal kits to go. St. Louis Magazine, 14 April 2020. https://www.stlmag.com/dining/restaurant-meal-kits-st-louis/
74. Sheng, H., Nah, F., Siau, K.: Strategic implications of mobile technology: a case study using value-focused thinking. J. Strat. Inf. Syst. **14**, 269–290 (2005)
75. Sheng, H., Nah, F., Siau, K.: An experimental study on U-commerce adoption: impact of personalization and privacy concerns. J. Assoc. Inf. Syst. **9**, 344–376 (2008)
76. Siau, K.: Health care informatics. IEEE Trans. Inf. Tech. Biomed. **7**, 1–7 (2003)
77. Siau, K.: Education in the age of artificial intelligence: how will technology shape learning? Global Analyst. **7**, 22–24 (2018)
78. Siau, K., et al.: FinTech empowerment: data science, artificial intelligence, and machine learning. Cutter Bus. Tech. J. **31**, 12–18 (2018)
79. Siau, K., Shen, Z.: Mobile commerce applications in supply chain management. J. Internet Commer. **1**, 3–14 (2002)
80. Siau, K., Shen, Z.: Building customer trust in mobile commerce. Commun. ACM **46**, 91–94 (2003)
81. Siau, K., Shen, Z.: Mobile healthcare informatics. Med. Inf. Internet Med. **3**, 89–99 (2006)
82. Siau, K., Sheng, H., Nah, F.: The value of mobile commerce to customers. In: Proceedings of the Third Annual Annual Workshop on HCI Research in MIS, pp. 65–69 (2004)
83. Siau, K., Southard, P., Hong, S.: e-healthcare strategies and implementation. Int. J. Healthcare Tech. Manage. **4**, 118–131 (2002)
84. Siau, K., Tian, Y.: Supply chains integration: architecture and enabling technologies. J. Comput. Inf. Syst. **44**, 67–72 (2004)
85. Siau, K., Wang, W.: Building trust in artificial intelligence, machine learning, and robotics. Cutter Bus. Tech. J. **31**, 47–53 (2018)
86. Sivasubramanian, S.: How AI and machine learning are helping to figure COVID-19, 28 May 2020. https://www.weforum.org/agenda/2020/05/how-ai-and-machine-learning-are-helping-to-fight-covid-19/
87. Somers, S.: Measuring resilience potential: an adaptive strategy for organizational crisis planning. J. Conting. Crisis Manage. **17**, 12–23 (2009)
88. Stanton, D., Smith, C.: Experience-driven engineering in IoT: the importance of user experience for developing connected products people love. AIS Trans. Hum. Comput. Interact. **11**, 232–243 (2019)
89. Statt, N.: Boston Dynamics' spot robot is helping hospitals remotely treat coronavirus patients. The Verge, 23 April 2020. . https://www.theverge.com/2020/4/23/21231855/boston-dynamics-spot-robot-covid-19-coronavirus-telemedicine

90. Stephanidis, C., et al.: Seven HCI grand challenges. Int. J. Hum. Comput. Interact. **35**, 1229–1269 (2019)
91. Stieg, C.: How this Canadian start-up spotted coronavirus before everyone else knew about it. CNBC, 3 March 2020. https://www.cnbc.com/2020/03/03/bluedot-used-artificial-intelligence-to-predict-coronavirus-spread.html
92. Tarasewich, P., Gong, J., Nah, F., DeWester, D.: Mobile interaction design: integrating individual and organizational perspectives. Inf. Knowl. Syst. Manage. **7**, 121–144 (2008)
93. Treiblmaier, H., Putz, L., Lowry, P.: Research commentary: setting a definition, context, and theory-based research agenda for the gamification of non-gaming applications. AIS Trans. Hum. Comput. Interact. **10**, 129–163 (2018)
94. Vincent, J.: Welcome to the automated warehouse of the future. The Verge, 8 May 2018. https://www.theverge.com/2018/5/8/17331250/automated-warehouses-jobs-ocado-andover-amazon
95. Vincent, J.: Spot the robot is reminding parkgoers in Singapore to keep their distance from one another. The Verge, 8 May 2020. https://www.theverge.com/2020/5/8/21251788/spot-boston-dynamics-robot-singapore-park-social-distancing
96. Vincent, J.: After the pandemic, doctors want their new robot helpers to stay. The Verge, 9 July 2020. https://www.theverge.com/21317055/robot-coronavirus-hospital-pandemic-help-automation
97. Vogus, T.J., Sutcliffe, K.M.: Organizational resilience: towards a theory and research agenda. In: Proceedings of the International Conference on IEEE Systems, Man, and Cybernetics, pp. 3418–3422 (2007)
98. Wang, W., Siau, K.: Artificial intelligence, machine learning, automation, robotics, future of work, and future of humanity – a review and research agenda. J. Database Manage. **30**, 61–79 (2019)
99. Watson, J., Builta, J.: IoT set to play a growing role in COVID-19 response. OMDIA Technology, 1 April 2020. https://technology.informa.com/622426/iot-set-to-play-a-growing-role-in-the-covid-19-response
100. Weick, K.E., Sutcliffe, K.M., Obstfeld, D.: Organizing and the process of sensemaking. Organ. Sci. **16**, 409–421 (2005)
101. Wyatt, J., Piggott, A.: The design of not-so-everyday things: designing for emerging experiences. AIS Trans. Hum. Comput. Interact. **11**, 244–252 (2019)
102. Zhang, X., Venkatesh, V.: From design principles to impacts: a theoretical framework and research agenda. AIS Trans. Hum. Comput. Interact. **10**, 105–128 (2018)
103. Zou, C., Zhao, W., Siau, K.: COVID-19 calls for remote reskilling and retraining. Cutter Bus. Tech. J. **33**, 21–25 (2020)

Numerical Analysis of Bio-signal Using Generative Adversarial Networks

Koki Nakane[1], Hiroki Takada[1(✉)], Shota Yamamoto[1], Rentarou Ono[1], and Masumi Takada[2]

[1] Department of Human and Artificial Intelligence Systems, Graduate School of Engineering, University of Fukui, Fukui 910-8507, Japan
takada@u-fukui.ac.jp
[2] School of Nursing, Yokkaichi Nursing and Medical Care University, Mie 512-8043, Japan

Abstract. In this decade, it is not necessary to have technical knowledge for the investment since the automatic algorithms to sell/buy investment destination have been developed with artificial intelligence (AI). In our previous study, measuring similarity (stationarity, fractality, and degree of determinism) of variations in the exchange rates to the pseudo-exchange rates generated by Generative Adversarial Networks (GANs), we compared Winner processes with the GANs. From the viewpoint of stationarity, the similarity of sequences in the pseudo exchange rates were higher than those generated by the Winner processes, and high scores in the similarity were resulted from both sequences in terms of degree of determinism. In this study, we have applied this AI system to numerical simulations of the stabilogram whose randomness is remarkably greater than that of the other bio-signal in accordance with the nonlinear analysis.

Keywords: GAN · Double-Wayland algorithm · Stochastic process · Stabilogram

1 Introduction

In this decade, it is not necessary to have technical knowledge for the investment since the automatic algorithms to sell/buy investment destination have been developed with artificial intelligence (AI). However, these kinds of mechanical trading systems may not support variations realized in future because the systems were developed with use of time series data of the investment outlets in the past, or time sequences generated by stochastic processes to verify the systems. Therefore, we considered applying the generative adversarial network (GAN), which has attracted attention in the field of image generation, to time series of the exchange rate. Learning the properties of variations in the exchange rate whose factors are not elucidated in detail by GAN, the pseudo exchange rates were generated to use as the data for verification of the mechanical trading system.

In previous study, measuring similarity (stationarity, fractality, and degree of determinism) of variations in the exchange rates to the pseudo-exchange rates

C. Stephanidis et al. (Eds.): HCII 2020, LNCS 12427, pp. 601–613, 2020.
https://doi.org/10.1007/978-3-030-60152-2_44

generated by GANs, we compared Winner processes with the GANs [1]. From the viewpoint of stationarity, the similarity of sequences in the pseudo exchange rates were higher than those generated by the Winner processes, and high scores in the similarity were resulted from both sequences in terms of degree of determinism.

February 2, 2018 marked the day of the largest Dow Jones Industrial Average decline [2]. The exact reasons for this decline remain unknown, but one is thought to be the use of artificial intelligence (AI), or automated trading algorithms, to continuously sell shares (mechanical trading) [3]. In recent years, mechanical trading has not been limited to major hedge funds. For example, many financial institutions (such as securities firms) provide mechanical investment services to general consumers. Mechanical investment offers the advantage of preventing the consumer from having to manage the question of what and when to buy. However, there is the risk that mechanical investment may not be able to predict future fluctuations. In general, automated trading algorithms are tested against past fluctuations to evaluate their effectiveness, but there is the possibility of overfitting if the system solely relies on past fluctuations. Overfitting is a common problem with machine learning. Thus, to prevent this problem, a test may be performed on a large number of time series generated by using a stochastic process. Many researchers have performed studies in which stock prices are considered stochastic processes [4, 5]. However, a time series generated by a stochastic process does not reflect actual stock prices and exchange rate fluctuations, making it unreliable. Therefore, in this study, we used stochastic process-generated time series data.

Current research also includes the use of neural networks to classify images [6]. In 2014, the concept of generative adversarial networks (GAN) was proposed by Goodfellow et al. [7]. Additionally, the amount of research that entails using neural networks to generate images has been increasing. In this study, G is defined as a network that generates simulative sequences (SSs, i.e., fake data) from input noise, and Network D distinguishes whether the data generated by the generator G is the desired real data (i.e., true data). The generator G learns as the discriminator D mistakes the SS with the true data, and D is trained to correctly distinguish between the true and SS. Repeated training of Networks G and D (Fig. 1) can result in the generation of a large amount of SSs if the output of G can generate data that are very similar to the true data.

Therefore, we thought that we could improve the reliability of automated trading algorithms by examining various fluctuation patterns using a GAN-generated pseudo time series.

This study was purposed with designing an artificially intelligent GAN by generating one-dimensional time series data that are similar to the true exchange rate data. We evaluated the pseudo exchange rates, as generated by a model based on stationarity [8], fractality, and the degree of determinism [9], in comparison with the actual exchange rate and time series data generated by a stochastic process.

In this study, we apply this AI system to numerical simulations of bio-signals such as stabilograms.

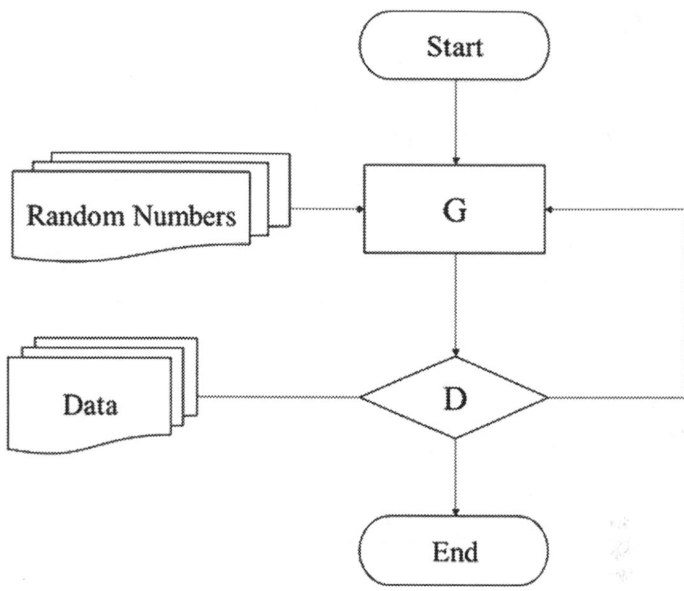

Fig. 1. GAN model.

2 Model Design

GAN learning is reported to be difficult to stabilise. In fact, we designed and trained several models in an effort to generate pseudo exchange rates, but there were few models that exhibited stable learning.

Alec Radford et al. provided some suggestions regarding how to stabilise GAN learning [10]. In this study, we used the hyperbolic tangent function (i.e., tanh) as the activation function in the output layer of the generator. With the exception of the output layer/fully connected layer, LeakyReLU [11] was used to design the discriminators, each of which act as an activation function. To design a model, various parameters need to be set. Moreover, because the accuracy of the generated pseudo exchange rate is dependent on the parameters, it is necessary to optimise various parameter settings. Neural network model optimisation often entails using the accuracy rate as an objective variable for classification and prediction [12]. However, the purpose of this study was to identify the characteristics of exchange rate fluctuation, and generate a pseudo time series that simulates the fluctuations. Therefore, GAN learning cannot be evaluated based on the degree of similarity between the actual exchange rate and generated pseudo exchange rate. In addition, it is difficult to describe the characteristics of exchange rates because the factors and systems of exchange rates are not clearly understood. Therefore, an objective variable must be defined. In this study, we discovered two metrics that would facilitate network training. These metrics are as follows: 1) the total output errors for the generator and discriminator training data must be small, and 2) lower output error values for the generator and discriminator training data indicates stable learning. Thus, we developed and optimised the optimisation function (1), which takes into account these two metrics:

$$OptimisationFunction(G_{Loss}, D_{Loss}) = ln\frac{G_{Loss} + D_{Loss}}{D_{Loss}/G_{Loss}} \tag{1}$$

G_{Loss} is the training error for the generator network, and D_{Loss} is the training error for the discriminator network.

In this study, the number of convolutional layers (i.e., four), number of filters per layer (16, 32, 64, 128), and filter sizes (1–10) were set to minimise the value of the optimisation function for the generator model (Fig. 2). The value was evaluated after the parameters were optimised. It should be noted that, because of the high computational expense, the discriminator model was not optimised (Fig. 3).

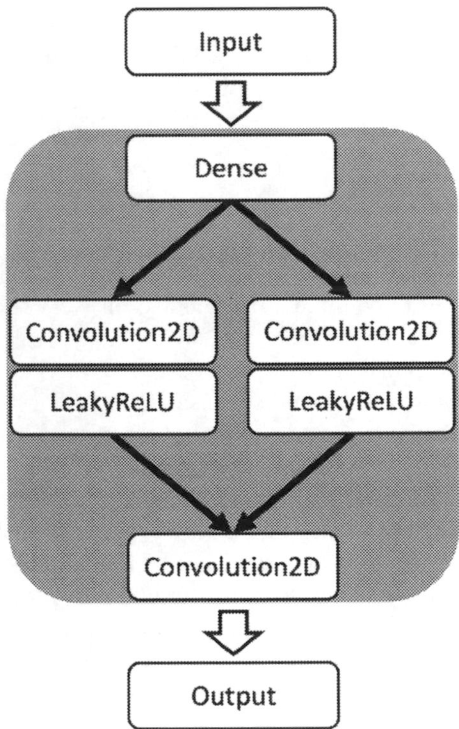

Fig. 2. Generator

Fig. 3. Discriminator

3 Empirical Study in Stabilometry

We studied the effects of stereoscopic video clips on the elderly. In addition to radial motion, body sway was simultaneously measured while the young and the elderly viewed stereoscopic video clips. The results showed that, in the elderly, the equilibrium function is affected by tracking the visual target in stereoscopic video clips.

3.1 Experiment 1

As a basic study, the stabilometry was conducted for 238 elderly people that stood with Romberg posture on a gravicorder GS3000 (Anima Corp. Ltd., Tokyo). Stabilograms were recorded at 20 Hz sampling with their eyes open/closed for 60 s, respectively.

This experiment was approved by the Ethics Committee of Graduate School of Information Science, Nagoya University.

3.2 Experiment 2

In this experiment, the experiment was conducted for twelve healthy volunteers (6 young and 6 elderly) that were 22.5 ± 1.0 yrs. (mean ± standard deviation) and 75.0 ± 8.2 yrs. of age, respectively. Beforehand, the experiment was fully explained to the subjects that could view stereoscopically, and written consent was obtained. The experiment was also approved by the Ethics Committee of the Department of Human and Artificial Intelligent Systems in the Graduate School of the Engineering University of Fukui (No. H2019003). Stabilometry and radial movements were simultaneously measured and recorded at 100 Hz and 60 Hz sampling rates, respectively.

Stereoscopic images used for this experiment was recreated based on Sky Crystal (Olympus Memory Works Corp, Tokyo) with permission (Figs. 4). The 3D video clips:

VC1) A normal 3D video clip with full backgrounds (Fig. 4a)
VC2) A 3D video clip with the static regulation of backgrounds (Fig. 4b)

On a liquid crystal display (LCD), 55UF8500-JB (LG Electronics, Seoul), were played in visual pursuit for 60 s or in the peripheral vision for 60 s in a dark room. In the VC1, the peripheral visual field was compulsory narrowed. An order effect was herein excluded in the protocol of this experiment in which a test with the subjects' eyes closed was conducted after each simultaneous measurement.

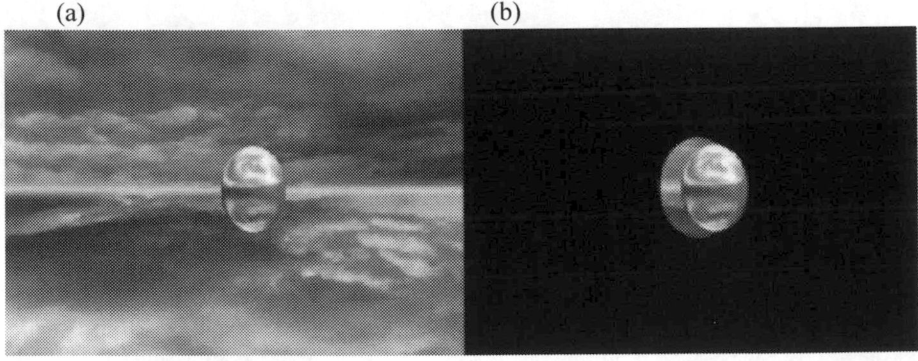

Fig. 4. Visual stimulus; a normal image (a), an image with static regulation of backgrounds (b)

In this experiment, the stabilometry was conducted by using a Wii balance board (Nintendo, Kyoto). Typical example of stabilograms were shown in Fig. 5a.

Also, we used an eye mark recorder, EMR-9 (Nac Image Technology, Tokyo) to measure the radial movement. The position of the viewpoint for each sampling time is composed of *x*-*y* coordinate [pix]. Total locus length, area of radial sway, and total locus length per unit area were evaluated as well as the analysis of the body sway. Also, we performed statistical tests for each analytical index. The significance level was set to be 0.05.

4 Results and Consideration

The elderly voluntary participated in this study. Their stabilograms were recorded while standing with Romberg posture (Fig. 5a). In previous studies, the mathematical models of the body sway have been described by stochastic processes on the basis of the following properties for each component;

(i) Markov property.
(ii) non-anomalous diffusion.

In stabilograms, variables x (right designated as positive) and y (anterior designated as positive) are regarded to be independent [13]. A linear stochastic differential equation (Brownian motion process) has been proposed as a mathematical model to describe body sway [14–16]. To describe the individual body sway, we show that it is necessary to extend the following nonlinear stochastic differential equations:

$$\frac{\partial x}{dt} = -\frac{\partial}{\partial x} U_x(x) + \mu_x w_x(t), \tag{2}$$

$$\frac{\partial y}{dt} = -\frac{\partial}{\partial y} U_y(y) + \mu_y w_y(t), \tag{3}$$

where $w_x(t)$ and $w_y(t)$ express the white noise [17]. The following formulas describes the relationship between the distribution in each direction, $G_x(x)$ and $G_y(y)$, and the temporal averaged potential constituting the stochastic differential equations (SDEs):

$$U_x(x) = -\frac{\mu_x^2}{2} \ln G_x(x) + const., \tag{4}$$

$$U_y(y) = -\frac{\mu_y^2}{2} \ln G_y(y) + const. \tag{5}$$

The variance of stabilograms depends on the temporal averaged potential function (TAPF) with several minimum values when it follows the Markov process (i) without abnormal dispersion (ii). SDEs can represent movements within local stability with a high-frequency component near the minimal potential surface, where a high density at the measurement point is expected. In the numerical analysis of Eqs. (2)–(3), the SDE was rewritten to the difference equation in which the term $w_x(t)$ or $w_y(t)$ was substituted into pseudorandom numbers produced by the white Gaussian noise [17] or the 1/f noise [18].

(a)

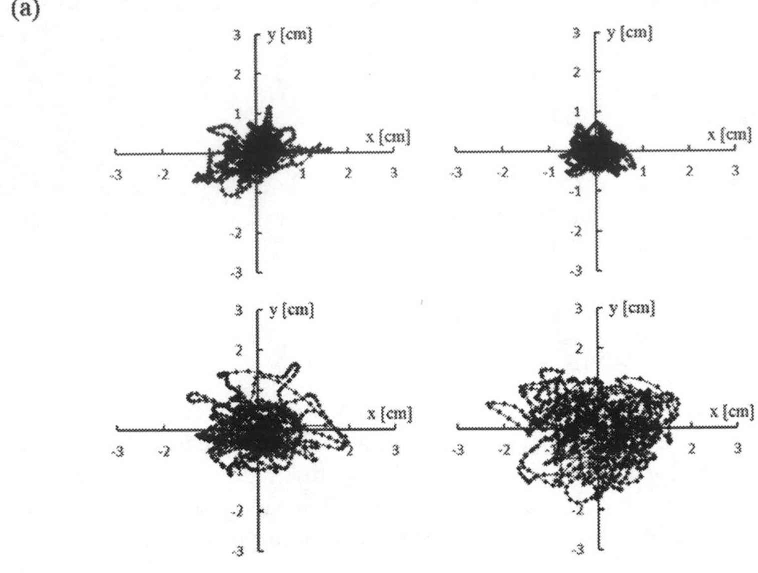

(b)

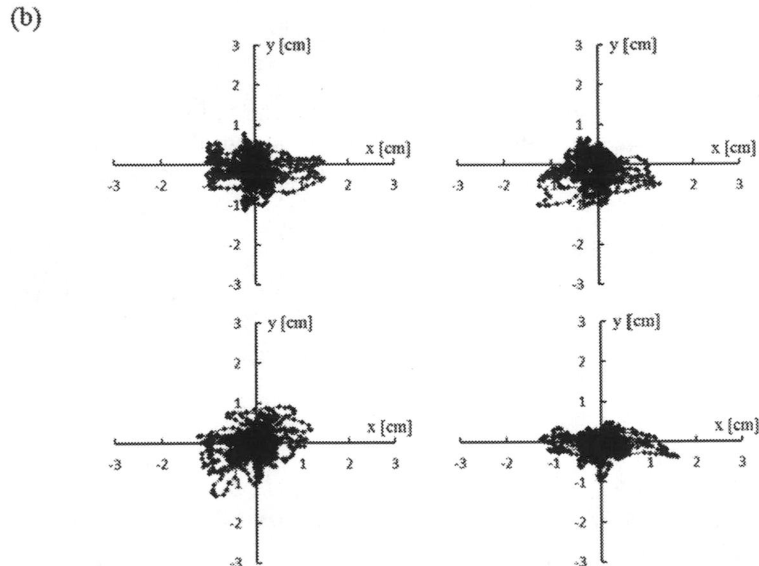

Fig. 5. Typical Stabilograms; data for 1 min with eyes open (a), simulation patterns after 10,000 step (b), simulation patterns after 20,000 step (c), simulation patterns after 30,000 step (d).

(c)

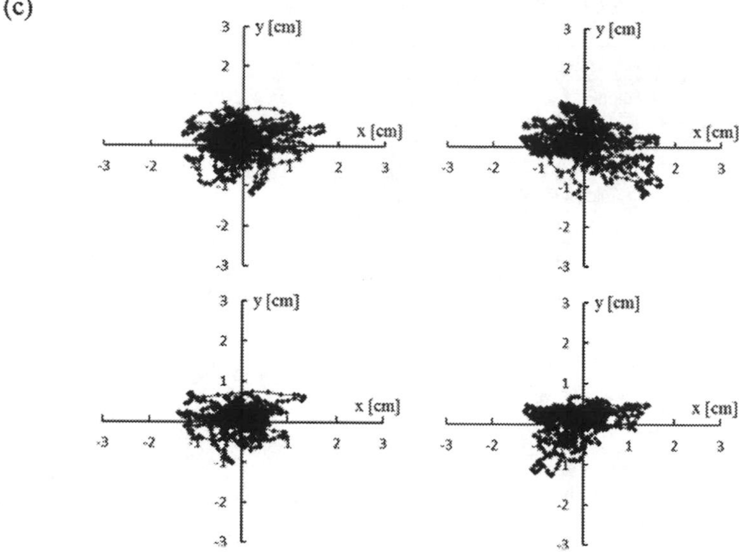

(d)

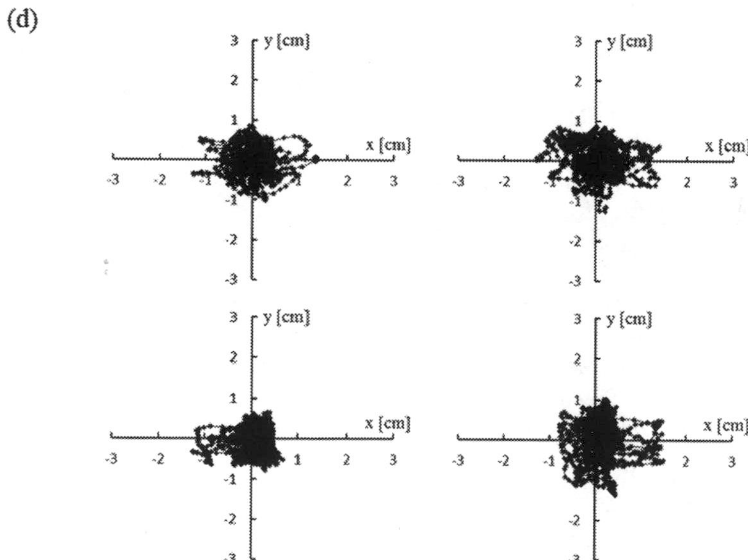

Fig. 5. (*continued*)

Table 1. Optimal parameter of the neural network as a generator of the GANs

Layers	Layer Name	Units	Kernel Size	Filters	Output Shape	Activation
0	Input	-	-	-	-	-
1	Dense	150	-	-	150	LeakyReLU
2-1	Dense	150	-	-	150	LeakyReLU
3-1	BatchNormalization	-	-	-	150	-
4-1	Convolution	-	1×3	128	1×300	LeakyReLU
5-1	BatchNormalization	-	-	-	1×300	-
6-1	Convolution	-	1×3	128	1×300	LeakyReLU
7-1	BatchNormalization	-	-	-	1×300	-
8-1	Convolution	-	1×3	64	1×600	LeakyReLU
9-1	BatchNormalization	-	-	-	1×600	-
10-1	Convolution	-	1×3	64	1×600	LeakyReLU
11-1	BatchNormalization	-	-	-	1×600	-
12-1	Convolution	-	1×3	32	1×1200	LeakyReLU
13-1	BatchNormalization	-	-	-	1×1200	-
14-1	Convolution	-	1×3	32	1×1200	LeakyReLU
15-1	BatchNormalization	-	-	-	1×1200	-
16-1	Convolution	-	1×1	1	1×1200	Tanh
2-2	Dense	150	-	-	1×150	LeakyReLU
3-2	BatchNormalization	-	-	-	1×150	-
4-2	Convolution	-	1×3	128	1×300	LeakyReLU
5-2	BatchNormalization	-	-	-	1×300	-
6-2	Convolution	-	1×3	128	1×300	LeakyReLU
7-2	BatchNormalization	-	-	-	1×300	-
8-2	Convolution	-	1×3	64	1×600	LeakyReLU
9-2	BatchNormalization	-	-	-	1×600	-
10-2	Convolution	-	1×3	64	1×600	LeakyReLU
11-2	BatchNormalization	-	-	-	1×600	-
12-2	Convolution	-	1×3	32	1×1200	LeakyReLU
13-2	BatchNormalization	-	-	-	1×1200	-
14-2	Convolution	-	1×3	32	1×1200	LeakyReLU
15-2	BatchNormalization	-	-	-	1×1200	-
16-2	Convolution	-	1×1	1	1×1200	Tanh
17	Concatenate	-	-	-	2×1200	-

In the experiment 1, we have succeeded in findings of the mathematical models of the body sway in the elderly with use of the GANs. It was confirmed that training was not stable in the created GAN-model because of the small number of real data. Therefore, two-dimensional noise was firstly generated by the independent Winer processes; 1,000,000 kinds of time sequences were provided for each component by the

Table 2. Optimal parameter of the neural network as a discriminator of the GANs

Layers	Layer Name	Units	Kernel Size	Filters	Output Shape	Activation
0	Input	-	-	-	-	-
1	Convolution	-	1×3	32	2×600	LeakyReLU
2	Convolution	-	1×3	64	2×300	LeakyReLU
3	Convolution	-	1×3	128	2×150	LeakyReLU
4	Convolution	-	1×3	256	2×75	LeakyReLU
5	Convolution	-	1×3	512	2×38	LeakyReLU
6	Convolution	-	1×3	32	1×38	LeakyReLU
7	Flatten	-	-	-	1216	-
8	Dense	32	-	-	32	LeakyReLU
9	Dense	1	-	-	1	Sigmoid

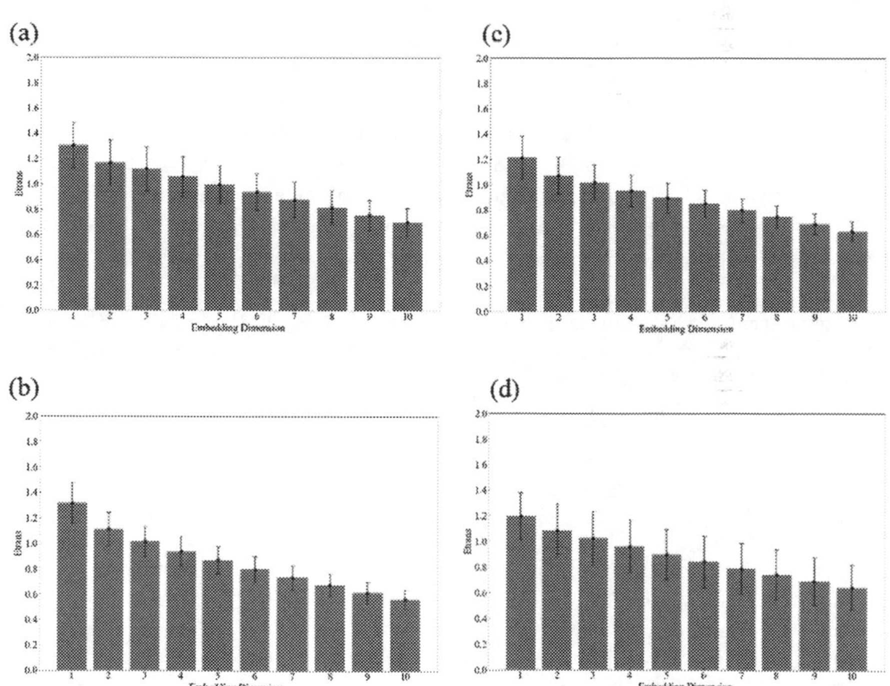

Fig. 6. Translation error estimated from the following *x*-component in each embedding space; measured stabilograms for 1 min (a), simulation patterns after 20,000 step (b), simulation patterns after 70,000 step (c), simulation patterns after 100,000 step (d).

Wiener processes. Substituting the noise into the generator G, fine-training has been secondly conducted for the optimal parameter of the neural network as a generator/discriminator of the GANs (Table 1 and 2). After the fine-training, training was performed using 204 stabilograms measured in the experiment 1–2.

Using the stabilograms stated in the last section, machine learning was thirdly conducted to generate simulative stabilograms by the generator G and to distinguish these fake data with measured stabilograms. Lastly, all stabilograms were evaluated by the translation error estimated in accordance with the Wayland algorithm [19].

According to the Wayland algorithm, values of the translation error were distributed around 0.7, which was resulted from both components of measured stabilograms (Fig. 5a). As shown in Figs. 6, these values were greater than those estimated from x-component sequences in the simulative stabilograms after iterations (<20,000 steps) on the deep-learning of the GANs (Fig. 5b–c). The values of simulative translation error were distributed around 0.7 after 60,000–70,000 iterations, which did not depend on the component. However, we could find decrease in the values of simulative translation error with increase of iterations (>70,000 steps).

In the experiment 2, we recorded the radial motion while viewing stereoscopic video clip. The radial motion of the elderly was quantitatively different from that of the young. We herein applied this AI system to numerical simulations of stabilograms. In the next step, this AI system can be applied to numerical simulations of figure patterns of the radial motion in order to evaluate the anomalous radial motion due to the deterioration of visual function with aging. Also, the AI system can be applied to numerical simulations of bio-signals such as time sequences measured by 3D motion capture, electrocardiograms (ECGs), and electrogastrograms (EGGs). This study shows an example of the application.

Acknowledgement. This work was supported in part by TAKEUCHI Scholarship Foundation, the Japan Society for the Promotion of Science, Grant-in-Aid for Research Activity Start-up Number 15H06711, Grant-in-Aid for Young Scientists (B) Number 16K16105, and that for Scientific Research (C) Number 17K00715.

References

1. Nakane, K., Takada, H., Hirata, T.: On Generation of Pseudo Exchange Rate Using GAN. IEICE Technical report, NLP2019-13 (2019-05). 71–76 (2019)
2. Wang, L., Wittenstein, J.: Dow Jones Industrial Average, worst decline in history - drops nearly $1600, Bloomberg (2018). https://www.bloomberg.co.jp/news/articles/2018-02-06/P3PHW06TTDS601
3. Ponczek, S., Popina, E., Wang, L.: Dow average, who dropped-machine criminal theory emerging, Bloomberg (2018). https://www.bloomberg.co.jp/news/articles/2018-02-06/P3PJGE6JIJVN01
4. Black, F., Scholes, M.: Pricing of options and corporate liabilities. J. Polit. Econ. **81**(3), 637–654 (1973)
5. Tasaki, T.: Motion equation of for stock prices. Bussei Kenkyu **81**(4), 518–519 (2004). (in Japanese)

6. Wakabayashi, K., Namatame, T.: Examination of applicability of deep learning to consumers' purchasing behavior. J Jpn. Soc. Data Sci. Soc. **1**(1), 48–57 (2017). (in Japanese)
7. Goodfellow, I.J., et al.: Generative adversarial Nets, pp. 1–9 (2014)
8. Matsuba, I.: Statistics of Long-Term Memory Processes: Self-Similar Time Series Theory and Method. Kyoritsu Shuppan, Tokyo, pp. 144–146 (2007). (in Japanese)
9. Aihara, K., Ikeguchi, T., Yamada, T., Komuro, M.: Basics and Applications of Chaos Time Series Analysis. Sangyo Tosho, Tokyo, p. 13 (2000). (in Japanese)
10. Radford, A., Metz, L.: Unsupervised representation learning with deep convolutional generative adversarial networks. In: Proceedings of the ICLR, pp. 1–16 (2016)
11. Maas, A.L., Hannun, A.Y., Ng, Y.A.Y.: Rectifier nonlinearities improve neural network acoustic models. In: Proceedings of the ICML, vol. 30, no. 1, p. 3 (2013)
12. Snoek, J., Larochelle, H., Adams, R.P.: Practical Bayesian optimization of machine learning algorithms. In: Proceedings of the NIPS, pp. 1–9 (2012)
13. Goldie, P.A., Bach, T.M., Evans, O.M.: Force platform measures for evaluating postural control: reliability and validity. Arch. Phys. Med. Rehabil. **70**, 510–517 (1986)
14. Emmerrik, R.E.A., Sprague, R.L.V., Newell, K.M.: Assessment of sway dynamics in tardive dyskinesia and developmental disability, sway profile orientation and stereotypy. Mov. Disord. **8**, 305–314 (1993)
15. Collins, J.J., Luca, C.J.D.: Open loop and closed-loop control of posture: a random-walk analysis of center of pressure trajectories. Exp. Brain Res. **95**, 308–318 (1993)
16. Newell, K.M., Slobounov, S.M., Slobounova, E.S., Molenaar, P.C.: Stochastic processes in postural center of pressure profiles. Exp. Brain Res. **113**, 158–164 (1997)
17. Takada, H., Miyao, M., Fateh, S. (eds.): Stereopsis and Hygiene. Springer, Singapore (2019). https://doi.org/10.1007/978-981-13-1601-2
18. Takada, H., Yokoyama, K. (eds.): Bio-Information for Hygiene. Springer, Singapore (2021, in Press)
19. Wayland, R., Bromley, D., Pickett, D., Passamante, A.: Recognizing determinism in a time series. Phys. Rev. Lett. **70**, 580–582 (1993)

The Effect of Mobile App Design Features on Student Buying Behavior for Online Food Ordering and Delivery

Narayan Prabhu[1,2](✉) and Vishal Soodan[3]

[1] Welcomgroup Graduate School of Hotel Administration, Manipal Academy of Higher Education, Manipal, India
narayan.prabhu@manipal.edu
[2] Lovely Professional University (LPU), Phagwara, Punjab, India
[3] Mittal School of Business, Lovely Professional University (LPU), Phagwara, Punjab, India

Abstract. The smartphone penetration in India is growing rapidly and online buying through mobile apps for products and services is growing at a faster pace. Restaurants in brick and mortar format in order to increase their customer base are trying to get their feet wet in an online mode. Though India has attracted food tech companies due to majority of population being millennial consumers, not many studies have focused on consumer behavior towards online food delivery and ordering through mobile apps especially pertaining to students' market. Hence this study in an initiative to study the effect of mobile app design features on decision making stages of consumers pertaining for online food ordering and delivery sector. The study adopted SEM analysis through Smart PLS-3 with sample being students studying in Manipal Academy of Higher Education campus from across the globe. The result of study reveals that design features of mobile apps has huge impact on the students' behavior design, with features, namely, "Security towards the retention of personal information of students", "Secure and accessible payment gateways", "Option of tracking real time order status of food and beverages ordered", "Access to customer service to respond to customers query without any hindrances", "Pop up of trending information of food and beverages", "Easy navigation of app features", "Search function to quickly access the information", and "Delivery services which are flexible". This study could be used as a base by organizations to enhance the design features of mobile apps in food ordering sector for engaging customers and enhancing their business.

Keywords: Mobile app design · Online food ordering · Students' behavior · India

1 Introduction

Smartphones have become an almost essential gadget in recent years providing consumers with an entertainment as well as yield through different mobile applications. Organizations have apprehended that consumers use a variety of application

C. Stephanidis et al. (Eds.): HCII 2020, LNCS 12427, pp. 614–623, 2020.
https://doi.org/10.1007/978-3-030-60152-2_45

(app) features to perform different tasks such as sharing, browsing, searching and purchasing, which in turn enrich brand image, awareness and experience, ensuing in higher sales [24]. According to eMarketer, global digital user update, the internet onlookers in 2019 grew at rapid pace in Asia-Pacific (5.8%), Middle East and Africa (5.3%) and also among individual countries, such as India (9.1%), China (5.0%), and Indonesia (9.4%). Currently with boom in m-commerce in India, development of food and beverage startups has been catching the limelight with growing demand for food ordering and delivery apps. In India, revenue from the food ordering and delivery segment is currently $7,092m and largest market segment is restaurant to consumer delivery market with business volume of $5,825m (Statista 2019). Restaurants in India have off late tied up with several food ordering and delivery apps as an option for market development and to retain frequent and loyal customer [20]. However, regardless of the enhanced interest in food ordering and delivery apps and potential for market and customers in India, there are few research studies pertaining to mobile apps for online food ordering and delivery. Though most of earlier studies have focused on acceptance and adoption of technology by consumers, there is dearth of research focusing on consumers' behavior towards design features especially mobile apps pertaining to online food ordering and delivery, Therefore this study is an initiative to study the effect of mobile design features pertaining to online food ordering and delivery platforms with decision making stages of students.

2 Literature Review

A vast majority of studies have inspected the acceptance and adoption of technology by consumers, such as the technology acceptance model (TAM) [5], theory of planned behavior (TPB) [1], and unified theory of acceptance and use of technology (UTAUT) [17]. All these research studies basically were drawn from domain of software usability, website usability [16]. Further there were studies in which it is empirically proved that product attributes in technology domain will enhance, product performance (navigation, ease of use, usefulness, innovation in technology), product appearance (visual and design) and product communication (self-expression, information), [10]. Further studies also revealed that mobile app interaction enhances consumer experience and brand value [2, 3, 12, 22]. [9] recognized that mobile user attractiveness features such as navigation and visual and graphic designs are important attributes which helps consumers to prefer and use particular mobile app. Although mobile food delivery apps are technology system adopted by the restaurant sector all over the world, scholars interest in understanding and addressing issues related to these apps are in nascent stage [13, 21, 23]. Few of the studies conducted in mobile food delivery apps have focused on motivation towards customer adoption [14], customer willingness to use mobile food delivery apps [13], customer's continued intention to use mobile food ordering apps [22], attitude of customer towards apps [4], and [8] focused on influence of mobile food delivery apps on online food ordering and delivery aggregators on collaboration and conversion. Though all these studies have focused on adoption, till date no study has explored on design attributes of mobile delivery apps and its influence on decision stages of consumer, hence this study is an initiative to understand the effect of mobile food delivery apps design features on buying behavior with specific focus on the students studying in the university campus.

3 Research Design

3.1 Design Attributes and Consumer Purchase Decision

Prior studies in area of web technology have discovered that design quality and features provides an adequate systematic support to consumers' during all the stages of consumer purchase decision. Also design attributes do enhance the shopping and purchase behavior in an online business or e commerce [18], and merchandising and showcasing number of visitors to your website as well as quick reviews will enhance awareness and business [11]. Further, providing consumers with suitable search provisions, specific information about product and services and state of art navigation features will enhance search and information capability of consumers [6, 20]. Features such as price comparison, access to consumer comments, buying history will help evaluation process of consumers [11]. Also delivery process, order que, payment options and gateway helps purchase decision [9]. Flexible delivery service, tracking of order, convenience to return back products and referral bonus with help post purchase [19].

Thus following hypothesis are being proposed for this study

H1. Design features will effect students' product awareness.
H2. Design features will effect students' information search.
H3. Design features will effect students' evaluation.
H4. Design features will effect students' purchase.
H5. Design features will effect students' post-purchase.

3.2 Research Model

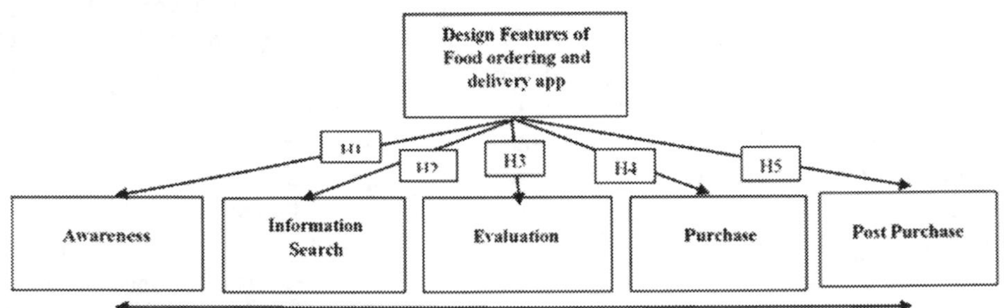

4 Methodology

4.1 Survey Instrument

A multi-stage approach was used to develop a robust survey instrument for this study. First stage involved development of structured questionnaire through systematic literature review. After shortlisting, following constructs and number of items were finalized, design features with fifteen items, awareness with five items, information Search with nine items, evaluation with six items, purchase with fourteen items and post purchase with six items. Second phase involved pretesting of survey instrument for content validity with experts from industry and academia. A total of seven experts three form academia with more than 15 years of experience and three from industry with more than 10 years of experience with food and beverage sector provided their feedback on questionnaire. The experts were requested to rate each of the items on importance level of 1 to 5, with 1 being least important and 5 being most important. Suitable modifications were incorporated as per the suggestions from the experts in the items of each construct. After shortlisting constructs and number of items through literature review, survey instrument consisted of design features with eleven items, awareness with five items, information search with seven items, evaluation with five items, purchase with twelve items and post purchase with four items. The instrument was then tested for content, item and scale validity with experts.

4.2 Sampling and Data Collection

The sampling unit for this study were students who are studying in Manipal Academy of Higher Education a leading private university with institute of eminence status with 33, 133 students in campus from 67 countries around the world. The campus has access to nearly 100 quick service restaurants with noted brands such as McDonald's, Starbucks, Pizza Hut, Dominos, Baskin Robins, Lavazza etc. having visibility in the leading food ordering and delivery mobile apps such as Zomato and Swiggy. The survey used purposive sampling method with respondents chosen who had the smartphone of and had used online food ordering and delivery apps at least once for ordering food through branded food apps like Zomato and Swiggy which are two major players in Manipal. The students were reached through hostel mess and food courts situated in the campus with data collected from time frame of two months from January to February, 2020. A total of 550 students had filled in and submitted the questionnaire during this tenure and after that the questionnaire which had errors and wrongly filled in were not taken into consideration for analysis. A total of 435 filled in questionnaire were considered for study out of which received from which 260 which justified the criteria set of study were considered for further analysis.

4.3 Data Analysis

The data which was collected from students through the questionnaire were fed into IBM SPSS V. 21 statistical software to calculate descriptive statistics and this data was imported into Smart PLS-3 for SEM (Structural equation modeling) and other analysis.

To test the projected hypotheses, SEM was adopted due to its ability in testing the theoretically supported and colorant causal models [7, 15]. The PLS-3 was opted over other methods available for analysis as PLS-SEM focuses more on the prediction and estimation, which is useful in research analysis for maximizing the described variance of the independent variables on the dependent variables [6, 15]. For this study, the value of significance (P) was set to less than 0.05 for convenience ($P < 0.05$). Design features were considered as dependent variable and consumer decision making stages were considered as an independent variable. Analysis of dependent variables against the independent variables was being processed through one-sample t-test.

5 Result Analysis and Discussion

5.1 Descriptive Statistics

The data was gathered from a total of 260 respondents i.e. students studying in the campus of Manipal Academy of Higher Education, Manipal. Male respondents were on higher number and the students from age group of 18–21 years (61%) were the majority as in a university set up the undergraduate students generally will be more in number compared to post graduates. Brunch and Lunch were mostly ordered time through food ordering and delivery apps as most the restaurants will be busy during this time. Further amount spend per order fell mostly in below ₹350 to ₹350 to ₹750 category (Table 1).

Table 1. Descriptive statistics

		Number of respondents	Percentage (%)
Gender	Male	156	60
	Female	104	40
Age (Yrs.)	18–21	159	61
	21–24	57	22
	24–27	32	12
	27 above	12	5
Time of ordering food	Breakfast	28	11
	Brunch	67	26
	Lunch	91	35
	Dinner	74	28
Amount spent per order (₹)	Less than 350	86	33
	350–750	78	30
	750–1500	66	25
	Above 1500	30	12

Table 2. Reliability of the constructs

Construct	No. of items	Cronbach's Alpha (α)	Composite Reliability (CR)	Average variance extracted (AVE)
Design features	13	0.816	0.857	0.532
Awareness	5	0.739	0.825	0.565
Information	7	0.832	0.874	0.500
Evaluation of alternatives	5	0.757	0.835	0.508
Purchase decision	12	0.896	0.914	0.575
Post purchase	4	0.909	0936	0.786

Table 3. Test on Discriminant Validity - Fornell-Larcker criterion

Constructs	DF	A	I	E	P	PPE
Design Features - DF	**0.639**					
Awareness - A	0.456	**0.631**				
Information - I	0.606	0.546	**0.712**			
Evaluation of alternatives – E	0.543	0.599	0.604	**0.730**		
Purchase decision - P	0.632	0.347	0.542	0.500	**0.877**	
Post purchase - PPE	0.545	0.601	0.675	0.707	0.566	**0.709**

5.2 Convergent Validity Test and Discriminant Validity Test

Before running PLS analysis through Smart PLS-3 for SEM and one sample t-test, the constructs of the study were assessed for the convergent validity test and discriminant validity test. Convergent validity is carried out to understand the relationship among items of the same construct. This is been done by accessing values of Cronbach's Alpha (α), Composite reliability (CR.) and average variance extracted (AVE.) of constructs. From Table 2, we can observe that values of Cronbach's Alpha (α), CR and AVE are all greater than 0.7, 0.8 and 0.5, which suggest that convergent validity test for constructs is reliable and accepted. Discriminant validity test were conducted by analysing correlations between the constructs. From Table 3 we can observe that, the square root of average variance extracted (AVE) for each of the construct of study is greater than its correlation value of all other constructs, thus we can conclude that discriminant validity test is accepted. Further Table 4 displays the correlation between the constructs. Strongest correlation were found between awareness and design feature (0.832) and lowest was found with Purchase and design feature (0.716) which specifies that design feature are very important to drive consumers into the app which will help engagement and exploring products and services being promoted by the organisation.

Table 4. Correlation between the constructs

Constructs	DF	A	IS	E	P	PPE
Design Features - DF	**1.000**					
Awareness - A	0.832	**1.000**				
Information - I	0.756	0.854	**1.000**			
Evaluation of alternatives -E	0.734	0.768	0.788	**1.000**		
Purchase decision - P	0.716	0.823	0.805	0.773	**1.000**	
Post purchase - PPE	0.788	0.734	0.771	0.845	0.855	**1.000**

5.3 Structural Equation Model (SEM)

Table 5 displays SEM analysis through Smart PLS 3. The analysis displays a strong relationship between the mobile app design features and purchase decision making stages of consumer. All the paths featured in Table 5 have shown that they are significant ($p < 0.05$), which proves that the research model and constructs are reliable. The path displays that the effects of mobile app design feature is strongly supportive on decision making stages of consumer under five percent significance level ($p < 0.05$). The result shows that there is significant correlations between design features of online food ordering delivery mobile apps and consumer purchase decision making stages ($P < 0.01$). These results authorize a strong relationship between the mobile app design features and the decision making stages of consumer. Also to quantify the effect of design feature of mobile app with every stage of purchase decision making, t test results with p value shows that "information search stage" has been more effected followed by might be due to the fact that most of online food ordering platform through their mobile app make sure that the students receive updates and detailed information into order to make sure the order is not missed out. The next stage which is effected is awareness and then subsequently by other stages which proves that mobile design features effects consumer buying during initial stages as they will be exploring the restaurant, the food and beverages menu, brand name and then will be interested in offers, payment gateway, timeliness of order delivery. Also focusing on design features, students have perceived following features as most important namely, "security towards the retention of personal information of students", "Secure and accessible payment gateways", "Option of tracking real time order status of food and beverages ordered", "Access to customer service to respond to customers query without any hindrances", "Pop up of trending information of food and beverages", "Easy navigation of app features", "Search function to quickly access the information", and "Delivery services which are flexible".

Table 5. SEM of hypotheses

	Original sample (O)	Sample Mean (M)	Standard Deviation (STDEV)	T statistics (O/STDEV)	P values
H1: Design features → Awareness	0.5768	0.6243	0.0673	8.57057	0.0002
H2: Design features → Information	0.5987	0.6104	0.0588	10.1835	0.0000
H3: Design features → Evaluation of alternatives	0.5463	0.5594	0.0893	6.1161	0.0000
H4: Design features → Purchase decision	0.6008	0.6168	0.1004	5.9870	0.0000
H5: Design features → Post purchase	0.3467	0.3642	0.0941	3.6854	0.0003

6 Conclusion

This study which is designed to study the effect of mobile app design features on consumer behavior especially of students studying in University campus contributes to existing literature by exploring the design features of mobile app that needs to consider as trending and important by food tech companies in order to build in more customers as well as maintain relationship. The reason of initiating this study is considering the fact that there is lack of empirical evidence pertaining to link connecting design features of mobile app of food ordering and delivery platforms and decision making stages of consumers especially of students. The outcomes of this study gained through SEM projects that, following features are termed important by student consumers namely, "security towards the retention of personal information of students", which is basically to make sure that students data is not been shared with other organization for marketing purpose, "Secure and accessible payment gateways", to make sure that payment gateway is safe and hackers do not take advantage of the same, "Option of tracking real time order status of food and beverages ordered", which is required to make sure the order is traceable and is been provided to right customer, "Access to customer service to respond to customers query without any hindrances", in Case the order need to be cancelled or there is issue with delivery of food and beverages or with wrong order been delivered and add additional items to order, "Pop up of trending information of food and beverages", for future orders or for referral and awareness, "Easy navigation of app features", for amateurs or students who are not much tech savvy and also for ones who do not have patience and need an easy process while ordering food, "Search function to quickly access the information", and "Delivery services which are flexible" to make sure that customization and ease of order is achieved. Thus having mentioned the these important features and aspects, the study concludes that, though food ordering and delivery market in India is in the nascent stage, this will help organization to understand importance of mobile app design features and how it can be used to enhance customer engagement and sales and also to build brand reputation for repeat sales.

References

1. Ajzen, I.: The theory of planned behavior. Organ. Behav. Hum. Decis. Process. **50**(2), 179–211 (1991)
2. Bellman, S., Potter, R.F., Treleaven-Hassard, S., Robinson, J.A., Varan, D.: The effectiveness of branded mobile phone apps. J. Interact. Market. **25**(4), 191–200 (2011)
3. Calder, B.J., Malthouse, E.C.: Media engagement and advertising effectiveness. In: Kellogg on Advertising and Media, pp. 1–36 (2008)
4. Cho, M., Bonn, M.A., Li, J.J.: Differences in perceptions about food delivery apps between single-person and multi-person households. Int. J. Hosp. Manage. **77**, 108–116 (2019)
5. Davis, F.D.: Perceived usefulness, perceived ease of use, and user acceptance of information technology. MIS Q. **13**, 319–340 (1989)
6. Gan, C.L., Balakrishnan, V.: Mobile technology in the classroom: what drives student-lecturer interactions? Int. J. Hum. Comput. Interact. **34**(7), 666–679 (2018)
7. Haenlein, M., Kaplan, A.M.: A beginner's guide to partial least squares analysis. Underst. Stat. **3**(4), 283–297 (2004)
8. Kapoor, A.P., Vij, M.: Technology at the dinner table: ordering food online through mobile apps. J. Retail. Consum. Serv. **43**, 342–351 (2018)
9. Kim, H.W., Gupta, S., Koh, J.: Investigating the intention to purchase digital items in social networking communities: a customer value perspective. Inf. Manag. **48**(6), 228–234 (2011)
10. Lee, S., Ha, S., Widdows, R.: Consumer responses to high-technology products: Product attributes, cognition, and emotions. J. Bus. Res. **64**(11), 1195–1200 (2011)
11. Liang, T.P., Lai, H.J.: Effect of store design on consumer purchases: an empirical study of on-line bookstores. Inf. Manag. **39**(6), 431–444 (2002)
12. Mollen, A., Wilson, H.: Engagement, telepresence and interactivity in online consumer experience: reconciling scholastic and managerial perspectives. J. Bus. Res. **63**(9–10), 919–925 (2010)
13. Okumus, B., Bilgihan, A.: Proposing a model to test smartphone users' intention to use smart applications when ordering food in restaurants. J. Hosp. Tour. Tech. **5**(1), 31–49 (2014)
14. Pigatto, G., Machado, J.G.d.C.F., Negreti, A.d.S., Machado, L.M.: Have you chosen your request? Analysis of online food delivery companies in Brazil. Br. Food J. **119**(3), 639–657 (2017)
15. Ramayah, T., Cheah, J., Chuah, F., Ting, H., Memon, M.A.: Partial Least Squares Structural Equation Modeling (PLS-SEM) Using SmartPLS 3.0: An Updated and Practical Guide to Statistical Analysis, 2nd edn. Kuala Lumpur, Pearson (2018)
16. Venkatesh, V., Ramesh, V.: Web and wireless site usability: Understanding differences and modeling use. MIS Q. **30**, 181–206 (2006)
17. Venkatesh, V., Morris, M.G., Davis, G.B., Davis, F.D.: User acceptance of information technology: toward a unified view. MIS Q. **27**, 425–478 (2003)
18. Verhagen, T., van Dolen, W.: The influence of online store beliefs on consumer online impulse buying: a model and empirical application. Inf. Manag. **48**(8), 320–327 (2011)
19. Vila, N., Kuster, I.: Consumer feelings and behaviours towards well designed websites. Inf. Manag. **48**(4–5), 166–177 (2011)

20. Wang, R.J.H., Malthouse, E.C., Krishnamurthi, L.: On the go: how mobile shopping affects customer purchase behavior. J. Retail. **91**(2), 217–234 (2015)
21. Wang, Y.S., Tseng, T.H., Wang, W.T., Shih, Y.W., Chan, P.Y.: Developing and validating a mobile catering app success model. Int. J. Hosp. Manag. **77**, 19–30 (2019)
22. Yadav, M.S., Varadarajan, R.: Interactivity in the electronic marketplace: an exposition of the concept and implications for research. J. Acad. Mark. Sci. **33**(4), 585–603 (2005)
23. Yeo, V.C.S., Goh, S.K., Rezaei, S.: Consumer experiences, attitude and behavioral intention toward online food delivery (OFD) services. J. Retail. Consum Serv. **35**, 150–162 (2017)
24. Zhou, T.: Understanding continuance usage intention of mobile internet sites. Univ. Access Inf. Soc. **13**(3), 329–337 (2013). https://doi.org/10.1007/s10209-013-0313-4

Digital Content Effects and Children as a Consumer

Uttam Kumar Roy[1(✉)] and Weining Tang[2]

[1] E-Commerce Technology, Information Engineering Department, Huzhou University, Huzhou, China
jonnlinn@outlook.com

[2] Business Department, Huzhou University, Huzhou, China
twnly@zjhu.edu.cn

Abstract. In recent years children are one of the biggest consumers. Many industries are targeting children to sell their products. Many individuals are creating full-time content for the kids. Children are attached to screen watching cartoons or playing games. There are many educational contents. After watching more content on the digital device, children's behaviors change dramatically. They become more stubborn. They become angry and addicted to digital materials. Industries are creating more content to attract kids. But those contents make the children addicted. It is a big question about the intention of the sector.

The E-commerce sector has boomed targeting, especially children, and spending thousands of dollars to reach the kids. Many platforms have risen, such as paid services for the children. Absent parents are happy to leave the kids to the digital device as the devices keep the kids happier. But it is a big question that how much helpful those digital content for children.

Keywords: Digital content · Digital device · Children · Consumer · E-commerce · Advertisement

1 Introduction

Recent booms of E-commerce where privacy is a big concern, but privacy neglected for the kids. Kids are playing games on the smart phone, watching multimedia, and also, they were a big target for many industries today. Kids industries are more interested in going inside the market but not investing enough to create awareness for the well-being of children.

The Digital content sector, where people upload a significant number of videos every day for kids, because of a lot of video content, famous online giants created a separate section for kids. Such as Youtube and Netflix have a different section for the kids.

And thousands of creators making dedicated digital content for the children. There's no doubt that many of the content is useful and enjoyable. But most of the materials are addictive.

We have talked to some parents from China, Bangladesh, the UK, South Korea, and USA. Some parents shared awful experiences. Some kids are so addicted that they don't want to eat without watching some particular content, especially videos. They are stubborn to buy something after viewing some advertisement [1].

C. Stephanidis et al. (Eds.): HCII 2020, LNCS 12427, pp. 624–629, 2020.
https://doi.org/10.1007/978-3-030-60152-2_46

Life is being comfortable with E-commerce development, so buying something from the internet is very easy now—just click-buy and pay. Parents have an extra budget for their children. China's children's wear market alone reached 160 billion yuan (24 billion US dollars) in 2018 and globally 169 billion US dollar. And global Children's wear market will reach more than $239 billion within 2023 [2].

But current digital markets' researches are not stopped only with children's wear market as the internet is now effortless to get. Software as a service has increased dramatically.

Many software companies created content for the kids commercially without advertisement. Children started using software for fun. But after using those software in a particular time every day, children are being addicted to those software's content. Enjoyment transforms into an addiction. They are attached to the screen like glue.

Companies are making free content to attract children and lock their premium content. And the companies selling the premium digital content monthly or annually basis.

And some creators share their affiliate link or products as an advertisement to the free versions. As there is no noticeable government policy for such ads, so there are big chances, children are getting adult content and inappropriate materials on the advertisements.

Children are joining social media from multiple locations and multiple devices. Some parents told us that they let their children use their social media account. Social media is not limited to certain things. Sometimes many hatred words, violent graphics, racism, debatable content, adult stuff going on social media.

Many children join social media giving the wrong date of birth information. For such cases, they are doing self-harm, learning discrimination, getting unwanted pornographic content, which causes sexual harassment [3].

So there is a significant risk for children using those contents. And the risk becomes harmful when the children make a decision such as suicide. The prominent example is the Blue Whale game, where many children believed that they deserve to die to save their loved ones. And many children commit suicide because of getting low score on the exam paper, so this increased because of using too much social media, depend on the reactions of social media and scared of being cyberbullied.

It does not stop within the cyberbullied, digital content effect on health as well; overuse of digital devices to play, excessive surf on the internet causes cognitive, social, emotional, attention problems [4].

Digital devices connected with the internet are filled with millions of digital content now. And updating new videos or content at any moment, and children who are surfing the device, their attention keep changing one direction to another direction. The overuse of digital materials keeps their concentration on different things at the same time, which causes attention problems later to learn something new. Children are not spending time playing or exercising because of using excessive digital device, which causes obesity for children. Significant issues arise with sleeping; there are big chances of having sleeping disturbance if the children watch digital content before sleeping. And it also causes a headache, irritated eyes, and computer vision problems.

Those digital devices also have a significant impact on emotional interaction.

2 Methodology

We talked with 30 parents (children age between 2–13) about using digital devices for their kids. We use qualitative methods to determine the problems. We asked the parents how long they allow their children to use digital devices. Are they well aware of the impact of digital devices/digital content? And what changes have they seen after giving the digital devices/digital contents for a long time?

We observe some kids after they use digital tools and watch digital content. We observe the concentration to remember something; kids who are learning from the digital devices and play for a minute in the digital device. And we observe some rural kids where chances are very low to get the digital device. We gave them some books, and they play some games after reading books.

3 Results

1 out of 10 parents said that their children are not much addicted to digital devices. Where 90% of parents admitted that they let their kids to use the digital device. But 90% of parents wish to stop giving the device to the kids often. But they are helpless because the kids are already addicted.

90% of parents said that if they stop giving the device, then their children start crying, being stubborn, don't sleep properly, and don't listen to them. And in some cases, some parents play some TV series to keep feeding their children because the children are giving them conditions not to eat if they don't play the device.

Parents said their kids use digital devices, taking them very near to eyes, and parents are afraid of getting a bad effect on their children's eyes.

The kids change very fast after using digital devices.

The kids from the rural area played a very significant role; their memory to remember something was better than the kids who are using digital devices. Only a few kids did well, even using digital tools, but the number is quite low.

Parents also admitted that they are not well aware of digital content. As the children learn faster from seeing, parents acknowledge that they often play with their smartphones in front of the kids where the kids get attracted to the digital devices first.

Some parents are aware of the risks, but still, they let their children play with digital devices because they are afraid that maybe the kid will do something wrong if they don't allow the kids to use digital tools. And 90% of the parents are not aware that children software may collect the data from their device to show some personalized advertising. Sometimes parents supervise their children on how to use digital tools and limit the usage time. But they don't pay enough attention to it what the children are doing on the devices.

As the children start using the device for a long time, joining many surveys unwillingly, privacy concern arises.

4 Detailed Discussion

In this economic world (research), children are a big part today. Children are lovable for a family. So all of the family members want their kids to be happiest, best. Children are also a part of some parent's/family's status. So the parents push so much pressure on kids, and in return, allow their kids use the internet and browse digital content. Mostly, in developing countries, parents often proud of their children if they do good homework in the class, getting a high score on the exam. Even some parents leave their job to take dedicated care for their children.

As now the kids are born in the Internet era, they got something extra. And we also don't have much idea about this generation. They are getting a ton of information quickly from the internet. They can watch something they like, just touching some screen buttons. So today's kid's life is different than who is born in the '90s or before the '90s. So the new parents who born 90's and before most of them are clueless for raising children in this internet era.

The industries are targeting children because it's a big market to make huge profits for a long time. And the market is solid. If a child likes a brand today, the possibility is very high that the brand taste will not change for a long time. So the market is quite profitable for the industries.

And day by day, online industries are growing up to catch this competitive market, and the market is not only about some cartoon contents or animation and gaming but also full of educational content and live teaching content.

Parents are also happy because their kids are learning something from the internet, from the small device. So the method is considered as a teacher. There is no doubt that these digital devices and digital content can make a significant change in society.

But Instead of creating social awareness, Industries are collecting information from the children and also from the parents and showing personalized advertise.

The question is why a lot of video content creators creating videos full time for the kids. The creators using some advertisements on the content and encourage the audience to browse it. The content editor offers free books, more free videos, and collecting email and age from the audience. The industry analyzes those data where to put the exact advertisement so that the parents willing to buy the products, and they put a banner there.

Children believe anything on the internet very quickly, which is a big privacy concern; they put much personal information on the website, which can create significant issues in the future.

Today's kids have the most significant footprints in online history. In some way, companies start collecting data, and even the kid is not born. Targeting the pregnant woman and collecting their buying nature, online industries are narrowing down their audience for specific products. CPA is calculated based on this.

Parents need to be aware of their privacy. Because when the online industries can know your taste and choices, they can easily target you. In some ways, it's a good thing that they know your preference, and they are recommending you the right products. But another way, Industries are trying to push you to buy certain things for the kids.

So the educational contents, enjoyable videos for the kids now a big question that is the online industries try to make something better for our kids? If yes, then why the enterprises are not also putting the same effort to raise the awareness to proper use of the digital devices and the contents. Online industries could quickly raise awareness of using digital devices and could make a massive change. If the parents are aware of the digital content, they could take extra care for the children. And many children can be saved to being stubborn, suicidal issues, social problems, and health problems.

Sexual abuse is a big concern. Children are joining the internet means they are not in a specific boundary. According to the Internet watch foundation, 105,047 webpages are showing sexual abuse in 2018 where 39% of victims are under ten years old; 1% aged two or under and 78% of images where victims were girls, 17% of images where victim were boys and 4% of images where victim were both genders.

To believe something easily of children's nature, they joined the scam very fast [5]. Children usually don't have much self-control; that's why many children enter the internet to bullying others, posting other images; and they are too much attached to the emotions of likes, comments, and share on social media.

However, expert estimates millions of children on social media Facebook using the false sign up information. And the same time, when children like, comment, share, react for any content, all their online footprint history saved on the net, and later on, they are getting ads according to the reactions and personal newsfeed/moments which kept very addictive.

Short videos App has gotten very famous because of the recommended videos—Tiktok, where people share short length videos and watch numerous videos according to the suggestion. And there are millions of children videos on this platform. If children use the TikTok without concern of their parents, Tiktok recommend video based on the likes, comment, subscribe, and how long the audience engaged on the videos. As the platform's AI understands that info and start recommending video analyzing the watching history; the platform can show more interesting videos as audience choice. The more the children spend time on a specific platform, the platform can keep the children for more time. And more time means more ads, and more ads mean more money for the platform. And the children were watching those contents without being aware of their health issues. They indulge in such platforms.

Many platforms use some age restriction to join their platform, but it's easy to bypass this kind of security. So the questions raised again about the intention for well-being of today's children from those online giants.

In the last 8–10 years, online industries made considerable changes to protect privacy and data, but about the age restriction, there is no noticeable progress. It's alarming questions that are they not capable of doing something in this regard, or don't they want to change their platform to collect data secretly. Today's demographic data mining technology can help efficiently to find the children who are bypassing those age restriction efficiently.

5 Conclusion

In this digital age, it's not possible to stop the children using digital devices or contents. It may slow down the development process of this century. So a better solution is to create awareness before the problem rises a massive way. We need to do more research on this field to make a better future. Because today's kids will be tomorrow's generation.

The biggest online industries have to come to a step forward to create awareness about the digital content; children are their most significant consumers. And it's not only a responsibility but also a moral duty to take steps quickly.

We need to do more research on how digital devices keep the children for a long time and change their behavior in a short amount of time. And raise voice for privacy, preserve the children's confidentiality, and not allowing to sell it to the third party.

World organizations can make a significant effort to do more research and create awareness on this field, and at the same time, digital device manufacturers should think about the children using their device. And make child-friendly devices.

And the most important how to use the internet properly and get relevant, useful data fast, big industries can think to create awareness about it and can take extra precautions.

Digital content creators can easily put a banner for awareness. The content can be videos, stories on digital devices, software, game, etc.

If the digital content creators start working on this as soon as possible, many problems will decrease very fast.

The children who are not much addicted to digital devices, their parents said that they seldom use digital tools in front of their kids, and instead of playing on digital devices, they prefer playing with the toys with their children.

We need further research on this field about raising the kids in this booming e-commerce time. And at the same time, when cryptocurrency, VR, block chain has already started an impact on people's life. So researching more about raising kids in this digital era is very necessary otherwise the enjoyment of the digital devices will be addictive, addiction will arise risk and risk will transform into harm.

References

1. Šramová, B., Pavelka, J.: The perception of media messages by preschool children. Young Consum. **18**, 121–140 (2017)
2. Children's wear market value worldwide in 2018 and 2023. https://www.statista.com/
3. Children in a Digital world. https://www.unicef.org/
4. Smahel, D., Wright, M.F., Cernikova, M.: The impact of digital media on health: children's perspectives. Int. J. Public Health **60**(2), 131–137 (2015). https://doi.org/10.1007/s00038-015-0649-z
5. Furnham, A., Gunter, B.: Children as Consumers. A Psychological Analysis of the Young People's Market. Routledge, London (2008)

A Review on Eye-Tracking Metrics for Sleepiness

Debasis Roy(✉) and Fiona Fui-Hoon Nah

Missouri University of Science and Technology, Rolla, MO, USA
{dr4b8, nahf}@mst.edu

Abstract. Sleepiness that can arise from sleep deprivation can increase human errors in task performance and create workplace hazards and accidents. Hence, it is critical to detect sleepiness to minimize hazards and human errors. This paper provides a review of the literature on eye tracking metrics that can be used to detect sleepiness. These metrics include blink duration, blink frequency, saccade latency, saccade peak velocity, saccade accuracy, smooth pursuit velocity gain, fixation rate, pupil size, and latency to pupil constriction.

Keywords: Cognitive performance · Sleep deprivation · Sleepiness · Eye tracking · Blink · Saccade · Fixation · Smooth pursuit

1 Introduction

Sleep deprivation affects our ability to respond to and perform well in tasks. It increases human errors, reaction times, and time to complete tasks [9, 30, 39]. Sleep deprivation can also increase the risks to our health, well-being, and longevity [6]. It can negatively affect performance in computer-based tasks and create hazards when safety is involved, such as in traffic control tasks [39]. Sleep deprivation can impact behavior in the following ways [30]: (i) it slows down reaction times; (ii) it increases errors on omission and commission; (iii) it increases time to complete tasks; (iv) its effect on vigilant attention is sensitive to circadian and homeostatic drives. The circadian drive promotes wakefulness according to one's biological clock while the homeostatic drive to sleep is affected by the duration of wakefulness.

Staying focused is important when performing mission critical tasks. The ability to stay focused or attentive can be affected by fatigue or sleepiness. Sleepiness reduces one's ability to respond to a stimulus in a timely manner and increases the likelihood of responding to a wrong stimulus [14]. The functioning of our cognitive process is affected when one is deprived of sleep because sleepiness induces general performance deficit due to impaired attention [51]. As per the theory of state instability, the effect of sleep deprivation on time on task is not strictly linear [52]. Rather, sleep deprivation creates a progressive increase in response variability over the duration of the task [4, 9].

C. Stephanidis et al. (Eds.): HCII 2020, LNCS 12427, pp. 630–640, 2020.
https://doi.org/10.1007/978-3-030-60152-2_47

2 Literature Review

Sleepiness can impact performance in computer-based tasks. It can increase the response time, the number of errors made, and the time taken to complete a task [30]. Popular scales that have been used to assess sleepiness include Multiple Sleep Latency Test (MSLT), Maintenance of Wakefulness Test (MWT), Stanford Sleepiness Scale (SSS), Epworth Sleepiness Scale (ESS), Karolinska Sleepiness Scale (KSS), Sleepiness Visual Analogue Scale (VAS), and Accumulated Time with Sleepiness (ATS). MSLT and MWT are laboratory-based methods to assess the ability to fall asleep and stay awake respectively. When focused attention is critical to a job that involves safety, MSLT and MWT can be used to assess the potential risk of work-related hazards or accidents. The SSS, ESS, KSS, VAS and ATS are self-reported measures of sleepiness.

MSLT is a validated measure of excessive daytime sleepiness that assesses the tendency to fall asleep [31]. It is assumed that the sleepier a person is, the quicker he or she falls asleep. MSLT measures how fast a person falls asleep in a laboratory setup (i.e., in a dark and quiet controlled environment) with electrodes and wires attached so sleep can be monitored [26, 41]. The testing for MSLT takes a full day as it involves assessments of sleep latency (i.e., the amount of time it takes to fall asleep) of five naps separated by two-hour breaks. Each nap lasts 15 min and each nap attempt will end after 20 min if sleep does not occur.

MWT is an assessment of the ability to stay awake for a defined period of time [31]. It assesses the wake tendency of a person while resisting falling asleep in a dull and quiet environment [8]. MWT has been used to assess responses to interventions for disorders associated with excessive sleep. External factors such as light, temperature and noise are isolated during the test.

SSS is a popular sleepiness scale that is widely used in sleep-related studies. SSS is a one-item self-reported assessment of sleepiness at a specific moment in time. Sleepiness is assessed on a scale of 1 to 7 (see https://web.stanford.edu/~dement/sss.html for the scale) to indicate the current level of sleepiness [21]. SSS has also been used to assess mood change following a phase shift of the sleep-wake schedule [32, 49].

Similar to SSS, KSS is also a one-item measure to assess sleepiness and has been validated against alpha and theta electroencephalographic (EEG) activity as well as slow eye movement electrooculographic (EOG) activity [1]. It is rated on a 9-point scale (1 = extremely alert, 2 = very alert, 3 = alert, 4 = rather alert, 5 = neither alert nor sleepy, 6 = some signs of sleepiness, 7 = sleepy, but no effort to keep awake, 8 = sleepy, some effort to keep awake, 9 = very sleepy, great effort keeping awake, fighting sleep). It exists in two versions. The first version only labels the odd points in the scale (1, 3, 5, 7, 9) whereas the second version adds labels to the even points and hence, labels all 9 points in the scale [37].

ESS is a self-administered questionnaire of eight items to assess general level of sleepiness (see https://epworthsleepinessscale.com/about-the-ess/ for the scale). It uses a 4-point (0–3) scale to assess the likelihood to doze off or fall asleep when engaged in eight different activities. The sum of the scores over the eight items or activities yields an overall score for average sleep propensity. The concept of ESS was derived from observations on the nature of the occurrence of daytime sleepiness [25].

VAS was developed to assess state or mood on a horizontal line with bipolar descriptions. It involves rating sleepiness along a line with both extremes labeled as 'very sleepy' and 'very alert' [27, 35]. VAS can be used for rating a variety of states or moods that also include fatigue, tension, anger, vigor, sadness and cheerfulness [34, 44].

ATS is an assessment of subjective sleepiness over long time periods. It involves asking a person to estimate the proportion of wake times that he or she experiences symptoms of sleepiness such as heavy eyelids, irresistible sleepiness, and reduced performance [53]. Ratings of ATS showed high correlations with those of KSS and VAS [16].

Sleepiness can also be assessed using eye-tracking metrics that involve saccadic movement (i.e., rapid eye movement between fixation points), smooth pursuit (i.e., eye movement following a moving target), blinks and pupil size, and fixation. The rest of this paper provides a review of eye-tracking metrics to detect sleepiness. The review focuses on synthesizing the literature on the different eye-tracking metrics in the literature. The following databases have been utilized for identifying relevant published articles: ACM Digital Library, Scopus, PsycINFO, ABI/Inform, IEEE, Medline, Oxford Academic, National Center for Biotechnology Information or NCBI, ScienceDirect, and PubMed. We used various combinations of the following keywords for the searches: 'eye tracking', 'sleepiness', 'sleep deprivation', 'blink', 'saccade', 'fixation', 'blink rate, 'pupil', 'eye tracking metrics'. The findings from the literature review are summarized in Table 1.

Table 1. Summary of findings

Reference	Research focus	Key findings	Eye-tracking metrics
Ahlstrom et al. [1]	Using eye movements and sleep-wake predictor to estimate sleepiness of drivers in a fit-for-duty test	Eye movement features and sleep-wake predictor model can be used to predict severe sleepiness	• Saccade curvature • Amplitude ratio of sinusoidal smooth pursuit
Barbato et al. [3]	Sleep deprivation affects spontaneous eye blinks	Blink rate increases with sleep deprivation, which increases central dopamine activity	Blink frequency
Bocca & Denise [5]	Saccadic eye movement was used to assess the effect of sleep deprivation	Saccadic eye movements can be used to assess alertness, sleep deprivation, and visuo-spatial attention	• Saccadic accuracy • Saccadic latency
Caffier et al. [6]	Measurement of eye-blink parameters for sleepiness	Blinks can be used to assess drowsiness or sleepiness	• Blink duration • Blink re-opening time • Blink frequency • Proportion of long-closure duration blinks

(*continued*)

Table 1. (*continued*)

Reference	Research focus	Key findings	Eye-tracking metrics
Crevits et al. [7]	Effect of sleep deprivation on saccades and eyelid blinking	Sleep deprivation affects blink rate but not saccade latency or number of saccade errors	• Blink frequency
De Gennaro et al. [10]	Total sleep deprivation: a dissociation between measures of speed and accuracy	Saccade latency increases, saccade peak velocity decreases, and smooth pursuit velocity gain decreases with sleep deprivation	• Saccade latency • Saccade peak velocity • Smooth pursuit velocity gain
Ferrara et al. [12]	Oculomotor performance after sleep deprivation	Saccade latency increases, saccade velocity decreases, and smooth pursuit velocity gain increases after sleep deprivation	• Saccade latency • Saccade velocity • Smooth pursuit velocity gain
Fransson et al. [14]	Smooth pursuit and saccadic eye movements in restricted sleep deprivation	Sleep deprivation decreases smooth pursuit gain, smooth pursuit accuracy and saccade velocity, and affects the ratio between saccade velocity and saccade amplitude	• Saccade velocity • Saccade amplitude • Smooth pursuit velocity gain • Smooth pursuit accuracy
Franzen et al. [15]	Pupillary reactivity on sleep deprived adults	Sleep-deprived individuals have larger pupil diameters when viewing negative pictures, which suggests greater reactions to negative emotional information	• Pupil diameter
Häkkänen et al. [18]	Detect sleepiness through blink duration	Sleepiness increases blink duration	• Blink duration
Heaton et al. [19]	Sleep deprivation decreases attention and visual tracking	Gaze instability and reaction time deterioration are found in sleep deprived individuals	• Gaze stability
Ingre et al. [22]	Blinking, lane drifting and KSS in sleep-deprived driving	Blink duration increases with sleep deprivation	• Blink duration
Jin et al. [24]	Eye movement variables to detect driver sleepiness	Blink frequency, gaze direction, fixation duration, and percent eye closure (PERCLOS) can detect driver sleepiness	• Blink frequency • Gaze direction • Fixation duration • Percent eye closure (PERCLOS)
Kurylyak et al. [28]	Infrared camera-based system to evaluate sleepiness	Blink time interval is shorter when sleepy than alert	• Blink frequency

(*continued*)

Table 1. (*continued*)

Reference	Research focus	Key findings	Eye-tracking metrics
Marshall [33]	Measurement of metrics to identify cognitive states of a sleep deprived subject	Assessed different cognitive states using eye metrics and corroborated with EEG measurements, and identified metrics that differentiate cognitive states including sleepiness	• Blink frequency • Pupil size • Saccade rate • Divergence (distance between horizontal locations of both cycs)
Miles et al. [36]	Extreme sleepiness can be detected by eye movement and visual fixation	Saccadic eye movement was slower and wavering of fixations was observed with sleepiness	• Saccade velocity • Gaze stability
Porcu et al. [40]	Nighttime sleepiness indicators	Saccadic performance and smooth pursuit are affected by sleepiness	• Saccade accuracy • Saccade latency • Rejected ("inappropriate") saccades • Smooth pursuit (velocity gain and phase)
Rowland et al. [42]	Responses of oculomotor during partial and total sleep deprivation	Sleep deprivation induces fatigue and results in delayed pupil light reflex and decreased saccadic velocity	• Saccadic velocity • Latency to pupil constriction
Russo et al. [43]	Oculomotor impairment during chronic partial sleep deprivation	Saccadic velocity decreases and latency to pupil constriction increases with sleep deprivation	• Saccadic velocity • Latency to pupil constriction
Schleicher et al. [45]	Eye movement indicators of fatigue for issuing sleepiness warnings	With increasing time in repetitive work, eye movements changed and by capturing different metrics, sleepiness warnings can be issued	• Blink duration • Delay of eyelid reopening • Blink interval • Eyelid closure speed • Saccadic duration • Fixation duration
Shiferaw et al. [46]	Sleep deprived drivers' behavior	Gaze behavior, blink rate, blink duration, fixation rate and saccade amplitude are affected by sleep deprivation	• Stationary gaze entropy • Gaze transition entropy • Fixation rate • Blink frequency • Blink duration • Saccade amplitude

(*continued*)

Table 1. (*continued*)

Reference	Research focus	Key findings	Eye-tracking metrics
Tong et al. [50]	Sleep deprivation and visual tracking synchronization	Acute sleep deprivation degrades visual tracking and prediction capability becomes less precise	• Saccade velocity • Anticipatory saccade amplitude • Gaze stability • Smooth pursuit velocity gain
Wilhelm et al. [55]	Pupil size measurement of sleep deprivation	Pupillary behavior is affected by sleep deprivation and pupillary oscillations quantify sleepiness	• Pupil size
Yang et al. [56]	Eyelid tracking for fatigue detection from sleep deprivation	An eye-tracking system was developed to predict fatigue from facial and eyelid tracking of sleep deprived and non-sleep deprived individuals	• Blink frequency • Percent [of] eye [lid] closure (PERCLOS)
Zils et al. [57]	Differential effects of sleep deprivation on saccadic eye movements	Saccadic eye movements can detect sleep deprivation	• Saccadic peak velocity • Saccadic accuracy • Saccadic latency

Sleepiness is regulated by three main factors: the circadian rhythm, time awake, and prior sleep [1]. The eye movements of an individual who is awake can take one of three states: blink, saccade, and fixation. Researchers have utilized eye-tracking metrics to study sleep deprivation and these metrics are discussed below.

Fixation: Agaze cluster constitutes a fixation when the eyes are locked toward an object. A fixation is defined as "a relatively stable eye-in-head position within some threshold of dispersion ($\sim 2°$) over some minimum duration (200 ms), and with a velocity threshold of 15–100°/s" [23, p. 581].

Sleep deprivation and fixations are closely related. With sleep deprivation, the corrective movement of a fixation becomes larger and less exact than in normal conditions [36]. Sleep deprivation also reduces the rate of fixations, increases the spatial dispersion of fixations, and increases the randomness of patterns of transitions between fixations [46]. Fixation rate decreases with sleep deprivation due to a decline in the amount of information sampled when one is sleepy or drowsy [36]. Long fixation durations may indicate difficulties extracting or processing information from a display element, which take place when one is sleepy [13, 17].

Gaze: A gaze refers to a grouping of fixations within a single area of interest. Its associated measures include the number of fixations within a single gaze, the total number of gazes, the frequency of a gaze, and the duration of a gaze, as well as the mean and statistics of those measures [20]. A large number of fixations around a gaze point signify instability of the eyes in positioning the gaze. This instability can be observed in sleep deprived individuals due to their inability to concentrate or focus on a display point or element [19, 36, 46, 50].

Smooth Pursuit: Smooth pursuit eye movements refer to the movements of the eye to follow a moving target. Smooth pursuit velocity gain refers to the ratio between the eye velocity and the target velocity during smooth pursuit, and smooth pursuit phase refers to the distance between the gaze and the target [1]. Sleep deprivation degrades the moment-to-moment synchronization between the gaze and the target [1, 10, 12, 14, 40, 50], and causes lapses in gaze-target synchronization sensitivity with increased attention loss [19]. Fransson et al. [14] found that as the length of sleep deprivation increases, only smooth pursuit velocity gain continues to degrade as compared to other metrics.

Saccade: A saccade is a rapid eye movement from one fixation to another. The characteristic feature of a saccade is that the movement is not smooth but is organized in distinct jumps. These jumps are rapid and their endpoint cannot be changed once the saccade is initiated.

Several metrics associated with saccadic eye movements that can be triggered by sleep deprivation include saccadic latency (i.e., the delay before the onset of a saccade), saccadic accuracy (i.e., ratio of the amplitude of the saccade and the amplitude of the target), saccadic amplitude (i.e., angular distance the eye travels during the saccade), and saccadic peak velocity (i.e., highest velocity in the saccade) [57]. These metrics are affected by the slowing of neural responsiveness and metabolism due to sleep deprivation. Sleep deprivation generally leads to increased saccadic latency, decreased saccadic accuracy, increased saccadic amplitude and curvature, and decreased saccadic peak velocity [1, 5, 10, 12, 14, 33, 36, 40, 42, 43, 50, 57].

Blink: Blink frequency (or rate) is the number of times a person blinks in a minute, and blink duration is the complete time from when the eyelid starts moving down until it is fully up again. With sleep deprivation, blinks become relatively slower for the same amplitude. Normal blink frequency is on the order of 9 to 13 per minute in the daytime, which is increased to 20 to 30/m under sleep deprivation [6, 43].

Blink duration and blink frequency are often used in sleep-related studies [3, 6, 7, 18, 22, 24, 29, 33, 45, 46]. We generally blink more frequently and at a slower speed when we are sleepy [47, 48]. Lapses due to sleepiness can increase errors and variability in cognitive performance [9]. A relation between subjective sleepiness measured using the KSS, objective indicators of sleepiness based on blinks, and driving performance has been demonstrated [22]. With higher KSS levels, the standard deviation of the lateral position is increased and eye blinks are longer in duration. Blink duration increases with sleep loss as found in several field studies on driving [28, 38] and in driving simulators [2, 6, 45, 54]. Sleepiness can also be assessed by the duration of eye closure (PERCLOS), which refers to the proportion of the time the eyes are closed (i.e., typically assessed as between 80% to 100% closed) [24, 56].

Pupil Size: Pupil size is the aperture in an optical system of a human eye. The normal pupil size varies from 2 to 4 mm in diameter in bright light and 4 to 8 mm in the dark.

The deeper the sleep, the more the pupil constricts. Pupil diameter is influenced by the arousal system and is a sensitive indicator of sleepiness [55].

Pupil diameter changes are a manifestation and representation of sleep deprivation [15, 33, 55]. Pupil changes in size and other pupillometry metrics can be used to deduce changes in cognitive processing [11]. Using an individual's pupil diameter as a baseline, relative changes in pupil size can be a consequence of sleepiness or sleep deprivation. Latency of pupil constrictions increases with sleep deprivation or drowsiness due to reduced neural responsiveness [42, 43].

3 Conclusion and Future Research

Eye-tracking metrics such as saccadic latency, saccadic accuracy, saccadic peak velocity, blink duration, blink frequency, and fixation rate can be used to assess sleepiness. In our future research, we plan to conduct an experimental study to assess the reliability and validity of these measures for assessing sleepiness of users in a computer-based environment. We will also analyze the relationship between sleepiness and performance based on task accuracy and response/completion time. We hope that our future research will contribute toward the body of research to better explain and understand the relationships between eye-tracking and performance metrics and the underlying mechanisms of sleepiness. Performance metrics such as lapses (i.e., response time >500 ms) [2] and error types in task completion will also be assessed on whether they are appropriate for detecting sleepiness. We hope to offer useful guidelines to detect sleepiness in the computer-based workplace environment and help enhance workplace productivity and quality of work.

References

1. Ahlstrom, C., et al.: Fit-for-duty test for estimation of drivers' sleepiness level: eye movements improve the sleep/wake predictor. Transp. Res. Part C Emerg. Tech. **26**, 20–32 (2013)
2. Anderson, C., Wales, A.W., Home, J.A.: PVT lapses differ according to eyes open, closed, or looking away. Sleep **33**, 197–204 (2010)
3. Barbato, G., Ficca, G., Beatrice, M., Casiello, M., Muscettola, G., Rinaldi, F.: Effects of sleep deprivation on spontaneous eye blink rate and alpha EEG power. Biol. Psychiatr. **38**, 340–341 (1995)
4. Bills, A.G.: Blocking: a new principle of mental fatigue. Am. J. Psychol. **43**, 230–245 (1931)
5. Bocca, M.L., Denise, P.: Total sleep deprivation effect on disengagement of spatial attention as assessed by saccadic eye movements. Clin. Neurophysiol. **117**, 894–899 (2006)
6. Caffier, P.P., Erdmann, U., Ullsperger, P.: The spontaneous eye-blink as sleepiness indicator in patients with obstructive sleep apnoea syndrome-a pilot study. Sleep Med. **6**, 155–162 (2005)
7. Crevits, L., Simons, B., Wildenbeest, J.: Effect of sleep deprivation on saccades and eyelid blinking. Eur. Neurol. **50**, 176–180 (2003)
8. Doghramji, K., et al.: A normative study of the maintenance of wakefulness test (MWT). Electroencephalogr. Clin. Neurophysiol. **103**, 554–562 (1997)

9. Doran, S.M., Van Dongen, H.P., Dinges, D.F.: Sustained attention performance during sleep deprivation: evidence of state instability. Arch. Ital. Biol. **139**, 253–267 (2001)
10. De Gennaro, L., Ferrara, M., Urbani, L., Bertini, M.: Oculomotor impairment after 1 night of total sleep deprivation: a dissociation between measures of speed and accuracy. Clin. Neurophysiol. **111**, 1771–1778 (2000)
11. Ellis, K.K.E.: Eye tracking metrics for workload estimation in flight deck operations. Theses and Dissertations, p. 288 (2009)
12. Ferrara, M., Gennaro, L.D., Bertini, M.: Voluntary oculomotor performance upon awakening after total sleep deprivation. Sleep **23**, 801–811 (2000)
13. Fitts, P.M., Jones, R.E., Milton, J.L.: Eye movements of aircraft pilots during instrument-landing approaches. Aeronaut. Eng. Rev. **9**, 24–29 (1950)
14. Fransson, P.A., Patel, M., Magnusson, M., Berg, S., Almbladh, P., Gomez, S.: Effects of 24-h and 36-h sleep deprivation on smooth pursuit and saccadic eye movements. J. Vestib. Res. **18**, 209–222 (2008)
15. Franzen, P.L., Buysse, D.J., Dahl, R.E., Thompson, W., Siegle, G.J.: Sleep deprivation alters pupillary reactivity to emotional stimuli in healthy young adults. Biol. Psychol. **80**, 300–305 (2009)
16. Gillberg, M., Kecklund, G., Åkerstedt, T.: Relations between performance and subjective ratings of sleepiness during a night awake. Sleep **17**, 236–241 (1994)
17. Goldberg, J.H., Kotval, X.P.: Computer interface evaluation using eye movements: methods and constructs. Int. J. Ind. Ergon. **24**, 631–645 (1999)
18. Häkkänen, H., Summala, H., Partinen, M., Tiihonen, M., Silvo, J.: Blink duration as an indicator of driver sleepiness in professional bus drivers. Sleep **22**, 798–802 (1999)
19. Heaton, K.J., Maule, A.L., Maruta, J., Kryskow, E.M., Ghajar, J.: Attention and visual tracking degradation during acute sleep deprivation in a military sample. Aviat. Space Environ. Med. **85**, 497–503 (2014)
20. Hendrickson, J.J.: Performance, preference, and visual scan patterns on a menu-based system: implications for interface design. In: Proceedings of the SIGCHI Conference on Human Factors in Computing Systems, pp. 217–222, March 1989
21. Hoddes, E., Dement, W.C., Zarcone, V.: Stanford sleepiness scale. Enzyklopädie der Schlafmedizin, p. 1184 (1972)
22. Ingre, M., Åkerstedt, T., Peters, B., Anund, A., Kecklund, G.: Subjective sleepiness, simulated driving performance and blink duration: examining individual differences. J. Sleep Res. **15**, 47–53 (2006)
23. Jacob, R.J., Karn, K.S.: Eye tracking in human-computer interaction and usability research: Ready to deliver the promises. In: The Mind's Eye, pp. 573–605. North-Holland (2003)
24. Jin, L., Niu, Q., Jiang, Y., Xian, H., Qin, Y., Xu, M.: Driver sleepiness detection system based on eye movements variables. Adv. Mech. Eng. **5**, 1–7 (2013)
25. Johns, M.W.: A new method for measuring daytime sleepiness: the Epworth sleepiness scale. Sleep **14**, 540–545 (1991)
26. Johns, M.W.: Sensitivity and specificity of the multiple sleep latency test (MSLT), the maintenance of wakefulness test and the Epworth sleepiness scale: failure of the MSLT as a gold standard. J. Sleep Res. **9**, 5–11 (2000)
27. Johnson, L.C., Spinweber, C.L., Gomez, S.A., Matteson, L.T.: Daytime sleepiness, performance, mood, nocturnal sleep: the effect of benzodiazepine and caffeine on their relationship. Sleep **13**, 121–135 (1990)
28. Klauer, S.G., Dingus, T.A., Neale, V.L., Sudweeks, J.D., Ramsey, D.J.: The impact of driver inattention on near-crash/crash risk: An analysis using the 100-car naturalistic driving study data. Virginia Tech Transportation Institute, Blacksburg, VA. (NTIS No. DOT HS 810 594) (2006)

29. Kurylyak, Y., Lamonaca, F., Mirabelli, G., Boumbarov, O., Panev, S.: The infrared camera-based system to evaluate the human sleepiness. In: IEEE International Symposium on Medical Measurements and Applications, pp. 253–256. IEEE, May 2011
30. Lim, J., Dinges, D.: Sleep deprivation and vigilant attention. Ann. New York Acad. Sci. **1129**, 305 (2008)
31. Littner, M.R., et al.: Practice parameters for clinical use of the multiple sleep latency test and the maintenance of wakefulness test. Sleep **28**, 113–121 (2005)
32. Maclean, A.W., Fekken, G.C., Saskin, P., Knowles, J.B.: Psychometric evaluation of the stanford sleepiness scale. J. Sleep Res. **1**, 35–39 (1992)
33. Marshall, S.P.: Identifying cognitive state from eye metrics. Aviat. Space Environ. Med. **78**, B165–B175 (2007)
34. McCormack, H.M., David, J.D.L., Sheather, S.: Clinical applications of visual analogue scales: a critical review. Psychol. Med. **18**, 1007–1019 (1988)
35. Monk, T.H., Fookson, J.E., Kream, J., Moline, M.L., Pollak, C.P., Weitzman, M.B.: Circadian factors during sustained performance: Background and methodology. Behav. Res. Methods Instrum. Comput. **17**, 19–26 (1985)
36. Miles, W.R., Laslett, H.R.: Eye movement and visual fixation during profound sleepiness. Psychol. Rev. **38**, 1–13 (1931)
37. Miley, A.Å., Kecklund, G., Åkerstedt, T.: Comparing two versions of the Karolinska sleepiness scale (KSS). Sleep Biol. Rhythms **14**(3), 257–260 (2016). https://doi.org/10.1007/s41105-016-0048-8
38. Mitler, M.M., Miller, J.C., Lipsitz, J.J., Walsh, J.K., Wylie, C.D.: The sleep of long-haul truck drivers. N. Engl. J. Med. **337**, 755–762 (1997)
39. Pilcher, J.J., Huffcutt, A.I.: Effects of sleep deprivation on performance: a meta-analysis. Sleep **19**, 318–326 (1996)
40. Porcu, S., Ferrara, M., Urbani, L., Bellatreccia, A., Casagrande, M.: Smooth pursuit and saccadic eye movements as possible indicators of nighttime sleepiness. Physiol. Behav. **65**, 437–443 (1998)
41. Richardson, G.S., Carskadon, M.A., Flagg, W., van den Hoed, J., Dement, W.C., Mitler, M. M.: Excessive daytime sleepiness in man: multiple sleep latency measurement in narcoleptic and control subjects. Electroencephalogr. Clin. Neurophysiol. **45**, 621–627 (1978)
42. Rowland, L.M., et al.: Oculomotor responses during partial and total sleep deprivation. Aviat. Space Environ. Med. **76**, C104–C113 (2005)
43. Russo, M., et al.: Oculomotor impairment during chronic partial sleep deprivation. Clin. Neurophysiol. **114**, 723–736 (2003)
44. Sanchez, A., Vanderhasselt, M.A., Baeken, C., De Raedt, R.: Effects of tDCS over the right DLPFC on attentional disengagement from positive and negative faces: an eye-tracking study. Cogn. Affect. Behav. Neurosci. **16**, 1027–1038 (2016)
45. Schleicher, R., Galley, N., Briest, S., Galley, L.: Blinks and saccades as indicators of fatigue in sleepiness warnings: looking tired? Ergonomics **51**, 982–1010 (2008)
46. Shiferaw, B.A., et al.: Stationary gaze entropy predicts lane departure events in sleep-deprived drivers. Sci. Rep. **8**, 1–10 (2018)
47. Stern, J.A.: Blink and you'll miss it. Hum. Factors Soc. Newsl. **2**, 14–15 (1990)
48. Stern, J.A., Walrath, L.C., Goldstein, R.: The endogenous eyeblink. Psychophysiology **21**, 22–33 (1984)
49. Surridge-David, M., MacLean, A.W., Coulter, M., Knowles, J.B.: Mood changes following an acute delay of sleep. Psychiatr. Res. **22**, 149–158 (1987)
50. Tong, J., Maruta, J., Heaton, K.J., Maule, A.L., Ghajar, J.: Adaptation of visual tracking synchronization after one night of sleep deprivation. Exp. Brain Res. **232**(1), 121–131 (2013). https://doi.org/10.1007/s00221-013-3725-8

51. Tucker, A.M., Whitney, P., Belenky, G., Hinson, J.M., Van Dongen, H.P.: Effects of sleep deprivation on dissociated components of executive functioning. Sleep **33**, 47–57 (2010)
52. Van Dongen, H.P.A., Belenky, G., Krueger, J.M.: A local, bottom-up perspective on sleep deprivation and neurobehavioral performance. Curr. Top. Med. Chem. **11**, 2414–2422 (2011)
53. Waage, S., Harris, A., Pallesen, S., Saksvik, I.B., Moen, B.E., Bjorvatn, B.: Subjective and objective sleepiness among oil rig workers during three different shift schedules. Sleep Med. **13**, 64–72 (2012)
54. Wierwille, W.W., Ellsworth, L.A.: Evaluation of driver drowsiness by trained raters. Accid. Anal. Prev. **26**, 571–581 (1994)
55. Wilhelm, B., Wilhelm, H., Lüdtke, H., Streicher, P., Adler, M.: Pupillographic assessment of sleepiness in sleep-deprived healthy subjects. Sleep **21**, 258–265 (1998)
56. Yang, F., Yu, X., Huang, J., Yang, P., Metaxas, D.: Robust eyelid tracking for fatigue detection. In: 19th IEEE International Conference on Image Processing, pp. 1829–1832. IEEE, September 2012
57. Zils, E., Sprenger, A., Heide, W., Born, J., Gais, S.: Differential effects of sleep deprivation on saccadic eye movements. Sleep **28**, 1109–1115 (2005)

Storytelling with Data in the Context of Industry 4.0: A Power BI-Based Case Study on the Shop Floor

Juliana Salvadorinho[1], Leonor Teixeira[2(✉)], and Beatriz Sousa Santos[3]

[1] Department of Economics, Management, Industrial Engineering and Tourism, University of Aveiro, 3010-193 Aveiro, Portugal
juliana.salvadorinho@ua.pt

[2] Institute of Electronics and Informatics Engineering of Aveiro, Department of Economics, Management, Industrial Engineering and Tourism, University of Aveiro, 3010-193 Aveiro, Portugal
lteixeira@ua.pt

[3] Institute of Electronics and Informatics Engineering of Aveiro, Department of Electronics, Telecommunications and Informatics, University of Aveiro, 3010-193 Aveiro, Portugal
bss@ua.pt

Abstract. Industry 4.0 (I4.0) is characterized by cyber physical systems (CFS) and connectivity, paving the way to an end-to-end value chain, using Internet of Things (IoT) platforms supported on a decentralized intelligence in manufacturing processes. In such environments, large amounts of data are produced and there is an urgent need for organizations to take advantage of this data, otherwise its value may be lost. Data needs to be treated to produce consistent and valuable information to support decision-making. In the context of a manufacturing industry, both data analysis and visualization methods can drastically improve understanding of what is being done on the shop floor, enabling easier decision-making, ultimately reducing resources and costs. Visualization and storytelling are powerful ways to take advantage of human visual and cognitive capacities to simplify the business universe. This paper addresses the concept of "Storytelling with Data" and presents an example carried out in the shop floor of a chemical industry company meant to produce a real-time story about the data gathered from one of the manufacturing cells. The result was a streaming dashboard implemented using Microsoft Power BI.

Keywords: Visualization · Storytelling · Industry 4.0 · Power BI

1 Introduction

The Third Industrial Revolution (3rd IR) brought computers and automation to the manufacturing system. The Fourth Industrial Revolution (4th IR) adds to these two mechanisms the concepts of cyber physical systems (CFS) and connectivity [1]. Industry 4.0 (I4.0) is characterized by CFS, preparing the way to an end-to-end value chain, using

C. Stephanidis et al. (Eds.): HCII 2020, LNCS 12427, pp. 641–651, 2020.
https://doi.org/10.1007/978-3-030-60152-2_48

Internet of Things (IoT) platforms supported on a decentralized intelligence in manufacturing processes. Connectivity is a key-factor in I4.0 environment, ensuring an automatic data collection, but in return responsible for the large amount of data present in most industrial environments that intend to embrace the challenge of I4.0 [2, 3]. In addition to these challenges, there is an urgent need for organizations to take advantage of this large volume of data; otherwise, the value of information will be lost. This data needs to be treated to produce consistent and valuable information to support decision making in organizations. Data science, a scientific approach that uses several mathematical and statistical techniques supported in computer tools for processing large amounts of data is becoming an invaluable area in I4.0 environments, since it can transform data into information and this in useful knowledge for the business. In addition, big data integrated with agile information systems can promote the solutions to convert those data in valuable information [2] improving at the same time the organization's capacity in response to internal, organizational and environmental changes in real-time [4].

According to Narayanan and Kp [5] "For a business to exist competently, the two things to keep up are: the management of time and better understanding of current status of the organization". Behind these issues is the importance of data visualization. In the context of a manufacturing industry, both data analysis methods and data visualization methods can drastically improve understanding of what is being done on the shop floor, thus enabling easier decision-making, ultimately reducing resources and perhaps costs. In fact, the human brain is an expert in memorizing data as images, so data visualization is just a clever idea to uncomplicated the business universe [5].

On the other hand, business intelligence (BI) is defined as "automatic data retrieving and processing systems that can help make intelligent decisions based on various data sources" [6]. Most of the BI solutions offer data analysis and data visualization which with the correct data capture technology should be able to treat data in real-time [7]. Some of the advantages of using BI tools are denoted by Stecyk [8] as the ability of linking to any data source, building up analyses in real time and having an intuitive and straightforward interface that helps in data visualization. However, to obtain this, some areas of knowledge need to be consistent and strong such as the ability to get data from a variety of sources, the aptitude to properly structure and relate the database and techniques about building key indicators (economic or performance) as well as dynamic reporting (visualization techniques) [8]. In addition, it is common sense that "the communication of information is an important capability of visualization" [9] and recently, literature has laid eyes on the new concept, more specifically the "Storytelling with Data" concept. This concept refers to a set of processes and mechanisms that help organizations to prepare multifaceted information, based on complex sets of data, with the purpose of communicating a story [9], including the arrangement of three elements: data, visualizations and narratives [7]. To address these issues, companies can use BI tools, but before it is important to choose the correct amount of information to deliver a message and adding to that the techniques that should be applied to produce story-like statements [9].

One of the open source BI tools referred to in the literature that allows achieving these objectives is the Microsoft Power BI, representing a tool gifted to create "shareable and customized visualizations to communicate data-based stories" [9], while providing visibility of the information flows [2].

In nutshell, despite the potential advantages in implementing the phenomenon of I4.0, organizations must be prepared to deal with the huge amount of data that IoT will bring. In addition, for that to happen it is essential that information flows be cleared and organized between all the departments in organizations. After that work done it is possible to implement BI tools in order to visualize what is going on in the shop floor. Considering these concerns, this paper intends to clarify the "Storytelling with Data" concept based on a literature review, and at the same time, pretends to describe methods and results carried out in a manufacturing company's shop floor, in order to implement the above concept. The study will be conducted in a chemical industry enterprise and the last goal set is to have a real-time story about the data that is gathered from one of the manufacturing cells. To tell this story we will have a streaming dashboard implemented by a BI tool, the Microsoft Power BI.

2 Background

2.1 Industry 4.0, Cyber Physical Systems and Digital Twin

Industry 4.0 principles are governed by the interconnection and transparency of information for decentralized decision making [10], requiring for that the combination of sensors, artificial intelligence, and data analytics [11]. This concept relies on the idea of combining optimized industrial processes with cutting-edge technology and digital skills and is the promotor of 'Smart Factory' or 'Factory of the future', concepts that are becoming the ambition of any enterprise. [12] Giving the concept behind smart factory and taking into account that it is still a utopia for many, it is important to understand the prerequisites to enable the smart in 'Digital Factory' [10]. According to [13], digital factory "refers to a new type of manufacturing production organization that simulates, evaluates and optimizes the production process and systems". While the Digital Factory provides tools for planning in Virtual Reality, the Smart one operates and optimizes the factory in real-time.

Information systems will be pivotal to achieve the vision of "real-time enterprise", remembering that they are "made of computers, software, people, processes and data" [10]. These components plan, organize, operate and control business processes [14], so they are pivotal in the integration of information flow.

Cyber physical systems are at the core foundation of Industry 4.0 and they intertwine physical and software components, each operating on different spatial and temporal scales. At the same time these components interact with each other in a multitude of ways that change with context [11].

The shop floor is the basic element of manufacturing, so the convergence between the physical and the virtual space becomes imperative [15].

As mentioned by Qi [16] "The digital twin paves the way to cyber-physical integration". This concept aims to create virtual models for physical objects in order to understand the state of these physical entities through sensing data (allowing predict, estimate and analyze dynamic changes). So it can be also assumed as a real-time representation of manufacturing systems or components [17].

This concept incorporates dynamic and static information, where data is transferred from the physical to the cyber part [18]. The data in digital twins are composed by physical world data as well as virtual models [16]. Digital twins combine and integrate data from multiple sources in order to achieve a more accurate and comprehensive information [15].

The digital twin is a prerequisite for the development of a Cyber-physical Production System although some difficulties must be overcome, such as data security concerns, standardization of data acquisitions, high costs for new IT-environments that inhibit the application of vertical industry 4.0 and the creation of a central information system which can be combined with decentralized data acquisition (taking into account that in-house implementation of industry 4.0 is frequently insufficient) [19].

2.2 Business Analytics, Visualization and Storytelling

Business analytics and business intelligence are assumed, in the Industry 4.0' context, as areas that can actually help to increase productivity, quality and flexibility. The importance of making quick and right decisions is even more fundamental for efficient and effective problem solving and process upgrading [20]. Today these two-knowledge fields have been valorizing the real-time production data, having influence in decision making [21].

Business analytics is a field which goal is to measure the company's performance, evaluating its position in the market and at the same time find where there is a need for improvement and what strategies should be carried out [22]. For that, statistical, mathematical and econometric analyses of business data need to be done in order to support operational and strategic decisions [23].

Visual analysis tools are assumed as technology products that combine information from complex and dynamic data in such a way that support evaluation, planning and decision making [24].

The understanding and the communication of information is supported by visualization that allows the abstraction of raw data and complex structure [25]. Therefore, data visualization is concerned with methods to obtain appropriate visual representations and interactions which accept users to understand complex data and confirm assumptions or even examine streaming data [26] Visualization is seen as a significant tool in many areas for clarifying and even perceive large and complex data [27] and affords the user to obtain more knowledge about the raw data which is gathered from a diversity of sources [22].

Although visualization plays an essential role in providing insights on real-time data, this may not be the exact solution for analyzing a large volume of data, as an adequate data extraction process must be carried out [22]. In the industry 4.0's context, where the big data concept carries a huge weight, the main goal of big visualization is to acknowledge patterns and correlations [28]. Newly, visual storytelling is receiving attention from the academic community, where authoring tools have been developed in order to create stories and provide visual support [9]. The entire process of modifying data into visually shared stories includes exploring the data, passing it into a narrative and then communicating it to an audience [29]. Stories offer an effective way of stowing information and knowledge and make it easy for people to perceive them [30].

There are already many communities that emphasize the importance of storytelling in data visualization [31] and this concept has also captivated significant interest in visual analytics. Texts and hyperlinks connecting to bookmarked visualizations can constitute a story which can embrace also graphical annotations [9, 29]. It is assumed that visual storytelling can be critical in contributing to a more intuitive and fast analysis of broad data resources [32]. Even in the scientific approach, the storytelling concept urges as scientific storytelling, which means telling stories using scientific data. Visualization is used in academia to validate experiments, explore datasets or even to transmit findings, so if properly done, such visualizations can be highly effective in conveying narratives [33].

The Microsoft Power BI software is a business intelligence tool where visualization seems interactive and rich, allowing the creation of dashboards in matter of minutes. Although there is the option of running a R script, the software doesn´t require programming skills. The program is able to connect to various data sources in order to extract and transform them, creating information [28].

Currently and as evidenced by Gartner's Magic Quadrant, the Power BI software has assumed the first position in the ranking, since February 2019, ahead of Tableau, which until then was recognized as the most used tool within the subject of business intelligence[1].

In the next sections an example developed using Power BI is presented that used a Drill-Down Story allowing the user to select among particular details, putting more attention on the reader-driven approach [32].

3 Construction of a Dashboard Reflecting a Manufacturing Cell

3.1 Context Goals and Methods

The case study presented in this article was carried out at a company whose production focuses on flush toilets. Belonging to the chemical industry, its production is divided into two sectors, injection (made up of several injection molding machines) and assembly (made up of several cells that cover different parts of the flush toilet). In the assembly area, there are numerous manufacturing cells where automation can effectively make a difference. The currently most automated one, having data capture through IoT mechanisms, is the tap cell. The data acquired in this cell do not have any meaning to the decision-maker; yet they may produce potentially relevant information.

The goal of this case study is the construction of a dashboard where it is possible to view the manufacturing cell data in a way capable to help understanding the actual production state and thus support decision-making. It uses data analysis and visualization techniques, as well as storytelling.

[1] https://powerbi.microsoft.com/en-us/blog/microsoft-a-leader-in-gartners-magic-quadrant-for-analytics-and-bi-platforms-for-12-consecutive-years (visited, Jan, 2020).

For the construction of the dashboard to be possible, firstly an analysis of problem, including the software, was carried out, through its modelling in Unified Modelling Language (UML). After this modelling and after understanding the data generated, exporting them to Excel was essential to better understand the problem. Through Power BI Software Power Query, several transformations were possible, obtaining a fact table of relevant information. This table was built using M and DAX language and in the end, the application of graphic elements was done, creating the final dashboard.

It is important to denote that the process of creating the dashboard application involved three representative company divisions (actors), namely, the data analyst, the IT technician and the head of continuous improvement. These three types of users contributed to understand the dashboard requirements, as well as what advantage would be derived from the use of this streaming data to assist in decision making.

3.2 Result with Some Software Dashboard Interfaces

Concerning the dashboard application, it was created following a user-centered approach. Figure 1 presents the first dashboard menu, where the user can choose among viewing station stops, cadences, actual production and efficiency levels of the station.

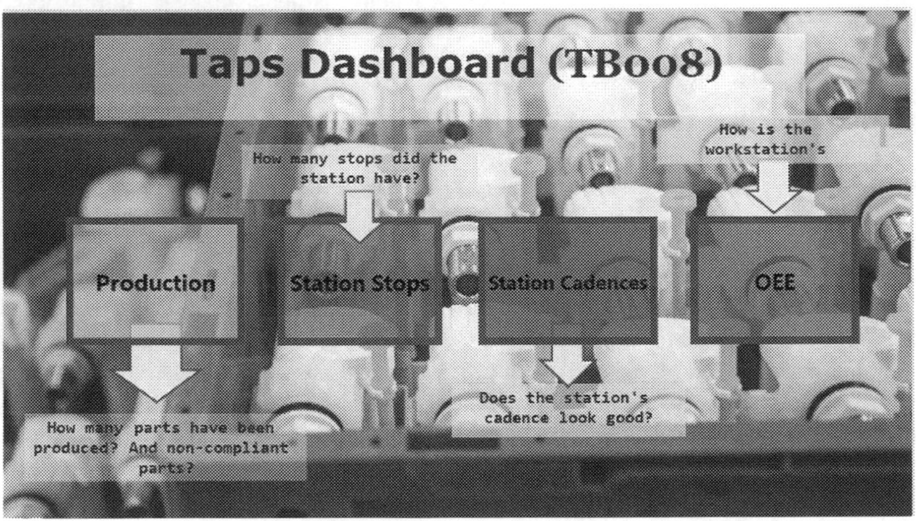

Fig. 1. Menu dashboard

In the "Production" (Fig. 2) bookmark the user can find the total amount of parts produced by the station and the total number of non-compliant parts. Once more, there are two filters, one is the date and the other is the product family.

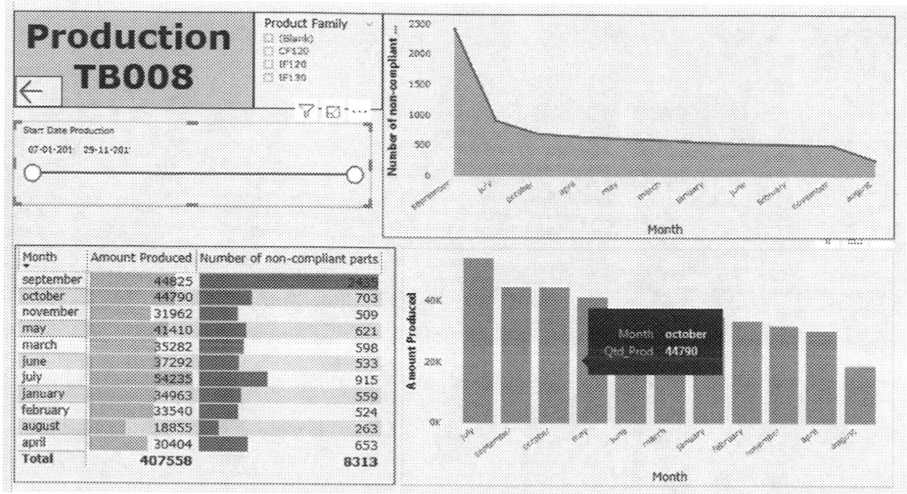

Fig. 2. Production bookmark

In the "Station Stops" (Fig. 3) bookmark is possible to visualize the total number of stops at the station, the total number of scheduled stops, micro stops and the total time available for production. The analysis can be filtered according to the date and product family chosen by the user.

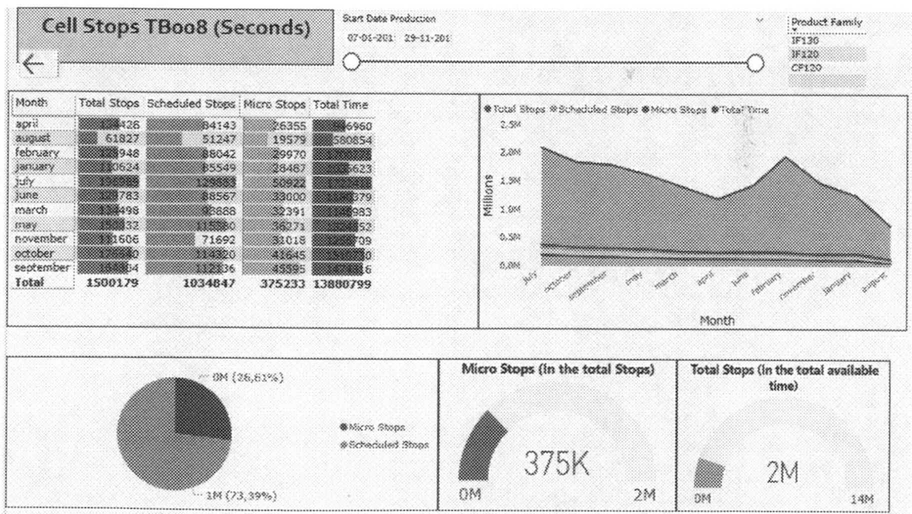

Fig. 3. Station Stops bookmark

The "Station Cadences" (Fig. 4) is another bookmark where the real cadence and the theoretical one can be compared.

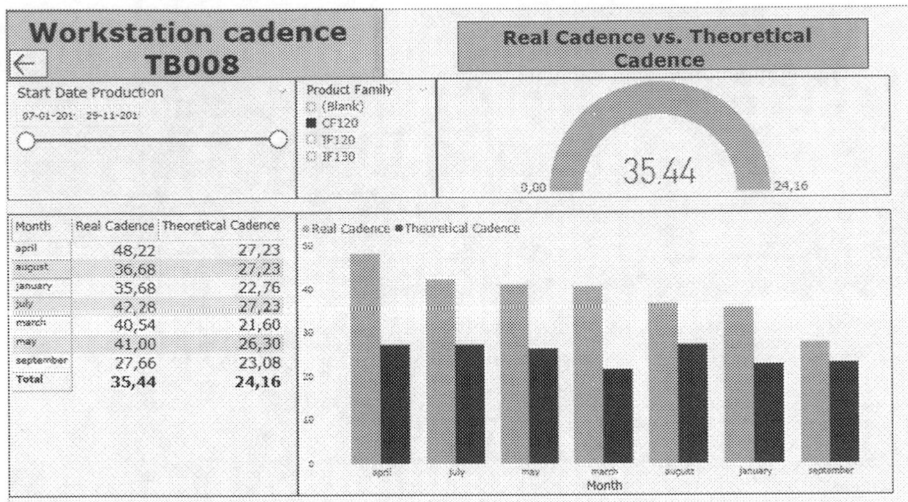

Fig. 4. Workstation Cadence bookmark

The final bookmark "OEE" (Fig. 5) displays the station efficiency levels, calculated by the concept of Overall Equipment Effectiveness. Here, the OEE Availability, OEE Operator/Performance and OEE Quality are calculated along time and the multiplication of the three allows us to obtain the global value (OEE Global). Filtering it is also possible using date and product family (Fig. 5).

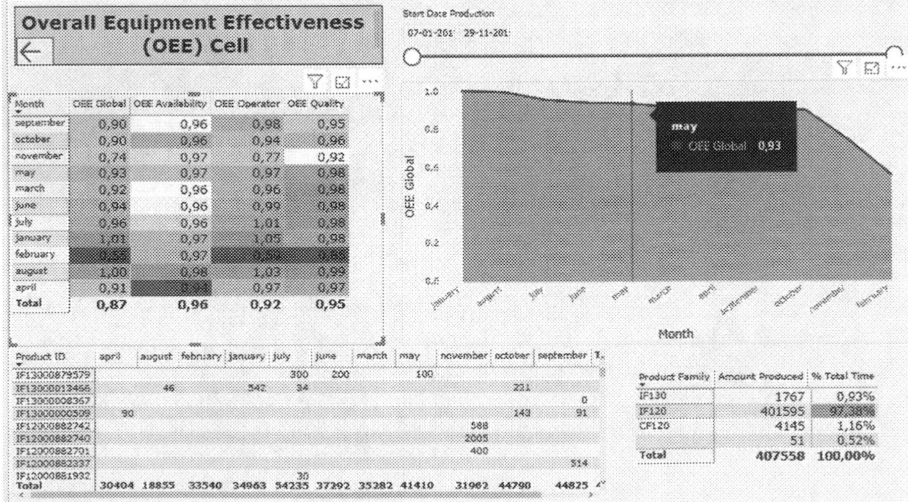

Fig. 5. Overall Equipment Effectiveness bookmark

4 Final Remarks and Future Work

The created dashboard allowed showing an informative overview to the user in order to facilitate the interpretation of the data resulting from the cell's production. Power BI proved to be a tool capable of representing data visually "telling a story" about how the cell is operating that can be easily understood by the user providing insights into the cell's activity.

More and more, particularly in the context of industry 4.0, the use of data becomes essential in order to bring value to the organization and easy decision making. The introduction of IoT mechanisms on the shop floor brings the need to take advantage of the data collected, using data visualization, data analytics and, more recently, storytelling tools. The ability to convey information in a more perceptible way has become a concern, considering the numerous resources and data sources scattered throughout the manufacturing space. Expertise in data processing and visualization is currently one of the foci of hiring companies and software such as Power BI facilitates these activities since they appear to be intuitive and accessible for people without advanced programming skills.

As future work, there is a need to test the dashboard with other types of users in order to evaluate it as a proof of concept, as well as to extend the dashboard to other manufacturing cells, so that operators in the shop floor can have better understanding of the data and of the complete manufacturing process. The application of the dashboard on the shop floor will allow acting in real time in the face of errors or discrepancies that may occur along the processes.

Acknowledgments. This research was supported by the Portuguese National Funding Agency for Science, Research and Technology (FCT), within the Institute of Electronics and Informatics Engineering of Aveiro (IEETA), project UIDB/00127/2020.

References

1. Hill, R., Devitt, J., Anjum, A., Ali, M.: Towards in-transit analytics for industry 4.0. In: Proceedings - IEEE International Conference on Internet Things, IEEE Green Computing and Communications IEEE Cyber, Physical and Social Computing, IEEE Smart Data, pp. 810–817 (2017). https://doi.org/10.1109/iThings-GreenCom-CPSCom-SmartData.2017.124
2. Arromba, A.R., Teixeira, L., Xambre, A.R.: Information flows improvement in production planning using lean concepts and BPMN an exploratory study in industrial context. In: 14th Iberian Conference on Information Systems and Technologies (CISTI), pp. 206–211 (2019)
3. Miragliotta, G., Sianesi, A., Convertini, E., Distante, R.: Data driven management in Industry 4.0: a method to measure Data Productivity. IFAC-PapersOnLine **51**, 19–24 (2018). https://doi.org/10.1016/j.ifacol.2018.08.228
4. Chaudhary, P., Hyde, M., Rodger, J.A.: Exploring the benefits of an agile information system. Intell. Inf. Manage. **09**, 133–155 (2017). https://doi.org/10.4236/iim.2017.95007
5. Narayanan, M., Sanil Shanker, K.P.: Data visualization method as the facilitator for business intelligence. Int. J. Eng. Adv. Technol. **8**, 2249–8958 (2019). https://doi.org/10.35940/ijeat.f9054.088619

6. Choi, T.M., Chan, H.K., Yue, X.: Recent development in big data analytics for business operations and risk management. IEEE Trans. Cybern. **47**, 81–92 (2017). https://doi.org/10.1109/TCYB.2015.2507599
7. Pribisalić, M., Jugo, I., Martinčić-Ipšić, S.: Selecting a business intelligence solution that is fit for business requirements. In: 32nd Bled eConference Humanizing Technology for a sustainable Society, pp 443–465 (2019)
8. Stecyk, A.: Business intelligence systems in SMEs. Eur. J. Serv. Manage. **27**, 409–413 (2018)
9. Chen, S., Li, J., Andrienko, G., et al.: Supporting story synthesis: bridging the gap between visual analytics and storytelling. IEEE Trans. Vis. Comput. Graph. **14**, 1077–2626 (2015). https://doi.org/10.1109/TVCG.2018.2889054
10. Mantravadi, S., Møller, C.: An overview of next-generation manufacturing execution systems: how important is MES for industry 4.0? Procedia Manuf. **30**, 588–595 (2019). https://doi.org/10.1016/j.promfg.2019.02.083
11. Da, X., Xu, E.L., Li, L.: Industry 4.0: state of the art and future trends. Int. J. Prod. Res. **56**, 2941–2962 (2018). https://doi.org/10.1080/00207543.2018.1444806
12. Savastano, M., Amendola, C., Bellini, F., D'Ascenzo, F.: Contextual impacts on industrial processes brought by the digital transformation of manufacturing: a systematic review. Sustainability **11**, 891 (2019)
13. Salierno, G., Cabri, G., Leonardi, L.: Different perspectives of a factory of the future: an overview. In: Proper, H., Stirna, J. (eds.) Advanced Information Systems Engineering Workshops. Lecture Notes in Business Information Processing, pp. 107–119. Springer, Cham (2019). https://doi.org/10.1007/978-3-030-20948-3_10
14. Qu, Y., Ming, X., Ni, Y., et al.: An integrated framework of enterprise information systems in smart manufacturing system via business process reengineering. Proc. Inst. Mech. Eng. Part B J. Eng. Manuf. (2018). https://doi.org/10.1177/0954405418816846
15. Tao, F., Zhang, M.: Digital twin shop-floor: a new shop-floor paradigm towards smart manufacturing. IEEE Access **5**, 20418–20427 (2017). https://doi.org/10.1109/ACCESS.2017.2756069
16. Qi, Q., Tao, F.: Digital twin and big data towards smart manufacturing and Industry 4.0: 360 degree comparison. IEEE Access **6**, 3585–3593 (2018). https://doi.org/10.1109/ACCESS.2018.2793265
17. Zhu, Z., Liu, C., Xu, X.: Visualisation of the digital twin data in manufacturing by using augmented reality. Procedia CIRP **81**, 898–903 (2019). https://doi.org/10.1016/j.procir.2019.03.223
18. Schroeder, G., Steinmetz, C., Pereira, C.E., et al.: Visualising the digital twin using web services and augmented reality. In: IEEE International Conference on Industrial Informatics, pp. 522–527 (2016). https://doi.org/10.1109/INDIN.2016.7819217
19. Uhlemann, T.H.J., Lehmann, C., Steinhilper, R.: The digital twin: realizing the cyber-physical production system for Industry 4.0. Procedia CIRP **61**, 335–340 (2017)
20. Schrefl, M., Neub, T., Schrefl, M., et al.: Modelling knowledge about data analysis modelling knowledge about data analysis in manufacturing processes. IFAC-PapersOnLine **48**, 277–282 (2015). https://doi.org/10.1016/j.ifacol.2015.06.094
21. Bordeleau, F.E., Mosconi, E., de Santa-Eulalia, L.A.: The management of operations Business intelligence and analytics value creation in Industry 4.0 : a multiple case study in manufacturing medium enterprises case study in manufacturing medium enterprises. Prod. Plann. Control 1–13 (2019). https://doi.org/10.1080/09537287.2019.1631458

22. Raghav, R.S., Pothula, S., Vengattaraman, T., Ponnurangam, D.: A survey of data visualization tools for analyzing large volume of data in big data platform. In: Proceedings of International Conference on Communication, Computing and Electronics Systems, ICCES 2016, pp. 1–6 (2016). https://doi.org/10.1109/CESYS.2016.7889976
23. Raffoni, A., Visani, F., Bartolini, M., Silvi, R.: Business Performance Analytics: exploring the potential for Performance Management Systems. Prod. Plann. Control **29**, 51–67 (2018). https://doi.org/10.1080/09537287.2017.1381887
24. Poleto, T., De Carvalho, V.D.H., Costa, A.P.C.S.: The full knowledge of big data in the integration of interorganizational information: an approach focused on decision making. Int. J. Decis. Support Syst. Technol. **9**, 16–31 (2017). https://doi.org/10.4018/IJDSST.2017010102
25. Morgan, R., Grossmann, G., Schrefl, M., Stumptner, M.: A model-driven approach for visualisation processes. In: ACM International Conference Proceeding Series (2019). https://doi.org/10.1145/3290688.3290698
26. Thalmann, S., Mangler, J., Schreck, T., et al.: Data analytics for industrial process improvement a vision paper. In: Proceeding - 2018 20th IEEE International Conference on Bus Informatics, CBI 2018, vol. 2, pp. 92–96 (2018). https://doi.org/10.1109/CBI.2018.10051
27. Zhou, F., Lin, X., Liu, C., et al.: A survey of visualization for smart manufacturing. J. Vis. **22**, 419–435 (2019). https://doi.org/10.1007/s12650-018-0530-2
28. Ali, S.M., Gupta, N., Nayak, G.K., Lenka, R.K.: Big data visualization: tools and challenges. In: Proceedings of 2016 2nd International Conference on Contemporary Computing and Informatics, IC3I 2016, pp. 656–660 (2016). https://doi.org/10.1109/IC3I.2016.7918044
29. Lee, B., Riche, N.H., Isenberg, P., Carpendale, S.: More than telling a story: transforming data into visually shared stories. IEEE Comput. Graph. Appl. **35**, 84–90 (2015). https://doi.org/10.1109/MCG.2015.99
30. Kosara, R., MacKinlay, J.: Storytelling: the next step for visualization. Computer (Long Beach Calif) **46**, 44–50 (2013). https://doi.org/10.1109/MC.2013.36
31. Tong, C., Roberts, R., Laramee, R.S., et al.: Storytelling and visualization: a survey. In: VISIGRAPP 2018 - Proceedings of the 13th International Joint Conference on Computer Vision, Imaging and Computer Graphics Theory and Applications, pp. 212–224. SciTePress (2018)
32. Segel, E., Heer, J.: Narrative visualization: telling stories with data. IEEE Trans. Vis. Comput. Graph. **16**, 1139–1148 (2010). https://doi.org/10.1109/TVCG.2010.179
33. Ma, K.-L., Liao, I., Frazier, J., et al.: Scientific storytelling using visualization. IEEE Comput. Graph. Appl. **32**, 12–15 (2012)

The Study on How Influencer Marketing Can Motivate Consumer Through Interaction-Based Mobile Communication

Kai-Shuan Shen(✉)

Fo Guang University, Jiosi, Yilan County 2624, Taiwan (R.O.C.)
ksshen319@gmail.com

Abstract. Compared to traditional sales model, influencer marketing motivates consumers in a totally different way in this era of mobile phone. This type of marketing uses various phenomenons hot online, such as a delicately designed image with high attention, so targeted merchandise can be concerned associatedly. The way which influencer marketing uses activates consumers' interests and improves the appeal of products. The success of influencer marketing also indicates the achievement of mobile communication and motivates my study.

Kansei Engineering was applied to study how influencer marketing can motivate consumer through mobile communication based on human emotions. Hence, EGM (Evaluation Grid Method) was used to determine the interdependent appeal factors and specific characteristics of influencer marketing based on experts' opinion. In addition, this study took advantage of Quantification Theory Type I to analyze the importance of each appeal factors and characteristics according to consumers' responses. This study combined in-depth interviews and questionnaire surveys, and can be individual analyzed through EGM and Quantification Theory Type I. Hence, both experts' and consumers' preferences to influencer marketing can be reveal. Then, the role of mobile phone can be positioned and clarified in influencer marketing & communication based on the results of this study.

The semantic structure of appeal of influencer marketing, as the results of EGM analysis, showed the hierarchy of the relationship among appeal factors, the reasons for consumers' preferences, and the specific characteristics. In addition, the results of Quantification Theory Type I analysis indicated that appeal factors will be affected in varying degrees by particular reasons and characteristics in this study. Then, the influencer marketing strategies could be used for mobile communication on the basis of the result of this study. Researchers who are concerned about the issues of influencer marketing can get useful information in this study. Furthermore, the achievements of this study can contribute to the field of marketing, mobile communication, social media, and media design.

1 Introduction

Influencer marketing are undoubtedly one of the most cost-effective way of promotion for sellers or brands. Many techniques of influencer marketing have been widely used by sellers or brands. For example, graffiti walls, action figures, and pop stores can be

C. Stephanidis et al. (Eds.): HCII 2020, LNCS 12427, pp. 652–660, 2020.
https://doi.org/10.1007/978-3-030-60152-2_49

set for consumers' taking a picture and uploading to social medial sites. In addition, celebrities or stars' participating activities, having experience in using or introducing products can be used for the increase of credibility through the communication of social media (see Fig. 1).

Fig. 1. The above pictures show how influencer marketing is practiced through a mobile phone and how to attract consumers to check in online with designed aesthetic landscape, such as a instagrammable wall.

I probe the appeal of interaction-based from the perspective of mobile communication in this study. In the following section, I interviewed related theories and documentation.

2 Review of the Literature

2.1 Influencer Marketing Through Social Media

Consumers' motivations to participate influential marketing are crucial to realize how this type of marketing are popular in recent years, especially for specific social media. As far as YouTube consumers were concerned, influential factors affecting consumers' perception of credibility were proved to be trustworthiness, social influence, argument quality, and information involvement [1]. Influencer marketing, as a model of innovative strategies, has become a powerful weapon to break the traditional advertising industry. In addition, the billion-dollar influencer marketing industry is worth exploring because it needs a change for effective implementation [2].

2.2 Interaction-Based Mobile Communication

Interactivity and mobility, as the characteristics of wireless devices, provide an innovative platform for marketing, makes commercial interaction happen anytime and anywhere [3]. In this study, the research focused on the interaction-based mobile communication and tried to find out how the technique of influencer marketing was used to interact with consumers and created the best cost effectiveness for commercial promotion.

2.3 Kansei Engineering

The Kansei concept come from consumers' feeling for products. Kansei engineering focuses on translating Kansei concepts into specific design concepts, such as product mechanical function [4]. It means that Kansei engineering attempts to merge consumers' preferences into product design. Hence, Kansei engineering is also famous for its consumer-oriented technology. This study would use Kansei engineering as the theory-based methodology to explore how influencer marketing can motivate consumers based on their emotions.

2.4 Evaluation Grid Method

Then, the EGM, functioned as visualization of users' requirements, was used as the first technique to conduct Kansei engineering in this study. The EGM adopts laddering [5], as leading questions to elicit higher/lower level of constructs so reviewees' original expression can be constructed systematically.

2.5 Quantification Theory Type I

Quantification Type I Method, as a quantitative tool, was used to analyze the importance of the appeal factors, reasons and characteristics of the interaction which was formed through influencer marketing in this study. More specifically, Quantification Type I Method could measure and quantify the upper-l and lower-level items using the importance levels from the original evaluation items. Quantification Theory Type I can statistically predict the relationship between a response value and the categorical values using multiple linear regression methods [6]. Moreover, in the field of design, the weights of the factors of users' preferences can be evaluate using Hayashi's Quantification Theory Type I [7, 8].

3 Research Objectives

This study focused on how influencer marketing can motivate consumer through mobile communication. Hence, the following critical issues would be probed. First, in order to comprehend why consumers preferred influencer marketing, the design characteristics and appeal factors of it could be determined through interviewing experts. Secondarily, in order to realize how consumers preferred influencer marketing, the importance of the appeal factors and design characteristics for influencer marketing could be quantified through statistical analysis.

4 Research Methods

In order to explore how influencer marketing can motivate consumer through mobile communication. Kansein Engineering, as emotion-based theory, was used for evaluating consumers' motivation in participating influencer marketing. Hence, this study hypothesized that influencer marketing could attract consumers. This means that influencer marketing was assumed to have appeal to consumers. Furthermore, the appeal of influencer marketing could be determined and quantified through Kansei Engineering. Hence, this study was divided into two stages to explore the appeal of influencer marketing as the following:

In the first stage, the design characteristics and appeal factors of influencer marketing could be determined through EGM (Evaluation Grid Method). 161 Kanssei words were selected from the articles of authoritative magazines, columns, and professional web sites. Then, 174 Kansei words were chosen from the interviews of five professional experts, which comprised three editors, one seller, and one experienced user with the ages who were ranging in age from 32 to 56 and were all good at the field of influencer marketing.

21 Kansei words were finally evolved from 355 ones which came from both experts' opinions and editors' perspectives in articles. The process evolution of these Kansei words were based on the method of the EGM. The core technique of the EGM could arrange the Kansei words from abstract to specific concepts in a hierarchy from. The abstract Kansei words were named as "upper-level" concepts in this study, such as "influential". The specific Kansei words were named as "lower-level" concepts in this study, such as "the increase of online exposure". The connection between "upper-level" and "lower-level" concepts lies in relation of subordination. For example, the corresponding concept of the above mentioned upper-level "influential" one is "the increased of online exposure".

5 Analysis and Results

The researcher conducted EGM for extracting Kansei words from two origins. First of all, 5 experts were interviewed for the collection of their professional opinions. Secondly, 9 selected articles were reviewed for the construction of the strong knowledge-based Kansei system. Finally, 21 Kansei words were evolved from 355 ones which came from both experts' opinions and editors' perspectives in articles. Then, "communicably-targeted" and "Interaction-based" were determined as the original evaluation items according to EGM. Then, each original evaluation item had its own corresponding "upper-level" and "lower-level" concepts. As far as "communicably-targeted" item was concerned, its corresponding "upper-level" concepts were "influential" and "exact", of which totally matched "lower-level" concepts were "easy to be communicated widely", "using hashtag reasonably", "the increase of online exposure", "a large number of followers or fans", "high cost effectiveness", "exact marketing" and "exact communication". As far as "interaction-based" item was concerned, its corresponding "upper-level" concepts were "social interactive" and "customer based", of which totally matched "lower-level" concepts were "online check-in", "taking a picture for consumers for getting their contact information", "opening a pop-up shop", "giving a consumer a gift for hashtag", "creating contents for a brand by customers", "realizing a customer's preferences", "realizing who are target customers" and "retaining customers".

This EGM structure composed of these selected evaluation items was also used for design of the questionnaire and could then be transferred to the following table, which shows their ranking by the number of accumulated times (Table 1). The upper-level reasons with the higher number of times were also determined using the EGM, as shown in Table 2. In order to quantify the selected items and to give a score to the correspondent appeal factors for interaction-based influencer marketing, questionnaires were created using a level-based construction, which is composed of upper and lower levels based on EGM, as shown in Table 3.

Table 1. The ranking from the hierarchical diagram by the number of times the descriptions appeared

Original images	Upper level (reasons)	Lower level (specific attributes)
Interaction-based 11	Interactive 13	Online check-in 15
Communicably-targeted 9	Influential 12	Opening a pop-up shop 13
Recommended 7	Consumer-based 10	Creating contents for a brand by customers 12
Cool 5	Exact 9	Easy to communicated widely 12
Fashionable 3	Sharing 7	Using hashtag reasonably 11

In addition, the contents of the interviews were recorded and summarized in the form of hierarchy diagrams (Figs. 2 and 3), which were then used as the foundation for the design of a questionnaire.

Note: The No.4 evaluation structure was constructed by retrieving the answers from an experienced user, who is 36 years old.

The first appeal factor categorized was "communicably-targeted", which included "influential" and "exact". The classifications indicated that users had the impression of

Table 2. The best four "original images and reasons" selected from the hierarchical diagram by the higher number of times they appeared

Classified	Original images	Reasons (upper level)
First	Interactive 13	Online check-in 15 Opening a pop-up shop 13 Creating contents for a brand by customers 12
Second	Influential 12	Easy to communicated widely 12 Using hashtag reasonably 11 The increase of online exposure 9
Third	Consumer-based 10	Realizing a consumer's preferences 10 Realizing who are target consumers 8 Retaining customers 7
Forth	Exact 9	Exact marketing 11 Exact communication 8 High cost effectiveness 7

Table 3. The setting of the level-based construction of questionnaire

Level of questionnaire	The first level	The second level	The third level
The type of question	Original evaluation item	Upper level	Lower level
The example of a question subject	Interaction-based	Interactive	Online check-in

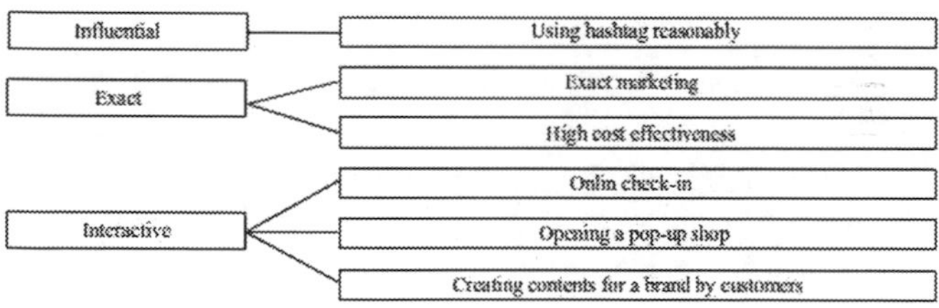

Fig. 2. An example of a participant's evaluation structure

"communicably-targeted" on influencer marketing because of the two above-mentioned reasons. The results of the Quantification Theory Type I shows the coefficient of determination (R2 = 0.686) in this study and means standard reliability for our survey results. The reason of "influential", with the highest partial correlation coefficient, had the most influence on the appeal factor "communicably-targeted". "easy to be communicated widely", "using hashtag reasonably", "the increase of online exposure", "A large number of followers or fans", and "meeting together offline" were included in this category. According to the partial correlation coefficients, "influential" had a stronger effect on "communicably-targeted". In addition, according to the category scores, "easy

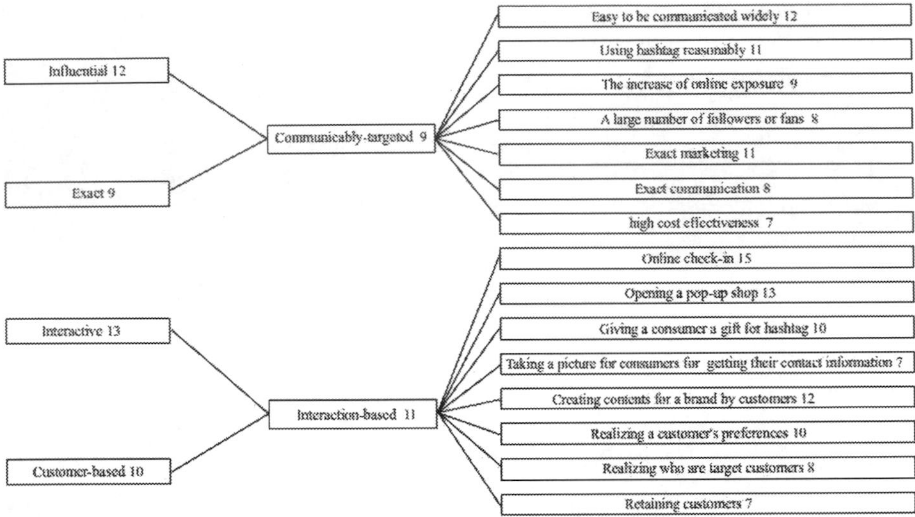

Fig. 3. The hierarchical diagram of preferences for users of influencer marketing, determined by the EGM

to be communicated widely" had the most positive influence on "influential" and using hashtag reasonably" had the most negative effect on "influential" (Table 4).

Table 4. The partial correlation coefficients, the category scores and the coefficient of determination for the factor of "communicably-targeted"

Items	Categories	Category scores	Partial correlation coefficients
Influential	Easy to be communicated widely	0.151	0.820
	Using hashtag reasonably	−0.154	
	The increase of online exposure	0.026	
	A large number of followers or fans	−0.104	
	Meeting together offline	0.016	
Exact	Exact marketing	−0.017	0.437
	Exact communication	−0.027	
	High cost effectiveness	0.053	
C	0.801		
R=	0.828		
R^2=	0.686		

The second factor to be analyzed was "interaction-based", which comprised "interactive" and "customer based" in the upper-level assessment. According to the results of the Quantification Type I, the coefficient of determination (R2) was 0.559 and shows standard reliability for the survey instrument. “Interactive” contributed most to the “interaction-based” factor because it had the highest correlation coefficient (see Table 5). The categories that belong to “interactive” included “online check-in”, “opening a pop-up shop”, “giving a consumer a gift for hashtag”, “taking a picture for consumers for getting their contact information”, and “creating contents for a brand by customers”. The partial correlation coefficient of Table 5 shows “interactive” had a stronger effect on “interaction-based” then “customer based”. Furthermore, according to the category scores, “creating contents for a brand by customers” had more positive effect on “interactive” of all the categories. Then, “realizing a customers’ preferences” had more negative influence on “customer based” of all the categories.

Table 5. The partial correlation coefficients, the category scores and the coefficient of determination for the factor of “interaction-based”

Items	Categories	Category scores	Partial correlation coefficients
Interactive	Online check-in	0.076	0.700
	Opening a pop-up shop	−0.069	
	Giving a consumer a gift for hashtag	−0.029	
	Taking a picture for consumers for getting their contact information	−0.089	
	Creating contents for a brand by customers	0.101	
Customer based	Realizing a customers’ preferences	−0.102	0.658
	Realizing who are target customers	0.048	
	Retaining customers	0.038	
C	0.801		
R=	0.748		
R^2=	0.559		

6 Discussion and Conclusions

The semantic structure of appeal of influencer marketing, as the results of EGM analysis, showed the hierarchy of the relationship among appeal factors, the reasons for consumers’ preferences, and the specific characteristics. This results showed that consumers were affected by influencer marketing based on the “communicably-targeted” and “interaction-based” factors. The two factors indicated that influencer marketing adopted innovative strategies different from traditional way. This type of strategies focused on how to interact with customers and how to expand influence through them. Furthermore, influencer marketing motivated consumers for the four reasons, including “influential”, “exact”, “social” and “customer-based” considerations. These four reasons disclosed customers’ motivations to participate influencer marketing. Customers enjoyed themselves in the process of influencer marketing because they can interact with friends and effort distribution of influence so a brand or a seller also can conduct a marketing strategy successful. In addition, all the "lower-level" concepts, such as "opening a pop-up shop", showed what specific means a brand or a seller used to attract consumers.

The limitations of this research are as follows. First, even though the interviews, including text and sound, were completely recorded for the process of generalising the results of the interviews for the EGM, decoding the implication of the words of the experts can be very difficult. Interpretation without bias involves semantics, especially for descriptive adjectives. Further, in the process of conducting the questionnaire survey, even though most of the responding individuals were well informed and motivated to participate, possible biases of cognition may exist due to personal differences in their comprehension of the descriptions in the questionnaire. Further, the influencer marketing in different countries or areas vary for many reasons, such as the influences of culture or society, so the current data collected from the specific institutions in Taiwan may not be representative of wider populations.

References

1. Xiao, M., Wang, R., Chan-Olmsted, S.: Factors affecting YouTube influencer marketing credibility: a heuristic-systematic model. J. Media Bus. Stud. **15**(3), 188–213 (2018)
2. Childers, C.C., Lemon, L.L., Hoy, M.G.: #Sponsored #Ad: agency perspective on influencer marketing campaigns. J. Curr. Issues Res. Advert. **40**(3), 258–274 (2019)
3. Barnes, S.J., Sornavacca, E.: Mobile marketing: the role of permission and acceptance. Int. J. Mob. Commun. **2**(2), 128–139 (2004)
4. Nagamachi, M.: Kansei engineering as a powerful consumer-oriented technology for product development. Appl. Ergon. **33**(3), 289–294 (2002)
5. Fransella, F., Bannister, D.: A Manual for Repertory Grid Technique. Academic Press, London (1977)
6. Hayashi, C.: On the quantification of qualitative data from the mathematico-statistical point of view. Ann. Inst. Stat. Math. **2**(1), 35–47 (1950)
7. Iwabuchi, C., et al.: Data Management and Analysis by Yourself, pp. 180–185. Humura Publishing, Tokyo (2001)
8. Sugiyama, K., Novel, K.: The Basic for Survey and Analysis by Excel: A Collection of Tool for Planning and Design, pp. 51–62. Kaibundo Publishing, Tokyo (1996)
9. Lee, W.Y.: How to attract consumers through influencer marketing. CommonWealth Mag., Taipei (2018)

A Visual Tracking Method to Explore the Effect of Presence on Online Consumers

Yu Sun(✉)

School of Business Administration, Shenyang University, Shenyang 110044, People's Republic of China
104549720@qq.com

Abstract. The online advertisements (ads) with presence design are widely proved to be more appealing for attracting users' attention. However, the different visual processing mechanism towards presence design and consumers' involvement has not been well investigated. This study combined subjective evaluation with eye-tracking data to investigate the effect of consumers' involvement on their visual attention to online ads with different sense of presence. Refined and re-calibrated presence scale and approaching-avoidance trend scale were constructed to respectively measure consumers' perceived sense of presence and behavioral intention to online ads. Two tasks were designed to simulate different sense of user involvement. In task one, participants just freely browsed two banner ads. Meanwhile binary choices were used in task two to ask participants to select one with higher purchase intention from two vertical search ads and then click the mouse. Both tasks were completed by each participant whose eye movement data was recorded by an eye-tracker. The perceived sense of presence and corresponding behavioral intention of each two ads were evaluated after each task. The results showed that online ads with high presence will significantly increase consumers' fixation counts, revisit counts, saccadic counts, blink counts in browsing-driven task which will significantly improve consumers' stickiness. As to purchasing-driven task, however, online ads with high presence will reduce fixation counts, saccadic counts, blink counts which indicates that the online ads with high sense of presence can reduce the difficulty of information extraction, thus helping the customer to make purchase decisions effectively.

Keywords: Presence · Users' involvement · Online advertising · Visual attention

1 Introduction

With the amount of online advertising on a steady rise, generic ads noticeably lose effectiveness. In order to break through the clutter, more and more experts and designers begin to explore how to stimulate the users' attention towards the online ads. However, as the feeling of remoteness between the internet and the field, the ads with presence design are proved to be more appealing, which can attract more attention and thus to cause purchase intentions. The effect of online ads with sense of presence has been a controversial issue, especially on the topic of how to effectively help users to

C. Stephanidis et al. (Eds.): HCII 2020, LNCS 12427, pp. 661–678, 2020.
https://doi.org/10.1007/978-3-030-60152-2_50

make decisions. This paper just uses field data to show that the effectiveness of online ads with presence design, no matter the involvement of the users, idling or purchasing, the presence design will considerably hinge on consumers' appreciation, which can be shown in eye movements obviously.

1.1 Presence and Current Research

Facing the fast increase of netizens and the ever-expanding advertising market, how to stand out, to attract attention and even to stimulate clicks from visitors in mass information is an important issue for advertisers to consider. In addition to focus on the product quality and price as the general advertising, online advertising also needs to take the virtual network environment into account. As an important influencing factor of virtual environment, the sense of presence not only helps products to achieve the stimulation effect of field sales in the network environment, but also stimulates consumers' attention, click rates and purchase intentions through subjective design. Presence can be illustrated as the individual's perception of the environment (Montes 1992) which can describe the true degree of an individual in a virtual environment, as a sense of "being there" (Matthew and Matthew 2007). In 2004, Lee divided this sense of presence into three levels, namely physical presence, social presence and self-presence according to the process of human perception. Throughout the current research, most scholars only discuss the sense of presence at one level.

At the physical level, scholars mostly study how products or objects from a remote distance can be vividly presented in front of them through media. As early as 2000, Jahng pointed out that physical presence would bring the "immediate" psychological and physical experience of products to consumers. Fiore (2001) also points out that physical presence can be expressed through rich media, allowing consumers to interact with products' images, so as to achieve users' perception of products' proximity. The presence at this level makes people realize that the vividness and verisimilitude of the product image display will have a positive impact on customers' perception and attitude.

At the level of social presence, many scholars have contributed a lot in their researches. Biocca et al. (2003), Steuer (1992) and Oh (2008) all take the social presence as the evaluation index of virtual reality, and adopt the method of questionnaire survey to measure the friendliness and accessibility of a media. Their research results make people realize that in the network, through the shaping of social situations, communicating and sharing with others, the customer's sense of social presence will be enhanced, thus influencing the customer's purchasing intentions.

In terms of self-presence, scholars mainly focus on the field of online games and e-commerce. Oatley (1994) pointed out that users can improve their self-presence, understand the purpose of the game and have more fun of the game by imagining themselves as the protagonist. Tan et al. (2012) and Namkee et al. (2010) have also verified through experiments that realistic avatar design and emotional background stories can improve users' sense of self-presence in online games, thus leading to a better task performance. Wang et al. (2007) combined the sense of self-presence with the sense of immersion and flow experience to illustrate the impact of customers' loyalty on purchasing intentions. The results of these studies showed that the customers who enhanced their self-presence were more immersed in the virtual environment and thus complete the tasks better. In short, presence, as an important factor influencing the interaction between users and virtual environment has been more and more recognized and valued.

1.2 Consumers' Involvement

The involvement refers to the degree of an individual devoting in a task. There have been some research findings about the degree of users' involvement and the communication effect of online advertising. Palanisamy and Wong (2003) discussed the impact of the inherent flexibility, expectation, involvement and perceived personal usefulness on the effectiveness of banner advertising. There is no gender difference in the influence of audience's search style and involvement level on advertising effect (Palanisamy 2004). Yang (2004) divided customers into two categories (high involvement and low involvement) and studied the role of customers' motivation in online search tasks using online advertisements. Kim et al. (2010) think that online shopping should emphasize the nature of advertising to attract customers with high degree of involvement, such as good service image and excellent service quality. For customers with low involvement, who may just choose gifts, the designers need to consider the entertainment information combined with product information and ensure smooth communication. Kim and Lee (2011) confirmed that visitors' searching style will affect their visual attention, and exploratory visitors are more likely to pay attention to advertisements than the targeted visitors. Butt and de Run (2011) studied the impact of users' purpose on advertising interactivity, and the results showed that targeted users had a positive attitude towards low-interactive ads, while aimless users had a positive attitude towards high-interactive ads.

Although it has been affirmed that the high degree of users' involvement has positive effects on advertising communication, it needs further discussion on how specific advertisements affect users' behavioral intention and how users' eye movements reflect it.

1.3 Eye-Tracking Method

In the past, questionnaire survey was used to mark the recall or recognition of advertising in order to evaluate the advertising effect, which lacked real-time measurement and had low ecological validity. AIDA sales model which is introduced by international marketing expert Haiimam Godman emphasizes the importance and sequence of attention, interest, desire and purchase. It can be seen that in most advertising modes, "attention" is placed in a very important position. Most advertising models agree that "visual attention" is a necessary precondition for advertising to be effective. Based on this, some scholars assert that "no matter how much the budget is, only the advertising that can capture consumers' attention can be regarded as a success". At present, scholars generally think that understanding when, where and how the consumers' attention will be directed to advertising, and what kind of factors will occupy consumers' attention distribution will be a very important research topic in the field. In an age of attention increasingly scarce, research on consumers' attention will become increasingly important.

As eye movement is a physiological analysis method that can be used to measure users' attention and information acquisition and often involves higher cognitive processes (Clifton et al. 2007), it can better reflect customers' information processing and perception experience of online advertisements. The users' onsite experience of different advertisements can be analyzed and understood from the physiological awakening of the subjects, especially changes in eye movements. Fixation counts, as early as 2003, Jacob and Karn pointed out that the fixation counts got related to the users' attention level and information relevance of them. Blink rate is associated with alertness, concentration, fatigue and anxiety (Qiao 2012). Saccades are rapid and sustained eye movements between fixations (Uzzaman and Joordens 2011). The higher the fixation counts, the higher the saccade counts. Browsing time is the user's overall time on the Advertising page, including fixation time and blink time. Clifton et al. (2007) and (Van der Laan et al. 2015 pointed out that fixation time and browsing time can reflect users' preferences and choices, and also reflect the importance of correlation between stimulus and users' interest and attention. Users tend to focus their attention on the interest areas, with a more fixation counts and a longer browsing stay, which we call stickiness today. The number of revisits is the number of times the user keeps returning to a certain area to watch. Generally speaking, the more revisit counts, the more attractiveness of the information (Li et al. 2009).

2 Materials/Apparatus

2.1 Banner Advertisings

Banner advertising is one of the main forms of online advertisings, and also one of the earliest and most commonly used online advertisings. It creates image files by GIF, JPG, Flash and etc. when users click on the images; it will be directly linked to the advertiser's website. Because of its low cost, fast transmission, wide coverage, and simple inputs with large quantity information, it has been received a great popularity by advertisers since its birthday in 1994 (Cho 2006). However, with the problem of information overloading on the whole network, more and more users start to turn a blind eye to banner advertisings, resulting in banner blindness (Bayles 2000). Pagendarm and Schaumburg (2006) pointed out that it was easy for users to ignore banner ads when looking for specific information on a website. Danaher and Mullarkey's (2003) research has also shown that, in visual search tasks, visitors pay little attention to ads and the re-recognition rate of ads is low. Facing this phenomenon, many scholars try to find the clues embedded in the ads which can attract consumers' attention, and made some progress. Wedel and Pieters (2000) found in the experiment that the attraction of advertisement elements from big to small was the title, picture and copywriter in turn. Yoo et al. (2004) showed the effect of moveability on the promotion of advertising through research. Xie et al. (2004) proved that presenting incentive information in advertisements significantly increased click-through rates. Bernard (2001) who found that banner advertisings are generally expected to be at the top of a web page emphasized the location design of the banner ads. Lohse and Wu (2001) also verified the ads locating on the top and in the middle of the news can gain long fixation duration and more fixation counts through experiments. Bayles (2002) demonstrated the important role of dynamic design on the effect of banner advertisings. Huhmann (2006) found the application effect of multimedia technology (color, photography and animation) on banner advertising. Although some design elements are explored, the combination consideration of the remote web environment and the ad design is still rare.

According to the overall development summary of China's online advertising released by I research and also a questionnaire of testing the most different two banner ads, the two major portals (Sohu and Tencent) were selected as the research objects based on their proportion, profitability and difference. As the layout of advertising isn't included in this experiment research, thus the location, size and the relationship with the web content were immobilized through technical processing, only choose the two major portal first screen banner ads as the research object. Screenshots of Sohu banner and Tencent banner are respectively shown in Fig. 1 and Fig. 2.

Fig. 1. Sohu banner (Ad.1)

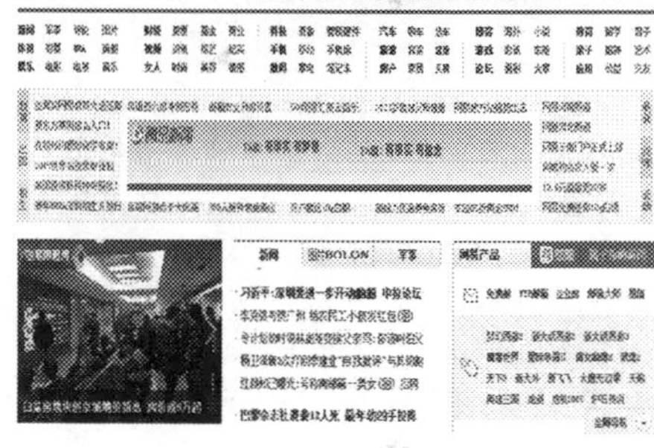

Fig. 2. Tencent banner (Ad.2)

2.2 Vertical Search Advertisings

Vertical search ads refer to the ads provided by vertical search websites, such as Taobao, Jingdong, Amozon, etc. From the perspective of interactive process, vertical search ads can help users in B2C business websites to obtain the information integrating product image, advertising copy, sales and purchase entrance by entering keywords, which thus can help online consumers to make purchase decisions and even complete online purchase. As this type of online ads has been seldom studied alone, the corresponding finding is very limited (Hassanein and Head 2007; Cyr et al. 2007), and this study wants to fill the blanks.

Take the clothes which account for 41.3% of the total online purchase as a keyword, select two ads with the largest differences by experts' investigation and cluster analysis among the top 10 B2C e-commerce sites. After the prices were unified, screenshots of Vancl ad and Taobao ad are respectively shown in Fig. 3 and Fig. 4.

Fig. 3. VANCL vertical search Ad (Ad.3)

Fig. 4. Taobao vertical search Ad (Ad.4)

2.3 Selection of Participants

Thirty volunteers (16 men and 14 women) took part in the experiment. All of participants were in good physical condition during the test, with normal vision or corrected visual acuity, right handed, Aged from 19 to 40, average age 24.8, standard deviation 5.16. In the six months before the experiment, most of the participants went online every day. All of them had seen the banner ads on the portal website. Among them, 27 subjects purchased online more than 10 times, accounting for 93.3% of the total sample. It can be seen that the participants of this study are familiar with online advertising and can make accurate evaluation.

2.4 Experiment Equipments

The experimental equipment mainly includes a red telemetry eye tracker, a laptop, a monitor and two sets of infrared light source and camera. The laptop has installed Center 2.5 experimental design and stimulus presentation software, iVew X 2.5 eye-tracking software and BeGaze 2.5 data analysis software.

2.5 Test Tasks

Participants are required to follow their normal surfing habits when browsing the information of the following portal websites and the banner advertisements on the website. They can decide their browsing time at their own discretion. If they do not want to browse, they can press space to exit. At the end of the task, the participants will be asked to fulfill a questionnaire including the recall of the presence experience in the process of completing the task, and an evaluation of their approach-avoidance tendency towards each advertising. The process of testing vertical search advertising is similar; the subjects will be asked to click the ad according to their purchase intentions.

The participants' eye movement data were derived by BeGaze2.5. The participants' questionnaire data were input and processed by SPSS20.0. The abnormal values were first eliminated according to the box diagram, then the normality test was conducted, and finally the abnormal data was eliminated by 3 times of standard deviation.

2.6 Data Analysis

The validity of the scale was validated by Amos7.0 software using the maximum likelihood estimation in confirmatory factor analysis. Then, SPSS20.0 was used for the following analysis:

Firstly, repetitive measurement deviation analysis was used to analyze the approaching trend data of four advertisements, with three items measure proximity, two items measure avoidance. The seven-point Likert scale was used, with scores increasing from 1 to 7, from strongly disagree to strongly agree. Therefore, the assessment score of avoidance trend will be converted first. In other words, when the avoidance trend score is 1, it is converted to approaching trend score of 7; when the avoidance trend score is 2, the conversion to approaching trend score is 6, and so on. Then the average score of the five items is calculated, which is the final approach trend of each advertisement. If the result of spherical test has correlation, Greenhouse-Geisser method is used to correct the result. Paired comparison was performed using the Bonferroni post hoc test.

Then, calculate the indicators of physical presence, social presence and self-presence respectively, and discuss the relationship between the three levels of presence. As the purpose of this study is to explore whether physical, social and self-presence have the same changing trend, Kendallτcorrelation coefficient is chosen. The measurement principle of Kendallτcorrelation coefficient is to pair all sample points together and then see whether the two observations in each pair increase or decrease at the same time. Kendallτcorrelation coefficient is a number between −1 and 1. The closer it is to −1 or 1, the more relevant it is, while the closer it is to 0, the less relevant it is (Wu 2007). Set the alpha level of all statistical analyses to 0.05.

2.7 The Reliability and Validity of the Scales

(1) Presence Scale. Presence is measured through three levels, including physical presence, social presence and self-presence according to 1.1. The physical presence refers to the user's perception of the authenticity of advertising products as if they were close at finger when users browse them. Vividness is considered as one of the perceptual characteristics of physical presence (Fortin and Dholakia 2005). In the field of human-computer interaction and e-commerce, vividness helps users to feel the advertising products as if they were in front of them. As long as the computer and screen can maximize user's sensory channels, the product will be shown vividly even though it is far away. Proximity, first proposed by Argyle and Dean in 1965, refers to geographical distance, eye contact, physical proximity, smile, facial expression and personal theme (Argyle and Dean 1965). In the fields of human-computer interaction and e-commerce, proximity emphasizes the geographical proximity, visual access and easy access of products. At present, many studies have confirmed that proximity is a perceptual dimension of the sense of presence (Jaggars 2014; Lee 2014; Kumar and Hart 2014). Therefore, the measurement dimension of physical presence can be summarized as vividness and proximity.

The social presence of online advertising refers to the warmth of interpersonal communication perceived by users through interaction with online advertising, and also a perception or recognition of social clues embedded in online advertising. As to the measurement of social presence, many scholars have made relevant explorations, and it is generally believed that warmth, friendliness and consideration of other people's emotions can be used as the measurement dimensions (Hassanein and Head 2007; Cyr et al. 2007).

Tamborini et al. (2004) proposed that users' perception of self-presence includes physical spatial perception (consciousness existing in a continuous spatial environment) and ecological realism (participants' consciousness of content credibility and authenticity). Physical space perception refers to the user's psychological acceptance of the online advertising context, and a perception of a continuous space environment between the product use atmosphere created by advertising and the user's reality. Realism refers to the user's perception of the authenticity of the product content provided by the advertising, where user regards online advertising viewing and purchasing are as the same as in reality. Huang and Liao (2014) found that augmented reality interaction technology (ARIT), which displays user's real avatar when they fit online, has a significant impact on user's sense of self-presence. It is shown that the more users' attention to virtual reality, the more they feel consistent in physical space between virtual reality and the real environment. Thus the more they believe that their virtual world is as real as reality, the more they can feel the sense of self-presence. So we measure self-presence by two dimensions, that is, whether users' idling or purchasing on ads are as they were in the advertising scenario or field purchasing.

Maximum likelihood estimation was performed to evaluate the effectiveness of the scale's factor structure through AMOS7.0. CFI (comparative fit index) of presence scale was 0.998, and SRMR (standard root mean-square residual) was 0.024, indicating that the model was well adapted (Hu and Bentler 1999). The scale's loading coefficient is significant, with a value between 0.806 and 0.952 and a combined reliability of 0.958, indicating that the factor reliability is at an acceptable level (Raykov 1997), shown in Table 1.

Table 1. Reliability and validity test results of presence scale

Latent variables	Questionnaire keywords	Factor loading	Cronbach's α	AVE
Physical presence	1. Vividness	0.952	0.943	0.893
	2. Proximity	0.938		
Social presence	3. Friendliness	0.921	0.937	0.837
	4. Warmth	0.917		
	5. Consideration	0.906		
Self-presence	6. As in the scenario	0.974	0.879	0.894
	7. As field purchasing	0.806		

(2) Approaching-avoidance Trends Scale. The approaching-avoidance trend scale was developed by Donovan and Rossiter (1982) to measure users' behavioral intention to the website. If users have an approaching tendency to the website, they may spend more time browsing the website. If users have an avoidance tendency to the website, they may leave the website and go to another website. Similarly, if users want to be close to the online advertising, they are likely to spend more time browsing the advertisement, the advertisement will be well spread. If the user has a tendency to avoid the advertising, it is possible to close the ad or ignore the ad, and the Ad communication effect is limited. In this paper, a preliminary experiment was made to optimize the scale, and the final scale contains three projects measuring the approaching trend (I like this Advertisement, I'm willing to take the time to browse the advertisement, I feel very enjoyable when browsing this ad), and two projects to measure the avoidance trend (I want to avoid to browse or ignore the advertisement, the advertisement can't interest me. Using the seven-level Likert scale, the scores increased from 1 to 7, from strongly disagree to strongly agree. CFI of the approaching-avoidance trend scale was 1.000 and SRMR was 0.010, indicating that the model was well adapted. The factor loading coefficient of the approaching-avoidance trend scale is significant, with a value between 0.900 and 0.975 and a combined reliability of 0.939, indicating that the factor reliability is at an acceptable level. Besides, the scales used in this paper all have good content validity.

3 Results

3.1 The Analysis of Banner Ads

Bonferroni showed that when the significance level was 0.05, there was a significant difference in the approaching trend between the two ads ($p < 0.001$). It can be understood that the subjects had an approaching trend towards the No.1 Sohu banner advertisement (hereinafter referred to as No.1 ad) and an avoiding trend to the No.2 Tecent banner advertisement (hereinafter referred to as No.2 ad).

(1) Comparison of Presence Indexes between the Two Banner Ads. Table 2 shows the results of the paired samples T test of the three levels of presence experience when browsing the two different advertisements. As can be seen from Table 2, when the significance level was 0.05, the subjects' physical, social and self-presence in No.1 ad were significantly higher than those in No.2 ad. From the perspective of variation range, physical presence changes the most, up to 120%, followed by self presence and social presence, which are respectively 100% and 96%.

Table 2. Comparison of presence experience of two banners

Subjective presence levels	No.1 ad		No.2 ad		t	df	p
	M	SD	M	SD			
Physical presence	5.40	1.05	2.45	1.07	11.125	29	<0.001
Social presence	4.98	1.25	2.54	1.18	8.367	29	<0.001
Self-presence	4.27	1.42	2.13	1.11	8.449	29	<0.001
Total presence	4.88	1.13	2.38	1.04	10.209	29	<0.001

(2) Comparison of the Eye Movements between the Two Banner Ads. Table 3 shows the results of the paired samples T test of eye-movement indicators when browsing the two different advertisements.

Table 3. Comparison of eye movements when browsing two banners

Eye movement indicators	No.1 ad		No.2 ad		t	df	p
	M	SD	M	SD			
Average pupil diameter/px	11.68	1.39	11.72	1.46	−.629	29	0.535
Fixation/count	43.59	25.23	24.62	13.52	5.90	29	**<0.001**
Average fixation/ms	196.51	48.83	194.48	46.74	−1.80	29	0.083
Blink/count	9.10	7.54	4.14	3.13	4.82	29	**<0.001**
Average blink/ms	185.15	234.68	172.14	140.95	0.50	29	0.686
Browsing duration/ms	13008.04	9035.94	7416.90	5096.47	5.13	29	**<0.001**
Saccade/count	74.72	32.87	38.72	17.87	8.27	29	**<0.001**
Average saccade/ms	66.57	52.63	60.31	53.82	1.23	29	0.229
Average saccade amplitude/。	8.51	6.48	7.36	6.79	1.42	29	0.168
Average saccade velocity/。/s	113.00	41.10	106.00	38.08	1.40	29	0.173
Entry time/ms	488.79	492.37	443.62	1115.09	0.24	29	0.814
Revisit/count	11.76	6.38	6.31	5.00	6.02	29	**<0.001**

As can be seen from Table 3, when the significance level was 0.05, there was a significant difference in the fixation counts, browsing duration, blink counts, saccade counts and revisit counts between the two advertisements. The fixation counts, saccade counts, browsing duration, blink counts and revisit counts on No.1 ad were significantly greater than the corresponding values on No.2 ad. At the same time, in terms of average fixation duration, there was an edge significant difference (sig = 0.083 < 0.01) between the two ads, while there was no significant differences in the other eye movement indexes between the two ads. The high presence banner will increase customers' fixation counts by 1.77 times (from 24.62 to 43.59), increase revisit counts by 1.86 times (from 6.31 to 11.76), increase saccadic counts by 1.93 times (from 38.72 to 74.72), increase blink counts by 2.20 times (from 4.14 to 9.10), and extend the customers' browsing duration by 2.55 times (from 7416.90 to 13008.04), which can be summarized as that high presence banner can nearly double users' browsing stay and visual attention compared with the low one.

3.2 The Analysis of Vertical Search Ads

Bonferroni showed that when the significance level was 0.05, there was a significant difference in the approaching trend between the two ads (p < 0.001). It can be understood that the subjects had an approaching trend towards the No.4 Taobao vertical search advertisement (hereinafter referred to as No.4 ad) and an avoiding trend to the No.3 Vancl vertical search advertisement (hereinafter referred to as No.3 ad).

(1) Comparison of Presence Indexes between the Two Vertical Search Ads. Table 4 shows the results of the paired samples T test of the three levels of presence experience after checking the two different vertical search advertisements and clicking for purchase. As can be seen from Table 4, when the significance level was 0.05, the subjects' physical, social and self-presence in No.4 ad were significantly higher than those in No.3 ad. From the perspective of variation range, self-presence changes the most, up to 116%, followed by social presence and physical presence, which are respectively 104% and 66%.

Table 4. Comparison of presence experience when shopping on websites

Subjective presence levels	No.3 ad		No.4 ad		*t*	*df*	*p*
	M	*SD*	*M*	*SD*			
Physical presence	3.00	1.10	4.97	1.00	7.339	29	**<0.001**
Social presence	2.70	1.08	5.52	0.69	12.720	29	**<0.001**
Self-presence	2.35	0.88	5.08	1.17	9.786	29	**<0.001**
Total presence	2.68	0.91	5.19	0.83	−11.094	29	**<0.001**

(2) Comparison of the Eye Movements between the Two Ads. Table 5 shows the results of the paired samples T test of eye-movement indicators when checking the two different advertisements.

Table 5. Comparison of eye movements when shopping on websites

Eye movement indicators	No.3 ad		No.4 ad		*t*	*df*	*p*
	M	*SD*	*M*	*SD*			
Average pupil diameter/px	**12.11**	**1.66**	**12.26**	**1.64**	**2.67**	**29**	**0.012**
Fixation/count	51.20	23.61	38.97	17.57	3.31	29	**0.002**
Average fixation/ms	242.60	110.89	278.80	143.01	−3.54	29	**0.001**
Blink/count	7.37	4.81	4.63	3.55	2.77	29	**0.010**
Average blink/ms	254.37	579.12	172.40	91.44	0.77	29	0.450
Browsing duration/ms	18154.20	7015.35	14753.03	4890.04	2.55	29	**0.016**
Saccade/count	51.47	24.40	39.03	18.36	3.23	29	**0.003**
Average saccade/ms	70.37	66.83	129.12	340.29	−1.07	29	0.293
Average saccade amplitude/。	9.44	10.47	24.28	7.38	−1.22	29	0.233
Average saccade velocity/。 /s	100.78	34.66	108.01	56.22	−0.90	29	0.375

As shown in Table 5, when the significance level was 0.05, there was a significant difference in fixation counts, average fixation duration, browsing duration, blink counts, saccade counts and average pupil diameter in the two ads. The fixation counts, blink counts and saccade counts are significantly more on No.3 ad than on No.4 ad, and the average fixation duration, browsing duration and average pupil diameter on No.3 ad were significantly shorter and smaller than on No.4 ad. Other eye movement indexes showed no significant difference between the two ads. The high presence vertical search ad will reduce users' fixation counts by 24% (from 51.20 to 38.97), reduce saccadic counts by 24% (from 51.47 to 39.03), reduce blink counts by 37% (from 7.37 to 4.63), shorten the browsing duration by 73% (from 18154.20 to 14753.03), increase average fixation duration by 15% (from 242.60 to 278.80), and enlarge average pupil diameter a little (from 12.11 to 12.26). In summary of these key numbers, it shows that high presence vertical search ad can reduce users' frequent eye movements (fixation counts, blink counts and saccadic counts) by nearly 1\4, and shorten users' browsing stay by 3\4 compared with the low one.

4 Discussion

4.1 Findings

The experimental results show that no matter the involvement degree of the users, customers tend to approach the high-presence online ads. But there is a great difference of the eye movements on the high-presence ads as the users have a different degree of involvement.

(1) The Same Users' Tendency to Approach the High-presence Online Ads Regardless the Degree of Involvement.

It can be seen from the analysis of the questionnaire that compared with the low-presence ads; users tend to approach the high-presence ads, no matter the degree of the involvement. When the involvement is low, the users just idle on internet, they would prefer not to turn a blind eye to a high-presence banner ad. When the involvement is high, the users want to make some purchase intentions, they also want to check a high-presence vertical search ad in order to have more information and reduce the difficulty of the information processing. This result is consistent with the existing finding by Hasseinin's (2007) research on clothing websites, which proved that the expression of products with characters and background can improve users' presence and attract users to buy. From the perspective of the design elements of the banners, No.1 ad and No.4 ad are all fresh and vivid with bright colors. The images of the stars or a beautiful woman in the picture are lively and charming. Besides, No.4 advertisement has a good character image, a friendly display background, a convenient shopping button and reasonable commodity details, all of which are conducive to stimulate the users' presence and to enhance the purchase intention (the questionnaire survey shows that 100% of the subjects are willing to buy No.4 advertising products).

Compared with the high-presence ads, the low-presence ones are not appealing. The single text of No.2 ad leaves large white space and the color is similar with the surrounding articles, so it is difficult to attract the attention and interest of the users, and not to mention bringing an immersive experience. No.3 ad shows products without models or detailed information, which cannot stimulate users' presence, nor arouse consumers' willingness to buy.

(2) Eye Movement Indexes Between the High-presence Ads and Low-presence Ads in the Same Degree of Involvement.

As to a low involvement, users just idled on internet and happened to see banner ads, the users' eye movements would make a significant difference on different degree of presence on banners as follows. The fixation counts, fixation duration, browsing duration, blink counts, saccade counts, and revisit counts of No.1 ad were significantly greater than the corresponding values on No.2 ad. The reason of a more fixation counts, a longer fixation duration and a longer browsing duration on No.1 advertisement can be interpreted as that when the subjects browse the banner advertisements, they found that there were many interesting areas on this banner, so the fixation counts, fixation duration and browsing duration are relatively high (Jacob and Karn 2003). Saccade is a rapid and sustained eye movement between fixations (Uzzaman and Joordens 2011). The more of the fixation counts, the more the saccades are. Therefore, the saccades counts in No.1 ad are significantly higher than that in No.2 ad. The more saccades in No.1 ad could be interpreted as that users were stimulated by multiple stimuli when browsing this advertisement, and they are always paying attention to and looking for interesting content. This advertisement is more attractive for the subjects to watch and browse. From advertising design elements, this may be related to advertising content. No.1 ad has several star images, and the content is a promotion of an entertainment program, which is more likely to attract the attention of the subjects than No.2 ad which is merely written words and mainly focusing on news. The revisit counts is determined by demarcating an interest area of the banner advertisement, and count the times of users' looking back at the advertisement again after leaving it in the whole process of browsing the webpage. Generally speaking, the more revisit counts are, the more attention the banner advertisement in the webpage has attracted. From the perspective of advertising design, the boring expression form of No.2 ad compared with No.1 ad is difficult to attract users' attention. After quick browsing, users hardly found any interest areas, so they choose to leave the website and rarely look back. The blink counts are associated with alertness, concentration, fatigue, and anxiety (Chen and Epps 2013). From the point of presence experience, No. 1 ad with rich color and lively characters can enhance participants' immersive experience more; and the stronger the participants' immersive experience, the more active they are to icon themselves to advertising in the role of the "running man". Thus their mood will be relatively excited; the blink counts will become more.

As to a high involvement, users would like to check the vertical search ads carefully in order to make a reasonable decision on purchase, where the users' eye movements would make a significant difference on different degree of presence on vertical search ads as follows. Users pay more on No.3 ad than No.4 ad in terms of fixation counts, blink counts, saccade counts and browsing duration, while the average fixation duration and average pupil diameter of No.4 ad were significantly longer and bigger than No.3 ad. The reason why low-presence ad stimulates more fixation counts, longer browsing duration but shorter fixation duration, indicates that the participants tend to search more information to support their purchase decision when they see the ad. Therefore it forms a lot of sight interest areas (shown as fixation counts), which cost users a longer time to check on the ad (shown as long browsing duration). But as the low-presence ad contains less information, users tend to spend little time on a single fixation point, which presents a frequent and intense gaze on No. 3 ad, but a lack of a sustained attention (shown as shorter average fixation duration). While No. 4 ad with high-presence provides more information for decision reference, including the well layout, rich content and image, clear and tidy information, as well as the convenient purchase button, which can help the users to check the ads easily, rather than to search the information in a mess. Thus users arrange less fixation points on this ad, but pay a longer time to read a single fixation point, thus formed an opposite result with less fixation counts but longer fixation duration compared with the No.3 ad. The main function of saccade is to change the fixation points in order to let the interest area fall in the central socket of the retina. Since No.3 ad has more fixation counts, the corresponding saccade counts will increase. Previous studies have shown that saccades are largely influenced by cognitive factors, including attention, memory, learning processing and decision-making (Gefen and Straub 2004). In this experiment, the more saccades on No. 3 ad, can be illustrated that users tend to search information or their interest areas when making purchase decisions, but No.3 ad with low presence could not supply with sufficient or valid information, which makes users to saccade frequently and leave finally. Most of the blink behaviors are endogenous blinking controlled by central nervous system, which is related to cognition (Blakney and Sekely 1994). The blink counts increase as the task difficulty increases (Jarvenpaa and Todd 1997). It is difficult to find the required information when viewing the No.3 ad with low presence for purchase decision, which will cause users' confusion, make it more difficult to complete the shopping task and thus increase the blink counts. From the perspective of design elements, the low presence No.3 ad with dull information, monotonous content and color can seldom call users' attention or cause visual fatigue, which will lead to a increased blink counts. Existed studies have confirmed that changes in pupil size are related to cognitive and emotional information processing (Wiener and Mehrabian 1968), and Loewenfeld (1966) also pointed out that increased light intensity can stimulate pupil changes. Hess and Polt (1960) supported the idea that the dilation of the pupil is related to the evaluation of potency. The pupil dilates when the user sees a pleasant stimulus. No.4 ad with high presence has rich colors and vivid images, which can stimulate the happy mood of the users, instantly attract the attention of the customers and the diameter of the pupil will also dilate significantly. This experiment conclusion further proves the previous research results.

(3) A Further Analysis of Users' Involvement on Their Eye Movements towards the High-presence Advertisings.

When the user is involved in a low degree and the task is designed to browse the banner ads, fixation counts can show user's attention and interests as the more fixation counts on the high presence ads. And when the user is involved in a high degree and the task is designed to purchase from the vertical search ads, the ads with high presence can better satisfy with the users' cognition, including information searching, processing, and storage, which will reduce users' fixation counts effectively, making a obvious show of less fixation counts on vertical search ads with high presence. In the low involvement, users tend to make a simple cognition. The banner ads with high presence will bring positive emotional responses with its abundant and vivid images. The more real the users feel on the banner ads, the more immersive the users are, which will be shown as a flexible visual processing on the behavioral level (blink more). Compared with banner ads, users are more involved in vertical search ads, and users' active information processing path can be opened. Previous studies have shown that the blink counts are related to task difficulty, the blink counts increase as task difficulty increases. Compared with low-presence ads, high-presence ads can effectively reduce users' difficulty in cognitive processing. Therefore, users can blink less on high-presence ads.

According to the study, saccades are largely influenced by cognitive factors, including attention, memory, learning and decision-making (Stasi et al. 2013). In the browsing task with low cognitive demands, the high-presence banner ad can attract more users' attention with its vivid expression technique, so it shows more saccade counts in the high-presence banner ads. However, in the task of purchasing, which requires high cognition; users need to make purchase decisions in combination of the ads information. It is no longer simply browsing ads, but to mobilize more mental participation to obtain the information to support his rational decision. When the vertical search ads with low presence show stiff, limited content, it is difficult to meet the needs of users. In order to support the decision, users often turn to constantly search for the interest areas in such ads, which makes a reflection of more saccade counts on the low presence ads.

From this study, when the involvement is low, the banner with high presence will induce high immersive cognition and experience, and stimulate users' interest and attention, so that the users are willing to allocate more time and attention to focus on high presence banner ads, which is shown in visual display as a longer browsing duration and a longer average fixation duration. While on the vertical search ads with higher degree of involvement, users have strong desire of purchase. When they see the information that can support their decision-making in the high-presence ads, they will stop searching information, which in turn forms a relatively short browsing duration and fixation duration.

The high presence banner ad with vivid product display can capture more users' visual attention, and attract users' repeated attention to the banner ads in the web page, forming more revisits counts. However, the low-presence banner ad information is monotonous, which makes it difficult for users to form an immersive positive experience. When users are unable to find the interest areas after a quick browsing, they will choose to leave the page and is unlikely to make any revisits.

4.2 Implications

No matter how involved the users are, or whether consumers want to buy goods or just idle online, compared with the low presence ads, high presence ads are more likely to attract users' attention, bring positive guidance and stimulate purchase. The eye movement data can be used to further analyze the specific cognition process at a microscopic level. The eye information processing methods of users are completely opposite towards the high presence ads in different degrees of involvement. This finding can be used for reference and guidance to design the online advertisements for different business platforms, and to help users to increase their online search efficiency, website stickiness and achieve purchase decision as soon as possible.

4.3 Limitations

The relationship between the three levels of presence and the main sense of presence that dominates users' behavior in different degrees of involvement has not been deeply discussed, and it is worth further exploration later.

4.4 Conclusion

Through the experimental research, the results showed that when the customer is driven by the task of browsing or idling, online advertising with high presence will nearly double the customer's fixation counts, revisit counts, saccadic counts, blink counts, and extend the customers' browsing duration, which will surely improve the customers' stickiness, a longer browsing stay and a more visual attention on the ads; When the customer is driven by the task of purchase, online advertising with high presence will reduce fixation counts, saccadic counts, blink counts by nearly 1\4, and shorten the browsing duration by 3\4, which indicates that the online advertising with high presence can reduce the difficulty of information extraction and help the customer to make purchase decisions effectively.

Author Contributions. Conceptualization & original writing, Yu Sun; methodology & review, Fu Guo; Review & edit, Vincent. G. Duffy.

Funding. This research was supported by the Natural Science Foundation of Liaoning Province, China (grant No. 20180550629).

References

Argyle, M., Dean, J.: Eye-contact, distance and affiliation. Sociometry **28**(3), 289–304 (1965)

Bernard, M.L.: Developing schemas for the location of common web objects. In: Proceedings of the Human Factors and Ergonomics Society, 45th Annual Meeting, Santa Monica, CA, pp. 1161–1165 (2001)

Biocca, F., Harms, C., Burgoon, J.K.: Toward a more robust theory and measure of social presence: review and suggested criteria. Presence Teleoperators Virtual Environ. **12**(5), 456–480 (2003)

Blakney, V.L., Sekely, W.: Retail attributes: influencing on shopping mode choice behavior. J. Manage. Issues **6**(1), 101–118 (1994)

Huhmann, B.A.: Visual complexity in banner ads: the role of color, photography and animation. Soc. Sci. Res. Netw. **19**, 10–17 (2006)

Butt, M.M., de Run, E.C.: Do target and non-target ethnic group adolescents process advertisements differently? Australas. Market. J. **19**(2), 77–84 (2011)

Chen, S., Epps, J.: Automatic classification of eye activity for cognitive load measurement with emotion interference. Comput. Methods Programs Biomed. **110**(2), 111–124 (2013)

Cho, C.: The effectiveness of banner advertisements: involvement and click-through. Journal. Mass Commun. Q. **80**(3), 623–645 (2006)

Tan, C.-H., Sutanto, J., Phang, C.W.: An empirical assessment of second life vis-a-vis chatroom on media perceptual assessment and actual task performance. IEEE Trans. Eng. Manage. **59** (3), 379–389 (2012)

Clifton, C., Staub, A., Rayner, K.: Eye movements in reading words and sentences. In: Eye Movements: A Window on Mind and Brain, pp. 341–372 (2007)

Cyr, D., Hassanein, K., Head, M., et al.: The role of social presence in establishing loyalty in e-service environments. Interact. Comput. **19**(1), 43–56 (2007)

Danaher, P.J., Mullarkey, G.W.: Factors affecting online advertising recall: a study of students. J. Advert. Res. **43**(3), 252–267 (2003)

Donovan, R.J., Rossiter, J.R.: Store atmosphere-an environmental psychology approach. J. Retail. **58**(1), 34–57 (1982)

Fiore, A.M., Yu, H.: Effects of imagery copy and product samples on responses toward the product. J. Interact. Market. **15**(2), 36–46 (2001)

Fortin, D.R., Dholakia, R.R.: Interactivity and vividness effects on social presence and involvement with a web-based advertisement. J. Bus. Res. **58**(3), 387–396 (2005)

Gefen, D., Straub, D.W.: Consumer trust in B2C e-Commerce and the importance of social presence: experiments in e-Products and e-Services. Omega **32**(6), 407–424 (2004)

Hassanein, K., Head, M.: Manipulating perceived social presence through the web interface and its impact on attitude towards online shopping. Int. J. Hum Comput Stud. **65**(8), 689–708 (2007)

Hu, L., Bentler, P.M.: Cutoff criteria for fit indexes in covariance structure analysis: conventional criteria versus new alternatives. Struct. Eqn. Model. Multi. J. **6**(1), 1–55 (1999)

Jacob, R.J.K., Karn, K.S.: Eye tracking in human-computer interaction and usability research: ready to deliver the promises. Mind **2**(3), 4–36 (2003)

Jahng, H., Jain, K.T.: Effective design of electronic commerce environments: a proposed theory of congruence and an illustration. IEEE Trans. Syst. Man Cybern. Part A **30**, 456–471 (2000)

Jaggars, S.S.: Choosing between online and face-to-face courses: community college student voices. Am. J. Distance Educ. **28**(1), 27–38 (2014)

Jarvenpaa, S.L., Todd, P.A.: Is there a future for retailing on the internet? In: Peterson, R.A. (ed.) Electronic Marketing and the Consumer. Sage, Thousand Oaks (1997)

Kim, G., Lee, J.: The effect of search condition and advertising type on visual attention to internet advertising. Cyberpsychol. Behav. Soc. Network. **14**(5), 323–325 (2011)

Kim, J.U., Kim, W.J., Park, S.C.: Consumer perceptions on web advertisements and motivation factors to purchase in the online shopping. Comput. Hum. Behav. **26**(5), 1208–1222 (2010)

Kumar, S., Hart, M.: Social presence in learner-driven social media environments. In: Society for Information Technology & Teacher Education International Conference, no. 1, pp. 73–78 (2014)

Lee, K.M.: Presence, explicated. Commun. Theory **14**(1), 27–50 (2004)

Li, Y., Jing, J., Zou, X.: A preliminary study on the cognitive features of facial emotion expression in autistic children. Chin. J. Evid. Based Pediatr. **4**(1), 23–28 (2009)

Lee, S.M.: The relationships between higher order thinking skills, cognitive density, and social presence in online learning. Internet High. Educ. **21**, 41–52 (2014)

Loewenfeld, I.E.: Comment on Hess' findings. Surv, Ophthalmol. **11**, 291–294 (1966)

Hess, E.H., Polt, J.M.: Pupil size as related to interest value of visual stimuli. Science **132**(3423), 349–350 (1960)

Lohse, G.L., Wu, D.J.: Eye movement patterns on Chinese yellow pages advertising. Electron. Markets **11**(2), 87–96 (2001)

Matthew, L., Matthew, T.J.: Identifying the (tele) presence literatur. Psychol. J. **5**(2), 197–206 (2007)

Bayles, M.E.: Designing online banner advertisements: should we animate? In: Conference on Human Factors in Computing Systems, pp. 363–366 (2002)

Bayles, M.E.: Just how 'blind' are we to advertising banners on the web? Usability News **6**(2), 520–541 (2000)

Montes, G.L.: Is interaction the message? The effect of democratizing and non-democratizing interaction in video-conferencing small groups on social presence and quality of outcome. Technol. Mediat. Commun. **1**, 187–223 (1992)

Namkee, P., Kwan, M.L., Seung-A, A.J., Sukhee, K.: Effects of pre-game stories on feelings of presence and evaluation of computer games. Int. J. Hum Comput. Stud. **68**, 822–833 (2010)

Oatley, K.: Taxonomy of the emotions of literary response and a theory of identification in fictional narrative. Poetics **23**, 53–74 (1994)

Oh, J., Fiorito, S.S., Choc, H., et al.: Effects of design factors on store image and expectation of merchandise quality in web-based stores. J. Retail. Consum. Serv. **15**(4), 237–249 (2008)

Palanisamy, R., Wong, S.A.: Impact of online consumer characteristics on web-based banner advertising effectiveness. Glob. J. Flex. Syst. Manage. **4**(1/2), 15–25 (2003)

Palanisamy, R.: Impact of gender differences on online consumer characteristics on web-based banner advertising effectiveness. J. Serv. Res. **4**(2), 45–75 (2004)

Pagendarm, M., Schaumburg, H.: Why are users banner-blind? The impact of navigation style on the perception of web banners. J. Digit. Inf. **2**(1), 37–38 (2006)

Qiao, F.: Research on the effects of highway tunnel sections on drivers' vision and heart rate. Jilin University, Changchun (2012)

Raykov, T.: Estimation of composite reliability for congeneric measures. Appl. Psychol. Meas. **21**(2), 173–184 (1997)

Stasi, L.L.D., Catena, A., Cañasc, J.J., Macknike, S.L., Martinez-Condo, S.: Saccadic velocity as an arousal index in naturalistic tasks. Neurosci. Biobehav. Rev. **37**(5), 968–975 (2013)

Steuer, J.: Defining virtual reality: Dimensions determining telepresence. J. Commun. **42**, 73–93 (1992)

Tamborini, R., Eastin, M.S., Skalski, P., et al.: Violent virtual video games and hostile thoughts. J. Broad. Electron. Media **48**(3), 335 (2004)

Huang, T., Liao, S.: A model of acceptance of augmented–reality interactive technology: the moderating role of cognitive innovativeness. Electron. Commer. Res. 1–27 (2014)

Uzzaman, S., Joordens, S.: The eyes know what you are thinking: eye movements as an objective measure of mind wandering. Conscious. Cogn. **20**(4), 1882–1886 (2011)

Van der Laan, L.N., Hooge, I.T.C., de Ridder, D.T.D., et al.: Do you like what you see? The role of first fixation and total fixation duration in consumer choice. Food Qual. Prefer. **39**, 46–55 (2015)

Wang, L., Baker, J., Wagner, J.A., et al.: Can a retail web site be social. J. Market. **71**(3), 143–157 (2007)

Wedel, M., Pieters, R.: Eye fixations on advertisements and memory for brands: a model and findings. Market. Sci. **19**(4), 297–312 (2000)

Wiener, M., Mehrabian, A.: Language within language: Immediacy, a channel in verbal communication. Ardent Media (1968)

Wu, X.: Statistics: From Data to Conclusion. China Statistics Press, Beijing (2007)

Xie, T., Donthu, N., Lohtia, R., et al.: Emotional appeal and incentive offering in banner hierarchy of effects model. J. Interact. Advert. **4**(2), 49–60 (2004)

Yang, K.: Effects of consumer motives on search behavior using internet advertising. Cyber Psychol. Behav. **7**(4), 430–442 (2004)

Yoo, C.Y., Kim, K., Stout, P.A.: Assessing the effects of animation in online banner advertising: advertisements. J. Interact. Advert. **4**(2), 30–37 (2004)

The Impact of Trust and Fairness on Information System's Resistance

Zoubeir Tkiouat(✉), Ryad Titah, and Pierre-Majorique Leger

HEC Montreal, Montreal, Canada
{zoubeir.tkiouat, ryad.titah, pierre-majorique.leger}@hec.ca

Abstract. The present study advances a new conceptual framework explaining and predicting individual resistance to information systems (IS). IS Resistance depends on how individual users perceive the implementation process, its outcome and their relationship with the managers and implementers of the system. We draw on the justice literature to conceptualize the characteristics of the implementation process and its expected outcomes, and from the trust literature to conceptualize the relationships between the users and their managers. In addition to the concept of distributive justice that refers to an individual's perceptions about the expected outcomes of an IS implementation process, our model includes the constructs of procedural justice, interpersonal justice, and informational justice to capture different aspects of the interaction between the users, the managers and the implementers. In line with previous conceptualizations developed in the justice literature, we also propose that trustworthiness towards managers and implementers will mediate the effect of procedural justice and interpersonal Justice on individual intention to resist an IS. We conclude with a discussion about the theoretical and practice related contributions to the HCI and IS fields.

Keywords: IS resistance · IT resistance · Justice · Fairness · Trust

1 Introduction

While past IS resistance research has produced rich and insightful knowledge, most of this research adopted a process theory perspective to investigate the phenomena. It has also mainly focused on explaining how resistance behaviours derived from individuals' perceptions about the negative outcomes of an IS implementation process such as loss of revenue, status, and power, perceived inequity, and change in routines [3, 4]. As such, we argue that the paucity of variance theories to explain IS resistance is a missed opportunity to a) capture the complex mechanisms that drive the "resistance process" and to b) to quantitatively test measure the effects of these mechanisms at the individual level [1]. We also argue that the perceived negative outcomes are not the only factors that trigger resistance behaviours.

Based on the assumption that resistance behaviours represent the outcome of individuals' sensemaking processes, we argue that these processes cannot be limited to the perceived negative outcomes of the IS but must also take into consideration the

C. Stephanidis et al. (Eds.): HCII 2020, LNCS 12427, pp. 679–692, 2020.
https://doi.org/10.1007/978-3-030-60152-2_51

implementation process itself, and the relationships between the different parties engaged in the implementation process i.e., the users, the implementers and the managers. Even though the effects of these relationships have been theorized in past studies [2], to the best of our knowledge, we believe that no study has yet investigated them in an integrated way which is what we propose in this study.

The purpose of this paper is to provide a conceptual model that explains how individual users process information relative to the implementation process and the information about their relationship with the management and implementers to supplement their expected outcomes. It proposes that individuals' sense-making of these different informations leads to their intention to resist or not the IS implementation. We draw on the justice literature to capture the characteristics of the implementation process and the expected outcomes and on the trust literature to qualify the relationships between the users and the management. We first review previous theories on IS resistance and discuss the relevance of looking at different factors such as individuals' perceived justice of the implementation and trustworthiness of the management. We then present our model by describing the relationships between the proposed constructs and conclude with a discussion about how this study contributes to both theory and practice in HCI.

2 IS Resistance

IS Resistance refers to *"the behaviours that are aimed at preventing the implementation or the use of a system or at preventing system designers from achieving their objectives"* [3]. These behaviours exist within a spectrum ranging from passive to more aggressive and active behaviours [4, 5]. This classification proposed by Coestsee [4] helps categorize resistance into four levels: Apathy, Passive resistance, Active resistance, and Aggressive resistance [4]. Apathy or indifference is a state where people demonstrate no interest towards the change introduced even though they are informed about it. In Passive resistance, people show weak forms of rejection such as voicing negative attitudes about the change. In Active resistance, people show strong non-destructive behaviour such as voicing strong opinions against the change, noncompliance, withdrawal, and peaceful boycotts. The Aggressive resistance is the highest form of rejection that manifests as sabotage, making errors on purpose, and destruction [4].

Resistance to information systems has mainly been studied from a process perspective. Lapointe and Rivard presented a model that theorized IS resistance as an outcome of perceived threats that are the result of the interaction between initial conditions and the object of resistance [5]. Initial conditions are contextual factors that impact the degree of threat that the users will perceive such as: the ability to perceive a threat, established routines and the distribution of power. The effect of the distribution of power on the decision to use or resist a system was first introduced in a model proposed by Markus [3] which advanced that if the system is seen as improving the position of power of a group, then the group will use the system. However, if the system is seen as detrimental to their position of power, then the group is more likely to resist it [3].

The model proposed by Joshi [6] extends equity theory [7] in the context of an IS implementation and conceptualizes the mental process by which individuals evaluate the fairness of the gains of the implementation into three stages. At first, individuals evaluate the change that is brought to them by comparing their personal outcomes to their input to the system, then individuals compare these outcomes to those of the employer/organization, finally individuals compare their outcomes to the outcomes that other users will receive [6]. Examples of such outcomes are the change in skills learned, control over the work, information control and ownership. Examples of inputs are the change in efforts needed to do the task, and efforts to adapt and learn the new system.

In the context of an IS implementation, most of the antecedents of IS resistance (including change in power, change in routines, and change in equity) are relative to the perceived outcomes of the implementation process. Drawing on the organizational justice literature, another important aspect that impacts employees' attitude and behaviour towards change and its outcomes is the fairness of the process of allocation [8]. To capture both of these aspects, we use the four-factors conceptualization of justice which is composed of the constructs of distributive justice, procedural justice, interpersonal justice, and informational justice [9, 10]. Distributive justice reflects the perceived fairness of the outcomes of the IS implementation including changes to the individuals' inputs and outcomes. Procedural justice, interpersonal justice, and informational justice capture the aspects relative to the implementation process including how the decision to implement was taken, how the users were treated and what information they were provided about the motives of the implementation.

Another relevant factor that impacts the reaction of the users towards the IS is the level of trust that the users have towards the implementers of the system and their management. The trust level impacts the response through which individuals appraise the change event. The higher the trust, the lower the likelihood that the change is seen as a threat. Along with empowerment, justice and work redesign, trust was found to impact the response of downsizing survivors to the perceived threat of the change [11]. Mishra and Spreitzer [11] argued that survivors that trusted management prior and during the implementation of downsizing were likely to exhibit constructive responses. Depending on the level of control they had, they either saw the change as benign and went along with it or they saw it as an opportunity to enhance the performance of the company.

3 Conceptual Development

Our conceptual model (Fig. 1) captures three aspects that individual users perceive during an implementation and how these different perceptions later affect their intentions to resist an IS. We draw on the justice and trust literatures to capture each of these aspects (Table 1). The first aspect relative to the perceived outcomes of an IS implementation is reflected in the construct of distributive justice. The second aspect is about the process of IS implementation itself and it is captured through the constructs of procedural, interpersonal and informational justice. The last one is the relationship between the users and the management and implementers which is captured by the construct of trustworthiness towards management. The conceptual model (Fig. 1) summarizes the relationships between these constructs and the users' intention to resist an IS. While it is outside the scope of this paper, we propose that higher intentions to resist will lead to higher levels of resistance.

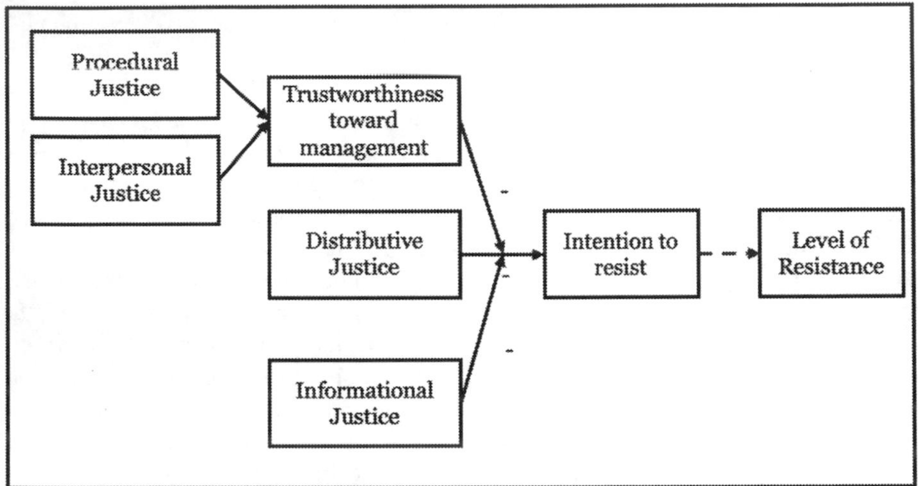

Fig. 1. Conceptual model

3.1 Conceptual Boundaries

The proposed model is about the reactions that individual users will exhibit following an IS implementation. These reactions are based on how they perceive elements from their environment. Our assumption about individuals is that when faced with uncertainty and perceived threats, they will attempt to make sense of their situation and will use cues from their environment which will include previous interactions with agents around them.

Due to our focus on intentions to resist, our model does not explicitly consider the level of control that the users have on information system use and assume that this use is mandatory. Based on the premise that the IS implementation is usually a team effort that involves different parties i.e., the management, the implementers and different groups of users, we consider that it is relevant and important to look at it from a multi-stakeholder point of view.

Table 1. Construct definitions

Construct	Definition	Dimensions	Source
Distributive justice	The degree of fairness of the outcomes or expected outcomes	• Information control • Expected outcomes • Expected adaptation efforts • Relative information control • Relative expected outcomes • Relative expected adaptation efforts	Adapted from [6]
Procedural justice	The individual's perception of the fairness of procedural components of the social system that regulates the allocative process	• Consistency of application across users time • Freedom from bias • Representation of the opinions of subgroups of users	Adapted from [12]
Interpersonal justice	The level of respect and dignity that the management exhibits in treating the involved users		[9]
Informational justice	The quality of explanations given to the users about the reason and the process for the IS implementation		[9]
Trustworthiness towards management	The competence and character of the management formed by its Ability, Integrity and Benevolence	• Ability • Integrity • Benevolence	[13]
Intention to resist	The individual's subjective probability of resisting the IS, indicating their readiness to resist it		[14]

3.2 Justice and Fairness

Organizational justice is composed of four constructs: distributive, procedural, interpersonal, and informational justice [9, 10]. **Distributive justice** reflects the fairness of the outcomes. Similar to Joshi's equity-implementation model [6], and based on the same theory i.e. Equity Theory [7] the person calculates a ratio of input to outcomes and compares it to a reference in order to evaluate fairness. **Procedural justice** is the individual's perception of the fairness of procedural components of the social system that regulates the allocative process [12]. It is concerned by what happens before the distribution of rewards (the outcome) takes place. The factors that drive this perception are: (a) the consistency of application across people and across time, (b) the freedom from bias and (c) the representation of the opinions of subgroups affected by the decision. Both interpersonal, and informational justice are aspects of the quality of the treatment people receive and sometimes referred to as "interactional justice". **Interpersonal justice** is the individual's perception of the degree of respect and dignity that the management and implementers display when treating involved users. **Informational justice** reflects the perceived quality of explanations given to users about the motives the IS implementation.

In their study of organizational retaliation behaviour (ORB) Skarlicki and Folger [15] proposed that employees that feel exploited by their companies will tend to "get even" and engage in overt retaliation behaviours such as theft and sabotage or go towards more covert behaviours such as psychological withdrawal or resistance behaviour depending on their perceived power relatively to the entity enacting the injustice [15]. Their argument is built on Inequity theory which states that perceived injustice in the distribution of outcomes leads to lower satisfaction in the relationship and goes beyond the dissatisfaction to create a desire to punish the wrongdoer (i.e., the entity enacting justice, the management). They argued that individuals do not only perceive the distribution of outcomes but take into consideration the process and the way by which they are treated to make their justice judgments. Using peer report to measure retaliation behaviour, Skarlicki and Folger obtained results that were similar to other empirical studies, e.g., [16–18]. They found that distributive justice, procedural justice and interactional justice explained 68% of the variance in retaliation behaviours [15].

Following Joshi's Equity-Implementation theory [6], and in accordance with the Rational economic man conceptualization of individuals, we posit that individuals seek to maximize their utility and will look at the impact that the implementation has on their equity. Joshi's model showed three levels of analyses that the individuals undertake to reach an equity judgment. The first step consists in assessing the change to their equity, the second is assessing their equity relative to the company's change in equity and lastly in assessing their equity in comparison to other users. We will only consider Joshi's first and third levels of analysis (i.e. the change in one's own equity and the difference in equities between a user and other users) as a basis for the dimensions of distributive justice for two reasons: first, because of the distinction between the enactor of justice and the recipients of justice in the justice literature. In this context the enactors are the management and the implementers and the recipients are the system's users. Second, because users have similar outcomes and inputs, it is easier for them to assess their own and compare them to other users in reference groups.

Looking at the main factors of outputs and inputs that individuals assess [6], we supplement these from existing theories of IS resistance [3, 5] which proposed that the main reasons for resistance are the perceived change in power and in work routines and synthesize them into three elements: expected outcomes, information control, and expected adaptation efforts. The construct of expected outcomes is [19, 20] defined as the likely consequences of using the system and is reflected in performance outcomes and personal outcomes. Performance outcomes include the time spent on the task, the quality of the task output, and the change in effectiveness. Personal outcomes are salient to the users such as the change in status, and the learning of a new marketable skill. Information control is the degree of ownership that the users have on the information produced in their business process. This was the main source of power in the study of Markus where users that felt loss of control over the information produced in their business process interpreted that as a loss of power and were more likely to resist [3]. The expected adaptation efforts represents the efforts that the users need to invest in the system after the implementation process due to discrepancies between individuals' abilities, task characteristics and technology characteristics. A user who need to invest a lot of time and effort into learning a new system will be more likely to resist it compared to a user who is familiar with it.

Proposition 1: Distributive justice will negatively impact intention to resist an IS.

Several theories were proposed to explain the effect of procedural justice on the perception and reaction to change. In their discussion about relationship models of procedural justice, Blader and Tyler [21] propose that fair treatment signals to the recipient of justice a positive message about their relationship with those that enact the justice in contrast to unfair treatment that sends a negative message about this relationship. This feedback signal reinforces the individual's attitudes and reactions to this treatment.

We will use Van den Bos's theorizing [22] about how the "Fair process effect" affects people's outcome fairness judgment, and their acceptance and satisfaction with the decisions. This author proposed that upon experiencing uncertainty provoking events/experiences, individuals engage in a sensemaking process to understand what is happening and construct future expectations from the information available to them.

This process of sense-making is similar to the appraisal done in coping with stress in which individuals try to evaluate the consequences of an event and how it impacts them [23]. While coping theory does not specify the information elements used in the appraisal process, the IS literature sheds light on factors contributing to the beliefs of the users [24]. The factors listed by Beaudry and Pinsonneault [24] can be classified into four main categories: technology characteristics, individual characteristics, the fit between the task, the technology, and the individual, and the performance expectancy of using the technology. All of these are relative to the outcomes (or expected outcomes) of an IS implementation.

The process of sense-making proposes *treatment fairness* as an essential source of information that individuals use to form their appraisal of the change in addition to the expected outcomes. Treatment fairness does not exclusively concern the perceived fairness of the formal procedures (procedural justice) but considers also the fairness of interactions between the enactors of justice and their recipients (interpersonal justice, and informational justice). The process of sensemaking is conceptualized as a cognitive *alarm system* triggered by uncertainty events. The process of sense making presented by Van den Bos [22] is based on the individual experiential-affective system from the Cognitive-experiential self-theory [25] and in situations of uncertainty, the role of the experiential-affective system is important. After individuals perceive these alarming events, they tend to process information about their experiences regarding this event. If they experience fair treatment, they are likely to calm down and exhibit positive reactions to this event. However, if they experience unfair treatment, the activation of their alarm system is likely to increase leading to negative reactions to the event [22].

From a rational system perspective, treatment fairness affects the individuals' evaluation of the amount of control they have over the outcomes they aspire to get [26].

3.3 Trust

Trust is defined as a "*psychological state comprising the intention to accept vulnerability based on positive expectations of the intentions or behaviours of another*" [27] and reflects a willingness to take risk. For it to be relevant in a relationship there needs to be dependence and risk in the relationship. In the context of an IS implementation, dependence and risk in the relationship exist between the users on one side and the management and implementers on the other side. This dependence stems from the users' ability to affect the success or failure of the implementation and from the power that the implementers and management have on the users' work and tasks. They are both dependent on each other for the success of the IS implementation and for the improvement of the users' work conditions. The risk factor comes from the uncertainty of the outcomes of the venture for both sides of the relationship.

Since the aim of the study is to explain the intentions and behaviours of the users, we will focus on the trust that the users have towards the implementers and management. While trust is the intention to accept vulnerability, trustworthiness is the perceived competence and character of the management and is formed by its Ability, Integrity and Benevolence [13]. Thus, following Mayer and colleagues' proposed model of trust [13], we consider perceived trustworthiness as the main antecedent of trust [13].

Ability refers to the perceived skills and competences that the trustee (i.e., management) has in a specific domain. Benevolence refers to the degree to which the trustee (i.e., management) is believed to behave in the benefit of the trustor (i.e., user) aside from a profit motive. Integrity refers to the perception that the trustee (management) adheres to a set of principles that the trustor (i.e., user) finds acceptable [13].

To study the effect of justice on trustworthiness, Colquitt and Rodell [28] draw on Fairness heuristic theory which proposes that individuals are faced with contradicting choices of either cooperating with the authorities (in our context, the implementers and management) which hence leads to positive outcomes or instead decrease the risk of being exploited by not cooperating [28]. These individuals go through two main phases: a "judgment phase" and a "use phase". During the "judgment phase" individuals will gather the information that is most available to them and which is most interpretable. Even though prior formulations of this theory do not make predictions about trustworthiness, Colquitt and Rodell [28] argue that the judgment of trustworthiness, especially integrity and benevolence, is one of the processes that individuals engage in using information they received about distributive, procedural, interpersonal and informational justice [28]. Their empirical study found that both interpersonal justice and procedural justice are the main justice components that drive the perception of trustworthiness. In their study about the impact of change on trust in management, Morgan and Zeffane [29] found that a significant fall in trust is expected when workers were not consulted about the change [29]. This lack of consultation is reflected in the low representation of their opinions in the decision making process which is one key factor of procedural justice.

Proposition 2: Perceived procedural justice and interpersonal justice will positively influence trustworthiness towards management.

The effect of trustworthiness on intention to resist operates through two causal mechanisms. The first is the impact that trustworthiness has on the appraisal of the individuals about the threat level. The second is the impact on the reaction to the threat after its appraisal. When management is perceived as trustworthy, it is seen as high in benevolence, integrity and ability. The higher the management's ability, benevolence and integrity, the more the users will believe the information provided by the management about the implementation and the more likely they are to perceive it to be true since the management is seen as "knowing what they are doing" and not having the intention to lie. A management that is perceived as benevolent will be expected to act in the benefit of the trustor (i.e. the users) as its intentions and motives are not driven by self-interest but instead by the interests of the users. If the users assume that the motives of the management is their benefit they are more likely to perceive the IS implementation as an opportunity instead of a threat. When the integrity of the management is high, the users perceive the management to have a set of principles that they find acceptable. These principles lower the perception that the management will act in an opportunistic way taking advantage of the users. Thus reducing the perception of the IS implementation as a threat.

The second mechanism by which trustworthiness affects intention to resist occurs after the appraisal of the level of threat. In Mayer and colleagues' model of trust, the outcome that trust explains is the level of risk taking in a relationship. Thus, the higher the trust, the higher the level of risk the trustor is willing to take. In this context risk taking translates into the acceptance of the IS by the users despite having perceived a level of threat. For the same level of threat perceived in the IS a user that sees the management as trustworthy will be more willing to accept the IS than a user that sees the management as untrustworthy.

When looking at the evolution of IT resistance, Lapointe and Rivard [5] highlighted the importance of the object of resistance since the latter is shaped by the characteristics of the object of resistance. They found a pattern by which resistance evolves and changes its object following a sequence that starts with resisting the system and its characteristics, then the system's significance, and later the implementers and the system's supporters. With resistance directed at the implementers and the system's supporters having higher levels of disruption. If the users perceive the management as being trustworthy, they are less likely to resist because they are more likely to attribute early failures in the implementation process to external factors rather than to the management. Also, trust was found to have a big impact on the communication between employees and management [30, 31]. In their study, Dirks and Ferrin [31] found that higher levels of trust lead to higher levels in the accuracy and amount of information shared and sent to the management. Thus, trust improves the communication between the management and the users during the implementation.

Proposition 3: Perceived procedural justice and interpersonal justice will influence individual intention to resist an IS and will be mediated by trustworthiness towards management.

Several studies found that the different components of justice interacted to predict employees reactions to change (e.g., organizational retaliation behaviour [15, 18], reactions to decisions [32]). However, Brockner and Siegel theorized that this interaction could be between the degree of trust resulting from procedural justice and distributive justice [33]. The reason for the interaction between the components of justice is that individuals engage in sense-making and analyse their interactions after events that entail uncertainty or are experienced as negative. The uncertainty and negative aspect of either unfair components of justice will heighten the receptivity of information about other components of justice. In other words, a negative experience about one aspect of justice will have a positive impact on the intention to resist directly. However, it will also trigger a sense making process by which individuals re-evaluate the impact of the other aspects of justice that they might have otherwise neglected. Therefore, we propose that the lower the resulting trustworthiness towards management the stronger the negative impact of distributive and informational justice will be. Also, the lower the distributive justice, the stronger the negative impact of informational justice will be.

Proposition 4: Perceived informational and Distributive justice will interact with trustworthiness towards management.

4 Discussion and Conclusion

The conceptual model proposed in this study aims at explaining individuals' intention to resist an IS. However the relationship between intention to resist and the behaviour of resistance itself will be affected by other factors. The well-established theories of planned behaviour and reasoned action propose that the link between intentions and behaviour will be affected by perceived behavioural control [14]. Perceived behavioural control could be internal or external to the individual user. An interesting concept that could shed light into this relationship is the *Punitive capability* which is the degree to which one party can affect the outcomes of another in the context of a conflict [34]. We can expect users that have higher punitive capability to resist more as a consequence to their intentions because they have more leverage on management and they would expect management's response to be in their favour since they have the power to affect its bottom-line.

If we look at the effect of justice and trustworthiness on intention to resist from the lens of the theory of planned behaviour, one relevant attitude that explains this effect on intention to resist is the user's change cynicism. Which is *"a pessimistic viewpoint about change efforts being successful because those responsible for making changes are blamed for being unmotivated, incompetent, or both"* [35]. This attitude was found to be shaped by the fairness of procedures and outcomes as well as the trust in management [36].

4.1 Research Design

In order to test our theoretical model, we propose a survey based study targeting users of a new mandatory information system in an organizational context. To improve the internal validity of the measurement model, this data collection will be longitudinal with measurements taken at two points in time. The first during the implementation process of an information system and the second six months after implementation.

Measurements for the constructs of intention to resist, trustworthiness towards management, and justice with the exception of distributive justice will be adapted to an IS implementation from existing validated scales [10, 14, 37]. Measurement for the distributive justice construct will be developed starting from the definition and dimensions provided in this paper and following established guidelines for measurement validation [38].

Since all of the measures are perceptual, it could be beneficial to assess them using implicit measures to control for a potential monomethod bias [39]. The construct in our model that could benefit from such measures the trustworthiness towards management. Similarly to Dimoka' study of trust and distrust's neural corollate [40], we could use neurophysiological measures to capture trustworthiness towards management.

For data analysis, Partial least squares (PLS) will be used as it is appropriate for theory development and have lower constraints in terms of sample size compared to covariance based SEM [41].

4.2 Contributions for Theory and Practice

Compared to previous theories in the IS literature, we propose one of the few variance models that aims at explaining and predicting users' intentions to resist an IS implementation.

To achieve this goal we introduced the constructs of justice and trust to the study of IS resistance. Namely procedural, interpersonal, and informational justice and trust represented by trustworthiness towards management. Even though the theory underlying the construct of Distributive Justice (i.e., equity theory) was used before in the IS literature [6], we have clarified this construct delineating its dimensions. The study's constructs cover individuals' perceptions not only exclusive to their expected outcomes about the system's characteristics but include those relative to the implementation process and to the relationships they have with the implementers and the management. This last relationship was investigated by Rivard and Lapointe [2] in their study about the effect of implementers' responses to resistance [2]. Also, we proposed relationships that capture patterns linking the justice components and trustworthiness towards management. Additionally, we provided explanations based on the sense-making process of individuals that gather information cues from their environment to shape their later reactions based on the level of threat that they perceive. For the contextual boundaries, the model considers the IS implementation process within a continued relationship between users, implementers and managers. This is congruent with previous IS resistance studies that had found that resistance behaviours manifested during the implementation process, e.g., [3] and long after the implementation had finished, e.g., [42].

For managers, previous studies suggest that early in the implementation the users tend to resist the system then later shift their resistance to the system's significance and its advocates depending on their response to users' resistance. However, we argue that intention to resist is not only shaped by the response of management to earlier resistance behaviours but also by the formal and informal interactions they engage in with the users. With this in mind, the process of managing resistance starts before the first signs of resistance appear. An effective management of resistance should be done through a better treatment of users, considering their needs and opinions in the decision making process and clearly explaining the motives of the IS implementation.

4.3 Limitations and Future Research

The model proposed in this study is an individual level model of IS resistance. However, since a) IS resistance has been studied from a multilevel perspective and b) as it has been shown that its manifestation can evolve from an individual to a group level [5], it would be useful to look at the effect of other multilevel factors on individual intentions to resist and how individuals' intentions to resist affect group level behaviours. One causal mechanism that could be investigated is the need to increase one's *Punitive capability* [34, 43] in the relationship between users and implementers and management under the condition that these users have similar desired outcomes.

The second limitation of the paper is that it considers the implementers and the management as one entity in their relationships with the users. However, this entity is not homogeneous as during the implementation, organizations can rely on the capabilities of their internal IT department, they can also use outside consultants to help them with the implementation process or they can even outsource the implementation project. This adds a level of complexity to the relationships between the users and the entity that drives the implementation and thus requires additional theorizing and research.

References

1. Van de Ven, A.H., Poole, M.S.: Alternative approaches for studying organizational change. Organ. stud. **26**(9), 1377–1404 (2005)
2. Rivard, S., Lapointe, L.: Information technology implementers' responses to user resistance: nature and effects. MIS Q. **36**, 897–920 (2012)
3. Markus, M.L.: Power, politics, and MIS implementation. Commun. ACM **26**(6), 430–444 (1983)
4. Coetsee, L.: From resistance to commitment. Public Adm. Q. **23**, 204–222 (1999)
5. Lapointe, L., Rivard, S.: A multilevel model of resistance to information technology implementation. MIS Q. **29**, 461–491 (2005)
6. Joshi, K.: A model of users' perspective on change: the case of information systems technology implementation. MIS Q. **15**, 229–242 (1991)
7. Adams, J.S.: Inequity in social exchange. Adv. Exp. Soc. Psychol. **2**, 267 (1965)
8. Greenberg, J.: A taxonomy of organizational justice theories. Acad. Manag. Rev. **12**(1), 9–22 (1987)

9. Colquitt, J.A., et al.: Justice at the millennium: a meta-analytic review of 25 years of organizational justice research. J. Appl. Psychol. **86**(3), 425 (2001)
10. Colquitt, J.A.: On the dimensionality of organizational justice: a construct validation of a measure. J. Appl. Psychol. **86**(3), 386 (2001)
11. Mishra, A.K., Spreitzer, G.M.: Explaining how survivors respond to downsizing: the roles of trust, empowerment, justice, and work redesign. Acad. Manag. Rev. **23**(3), 567–588 (1998)
12. Leventhal, G.S.: What should be done with equity theory? In: Gergen, K.J., Greenberg, M. S., Willis, R.H. (eds.) Social Exchange, pp. 27–55. Springer, Boston (1980). https://doi.org/10.1007/978-1-4613-3087-5_2
13. Mayer, R.C., Davis, J.H., Schoorman, F.D.: An integrative model of organizational trust. Acad. Manage. Rev. **20**(3), 709–734 (1995)
14. Fishbein, M., Ajzen, I.: Predicting and Changing Behavior: The Reasoned Action Approach. Psychology Press, New York (2011)
15. Skarlicki, D.P., Folger, R.: Retaliation in the workplace: the roles of distributive, procedural, and interactional justice. J. Appl. Psychol. **82**(3), 434 (1997)
16. Greenberg, J.: Employee theft as a reaction to underpayment inequity: the hidden cost of pay cuts. J. Appl. Psychol. **75**(5), 561 (1990)
17. Greenberg, J.: Using socially fair treatment to promote acceptance of a work site smoking ban. J. Appl. Psychol. **79**(2), 288 (1994)
18. Skarlicki, D.P., Folger, R., Tesluk, P.: Personality as a moderator in the relationship between fairness and retaliation. Acad. Manag. J. **42**(1), 100–108 (1999)
19. Compeau, D.R., Higgins, C.A.: Computer self-efficacy: Development of a measure and initial test. MIS Q. **19**, 189–211 (1995)
20. Compeau, D., Higgins, C.A., Huff, S.: Social cognitive theory and individual reactions to computing technology: a longitudinal study. MIS Q. **23**, 145–158 (1999)
21. Blader, S.L., Tyler, T.R.: Relational models of procedural justice. In: The Oxford Handbook of Justice in the Workplace, vol. 351, pp. 370 (2015)
22. Van den Bos, K.: Humans making sense of alarming conditions: psychological insight into the fair process effect. In: Oxford Handbook of Justice in Work Organizations, pp. 403–417 (2015)
23. Lazarus, R.S., Folkman, S.: Stress, Appraisal, and Coping. Springer, New York (1984)
24. Beaudry, A., Pinsonneault, A.: Understanding user responses to information technology: a coping model of user adaptation. MIS Q. **29**, 493–524 (2005)
25. Epstein, S., Pacini, R.: Some basic issues regarding dual-process theories from the perspective of cognitive-experiential self-theory. In: Dual-Process Theories in Social Psychology, pp. 462–482 (1999)
26. Thibaut, J.W., Walker, L.: Procedural Justice: A Psychological Analysis. Lawrence Erlbaum Associates, Mahwah (1975)
27. Rousseau, D.M., et al.: Not so different after all: a cross-discipline view of trust. Acad. Manag. Rev. **23**(3), 393–404 (1998)
28. Colquitt, J.A., Rodell, J.B.: Justice, trust, and trustworthiness: a longitudinal analysis integrating three theoretical perspectives. Acad. Manag. J. **54**(6), 1183–1206 (2011)
29. Morgan, D., Zeffane, R.: Employee involvement, organizational change and trust in management. Int. J. Hum. Resour. Manage. **14**(1), 55–75 (2003)
30. Stanley, D.J., Meyer, J.P., Topolnytsky, L.: Employee cynicism and resistance to organizational change. J. Bus. Psychol. **19**(4), 429–459 (2005)
31. Dirks, K.T., Ferrin, D.L.: The role of trust in organizational settings. Organ. Sci. **12**(4), 450–467 (2001)
32. Brockner, J., Wiesenfeld, B.M.: An integrative framework for explaining reactions to decisions: interactive effects of outcomes and procedures. Psychol. Bull. **120**(2), 189 (1996)

33. Brockner, J., Siegel, P.: Understanding the interaction between procedural and distributive justice: the role of trust (1996)
34. De Dreu, C.K., Giebels, E., Van de Vliert, E.: Social motives and trust in integrative negotiation: the disruptive effects of punitive capability. J. Appl. Psychol. **83**(3), 408 (1998)
35. Wanous, J.P., Reichers, A.E., Austin, J.T.: Cynicism about organizational change: measurement, antecedents, and correlates. Group Org. Manage. **25**(2), 132–153 (2000)
36. Choi, M.: Employees' attitudes toward organizational change: a literature review. Hum. Resour. Manage. **50**(4), 479–500 (2011)
37. Mayer, R.C., Davis, J.H.: The effect of the performance appraisal system on trust for management: a field quasi-experiment. J. Appl. Psychol. **84**(1), 123 (1999)
38. MacKenzie, S.B., Podsakoff, P.M., Podsakoff, N.P.: Construct measurement and validation procedures in MIS and behavioral research: integrating new and existing techniques. MIS Q. **35**(2), 293–334 (2011)
39. De Guinea, A.O., Titah, R., Léger, P.-M.: Measure for measure: a two study multi-trait multi-method investigation of construct validity in IS research. Comput. Hum. Behav. **29**(3), 833–844 (2013)
40. Dimoka, A.: What does the brain tell us about trust and distrust? Evidence from a functional neuroimaging study. MIS Q. **34**, 373–396 (2010)
41. Gefen, D., Straub, D., Boudreau, M.-C.: Structural equation modeling and regression: guidelines for research practice. Commun. Assoc. Inf. Syst. **4**(1), 7 (2000)
42. Ferneley, E.H., Sobreperez, P.: Resist, comply or workaround? An examination of different facets of user engagement with information systems. Eur. J. Inf. Syst. **15**(4), 345–356 (2006)
43. Carnevale, P.J., Pruitt, D.G.: Negotiation and mediation. Ann. Rev. Psychol. **43**(1), 531–582 (1992)

The Impact of Work from Home (WFH) on Workload and Productivity in Terms of Different Tasks and Occupations

Hongyue Wu[1] and Yunfeng Chen[2(✉)]

[1] Construction Animation, Robotics, and Ergonomics (CARE) Lab, School of Construction Management Technology (SCMT), Purdue University, West Lafayette, IN 47906, USA
wu1513@purdue.edu
[2] CARE Lab, SCMT, Purdue University, West Lafayette, IN 47906, USA
chen428@purdue.edu

Abstract. Most people have to work from home (WFH) due to stay-at-home orders in response to COVID-19 pandemic. The shifting of work environment from regular office to home has caused changes in workload and productivity, which may lead to reduced salaries for employees, economic loss of the companies, and impacts on the national economy. Thus, it is urgent to explore the impacts of WFH on workload and productivity of different employees. A nationwide survey was distributed to collect data about the workload and productivity of regular work and WFH considering different types of tasks and occupations. The findings indicate that WFH causes an increase in workload for all participants by three hours per week and a loss of productivity for 38% participants. Moreover, the technical issues, such as less efficiency of online communication technologies, are the core reasons for the decrease in productivity. For different occupations, employees who regularly work in an office or workstation show a higher workload because their major work can be done at home but require more time due to the technical issues, while on-site occupations and many researchers have less workload because their major work cannot be done at home, such as onsite work and experiments. Then, the workload and complexity of tasks lead to the differences in productivity. In the future, the key problems are how to address the technical issues and strengthen the human-computer interaction to improve the productivity, and support on-site work and lab-based tasks to improve the feasibility of WFH.

Keywords: Work from home (WFH) · Productivity · Workload · Home-based work

1 Introduction

Due to the COVID-19, the United Stated (U.S.) was under stay-at-home orders. At the end of March 2020, more than 308 million people, which accounted for 94% of the U. S. population, had to stay at home [1]. Therefore, many workers have to work from home. It was indicated that over 34% of employees who were commuting are now

C. Stephanidis et al. (Eds.): HCII 2020, LNCS 12427, pp. 693–706, 2020.
https://doi.org/10.1007/978-3-030-60152-2_52

working from home in the U.S. [2]. Also, over 16 million knowledge workers have started home-based work since March 27, 2020 [3]. This rapid change has caused many problems. Many employees reported the change of workload at home and suffered from a productivity loss, especially the new home-based workers [3]. The loss of productivity will not only lead to a decrease in salaries of employees [4], but also the economic loss of companies and even the whole industry and national economy. Besides, many companies provide or consider the options of partial or complete home-based work in a long-term [5]. It was reported that previous experiences of Work from Home (WFH) are helpful for employees to adapt to future remote work [3]. There are studies investigating the impacts of WFH on quality of life and the online content contributions of workers [6, 7]. Also, many researchers investigated the feasibility of WFH due to the impacts of COVID-19 considering different types of work, demographic information, different areas, etc. [8–10]. However, there is still a lack of research focusing on the impact of WFH on workload and productivity using quantitative evidence. Thus, this research aims to fill the gaps to analyze the differences of workload and productivity between regular work and WFH, and identify the reasons leading to these differences.

To achieve this goal, a nationwide survey was conducted to collect data about the workload and productivity of both regular work and home-based work for employees with different occupations. Then, the workload and productivity of different tasks were compared to show the impacts of WFH. Finally, the reasons causing the change in workload and productivity were discussed to sheds light on how to support future home-based work based on current experiences.

2 Literature Review

It was indicated that 37% of U.S. jobs can be conducted at home, including educational services, professional and technical services, management, finance and insurance, and information [11]. Therefore, WFH is a popular trend for current employees. The advantages of WFH are the flexible schedule, cost-saving for transportation, and better work-life balance, while there are also disadvantages including the loss of work motivation and productivity, possible data security problems, etc. [12]. For example, a study showed that 56% of 1014 respondents reported less productivity and less workload when working from home [13].

Furthermore, there are many researches exploring different factors that influence productivity of WFH. First, the distraction from family members is a core problem impacting productivity for parents, especially women [14]. Also, the lack of in-person collaboration caused by home-based work may lead to a decrease in creativity and innovation [14]. Meanwhile, it was indicated that WFH could lead to the increase in mental health disorders because of the workplace transmission and limitation of work relationships [8, 15], which also drain the productivity. Besides, many organizations lack proper plans and resources to support home-based work [4], which is a major reason for the loss of productivity for the employees.

However, how WFH impacts the workload and productivity are still unclear due to the lack of quantitative evidences. Therefore, this paper aims to explore the differences in workload and productivity between regular work and WFH, and identify the problems causing the change of workload and productivity to support current and future home-based work.

3 Research Methodology

A nationwide survey about WFH were distributed this May in the U.S. More than 13000 people were reached individually by email, LinkedIn, or other social media. Meanwhile, 15 different professional associations were reached out to help distribute the survey. Total responses are 774, with a response rate of 6%. For the workload and productivity of employees with different tasks, there are 200 complete responses, which were used for further analysis.

3.1 Participants

The participants are employees in the U.S. from both industry and education areas, covering seven different occupations [16, 17], which are listed in Table 1. Also, the participants cover 26 different states from all regions in the U.S.

Table 1. Occupations of participants in the survey.

Categories	Occupations	Definitions
Education	Teacher/Instructor	Major work is teaching
	Researcher	Major work is research
	Professor	Need to do both teaching and research
	Staff	Major work is administration
Industry	On-site occupations	Usually conduct your work outside office
	Project management occupations in office	Usually conduct project-specific work in office
	Staff	Major work is company administration and support

3.2 Measurements

Different tasks cover listening, speaking, reading, and writing using both traditional ways and electronic devices, which are shown in Table 2. The tasks were compiled considering both the major skills and the devices/tools that will be used [18, 19]. In other words, these tasks indicate different types of information that employees can obtain and communicate with others. Based on the following tasks, the hours spent on each task per week were used to measure the workload, while the productivity was evaluated by a five-point Likert scale.

Table 2. Tasks for employees considering listening, speaking, reading and writing.

Category	Tasks
Listening & Speaking	Phone call
	In-person meeting
	Online meeting
	Other communication (text/chat/etc.)
	Presentation
Reading & Writing	Email
	Review documents in print
	Prepare documents in print
	Review documents on-screen (computers/iPads/etc.)
	Prepare documents on-screen (computers/iPads/etc.)
	Other tasks on paper (calculation/drawing/etc.)
	Other tasks on electronic devices (calculation/drawing/coding/etc.)

4 Data Analysis and Results

4.1 Demographic Information

Occupation. The distribution of different occupations is shown in Fig. 1. The project management occupation in office in the industry has the highest proportion of 32% (64 responses). At the same time, professors in the education area account for 31% (61 responses). Then, staff in the education area accounts for 9% (18 responses). After that, the number of responses from researchers in the education area is 16 (8%). Also, the percentage of staff in the industry is 7% (14 responses), which is the same as on-site employees in the industry. Finally, teacher/instructor in the education area accounts for 6% (13 responses). Overall, there are 108 responses from the education area, while 92 responses from the industry. Meanwhile, the respondents cover different majors, including architecture, engineering, construction, and operation, management, public administration, etc. Therefore, the data can represent the workload and productivity of diverse occupations and fields.

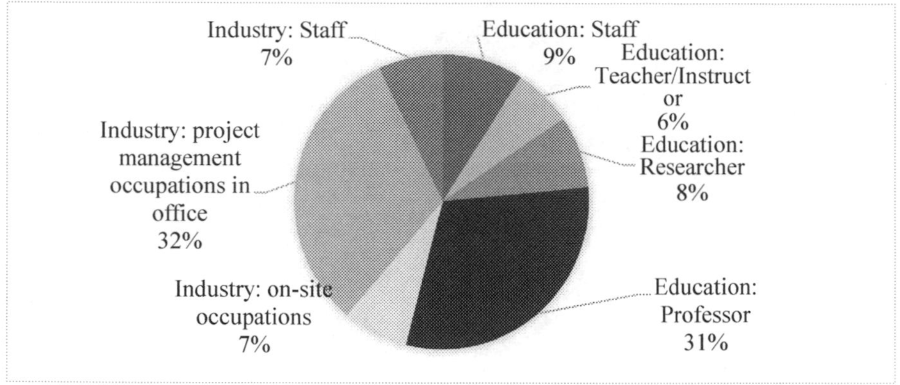

Fig. 1. The distribution of occupations for 200 responses.

Age and Gender. The participants cover different ages from 23 to 79 with only one participant who didn't want to provide this information. The distribution of age is shown in Fig. 2. The major part is from 31 to 39 years old, which accounts for 26.63%. Also, there are 43 respondents who are 23 to 31 years old. Then, the third area is from 39 to 47 years old, which accounts for 17.59%. After that, 29 participants whose ages range from 47 to 55. Besides, the number of participants who are older than 55 years old is 39, which accounts for 19.60%. It shows that the distribution of age covers a large range and mainly focuses on the range from 23 to 55, which is also the common age of current employees [20].

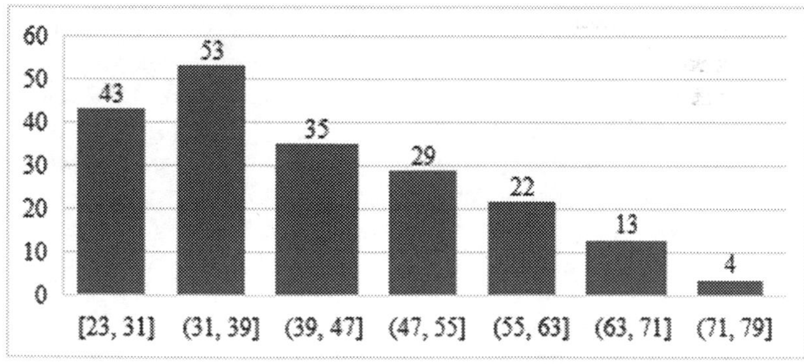

Fig. 2. The distribution of ages for participants.

As for gender, about 78% of participants are male, while only 20% of them are female. In addition, there are 2% (3 respondents) chose the Non-binary, Not listed, or Prefer not to disclose. The result is shown in Fig. 3.

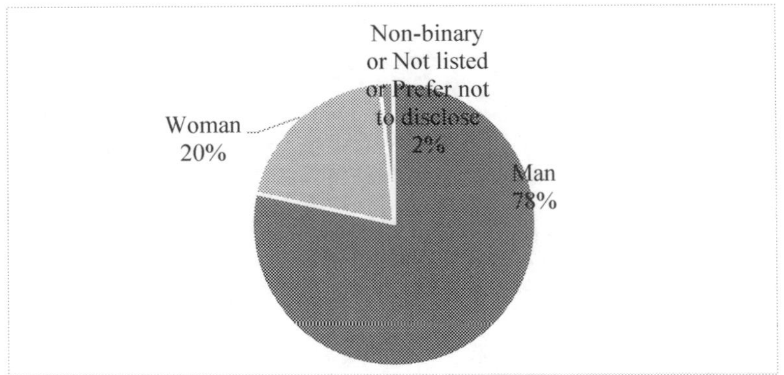

Fig. 3. The distribution of genders for participants.

4.2 Analysis of Workload

Different Tasks. The mean and standard deviation of hours that the participants spent on each task for both regular work and WFH per week are shown in Table 3. After data screening, four responses were not used for the analysis of workload because their total working hours per week is higher than 165 h, which is unreasonable. Therefore, there are 196 responses that were used for the following analysis.

For regular work, email and in-person meeting are the tasks that require most hours to finish (6.7449 h/week and 6.4786 h/week on average), while prepare documents in print and other tasks on paper (such as calculation/drawing/etc.) account for the smallest part (1.7474 h/week and 1.5638 h/week on average) of total workload. Considering WFH, email and online meeting show more working hours per week (7.6097 h/week and 7.0536 h/week on average), while in-person meeting and prepare documents in print have less workload (0.8699 h/week and 0.8827 h/week on average).

For the differences between regular work and home-based work, seven out of 12 tasks show higher workload when working from home. For the category of listening and speaking, in-person meeting decreases most (5.6087 h/week) and presentation reduces by 0.2474 h/week for home-based work, while the workload of the other three tasks increases from 0.6071 to 5.1505 h/week. As for the category of reading and writing tasks, all the tasks relevant to printed paper and documents (prepare documents in print, review documents in print, and other tasks on paper) decrease ranging from 0.1913 to 0.8648 h/week, while the other four tasks relating to electronic devices increase ranging from 0.4974 to 1.3750 h/week.

In addition to the tasks listed above, the respondents added some other major tasks or specific tasks that they wanted to mention. The answers include webinars, brainstorming, other technical tasks, individual work, etc. In total, the overall workload of regular work is 43.0268 h/week on average with a standard deviation of 26.3376. For WFH, the average workload per week is 46.0967 h with a standard deviation of 29.1805. Therefore, WFH lead to an increase of overall workload by 3.0699 h/week.

However, the standard deviation is 23.2634, which means there are significant individual differences.

Table 3. Workload of different tasks for regular work and WFH.

Category	Tasks	Regular work (hours/week)		Work from home (hours/week)		Differences (regular work - WFH)	
		Mean	SD	Mean	SD	Mean	SD
Listening & Speaking	Phone call	3.4043	6.8023	4.3431	5.6054	−0.9388	6.6507
	In-person meeting	6.4786	6.1122	0.8699	3.0314	5.6087	5.8809
	Online meeting	1.9031	2.6118	7.0536	6.3974	−5.1505	5.9521
	Other communication (text/chat/etc.)	2.6250	4.5464	3.2321	4.5883	−0.6071	3.3853
	Presentation	1.8148	2.2511	1.5673	2.2187	0.2474	1.6144
Reading & Writing	Email	6.7449	5.7678	7.6097	6.1523	−0.8648	4.0371
	Review documents in print	2.3929	2.9218	1.6020	2.9587	0.7908	2.5665
	Prepare documents in print	1.7474	2.7021	0.8827	1.9139	0.8648	2.4058
	Review documents on-screen (computers/iPads/etc.)	4.6939	4.2545	6.0689	5.1790	−1.3750	4.3915
	Prepare documents on-screen (computers/iPads/etc.)	5.0459	5.4418	5.9770	6.5332	−0.9311	4.0296
	Other tasks on paper (calculation/drawing/etc.)	1.5638	3.4869	1.3724	3.6132	0.1913	1.5057
	Other tasks on electronic devices (calculation/drawing/coding/etc.)	3.3495	6.6004	3.8469	7.1437	−0.4974	3.0066

Different Occupations. To better understand the differences of workload between different occupations, the total working hours for each occupation, which are the summation of all the hours spent on each task, were listed in Table 4.

In the education area, the workload of staff is the highest in both regular work and WFH (45.8889 h/week and 53.1111 h/week on average), while teacher/instructor shows the least workload (26.0385 h/week for regular work and 33.8846 h/week for WFH). Considering the differences between regular work and WFH, only the workload of researchers decreases when working from home by 5.2 h/week, while the other three occupations have more workload of home-based work, ranging from 6.1918 to 7.8462 h/week.

For industry, the average workload of regular work is higher than the average in the education area. On-site occupations have the most working hours per week for regular work (53.0769), while project management occupations in office shows the highest workload when working from home (51.5726 h/week). As for the differences, only project management occupations in office have more work to do when working from home, increasing by 3.371 h/week. However, the on-site occupations and staff shows less workload by 6.6154 and 3.7857 h/week.

Table 4. Total workload for different occupations.

Category	Occupations	Regular work (hours/week)		Work from home (hours/week)		Differences (regular work - WFH)	
		Mean	SD	Mean	SD	Mean	SD
Education	Staff	45.8889	13.0964	53.1111	19.8906	−7.2222	13.2023
	Teacher/Instructor	26.0385	22.5725	33.8846	27.5999	−7.8462	14.1604
	Researcher	39.0667	31.7630	33.8667	24.1753	5.2000	29.4666
	Professor	37.5123	18.1280	43.7041	27.4931	−6.1918	21.4662
Industry	On-site occupations	53.0769	45.8411	46.4615	34.0260	6.6154	54.9144
	Project management occupations in office	48.2016	26.7178	51.5726	29.2643	−3.3710	16.7057
	Staff	51.1429	32.5195	47.3571	42.2513	3.7857	15.2829

4.3 Analysis of Productivity

Different Tasks. The mean and standard deviation of productivity for both regular work and WFH are shown in Table 5. Productivity is defined as a real output per hour, which was measured by a five-point Likert scale with 1 indicating the lowest productivity and 5 indicating the highest productivity.

For regular work, in-person meeting shows the highest productivity, which is 4.1640 on average. Then, prepare documents on-screen and presentation also have high productivities, which are 4.0757 and 4.0571 on average respectively. However, online meeting and other communication through text/chat/etc. show lower productivity than other tasks (3.7257 and 3.7457). It is interesting to see that the differences in productivity among tasks are not significant because all the results are between 3.7 and 4.2 out of 5. Moreover, the overall productivity of all the listening and speaking tasks are 4.0311 on average, while it is 4.1058 for the reading and writing tasks. Besides, the standard deviations of all the scores range from 1.1457 to 0.7851.

As for WFH, email and review documents on screen show the highest productivities, which are 4.0885 and 4 on average. But the in-person meeting had a significant lowest productivity, which is only 2.7835 out of 5. In this part, there is a significant difference in productivity among all tasks, which is different from regular work. It indicates that WFH leads to the variance of productivity among tasks. For the overall productivity, the listening and speaking tasks (3.7268) have lower productivity than the reading and writing tasks (3.9683), which is the same as regular work. In addition, the standard deviations range from 1.4806 to 0.9104.

Then, for the differences between regular work and home-based work, five out of 12 tasks show higher productivity when working from home, which are phone call, review documents on-screen, other communication, email, and online meeting. Online meeting has the highest improvement in productivity when changing from regular work to WFH (0.53 on average). On the other hand, in-person meeting shows a significant decrease in productivity of home-based work (2.585 on average). Also, all the tasks relevant to documents and paper in print represent lower productivity when working from home.

Table 5. Productivity of different tasks for regular work and WFH.

Category	Tasks	Regular work		Work from home		Differences (regular work - WFH)	
		Mean	SD	Mean	SD	Mean	SD
Listening & Speaking	Phone call	3.8033	1.1457	3.7234	1.1646	−0.0200	1.4317
	In-person meeting	4.1640	0.9891	2.7835	1.4806	2.5850	2.1202
	Online meeting	3.7257	1.0138	3.9479	0.9907	−0.5300	1.7361
	Other communication (text/chat/etc.)	3.7457	1.0196	3.7740	1.0251	−0.1000	1.4035
	Presentation	4.0671	0.9791	3.4658	1.2847	0.5450	1.5749
	Overall productivity	4.0311	0.8472	3.7268	0.9988	0.2750	1.4246
Reading & Writing	Email	3.9635	0.9835	4.0885	0.9420	−0.1200	1.2977
	Review documents in print	3.8571	1.0513	3.5865	1.2378	0.8550	1.9909
	Prepare documents in print	3.7484	1.0787	3.4173	1.2500	0.7350	1.7493
	Review documents on-screen (computers/iPads/etc.)	3.9202	0.9244	4.0000	1.0000	−0.0950	1.3285
	Prepare documents on-screen (computers/iPads/etc.)	4.0757	0.8875	3.9462	1.0792	0.1000	1.4284
	Other tasks on paper (calculation/drawing/etc.)	3.9291	0.9830	3.7259	1.1292	0.2550	1.3148
	Other tasks on electronic devices (calculation/drawing/coding/etc.)	3.9530	1.0090	3.8725	1.1107	0.0600	1.2467
	Overall productivity	4.1058	0.7851	3.9683	0.9104	0.1300	1.3121

Considering the overall productivity of all the participants, the average overall productivity of the two categories is used. For regular work, the overall productivity is 3.8850 with a standard deviation of 1.0724. Then, for WFH, the overall productivity is 3.6825 with a standard deviation of 1.0706. Therefore, the difference in productivity between regular work and home-based work is 0.2025 with a standard deviation of 1.2820. It shows that WFH has caused the loss of productivity. As for the differences in productivity for each participant, 38% of the participants experienced less productivity for WFH, while 37.5% of them showed an increase in productivity. 24.5% of the respondents indicated there are no obvious differences in productivity between regular work and home-based work.

Different Occupations. For the differences between seven occupations, the overall productivity of each occupation, which was calculated by the average of overall productivities of two categories, is shown in Table 6.

In the education area, for regular work, professor shows the highest productivity (4.082 out of 5), while the teacher/instructor has the lowest productivity (3.4615). Then, considering home-based work, the productivity of teacher/instructor becomes the highest (3.7308), whereas researcher shows the lowest productivity (3.2813). In addition, considering the differences between regular work and WFH, only teacher/instructor reports the increased productivity when working from home (0.2692), while the other three occupations have reduced productivities ranging from 0.1389 to 0.4089.

For industry, project management occupations in office shows the highest productivity for regular work (3.9844), while on-site occupations have the lowest productivity (3.3571) when they work on-site due to the complexity of their work contents and environment. Meanwhile, for WFH, project management occupations in office still have the highest productivity (3.8672), but it decreased comparing with regular work (0.1172). The productivity of staff decreases the most (0.5) when working from home among all the seven occupations. However, on-site occupations shows an increase in productivity for home-based work (0.2857).

Table 6. Overall productivity for different occupations.

Category	Tasks	Regular work (hours/week)		Work from home (hours/week)		Differences (regular work - WFH)	
		Mean	SD	Mean	SD	Mean	SD
Education	Staff	3.8611	0.8368	3.7222	0.8613	0.1389	0.8712
	Teacher/Instructor	3.4615	1.3301	3.7308	0.6651	−0.2692	1.3168
	Researcher	3.6563	1.4574	3.2813	1.4602	0.3750	0.9747
	Professor	4.0820	0.7862	3.6721	0.8750	0.4098	0.8921
Industry	On-site occupations	3.3571	1.8855	3.6429	1.5495	−0.2857	2.8603
	Project management occupations in office	3.9844	0.9999	3.8672	1.0437	0.1172	1.2432
	Staff	3.7857	0.7523	3.2857	1.3966	0.5000	1.1266

5 Discussion

Overall, the results indicate that WFH leads to an increase in workload and a loss of productivity for current employees. The workload increases by 3 hours per week on average (from 43.0268 h/week to 46.0967 h/week), while over 38% of participants report a decrease in productivity.

The major difference in workload is that the working hours for all the tasks on-screen and using electronic devices are increased when working from home, while the workload of reading and writing tasks based on printed papers and documents is reduced. It shows that tasks relying on printed papers and documents or in-person communication have been changed to online communication and work based on electronic devices. Many participants reported that “I don’t have access to a printer”, “I cannot have face-to-face meetings with my colleagues”, and “There is a lack of access to resources and documents in office”, which cause the change of workload. In addition, the results indicate that email showed the highest percentage of workload for both regular work and WFH. It has become the most commonly used way for people to conduct professional communications. Therefore, WFH makes people use more technologies to assist their work, such as electronic devices and online communication tools, which means the human-computer interaction becomes more important.

Then, considering different occupations, all the occupations in the industry had higher working hours per week than occupations in the education area. Professionals in the industry have a more stressful schedule than employees in the education area. Then, after shifting to WFH, staff in the education area, teacher/instructor, professor, project management occupations in office show increased workload. It may be because of their regular work in an indoor work environment, which means their major tasks can be done at home computers or other electronic devices [11]. However, WFH makes the workload become heavier from several aspects. Many respondents indicated that "more time to prepare online courses", "it takes longer to complete documentation virtually", and "more hours to contact and communicate with co-workers".

On the contrary, the researcher in the education area, on-site occupations, and staff in the industry show less workload when working from home. There are several possible reasons for the difference. One of the major reasons may be that many of their work contents cannot be done at home. For on-site occupations, many participants indicated that "safety management requires direct observation and interaction with jobsite staff" and "there are some quality and safety tasks I cannot do as well even with the drone and camera". Also, many researchers showed that "our research is mainly lab-based experiments for which lab facilities are required" and "various events were canceled because of the pandemic". Therefore, there is less work that they can finish at home. However, for staff in the industry, their workload is reduced although their regular work is conducted in the office. The respondents proposed some explanations including "fewer projects, fewer employees, fewer payables and receivables" and "some documentation must be completed when we are allowed to return to the office". Therefore, the change of workload is relevant to the status of the industry and requirement of their companies.

Then, for the decrease in productivity, the technical issue is the main reason. First, the reduced productivity of remote communication leads to lower productivity. The results show that in-person meeting has the highest productivity during regular work, while it has the lowest productivity of WFH because of the limitation of face-to-face meeting. Therefore, the major part of communication should be finished by online meeting, whose productivity increases. However, when changing to online meetings at home, their productivity is lower than in-person meetings during regular time, which indicates that remote communication by technologies cannot have the same high efficiency as face-to-face communication. Majority of the respondents complained about the less efficiency of communication when working from home, such as "face to face communication is more efficient", "working from home has greatly reduced the communication efficiency", and "collaboration with team members is more difficult. Communicating via email and conference calls, while helpful, are not as productive as typical in-person communication". Therefore, there is a need to improve the efficiency of virtual meetings and communication based on technologies to better support home-based work.

Also, when working from home, the tasks on electronic devices, such as review documents on-screen, email, online meeting, etc., have higher productivity, while the tasks based on printed papers or in-person communication, such as review documents in print, other tasks on paper, in-person meeting, etc., have lower productivity. The main reason may be the lack of access to resources as mentioned in previous part. Online meeting shows the highest improvement in productivity because most people mentioned that "communication with colleagues is critical for collaboration". But they cannot have face-to-face meeting at home, which means the in-person meeting shows a significant lower productivity. However, current technologies allow them to communicate with each other online. Therefore, tasks based on electronic devices have become a common and more productive way to finish work when working from home. However, WFH still causes a loss of productivity because of the technical issues mentioned by the respondents, such as "Slower technology at home, wi-fi issues, only one monitor", "Less efficient due to technology available", "There have been WIFI issues working from home that I don't experience at work", etc. Therefore, future studies should focus on how to address these problems and enhance human-computer interactions to improve the productivity of WFH.

Besides, another key reason for the loss of productivity is the distractions from family, especially childcare and housework. Many respondents made comments about it, such as "More distractions/responsibilities at home", "As the schools and daycare are closed, I need to do parenting, teaching, and spend more time cleaning and sanitizing", "Household demands, childcare, child education", "Interference with private life (household, family matters, etc.)", and "I'd rather work at the office. That's where I work. No kids, dog, laundry, but big screen, etc." Therefore, future research should pay attention to the work-life balance when WFH to improve work performance.

Finally, for different occupations, only teacher/instructor in the education area and on-site occupations in the industry show higher productivity, while the other five occupations have reduced productivity when working from home. The possible reason for the difference is that teacher/instructor only needs to work for the online courses at home. The single type of work makes it easier to adapt to WFH and have high productivity, while professor, staff, project management occupations in the office all have various types of work and need to spend much time communicating with others. The task complexity impacts the productivity [21]. Meanwhile, for on-site occupations, their productivity increases potentially because of the lower workload. However, although the researchers have less workload, their productivity decreased possibly because it takes them more time to adjust some lab-based tasks to home.

6 Conclusion

Due to stay-at-home orders under this pandemic, most employees have to shift from regular work to WFH. It has caused changes in workload and productivity, which impact employees, companies, and the national economy. Therefore, there is an urgent need to understand the differences between regular work and WFH. A nationwide survey was distributed to explore the workload and productivity considering different types of tasks and occupations. The results indicated that WFH leads to an increase in workload by 3 hours per week on average, and over 38% of respondents show a loss of productivity when working from home.

Moreover, there are differences in workload and productivity among different tasks and occupations. For the workload, WFH needs to use more technologies, including computers, online communication tools, etc., which require a high-level of human-computer interaction. However, the technical issues, including less efficiency of online communication, difficulties of technology accessibility, WIFI issues, etc. are major reasons for the reduced productivity. Also, the distraction from family is another key factor impacting productivity when working from home. Considering different occupations, employees whose tasks are regularly finished in an office or using computers show a higher workload because most of their work can be done remotely. However, on-site occupations and researchers, whose major work needs to be done onsite or in the lab, have less workload at home. Then, teacher/instructor and on-site occupations show higher productivity due to their single type of work or less workload. However, the other five occupations suffer a loss of productivity due to the increased workload, complexity of tasks, high-level of communication with others, and difficulty of shifting work from regular place to home.

The paper contributes to the theoretical understanding of WFH considering both workload and productivity from the new perspectives of different tasks and occupations. Also, the findings can provide insight for both individuals and institutions or companies on WFH and how to improve their remote work efficiency in practice. According to the findings, future studies should focus on the development of technology to support better online communications and reduce problems caused by technology, which can improve the productivity of WFH. Also, it is critical to enhance the human-computer interaction and develop innovative tools that allow people to finish on-site tasks, lab-based work or other things that cannot be done remotely now at home in the future.

References

1. Secon, H.: An interactive map of the US cities and states still under lockdown - and those that are reopening. Business Insider. https://www.businessinsider.com/us-map-stay-at-home-orders-lockdowns-2020-3. Accessed 30 May 2020
2. Brynjolfsson, E., Horton, J., Ozimek, A., Rock, D., Sharma, G., Ye, H.Y.T.: COVID-19 and Remote Work: An Early Look at US Data. Unpublished work (2020)
3. Team at Slack: Report: Remote work in the age of Covid-19. https://slackhq.com/report-remote-work-during-coronavirus. Accessed 21 Apr 2020
4. Ralph, P., Baltes, S., Adisaputri, G., Torkar, R., Kovalenko, V., et al.: Pandemic Programming: How COVID-19 affects software developers and how their organizations can help. https://arxiv.org/abs/2005.01127. Accessed 30 May 2020
5. Shamir, B., Salomon, I.: Work-at-home and the quality of working life. Acad. Manage. Rev. **10**(3), 455 (1985)
6. Choudhury, P., Koo, W., Li., X.: Working from Home Under Social Isolation: Online Content Contributions During the Coronavirus Shock. Harvard Business School Technology & Operations Mgt. Unit Working Paper, 20–096 (2020)
7. Baker, M.G.: Who cannot work from home? Characterizing occupations facing increased risk during the COVID-19 pandemic using 2018 BLS data. https://www.medrxiv.org/content/10.1101/2020.03.21.20031336v2. Accessed 30 May 2020

8. Gottlieb, C., Grobovšek, J., Poschke, M.: Working from home across countries. Covid Econ. **8**, 71 (2020)
9. Hensvik, L., Le Barbanchon, T., Rathelot, R.: Which jobs are done from home? Evidence from the American Time Use Survey. https://www.iza.org/publications/dp/13138. Accessed 29 May 2020
10. Gupta, A.: Accelerating Remote Work After COVID-19. https://www.thecgo.org/research/accelerating-remote-work-after-covid-19/. Accessed 29 Apr 2020
11. Dingel, J.I.: How Many Jobs Can be Done at Home? National Bureau of Economic Research, NBER Working Paper, (2020)
12. Purwanto, A., Asbari, M., Fahlevi, M., Mufid, A., Agistiawatti, E., et al.: Impact of work from home (WFH) on indonesian teachers performance during the Covid-19 pandemic: an exploratory study. Int. J. Adv. Sci. Technol. **29**(5), 6235–6244 (2020)
13. Rubin, O., Nikolaeva, A., Nello-Deakin, S., te Brömmelstroet, M.: What can we learn from the COVID-19 pandemic about how people experience working from home and commuting?. Centre for Urban Studies, University of Amsterdam. https://urbanstudies.uva.nl/content/blog-series/covid-19-pandemic-working-from-home-and-commuting.html. Accessed 06 May 2020
14. Gorlick, A.: The productivity pitfalls of working from home in the age of COVID-19. Stanford News, Stanford University Communications, California, USA (2020)
15. Lam, H.W., Giessner, S., Shemla, M.: Tips: loneliness and working from home during the COVID-19 crisis. https://discovery.rsm.nl/articles/431-tips-loneliness-and-working-from-home-during-the-covid-19-crisis/. Accessed 27 Mar 2020
16. King, W.R., Cleland, D.I.: Project Management Handbook. Van Nostrand Reinhold, New York (1988)
17. Houston, D., Meyer, L.H., Paewai, S.: Academic staff workloads and job satisfaction: expectations and values in academe. J. High. Educ. Pol. Manage. **28**(1), 17–30 (2006)
18. Czerwinski, M., Horvitz, E., Wilhite, S.: A diary study of task switching and interruptions. In: Proceedings of the SIGCHI Conference on Human Factors in Computing Systems, Vienna, Austria, CHI 2004, vol. 6, pp. 175–182 (2004)
19. Reddy, K.J.: Relevance of listening and speaking skills for engineering students in their professional career. Strength Today Bright Hope Tomorrow **19**(11), 150–159 (2019)
20. U.S. Bureau of Labor Statistics: Labor Force Statistics from the Current Population Survey. https://www.bls.gov/cps/demographics.htm#age. Accessed 15 June 2020
21. Goparaju Purna, S., Ayesha, F., Sanghamitra, P.: Measuring productivity of software development teams. Serb. J. Manage. **7**(1), 65–75 (2012)

Examining Emerging Technology Awareness in the Accounting and Finance Industries Through Twitter Data

Jiawei Xing[1(✉)], Jiayang (Jocelyn) Lin[2], Manlu Liu[1], and Jennifer Xu[3]

[1] Rochester Institute of Technology, Rochester, NY 14623, USA
jxx9924@rit.edu
[2] Pittsford Sutherland High School, Pittsford, NY 14534, USA
[3] Bentley University, Waltham, MA 02452, USA

Abstract. The accounting and finance industries have been increasingly adopting new technologies in recent years. In this exploratory study, we examine the degree of technology awareness in the accounting and finance industries over the past ten years. By identifying all the twitter accounts of the top 100 accounting firms and top 100 finance institutions and examining how they have discussed about technologies. We are able to conduct a series of analyses to help the audience gain a better understanding of the patterns and changes of technology awareness in the accounting and finance industries.

Keywords: Technology awareness · Text mining · Twitter · Accounting · Finance

1 Introduction

With rapid advancements in technology, the accounting and finance industries have experienced drastic changes. More and more accounting and financial firms are adopting new technologies and deploying them in their operations. Sutton and Arnold (2013) found an emerging technology-driven phenomenon in accounting and information systems. Liu, Hsu, and Yen (2018) also argued that the Chief Information Officers (CIO) face dynamic challenges in a constantly changing accounting environment, triggered by the significant impact of the adoption of information technology. However, there is little research studying the awareness of technologies and applications in the accounting and finance industries and how this awareness changes from time to time in recent years.

Our study focuses on examining the technology awareness in the accounting and finance industries in the recent ten years through social media. More specifically, this study is intended to address the following research questions: (1) What are the main technologies that accounting and finance industries have considered relevant in recent years? (2) What is the difference between accounting and finance industries in terms of technology awareness? (3) What is the difference between the Big Four and other accounting firms in terms of technology awareness? (4) How has this degree of

C. Stephanidis et al. (Eds.): HCII 2020, LNCS 12427, pp. 707–719, 2020.
https://doi.org/10.1007/978-3-030-60152-2_53

technology awareness changed in recent years? Through this study, we hope to show the big picture of the technology awareness in these two industries. Our findings may assist decision-makers of accounting and financial firms design better technology adoption strategies, thereby increasing their competitive advantage. In addition, technology firms may also find more opportunities for collaboration with accounting and financial firms. Our study may also motivate more research on the topic of technology awareness in industries that have traditionally been considered non-technical.

Social media have served as platforms for individuals and organizations to interact and discuss various topics and issues, providing research with rich databases to study many interesting social, political, and business phenomena. In this study, we chose to use Twitter data as it is the most common data source for researchers due to its accessibility. In addition, it is also one of the most popular social media platforms that many firms choose to interact with their customers, partners, or other stakeholders. We perform a series of analyses using Twitter text data analysis techniques to help the audience better understand how technology awareness changes in the accounting and finance industries.

The rest of the paper is organized as follows: The next section reviews the literature regarding the social media analytics and the changes in technology awareness in the finance and accounting industries. The two sections that follow present the data collection, analysis methods, results, and discussion. The last section concludes the paper and discusses plans for future work.

2 Literature Review

2.1 Social Media Analytics

Social media are electronic communication platforms, where users create or join online communities to share information or ideas (Kaplan and Haenlein 2010). Social media analytics is to discover patterns and knowledge by analyzing user generated contents on social media platforms. User generated contents on social media are usually unstructured and in diverse formats, including audio, image, text and video. Among the different types of social media analytics, text analysis is the most common and widely employed. Table 1 summarizes the techniques used in unstructured textual data analysis (Batrinca and Treleaven 2015):

Li (2010) categorized social media text analysis approaches into two general types: (1) rule-based approaches that process and classify the text based on pre-defined rules, and (2) statistical approaches which employ statistical techniques to analyze text. Guo, Shi, and Tu (2016) found that neural network outperforms other machine learning techniques in news categorization.

2.2 Technology Changes in Accounting and Finance Industry

Suddaby, Saxton, and Gunz (2015) discussed the domain of accounting expertise is reconstituted in new social media – Facebook, LinkedIn, and Twitter in the Big Four accounting firms. They also used content analysis and interview data to generate the

Table 1. Textual data analysis techniques.

Techniques	Explanations
Natural language processing	A series of techniques to process natural language using computer science, artificial intelligence, and linguistics
News analytics	The measurement of the various qualitative and quantitative attributes of textual data
Opinion mining	A process to determine people's opinions from natural language text
Scraping	A technique to collect online data from social media or other online textual data by programming
Tokenization	A technique under Natural language processing that takes a text or sets of text and separates them into individual words
Sentiment analysis	A technique to extract subjective information for certain textual data
Text analytics	A series of techniques to analyze text from a word frequency distribution, pattern recognition, visualization, and predictive analytics

change in the professional domain of the accounting industry. Cobbin, et al. (2013) argued that the application of advanced digitization technologies to accounting and business archives has created new opportunities for accounting and business professions. Additionally, Arnold (2018) discussed the impact technology changes have had on audit, financial reporting, and management accounting (or control). The technology changes have a significant impact on the firm's profitability and operational efficiency. PricewaterhouseCoopers (PwC 2015) posted the white paper about how information technology and data analytics are changing the practices of business, and identified technical skills that will be highly sought after by employers. Other Big Four accounting firms have also published a number of white papers to demonstrate the significant impacts of information technology on business transformation.

3 Data and Methods

We started by generating the technology-related keyword list. After a series of Google search within the accounting and finance industries, we identified 80 keywords related to emerging technology. We manually reviewed these keywords and compared them against elite information systems journal databases and keyword classification systems of professional societies (e.g., IEEE and ACM). These keywords were then consolidated and standardized. The final list contains 43 technology-related keywords (see Table 2). The keywords were then used to identify the posts in accounting and financial firms' tweets to explore the domain knowledge and changes in technology awareness.

Next we identified top 100 accounting firms and 100 financial firms based on 2018 revenue (Data source: Statista). We use the two samples to represent the accounting and financial industries, respectively. We collected the tweets posted by these 200 firms from August 2008 to May 2019. For those firms owning multiple twitter accounts, we collected the data from all their accounts. For example, each of the Big Four accounting

Table 2. Technology-related keyword list.

Keywords	
AI - Artificial Intelligence	Implementation
AML - Anti-Money Laundering	Insurtech
Analysis	Information Technology
API - Application Programming Interface	IoT - Internet of things
BI - Business Intelligence	IT - Information Technology
Bitcoin	Machine Learning
Blockchain	Mobile Banking
Crowdfunding	Mobile payment
Cryptocurrency	Neobank
Data Source	P2P - Peer to Peer
Data Stewardship	POS - Point of Sale
Database	Predictive Analytics
Development	Programming
Digital Banking	RegTech - Regulation Technology
Digital Processing	Saas - Software as a Service
Distributed ledger	Smart Contract
DLT - Distributed Ledger Technology	Spam
Ethereum	TES -Technology-Enabled Service
Fraud	Tokenization
Geolocation	Validation
Hyperledger	Wealth Management
ICO - Initial Coin Offering	Implementation

firms has many Twitter accounts. The number of the Big Four firms' Twitter accounts accounted for almost 50% of the total tweets in the accounting sample. Thus, we split the accounting sample into two parts: the Big Four and the rest accounting firms. Table 3 presents a summary of the data we collected.

Table 3. Summary of Twitter samples.

	Accounting firms (Big four)	Accounting firms (others)
Number of Twitter Accounts	85	99
Number of Tweet records	253503	234933

We used the text wrangling and pre-processing techniques in Natural Language Processing (Farzindar and Inkpen 2015) to conduct data cleansing for the twitter data. As a social media platform, Twitter has a huge constant stream of short text generated by users in an informal environment. There are a few "noises" affecting the accuracy of the analysis result. Natural Language Processing (NLP) package in R presents a useful tool to eliminate these "noises" through data cleansing. Through the data cleansing process, we removed the hyperlink, numbers, punctuations, special symbols, and additional white space and checked for typos. We also formatted the date when the tweets were posts, making it readable for the analytical tools we used in the following analysis.

We use a rule-based ("dictionary") approach (Li 2010) in text mining to obtain keyword frequency. A rule-based approach uses a "mapping" algorithm to read the text and look for the keywords based on a pre-defined rule (Li 2010). Specifically, we matched the text of the tweets with our keyword list and counted their occurrence frequency using R programming. In our keyword list, there are abbreviations (such as IT, POS, AI) and full names (such as Information Technology, Point of Sale, and Artificial Intelligence). We assigned different matching rules for them. For keywords in the abbreviation form, we directly matched the text of the tweets with our abbreviation keyword list. As R is a case sensitive language, we can differentiate "IT" (Information Technology) with "it" (pronoun) by directly matching. For keywords in full name form, we converted all the tweets text and keywords into lower case, and then matched them. Finally, we combine the results of the abbreviation and full name keyword frequencies. For instance, AI and Artificial Intelligence are the same keywords so we summed up their results.

4 Analysis and Results

Based on the method discussed above, we obtained the keyword frequency data in a time series for the Big Four accounting firms, other accounting firms, and financial firms. We analyzed the data from these three perspectives. First, we generated the big picture of the popularity of technology in the accounting and finance industries by summarizing the individual keyword frequency data and observed them overall. Second, we compared the differences in individual keyword frequencies between accounting firms and financial institutions through parallel analysis. Third, within accounting firms, we also compared the differences in individual keyword frequencies between the Big Four accounting firms and other accounting firms. Fourth, we added the time dimension and conducted longitudinal analysis to examine the change in the keywords' frequencies over time.

From the overall level of the analyses, we used an outer join to merge the accounting and financial firms' data by date and aggregate for all the dates. Figure 1 shows the technology keywords mentioned in the accounting and finance industries in the past ten years in descending order. We can tell Artificial Intelligence, Information Technology, Development, Blockchain, Fraud and Internet of things show high frequencies. Figure 2 shows the popularity of the technology keywords vividly using Text Cloud.

In the second level of analysis, we compared the differences between accounting firms and financial firms' keyword frequencies aggregated by date in Fig. 3 and 4. Generally, the number of keywords mentioned in accounting firms are exceed those mentioned in financial firms. In particular, Artificial Intelligence, Information Technology, and Blockchain are the most used frequent terms in accounting firms. For

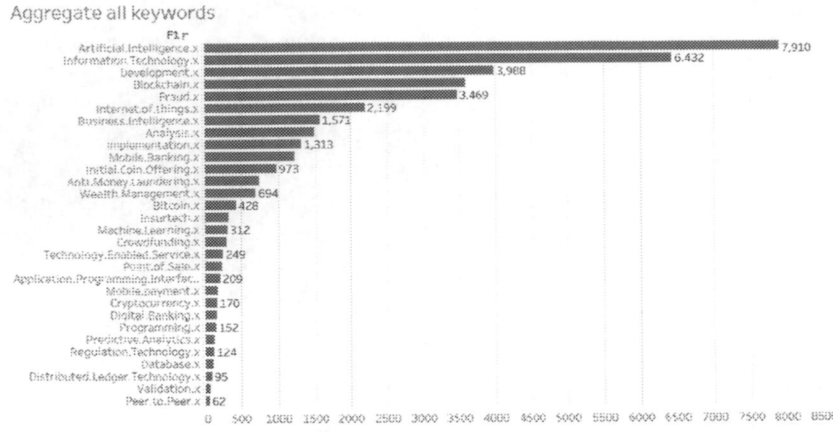

Fig. 1. Technology keyword frequency in the accounting and finance Industry.

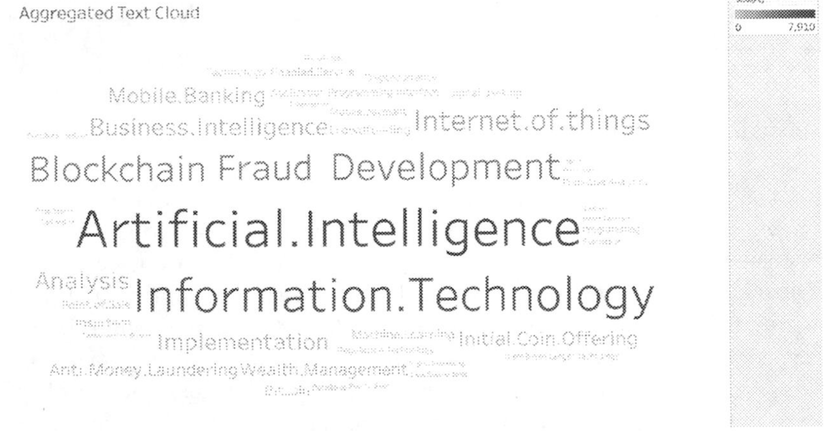

Fig. 2. Technology word cloud for accounting and finance industry.

financial firms, Information Technology, Mobile Banking, and Fraud are the most frequent terms. Both accounting and financial firms attach importance to Information Technology, Fraud, and Artificial Intelligence.

In the third level, within the accounting firms, we compared the differences between the Big Four accounting firms and the financial firms in Fig. 5. We can see that the Big Four accounting firms mention the technology keywords more frequently than other accounting firms. In the Big Four accounting firms, Artificial Intelligence is mentioned more than 5000 times and Blockchain and Information Technology are mentioned around 3000 times. Therefore, it can be can concluded that all the accounting firms pay attention to their development in Artificial Intelligence and Information Technology.

Next, we added the time dimension to track the changes in technology awareness changes occurring in the accounting and finance industries through longitudinal

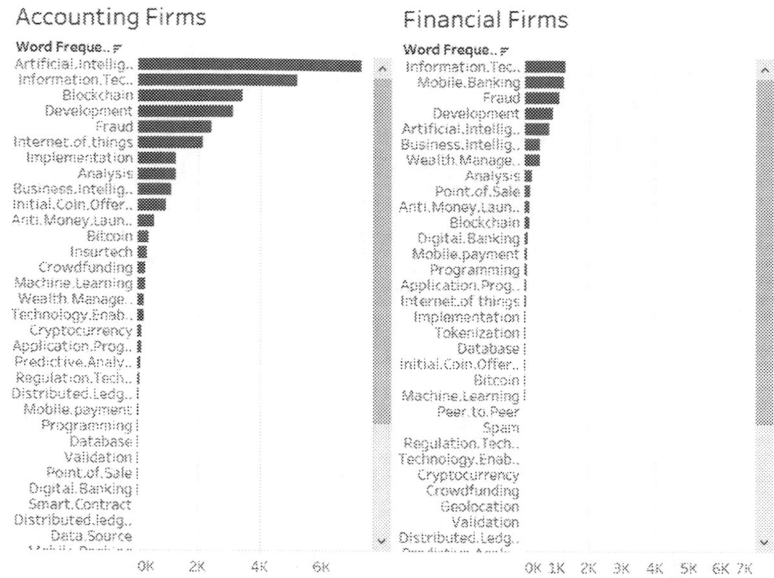

Fig. 3. Comparisons between accounting and financial firms.

Fig. 4. Comparisons between accounting and financial firms in word cloud.

analysis. We chose to cut off the collected data from 2018 as we just have data for the first half of 2019. Figure 6 shows the combined keyword occurrence frequencies for accounting and financial firms in a time series. We can see that there is a peak in 2010 for most of the keywords. The most frequent keyword is Artificial Intelligence. However, after 2015, Information Technology replaces Artificial Intelligences as the most frequent keyword. From 2010 to 2018, all the technology-related terms continue the downward trend.

Excluding the highlighted keywords in Fig. 6, we used area chart to zoom in on the changes of other keywords that do not appear in a high frequency in Fig. 7 and also to view accumulated totals overtime. In the area chart, we can see Wealth Management and Machine Learning are topics given constant attention. After 2018, attention to Peer-to-peer, Digital Banking, and Wealth Management increases.

When we view the accounting firms independently in Fig. 8, we can see trends similar to Fig. 6. In the accounting industry, the occurrence of these technology

Fig. 5. Comparisons between big four accounting firms and other accounting firms.

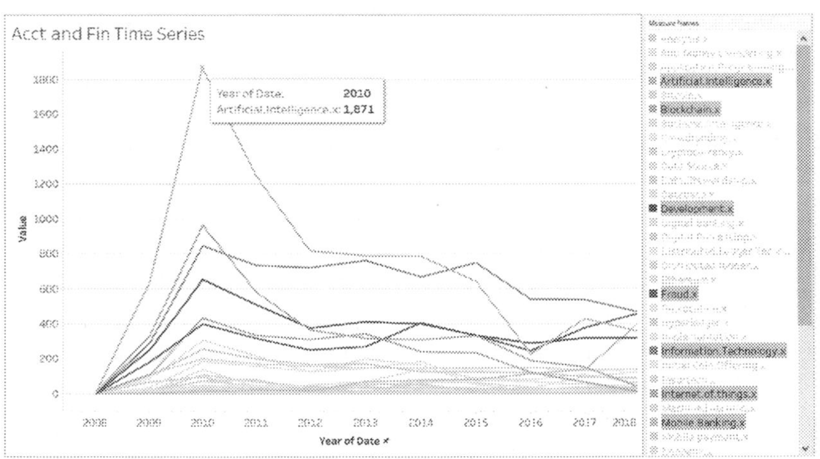

Fig. 6. Keyword occurrence frequency in time series for accounting and financial firms.

keywords is decreasing. The change of the most frequent keyword from Artificial Intelligence to Information Technology occurs in 2016. Figure 8 highlights the top five most frequently occurring keywords. Figure 9 shows the area chart for keyword occurrence frequencies in a time series for accounting firms after excluding the keywords highlighted in Fig. 8. We can see that almost all the keyword frequencies are decreasing except Mobile Banking.

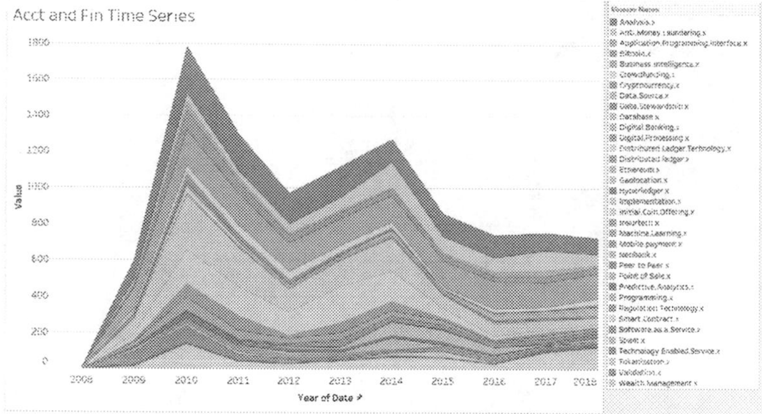

Fig. 7. Area chart for keyword occurrence frequency in time series for accounting and financial firms (excluding the keywords highlighted in Fig. 5).

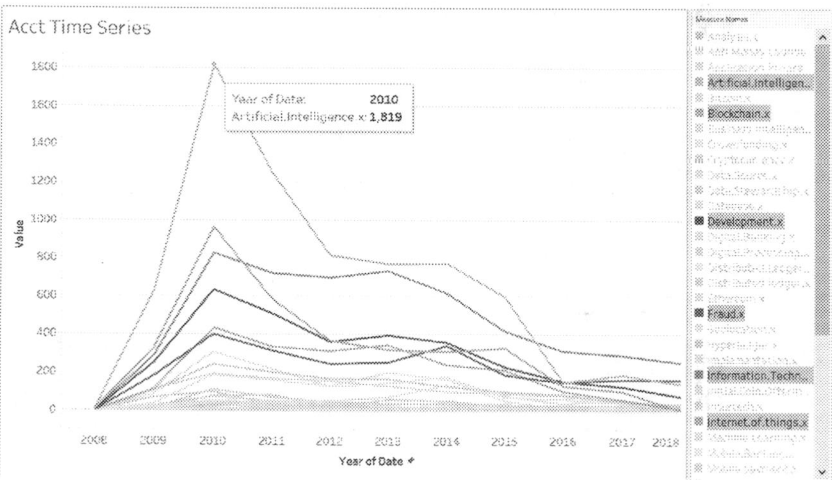

Fig. 8. Keyword occurrence frequency in time series for accounting firms.

Figure 10 shows the changing trends for keywords in financial firms. There are three peaks, Wealth Management in 2010, Information Technology in 2015, and Mobile Banking in 2018. Similarly, we obtain the area chart in Fig. 11 for keyword occurrence frequencies in a time series for financial firms after excluding the keywords highlighted in Fig. 10. We can see that the financial firms started to mention Blockchain, Internet of Things, and Digital Banking in 2014, and Crowdfunding in 2015. Compared with accounting firms, the technology keywords for financial firms show an upward trend from a longitudinal analysis perspective. In parallel comparison with accounting firms, we see that there is a hysteresis phenomenon-a delay in the overall trend- for financial firms.

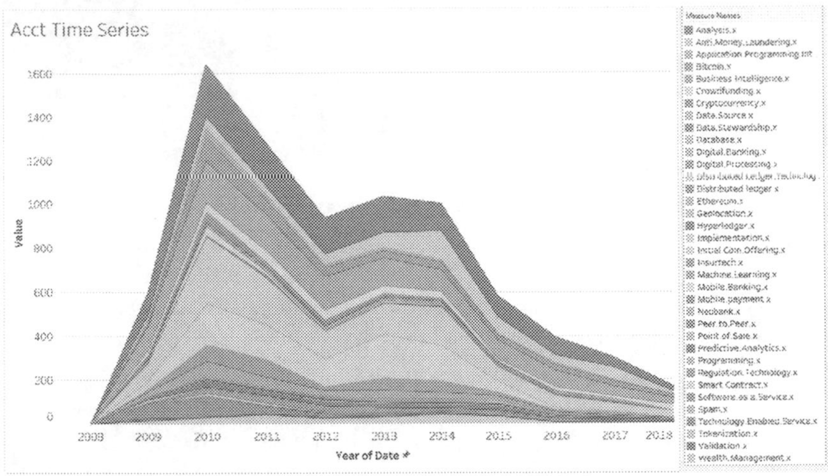

Fig. 9. Area chart for keyword occurrence frequency in time series for accounting firms (excluding the keywords highlighted in Fig. 7).

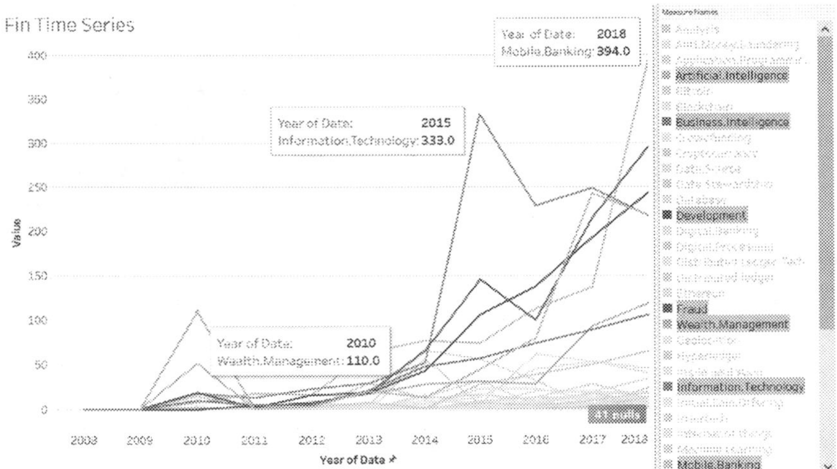

Fig. 10. Keyword occurrence frequency in time series for financial firms.

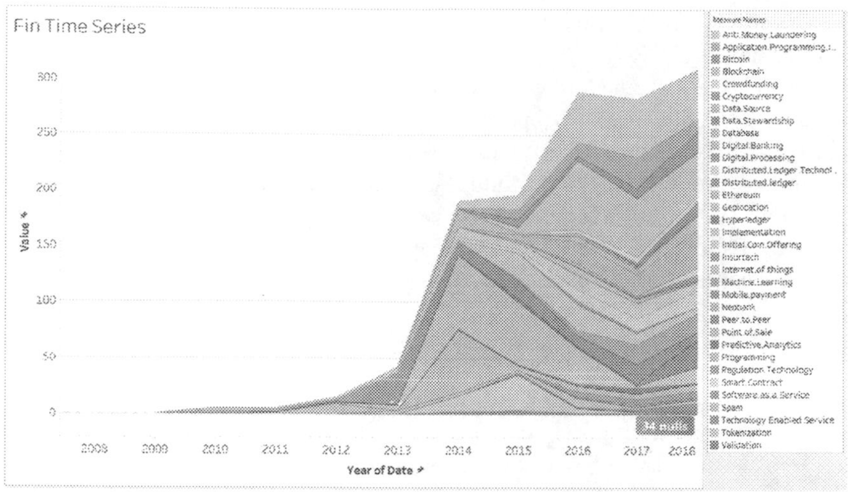

Fig. 11. Area chart for keyword occurrence frequency in time series for financial firms (excluding the keywords highlighted in Fig. 9).

5 Discussions

Our analysis describes a process for examining the technology changes in the accounting and finance industries through social media. Based on our twitter text data mining, we achieved the results and findings of four levels of analysis. The findings show the importance attached to technology and the changes in technology in the accounting and finance industries intuitively. These findings also provoke our reflection of the reasons behind the results.

1) Among all the technology-related keywords, Artificial Intelligence (AI) and Information Technology (IT) are the top two emerging technologies the accounting and finance industries are aware of. However, there are still many questions we need to care about. For example, do people show a positive attitude towards these technologies? How will these technologies evolve in the following five years or even ten years?

2) Compared with accounting firms, the keyword occurrence frequencies in financial firms are much lower and also show a hysteresis phenomenon. The overall trend for accounting firms is downward while that for financial firms is the upward. Consequently, the peak time of the accounting industry and the finance industry is different. Also in our data collection, we found that the twitter accounts for accounting firms are more numerous than that for financial firms. One assumption to explain this difference between accounting and financial firms is that the Big Four accounting firms demonstrate strong leadership in guiding the technology development in accounting industry. They explore and adopt these technologies earlier. After they adopt or deploy these technologies, they might decrease mention of these keywords. Instead, they will pay attention to the more detailed branch of these technologies or other new technologies. Meanwhile, the inherent differences between the accounting and finance industries can also contribute to the differences in the adoption of technology for

accounting and financial firms. Further research is required to understand the differences in depth.

3) In 2010, the awareness of emerging technologies reaches its peak. As we know, the world-wide financial crisis happened in 2008. One possible explanation is that after the financial crisis, companies are trying to find the area that can boost their performance. Investing in information technology might be one of the strategies to help with this situation. To test this assumption, we can collect accounting data from the annual financial reports for these finance and accounting firms from 2008 to 2010 and test the correlation between the accounting data from financial reports and these keywords frequencies.

6 Future Research

In this research, we proposed a new perspective to examine the changes in technological awareness in the accounting and finance industries, particularly, in firms' twitter text data. We applied the Natural Language Processing and text mining methodology in this analysis. Finally, we obtained the findings of the current situation of the popularity of technology in the accounting and finance industries as well as the differences in the overall trends concerning the popularity of technology within the accounting and finance industries.

Based on our research, we suggest several directions for future study. First, in our keyword list, we would have the keywords on different levels. For example, Information Technology, Blockchain, and Internet of Things are at different levels. We will use topic modeling and topic clustering to group our keywords and examine the differences between different keyword groups. Also, we will rephrase our keyword list. Second, in our study, the occurrence frequency of technology-related keywords exhibits a downward trend in the past ten years. There might be new technological keywords mentioned more in recent years in accounting firms. Currently, we only count the occurrence of the technology keywords but we have not no knowledge of people's attitudes towards them. We will add sentiment analysis to examine these firms' attitudes towards these technologies by detecting the mood words near the technology-related keywords. Third, we will also include other social media data (Facebook, LinkedIn), firms' official announcements and firms' interview data in our data source to increase the accuracy and validity of our research. Last, we will discuss and test the influence that these technologies bring to the firms in terms of profitability, brand images and other aspects.

References

Arnold, V.: The changing technological environment and the future of behavioral research in accounting. Account. Finan. **58**(2), 315–339 (2018)

Batrinca, B., et al.: Social media analytics: a survey of techniques, tools and platforms. AI Soc. **30**(1), 89–116 (2015)

Cobbin, P., et al.: Enhancing the Accessibility of accounting and business archives: the role of technology in informing research in accounting and business. Abacus **49**(3), 396–422 (2013)

Farzindar, A., Inkpen, D.: Natural language processing for social media. Synthesis Lect. Hum. Lang. Technol. **8**(2), 1–166 (2015)

Guo, L., Shi, F., Jun, T.: Textual analysis and machine learning: crack unstructured data in finance and accounting. J. Finan. Data Sci. **2**(3), 153–170 (2016)

Kaplan, A.M., Haenlein, M.: Users of the world, unite! The challenges and opportunities of social media. Bus. Horiz. **53**(1), 59–68 (2010)

Li, F.: Textual analysis of corporate disclosures: a survey of the literature. J. Account. Literat. **29**, 143 (2010)

Liu, F.-C., Hsu, H.-T., Yen, D.C.: Technology executives in the changing accounting information environment: impact of IFRS adoption on CIO compensation. Inf. Manag. **55**(7), 877–889 (2018)

Loughran, T., Mcdonald, B.: Textual analysis in accounting and finance: a survey. J. Account. Res. **54**(4), 1187–1230 (2016)

PricewaterhouseCoopers. Data Driven White Paper (2015)

Suddaby, R., Saxton, G.D., Gunz, S.: Twittering change: the institutional work of domain change in accounting expertise. Account. Organizat. Soc. **45**, 52–68 (2015)

Sutton, S.G., Arnold, V.: Focus group methods: using interactive and nominal groups to explore emerging technology-driven phenomena in accounting and information systems. Int. J. Account. Inf. Syst. **14**(2), 81 88 (2013)

A Study on the Influence of E-commerce Live Streaming on Consumer's Purchase Intentions in Mobile Internet

Shi Yin[1,2(✉)]

[1] Research Center of Resource Recycling Science and Engineering, Shanghai Polytechnic University, Shanghai 201209, China
yinshi@sspu.edu.cn
[2] School of Economics and Management, Shanghai Polytechnic University, Shanghai 201209, China

Abstract. With flow dividend of traditional e-commerce is diminishing gradually, the growth rate of e-commerce users is slowing down, the major e-commerce platforms are in great need of content innovation and new flow portals. E-commerce live streaming has become one of the breakthroughs for the platforms. Its growth prospects depend on whether the live streaming users can turn into orders. Based on the characteristics of China's e-commerce live streaming platforms and users, this paper explores the impact of e-commerce live streaming on consumers' purchase intentions under the mobile Internet environment. During e-commerce live streaming, consumers are likely to produce a herd effect, thus influencing their purchase intention. Results show that perceived ease of use, situational factors and follow others' behavior significant impact on purchase intention, perceived usefulness of influence on purchase intention is lesser, incomplete information will not directly affect consumer's purchase intention, but by following others it might affect their purchase attitude when watching live.

Keywords: E-commerce live streaming · Mobile internet · Consumer's purchase intentions

1 Introduction

With the development of e-commerce industry growing faster and faster. Low cost and low threshold also attract a lot of people to enter the e-commerce industry, the competition is increasingly fierce. By June 2019, the number of Internet users in China had reached 854 million, and the growth rate of Internet users had slowed down or even stagnated. Various Internet businesses had entered a state of stock competition. The flow bonus period of traditional e-commerce has passed, and the major e-commerce platforms are in urgent need of content innovation and new flow portals. E-commerce live streaming has become one of the breakthroughs for the platforms.

According to the latest figures released by Ministry of Industry and Information Technology of China, by 2019, China had a net increase of 149 million mobile phone

C. Stephanidis et al. (Eds.): HCII 2020, LNCS 12427, pp. 720–732, 2020.
https://doi.org/10.1007/978-3-030-60152-2_54

users, bringing the total to 1.57 billion. This shows that China has entered the era of mobile Internet. In the era of mobile Internet, consumers gradually get used to doing everything through smart phones, and network streaming has become an indispensable way of information transmission for e-commerce enterprises.

The development prospect of the new model depends largely on whether the users watching the live streaming can be converted into orders, that is, consumers' purchase intention. It is meaningful to study the influence of e-commerce live streaming on customers' purchase intention.

2 Relevant Theories and Literature Review

2.1 E-commerce Live Streaming

The characteristics of mobile Internet are not limited by time and space. While providing convenience for customers, mobile Internet breaks the original mode of flat interaction and time delay interaction of live streaming and has better transmission effect. In addition, compared with offline shopping, traditional e-commerce is mainly through text description and picture display, so that consumers have a better understanding of products. Consumers may not havc cnough information from thc Internet to make a purchase decision, especially for products that need to know the actual effect. Some e-commerce platforms have tried to overcome this pain point, such as 3D fitting rooms and inviting specific buyers to film their shows, but these appear to have little effect. Compared with pictures and texts, video has a richer information dimension and better information transmission effect, which enables consumers to have a more intuitive and comprehensive understanding of product and service information, which could reduce the cost of trial and error, and enables consumers to integrate into the shopping scene [1]. Thus, live streaming marketing comes into being.

In live streaming, users sitting in front of the mobile phones can feel directly facing with the anchors. Anchors and consumers can also greatly improve the shopping experience through the interactive method of real-time question answering and can also arouse consumers' interest in the purchase of goods, which can better attract customers to purchase and make the purchase decision easier. Live streaming can also be implemented like the shopping mall scene guide. People watching the live streaming together can communicate with each other and influence each other. Therefore, the live streaming e-commerce has certain social attributes. Seeing that everyone in the same live streaming room is buying commodities also helps users to make purchase decisions.

E-commerce live streaming mode is a new marketing communication mode integrating digital technology, Internet/mobile Internet, communication technology and other technologies. It involves consumers' use of mobile phones/computers/tablets, e-commerce websites or apps, which are integrated into a new information system to some extent. This new model can promote users' understanding and interest in commodities, spread brand value, enhance users' demand for stickiness, or promote the leap of traditional e-commerce.

2.2 Herd Behavior

Herd behavior comes from the study of animal cluster behavior, which means you don't need a centralized coordination within a group only through learning and imitation, ideas or behavior show the degree of consistency between the individual. It is now extended to under the condition of asymmetric information, the behavior subject is influenced by the behavior of others, choose to give up their original decisions, take actions consistent with most of the individuals.

The origin of herd behavior is information asymmetry. In real life, due to the cognitive cost, people can hardly get all the information about a certain transaction, so they tend to blindly follow others' decisions and make choices against their own will when making decisions.

At the beginning of the rise of E-commerce live streaming, a large number of live streaming platforms emerged, and various websites were also connected to these platforms. In order to experience new and interesting ways, make themselves not so backward or not to miss the various preferential activities provided by businesses, more and more consumers start to use network broadcast, "herd behavior" has come into being.

At present, the research on herd behavior also focuses on the fields related to stock, investment and securities. Yuan [2] and Chen [3] used Agent imitation strategy to simulate the herd effect caused by stock market price fluctuation, and proved that the herd behavior of investors under heterogeneous hypothesis would in turn significantly affect stock price. Hu [4] improved the information waterfall model, studied the effects of various factors on the generation of herd behavior, and verified the existence of herd behavior in e-commerce by regression calculation of the fixed effect model. E-commerce Live streaming is a popular element integrating ideas, behaviors and materials. When people choose to shop through live streaming, herd mentality plays a greater role, and herd behavior may exist in the adoption process of e-commerce.

2.3 Technology Acceptance Model

The technology acceptance model studies the perception of the adoption of new technologies by relevant individuals within the organization. Through the superposition of the perception of risks and interests, the attitude and cognition of the organization towards the adoption of new technologies are formed. Technology acceptance model was first proposed by Davis in 1985, which is of great significance to the study of individual behavior in technological innovation. The theory is that, under normal circumstances, factors influencing the people to accept and adopt new technology are: Perceived Usefulness, Perceived Ease of Use and Attitude toward using. Perceived usefulness reflects the user's perception of the possibility and extent to which the adoption of a new technology improves their productivity; Perceived ease of use reflects the user's perception of how easy it is to adopt a new technology.

Based on the technology acceptance model, Singh [5] constructed a model to study the factors influencing customers' willingness to purchase goods or services online. The results showed that the perceived value, usefulness, ease of use, protection of consumer information, and credibility of websites all positively affected consumers' willingness to buy. The determinants of customers' continuous online shopping intention through the improved technology acceptance model, and the results showed that perceived usefulness, entertainment and social pressure had a positive impact on online shopping intention of customers.

3 Model Specification and Research Hypothesis

3.1 Model Specification

Perceived Usefulness. Technology acceptance model has been applied and verified many times in the field of e-commerce. Perceived usefulness and perceived ease of use are its main determinants. It is believed that in the technology acceptance model, perceived usefulness is the key driving factor, which affects users' acceptance behavior by improving the effect of using information technology. If mobile e-commerce live streaming can provide services to meet customers' needs and improve the efficiency and effect of customers' purchase, customers will think it is useful. The following hypothesis is proposed:
Hypothesis 1: Perceived usefulness has a significant positive impact on consumers' purchase intention.

Perceived Ease of Use. It reflects customers' perception of the ease of using mobile live streaming, which generally includes whether it is simple, clear and convenient to use. Kasiri [6] think good service and convenience can improve the user's satisfaction, in turn, affects the user use motivation. These two factors as perceived convenience and believe that the perceived convenience of use directly affects users' use of the technology.

The screen of the mobile terminal is usually more than ten times larger than that of the PC terminal. In limited space, the mobile shopping terminal needs to display useful information to customers. At the same time, the operation is relatively simple, and the process is not too complicated. The following hypothesis is proposed:

Hypothesis 2: Perceived ease of use has a significant positive impact on consumers' purchase intention.

3.2 Herd Effect

Incomplete Information. The collection of data related to behavior is the basis for people to make action decisions. The completely rational behavior requires a high level of behavioral data, which requires not only quality and quantity of information, but also timely access to time.
With the development of science and technology, especially the popularity of the Internet, people have made a qualitative leap in the ability to collect and process

information, but this is still far from the requirements of rational behavior. It's showed by two points: first, the uncertainty of behavioral information. There are various factors influencing behavioral decision, and some of them will change under certain conditions, which makes it difficult to obtain certain information. Second, information collection cost accessibility. Collecting behavioral information requires a certain time cost and economic cost. If too much information needs to be collected, which exceeds the range that users can bear, even if such information can be collected, it cannot be achieved from the perspective of cost [7]. Because people do not have access to complete information, the act of following others is inevitable.

The incompleteness of information leads to great uncertainty in the decision-making process. Consumers usually adopt heuristic strategies based on the principle of investing least cognitive resources, that is, they trust other consumers based on cognitive judgment, which leads to herd behavior. When consumers make shopping decisions through e-commerce live streaming, each consumer will obtain relevant information through personal and public channels. While knowing the decisions made by others, their own decisions also affect the choices of other consumers. The majority of consumers will be due to other people's evaluation and purchase behavior of the network trust to promote the adoption of herd behavior. The following hypothesis is proposed:

Hypothesis 3: incomplete information has a significant positive impact on following others' behavior.

Hypothesis 4: incomplete information has a significant positive impact on consumers' purchase intention.

Situational Factors. In the mobile Internet environment, different from the traditional consumption scenario, consumers can't get access to goods and can only get information through their eyes. The traditional e-commerce mainly carries on the information transmission through photos, but the disadvantage is the relative lack of human factors. After the emergence of e-commerce live streaming, merchants can improve the vividness of the shopping process through video and interaction to meet consumers' sensory experience and entertainment needs. In the live streaming process of e-commerce, consumers are easily stimulated by discount promotion and shopping environment and are prone to purchase impulse. In the process of watching, consumers are likely to feel the purchasing passion of other consumers, so as to follow others and generate purchasing intention [8]. The following hypothesis is proposed:

Hypothesis 5: situational factors have significant positive impact on following others' behavior.

Hypothesis 6: situational factors have significant positive impact on consumers' purchase intention.

Follow Others' Behavior. The essence of herd behavior is that individuals follow the behavior and decisions of others. When making decisions, users will ignore the collection of information and reduce the estimation of risks, which will seriously affect users' rational judgment and may lead to the situation of following others, thus resulting in purchase behavior.

In general, people's self-protection mechanism is to avoid damage and reap the benefits and imitation strategy can reduce the various risks brought by the uncertain situation, make the organism to avoid mental or physical damage [9]. Therefore, follow the actions of others to individual better adapt to the environment.

As a typical uncertain situation, online shopping has a large amount of information. When personal knowledge or experience is insufficient to support independent decision-making, recommendation or review by others is the dominant strategy to generate trust in online shopping [10]. In order to obtain higher value or better avoid risks, following others, as a simple and quick decision rule, is effective and reasonable, especially for inexperienced consumers. Consumers adopt imitation strategies to reduce the threat of uncertain situations to themselves, resulting in the herd effect of online shopping. The following hypothesis is proposed:

Hypothesis 7: follow others' behavior has a significant positive impact on consumers' purchase intention.

4 Research Design and Empirical Analysis

4.1 Questionnaire Design

Based on the mutual influence of perceived ease of use, perceived usefulness, incomplete information, situational factors and follow others' behavior mentioned above, this paper designs a questionnaire based on the characteristics of mobile e-commerce live streaming by referring to the questionnaires in existing studies.

The first part of the questionnaire is general selection, including gender, age group, frequency of live shopping through e-commerce and other basic information of respondents. The second part is used to measure consumers' purchase intention when watching live stream of e-commerce. It is designed according to the five dimensions, and the items are measured by Likert seven-point scale, from 1 indicating "strongly disagree" to 7 indicating "strongly agree" (Table 1).

Table 1. Measurement items for each variable.

Measurable variable	Item number	Measuring item	Measurable variable	Item number	Measuring item
Perceived usefulness	PU1	Can be used for shopping	Perceived ease of use	PEU1	Simple operation interface
	PU2	Get more deals		PEU2	Smooth operation
	PU3	Convenient for daily life		PEU3	Convenient purchase process
	PU4	Help to make better decisions		PEU4	Services offered are acceptable
	PU5	More efficient		PEU5	Interaction operation

(*continued*)

Table 1. (*continued*)

Measurable variable	Item number	Measuring item	Measurable variable	Item number	Measuring item
Incomplete information	INI1	Price of goods	Situational factors	SF1	Recommendation from people around
	INI2	Compare prices		SF2	Help integrate into society
	INI3	Multiple purchase channcls		SF3	Number of followers
	INI4	Public evaluation		SF4	Interactivity
	INI5	Previous experience		SF5	Intense and stimulative environment
Follow others' behavior	FOB1	Trust public evaluation	Purchase intention	PI1	Learn more about the features
	FOB2	Follow others' decision		PI2	Used in the future
	FOB3	Trust live streaming information		PI3	May be used frequently
	FOB4	Trust public information		PI4	Be willing to share advantage
	FOB5	Mitigate risks		PI5	Recommend it to others

4.2 Data Collection and Statistical Analysis

The questionnaires were distributed online, and the survey period was from 2019 November 1 to December 1. During the effective survey period, a total of 423 questionnaires were collected, and those that had not watched the live stream and had answered contradictory questions were deleted. In the end, 377 questionnaires were valid, with an effective rate of 89.1%.

After the questionnaire is collected, the data are collected and sorted out. First, descriptive statistical analysis was carried out on the survey samples. Then, SPSS 20.0 and AMOS 20.0 software were used for reliability and validity analysis. Finally, the influence of various factors on consumers' purchase intention is discussed.

Among the 46 invalid questionnaires, 37 were respondents who had never shopped through live streaming, among which 31 were not using but intended to use, and only 6 said they would not use in the short term. It indicates that most of the respondents are interested in e-commerce live shopping.

There are 201 males and 176 females in this survey. The sample proportion basically conforms to the gender structure of Internet users [11]. Young people like to try new things and the respondents who took the initiative to participate in the examination paper were mainly aged between 18 and 24 and between 25 and 35. According to the survey, most are between the ages of 18 to 24, accounting for 34.48% of the overall sample. In terms of level of education, samples are mainly concentrated in colleges and universities level, accounting for 53.05% of the overall sample, followed by a high school education accounted for 24.93%, master degree or above accounted for 17.24%, respondents high level of education is the goal of live electrical business customers. In terms of frequency of live streaming, daily use accounts for the largest proportion (42.71%), indicating that most of the respondents in this study are familiar with mobile Internet live streaming and have rich experience in using it (Table 2).

Table 2. Basic information of the survey samples.

Gender	Proportion	Age	Proportion	Education	Proportion	Frequency of use	Proportion
Male	53.32%	Under 18	10.88%	Junior high school and below	4.77%	Per day	42.71%
Female	46.68%	18–24	34.48%	High school or equivalent	24.93%	Per week	37.40%
		25–35	31.03%	Colleges and universities	53.05%	Per month and above	19.89%
		Above 35	23.61%	Master or above	17.24%		

4.3 Reliability and Validity Test

First, the maximum principal component analysis method was used to extract the factors, and exploratory factor analysis was conducted on the purchase intention of e-commerce live streaming customers. It was found that the variance estimators of the three indicators PU4, INI3 and FOB4 explained by factors were 0.231, 0.145 and 0.351, respectively, less than 0.5, indicating that these three variables should be deleted. Factor analysis was conducted on the modified model, and the results showed that KMO = 0.872, Bartlett's spherical value was 7289.976, df = 336.

Reliability analysis is generally used to verify the internal consistency of data. This study does not involve external reliability measurement, so only internal reliability is tested. Validity analysis reflects that the measurement tool can accurately measure the trait degree to be measured.

The reliability and validity tests were conducted by SPSS and AMOS software respectively, and the results showed that the project-overall correlation coefficient (CITC) was at least 0.608. According to the standard of CITC $\geq$ 0.3, the correlation between the measured items was relatively high. Cronbach's α coefficients are above 0.8, which conforms to the general standard that the Cronbach's α coefficient is greater than 0.7, so the scale has a high internal consistency. Cronbach's α if Item Deleted represents the Cronbach's α coefficient of the questionnaire after removing the current question, which can generally be used as an important reference basis for adjusting the question. The values of all questions in this questionnaire are all greater than the standard of 0.6, and the range of change is small. Therefore, the test value of this index indicates that the design after removing the index is more reasonable. The combined reliability (CR) of all indicators was between 0.810 − 0.923, all of which exceeded the level of 0.7. The overall reliability and reliability of the questionnaire data in this study are good. The validity test of measurement variables is mainly verified by factor load and average variance extracted (AVE) values.

This paper has established the initial model of mobile e-commerce live shopping intention above, and the calculation results of the model show that the fitting index is not very good. After modification, AVE values were all between 0.501 − 0.691, both higher than the critical value standard of 0.5, indicating that the model had good aggregation validity (Table 3).

Table 3. Model verification table.

Measurable variable	Indicator variables	Load factor	CITC	Cronbach's α if item deleted	Reliability	CR	AVE	Cronbach's α
Perceived usefulness	PU1	0.651	0.613	0.833	0.921	0.810	0.501	0.844
	PU2	0.672	0.651	0.737				
	PU3	0.678	0.640	0.829				
	PU4	–	–	–				
	PU5	0.823	0.866	0.812				
Perceived ease of use	PEU1	0.766	0.764	0.789	0.959	0.875	0.585	0.856
	PEU2	0.642	0.682	0.798				
	PEU3	0.772	0.778	0.840				
	PEU4	0.869	0.857	0.812				
	PEU5	0.704	0.765	0.831				
Incomplete information	INI1	0.735	0.778	0.747	0.903	0.846	0.618	0.922
	INI2	0.881	0.889	0.872				
	INI3	–	–	–				
	INI4	0.799	0.671	0.874				
	INI5	0.745	0.722	0.780				
Situational factors	SF1	0.796	0.697	0.782	0.916	0.923	0.691	0.923
	SF2	0.844	0.818	0.821				
	SF3	0.895	0.638	0.914				
	SF4	0.867	0.727	0.871				
	SF5	0.824	0.892	0.864				

(*continued*)

Table 3. *(continued)*

Measurable variable	Indicator variables	Load factor	CITC	Cronbach's α if item deleted	Reliability	CR	AVE	Cronbach's α
Follow others' behavior	FOB1	0.829	0.726	0.762	0.918	0.897	0.689	0.873
	FOB2	0.814	0.827	0.781				
	FOB3	0.818	0.695	0.871				
	FOB4	–	–	–				
	FOB5	0.771	0.841	0.911				
Purchase intention	PI1	0.772	0.727	0.827	0.908	0.835	0.524	0.911
	PI2	0.776	0.714	0.622				
	PI3	0.669	0.669	0.759				
	PI4	0.654	0.648	0.825				
	PI5	0.768	0.763	0.754				

4.4 Hypothesis Testing

After the overall fitting degree of the model reaches an acceptable level, the structural equation model can be tested. According to the test, the χ^2/df of the model was 1.676, less than 3, indicating that the adaptability of the model is good. The GFI value is 0.890, exceeding the adaptation standard of 0.8;R MSEA is 0.071, which meets the standard less than 0.08; CFI, TLI, and IFI were 0.91, 0.912, and 0.924, respectively, which were both greater than 0.9, indicating that the fitting degree of each index was good and the model and data were relatively adaptable.

The preliminary model has been established above, and it is believed that perceived usefulness has a certain influence on whether consumers are willing to choose mobile live streaming for shopping, but to what extent? According to the results in Table 4, the indicators of incomplete information, situational factors, following others and purchase intention better explain the respective hidden variables. The first three measures of perceived usefulness explain the latent variables, the correlation coefficient of "live streaming through e-commerce can make online shopping more efficient" is only 0.31.

Table 4. Parameter estimation table.

Measurable variable	Indicator variables	Parameter estimate	Measurable variable	Indicator variables	Parameter estimate
Perceived usefulness	PU1	0.52***	Incomplete information	CT1	0.52**
	PU2	0.68**		CT2	0.71***
	PU3	0.78***		CT4	0.59***
	PU5	0.31***		CT5	0.66***
Perceived ease of use	PEU1	0.78***	Situational factors	CS1	0.72***
	PEU2	0.80***		CS2	0.68***
	PEU3	0.67**		CS3	0.64***
	PEU4	0.49***		CS4	0.59***
	PEU5	0.65***		CS5	0.83***

(continued)

Table 4. (*continued*)

Measurable variable	Indicator variables	Parameter estimate	Measurable variable	Indicator variables	Parameter estimate
Follow others' behavior	CRC1	0.76**	Purchase intention	CRI1	0.55***
	CRC2	0.62***		CRI2	0.72***
	CRC3	0.58***		CRI3	0.64***
	CRC5	0.53**		CRI4	0.51***
				CRI5	0.71***

As can be seen from the results in Fig. 1, except for the low parameter estimation of H1 and the unsupported H4, the estimated values of the other hypotheses are all above 0.5, and there is a strong relationship between potential variables and the hypotheses are supported, which is basically consistent with the former research hypothesis.

The results show that perceived usefulness has a significant positive effect on purchase intention, and hypothesis 1 is accepted. Perceived ease of use has a significant positive effect on purchase intention, and hypothesis 2 is accepted. Incomplete information has a significant positive effect on following others' behavior, and hypothesis 3 is accepted. Incomplete information has a significant positive effect on purchase intention, and hypothesis 4 is rejected. Situational factors have a significant positive effect on following others' behavior, and hypothesis 5 is accepted. Situational factors have a significant positive effect on purchase intention, and hypothesis 6 is accepted. Following others' behavior has a significant positive effect on purchase intention, and hypothesis 7 is accepted.

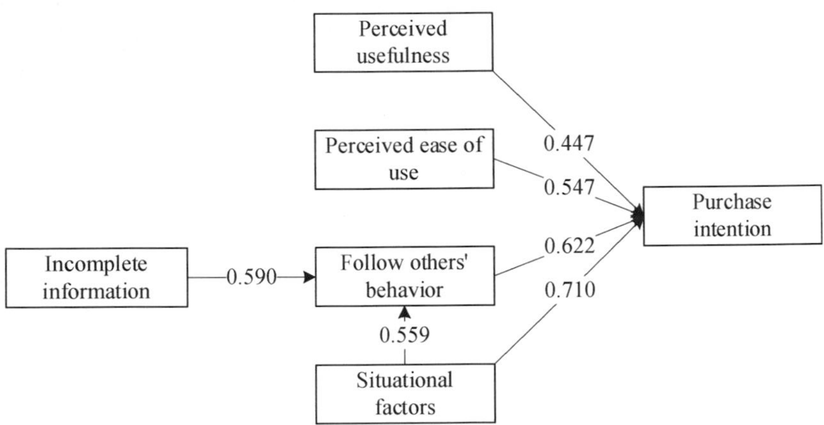

Fig. 1. Path coefficient diagram.

5 Conclusion

In this paper, perceived ease of use, perceived usefulness, incomplete information, situational factors, follow others' behavior and customers' willingness to use mobile e-commerce live streaming for shopping are integrated into a theoretical research framework to discuss the herding behavior under the mobile Internet environment on consumers' purchase intention. Then the statistical analysis method and structural equation model are used to modify and verify the theoretical model. It is believed that consumers will go through a series of complex psychological processes before making purchase decisions, and whether consumers will have purchase intention in the process of e-commerce live streaming will be affected by various aspects. Among them, perceived ease of use, situational factors and following others' behavior are more obvious. The perceived usefulness impacts turn out to be small, incomplete information does not directly affect purchase intention during live streaming, but it may affect their attitude during live streaming after following others.

For e-commerce shopping platforms, promoting consumers' buying behavior through live streaming is conducive to improving their sales. For consumers shopping through e-commerce live streaming, they are easily influenced by others. They may go shopping due to herd behavior, so consumers should avoid irrational following behaviors.

Acknowledgements. This work was supported by Gaoyuan Discipline of Shanghai–Environmental Science and Engineering (Resource Recycling Science and Engineering), Discipline of Management Science and Engineering of Shanghai Polytechnic University and Construction of Ideological Education System for E-commerce Majors.

References

1. Wu, B., Gong, C.Y.: Research on E-commerce live streaming based on the information system success model—Taobao E-commerce live streaming as an example. Bus. Globalization **5**(3), 37–45 (2017)
2. Yuan, J.H., Deng, R., Cao, G.X.: Computational experiment of herd behavior. Syst. Eng. Theory and Pract. **31**(5), 855–862 (2011)
3. Chen, Y., Yuan, J.H., Li, X.D., Xiao, B.Q.: Research on cooperative herding behavior and market fluctuation based on computational experiments. J. Manag. Sci. **13**(09), 119–128 (2010)
4. Hu, Q.: Herd Behavior in C2C E-commerce. Tianjin University, Tianjin (2011)
5. Singh, D.P.: Integration of TAM, TPB, and self-image to study online purchase intentions in an emerging economy. Int. J. Online Mark. **5**(1), 20–37 (2015)
6. Kasiri, L.A., Cheng, K.T.G., Sambasivan, M.: Integration of standardization and customization: impact on service quality, customer satisfaction, and loyalty. J. Retail. Consum. Serv. **35**, 91–97 (2017)
7. Astier, N.: Comparative feedbacks under incomplete information. Resour. Energy Econ. **54**, 90–108 (2018)
8. Sheehan, D., Hardesty, D.M., Ziegler, A.H.: Consumer reactions to price discounts across online shopping experiences. J. Retail. Consum. Serv. **51**, 129–138 (2019)

9. Giachetti, C., Torrisi, S.: Following or running away from the market leader? The influences of environmental uncertainty and market leadership. Eur. Manag. Rev. **15**(3), 445–463 (2018)
10. Xiao, B., Benbasat, I.: An empirical examination of the influence of biased personalized product recommendations on consumers' decision making outcomes. Decis. Support Syst. **110**, 46–57 (2018)
11. Statistical report on Internet development in China (44th). http://www.cac.gov.cn/2019-08/30/c_1124938750.htm. Accessed 10 Feb 2020

Author Index

Printed in the United States
By Bookmasters